International Conference on Digital Solutions for Automotive Industry, Roadway Maintenance and Traffic Control (DS ART 2020)

AA007530

IOP Conference Series: Materials Science and Engineering
Volume 1159

Moscow, Russia
14 - 16 December 2020

ISBN: 978-1-7138-4625-3
ISSN: 1757-8981

Printed from e-media with permission by:

Curran Associates, Inc.
57 Morehouse Lane
Red Hook, NY 12571

Some format issues inherent in the e-media version may also appear in this print version.

This work is licensed under a Creative Commons Attribution 3.0 International Licence.
Licence details: http://creativecommons.org/licenses/by/3.0/.

No changes have been made to the content of these proceedings. There may be changes to pagination and minor adjustments for aesthetics.

Printed with permission by Curran Associates, Inc. (2022)

For permission requests, please contact the Institute of Physics at the address below.

Institute of Physics
Dirac House, Temple Back
Bristol BS1 6BE UK

Phone: 44 1 17 929 7481
Fax: 44 1 17 920 0979

techtracking@iop.org

Additional copies of this publication are available from:

Curran Associates, Inc.
57 Morehouse Lane
Red Hook, NY 12571 USA
Phone: 845-758-0400
Fax: 845-758-2633
Email: curran@proceedings.com
Web: www.proceedings.com

TABLE OF CONTENTS

 doi:10.1088/1757-899X/1159/1/011001

PREFACE

International Conference on Digital Solutions for Automotive Industry, Roadway Maintenance and Traffic Control (DS ART 2020)

International Conference on Digital Solutions for Automotive Industry, Roadway Maintenance and Traffic Control was held virtually on 14–16 December 2020. Moscow Automobile and Road Construction State Technical University (MADI), Moscow, Russia organized DS ART 2020, with the promotional sponsorship of International Association for Automotive and Road Engineering Education (IAAREE).

Due to the pandemic and travel restrictions, the event was held in the virtual format via Microsoft Teams conferencing tool. The access to the meetings was provided for all speakers and preregistered staff. The plenary session is shown on the picture below.

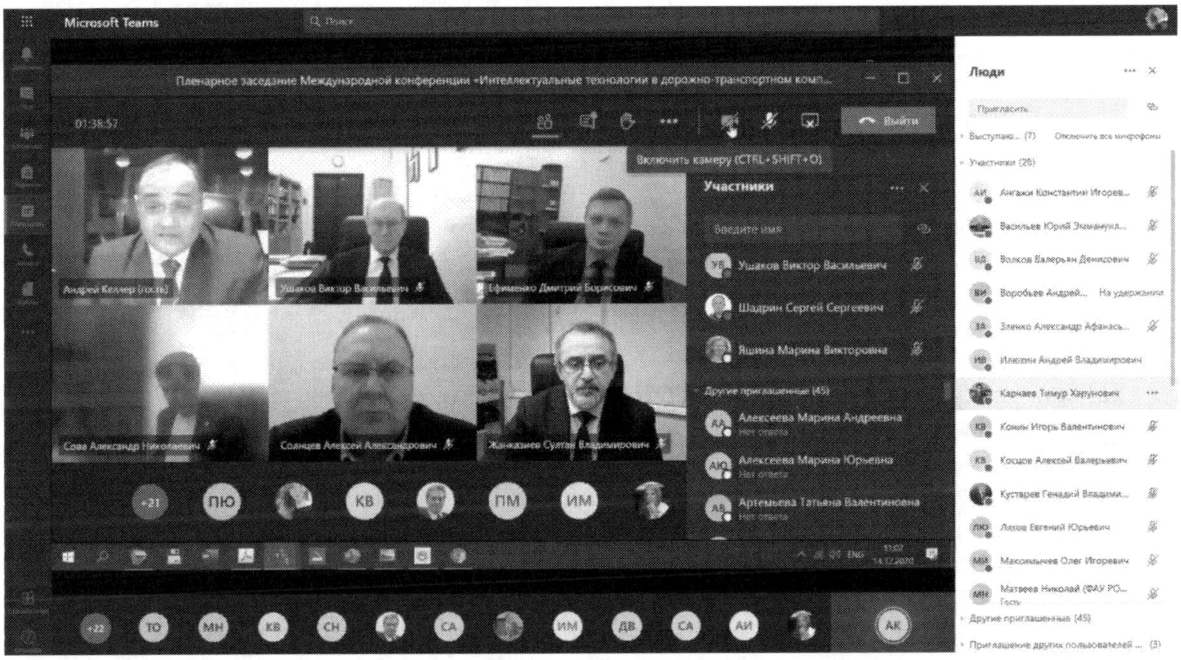

The participants came together to discuss the topical issues of automobile and roadway complex, automotive production, maintenance and repair as well as the acute problem of traffic control. The questions of personnel training in the automotive industry were also raised. The need to move towards creating digital solutions in the industry was emphasized.

All the papers submitted to DS ART 2020 were carefully reviewed and proofread by the editors to select the most valuable and significant ones. They cover the range of relevant research on the current state of the automotive industry and innovative ways to solve its problems.

We express our gratitude to all the participants and the organizing committee for their contribution to preparing the DS ART 2020 proceedings.

With the best regards,
Prof. Aleksei A. Solntsev, Chairman of the Organizing Commette

Content from this work may be used under the terms of the Creative Commons Attribution 3.0 licence. Any further distribution of this work must maintain attribution to the author(s) and the title of the work, journal citation and DOI.
Published under licence by IOP Publishing Ltd

IOP Conf. Series: Materials Science and Engineering 1159 (2021) 011001 doi:10.1088/1757-899X/1159/1/011001

Scientific Committee

1. Didmanidze O.N. – Chair – Academician of RAS, Dr., Professor, Head of the Department of Tractors and cars, Institute of Mechanics and Power Engineering named after V.P. Goryachkin, Russian State Agrarian University Moscow Agricultural Academy named after K.A. Timiryazeva
2. Karelina M.Ju. – Co-Chair – Dr., Professor, Vice-rector for scientific work, Head of the Department of Machine parts and the theory of mechanisms, MADI
3. Demijanushko I.V. – Co-Chair – Dr., Professor, Head of the Department of structural mechanics, MADI
4. Prihodko V.M. – Co-Chair – Corresponding Member of Russian Academy of Sciences, Dr., Professor, MADI
5. Gaidar S.M. – Dr., Professor, Head of the Department of Materials Science and Technology of Mechanical Engineering, Russian State Agrarian University-Moscow Agricultural Academy named after K.A. Timiryazeva
6. Ivanov A.M., Dr., Professor, Head of the Department «Vehicles» MADI
7. Kozlovskij V.N., Dr., Professor, Head of the Department of Theoretical and general electrical engineering, SamSTU
8. Ostrouh A.V., Dr., Professor of MADI
9. Siljanov V.V., Dr., Professor of MADI
10. Sova A.N., Dr., Professor, Head of the Department of Transport installations, MADI
11. Trofimenko Ju.V., Dr., Professor, Head of the Department of Technosphere safety, MADI
12. Shatrov M.G., Dr., Professor, Head of the Department of Heat engineering and automotive engines, MADI
13. Jashina M.V. – Dr., PhD, Head of the Department «Higher mathematics» MADI
14. Girma Gebresenbet – Professor of the Department of Energy and Technology, Swedish university of agricultural sciences, P.O. Box 75651 Uppsala, Sweden; https://www.slu.se/en/ew-cv/girma-gebresenbet/
15. Dzhumanov R. B., PhD, Engineer of the Federal Highway Administration Project (FHWA), USA
16. Matkerimov T.Y., Dr., Professor, Academician of the Engineering Academy of the Kyrgyz Republic, Academician of the International Academy of Pedagogical Education, Dean of the Faculty of Transport and Mechanical Engineering of KSTU named after I. Razzakova, Kyrgyzstan
17. Makovski C. – Coordinator of Cooperation Programs, Associate Professor at the Gdansk Polytechnic University (Sławomir Makowski – doktor nauk technicznych, Politechnika Gdańska, Wydział BM (https://pg.edu.pl/3837a451cd_slawomir.makowski))
18. Hoang D.K. – PhD, South Branch of the Russian-Vietnamese Tropical Scientific Research Technological Center, Vietnam, Ho Chi Minh

IOP Conf. Series: Materials Science and Engineering 1159 (2021) 011001 doi:10.1088/1757-899X/1159/1/011001

Organizing Committee

1. Keller A.V., Dr., Professor, vice-rector of MADI – Chair
2. Nizamov R.K., Dr., Professor, rector of KGASU – Co-Chair
3. Vasilev Ju.V., Dr., Professor, Head of the Department of Road construction materials, MADI
4. Vdovin E.A., PhD, assistant Professor, Vice-rector for research work, Head of the Department of Highways, bridges and tunnels, KGASU
5. Evtjukov S.A., Dr., Professor, Head of the Department, Legal and customs regulation in transport, SPbGASU
6. Efimenko D.B., Dr., Professor, Vice Rector for Academic Affairs, Head of the Department of Legal and customs regulation in transport, MADI
7. Lazarev Ju.G., Dr., Professor of SPbPU
8. Novikov A.N., Dr., Professor, Director of the Polytechnic Institute named after N.N. Polikarpova, Head of the Department "Service and repair of machines, Orlov State University, named after I. S. Turgenev"
9. Novikov I.A., Dr., Director of the Transport and Technological Institute of BSTU named after V.G. Shukhov
10. Pospelov P.I., Dr., Professor, Head of the Department of Roadways Design and Development, MADI
11. Solntsev A.A., Professor, Head of the Department of Car Operation and Service MADI.

Peer review declaration

All papers published in this volume of *IOP Conference Series: Materials Science and Engineering* have been peer reviewed through processes administered by the Editors. Reviews were conducted by expert referees to the professional and scientific standards expected of a proceedings journal published by IOP Publishing.

- **Type of peer review:** Double-blind peer review

- **Review criteria:** Consistency with the conference scope; Academic value of the paper; Presentation style of the content. Each criterion was assessed with the following scale: Poor, Satisfactory, Good, Excellent. The majority rule was applied to determine the outcome of the review. If all the referees give different marks for a single position, that the intermediate mark used as a consensus. All the manuscripts with Good and Excellent marks on each position were accepted as is. The manuscripts with Poor marks on any position were declined. Other manuscripts were sent to the authors for revision. After revision, an editor performed a check if the revision was successful. If necessary, the editor could submit the updated manuscript to the reviewers to get their confirmation of revision.

- **Conference submission management system:** submissions were received and handled via e-mail

- **Number of submissions received:** 190

- **Number of submissions sent for review:** 139

- **Number of submissions accepted:** 109

- **Acceptance Rate (Number of Submissions Accepted / Number of Submissions Received X 100):** 60%

- **Average number of reviews per paper:** 3

- **Total number of reviewers involved:** 30

- **Any additional info on review process:** three independent referees one of which invited from outside the conference committee reviewed all the papers. The review process was held under the supervision of the organizing commitee.

- **Contact person for queries:** Aleksey Solntsev, chairman of the organizing comittee, professor, dean of the faculty of Automotive Vehicles, Moscow Automobile and Road Construction State Technical University (MADI), Moscow, Russia, solntsev@madi.ru

Content from this work may be used under the terms of the Creative Commons Attribution 3.0 licence. Any further distribution of this work must maintain attribution to the author(s) and the title of the work, journal citation and DOI.
Published under licence by IOP Publishing Ltd

IOP Conf. Series: Materials Science and Engineering 1159 (2021) 012001 doi:10.1088/1757-899X/1159/1/012001

Improvement of the technology of cold reclamation of asphalt concrete in the asphalt plant and the possibility of its application in the republic of Iraq

M M Al-Karaguli[1], A V Lupanov[1], V V Silkin[1], M M Haba[2]

[1] MADI, Leningradsky pr. 64, Moscow, 125319, Russia
[2] BME, Technological University, Budapest,31111, Hungary

E-mail: mmg_1989m@yahoo.com

Abstract. This article presents the state of the highways in the Republic of Iraq and studying the possibility of using the regeneration technology of asphalt concrete as a base for road pavements strengthen. This technology has not yet used in Iraq. Therefore it proposed to use for the maintenance and reconstruction of roads the technology of cold regeneration of asphalt concrete in the asphalt plant, which ensures the improvement of the properties of asphalt granules due to its preliminary crushing, the introduction of cement additives as a mineral binder. Compositions of asphalt-granular-concrete mixtures have developed, and data from studies of the properties of asphalt-granular-concrete of various documents presented. The most common pavement defects in Iraq are plastic deformations, which are caused by a combination of high temperatures and low strength characteristics of asphalt concrete. Pregame compositions used to strengthen the construction of road pavements with increased strength and shear stability. A cost-effectiveness calculation of using the cold technique is also presented moreover compare the cost this application in Russia and Iraq also.

1.Introduction

Asphalt concrete mixtures is the main material used in repairing the pavement. This material is gradually increasing the price of it due to the increasing in the price of the resources, the transfer and the shipping costs etc.

Many studies and researches are working on reducing the time and the cost of the pavements repairing by studying different technologies based on the processing of old asphalt concrete, which are implemented both at asphalt concrete plants and directly on the road. Nowadays, these technologies are widely used around the World, where the volume of used old asphalt concrete is 20-30% of the total amount of produced asphalt concrete mixtures. Cold regeneration (CR) technology is the most economical maintenance technology which ensures restoring the original strength of the pavement.

Currently, Iraq and other middle east countries resorting to economic solutions in light of financial crises. Therefore, using cold regeneration (CR) technology assumed to be good Choice to improve the properties of old asphalt concrete, reduce the harmful effects of emissions on the environment and minimize the costs associated with the construction and reconstruction of highways [1].

The analysis of the weather and climatic conditions of Iraq showed that in the summer, the heating temperature of the asphalt concrete pavement reaches 70 ° C [2,3]. That makes changes in the pavement structure and reducing the service life and causes an increase in the risk of the plastic deformation of the surface. The Figure below shows the main type of deformation occurs in the roads of the hot weather countries under a high traffic volume and heavy trucks, [4,5].

Content from this work may be used under the terms of the Creative Commons Attribution 3.0 licence. Any further distribution of this work must maintain attribution to the author(s) and the title of the work, journal citation and DOI.
Published under licence by IOP Publishing Ltd

a

b

Figure 1. Track on asphalt concrete pavement on the road: a-Baghdad-Diyala; b-Baghdad-Kut.

The proposed repair should be increasing structures of road pavements and increasing the strength and shear resistance of pavements in which we can get a smooth surface with a high rate of.

2. Materials and methods

In Russia, research in the field of asphalt concrete reuse was carried out by V.M. Goglidze, A.M. Aliev, B.A. Asmatulaev, G.S. Bahrach, JI.B. Bilay, S.F. Balashov, E.B. Ilyev, N.V. Gladyshev, A.P. Lupanov, A.V. Rudensky, B.S. Gmyrya, V.F. Poloiko, G.N. Kiryukhin, V.V. Silkin et al. Despite a wide range of domestic studies, technologies for the regeneration of old asphalt concrete in Russia are still at the initial stage of development and have not received such widespread use as abroad, where the volume of use of this material is up to 20% of the total amount of asphalt concrete mixtures produced. In Iraq, these technologies are still practically not used.

A comparison between the CR of asphalt concrete pavements and the traditional repair methods has been made, and cold repair technique showed better results based on cost and time of maintenance. The analysis showed that for the conditions of Iraq, taking into account the main types of road damage, It is more convenient to use cold regeneration technology where this technique does not require additional equipment and can be implemented in a short time. In addition, carrying out work without heating the material leads to minimal damage to the environment and organometallic materials.

As a result of the cold processing of asphalt concrete, an asphalt-granule-concrete mixture (AGC-mixture) is formed, which has its own structural features.

Theoretical studies of the structure formation of asphalt concrete are devoted to the works of I.A. Ryb'sva, B.I. Ladygin, L.B. Gesenzvey, I. B. Gorelsheva. I.V. Koroleva, V.A. Zolotareva, G.K. Xunyi, A.V. Rudensky and others. The results of the reaches showing that the properties of asphalt concrete as a highly concentrated dispersed system are highly determined by the properties of the liquid phase, represented by bitumen.

Unlike hot a / b mixtures, where the dispersed medium is represented by bitumen, in the AGC-mixture, the dispersed phase is represented by cement paste.

Crystallization contacts, which are formed during hardening of cement, which are introduced together with water and form a strong framework or paste, play an important role in the structure formation of AGC (Figure. 2). The required amount of cement paste depends on the total surface of the granular and the thickness of the paste film, which also depends on the size of the granules.

IOP Conf. Series: Materials Science and Engineering 1159 (2021) 012001 doi:10.1088/1757-899X/1159/1/012001

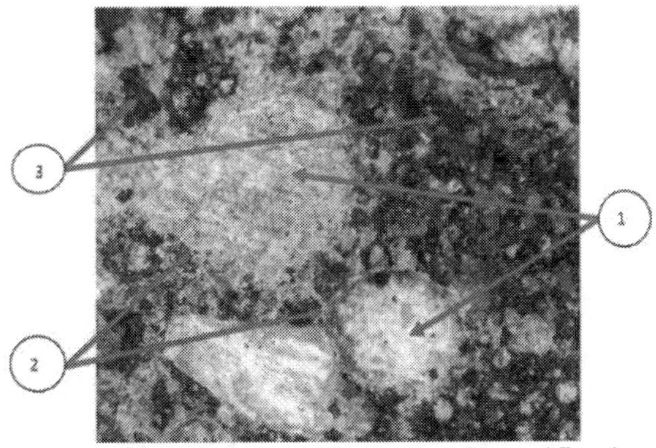

Figure 2. Structure of asphalt granular concrete with cement (magnification 30x): 1-grain AG; 3– a film from cement hardening; 3 – bitumen.

The volume of cement paste for the AGC mixture can be calculated with the following expression [6]:

- P = ∮ ∑ δi * Si * Pi

Where:

P: Paste volume,

∮: Density of cement paste kg/m^3.

δi: Thickness of the paste layer, m.

Si: The specific surface area of the fraction, m^2/kg.

Pi: the content of grains of the fraction,%.

To determine the value of Si, one can use the data of I.V. Koroleva [7] on the specific surface area of particles of different fractional composition. See figure 3.

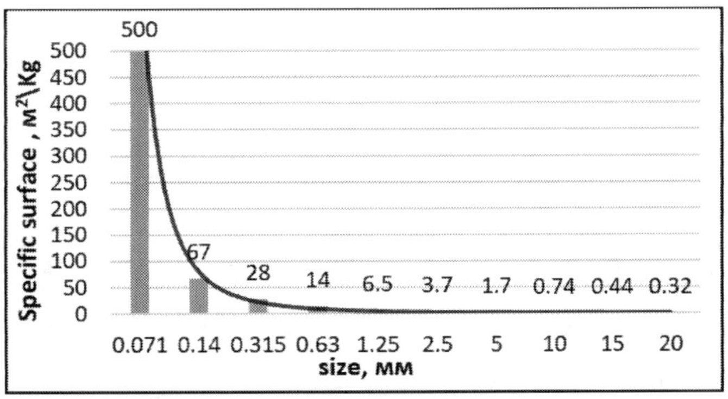

Figure 3. Dependence of the specific surface area of materials on the size of the fraction.

The performed calculations showed that the volume of cement paste for the formation of a uniform technological layer in the AGC mixture is 0.05 kg / m3 or 100 kg / t. With a water-cement ratio w / c = 0.5, the cement content per 1 ton of granules is 0.03 kg or 3% of the mass of the granules.

3. Results and Discussion

Based on the results of the theoretical studies, the experimental work aimed at selecting the compositions of asphalt granular concrete with the highest density and strength characteristics with an optimal amount of cement.

The experimental works were performed in the laboratory of Dorexpert LLC.

As a starting material, granules of old asphalt concrete was used, obtained as a result of milling of asphalt concrete pavements with a WR 2500 cutter from Wirtgen during repair work at facilities in Moscow. Granulometric curves are shown in Fig. 4.

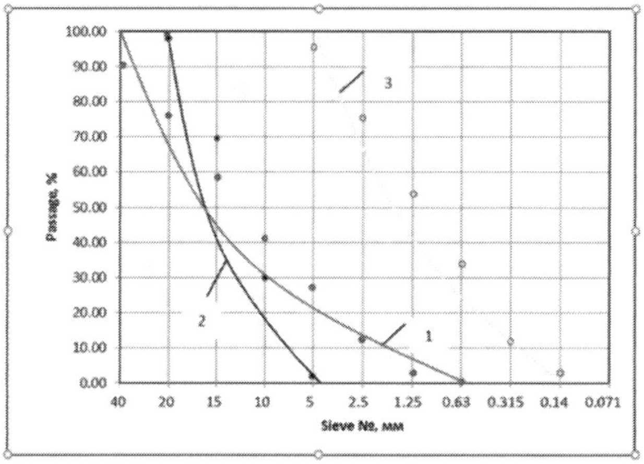

Figure 4. Aggregate composition of old asphalt concrete granules:1-fraction 0/40; 2-fraction 5/20; 3-fraction 0/5.

The work considered various options for the composition of AGC-mixtures based on granules of old asphalt concrete with the addition of cement in the range from 0.75 to 3%. Two types of granules were used to prepare AGC mixtures. Granules with a size of 0-40 mm (100%) obtained after milling without crushing and fractionation and granules with a size of 0-20 mm obtained after crushing, screening and mixing fractions of 5-20 mm and 0-5 mm, which ensured the production of denser AGC- mixtures. Table .1. is Illustrates the compositions of the mixtures.

Table 1. Compositions of AGC mixtures with granules 0-40 mm and granules 0-20 mm.

Materials	Comp. № 1	Comp. № 2	Comp. № 3	Comp. № 4	Comp. № 5
A) Granules of old asphalt concrete (100% fraction 0-40 mm) B) Granules of old asphalt concrete (60% fraction 5-20 mm and 40% fraction 0-5 mm)	100%	97.75%	95.5%	93.25%	91.0%
Cement м500	-	1.5%	3%	4.5%	6%
Water	-	0.75%	1.5%	2.25%	3.0%
Total:	100%	100%	100%	100%	100%

The test results of the main indicators of the properties of AGC concrete from uncrushed and crushed granules at the age of 7 days are shown in Figures, 5,6,7 and 8.

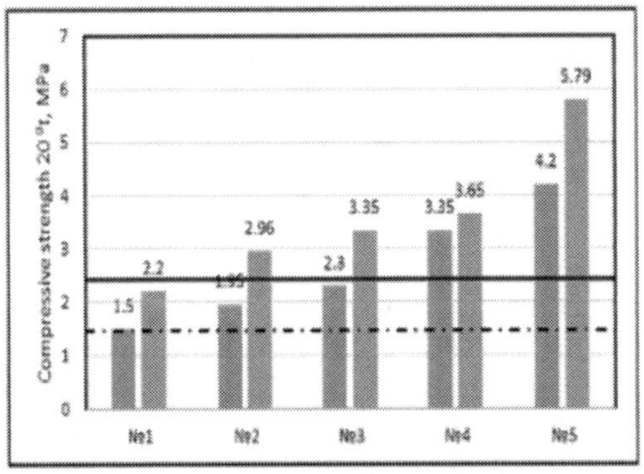

(a)

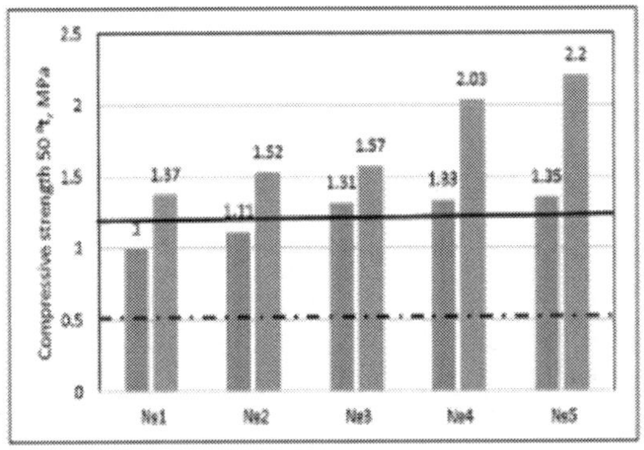

(b)

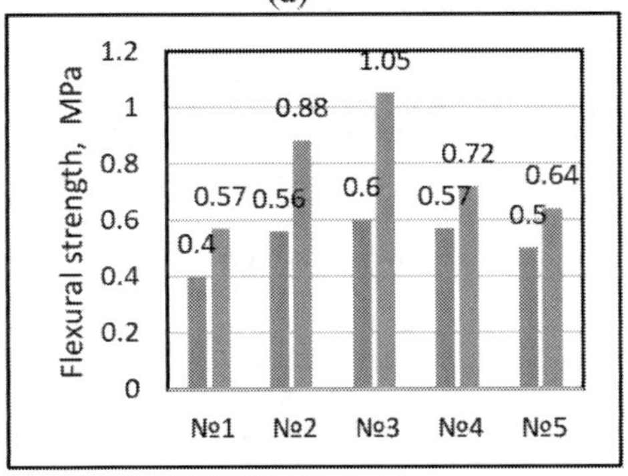

Figure 5. Value of the flexural strength at a temperature of 20 ° C after 7 days.
No crushing; After crushing and fractionation.

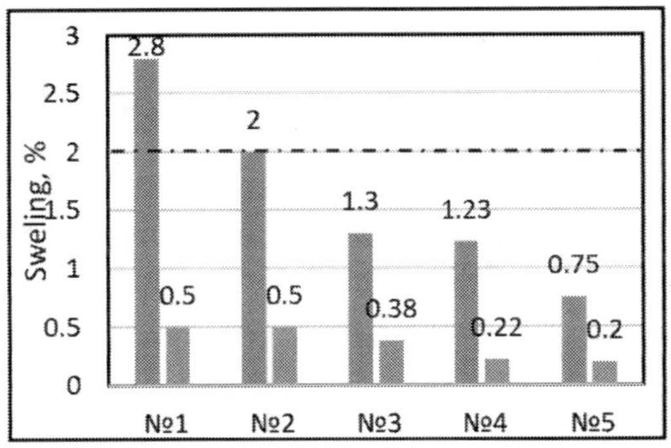

Figure 6. Swelling of samples at the age of 7 days.
No crushing; After crushing and fractionation.
The requirements of (STO NOSTROY 2.25.35-2011).

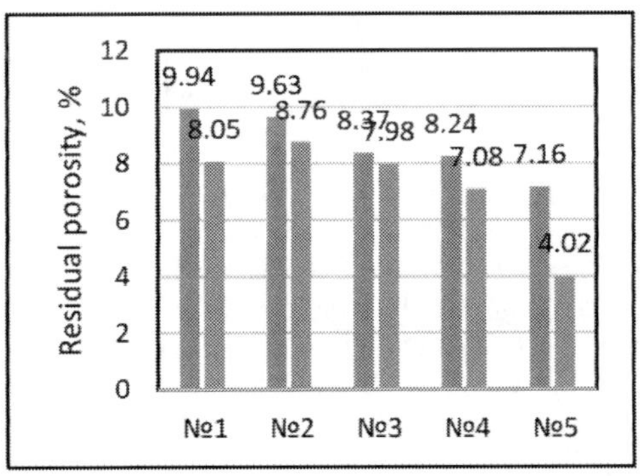

a

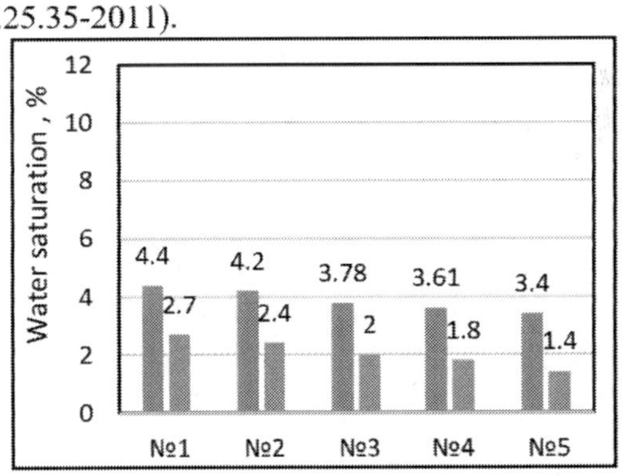

b

Figure 7. Comparison of the main indicators of the properties of AGC at 7 days: a- Residual porosity of samples; b- Water saturation of samples.
No crushing; After crushing and fractionation.
‒ · ‒ · The requirements of (STO NOSTROY 2.25.35-2011).

Summarizing the data obtained, we can conclude that, based on granules, it is possible to get mixtures for the construction of bases and lower layers of coatings, providing the required properties according to STO NOSTROY 2.25.35-2011 [9]. The properties of granular asphalt concrete can be modified by adding cement and aqueous Additives plasticizer through the grinding and the crashing the mixture to fractionated granules. Studies have shown that asphalt granule concrete at the age of 7 days using fractionated granules of 0-5 and 5-20 mm has strength at different temperatures higher than granular asphalt concrete with granules 0-40 mm without crushing and screening.

Moreover, with similar values of the properties of the studied compositions of granular asphalt concrete, the highest strength at 20 and 50 ° C, which characterizes its shear resistance, as shown by samples with the addition of cement. The use of AGC from crushed granules with the addition of 1.5-3% cement [8], in terms of its indicators, provides, and in terms of strength indicators, exceeds the requirements of GOST 9128-2013[10] for hot coarse-grained dense asphalt concrete.

From the given data, it follows that preliminary crushing and dosed granulate by fractions significantly increases the density of AGC. These reduce the water saturation and porosity of the samples. Crushing the granulate and increasing the amount of cement increases the flexural strength. However, increasing the cement is slightly decreasing the deformation due to an increase in the rigidity of the material. At the same time, the obtained values fit into the requirements of GOST 9128-2013 concerning coarse-grained asphalt concrete.

The use of fractionated granules significantly reduces the swelling rate. As for the indicator of water resistance and long-term water resistance, as can be seen from the data obtained, crushing and fractionation of the granules does not have a noticeable effect on this indicator.

The calculations help to assess the economic efficiency of the construction work using the cold regeneration technology in the asphalt plant for the conditions of Russia and Iraq. The following Figure 9 illustrates the analysis comparing the cost of coating (1m²) with a thickness of 10 cm from AGC mixtures for the conditions of Russia and Iraq.

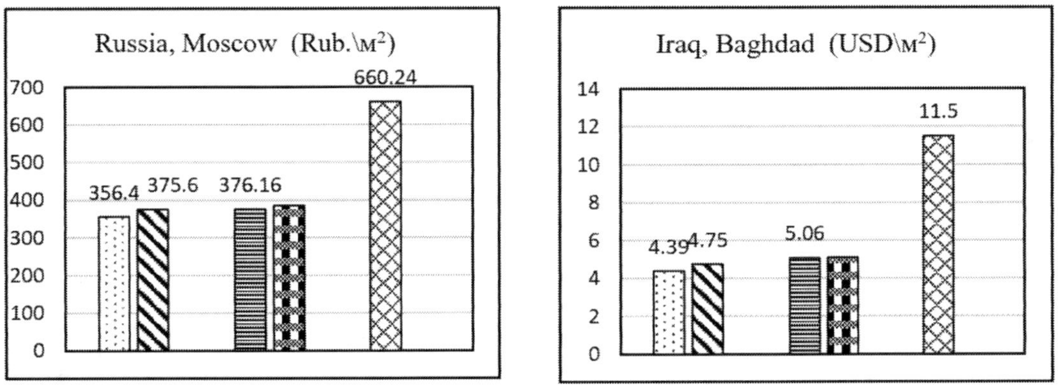

a b

Figure 8. Comparison of the cost of coverage by options: a-cost in Russia, Moscow; b-cost in Iraq, Baghdad. granules (0-40mm) without cement; fractionated granules (0-20mm); granules (0-40mm), 3% cement; fractionated granules (0-20mm) with cement; hot coarse-grain mixture (type B).

The calculations showed that the most economical option is a technology that provides for the device of a coating or base from crushed granules with fractions of 0-20 mm with the addition of cement and a plasticizer. For a layer with a thickness of 10 cm, savings according to the selected option, in comparison with a hot coarse-grained mixture (type B). is 284 rubles. Per 1 m². For Moscow and USD 6.45 for Baghdad.

4. Conclusion

- An analysis of the climatic conditions of Iraq showed that the primary defects of the pavements are associated with the low strength of the pavement structures and insufficient shear stability of the pavements, which requires appropriate adjustments in terms of shear resistance requirements.
- Studied modern methods of road surface repair and various technologies for the regeneration of old asphalt concrete and studying the techniques of hot and cold rejuvenation. The results show that the

cold method with the preparation of mixtures in the manufacture is the most suitable for the conditions of Iraq.

- It is possible to modify the structure of AGC mixtures with adding a certain amount of cement and choosing the appropriate granular size in order getting the optimal AGC
- The experimental studies obtained shows the possibility of improvement in the properties of AGC by optimizing the granules composition of the granules and the selection of the optimal amount of cement.
- Economic calculations performed both for the conditions of the Russian Federation and for the conditions of Iraq have confirmed the effectiveness of the technology of cold processing of asphalt concrete with the addition of cement, Compared to a 10 cm thick hot asphalt concrete pavement, the savings for the selected option is 284 rubles. Per 1 m2. For Moscow and USD 6.45 for Baghdad.

References

[1] Sawsan S 2016 Traffic density and its impact on the high pollution levels in the city of Baghdad *Al-Mostansiryah journal for Arab and international studies* **54(10)** 217-243
[2] Shoman S S ans Al-Karaguli M M 2016 Climatic conditions and the main types of road destruction in the Republic of Iraq. *Jurnal ADI* **4 (10)** 93-101
[3] Al-Karaguli M M 2019 Efficiency of cold regeneration of road surfaces in the Republic of Iraq actual problems of construction, utilities and technosphere safety, *Materials of the VI All-Russian Scientific and Technical Conference of Young Researchers* 22-27
[4] Babaev M G 1984 *Asphalt materials in hot climates* (Moscow) p 192
[5] Nosov V P, ans Rozov Yu N 2001 *Practical guidance for the current repair of asphalt concrete pavements of the urban road network* p 128
[6] Akhverdov I N 1981 *Fundamentals of concrete physics* p 464
[7] Korolev I V, Ageev E N, Golovko V A and Fomenko G R 1984 *Road warm asphalt concrete* (Moscow) p 200
[8] Bakhrakh G S 2020 Influence of powdered additives on the strength of asphalt granular concrete *World of roads* **1(26)** 38-40
[9] STO NOSTROY 2012 Device of bases for road clothing, Part 7: *Construction of foundations using asphalt concrete granules* p 16
[10] GOST 9128-2013. *1998 airfield and asphalt concrete mixtures* p 24

Study of material strength properties for horizontal road marking using innovation equipment

V I Bochkarev[1], **E V Zhustareva**[1], **S V Fedorov**[1]

[1]Moscow Automobile and road construction state technical university (MADI), Moscow, Russia

E-mail: joustareva@mail.ru

Abstract. The article deals with the study of strength properties of materials used for horizontal road marking. The investigations were carried out on the basis of the chair "Construction and Maintenance of Roads" of MADI and NTTs "Katafot". The tests were performed on marking materials that are most common and frequently used in the Russian Federation. Thermoplastics and cold plastics were selected for research. For the tests, up-to-date high-precision equipment (AMPT / SPT installation) was used, which made it possible to take measurements in automatic mode as well as to save and process data. During the experiments, the dependences of the dynamic modulus of thermoplastics and cold plastics on the frequency of load application (loading frequency) and temperature were investigated. The researches resulted in recommendations for the use of materials to arrange the horizontal road marking under various operating conditions. The aim of the study is to improve the quality and durability of road markings as well as to ensure road traffic safety. The current scientific work has relevance to the implementation of the national project "Safe and High-Quality Roads".

1. Introduction

At present, the national project "Safe and High-Quality Roads" is being implemented. The completion date of the project is 2024. The project includes four federal projects: "Road network", "System-wide measures for the development of road sector", "Traffic safety" and "Highways of the Russian Ministry of Defense". One of the main objectives of the federal projects "Road Network" and "System-Wide Measures for the Development of the Road Sector" is to decrease the number of places, where traffic accidents (TA) are concentrated, as well as the number of federal and regional roads at heavy traffic.

To resolve these tasks there is continuous effort to improve the methods and means of traffic organization. Road marking is one of three kinds of technical means to organize information traffic. World experience shows that the cost of road marking is at least an order of magnitude lower than the damage caused by traffic accidents.

In recent years in Russia, the areas of thermoplastic and cold plastic markings on the federal public road network have dramatically increased [1]. As a rule, thermoplastics are used for applying longitudinal lines of road markings. In so doing, marking machines to apply materials by a mechanized mode are used. From a technological point of view, cold plastic is traditionally a more convenient material for performing "manual works", i.e., applying "symbolic" markings using the scotch-stencil method. However, recently, the technologies for cold plastic longitudinal marking by a mechanized mode, which use special aggregates incorporated in marking machines, have been intensively developed.

In spite of the fact that the cost of 1 kg of cold plastic is approximately 3 times higher than the cost of thermoplastic, the share of the cold plastic road marking is increasing. Nevertheless, even if taking into account the facts cited, the cost of material per unit of the marking area for cold plastic is 1.3-2.0 times higher as compared with that of thermoplastic.

In foreign countries, cold plastic is believed to be a high-quality material for road marking. Firstly, cold plastic is a more stable and durable material; secondly, its physical and mechanical features do not

Content from this work may be used under the terms of the Creative Commons Attribution 3.0 licence. Any further distribution of this work must maintain attribution to the author(s) and the title of the work, journal citation and DOI.
Published under licence by IOP Publishing Ltd

depend on temperature. The Federal Road Agency also agrees with this point of view, considering that the use of more durable, but more expensive types of materials will be economically beneficial.

The scientific and normative literature, however, does not provide sufficient information about strength indices of these materials, as well as about the change in these indices depending on temperature. Data on comparative tests of thermoplastics and cold plastics are utterly lacking. So, there is a need to carry out researches on changes in strength properties of marking materials. Reliable information allows selecting properly these materials for the use under different operating conditions. Thus, this will provide durability of road marking and, as a result, highway traffic safety as well.

2. Materials and methods

Comparative tests of thermoplastics and cold plastics were performed on the basis of the MADI chair "Construction and Maintenance of Road" and NTTs "Katafot". Following tasks were set:

- to measure the strength of thermoplastics and cold plastics used for road marking depending on temperature;
- to define the field of application and to give recommendations for using these materials under different operating conditions.

The methods to estimate and compare strength characteristics of thermoplastics and cold plastics for road marking are lacking both in domestic and foreign standard test methodologies [2, 3, 4].

The AMPT/SPT (Asphalt Mixture Performance Tester) plant that is usually applied for asphalt concrete testing under various temperature conditions was used for tests. Thermoplastics and cold plastics are elastic-viscous-plastic materials, so the plant in question was used as a system that made it possible to determine similar properties of materials.

The AMPT/SPT plant (Figure 1) takes measurements of the dynamic modulus that characterizes a material hardness [5].

Three labels of thermoplastics and two labels of cold plastics, which are widely applied for road marking in Moscow as well as on Russian federal roads, were selected for experimental studies. Each material was assigned its code in order to avoid advertising - anti-advertising of manufacturers:

- thermoplastics TP 1, TP 2, TP 3;
- cold plastics CP 1, CP 2.

The AMPT / SPT test requires cylindrical samples 100 mm in diameter and 150 mm in height (Figure 2).

Thermoplastics were heated to the operating temperature advised by the material manufacturer. The material was stirred at operating temperature within the time also recommended by the manufacturer. To avoid strong deformation of samples owing to thermal shrinkage during cooling, the mold was poured layer-by-layer. The pouring was performed in three steps, until each layer cooled down to a temperature of 60-80 °C.

During making cold plastic samples the ratio of the base material to the hardener advised by the manufacture was strictly observed. The material was thoroughly stirred and poured into forms before hardening.

Figure 1. General view of AMPT/SPT plant.

Figure 2. Test samples.

3. Sample testing

Before testing, 6 anchors were glued on the sample to set sensors for reading the sample deformation during loading. Then the sample was placed in a climatic chamber (Figure 3), where it was kept for two hours at different temperatures (+ 4 °C, + 10 °C, + 20 °C, + 30 °C, + 35 °C). After that, the dynamic loading process was started (Figure 4). The test was performed in automatic mode. The sample was loaded at different frequencies (from 0.1 to 25 Hz). Upon completion of the tests, a summarized table with the results of measured values was displayed on the computer. The process was repeated for each temperature level.

The tests resulted in obtaining the values of the dynamic modulus at various temperatures and loading frequencies. Figures 5-7 show the data for thermoplastics. Figures 8-9 demonstrate the data for cold plastics.

Figure 3. Placing of the sample into a climatic chamber.

Figure 4. Startup of the test program.

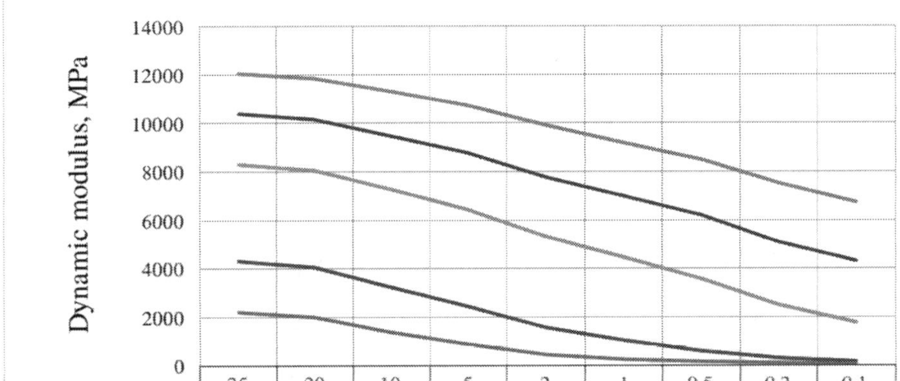

	25	20	10	5	2	1	0.5	0.2	0.1
4°C	12065	11872	11320	10719	9888	9195	8490	7501	6745
10°C	10399	10156	9482	8762	7775	6990	6179	5085	4271
20°C	8309	8035	7250	6425	5301	4434	3582	2509	1771
30°C	4311	4042	3237	2462	1553	1003	609	296	167
35°C	2215	1993	1364	869	441	256	151	76	45

Frequency, Hz

Figure 5. Diagram of the dynamic modulus change in TP 1.

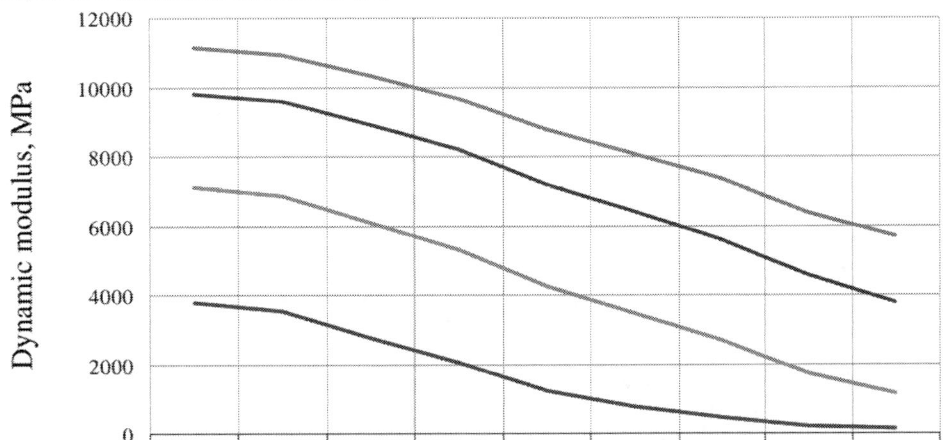

	25	20	10	5	2	1	0.5	0.2	0.1
4°C	11149	10925	10323	9677	8792	8079	7352	6382	5688
10°C	9812	9594	8934	8204	7197	6403	5605	4556	3785
20°C	7120	6858	6098	5312	4260	3458	2677	1744	1161
30°C	3804	3536	2762	2045	1245	784	471	229	135

Frequency, Hz

Figure 6. Diagram of the dynamic modulus change in TP 2.

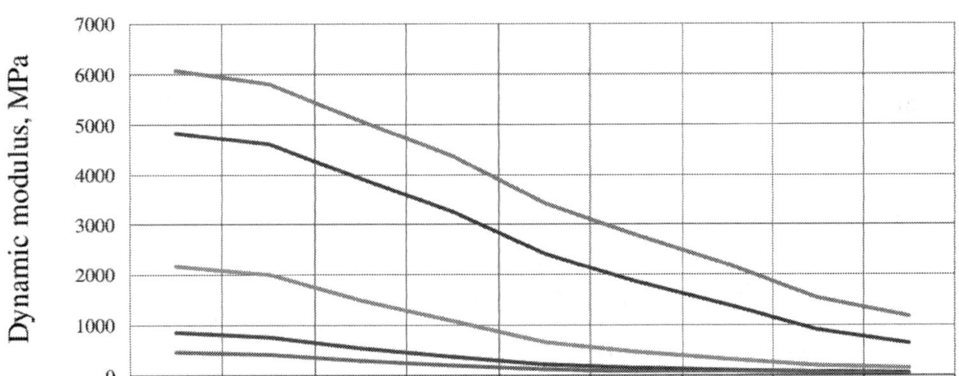

	25	20	10	5	2	1	0.5	0,2	0.1
4°C	6059	5805	5069	4349	3436	2790	2199	1544	1179
10°C	4831	4611	3919	3249	2421	1861	1386	901	646
20°C	2187	2007	1500	1080	671	462	320	202	147
30°C	865	770	527	358	217	151	108	73	55
35°C	473	419	285	194	120	85	63	44	33

Frequency, Hz

Figure 7. Diagram of the dynamic modulus change in TP 3.

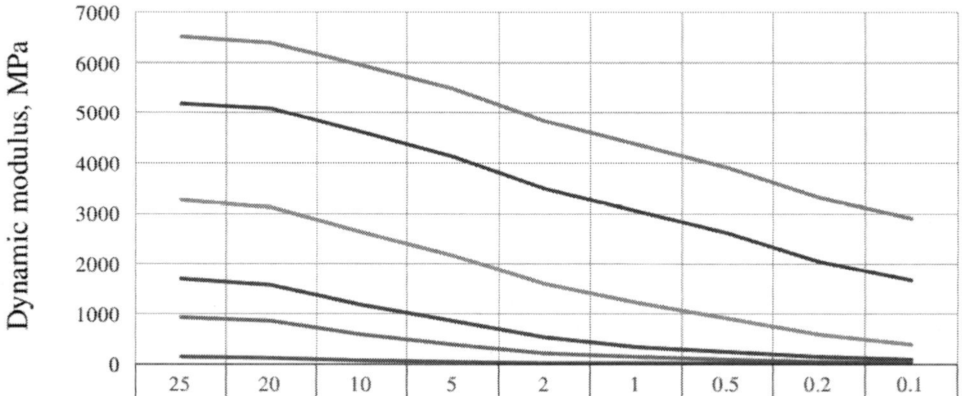

Frequency, Hz	25	20	10	5	2	1	0.5	0.2	0.1
4°C	6521	6397	5937	5469	4841	4374	3906	3315	2903
10°C	5194	5082	4611	4130	3499	3036	2595	2045	1676
20°C	3268	3125	2633	2162	1598	1225	910	585	404
30°C	1699	1570	1183	858	530	354	234	135	89
35°C	944	849	589	396	228	149	100	61	43
40°C	147	123	81	57	35	26	20	15	12

Frequency, Hz

Figure 8. Diagram of the dynamic modulus change in CP 1.

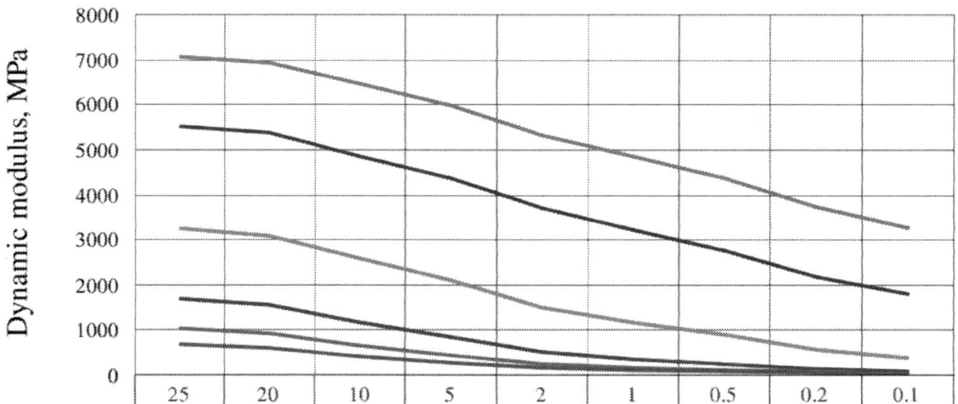

Frequency, Hz	25	20	10	5	2	1	0.5	0.2	0.1
4°C	7064	6945	6467	5981	5331	4860	4370	3737	3288
10°C	5519	5369	4868	4367	3709	3227	2761	2191	1808
20°C	3245	3078	2589	2103	1499	1185	890	560	385
30°C	1693	1561	1171	848	524	351	233	137	91
35°C	1042	938	656	445	259	170	115	70	49
40°C	681	603	406	268	154	103	71	45	33

Frequency, Hz

Figure 9. Diagram of the dynamic modulus change in CP 2.

4. Analysis of test results

The analysis of diagrams of changes in the dynamic modulus depending on loading frequency showed that thermoplastics and cold plastics had the properties of classic elastic-viscous-plastic materials. The

IOP Conf. Series: Materials Science and Engineering 1159 (2021) 012002 doi:10.1088/1757-899X/1159/1/012002

values of the dynamic modulus of such materials depended on the frequency of loading impact: the more frequent the load is, the higher the values of the dynamic modulus will be.

To compare and analyze the properties of test materials the diagrams of the dynamic modulus changes depending on the temperature at a frequency of 10 Hz), which corresponded to the duration of a single load of 0.1 s, were plotted (Figure 10). According to the diagrams, some thermoplastics showed higher values of the dynamic modulus than cold plastics. Thus, it cannot be argued that cold plastics have higher strength properties than thermoplastics.

The tests have demonstrated that there was a significant decrease in the values of the dynamic modulus with the increase in temperature both for cold plastics and thermoplastics. Thus, a generally accepted statement that thermoplastics and cold plastics have a different nature of changes in the strength properties at temperature change has not been confirmed.

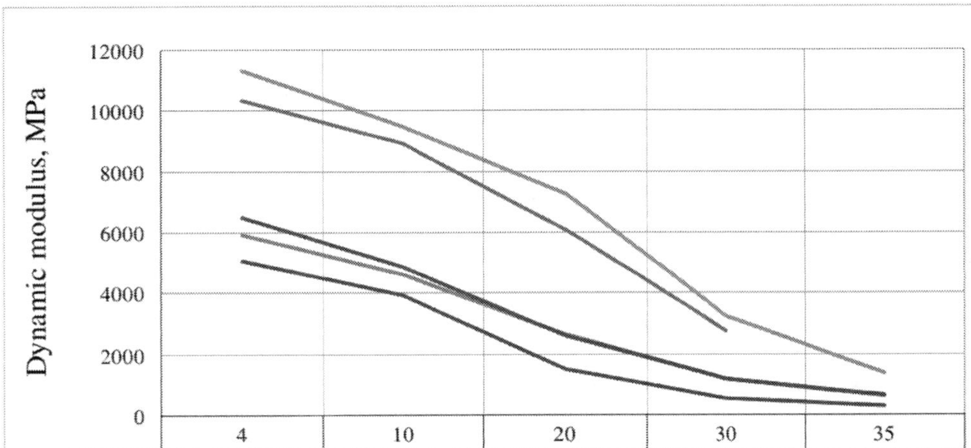

	4	10	20	30	35
CP 1	5937	4611	2633	1183	589
CP 2	6467	4868	2589	1171	656
TP 1	11320	9482	7250	3237	1364
TP 2	10323	8934	6098	2762	
TP 3	5069	3919	1500	527	285

Figure 10. Cumulative diagram of test results at 10 Hz loading frequency.

5. Conclusion

Researches resulted in the following conclusions:

1) there is no significant difference between the properties of cold plastics and thermoplastics under short-term loading of the samples (determination of the dynamic modulus);
2) contrary to well-established opinion, the properties of cold plastics change at temperature change as well: the dynamic modulus decreases 5 times when the temperature changes from +10 to + 30°C;
3) if there is no difference in the properties of materials studied under short-term loading, it makes no sense to use more expensive cold plastic for those lines of road marking where continuous static loads or significant shear loads are hardly probable.

References

[1] ODM 218.6.020-2016 *Guidelines for road marking* (Moscow: Rosavtodor [Federal road agency]), 2016

[2] GOST 32829-2014 *Automobile roads of general use. Road marking materials. Methods of testing* (Moscow: Standartinform), 2015

[3] GOST R 52576-2006 *General use highways. Road marking materials. Test methods* (Moscow: Standartinform), 2007

[4] EN 1871:2000 *Road Marking materials. Physical properties* (Brussels: European Committee for Standardization), 2012

[5] AASHTO T 378 *Standard Method of Test for Determining the Dynamic Modulus and Flow Number for Asphalt Mixtures Using the Asphalt Mixture Performance Tester (AMPT)* (Washington: American Association of State Highway and Transportation Officials), 2017

IOP Conf. Series: Materials Science and Engineering 1159 (2021) 012003 doi:10.1088/1757-899X/1159/1/012003

Technology and principles of asphalt paving

N M Andryukhov[1], **S A Pavlov**[1]

[1.] Moscow automobile and road construction state technical university, (MADI), 64 Leningradsky Prospekt, Moscow, 125319, Russia

E-mail: Andryukhov_nm@mail.ru

Abstract. The use of high quality equipment in the construction of asphalt pavements is one of the ways to get financial profit from the further use of the developed roadbed. Taking into account the peculiarities of delivery of asphalt mixture to the object of construction, its storage, moving into the hopper of the paver, can significantly reduce the phenomenon of segregation of the mixture, both fractional and temperature. Attention paid to the process of surface formation by the paver, docking with arriving dump trucks, prevents the appearance of defects in the laid layer: joints, potholes, irregularities. The analysis of negative phenomena that may arise in the technology of paving, made by the authors, gives the reader an opportunity to understand how technologically precise the process of building roads should be.

1. Introduction

The topics related to the paving of asphalt concrete mixtures are the most relevant in countries with developing road infrastructure. Such countries include the Russian Federation, Canada, Australia. Paving asphalt mixture is not only the working principles of the paver, but also the planning of works, solving the logistical problems, overloading, pre-compaction, fractional and temperature-related segregation.

2. Problems of laying asphalt concrete mixture

The main task of an asphalt paver is to form a layer of asphalt mixture according to the technical specification on the thickness of the layer, texture, smoothness and surface profile.

During the paving operation, it is important to ensure that the four basic criteria are met: mixture tonnage, paving width and thickness, control of parameters.

One of the problems of laying is to ensure the continuous delivery of the mixture to the construction site. This is affected by the productivity of the asphalt plant, the asphalt mixture transport routes, road conditions and the number of dump trucks available.

Asphalt plants can have different output volumes. The main task is to find out how many dump trucks it will take to ensure the constant availability of the mixture in the hopper of a working asphalt paver, taking into account the lack of equipment downtime.

In case of incorrect calculations or other factors, such as equipment failure, there will be a problem with the lack of asphalt mixture. In this case, it will be necessary to increase the time of paving the existing volume of the mixture, either by reducing the speed of paving, or be forced to stop the work in an emergency situation. But the stoppage should occur only when the hopper is filled with the mixture, this is done in order to preserve the temperature of the contact zone in which the next layers of the mixture are connected, otherwise, when the paver stops, a seam is formed (Fig. 1).

Content from this work may be used under the terms of the Creative Commons Attribution 3.0 licence. Any further distribution of this work must maintain attribution to the author(s) and the title of the work, journal citation and DOI.
Published under licence by IOP Publishing Ltd

Figure 1. Seam that occurs when the paver stops

To minimize the formation of seams, an efficient paving mode is used, which uses the next cycle:

1- Paving of a certain length of road;

2- Paver stoppage;

3- Pausing for the maximum time interval, which will not lead to cooling of the mixture;

4- Paving the next length of road.

This mode of operation is maintained until a dump truck with asphalt mixture arrives or until the mixture is finished.

The second unfavourable option is the overabundance of dump trucks. In this case, the problem of the asphalt mixture cooing also occurs. The duration of dump truck downtime with the mixture inside is influenced by weather factors, which significantly reduce the temperature of the mixture: rain, snow and cold weather [3].

Permissible transport range depends on the type of mixture, climatic conditions, the state of the delivery routes to the jobsite. The temperature of hot asphalt mixture at delivery should be within 115 - 155 ° C. For indicative calculations, it is considered that the mixture cools down by 1 ° C for each kilometer of the way or 20 ° C for each hour of the way. Based on experience, in dry hot weather, hot mixture can be transported up to 40 ... 50 km, and in cool weather - up to 20 ... 30 km.

An important problem associated with the delivery of the mixture is segregation, which can be of two types (Fig. 2.):

- Temperature segregation
- Fractional segregation

Figure 2. Segregation of asphalt mixture

Fractional segregation of the mixture is a change in the particle size distribution of aggregates and a change in the binder content in the originally homogeneous asphalt mixture due to separate movements of particles of coarse and fine aggregate in the process of working with the mixture. In the process of transporting the dump trucks move on uneven surfaces, in urban traffic constant braking and acceleration lead to vibration in the body of the dump truck. Denser particles such as gravel, crushed stone, fall to the bottom, and lighter particles such as bitumen rise to the top. Mixture in which segregation has occured needs to be mixed because if you pave with such a mixture, the entire asphalt surface will be heterogeneous, because the lighter particles will be on top of the entire mixture and they will get to the hopper of the paver first. As a result of paving, the mixture will be very fine at first, but with each meter of coating, the granules will become larger and larger.

Temperature segregation is the uneven temperature distribution of the mixture over its volume. When paving the mixture, it is necessary to achieve temperature uniformity, because if the mixture cools down earlier, compaction with a roller will not be effective.

To solve the problem of segregation special loaders are used, these machines are used only on large volumes of work, as they are quite expensive.

3. Asphalt paving planning

Paving parameters: width, length, thickness. Based on these parameters, the volume of mixture per square meter is determined. The thickness of the layer to be paved is adjusted with the paver's slab, and it is attached to a rocker, which is held on two hinges on each side of the machine. A hydraulic cylinder is attached to each hinge and they adjust the thickness of the layer and the angle of attack when paving the mixture.

Figure 3. Asphalt paver: 1-hydraulic cylinder for tamping plate angle adjustment; 2-hydraulic cylinder for tamping plate lifting and lowering

Hydraulic cylinders 1 and 2 are responsible for the plate height above the surface and the angle of attack. When hydraulic cylinders 1 and 2 work in parallel, the lifting of the plate above the surface is adjusted, when one of the hydraulic cylinders is adjusted, the angle of attack is adjusted.

The angle of attack is the angle that determines the change in thickness of the layer to be laid. Usually, the angle of attack is 3-6 mm (Fig. 4).

IOP Conf. Series: Materials Science and Engineering 1159 (2021) 012003 doi:10.1088/1757-899X/1159/1/012003

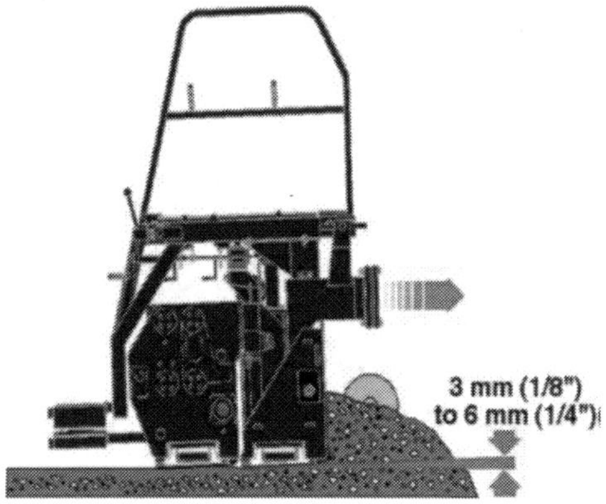

Figure 4. Tamping plate angle of attack

Because the plate is a very massive, strong stresses can occur on the paver during transport on the rocker arm. Retractable rods are used on which the rocker arm is lowered to relieve the load from hydraulic cylinders and the entire structure (Fig. 5).

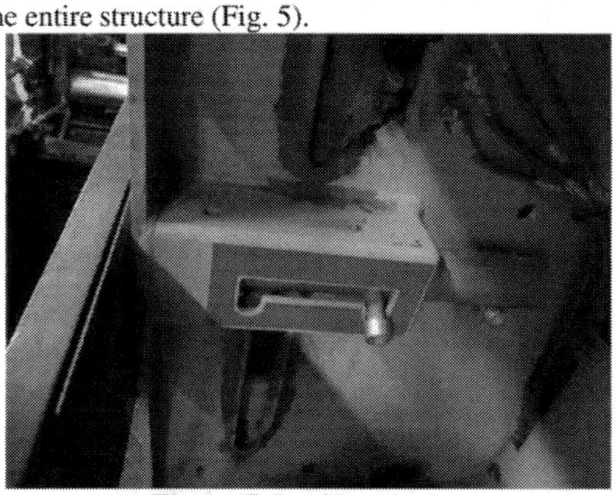

Figure 5. Locking device

The paving length is determined by the distance that the paver has passed.

In contrast to the thickness, the width of the layer is continuously adjustable, as it is necessary to bypass different barriers. There are two consoles installed on the paver's slab, which are responsible for its control; the consoles regulate the screws, the height of the slab's elevation, the angles of attack and the paving width. Paving in difficult places requires two operators on the right and left sides of the machine, each of them is responsible for the left and right sides respectively. The paving width of the mix can be adjusted using the joystick at the bottom of the control panel.

The standard paving widths are 2 - 4.5 meters. At the maximum paving width, screw extensions must be used, which are attached mechanically to ensure that the mixture is distributed over the entire width of the intended paving strip.

The next step is to determine the method of tracking the thickness and direction of travel, using sensors, the simplest of which is the tensioned stringline. Nowadays, both sound and light rangefinders are used; setting up the sensors of these devices is one of the most difficult stages. To improve the paving accuracy, the number of sensors installed on the left and right sides of the special supports is usually increased. When installing a large number of sensors also use a rod, which is usually attached to the sensor on the edges and center, all these measures are aimed at improving the accuracy of the paving.

4. Material feed system

The material feed system consists of a hopper, feeders, distributors and material feed sensors (Fig. 6).

The hopper plays the role of a storage for the mixture, its main property is to increase the cooling time of the mixture, the feed comes from chute conveyors, they are designed to move the mixture from the hopper to the screws. The spreader is a screw and is responsible for distributing the mixture over the width of the layer to be paved.

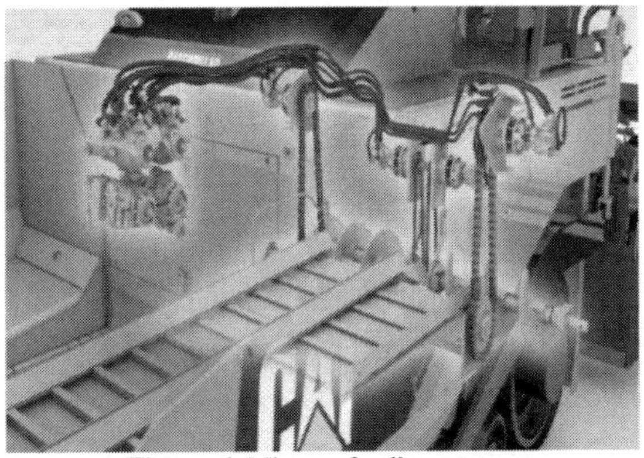

Figure 6. Mixture feeding system

Feed sensors are responsible for controlling the amount of mixture fed to the plate.

Also, on the front side of the paver, in front of the hopper, there are two iron rollers designed to dock the paver to the dump truck.

To ensure the smoothness of the layer to be paved, you must ensure continuous operation of the paver. Any stoppages or impacts may affect the smoothness of the paved surface.

An unfavorable point at which a stop or impact can occur is the discharge of the mixture from the dump truck into the paver's hopper. If the unloading technology is wrong, the impact will occur because most drivers will reverse at high speed until they dock with the paver. This impact causes the paver to oscillate, which in turn causes a seam to form on the paved surface.

The correct method of paving - when the dump truck stops at a short distance from the paver and turns on a neutral gear, and the paver itself approaches the dump truck and starts pushing it. Only at this moment the mix should be unloaded from the dump truck. The paver pushes the dump truck until the mixture in the dump truck body is finished. The first dump truck drives off and the full truck arrives.

5. Conclusion

Asphalt paving is one of the most expensive and technically demanding works, which requires qualified personnel. Savings on repairs and maintenance of the road network can reduce costs by up to 20 percent. Compliance with all the rules of paving technology will reduce the cost of further operation of the road by another 10-15 percent. Analysis of undesirable phenomena arising in the process of paving asphalt concrete pavement with the help of asphalt paver is one of the ways to form a rational approach in road construction.

References

[1] Andryukhov N M, Pavlov S A and Mikhailovskaya V A 2020 Analysis of the destruction of asphalt concrete pavement in the areas of cutting down core samples *Construction and road machines* **9** 8–12

[2] Shestopalov K K, Morozov R V 2020 Coordination of parameters of the dump truck with bottom unloading and the reloader *Truck* **7** 28–31

[3] Kazakov O Yu 2020 Improvement of the roller compaction technologies in the squad for high-speed asphalt laying *21st century: basic science and technology* **2** 19–25

[4] Antonova V V and Ushkov A V 2019 Influence of the temperature regime on the quality indicators of asphalt concrete mix laying *Car. road. infrastructure* **2 (20)** 5

[5] Kustarev G V, Ushkov A V, Pavlov S A 2017 High-speed construction of road surfaces *World of roads* **102** 101–102

The influence of the water flow on the stability of the coastal slope and adjacent transport constructions

O Ya Maslikova[1], **I I Gritsuk**[1,2,3]

[1]Russian Academy of Sciences Water Problems Institute (IWP RAS), 3, Gubkina Str., Moscow, 119333, Russia
[2]Peoples' Friendship University of Russia (RUDN University), 6, Miklukho-Maklaya Str., Moscow, 117198, Russia
[3]Moscow Automobile and Road Construction State Technical University (MADI), Leningradsky prospect, 64, Moscow, 125319, Russia

E-mail: gritsuk_ii@pfur.ru

Abstract. The mechanical behavior of the frozen soil is dynamic and complex. When formulating the strength-to-deformability ratio, it is important to understand the characteristics of the soil, which is being estimated. At the same time different soil samples that have analogical physical characteristics in the unfrozen state may behave differently after freezing and during thawing, as frozen relict and seasonally thawed soils behave differently. All this requires a number of laboratory researches, which will then allow to determine the soil parameters for a specific project. In the hydraulic laboratory, experiments were conducted to study the strength of the different samples of frozen soil under the influence of the water flow and the same load.

1. Introduction

In recent years, attempts have been made to create classifications based on the division of the hydraulic structures located in riverbeds and floodplains into active and passive ones and the degree of their impact on the channel processes. An example is the classification of Snischenko [1]. The first category includes: dams, bridge crossings that overlap floodplains, weaning of runoff from rivers and flooding of rivers during the transfer of runoff, and the second — river embankments hundreds of kilometers long, which completely separate the riverbed from the floodplain and radically change the channel processes of these river sections. Objects such as bridge crossings that overlap floodplains change channel processes on long river sections.

It should be emphasized that channel processes react most quickly and acutely to anthropogenic impact, and the degree of this impact can currently be so great that it radically changes the channel process on a water object.

Unfortunately, the methods of calculations and forecasts of channel processes have not yet been developed sufficiently.

All types of anthropogenic impacts can be conditionally divided into three groups:

- the impact on the climate;

Content from this work may be used under the terms of the Creative Commons Attribution 3.0 licence. Any further distribution of this work must maintain attribution to the author(s) and the title of the work, journal citation and DOI.
Published under licence by IOP Publishing Ltd

IOP Conf. Series: Materials Science and Engineering 1159 (2021) 012004 doi:10.1088/1757-899X/1159/1/012004

- a set of activities in the river basin;
- various types of constructions and activities in riverbeds and floodplains.

The first group has an indirect effect on runoff and other hydrological parameters through precipitation, sediment flow to streams and rivers, and other factors. More specific is the anthropogenic impact on river basins, as well as the impact of hydraulic constructions and water management measures in riverbeds and floodplains on the hydrological characteristics of a water object.

Various researchers have repeatedly noted that a water object is a self-regulating system. That means that a change in some hydrological characteristics in one place (for example, embankments, ditches in the floodplain, riverbed constraint due to the building of bridges) leads to a change in the balance of flow and sediment downstream, or even to significant washouts and niche formations, which clearly affects the stability of the nearby constructions. Traditionally, frozen rocks in the cryolithozone were considered to have stability comparable to hard rock, and such assumptions were used in the building of the hydraulic constructions and bridge crossings. However, the climate warming and the associated with it processes of the degradation of frozen soils have forced to review the basic construction standards. In [2] it was noted that the frozen soil during thawing loses its strength properties to a degree even less than the similar structure soil in the middle latitudes.

In recent years, new problems related to the degradation of permafrost initiated by an increase in soil temperature have emerged. Slope stability problems have been reported in mountain environments [3], and a new impact on the Arctic infrastructure is expected [4] and coastal zones [5] with the consequences of climate change [6].

2. The purpose of the work

The observed climate warming requires a review of the standards of construction near rivers and other water objects, since at high loads in low-strength soils there is a high risk of mobility of thawing rock and collapse of construction structures, which is fraught with catastrophic consequences. For the tasks of design in the conditions of the cryolithozone, it is necessary to know the behavior of frozen soils under various changing conditions. In this regard, new requirements related to the degradation of permafrost arise and, as a result, the criteria for strength and stability of soils suitable for construction are changing. The purpose of this work is to study various soil samples for stress resistance under thermal (natural thawing) and mechanical (water flow influence) influences.

3. Materials and methods

Physical modeling was performed on a hydraulic circulation tray with the following parameters:
- length – 4.5 m, of which 1.5 m was selected as the working area for installing models;
- width – 0.6 m;
- depth – 0.2 m.

To create a model close to natural conditions, 0.2 mm size sand was placed on the bottom of the tray. The samples of beforehand (several days in the chamber at a temperature of -18°(C)) frozen soil of 5 different physical characteristics were laid in such a way that they were embedded in the river bank of the water stream. On the each sample the load of the same shape and weight (6.21 kg each) was put. This is shown in Figure 1.

The width of the stream after the model was laid was 0.35 m. The depth was 0.04 m, which was equal to half the height of the models. Herewith the surface speed of the flow was 0.2 m/s.

The shapes for all samples were identical and had the following dimensions – 0.28 m x 0.18 m x 0.08 m (L x W x H). Before laying the models, they were kept for several days in the freezer at a temperature of -18°C until completely frozen. The soil models differed in the following characteristics:

1. Moistened sand (particle size 0.2 mm.);
2. Moistened peat in equal parts with sand (particle size 0.2 mm.);
3. Moistened peat;
4. Peat with sand (particle size 0.2 mm.) of natural humidity;
5. Peat of natural humidity

Figure 1. The laboratory experiment.

4. The laboratory experiment

Within a few minutes after the experiment started, the direct influence of the water flow could be noticed, which was expressed in the erosion of the submerged in water part of the sample (Figure 1). In [7] it was shown that thawing occurs slightly faster than the erosion itself. For different types of soil, the value of the thermal conductivity coefficient increases with the increasing of the ice content. All things being equal, the rate of thawing increases in the direction: clay – sandy loam – sand.

Also in [7] the measurements of thawing and erosion for a time were made, and the graphics of various freezes soil thawing and the erosion graphics were plotted for each experiment depending on time. The resulting graphs are uniquely interpolated by a logarithmic dependence (Figure 2).

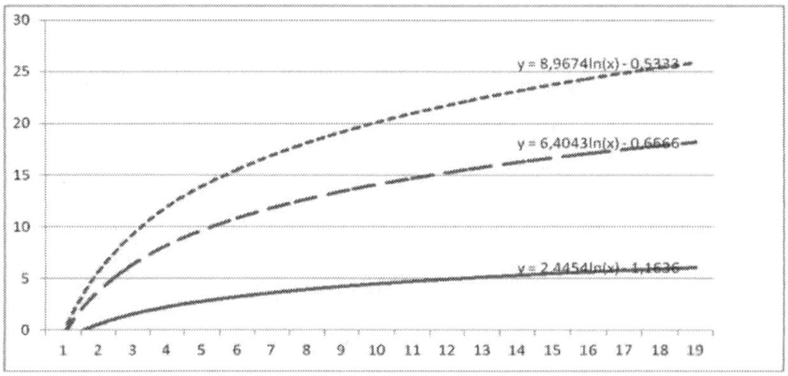

Figure 2. The movement of the thawing front (logarithmic approximation) – dotted line – for wetted soil, point – for natural wet soil; solid line – the movement of the erosion front.

IOP Conf. Series: Materials Science and Engineering 1159 (2021) 012004 doi:10.1088/1757-899X/1159/1/012004

The graphs show that the time of the beginning of erosion does not coincide with the time of the beginning of thawing, thawing begins earlier, and at the same time thawing is about twice as fast. Thawing of various soils, such as frozen natural soil (analogous to the above-water slope) and frozen wetted soil (analogous to the underwater part of the slope), differ by about one and a half times.

Only the area in the water is subject to erosion. After the start of erosion, the upper, above-water part of the slope begins to sink and collapse. In our experiments, the load-bearing capacity of the soil models during thawing and the influence of water flow gradually decreased, which resulted in a collapse in the following sequence: 4→5→1→2→3. Figure 3 shows the load on the sample that showed the highest strength characteristics, and Figure 4 shows the state of the riverbed at the end of the experiment after the impact of the flow and loads on the thawed areas.

Figure 3. The gradual collapse of the load in the thawing of the framework process.

Figure 4. Deformations of the coastal slope under the influence of the flow and load.

5. Results and discussions

As a result of the experiment, moistened peat showed the greatest strength characteristics. This is due to the fact that the thawing of the contained in it ice occurs much slower than the thawing of the ordinary peat with high porosity. When such a sample is placed in the stream, water fills the pores and washes out the substance, and the pores of the above-water part are filled with ambient air, which also contributes to rapid defrosting of the sample.

In second place in terms of stability was moistened peat in equal parts with sand, which also can be explained by the cementing effect of frozen moisture. This sample thawed more slowly than a similar sample of natural humidity, which is also explained by filling the pore space with water or air. Moistened sand was held only by the adhesion of ice, and at thawing it immediately crumbled, which is quite obviously due to its loose properties.

Thus, the strength properties of various soils are described by many factors, including porosity and ice content. In our case, it was not possible to obtain relict ice, which is found in the cryolithozone, but we used an analog of the seasonally thawed layer of soil. This work is a continuation of previously started studies of deformations of the coastal slopes of Northern rivers, where the dependence of collapses on flow rates, the presence of snow cover, rain, and the direction of the coastal slant was established [8] as well as studies of loads on the frozen ground during thawing [9].

According to [10], the degree of thermal shrinkage of perennially frozen rock during thawing is determined by the formula:

$$\delta = (\rho_{d,t} - \rho_{d,f})/\rho_d \,,$$

where $\rho_{d,t}$ is density of dry (skeleton) thawed soil that thawed under the design load, g/cm^3;
$\rho_{d,f}$ is density of dry (skeleton) soil in the frozen state, g/cm^3.
In the classification of permafrost soils they are divided into four categories depending on the degree of thermal shrinkage in the base:

1. non-shrinking, durable $\delta < 0.03$;
2. low-shrinking, not durable enough $0.03 < 8 < 0.1$;
3. shrinking, soft $0.1 < \delta < 0.4$;
4. high-shrinking $\delta > 0.4$, and also containing underground ice in the upper horizons.

The condition of thermal stability of permafrost: the depth of freezing should exceed the depth of seasonal thawing by an average of 10-30% or more, where K_{ts} is the coefficient of thermal stability:

$$K_{ts} = \frac{H_{fr}}{H_{th}} \geq 1.1 - 1.3$$

Load calculation during hydrotechnical construction in the cryolithozone is a complex task and requires a special approach in each case, since it is impossible to make a single model due to the variety of soils and their freezing history. Soil typification does not provide a clear picture at subzero temperatures, since the water-ice-water transition in the soil changes the behavior of rocks, and the unpredictability of the climate changes affecting this process significantly complicates the task. However, this issue requires further research.

6. Conclusion

- The main suggestion is not to build roads on floodplains (the part of the river valley that is above the riverbed and is flooded during high water or during floods). This also relates to wetlands.
- When designing bridges, observed flood data should be taken into account, meaning that bridges should not be turned into dams. Take into account changes in the type of the riverbed processes and general erosion that occurs during the periods of flood runoff.
- Apply a method for calculations that takes into account both the mass transfer between the riverbed and the floodplain, and the effect of interaction between the riverbed and floodplain flows.
- Keep in mind that the accumulation of the sediment in front of the bridge leads to an increase in the erosion of the riverbed outside the bridge. This leads to coastal deformations and destruction of the coastal anchorages.

Acknowledgments
This study was supported by state program № 0147-2019-0001 (registration № AAAA-A18-118022090056-0). The experiments with frozen grounds in laboratory flume were supported in part by RFBR, projects 18-05-00178, 20-08-00807.

References

[1] Snischenko D V, Mecerlyanc G G 1983 Development of the channel process in the areas of excavation of river alluvium *Channel flow dynamics, Gidrometeoizdat.* Vol 83

[2] Scott K M 1978 Effect of Permafrost on Stream Channel Behavior in Arctic Alaska *Geological Survey Proffessional* Paper 1068, Library of Congress Catalog – card No 78-600058

[3] Gruber S, Hoelzle M, Haeberli W 2004 Permafrost thaw and destabilization of Alpine rock walls in the hot summer of 2003 *Geophysical Research Letters* **31** Art. No. L13504 12004

[4] Phillips M, Springman S M, Arenson L U, Balkema A A 2003 Climate change and possible impact on Arctic infrastructure *Eighth International Conference on Permafrost* (Zurich, Switzerland) 461–466

[5] Demers D, Leahy D et al. 2004 Forbes D L: Climate-change impacts in the coastal zone: Implications for engineering practice *Géo Québec 2004: 57th Canadian Geotechnical Conference* (Québec City, QC, Canada) 7A 10–17

[6] Nelson F E, Anisimov O A, Shiklomanov N I 2002 Climate change and hazard zonation in the circumarctic permafrost region *Natural Hazards* **26** 203–225

[7] Debolsky V K, Maslikova O Ya, Ionov D N, Gritsuk I I, Dzhumagulova N T 2017 Laboratory study of the influence of the flow on thawing of the underwater slope and the rate of coastal erosion in permafrost conditions *RUDN Journal of Engineering researches* **2** 182–192

[8] Debolsky V K, Gritsuk I I, Ionov D N, Maslikova O Ya 2018 Blurring of frozen banks of northern rivers depending on the direction of the coastal slope *Vestnik MGSU* **13** Vol 9 1112–1124

[9] Maslikova O Ya, Debolsky V K, Ionov D N, Gritsuk I I, Sinichenko E K 2020 Laboratory research of thawing of frozen soils composing a riverbed *International conference "Engineering Systems - 2020"* RUDN Vol 1 66–74

[10] SNiP 2.02.04-88 *"Bases and foundations on permafrost soils"*

Semi-Markov modeling for assessing reliability of road construction machines in the process of their operation

V K Fyodorov

Moscow Automobile and Road State Technical University (MADI) Leningradsky ave., 64, Moscow, 125319, Russia

e-mail: kaskadf@yandex.ru

Abstract. The article provides a theoretical description of the processes of changing the technical state of road-building machines to assess their reliability at the stage of operation based on semi-Markov models. Reliability is the most important property of machines and characterizes their ability to perform their functions during normal operation. Reliability covers all stages of the life cycle of equipment-its design, manufacture, and operation. The operation stage is the longest in time and the costliest to ensure reliability. This explains the relevance of developing a mathematical model of operational reliability of road-building machines. During the operation of technical objects, their technical condition changes under the influence of objective and subjective factors. As a result, quantitative indicators that characterize the technical condition of machines vary widely and are random in nature. Road-building machines may be in one of the following technical states during operation: perfect, imperfect, upstate, down state. At the same time, a perfect machine is always upstate, and an imperfect one can be either upstate or down state. The transition from one technical state to another is random and can be considered instantaneous with a sufficient degree of accuracy. The probability of any technical state in the future depends only on the state of the process in the present and does not depend on when and how the process was in this state. This makes it possible to use the theory of semi-Markov processes when formalizing the process and developing a mathematical model. Based on semi-Markov modeling, the article shows the sequence of constructing a graph of states and transitions from one state to another for a road construction machine during operation; matrices of transitions and probabilities of transitions from one state to another; systems of differential equations for determining the limiting probabilities of finding a machine in a particular technical state during operation. The developed method of semi-Markov modeling allows estimating the operational reliability of equipment with the reliability allowed by probability theory.

1. Introduction

The complex property of reliability [1] characterizes the ability of machines to maintain their functional properties during normal operation. It is laid down in the design of equipment, ensured during its production and supported during operation. Thus, reliability covers all stages of the machine life cycle.

The longest stage of the life cycle is the stage of operation, including the use of machines for their intended purpose (main purpose), transportation, storage, installation and adjustment (if necessary), as well as maintenance and repair (M&R) [2]. The operating stage is not only the longest in time, but also the costliest to ensure dependability. So, over the entire period of operation, the costs of M&R are several times higher than the cost of a new machine: for cars and construction machines - 5-6 times, for aircraft - 4-5 times, for machine tools - 3 times.

Content from this work may be used under the terms of the Creative Commons Attribution 3.0 licence. Any further distribution of this work must maintain attribution to the author(s) and the title of the work, journal citation and DOI.
Published under licence by IOP Publishing Ltd

IOP Conf. Series: Materials Science and Engineering 1159 (2021) 012005 doi:10.1088/1757-899X/1159/1/012005

Therefore, the development of a mathematical model for assessing the reliability of road construction machines at the operation stage is an urgent task. The solution to this problem allows us to formalize the processes of changing the technical state of machines, to develop organizational and technical measures to maintain their operability and to ensure the development of proposals for improving dependability in the design and manufacture of equipment.

2. Materials and methods

During operation, the technical condition of machines changes under the influence of objective and subjective factors [3]. Objective factors include: environmental impact (temperature, humidity, dustiness), natural wear and tear, corrosion and aging of structural materials, their durability etc. Subjective factors are the qualifications of drivers, the choice of operating modes etc. As a result of the impact of a combination of factors, the quantitative parameters characterizing the technical condition of a machine change over a wide range, and these changes are random.

Road construction machines belong to the class of recoverable technical objects, and during operation they can be in one of the following technical states [1] – perfect S_p, imperfect S_{ip}, up state S_{us}, down state S_{ds}.

The perfect state is a state of an object in which it meets all the requirements established in the documentation for it. The imperfect state means that the object does not meet at least one of the requirements established in the regulatory and technical documentation. The up state is the kind of state, in which the machine is capable of performing its functions. Being down state, the machine is not capable of performing at least one of the required functions.

Thus, the set of technical conditions S can be divided into two subsets S_p and S_{ip}, S_{us} and S_{ds}. The following relations exist between these subsets: $S_p \subset S_{us}$, $S_{ds} \subset S_{ip}$, i.e. the subsets S_p and S_{ds} are included into the subsets S_{us} and S_{ip}.

The subsets S_{us} and S_{ip} intersect (the machine may be up state, but imperfect, for example, with a slight oil leak from the unit): $S_{us} \cap S_{ip}$, and they have a common part. The subsets S_{us} and S_{ds}; S_p and S_{ip} are incompatible-orthogonal. That is, a perfect machine is always up state and an imperfect machine is always down state. However, an imperfect machine can be both up state and down state.

In practical operation, it is enough to talk about the up state of the object S_{us}. It is it that determines the ability of the machine to perform the required functions. This allows, with a sufficient degree of error, not to take into account the technical condition S_{ds} separately in the future, by including it in the subset of states S_{us}. In this case, a diagram of possible technical states of a road construction machine during operation and transitions from one state to another can be represented in the following form (Fig. 1).

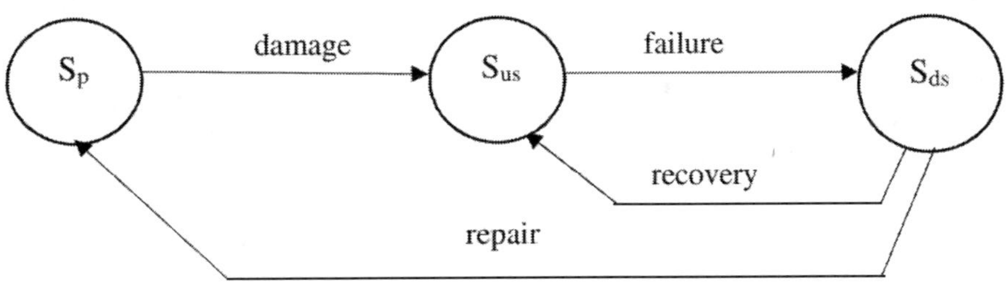

Figure 1. Scheme of possible technical conditions of a road construction machine

Markov processes are widely used in reliability theory. A random process is called a Markov process if the probability of any state in the future depends only on its state in the present and does not depend on when and how the process ended up in this state [4]. If in this case the time spent in any of the states is described by an arbitrary (except for exponential) distribution function, then such a process is defined as a semi-Markov - SM-process. SM- processes are the development and generalization of Markov processes [5].

3. Semi-Markov modeling of the operational reliability of a road construction machine

An SM- process is given both by the matrix function $Q(t)=(Q_{ij}(t))_{m\times m}$, the elements of which are the probabilities of transition from the state i to the state j during the time t, and by the initial distribution vector P(0).

The graph of states and transitions from one technical state to another during the operation of a road construction machine, taking into account Fig. 1, will have the following form (Fig.2).

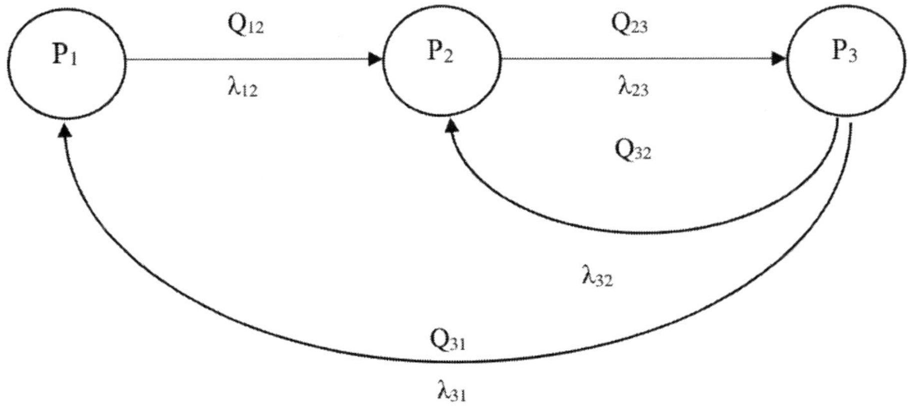

Figure 2. State and transition graph:

P_1 – probability of perfect state; P_2 – probability of up state; P_3 – probability of down state; Q_{12} (λ_{12}) – probability (intensity) of damage; Q_{23} (λ_{23}) - probability (intensity) of failure; Q_{32} (λ_{32}) - probability (intensity) of recovery; Q_{31} (λ_{31}) - probability (intensity) of repair

The evolution of the SM-process with respect to the transition times is described by the following equation of state

$$P_{i+t}=Q_{ij}(t) \times P_j \qquad (1)$$

The moment in time corresponding to the beginning of a certain state is denoted as t, to the end - t_{i+1}. Obviously, the time interval t is a characteristic of the state, since it determines its duration. The random variable t_i can be characterized by the mathematical expectation of the time spent in the state $M[t_i]$.

The process of changing the technical state of road construction machines is characterized not only by the set of states – S - and by the set of random variables of the time of stay in the states - t. These states arise in the process under consideration continuously and sequentially, i.e. at a certain moment of time t_1, t_2, t_3 are "stitched together" and form a process. At these moments, the machine makes transitions from one technical state to another. Assuming that in SM-processes such transitions are carried out instantaneously, one can consider as a quantitative characteristic of transitions not the time of transition to an adjacent state, but the probabilities or intensities (relative frequencies) of transitions λ_{ij} from the i-th to the j-th state.

$$Q_{ij} = \frac{n_{ij}}{\sum_{i=1}^{N} n_{ij}}, \quad \lambda_{ij} = \frac{n_{ij}}{\sum_{i=1}^{N} n_{ij}}, \qquad (2)$$

where n_{ij} is the statistical number of direct transitions from the i-th to the j-th state at a sufficiently large fixed time interval for observing the object – T_o.

N is the total number of states observed in the interval T_o

The matrices of transitions and transition probabilities, taking into account Fig. 2, are presented in Tables 1, 2.

Table 1. Transition matrix

Technical state S	Perfect S_p	Up state S_{us}	Down state S_{ds}	$\sum_{i=1}^{N} n_{ij}$
Perfect S_p	0	n_{ij}	n_{ij}	$\Sigma\ n_{ij}$
Up state S_{us}	0	n_{ij}	0	$\Sigma\ n_{ij}$
Down state S_{ds}	n_{ij}	n_{ij}	0	$\Sigma\ n_{ij}$

Table 2. Transition probability matrix $Q_{ij}(t)$

Technical state S	Perfect S_p	Up state S_{us}	Down state S_{ds}
Perfect S_p	0	Q_{ij}	Q_{ij}
Up state S_{us}	0	Q_{ij}	0
Down state S_{ds}	Q_{ij}	Q_{ij}	0

The matrix $Q_{ij}(t)$ $_{3\times3}$ (Table 2) determines the structure of the local process of transitions of a road construction machine from one technical state to another. In order for this matrix to determine the structure of all transitions to the technical state, it is necessary to show the ergodicity of the process in the observation time T_o section.

An ergodic random process [6] is a special case of a stochastic stationary process. A stationary process is considered ergodic if the normalized correlation function characterizing in this case the transition time tends to 0 (instantaneous transition).

The absolute frequencies or probabilities that the machine will get into one or another technical condition during operation during the observation time T_o are determined by the formula with normalization to 1.

$$P_{ij} = \frac{\sum_{i=1}^{N} n_{ij}}{\sum_{i=1}^{N} \sum_{j=0}^{N} n_{ij}} . \quad (3)$$

The resulting one row table $P=(P_1; P_2; P_3)$ is called an embedded row vector of state frequencies and is the vector P(o) of the initial probability distribution.

The set of the matrix $Q_{ij}(t)$ and the vector P(0) defines an embedded chain of a semi-Markov process. The set of sets {S}, {t}, the matrix $Q_{ij}(t)$ and the vector P(0) determines the whole process of changing the technical conditions of a road construction machine during operation. This process belongs to SM-processes, since the following conditions are met [7]:

The transition matrix $Q_{ij}(t)$ satisfies the conditions of the Markov matrix; the matrix $Q_{ij}(t)$ is square and has a finite order m×m=3×3=9, all elements of the matrix are non-negative;

The values t_i are random and characterized by mathematical expectation $M[t_i]$;

the vector P (0) has only stationary states.

In practice, it is advisable to use the transition intensities λ_{ij} as consistent estimates of the probabilities of transitions from one technical state to another. The limiting probabilities of finding a machine in a particular state are determined using the system of differential equations by A.N. Kolmogorov [8].

To construct differential equations, we use the following mnemonic rule. On the left side, the time derivative of the probability of finding the road construction machine in the i-th technical state at the moment t_i. The right-hand side contains the products of the transition intensity λ_{ij} by the probability of the state from which the arrow leaves. The sign of the product is positive if the arrow enters the state under consideration, and negative if the arrow leaves the state under consideration (Fig.2).

Thus, the number of terms on the right side of the equation is equal to the number of arrows connecting the state under consideration with other states. The system of differential equations is supplemented by the normalization condition $\sum_{i=1}^{N} P_i(t)=1$.

Then the system of differential equations will have the following form.

For the perfect state S_p

$$\frac{dP_1(t)}{dt} = \lambda_{31} P_3(t) - \lambda_{12}P_2(t). \quad (4)$$

For the up state S_{us}

$$\frac{dP_2(t)}{dt} = \lambda_{12} P_1(t) - \lambda_{23}P_3(t)+ \lambda_{32}P_3(t) \quad (5)$$

For the down state S_{ds}

$$\frac{dP_3(t)}{dt} = \lambda_{23} P_2(t) - \lambda_{32}P_2(t)+ \lambda_{31}P_1(t) \quad (6)$$

For a stationary steady state operation, all derivatives $P_i(t)=0$, and the system of differential equations transforms into a system of algebraic equations and with the addition of the normalization condition has the following form:

$$\lambda_{31} P_3 - \lambda_{12}P_2=0; \quad (7)$$

$$\lambda_{12} P_1 - \lambda_{23}P_3+ \lambda_{32}P_3=0; \quad (8)$$

$$\Lambda_{23} P_2 - \lambda_{32}P_2- \lambda_{31}P_1=0; \quad (9)$$

$$\sum_{i=1}^{3} P_i(t) =1. \quad (10)$$

By solving this system of equations, the limiting probabilities of finding a road construction machine in various technical states during operation are determined.

Statistical values for determining the P_{ij} and λ_{ij} values can be obtained quite simply by timely and competent processing of the currently existing operational documentation, such as the "Machine maintenance and repair log", "Incident log" for elevator equipment, etc. With the statistical generalization and accumulation of the information obtained, the accuracy of the calculations will increase.

4. Conclusion

The proposed method of semi-Markov modeling allows, with the reliability admitted by the theory of probability and the apparatus of mathematical statistics, to assess the operational reliability of equipment and to determine the limiting probability of finding a road-building machine in serviceable, efficient and faulty technical states, as well as the probability of transition from one state to another throughout the entire machine life cycle operation. In turn, this makes it possible to solve the following tasks: determination of the readiness of equipment to fulfill its functional purpose; planning stocks of human and material resources for maintenance and repair; formation of car parks taking into account reservation; provision of feedback between the operator-manufacturer-designer to increase the reliability of newly created equipment.

References

[1] GOST 27.002-2015 *Reliability in Engineering (SSNT). Terms and Definitions*
[2] GOST 25646-95 *Operation of construction machines. General requirements*
[3] Zorin V A 2015 *Reliability of mechanical systems* (Moscow: Tsifra) p 380

[4] Dynkin E B 1984 *Probability theory and Markov processes and their applications* (Moscow: Mir) p 238
[5] Korolyuk V S and Turbin A F 1976 *Semi-Markov processes and their applications* (Kiev: Naukova dumka) p 184
[6] Ventsel E S and Ovcharov L A 2016 *The theory of stochastic processes and its engineering applications* (Moscow: Knorus) p 448
[7] Janssen J and Manca R 2006 *Applied semi-Markov processes* (New York: Springer) p 310
[8] Kolmogorov A N 1986 *Theory of Probability and Mathematical Statistics* (Moscow: Nauka) p 534

Hydraulic modeling of surface vortex funnels

T V Artemyeva[1], **A L Zuikov**[1, 2]

[1] Moscow Automobile and Road Construction State Technical University, 64, Leningradsky prospect, Moscow, 125319, Russia
[2] Moscow State University of Civil Engineering, 26, Jaroslavskoe Shosse, Moscow, 129337, Russia

E-mail: zuykov54@mail.ru

Abstract. The peculiarities of the operation of entrance portals for concrete and corrugated metal road pipes, water intake pipelines, spillway structures and hydroelectric power plants include the formation of surface vortex funnels. The formation of surface funnels reduces the water consumption in the pipeline, provokes the destruction of the fastening of the slopes at the inlet, the suction of sludge, ice and debris into the pipeline. It is desirable to avoid these phenomena. The article is devoted to the issues of physical modeling of surface funnels. The aim of the research is to determine the criteria for the breakthrough of the air core of the surface funnel into the pressure pipeline. As a result of research, it was found that the formation of vortex funnels in the upper basin is determined by the structure of the flow in the local area adjacent to the water intake. It is shown that the profile of the free surface of the vortex funnel and its depth on the axis of rotation are determined by the intensity of the circulation generating the funnel. A formula is obtained for calculating the limiting depth of a surface vortex funnel with an accuracy sufficient for engineering calculations. It is shown that in physical modeling according to Froude's defining criterion, the depth of the funnel obtained on the model should be recalculated for a full-scale object with a scale factor $m^{-1.5}$, where m is the linear scale of the model. Another method for obtaining the depth of the funnel on the model, corresponding to its conversion to a full-scale object on a linear scale, is forcing the current velocities by $m^{-3/14}$ times in relation to their value according to the Froude criterion.

1. Introduction

The peculiarities of the operation of the entrance portals of tubular structures (concrete and corrugated metal road pipes, water intake and spillway structures, pipelines of hydroelectric power plants, etc.) include the formation of surface vortex funnels. The formation of surface funnels reduces the flow rate of water in the pipeline, provokes the destruction of the fastening of slopes at the inlet, suction of slush, ice, debris, etc. into the pipeline [1].

The circulating flow in the surface vortex funnel in front of the pipeline water intake (Figure 1) is a widespread phenomenon in nature. It is desirable to avoid it or prevent the burst of the funnel air core into the pressure water conduit, which occurs when the depth of the funnel exceeds the depth of the intake wall of the water intake ($Z_0 > h$).

Today, the methods for calculating this flow are far from perfect, therefore, the conditions for the formation of vortex funnels at the designed objects are studied on large-scale hydraulic models. But the physical modeling of vortex funnels is not an easy task, because such a flow is a multifactorial

Content from this work may be used under the terms of the Creative Commons Attribution 3.0 licence. Any further distribution of this work must maintain attribution to the author(s) and the title of the work, journal citation and DOI.
Published under licence by IOP Publishing Ltd

process determined by a number of dynamic similarity criteria, which cannot be simultaneously ensured for a model and a full-scale object due to their scale incompatibility.

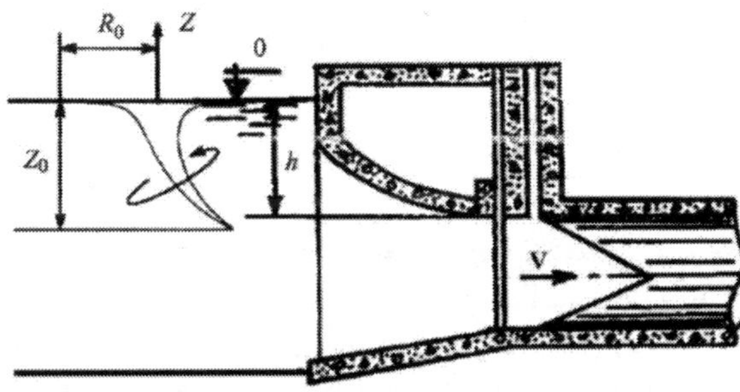

Figure 1. Diagram showing surface vortex funnel.

2. Results and Discussion

In the general case, the hydrodynamics of a steady flow of a viscous fluid in a surface vortex funnel is described by the equation

$$\text{rot}\, V \times V = -\text{grad}\left(\frac{P}{\rho} + \frac{V^2}{2} - \Pi\right) - \varepsilon\, \text{rot}(\text{rot}\, V), \tag{1}$$

where ρ and ε are density and molecular kinematic viscosity of the liquid; *P*, *V* and Π are pressure, local velocity vector and potential of external mass forces.

Moreover, the pressure on the surface of the air core of the vortex funnel can be assumed constant and equal to the atmospheric, then

$$\text{grad}(P/\rho) = 0.$$

The local velocity should be considered as the sum of time-averaged ($\overline{V}$) and pulsation (V') components

$$V = \overline{V} + V',$$

then, when averaged over a finite time interval

$$\text{rot}\, V \times V = \text{rot}\, \overline{V} \times \overline{V} + \text{rot}\, V' \times V',$$

and the potential of external mass forces on the surface of the air core, if the *Z* axis of the coordinate system is directed vertically upward, can be written as

$$\Pi = -gZ - K\sigma/\rho$$

where *g* is the gravitational acceleration; *Z* is current coordinate of the surface of the air core from the horizon of the pool in advance of the water intake; σ is the surface tension; *K* is the curvature of the surface of the air core.

Summarizing the above and taking in the first approximation a model of isotropic and homogeneous turbulence with constant vortex viscosity ε, as a result of normalization Eq. (1) with respect to the characteristic velocity V_0 and dimention L_0, we obtain [2]

$$\text{rot}\, \overline{v} \times \overline{v} = -\text{grad}\left(\frac{\overline{v}^2 + \overline{v'^2}}{2} + \frac{z}{\text{Fr}} + \frac{k}{\text{We}}\right) - \left(\frac{1}{\text{Re}} + \frac{1}{\text{Re}_t}\right)\text{rot}(\text{rot}\, \overline{v}), \tag{2}$$

where $\overline{v} = \overline{V}/V_0$ is the normalized value of the time-averaged local velocity; $\overline{v'^2}$ is the standard of pulsations; $z = Z/L_0$ and $k = KL_0$ are current relative (normalized) values of the surface coordinate and curvature of the vortex air core; Froude number

$$\text{Fr} = \frac{V_0^2}{gL_0}, \tag{3}$$

$\mathrm{We}=\dfrac{\rho V_0^2 L_0}{\sigma}$ is Weber number, Reynolds number and the turbulent analogue of the latter

$$\mathrm{Re}=\frac{V_0 L_0}{\varepsilon} \quad \text{and} \quad \mathrm{Re}_t=\frac{V_0 L_0}{\varepsilon_t}, \tag{4}$$

here ε_t is the vortical viscosity.

It can be assumed that for real hydraulic engineering objects (concrete and corrugated metal road pipes, water intake and spillway structures, pipelines of hydroelectric power plants, etc.), the Weber and Reynolds numbers are so large that they do not affect the formation of vortex funnels, that is, the k/We and 1/Re complexes in the equation motions (2) are negligible. It is also possible to consider these complexes to be insignificant if the flow simulation is performed in the zone of quadratic resistance, when the conditions of self-similarity are provided both by the Reynolds criterion and by the Weber criterion. However, this cannot be said about the $1/\mathrm{Re}_t$ complex, because, as is known [3], ε_t >> ε. As a result, the normalized equation of motion is reduced to the form

$$\mathrm{rot}\ \bar{v}\times\bar{v}=-\mathrm{grad}\left(\frac{\bar{v}^2+\bar{v}'^2}{2}+\frac{z}{\mathrm{Fr}}\right)-\frac{1}{\mathrm{Re}_t}\mathrm{rot}\left(\mathrm{rot}\ \bar{v}\right).$$

Consequently, the criterion equation in the physical modeling of vortex funnels is

$$f(\mathrm{Fr},\mathrm{Re}_t)=0. \tag{5}$$

The Froude and Reynolds dynamic similarity criteria (turbulent analogue) are incompatible in scale. Under these conditions, they usually go to an approximate similarity according to the determining Froude criterion. However, the fact of discrepancy between the model and full-scale surface vortex funnels in hydraulic modeling according to Froude is well known. This fact is described in almost all works devoted to this problem. As an example, let us refer to the fundamental work published under the editorship of Knauss [4].

In work [2], devoted to mathematical and physical modeling of surface vortex funnels, we obtained the equations for the profile of the surface of the air core of the funnel and its depth on the axis of rotation (Z_0), the latter is written in the form

$$\frac{Z_0}{R_0}=\frac{\mathrm{Fr}}{2}\left(1+\Gamma_0^2\right)+\frac{\mathrm{Fr}\,\mathrm{Re}_t\,\Gamma_0^2}{2[1-\exp(-\mathrm{Re}_t/2)]^2}\left[\mathrm{Ei}(-\mathrm{Re}_t)-\mathrm{Ei}\left(-\frac{\mathrm{Re}_t}{2}\right)-\ln(2)\right], \tag{6}$$

where Ei(...) is integral exponential function; R_0 is the outer radius of the funnel, beyond which the free surface of the liquid can be assumed horizontal; the Froude and Reynolds numbers from (3) and (4) contain as the characteristic velocity the average radial flow velocity to the $V_0 = V_{R0}$ funnel at its outside radius R_0, the outside radius o the funnel $L_0 = R_0$ is taken as the characteristic dimension; Γ_0 is relative circulation

$$\Gamma_0=V_{\theta 0}/V_{R0}, \tag{7}$$

where $V_{\theta 0}$ is the circumferential (tangential) component of the fluid velocity at the radius R_0. The number (7) is also called the Rossby number.

It can be seen that in the analytical calculation, it is required to know the values of the radius R_0, runoff V_{R0} and tangential $V_{\theta 0}$ velocities, which are determined by the structure of the flow in the water intake zone. The values of these parameters can be found by the methods of computational hydromechanics, however, even in this case, the combination of numerical and analytical calculations, which makes it possible to close the problem and obtain the desired solution, does not exclude the need to check it on a physical model, just as projected hydraulic structures and hydraulic calculation methods cannot do without laboratory studies. This allows us to speak of physical modeling as the most important stage in substantiating the adopted technical solutions, at which the main problem is the reliable interpretation of the data obtained. There is a need to find such a combination of scale-independent criteria of dynamic similarity included in equation (5), which achieves a reliable recalculation of results from model to real objekt. Note that there are two of these criteria, since the appearance of the Rossby number (7) in equation (6), which determines the depth of penetration of the air core of the vortex funnel into the water column in front of the water intake, does not change the form of criterion equation (5), because it is not a dynamic, but a kinematic criterion, which, subject to the geometric similarity of the model and real objekt, must be performed unconditionally. But in (4), which is included in equations (5) and (6), the value of the vortex viscosity ε_t remains unknown. A very close correspondence to the conditions of turbulent flow in surface vortex funnels takes place in swirling submerged jets, and such flows have been studied in sufficient detail [5, 6]. In these works, the following relationship was recommended, linking the vortex and molecular viscosity

$$\frac{\varepsilon_t}{\varepsilon} = 0.185\left(\frac{V_{\theta 0} R_0}{\varepsilon}\right)^{2/3}. \tag{8}$$

Verification (6) showed that the use of formula (8) makes it possible to obtain the calculated values that most accurately agree with the observed experimental and field data.

In the process of modeling according to the Froude criterion, the velocity field on the approach to the vortex surface funnel is determined by the general structure of the flow in the water intake zone. If the geometric similarity of the model and the full-scale object is observed, as a consequence, we can assume the kinematic similarity of the model and full-scale flows. Thus, if the Froude number and the normalized value of circulation on the model and on a full-scale object are equal, then the turbulent Reynolds number on the model will not correspond to its value on a full-scale object. This will distort the scale conversion according to the Froude criterion of the model dimensions of the surface funnel to the dimensions of the surface funnel on a natural object. To clarify this position, we substitute the value of the eddy viscosity according to (8) into the turbulent Reynolds number according to (4), as a result, we obtain

$$\mathrm{Re}_t = \frac{1}{C}\mathrm{Re}^{1/3}, \tag{9}$$

where $\mathrm{Re} = V_{R0}R_0/\varepsilon$ is the Reynolds number, and $C = 0.185(V_{\theta 0}/V_{R0})^{2/3}$ is a constant, the value of which will be equal to that of model and full-scale entity when the kinematic similitude $V_{\theta 0}/V_{R0}$ = idem is observed.

That is, the turbulent Reynolds number grows in proportion to the cubic root of the Reynolds number, calculated from the molecular kinematic viscosity. But Reynolds (molecular) and Froude's criterion are scaled incompatible, therefore, when modeling by the prevailing Froude criterion, the parameters of the model and natural flows in surface funnels will also be scaled incompatible. So, if the linear scale of the model

$$m = L_M / L_F, \tag{10}$$

where L_M and L_F are the characteristic linear dimensions of the model and full-scale entity, is equal, for example, m = 1:25, then when modeling according to Froude, the molecular Reynolds number in the transition from model to nature will increase by

$$\frac{\mathrm{Re}_F}{\mathrm{Re}_M} = \frac{1}{m\sqrt{m}} = 125$$

while the turbulent Reynolds number increases only by

$$\frac{\mathrm{Re}_{tF}}{\mathrm{Re}_{tM}} = \left(\frac{\mathrm{Re}_F}{\mathrm{Re}_M}\right)^{1/3} = \left(\frac{1}{m\sqrt{m}}\right)^{1/3} = \frac{1}{\sqrt{m}} = 5, \tag{11}$$

here, and below the subscripts "M" and "F" refer to the model and full-scale (natural) object, respectively.

To clarify the effect of the increase in the Reynolds number when switching from model to full-scale entity, let us analyze how this is reflected in the value of the main parameter of the funnel - its depth on the axis of rotation Z_0. The calculated turbulent Reynolds numbers, as practice shows, are quite high and even for small hydraulic models, for example, with a scale of m = 1: 100, are at a level not lower than $\mathrm{Re}_t = 25$. Taking this into account, it is possible to exclude from (6) small high orders

$$\exp(-\mathrm{Re}_t/2) \to 0,$$
$$\mathrm{Ei}(-\mathrm{Re}_t) \approx -\exp(-\mathrm{Re}_t)/\mathrm{Re}_t \to 0,$$
$$\mathrm{Ei}(-\mathrm{Re}_t/2) \approx -2\exp(-\mathrm{Re}_t/2)/\mathrm{Re}_t \to 0,$$

this will reduce it to the form

$$Z_0 = R_0 \frac{\mathrm{Fr}}{2}\left\{1 + \Gamma_0^2[1 - \mathrm{Re}_t \ln(2)]\right\} \approx -0.347 \cdot R_0 \cdot \mathrm{Fr} \cdot \mathrm{Re}_t \cdot \Gamma_0^2. \quad (12)$$

The formula gives reason to believe that it is circulation that has an overwhelming effect on the depth of the surface vortex funnel, and in general it allows us to conclude that the change in the velocity field in time, that is, the dynamics associated with an unsteady flow regime, or velocity pulsations, will affect the shape and the depth of the funnel to a degree greater than the square of the velocity. At a high level of pulsation, the depth of the funnel will be subject to significant fluctuations, which is often observed in experiments.

According to (12), for model and full-scale funnels, we can write

$$Z_{0\mathrm{M}} = -0.347 \cdot R_{0\mathrm{M}} \cdot \mathrm{Fr}_\mathrm{M} \cdot \mathrm{Re}_{t\mathrm{M}} \cdot \Gamma_{0\mathrm{M}}^2, \quad (13)$$

$$Z_{0\mathrm{F}} = -0.347 \cdot R_{0\mathrm{F}} \cdot \mathrm{Fr}_\mathrm{F} \cdot \mathrm{Re}_{t\mathrm{F}} \cdot \Gamma_{0\mathrm{F}}^2. \quad (14)$$

Since according to the conditions of Froude modeling $\mathrm{Fr}_\mathrm{M} = \mathrm{Fr}_\mathrm{F}$ and $\Gamma_{0\mathrm{M}} = \Gamma_{0\mathrm{F}}$ then, dividing (14) by (13), taking into account (10) and (11), we find

$$\frac{Z_{0\mathrm{F}}}{Z_{0\mathrm{M}}} \approx \frac{R_{0\mathrm{F}} \cdot \mathrm{Re}_{t\mathrm{F}}}{R_{0\mathrm{M}} \cdot \mathrm{Re}_{t\mathrm{M}}} = \frac{1}{m\sqrt{m}}. \quad (15)$$

Consequently, the ratio of the depths of the full-scale and model funnels in the physical modeling according to Froude is $1/(m\sqrt{m})$, and not $1/m$, as one would expect from a linear scale conversion. The graph of dependence (15) is shown in Figure 2. It can be seen that in our example, the depth of the funnel measured on the model under the Froude simulation should be increased for full-scale entity not $1/m = 25$, but $1/(m\sqrt{m}) = 125$ times; that is, 5 times higher.

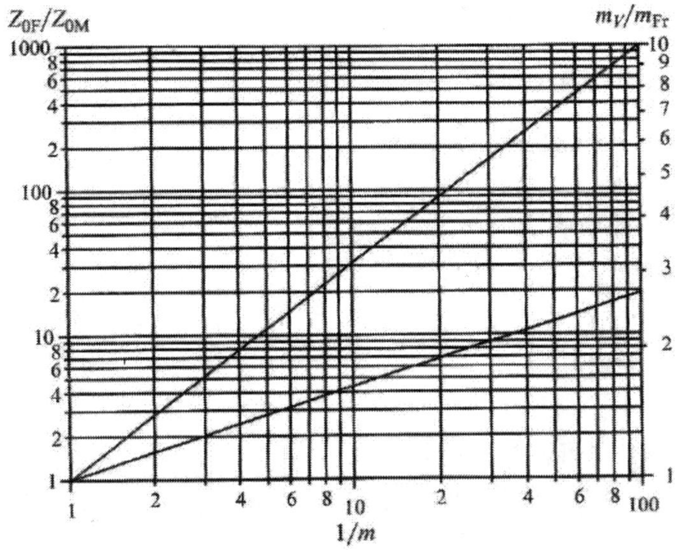

Figure 2. $Z_{0\mathrm{F}}/Z_{0\mathrm{M}} = m^{-1.5}$ and $m_V/m_{\mathrm{Fr}} = m^{-3/14}$ functions.

Considering the fact that the physical modeling of vortex funnels does not meet the Froude criterion, underestimating the values of their parameters by scale conversion from model to full-scale entity, many researchers use the method of increasing (forcing) the flow velocities on the model. Let us answer the question, what kind of velocity boost is necessary to obtain the depth of the funnel on the model, corresponding to the depth on the full-scale entity. Obviously, this is achieved subject to condition (10), according to which

$$\frac{Z_{0F}}{Z_{0M}}=\frac{R_{0F}}{R_{0M}}=\frac{1}{m}. \tag{16}$$

If the conditions of kinematic similarity ($\Gamma_0 = V$) are preserved, then (16), taking into account (13) - (14), is reduced to the equality

$$\mathrm{Fr_M}\cdot \mathrm{Re}_{t\mathrm{M}}=\mathrm{Fr_F}\cdot \mathrm{Re}_{t\mathrm{F}}. \tag{17}$$

where the turbulent Reynolds numbers and Froude numbers for the model and full-scale entity are written in accordance with (9) and (3) as

$$\mathrm{Re}_{t\mathrm{M}}=\frac{1}{C}(\mathrm{Re_M})^{1/3}=\frac{1}{C}\left(\frac{V_{R0\mathrm{M}}R_{0\mathrm{M}}}{\varepsilon}\right)^{1/3} \quad \text{and} \quad \mathrm{Fr_M}=\frac{V_{R0\mathrm{M}}^2}{gR_{0\mathrm{M}}},$$

$$\mathrm{Re}_{t\mathrm{F}}=\frac{1}{C}(\mathrm{Re_F})^{1/3}=\frac{1}{C}\left(\frac{V_{R0\mathrm{F}}R_{0\mathrm{F}}}{\varepsilon}\right)^{1/3} \quad \text{and} \quad \mathrm{Fr_F}=\frac{V_{R0\mathrm{F}}^2}{gR_{0\mathrm{F}}}.$$

Substituting these equalities into (17), as a result, we obtain

$$\left(\frac{V_{R0\mathrm{M}}}{V_{R0\mathrm{F}}}\right)^{7/3}=\left(\frac{R_{0\mathrm{M}}}{R_{0\mathrm{F}}}\right)^{2/3},$$

but $V_{R0\mathrm{M}}/V_{R0\mathrm{F}}=m_V$ is the scale of the velocity, and $R_{0\mathrm{M}}/R_{0\mathrm{F}}=m$ is the linear scale of the model, therefore,

$$m_V=m^{2/7}. \tag{18}$$

When simulating by Froude's criterion, the velocity scale is equal to $m_{\mathrm{Fr}}=\sqrt{m}$; dividing expression (18) by it, we find

$$m_V/m_{\mathrm{Fr}}=m^{-3/14}. \tag{19}$$

This is the necessary velocity boost when simulating by Froude's criterion. Function (19) of the required velocity boost depending on the linear scale of the model is shown in Figure 2. Thus, in our example, with a linear scale of the model m = 1:25, the velocity boost must be taken equal to $m_V/m_{\mathrm{Fr}}=m^{-3/14}=(1/25)^{-3/14}=1.99$, that is, the flow velocity should be increased in relation to its value by a large-scale recalculation according to Froude's rule is 2 times. This almost exactly corresponds to the calculation of the forcing according to the empirical version described in [7], but refutes the popular opinion that the velocity should be forced more significantly - up to full-scale values, for our example it is $m_V/m_{\mathrm{Fr}}=1/\sqrt{m}=5$ times.

3. Conclusion

The circulating flow in the surface vortex funnel in front of the pipeline water intake (Figure 1) is a widespread phenomenon in nature. It is desirable to avoid it or prevent the burst of the funnel air core into the pressure water conduit, which occurs when the depth of the funnel exceeds the depth of the intake wall of the water intake ($Z_0 > h$).

It is shown that the profile of the free surface of the vortex funnel and its depth Z_0 on the axis of rotation are determined by the intensity of the circulation $\Gamma_0 = V_{\theta 0}/V_{R0}$ generating the funnel, the values of the Reynolds numbers $\mathrm{Re}_t = V_{R0}R_0/\varepsilon_t$ and Froude $\mathrm{Fr}=V_{R0}^2/gR_0$.

It is established that, with an accuracy sufficient for engineering calculations, the depth of the surface funnel on the axis of its rotation can be found by the formula

$$Z_0 = -0.347 \cdot R_0 \cdot \mathrm{Fr} \cdot \mathrm{Re}_t \cdot \Gamma_0^2,$$

showing that circulation has the most significant effect on the depth of the surface vortex funnel.

A change in the velocity field in time, that is, the dynamics associated with an unsteady flow regime, or velocity pulsations will be reflected in the shape and depth of the vortex funnel to a degree exceeding the square of the velocity, and for example, at a high level of pulsations, the depth of the funnel will be subject to significant fluctuations, which in practice, it is observed quite often.

Comparison of the funnels obtained on the models and taking place in natural conditions shows that the recalculation of the model parameters according to the Froude criterion gives significantly underestimated values of the depth of the full-scale surface funnel. Such a discrepancy between the model and full-scale funnel is determined by the scale independence of the Froude and turbulent Reynolds criteria.

In hydraulic modeling according to Froude's defining criterion, the depth of the funnel obtained on the model should be recalculated to a full-scale object with a scale factor $1/(m\sqrt{m})$, where m is the linear scale of the model.

Another method for obtaining the depth of the funnel on a model corresponding to a linear recalculation in scale for a full-scale object is forcing the flow velocities in $m^{-3/14}$ times in relation to their value according to the Froude criterion.

References

[1] Volosukhin V A, Belokonev E N and Volynov M A 2012 *Proc. VIII Inter. scient. – pract. conf. "DNY VĚDY-2012"* 60-74
[2] Zuikov A L 2011 *Moscow Univ. of Civ. Eng. Press* 1-72
[3] Zuikov A L 2019 *Moscow Univ. of Civ. Eng. Press* **1** 1-544
[4] Swirling flow problems at intakes 1987 *A. A. Balkema* **XIV** 1-165
[5] Delery J M 1994 *Prog. Aerospace Sci.* **30** 1-59
[6] Fernandes F L and Lubard S C 1971 *AIAA Pap.* **No. 0615**
[7] Baum M R and Cook M E 1975 *Nucl. Engin. Design* **32**(2) 239-245

IOP Conf. Series: Materials Science and Engineering 1159 (2021) 012007 doi:10.1088/1757-899X/1159/1/012007

Transformation of the Couette vortex along the length of the tubular structure

E Yu Zhazha[1], **A L Zuikov**[1, 2]

[1] Moscow Automobile and Road Construction State Technical University, 64, Leningradsky prospect, Moscow, 125319, Russia
[2] Moscow State University of Civil Engineering, 26, Jaroslavskoe Shosse, Moscow, 129337, Russia

E-mail: zuykov54@mail.ru

Abstract. The article is a further research of a circular-longitudinal flow created in a cylindrical pipe by a continuous swirler called Couette vortex, which the author started to study in his previous works. The key question is how Couette modified vortex is transformed along the channel (pipe). The author regards variation of azimuthal velocities (u) and the Heeger – Baer swirl number (Sn) in turbulent irregular circular-longitudinal flow, which is described by the model of modified Couette vortex along the cylindrical channel. It is confirmed that the model of the modified Couette vortex and free-forced Burgers – Batchelor vortex show almost similar results in calculations and both vortex models can be equally used in engineering practice in calculations and the analysis of circulating and longitudinal flow operating modes (swirl flows).

1. Introduction

Quite often tubular structures: small tubular culverts in road construction, corrugated metal road pipes, pipelines for water supply and drainage systems, pipelines for hydroelectric power plants operate with swirling flows. The underestimation of the peculiarities of the movement of swirling flows leads to unjustified construction costs, to an underestimation of the volume of water passed through, and often to accidents associated with an incorrect determination of the throughput of the designed pipelines. This article is devoted to accounting for the peculiarities of swirling flows hydraulics.

In one of my previous articles [1] I regarded steady regular circular-longitudinal flow created in a cylindrical pipe by a continuous swirler (strap, tape or other). The flow is called Couette vortex. I also showed that it is possible to modify Couette vortex and find azimuthal velocities radial distribution formula, corresponding to irregular motion profiles of non-uniform motion of circular-longitudinal flows in a pipe, formed by local swirlers (bladed or tangential). The classical example of such profile is given in [2] (Figure 1).

Content from this work may be used under the terms of the Creative Commons Attribution 3.0 licence. Any further distribution of this work must maintain attribution to the author(s) and the title of the work, journal citation and DOI.
Published under licence by IOP Publishing Ltd

2. Results and Discussion

The modified Couette profile is described by the formula

$$u = u_R \frac{r(R^2 + r_m^2)}{R(r_m^2 + r^2)}. \tag{1}$$

The decoding of the designations in the formula (1) is given in the figure caption to Figure 1.

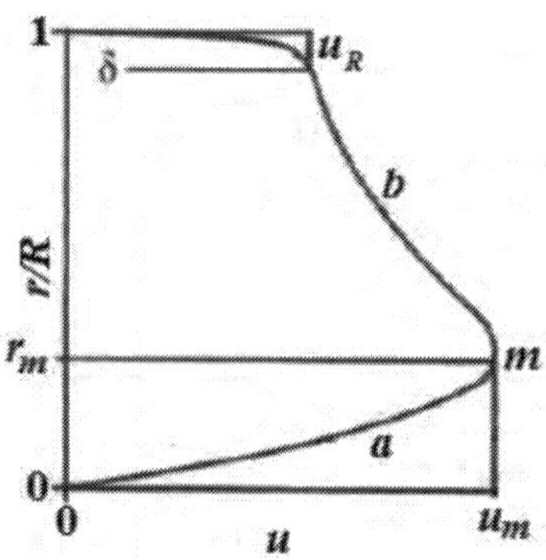

Figure 1. Azimuthal (tangential) velocity profile in turbulent circular-longitudinal flow: where u and u_R – azimuthal velocity on the reference radius r and velocity on the border of turbulent flow core with a boundary layer thick δ; r, r_m and R are reference radius, radius of peak azimuthal (tangential) velocities (u_m) and cylindrical pipe radius; line b – free (potential) rotation; line a – forced vortex (rigid-body rotation); point m – maximum tangential velocity point.

However, in [1] the key question remained open. How is the modified Couette vortex transformed along the length of the channel (pipe)? To answer it, we turn to articles [3, 4], in which, along with the Couette vortex, the Burgers – Batchelor vortex is considered

$$u = u_{R0} \frac{R}{r}\left[1 - \exp\left(-\eta \frac{r^2}{r_m^2}\right)\right], \tag{2}$$

where u_{R0} – azimuthal velocity on the border of a boundary layer at the channel inlet, i.e. directly behind the local swirler forming irregular (fading along the cylindrical channel) circular-longitudinal flow; η – constant, equal to $\eta = 1.256$.

The free-forced Burgers – Batchelor vortex (2) in a turbulent regime of fluid motion, as is known, is obtained as a result of the transformation of a free vortex along the length of a cylindrical tube with the distribution of azimuthal velocities according to the law of dynamic rotation

$$ur = u_{R0}R = \text{const}.$$

It is also known that the modified Couette vortex (1) and the free-forced Burgers – Batchelor vortex (2) give practically the same radial profile of azimuthal velocities in the calculations. In this case, the following relations are characteristic of the Burgers – Betchelor vortex [5]

$$\frac{r_m^2}{\eta R^2} = \chi\sqrt{2\lambda} \cdot \frac{z}{R}, \tag{3}$$

$$u_m r_m = u_{R0}R[1 - \exp(-\eta)] = 0.7152 u_{R0} R = \text{const}, \tag{4}$$

where χ is a universal constant equal to $\chi = 0.2$ for water; λ is flow frictional resistance.

But, according to (1), for the modified Couette profile, we have

$$u_m r_m = u_R \frac{R^2 + r_m^2}{2R}$$

and

$$u = u_m \frac{2r_m r}{r_m^2 + r^2}. \tag{5}$$

Then, taking equalities (3) and (4) valid not only for the Burgers – Batchelor vortex, but also for the modified Couette vortex, we obtain

$$u = u_{R0} \frac{2rR[1-\exp(-\eta)]}{\eta\chi\sqrt{2\lambda zR} + r^2}. \tag{6}$$

Formula (6) describes the radial distribution of azimuthal velocities in a turbulent circulation-longitudinal flow in an arbitrary pipe section. For the calculation, you only need to know the value of the hydraulic resistance coefficient along the length of λ. As practice has shown, for a turbulent circulation-longitudinal flow, the coefficient λ can be assigned in accordance with the equivalent equal-grained absolute roughness of the pipe walls (k_e) according to the well-known formulas Prandtl – Nikuradze

$$\frac{1}{\sqrt{\lambda}} = 2\lg\left(\frac{D}{k_e}\right) + 1.14$$

or Shifrinson

$$\lambda = 0.11\left(\frac{k_e}{D}\right)^{0.25},$$

where D is a pipe diameter, $D = 2R$.

In this case, hydraulic friction will occur not only in the axial, but also in the azimuthal direction.

Figure 2,a shows the calculated radial profiles of normalized azimuthal velocities (u/u_{R0}) in a turbulent circulation-longitudinal flow in seven sections along the pipe length at a distance of z = 25, 50, 100, 200, 400, 800, and 1600 mm from the local swirler. The pipe has a diameter D = 50 mm, the material of the pipe walls is polished organic glass with an equivalent equal-grained absolute roughness equal to k_e = 0.02 mm.

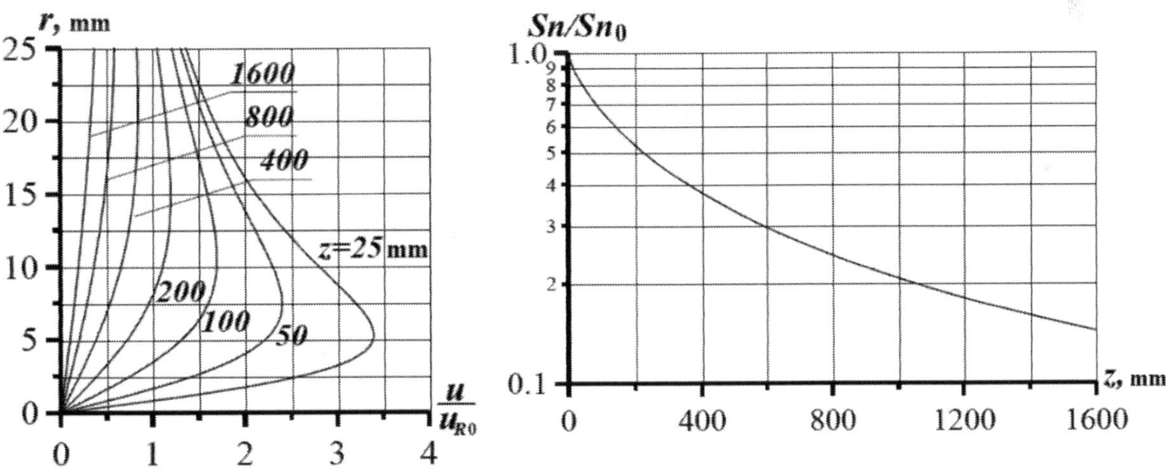

a b

Figure 2. Radial profiles of azimuthal velocities (a) and reduction of Heeger – Baer swirl number along the pipe length (b).

Let us consider the change along the length of the cylindrical channel of the integral characteristic of the circulation-longitudinal flow - the Heeger – Baer swirl number [6]

$$Sn = \frac{M}{RI}, \tag{7}$$

here M and I are the angular momentum and the momentum of the circulation-longitudinal flow, respectively.

$$M = \int_0^R \rho ruv 2\pi r\, dr, \tag{8}$$

$$I = \int_0^R \rho v^2 2\pi r\, dr = \rho\alpha_0 QV, \tag{9}$$

where ρ is the density of the liquid; v is the axial flow velocity at the current radius r; Q is flow rate; V is average flow velocity, $V = Q / \pi R^2$; α_0 is the Boussinesq coefficient.

According to (7) and (9), the change in the swirl number Sn to within the Boussinesq correction α_0 is associated exclusively with the drop in the angular momentum of the circulation-longitudinal flow along the channel length caused by hydraulic friction. In this case, the ratio of the angular momentum of the circulation-longitudinal flow in an arbitrary section of the pipe to its initial value at the entrance to the pipe behind the local swirler is

$$\frac{Sn}{Sn_0} = \frac{M}{M_0}. \tag{10}$$

Let us find the angular momentum of the circulation-longitudinal flow in an arbitrary section along the length of the pipe, taking integral (8) with the radial distribution of azimuthal velocities according to formula (5) and taking the radial distribution of axial velocities is uniform $v = V$

$$M = \int_0^R \rho ruv 2\pi r\, dr = 2\rho u_m r_m Q\left[1 - \frac{r_m^2}{R^2}\ln\left(1 + \frac{R^2}{r_m^2}\right)\right]. \tag{11}$$

But, the density ρ, the flow rate Q and the product $u_m r_m$ are constants that are the same in any section of the conduit, including in its initial section, where r_m tends to zero, u_m tends to infinity, and their product is a finite number equal to $u_m r_m = u_{R0} R[1 - \exp(-\eta)]$. Then the initial moment of momentum of the circulation-longitudinal flow in the initial section of the pipe behind the local swirler according to (11) will be equal

$$M = 2\rho u_m r_m Q = 2\rho u_{R0} RQ[1 - \exp(-\eta)].$$

From this, using (10), we find

$$\frac{Sn}{Sn_0} = 1 - \frac{r_m^2}{R^2}\ln\left(1 + \frac{R^2}{r_m^2}\right) \tag{12}$$

or taking into account (3)

$$\frac{Sn}{Sn_0} = 1 - \eta\chi\sqrt{2\lambda}\cdot\frac{z}{R}\ln\left(1 + \frac{R}{\eta\chi\sqrt{2\lambda} z}\right). \tag{13}$$

The graph of the change the Heeger – Baer swirl number along the length of the cylindrical channel for the design case is shown in Figure 2,b.

Analysis of the results obtained shows that as the flow follows the pipe along the axial coordinate z, the azimuthal velocities u and the swirl number Sn decrease. Moreover, at $z = R/\eta\chi\sqrt{2\lambda}$, the maximum tangential velocities reach the pipe walls, then, according to (12) and (13), we have $r_m = R$ and

$$\frac{Sn}{Sn_0} = 1 - \ln 2 = 0.3069.$$

Consequently, the ratio $Sn/Sn_0 < 0.3069$, which corresponds to the value of the parameter $z = R/\eta\chi\sqrt{2\lambda}$, characterizes the rotation of the liquid according to the law of a rigid body, that is, the stage of degeneration of the circulation flow.

In the design example, quasi-rigid rotation occurs at $z > 564$ mm, which is more than 11 pipe diameters.

3. Conclusion

Complete damping of the flow swirl occurs at $Sn = 0$, which is achieved only at $z \to \infty$. Thus, the damping of the azimuthal velocities and the Heeger – Beer swirl number along the length of the channel in the flow that had free (dynamic) rotation at the inlet is impossible except by gradual transition to quasi-rigid rotation.

Note that the results obtained on the transformation of the modified Couette vortex along the length of the cylindrical channel are close to the analogous characteristics of the change in the parameters of the free-forced Burgers – Batchelor vortex. Both the first and the second vortex model can be equally used in engineering practice when calculating and analyzing the modes of motion of circulation-longitudinal flows (swirling flows).

References

[1] Zuikov A L 2010 *Proc. Moscow State Univ. of Civ. Eng.* **4** 66-71
[2] Fernandes-Feria R, Fernandes de la Mora J, Barrero A 1995 *J. Fluid Mech.* **305** 77-91
[3] Burgers J M 1948 *Adv. in Appl. Mech.* **1** 171–199
[4] Batchelor G K 2002 *Cambridge Univ. Press* 1-631
[5] Zuikov A L 2012 *Proc. Moscow State Univ. of Civ. Eng.* **5** 23-28
[6] Gupta A K, Lilley D and Syred N 1984 *Abacus Press* 1-475

IOP Conf. Series: Materials Science and Engineering 1159 (2021) 012008 doi:10.1088/1757-899X/1159/1/012008

Implementation of FPGA-based camera video adapter for vehicle identification tasks based on EUR 13 classification

A V Ilyukhin[1], V S Seleznev[1], E O Antonova[1], M Y Abdulkhanova[1], E V Marsova[1]

[1] Department of Automation of Production Processes, Moscow Automobile and Road Construction State Technical University, 64, Leningradsky Avenue, Moscow, 125319, Russia

Email: unit.ussr@gmail.com

Abstract. This article describes the first stage of the research work on the development of FPGA-based cameras for vehicle identification tasks being widespread in automated points of weight and dimensional control. Since FPGA represents an alternative to conventional processors, which features the ability to perform multiple tasks in parallel, a camera equipped with FPGA will be able to perform vehicle detection and identification functions simultaneously. Thus, the camera will perform not only the transmission of the image but also the transmission of the processing result, after which the server will only need to confirm or deny the results of the camera operation, which will significantly reduce the processing time of images from all automated points of weight and dimensional control. A simple VGA port board, a program of a static image for displaying it on a monitor in 640x480 resolution, and a program for a pixel counter were implemented in the development phase. EP4CE6E22C8 is used as FPGA, which power is more than enough to achieve the result.

1. Introduction

Using the example of automated points of weight and dimensional control in the road industry, it may be noted that conventional cameras of photo and video fixing are used instead of "smart" cameras. All stages of processing images received from cameras, including determining the registration number of the vehicle, identifying the vehicle according to the EUR 13 classification, etc., are performed on the servers of information processing centers, as well as in traffic management centers. The use of cameras equipped with a field-programmable gate array (FPGA) will reduce the number of computing operations on the server, reducing the load on it by implementing primary image processing directly on the camera itself [1]. The latest generation of gate arrays called FPGA (Field-Programmable Gate Array) has great potential for performing various image processing methods, such as threshold or adaptive binarization, and others [2].

Since for the task of identifying vehicles according to EUR 13 classification, it is necessary to perform all the stages of image processing, such as preprocessing, segmentation, detection, and post-processing on FPGAs, the first stage in the development of a camera equipped with FPGAs is to implement a VGA controller on FPGA.

2. Implementing VGA on FPGA

To visualize the image processing process, a VGA monitor is used, which is a matrix of pixels. The screen resolution is 640 horizontal by 480 vertical. The monitor operates through the sequential transmission of information on each pixel in RGB format, where each color (red, green, blue) is

Content from this work may be used under the terms of the Creative Commons Attribution 3.0 licence. Any further distribution of this work must maintain attribution to the author(s) and the title of the work, journal citation and DOI.
Published under licence by IOP Publishing Ltd

transmitted as an analog signal. Since the monitor is not equipped with video memory and cannot remember information about each pixel, a continuous stream of information is required to monitor to save the image. The VGA sync frequencies are shown in Table 1.

Table 1. VGA sync frequencies

Name	Frequency
Vertical	60 Hz
Horizontal	31.46875 KHz
Pixel	25.175 MHz

The debug board on which the project is implemented is equipped with a quartz resonator with a vibration frequency of 50 MHz. Hence the first task arises: implementing a frequency divider to obtain the pixel frequency. Based on the value of the reference frequency, to get the pixel frequency equal to 25.175 MHz, it is necessary to multiply it by the division factor equal to 0.5035. The coefficient can be represented as a fraction, where the numerator is 5035, and the denominator is 10000. Reducing this fraction, the numerators are 1007, and the denominator is 2000. Since a meander is needed, the duration of the active level should be equal to the duration of the pause between individual pulses. The part of the VHDL code ensuring the frequency divider implementation is shown in Figure 1.

```
clk0_divide_by => 1,
clk0_duty_cycle => 50,
clk0_multiply_by => 2,
clk0_phase_shift => "0",
```

Figure 1. Frequency divider code

The next task is to get horizontal and vertical sweeps from the pixel frequency. Horizontal and vertical sweep time delays are presented in Tables 2 and 3.

Table 2. Horizontal delays

Pixel plot	Number of pixels	Time
Horizontal visible area	640	25.422045680238 μs
Leading edge time	16	0.63555114200596 μs
Synchronizing pulse	96	3.8133068520357 μs
Trailing edge time	48	1.9066534260179 μs

Table 3. Vertical delays

Pixel plot	Number of pixels	Time
Horizontal visible area	480	15.253227408143 ms
Leading edge time	10	0.31777557100298 ms
Synchronizing pulse	2	0.063555114200596 ms
Trailing edge time	33	1.0486593843098 ms

The total number of pixels for horizontal scanning should therefore correspond to 800 for a time interval of 31.777557100298 μs, and for 525 pixels vertical in 16.683217477656 ms [3]. To simplify the task, horizontal pixels can be represented as an abscissa axis, while the vertical ones can be shown as an ordinate axis. Part of the Verilog code for setting the pixel count is shown in Figure 2.

```
wire CounterXmaxed = (CounterX == 800); // 16 + 48 + 96 + 640
wire CounterYmaxed = (CounterY == 525); // 10 + 2 + 33 + 480
```

Figure 2. Pixel counter code

It is possible to check the performance of the programs by connecting the oscilloscope to the output contacts (pins) of the horizontal and vertical sweep on the board to evaluate the frequency characteristics. The measurement results are shown in Figures 3 and 4.

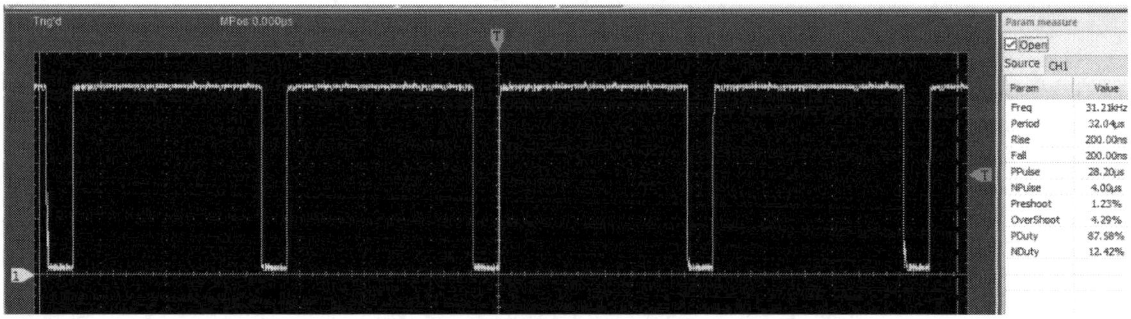

Figure 3. Horizontal scan

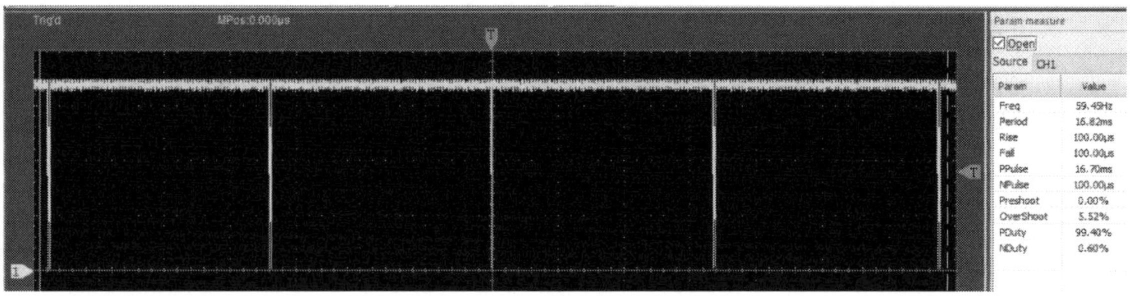

Figure 4. Vertical scan

Based on the oscilloscope readings, it turns out that the vertical and horizontal sweep often do not fall within the acceptable range. Consequently, in order to check the functionality of the program, a more accurate frequency counter is required, or it is required to generate an image and then transfer it directly to the monitor.

Based on this, the last task arises, which consists in generating a static image on board the FPGA with the subsequent transfer of the color scheme to the monitor. Since, according to the VGA standard, color is transmitted by an analog signal, and the FPGA operates on digital signals, it is necessary to develop the following electrical circuit through which it will be possible to obtain the analog values required for the VGA monitor operation from parallel digital signals [4]. The electrical circuit is shown in Figure 5, the computer model in Figure 6.

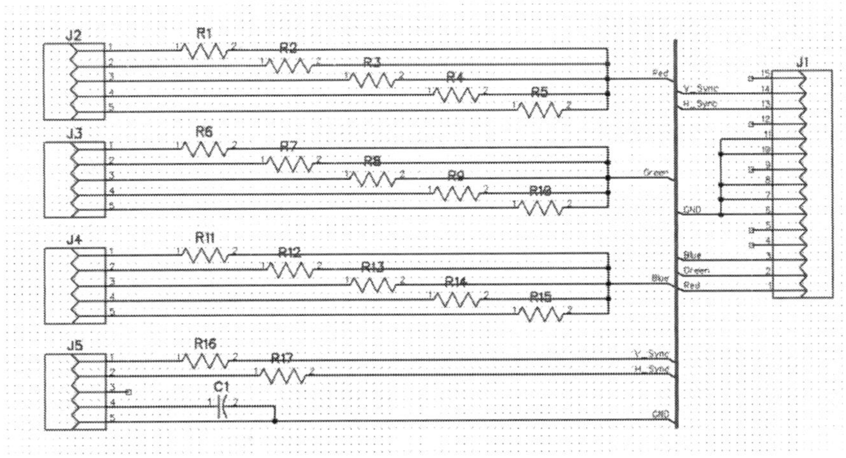

Figure 5. VGA port wiring diagram

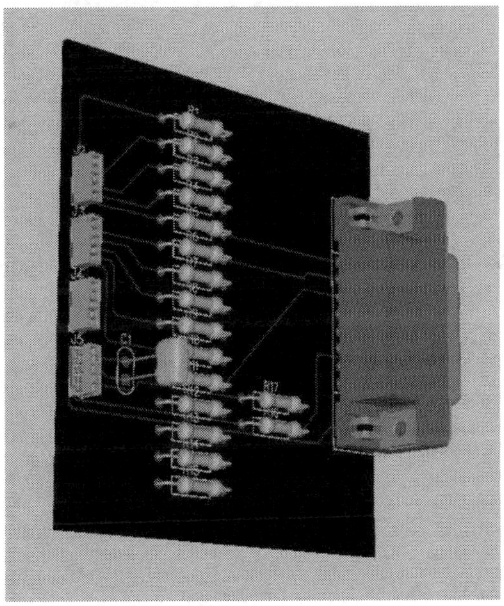

Figure 6. Computer model of VGA port

Now it is necessary to generate a static image in the form of three squares of different colors. Since the VGA monitor uses RGB additive color, it is necessary to define the areas of these squares. Since the visible area of the image is known to be 640 pixels horizontally and 480 pixels vertically, it is possible to calculate the vertices of three squares so that the result is represented by the coordinates of the beginning of the square for each color and its end. When implementing the electrical circuit shown in Figure 5, the RGB555 color transmission is used, which in turn means that 5 bits transmit the value of one color. The implementation procedure of the static image program is shown in Figure 7 and the result of the program is shown in Figure 8.

```
module Static_image(
    input wire CLK,              // pixel frequency: 25.175 MHz
    output wire [4:0] VGA_R,     // 5-bit VGA red output
    output wire [4:0] VGA_G,     // 5-bit VGA green output
    output wire [4:0] VGA_B      // 5-bit VGA blue output
    );

    wire A, B, C;

    assign A = ((x > 140) & (y >  60) & (x < 300) & (y < 220)) ? 1 : 0;
    assign B = ((x > 200) & (y > 120) & (x < 360) & (y < 280)) ? 1 : 0;
    assign C = ((x > 260) & (y > 180) & (x < 420) & (y < 340)) ? 1 : 0;

    assign VGA_R[4] = sq_A;   // Area A is Red
    assign VGA_G[4] = sq_B;   // Area B is Green
    assign VGA_B[4] = sq_C;   // Area C is Blue

endmodule
```

Figure 7. Static image program

Figure 8. Result of the program operation

The device is designed using Intel's Quartus software. To simplify the program, it is better to use PLL (Phase-Locked Loop) as the main frequency divider. The number of used logic gates will therefore be reduced by an order of magnitude and will amount to only 91 logic gates, which is about 1% of the total number of FPGA Cyclone IV E (EP4CE6) FPGA gates.

3. Conclusion

Having performed the first stage of the research work on the development of an FPGA-based camera for vehicle identification tasks, which consists in implementing a VGA port board and writing a program for displaying a static image, as well as writing a pixel counter program, the following feature of connecting an FPGA to a VGA monitor was identified: since the monitor is not equipped with built-in video memory, there is a need for a continuous flow of information to preserve the image.

To obtain the required pixel frequency with a value of 25.175 MHz, a code was written that describes the reference frequency divider with a factor of 0.5035. Then, using the obtained frequency values, horizontal and vertical sweeps were extracted with a total number of pixels of 800 and 525 respectively.

The last work stage comprised the development of an electrical circuit converting a parallel digital signal into an analog signal, which is required for the operation of a VGA monitor and displaying a static image.

Acknowledgments
This research work was performed with the financial support of the Fund for Assistance to the Development of Small Forms of Enterprises in the Scientific and Technical Sphere within the framework of the UMNIK program in accordance with research project No. 15411GU/2020.

References

[1] Palagin A V, Yakovlev Yu S, Eliseeva E V 2012 Features of the approach to the choice of FPGAs for the design of PIM-systems *Mathematical machines and systems* **3** p 19-28
[2] Ilyukhin A V, Zaripova I I, Marsova E V, Seleznev V S 2018 *Digital automation devices* (Moscow: Moscow Automobile and Road Construction State Technical University (MADI))
[3] Ivashinnikov B A, Kovalenko M V, Leskov O M 2016 Implementation of VGA-interface based on FPGA of ALTERA Company *"Young Scientist Publishing House" Limited Liability Company* **17** p 121
[4] Strogonov A V, Babich D A 2012 Development of a VGA controller in the VHDL language in the QUARTUS II environment *Bulletin of Voronezh State Technical University* **6**
[5] Gorodnichev M G, Marsova E V, Gematudinov R A, Dzhabrailov K A 2020 Technical vision for monitoring and diagnostics of the road surface quality in the smart city program *E3S Web of Conferences 2019 Topical Problems of Green Architecture, Civil and Environmental Engineering* Volume 164 03013
[6] Dzhabrailov K A, Gorodnichev M G, Gematudinov R A, Chantieva M E 2020 Development of a Control System for the Transportation of Asphalt Mix with the Maintenance of the Required Temperature *VIII International Scientific Siberian Transport Forum. TransSiberia 2019. Advances in Intelligent Systems and Computing* Volume 1116 Springer
[7] Ilyukhin A V, Marsov V I, Marsova E V, Syetina T A, Dzhabrailov K A 2019 Improving the efficiency of frozen soils excavation using high-frequency vibration *IOP Conference Series: Materials Science and Engineering* Volume 643
[8] Gorodnichev M G, Gematudinov R A, Dzhabrailov K A, Potapchenko T D 2019 The concept of an automated weight and size control system for measuring the mass of freight vehicles in a traffic flow (Weight-In-Motion) *Systems of Signal Synchronization, Generating and Processing in Telecommunications, SYNCHROINFO* IEEE
[9] Marsov V I, Gematudinov R A, Seleznev V S, Dzhabrailov K A 2019 Dependence of the Efficiency of the Distribution of Anti-icing Material on the Type of Suspension of the Weighing Conveyor *Systems of Signals Generating and Processing in the Field of on Board Communications* IEEE
[10] Baytesh K, Gematudinov R G, Chantieva M E, Dzhabrailov K A, Gorodnichev M G 2019 Development of an automatic method for determining vehicle dimensions through computer vision *International Scientific Conference Energy Management of Municipal Facilities and Sustainable Energy Technologies* Volume 1614 Journal of Physics: Conference Series

IOP Conf. Series: Materials Science and Engineering 1159 (2021) 012009 doi:10.1088/1757-899X/1159/1/012009

Economic efficiency of road surface construction using the cold technology of recycling the asphalt-concrete in Iraq and Russian Federation

M M Al-Karaguli[1], A V Lupanov[1], V V Silkin[1], B G Zumbadze[1], S S Sarmad[1]

[1] Moscow Automobile and Road Construction State Technical University (MADI), Leningradsky pr.64, Moscow, 125319, Russia

E-mail: mmg_1989m@yahoo.com

Abstract. This article illustrates the state of roads in the Republic of Iraq and studies the possibility of using the cold regeneration technology of asphalt concrete as the base of the pavement. Moreover, it shows the economic efficiency of using the technology of cold regeneration of asphalt concrete in the asphalt plants and the roads, including crushing of granulate, cement and an aqueous solution of an emulsifier. As compared with the traditional methods of hot asphalt paving, the use of cold technology of recycling the asphalt concrete in (Iraq and the Russian Federation), shows remarkable advantages in both economic and environmental effects.

1. Introduction

For the last three decades, Iraq's transport infrastructure has suffered from low funding. An extended period of economic sanctions and the lack of repair work contributed to the annual accumulation of wear on coatings and an overall decrease in the quality of transport services [1]. Highways provide 80% of all traffic in Iraq. The total length of the country's road network is over 60 thousand km. 80% of roads have a non-rigid surface, 10% - Rigid pavement, 10% without hard surface [2,3]. Over 80% of the length of the road network in Iraq does not meet regulatory requirements because of untimely repairs [4,5].

Currently, the state of Iraq is going through an economic crisis and is unable to allocate a large number of funds for the construction and reconstruction of highways. It is assumed that the use of cold reclaim technology (CR) improves the properties of old asphalt concrete and reduce the cost of repairing road surfaces.

The cost of the road surfaces maintenance is determined from the cost of materials, energy resources, equipment repair costs, etcetera. In the past three years, the cost of materials Iraq has been increasing and fluctuates noticeably during the construction season (Fig. 1) [6,7,8].

Content from this work may be used under the terms of the Creative Commons Attribution 3.0 licence. Any further distribution of this work must maintain attribution to the author(s) and the title of the work, journal citation and DOI.
Published under licence by IOP Publishing Ltd

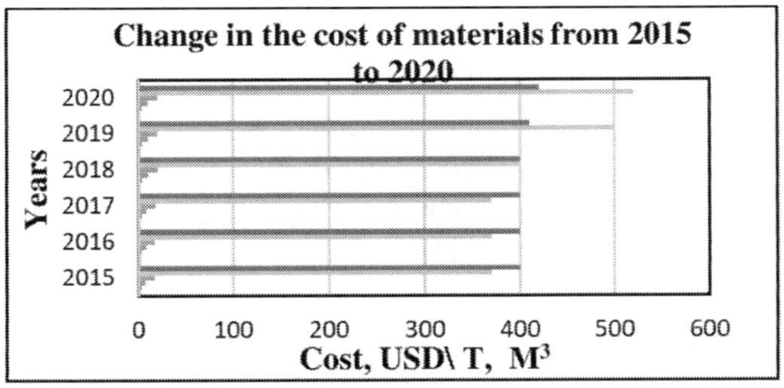

Figure 1. Change in the cost of materials in the Republic of Iraq:
Energy resources,
Gravel,
Sand,
Bitumen,
Workforce.

Figure 1 above shows the costs increase in the materials for the previous two years due to the increases demand and decreases in their production.

The main costs for the production of mixtures include costs of raw materials and materials consumed in the production process; fuel and energy costs spent on technological purposes; the cost of purchased products and semi-finished products that undergo further additional processing and the cost of spare parts for repairs and specialised equipment.

The energy savings in road rebuilding depends on:

- The amount of additional new material and the distance of its transportation;
- The method of removing the old coating and the distance of its transportation;
- The method of crushing the asphalt concrete pavement and the distance of its transportation;
- Type of regeneration technology and location from the project;
- Amounts of moisture content of asphalt concrete and stone materials.

According to UNESCO data for 2018, the level of environmental pollution in the republic continues to grow (Fig. 2). That is primarily due to military events and the growth of road transport, especially in Baghdad and the surrounding area.

Figure 2. The level of environmental pollution according to UNESCO data.

Over six million vehicles are in Iraq. 33% of the vehicles are in Baghdad. The daily fuel consumption is 7 million litres. Environmental pollution by lead oxides exceeds the permissible level by six times, and the carbon dioxide (CO2) exceeds the permissible level by 70% [2]. The second factor of pollution is industrial waste and emissions, including emissions released during the production of asphalt concrete in the asphalt plants manufactures.

Pollution arising during the asphalt plants operations of Hot mixture affect the quality of atmospheric air, soil, groundwater. The production activity of factories leads to the various releasing pollutants into the environment, such as carbon and nitrogen oxides, hydrocarbons, soot, sulfur oxides, resinous substances, benzopyrene, vanadium oxide, phenol, formaldehyde. Therefore inorganic dust is the most dangerous (Fig. 3) [10,11].

a b

Figure 3. Emissions from asphalt factory: a-Russian Federation; b-Iraq.

To reduce the level of environmental pollution, the cost and timing of road repair work, it seems advisable to restore worn-out road surfaces in Iraq using the (CR) method. Cold mixes are prepared on the road (cast in place) or in the factory.

2. Results and Discussion

About 70% of the costs of maintenance pavements are the costs of producing asphalt concrete mixtures. Table 1 below shows the composition of the costs of the mixtures for the conditions in Iraq.

Table 1. Composition of the costs for the production of mixtures, USD.

Cost of materials	Energy costs	Labour and equipment repair	Profit	Total
21	3	4,2	1,8	30

The analysis of the technological process of production and paving of asphalt concrete mixtures shows that the base energy consumption spent on the preparation of mixtures. In this case, more than 80% of energy resources (electricity, gas or diesel fuel) in the process of preparing mixtures are for the drying and heating materials in the drying drum of an asphalt mixing plant [9].

One of the main ways to reduce costs when introducing (CR) technology in the Republic of Iraq is the optimal selection of used road building materials with a preliminary study of their properties.

Table 2 shows the materials costs in Russia (Moscow/Rubles) and Iraq (Baghdad/USDollar).

The mixture obtained as a result of cold reclamation is called asphalt granule concrete (AGC).

Table 2. The cost of materials for the preparation of AGC mixture.

Materials	Unit	Cost in RUB. excluding VAT	Cost in USD. excluding VAT
Cement м500	T.	3700	100
Water	T.	100	8
Plasticiser	T.	12700	240
Uncrushed old asphalt concrete granulate	T.	670	6,5
crushed granulate of old asphalt concrete	T.	750	8

Table 3 shows The cost of the main types of work on the construction of asphalt granules concrete (AGC) coatings for the conditions of Russia (Moscow) in rubles and Iraq (Baghdad) in dollar terms.

Table 3. The cost of work on the construction of an asphalt-granular concrete pavement.

Name of works	Unit	Cost in RUB.	Cost in USD.
Milling old asphalt concrete (1t, a / b is obtained by milling 4 m^2 of coating 0.1m thick)	T.	520	4
Transportation of granules to the plant at a distance of up to 10 km	T.	150	2,5
Crushing granules at the plant in a crusher	T.	80	1,5
Cold mix preparation	T.	165	2
Transportation of AGC mixture from the plant by dump truck up to 10 km	T.	150	2.5
Paving of AGC mixtures with compaction with rollers.	$м^2$	120	1,75

Cold regeneration technology can significantly reduce the cost of mixtures, as well as reduce harmful emissions and reduce energy costs.

The calculations for four variants mixtures of (AGC) for the conditions of the Russian Federation in rubles, and the Republic of Iraq in dollars have been done, see (Table 4).

Table 4. Compositions of AGC mixtures with granulate 0-40 mm and granulate 0-20 mm.

Materials	Before crushing		After crushing	
	Comp.№ 1	Comp.№ 3	Comp.№ 1	Comp.№ 6
A) Granulate of old asphalt concrete (100% fraction 0-40 mm) B) Granulate of old asphalt concrete (60% fraction 5-20 mm and 40% fraction 0-5 mm)	100%	95,5%	100%	97,73%
Cement м500	-	3%	-	1,5%
Water	-	1,5%	-	0,75%
Plasticizer	-	-	-	0,018%
Total:	100%	100%	100%	100%

As a result, the average cost of 1 ton of hot fine-grained dense asphalt concrete mixture (type B) is 2500 rubles for Moscow, Excluding VAT, however, for Baghdad, the cost of a similar mixture is 30.00 USD.

For the comparison of the cost of materials by options in the factory, see Figure 4.

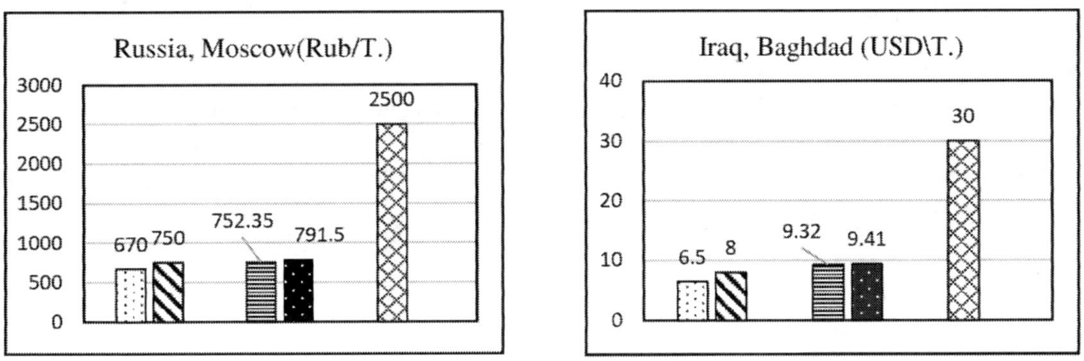

a b

Figure 4. Comparison of the cost of coverage by options in the asphalt plant:
a-Russia (Moscow); b-Iraq (Baghdad).
granules (0-40mm) without cement; fractionated granules (0-20mm);
granules (0-40mm), 3% cement; fractionated granules (0-20mm) with cement;
hot coarse-grain mixture (type B).

From the data obtained, it follows that the cost of (AGC) mixtures with crushed granules, plasticiser and cement (1.5%) obtained in the asphalt plant does not significantly differ from the cost of a similar mixture without crushing with cement (3%) used for cold regeneration on-site, while the cost AGC is 2.5 times less than the cost of hot asphalt concrete mix (type B).

For Moscow, the cost of 1m^2 of a hot coarse-grained dense asphalt-concrete mixture is 660.24 rubles. Excluding VAT, for Baghdad, the cost of coverage from a similar mixture is 11.51 USD.

Figure 5 shows a comparison of the cost of asphalt concrete pavement (1m^2) with a thickness of 10 cm in the Asphalt plant between both countries.

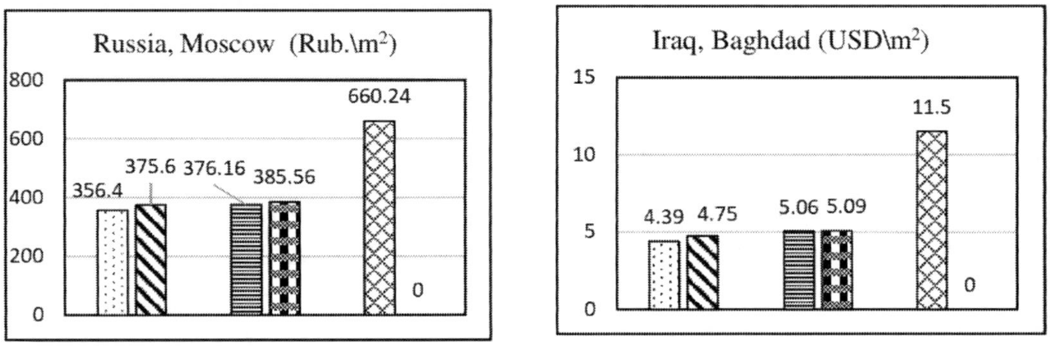

a b

Figure 5. Comparison of the cost of coverage by options in the asphalt plant:
a-Russia (Moscow); b-Iraq (Baghdad).
granules (0-40mm) without cement; fractionated granulate (0-20mm);
granules (0-40mm), 3% cement; fractionated granulate (0-20mm) with cement;
hot coarse-grain mixture (type B).

Cold regeneration technology in the road is not used yet in Iraq. Therefore the calculations are performed for the conditions of the Russian Federation.

For our calculation, we selected the most commonly used technology using a recycler (WR 2500).

The milling asphalt width is 2 m, but in some specific cases, the milling width can increase to 2.5 m. and the milling depth reaches to 30 cm.

The speed of the machine significantly depends on the milling depth and averages 5 - 7 m / min.

Average working productivity per shift is 1.5 km/day depending on the depth and the width of the milling and the current situation on the site.

The proposed width of the milling is 2.5 m, and the length of the section is 1500 m with a thickness of 0.1 m. The productivity per shift is 3750 m2 or 937.5 T / shift.

The duration of the shift is 8 hours, which means 117 T / hour. and as a result, the cost of a water tanker is 12.8 rubles / T.

Figure 6, summarizes the comparison of the cost of AGC pavement (1m^2) 10 cm thick in the asphalt plant and the site.

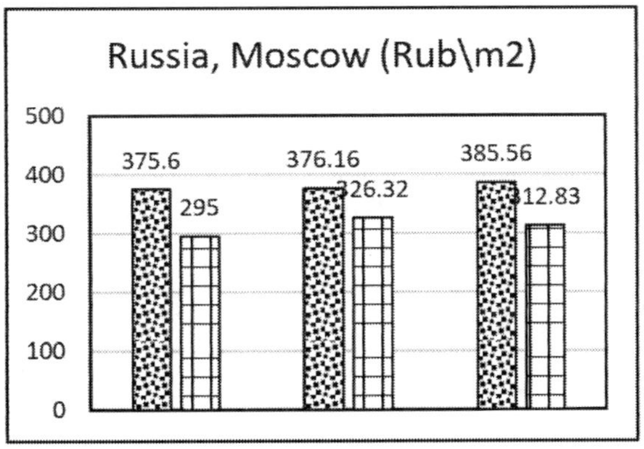

Figure 6. Comparison of the cost of coverage in the asphalt plant manufacture and on the site by options (Comp.№ 1, Comp.№ 3 and Comp.№ 6) respectively:
▒ In the asphalt plant; ▦ in site.

3. Conclusion

- The use of cold regeneration technology has a significant economic effect in comparison with the traditional hot mixture technologies.
- The performed calculations for comparative variants of mixtures and coatings for the conditions of Moscow and the requirements of the state of Iraq showed that the use of old asphalt concrete granulate could significantly reduce the cost of AGC mixture.
- The calculations showed that the most economical technology for improving the physical and mechanical characteristics of the mixture is the technology that provides for the device of a coating or base from crushed fractionated granulate of fractions 0-20 mm with the addition of cement and a plasticiser. For 1 m^2 of coating 10 cm thick, saving according to the selected option No. 6, in comparison with the hot coarse-grained mixture (type B). is 284.08 rubles for Moscow and USD 6.45 for Baghdad.
- The cost of regeneration in the plant is slightly higher if compared to CR in the road site due to transportation costs and additional crushing and sieving operations.

At the same time, the characteristics of the AGC mixture prepared in the asphalt plant are higher, also higher than the calculated features of the AGC. That makes it possible to obtain additional economic efficiency by reducing the thickness of the coating layer.

4.References

[1] Babaev M G 1984 *Asphalt materials in hot climates* (Moscow) p 192
[2] Sawsan S 2016 Traffic density and its impact on the high pollution levels in the city of Baghdad *Al-Mostansiryah journal for Arab and international studies* **54(10)** 217–243
[3] Al-Jumaily M A 2010 Adopting of Performance Grading System for Local Asphalt Cement *Kufa Journal of Engineering* **2(1)** 11–17
[4] Lupanov A P, Sukhanov A S, Al-Karaguli M M, Shoman S S, Shakir Y A 2020 The Rating of the Roadway Network in the Republic of Iraq *East African Scholars Multidiscip Bull* **3(7)** 5–9
[5] Lupanov A P 2017 Processing of asphalt granulate using cold technology at the ABZ *Science and technology in the road industry* **4(82)** 39–41
[6] Lupanov A P, Gladyshev N V 2013 Energy consumption in the production of asphalt concrete mixtures *Science and technology in the road industry* **2(1)** 36–37
[7] Silkin V V, Lupanov A P, Avseenko A A 2012 *Industrial road construction enterprises. Reference encyclopedia of the road builder* (Moscow) p 443
[8] Silkin V V, Lupanov A P 2015 Why are asphalt concrete mixtures becoming more expensive *Automobile roads* **11(2)** 49–51

[9] Al-Karaguli M M 2019 Reconstruction of road surfaces by the method of cold regeneration of asphalt concrete in a hot climate from the example of the Republic of Iraq *International Scientific and Practical Conference* **2(2)** 36–41
[10] Nemchinov M V, Sister V G, Silkin V V, Rudakova V V 2009 *Environmental protection during the design and construction of highways* (Moscow) p 280
[11] Donchenko V V, Manusadjiants Z G, Samoilova L G 1998 *Inventory of pollutant emissions into the atmosphere for asphalt concrete plants* (Moscow) p 24

Assessment of the possibility of local object to be flooded in megalopolis as a result of a hydraulic system breakage in the area of a transport tunnel

T A Suetina[1], **O N Chernykh**[2] **and A V Burlachenko** [3]

[1]Moscow Automobile & Road Construction State Technical University (MADI), Moscow, Russia
[2] Russian State Agricultural University, Moscow, Russia
[3] Mercedes Benz RUS JSC, Moscow, Russia

E-mail: t_a_souetina@mail.ru

Abstract. In the article is observed the approach to assessing the safety of the downstream of retaining waterworks in the conditions of the existing urban infrastructure of the metropolis and the transport tunnel built on the border of the water system. Also are described the peculiarities of the methodology for determining the size of probable harm in the event of a hydrological accident of a low-pressure hydroelectric megalopolis complex, the lower pool of which is filled up, the watercourse is taken into the collector. The position of the groundwater level causes flooding of the territory, increasing the hazard indicator due to the activation of included processes (change in soil properties, suffusion, appearance a source of local flooding of an urbanized area, etc.). The results of a numerical experiment to assess changes in the position of the ground water level in the territory in contact with the Mikhalkovsky tunnel in Moscow are presented. It is shown that if the drainage system fails near the tunnel, a significant rise in the ground water level (more than 1 m) may occur, which will cause flooding of the surrounding area and residential buildings. To improve the efficiency of drainage in the lower reaches of the hydroelectric system, it is proposed to create an additional drainage system, including from the collector of the Zhabenka river. Calculation also shows that the emergency situation which may occur due to hydrological crash with implementation of the most difficult and simultaneously the most likely of several possible crash scenario with overtopping of the dam and the formation of the closure channel on the waterworks of Moscow is not likely to occur. The damage in the probable and most severe scenario of a hypothetical hydrological accident is close to zero. Therefore, there will be no damage on most ponds in the megalopolis where the downstream is filled in, planned, fortified, or the diverting section of the watercourse is taken into the collector.

1. Introduction

Suffusion associated with both natural and technogenic factors is currently spread over an area of more than 13% of the territory of the Russian Federation. Natural and technogenic suffusion processes in ground bases affect the majority of urban industrial territories, where technogenic filtration suffusion dominates the natural one. For such large megacities as Moscow, the problem has become relevant in the last decade. It is associated with a number of factors: an increase in the level of ground water that

Content from this work may be used under the terms of the Creative Commons Attribution 3.0 licence. Any further distribution of this work must maintain attribution to the author(s) and the title of the work, journal citation and DOI.
Published under licence by IOP Publishing Ltd

causes flooding of the territory (in Moscow, more than 40% of the territory is flooded); intensive construction, which encapsulate pre-existing water bodies and disturbed the prevailing water balance in the territory (in the last 10 years in "old Moscow" borders were buried about 20 ponds and small watercourses, was newly constructed only 1 pond-dugout); underground urbanization; changing climatic indicators territory; leakage from water-bearing communications, slurry tanks and other city water facilities; development and condition of water supply and Sewerage networks (wear of water supply and sanitation networks in Russia as a whole is 60...80%, and water losses are 2.5...3 times higher than the standard permissible level [1]); the service life of water-carrying networks and water systems, etc. 42 karst-suffusion sinkholes with a funnel diameter of up to 40 m and a depth of up to 5...8 m have been formed in the North-West of Moscow over the past 30 years. This has a negative impact on traffic on certain main sections of the capital, the condition of vehicles and 3-and 5-floor buildings.

One of the main reasons that initiates the development of suffusion processes is the high position of the ground water level (GWL). The development of suffusion manifestations and deformations, along with the engineering and geological structure of the territory, is influenced by both one-time rises of the GWL, and the characteristics of fluctuations, their amplitude and frequency. The intensity of suffusion development in a local area of an urbanized territory is directly related to the production of earthworks, drilling, and sinking operations, the number of floors in a building, the destruction of underground water-bearing communications, and so on.

Mikhalkovsky tunnel is section of the North – Western chord in Moscow (Figure 1a) held in Koptevo district in complex engineering and hydrogeological conditions in the thickness of the heterogeneous meso-Cenozoic sediments (water-saturated Sands and fluid sandy loam), close to a drainage system associated with the river Zhabenka collector, which runs almost parallel to the tunnel and in close proximity to a Large Garden pond. The presence of drainage devices on certain sections of the route and their condition are not known. Total length of the tunnel section on Bolshaya Akademicheskaya street is 410 m: the closed part is 110 m, the southern ramp is 161.5 m, and the Northern ramp is 138.5 m. The tunnel width along the outer contour is 29,86 m. During construction, there were concerns that the construction of an underground tunnel in the immediate vicinity could dramatically affect the ecological environment of the reservoir and surrounding areas, and construction work near the border of the Petrovskoe-razumovskoe estate Park would lead to waterlogging of the Park and the death of valuable trees. The reservoir in which the Zhabenka river is enclosed is in a satisfactory condition (Figure 1b). The underground channel of a fairly deep river has a small historical section made of large stones, but the main part of it is from reinforced concrete. The average height of the collector is 3.2 m, and the width is 2.7 m.

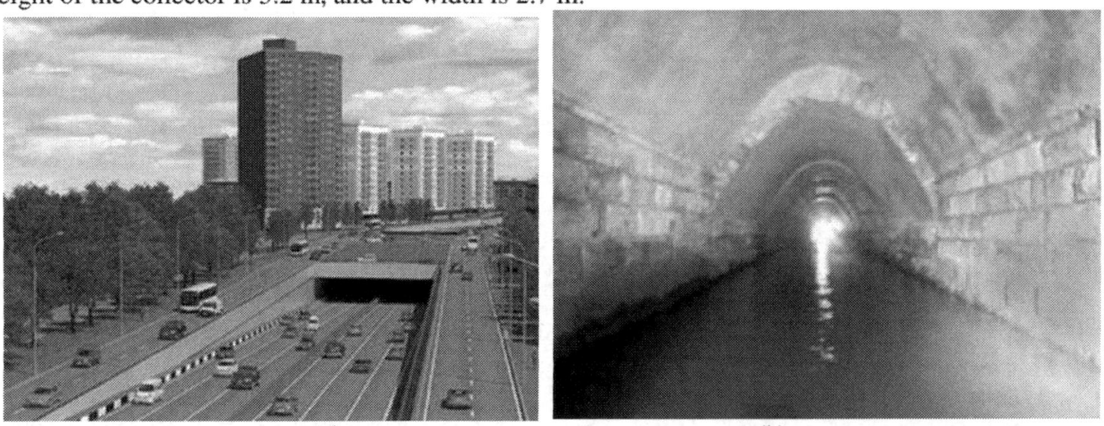

(a) (b)

Figure 1. General view: (a) - Mikhalkovsky tunnel; (b) - underground collector of r. Zhabenka

As long ago as in the 50s and 60s of the twentieth century, the entire Koptev plain was marked by increased GWL and wetlands. Therefore, a special horizontal drainage network was designed. Then the construction of 5-and 9 - floor residential buildings began. In the 90s, the street was built up with multi-storey modern buildings. However, drainage is not laid everywhere, so to this day there is flooding of basements of individual residential buildings, sinkholes due to seeps and springs beating from underground. The area is full of utilities; historically, it has a complex water system that has been repeatedly reconstructed, but despite this, the area in the lower reaches of the hydroelectric complex, which was filled in last century, is now very swampy. Knowing that the neighborhood is located on water-saturated ground, any movement can lead to unpredictable consequences.

2. Materials and methods

Many works are devoted to identifying the causes of accidents and assessing the risk of accidents in urbanized areas. In particular, Dar/VODGEO has developed an integrated risk assessment method based on the analysis of factors and types of harmful effects of natural and man-made filtration suffusion on the infrastructure of the urban industrial territory and its degree of susceptibility to dangerous impacts and possible consequences. The assessment of the risk of flooding and damage, both for the entire territory as a whole and for local objects, is based on the calculation of the doses of harmful effects D caused by it.[1,2]

$$D = \lambda_{оп} \cdot v_{уязв} \quad (1)$$

where $\lambda_{оп}$, $v_{уязв}$ – coefficients correspond to the risk of flooding and vulnerability of the territory. For urban-industrial territories, taking into account the features of development, their values are determined for each indicator of the risk of flooding according to the corresponding tables [1, 2]. For local objects, the calculation mechanism is integral.

$$\lambda_{оп} = \Sigma^{4}_{i=1}\, \delta_i\, a_i\, \lambda_o, \qquad v_{уязв} = \Sigma^{4}_{i=1}\, \varphi_i b_i\, v_o \, , \quad (2)$$

where: δ_i and φ_i - significance coefficients of the i–th hazard and vulnerability indicator; a_i , b_i – the value of the score for the i–th hazard and vulnerability indicator; λ_o, v_o - normalizing multipliers.

The algorithm for determining the values of all coefficients is described in detail in [1,2]. With a known dose of harmful effects D, the method called "probit analysis" in risk theory is used to find the annual risk from flooding, taking into account the cost of the object before the damage, i.e. without taking into account the flooding of the territory. To get correct results, you need to know some constants that characterize the specifics of harmful effects on this object. As a first approximation for the calculation to estimate the annual damage from flooding R object value C_e can be taken dependence (3), recommended in [2] for any value of the dose D, which should further take into consideration the nonstationary nature of flooding and the design features of the object under construction and the surrounding area

$$R = 6{,}5 \, .10^{-3}\, erfc[-2{,}36 lnD - 3].C_e \quad (3)$$

If there is no harmful effect of flooding $D \rightarrow 0$. In accordance with (3) and $R \rightarrow 0$, i.e. there is no risk (damage) from flooding.

One of the main factors in the list of risk factors from flooding, when calculating the doses of harmful effects on individual local objects, is the depth of groundwater, which is included in the first hazard indicator, the degree of which (large, medium or small) is characterized by the capacity of the flooding zone and the coefficient of flooding of the object. In this regard, to assess the degree of danger for the first hazard indicator, it is necessary to make a forecast of flooding on the territory of the object being designed, built or operated. In this area of the capital, flooding of the territory can cause activation of landslide processes with an average degree of danger. The neighborhood is characterized by relatively high position of the GWL. The first horizon from the surface lies at a depth of 0.5 to 5 m, with an oscillation amplitude in the range of 1...1.5 m.

3. Results and Discussion

The extraordinary nature of the situation has led to the need for numerical modeling to assess the impact of the tunnel on the groundwater of the adjacent territory, one part of which belongs to the protected areas, and the other to the microdistrict of dense residential development. Materials from surveys conducted by MOSINZHPROEKT and MOSGEOTREST were used as the basis for modeling, and the position and diameter of drains were taken from the geo-base. The paper presents the calculation results and the influence on the degree of danger according to the first indicator of the risk of flooding of territories adjacent to the highway Mikhalkovskaya tunnel on Bolshaya Akademicheskaya street in the northwestern administrative area of Moscow [3]. The aquifer complex has a single level surface, which is characterized by the hydroisohypses map (Figure 2a). The site is composed of upper Jurassic sand, sandy loam and layered loam. The underground water level of the complex lies at a depth of 0.5...5 m. In unfavorable years, levels may rise by 1...1.5 m. Oxford clays serve as a water barrier for the aquifer complex at the site of the tunnel passage. The assessment of the impact of the Mikhalkov tunnel on groundwater was based on a comparison of the results of modeling the position of the GWL before and after its commissioning.

Calculations of changes in GWL on the site territory, including adjacent territories, to assess the degree of danger for the first hazard indicator were performed using mathematical modeling on a three-dimensional model [3, 4]. Since the tunnel does not completely cover the nadyursky aquifer complex, the US Geological survey MODFLOW program and a preprocessor for preparing PMWIN data were used to solve the problem [4, 5]. Vertically, for numerical calculations, the filtration area was divided into 4 layers. In the numerical solution, the spatial steady-state filtration was described by the equation

$$\frac{\partial}{\partial x}\left(k_x\frac{\partial h}{\partial x}\right)+\frac{\partial}{\partial y}\left(k_y\frac{\partial h}{\partial y}\right)+\frac{\partial}{\partial z}\left(k_z\frac{\partial h}{\partial z}\right)=0 \tag{4}$$

where x, y and z - spatial coordinates, m; h - ground water pressure, m; kx , ky and kz - filtration coefficients in the direction of the coordinate axes, m/day.

The solution of equation (4), taking into account changes in the values of the filtration coefficients, is carried out numerically, based on the integral finite difference method. At the same time, the filtration area is divided into blocks-parallelepipeds of different sizes: near water intake wells, the rectangular bases of these parallelepipeds are smaller in size compared to those located in remote areas. In other words, a rectangular grid that covers the entire filter area in the plan has smaller sections in areas that require more detailed analysis. The solution of equation (4) can be represented as a volume integral

$$\int_v\left[\frac{\partial}{\partial x}\left(k_x\frac{\partial h}{\partial x}\right)+\frac{\partial}{\partial y}\left(k_y\frac{\partial h}{\partial y}\right)+\frac{\partial}{\partial z}\left(k_z\frac{\partial h}{\partial z}\right)\right]dV = \\ = F_{yz}\left(V_{x_{i-1/2}}-V_{x_{i+1/2}}\right)+F_{xz}\left(V_{y_{j-1/2}}-V_{y_{j+1/2}}\right)+F_{yx}\left(V_{z_{k-1/2}}-V_{z_{k+1/2}}\right) \tag{5}$$

The right side of equation (5) is the balance for a parallelepiped with integer coordinates $i = 1,...N_x$; $j = 1,...N_y$; $k = 1,...N_z$, corresponding Cartesian coordinates x, y, and z, respectively; N_x, N_y, N_z - the maximum values of integer coordinates. F_{yz}, F_{xz} и F_{xy} — sectional areas of the parallelepiped along the yz, xz and xy planes, respectively. The cross-sectional areas are determined based on the size of the parallelepiped block Δx_i, Δy_j, Δz_k :

$$F_{yz} = \Delta y_j \text{ x } \Delta z_k ;\quad F_{xz} = \Delta x_i \text{ x } \Delta z_k ;\quad F_{xy} = \Delta x_i \text{ x } \Delta y_j \ , \tag{6}$$

V_x, V_y, V_z - filtration rates, which are calculated for the indoor units according to the formulas:

$$V_{x_{i-1/2}} = k_{x_{i-1/2}}\frac{2\left(H_i-H_{i-1}\right)}{\left(\Delta x_{i-1}+\Delta x_i\right)},$$

$$V_{x_{i+1/2}} = k_{x_{i+1/2}}\frac{2\left(H_{i+1}-H_i\right)}{\left(\Delta x_i+\Delta x_{i+1}\right)}, \tag{7}$$

where H_{i-1}, H_i , H_{i+1} - pressures in the centers of gravity of the blocks $i-1, j, k$; i, j, k; $i+1, j, k$ respectively; Δx_{i-1}, Δx_i , Δx_{i+1} – size of blocks $i-1, j, k$; i, j, k; $i+1, j, k$ along axis x; $k_{x_{i-1/2}}$ - average

filtration coefficient along the x axis between blocks $i-1, j, k$ and i, j, k ; $k_{x_{i+1/2}}$ - average filtration coefficient along the x axis between blocks i, j, k and $i+1, j, k$.
Speeds are calculated similarly for indoor blocks $V_{y_{j-1/2}}$; $V_{y_{j+1/2}}$; $V_{z_{k-1/2}}$ and $V_{z_{k+1/2}}$.
Integration (5) performed for each block of the model allowed us to obtain a system of algebraic equations of the next type:

$$A_{i,j,k} H_{i-1,j,k} + B_{i,j,k} H_{i,j-1,k} + C_{i,j,k} H_{i,j,k-1} + P_{i,j,k} H_{i,j,k} + E_{i,j,k} H_{i+1,j,k} + F_{i,j,k} H_{i,j+1,k} + G_{i,j,k} H_{i,j,k+1} = I_{i,j,k}, \quad (8)$$

where H - unknown pressures or levels in the centers of gravity of the model blocks;

$A_{i,j,k}$ and etc. — coefficients determined based on filtration parameters and the geometry of the aquifer;

$I_{i,j,k}$ - the free term of equation (8), which mainly includes the source-stock components.
The total number of equations and unknowns is equal to the number of parallelepiped blocks $Nx \times Ny \times Nz$.

As a result of verification of the model and solving the inverse problem, a design map of hydroisohypses was obtained (Figure 2b), which characterizes the position of levels even before construction, which was then used for comparison with forecasts based on the results of modeling (before and after construction) and obtaining maps of changes in GWL depending on the location and operation of drainage (Figure 3a). When calculating changes in the distribution of GWL after the construction of the Mikhalkovsky tunnel, it was assumed that the tunnel overlaps underground water.

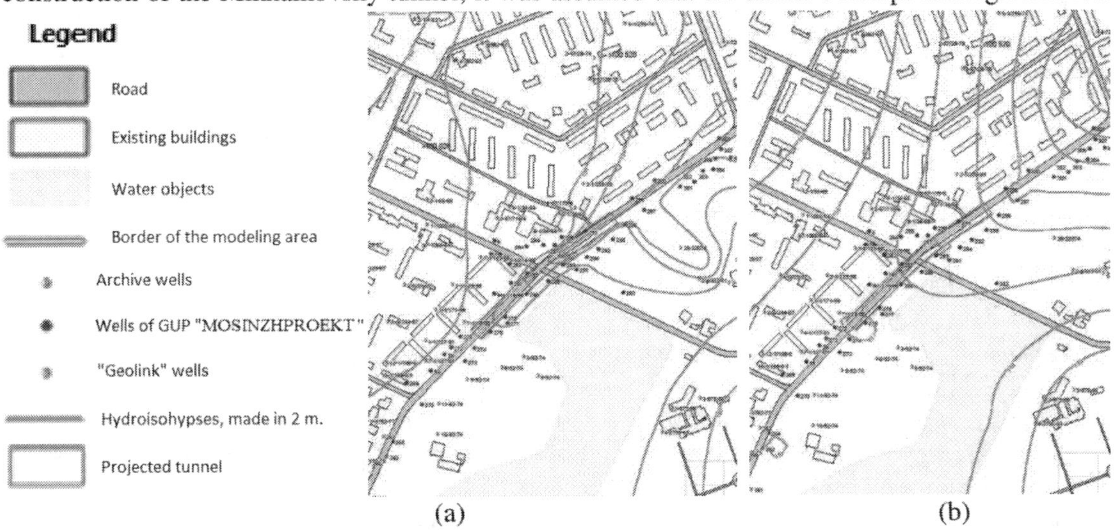

Figure 2. Map of isohypses GWL: (a) according to the survey; (b) the results of solving the inverse problem

The forecast of changes in the GWL is a result of tunnel construction (Figure 3b) showed that the rise of GWL in the entire territory where the object is being built, including the adjacent territories on both sides of the tunnel, cannot be assumed to be the same, including the area of the object. However, the change in the position of the GWL caused only by the tunnel was predicted to be very small, the first centimeters. This value is comparable to the accuracy of the measurement of GWL and can be practically ignored. This slight rise can be explained by two reasons: the tunnel does not completely block the flow of underground water (the nadyursky aquifer complex) and the smooth operation of existing drains that stabilize the GWL. Therefore, calculations were made to analyze changes in the position of the GWL under the combined action of two factors: the impact of the tunnel and the destruction of drains near it. The results of the forecast of changes in the GWL for this option showed that the rise of GWL when drains fail can exceed 1 m, which, taking into account the amplitude of fluctuations and the unsteadiness of the flooding process, can cause dangerous flooding of the territory and residential buildings located on it, and in particular the house No. 57 on Bolshaya Akademicheskaya street, whose residents constantly and previously complained about flooding of basements. To prevent the situation of cessation of operation of the drainage recommended to develop additional measures to preserve existing drainage and/or create a new one, i.e. the creation of additional support drainage system to ensure a stable position of the groundwater table at the existing marks and at least two observation wells to a depth of 10...15 m, to upgrade stormwater and urban sewer, water and heating in the district. Analysis of the map obtained for this case of changes in the position of the GWL caused by the impact of the tunnel, partial failure of the existing drainage and the construction of additional drainage showed that changes in the GWL in this case are insignificant.

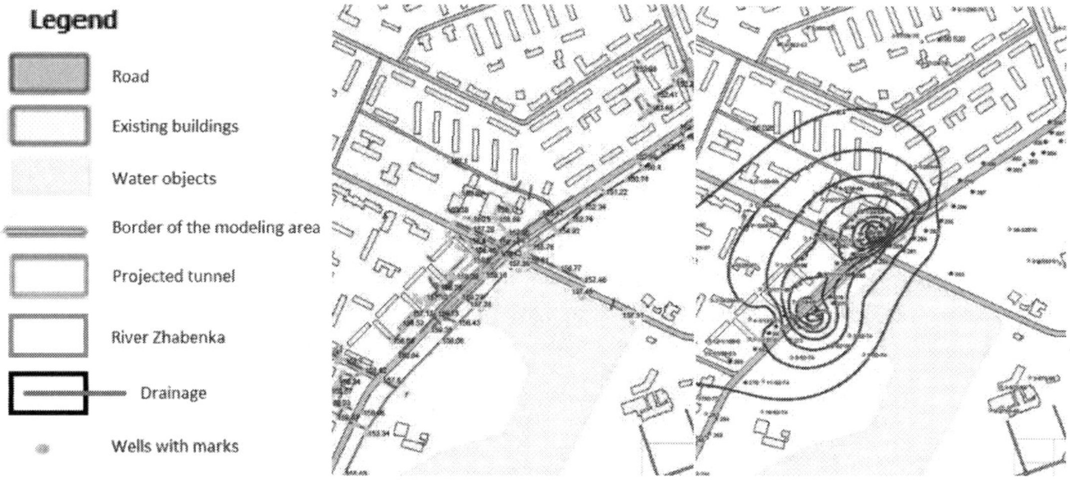

Figure 3. Maps: (a) - placement of drains and water bodies in the area of construction of the Mikhalkovsky tunnel; (b) - isolines of changes in the GWL during operation of the Mikhalkovsky tunnel in case of failure of drains

For a more complete assessment of the integral degree of danger λ_{on} of flooding of the tunnel adjacent territory and local objects on it, in addition to the depth of groundwater occurrence, it is necessary to more correctly determine the degree of danger for each indicator separately: corrosion aggressiveness of groundwater, changes in the strength and deformation properties of soils during flooding and drainage, activation of dangerous processes. To estimate the dose of the harmful impact of flooding on an object using (1), using the dependencies (2) and (3), it is also necessary to collect additional data and find all four indicators of vulnerability: urban, geotechnical, environmental and operational. Only then will it be possible to fully assess the maximum annual damage from possible flooding of the object.

However it should be noted that in modern conditions for the water system of the Big Garden pond and the territory, adjacent to tunnel, also assessment of emergence and development of number of possible scenarios of hydrological breakage is relevant. Such scenarios are connected with changes in work of the collector of the Zhabenki River and the system of drainage, its interaction its developed, but partially destroyed with the tunnel, and also with impossibility of the admission water throughput constructions of the water-engineering system of high-water flows and the water flow rate which is specially arriving from Golovinsky ponds for watering of the Yauza River a natural channel of the Likhoborki River. This analysis was carried out under an agreement with Mosvodostok [6] and showed that when the maximum calculated flow rate of the rare supply of the Zhabenka river (approximately 1 m3/s) and water discharge from Golovinsky ponds are in transit, it will take more than 1.4 days to create a normal oxygen balance in the Yauza river (the water supply flow rate is 5 m^3/s) even with the spillway culverts are not working (assumption) to raise the water level in the

IOP Conf. Series: Materials Science and Engineering 1159 (2021) 012010 doi:10.1088/1757-899X/1159/1/012010

Bolshoy Sadovoye pond with its current dimensions to the dam crest mark. If hypothetically such filling occurs, and the operation service cannot eliminate the emergency situation during this time, then the prerequisites for the operation of the hydroelectric power plant will be created in 5 possible scenarios [7]. However, a preliminary assessment shows [6, 7] that in the most severe scenario, the probability of a hole forming in the dam body is unlikely, since the lower slope of the dam is filled in, planned, reinforced, and partially paved. It is more likely that a dam crest of limited width will overflow than that a significant mass of water will flow through the resulting gap in the form of a breakout wave. In this case, the height of the overflow layer when the culverts are not working (assumption) it will not exceed 6 cm when spilling over the crest of the dam on a section equal to the length of the dam along the crest (about 310 m). The water level in the pond towards the edge of the shore from the Timiryazevsky Park will rise slightly, the flooding will be unnoticed, and environment of the coastal Park area at the end of the pond and the left shore, where is the stadium "Science" with number of areas (ex: dealership, body shop, OOO Elkom-yuna, etc.) and Bolshaya Akademicheskaya street. Only the beach and boat dock can be briefly under a layer of water from 1.5 to 0.1 m deep and the width of the coastal strip is less than 3 m.

Pass of the verification flow, occuring in this scenario, will cause the breakdown through the manifold R. Zhabenki laid along Bolshaya Akademicheskaya St., will happen in free-flow mode at the maximum depth of 0.85 m at the height of the cross section of the reservoir is 3.2 m. Thus, flooding of areas downstream of Large Garden pond with the big share of probability will not be. At the same time, it should be noted that in any of the other considered accident scenarios, is affected the territory of the RGAU Park Timiryazev (Forest experimental dacha, arboretum garden named after R. I. Schroeder, Botanical garden of the Moscow agricultural Academy named after S. I. Rostovtseva and estate Petrovsko-Razumovskaya adjacent to the waters of the pond and having the status of nature reserve protected areas), and the area Bolshaya Academic, a distance of from 40 to 250 m from the water practically will not be flooded above the corresponding regular operation of the main spillway of the pond when you pass the waterworks flow calibration repeatability 1 time in 1000 years. Thus, the area of hypothetical flooding does not include settlements, industrial facilities, valuable land, forests, flooding of which may cause damage to third parties, therefore, there will be no damage to all components of the damage as a result of the accident. The damage from the consequences of all accident scenarios is approximately equal to zero. An emergency situation due to a hydrodynamic breakdown will not occur if the most severe and at the same time the most probable accident scenario is realized.

4 Conclusions

Analyzing the materials obtained, we can conclude that the problem of improvement, correct assessment of the position of the GWL and stabilization of its position after the construction of any transport tunnel in the metropolis should be solved comprehensively. The formation of a groundwater table in the area of the tunnel with fixation of its provisions should be adapted to scientific requirements for the preservation of ecological security and state waters adjacent thereto of a water body, in the case of water systems of district and historical monument of the estate "Petrovsky-Razumovsky" with a Large Garden pond. Calculations have shown that the creation of additional drainage will allow maintaining the GWL at existing levels and, if necessary, improve the situation. Special attention should be paid to the organization of operation of existing drainage systems and structures that are on the balance of various organizations. Since one of the reasons for the rise of GWL may be a violation of water-bearing communications (water supply, Sewerage, etc.), it is necessary to take measures to preserve them. It is quite possible that the results of monitoring will require the reconstruction of the existing old drainage system of the entire microdistrict, including the main collector of the Zhabenka river and household sewer collectors.

In addition to performing the recommended actions is necessary operational evaluation of the quantitative and qualitative condition of groundwater at the ebb side of a dam of waterworks facility, which, coupled with the assessment of the harmful effects of groundwater flooding at the site will allow you to obtain the correct value of the likely damage resulting from hydrological accident, hydraulic and transport facilities in urban areas. The analysis of the probability of occurrence of hydrological hypothetical accident on a water body adjacent to Mihalkovskiy tunnel, and location of residential and economic facilities downstream of the dam showed that the emergency situation as a result of hydrodynamic accidents during the implementation of the most difficult and simultaneously the most likely scenario accident will arise. A potential accident of the pond structures cannot create a direct threat to the surrounding area. The probable damage for the most severe and at the same time the most probable accident scenario turned out to be the same and equal to zero. There will be no damage to third parties for which liability arises when the Bolshoy Sadoviy pond dam breaks, as well as for most small ponds in the territories of most administrative districts of the megalopolis, the diverting channel of watercourses in the lower reaches of which is taken into the collector.

References

[1] Kuranov P N Koroteev D G 2012 Calculations of risk from flooding of local facilities *Water supply, sanitation, environmental safety of construction and urban economy* **14** 25

[2] Senatorova N A Timofeeva E A Korostelev A D Kuranov N P 2012 Risk factors of natural and technogenic filtration suffusion processes in urban industrial territories *Water supply, sanitation, environmental safety of construction and urban economy* **14** 12

[3] Ushakov V V Rastorguev A V Altunin V I and Chernykh O N 2013 Impact of tunnel construction on ground water level *Science and Technology in the road industry* **4** 35

[4] McDonald M C and A W Harbaugh 1988 MODFLOW A modular three-dimensional finite difference ground-water flow model *U S Geological Survey Open-file report* 83-875

[5] Chiang W H and W Kinzelbach 1998 Processing Modflow: A Simulation System for Modeling Groundwater Flow and Pollution Hamburg–Zürich

[6] Chernykh O N Khanov N V Burlachenko A V 2019 Ways to solve the problems of integrated environmental rehabilitation and natural restoration of small rivers of Russian estates in Moscow *Environmental Engineering* **1** 66

[7] Chernykh O N Volkov V I Sabitov M A Altunin V I 2014 About some aspects of assessing the amount of probable damage as a result of an accident of hydraulic structures *Environmental Engineering* **4** 46

IOP Conf. Series: Materials Science and Engineering 1159 (2021) 012011 doi:10.1088/1757-899X/1159/1/012011

A method for estimating the potential of the physiological and emotional characteristics of a driver in a changing traffic situation using the ADES information system

A S Dotkulova[1,2]

[1] Department of Higher Mathematics, Moscow Automobile and Road Construction State Technical University (MADI), Leningradsky Prospect 64, Moscow, 125319, Russia
[2] Department of Mathematical Cybernetics and Information Technologies, Moscow Technical University of Communications and Informatics (MTUCI) Aviamotornaya 8-a, Moscow, 111024, Russia

E-mail: asoll05@inbox.ru

Abstract. Road traffic accidents are one of the problems of organizing road safety. According to statistics, every fifth road traffic accident in Russia occurs due to the unsatisfactory condition of streets and roads. The driver must react quickly to changes in the traffic situation while driving. Analysis of the human factor and identification of patterns in the driver's behavior on the road remains an urgent issue. This article is devoted to the design of an information system that helps to assess the emotional and psychophysiological state of the driver, for further forecasting his actions on the road. The method of processing the research results using the Wilcoxon test is presented.

1. Introduction

The road transport industry is actively developing. There are more and more cars on the roads, which provokes the appearance of road defects. While driving, the driver must quickly respond to the appearance of holes in front of the hood, potholes, roadbed subsidence, deviations of the manhole cover, etc. It is especially difficult to identify them in the dark and winter conditions. Therefore, checking such driver characteristics as reaction time, attentiveness, concentration is highly relevant.

The structure of the flow itself is currently changing [1]. Different levels of autonomy of vehicles appear on the roads, making their own adjustments to the driver's behavior on the roads. The structure of cars becomes more complex, there are many electronic elements that, on the one hand, help the driver, on the other, they require a sequence of actions, for the implementation of which it is necessary to perform the actions without errors. In this regard, it is necessary to assess not only the driving experience, but also the condition and mood of the driver. Therefore, it becomes necessary to classify driver behavior on the road and develop methods for monitoring and assessing dangerous driving [2]. Emotional state largely influences driving style, concentration and decision-making time, so it is important to find a correlation between the behavior and state of the driver during driving. Such analysis can greatly assist in predicting the road behavior of fully autonomous vehicles. Thus, the development of system that helps to assess the emotional and psychophysiological state of the driver is relevant.

Content from this work may be used under the terms of the Creative Commons Attribution 3.0 licence. Any further distribution of this work must maintain attribution to the author(s) and the title of the work, journal citation and DOI.
Published under licence by IOP Publishing Ltd

IOP Conf. Series: Materials Science and Engineering 1159 (2021) 012011 doi:10.1088/1757-899X/1159/1/012011

2. Description of the development environment for game components

The developed information system (IS) Analyzer of Driver Emotional State (ADES) has a client-server architecture similar to work [3]. The main feature of the IS is a testing model using gaming technologies created with Unity.

Unity is a cross-platform computer game development environment developed by the American company Unity Technologies [4]. Unity lets you create applications that run on over 25 different platforms, including personal computers, game consoles, mobile devices, web applications and more. The main advantages of Unity are the presence of a visual development environment, cross-platform support and a modular component system.

The Unity editor has a simple Drag & Drop interface that is easy to customize, consisting of various windows, so you can debug the game right in the editor. The engine uses C # to write scripts. Unity also supports solid and tissue physics. The editor has an object inheritance system; child objects will repeat all changes to the position, rotation, and scale of the parent object. Scripts in the editor are attached to objects as separate components.

When importing a texture into Unity, you can generate an alpha-channel, mip-levels, normal-map, light-map, a reflection map, but you cannot attach a texture directly to the model - a material will be created to which a shader will be assigned, and then the material will be attached to the model. The Unity Editor supports writing and editing shaders. The Unity editor has a component for creating animations, but animations can also be pre-created in the 3D editor and imported with the model, and then split into files.

3. Filling the information system

Unity is a multi-platform 3D and 2D game development tool. This engine provides a fast and flexible way to port your application to almost any popular operating system. The developed information system ADES consists of 7 games aimed at determining the physiological characteristics of the driver:

- Test game for reaction and concentration.

The task is to keep the red square within the playing field for as long as possible and not collide with obstacles. Output data: number of attempts and elapsed time.

- Reaction speed.

The test game is designed to determine a simple reaction speed. The task is to press the button in the minimum amount of time when the screen color changes.

- Mindfulness and memory.

The "Matrix" test game is designed to test and train visual memory. The task is to open all pairs of pictures hidden behind a blue background in the shortest possible time. The goal of the "Matrix" game is to find identical images under the buttons, remembering the desired color and location of the button it needs.

- Test game "Ball" to test a simple reaction.

The game implements a dynamic drop of objects from the top of the screen. The player needs to collect all the balls, trying not to make mistakes. In addition to the "ball" object, other round objects fall from above.

- Test game "Difficult reaction".

To determine the difficult reaction of the driver, a game was created, the goal of which is to drive as far as possible and score the most points. The game contains objects that help the player and hinder the achievement of the goal. In the game, depending on the objects caught, the speed of the game, the direction, the level of health changes.

- Test game for switching and distribution of attention.

The game "Switching and Distribution of Attention" is to correctly determine the shape and the location (complex reaction) of a figure in a minimum amount of time by pressing a button (simple reaction). Description of the game: the screen of the driver's device is divided into two parts, in which random figures will appear; the driver must choose the side of the screen on which there is no figure in the minimum time; the game ends after 60 seconds. During the game, the program calculates the correct and incorrect answers, as well as the average response time, this data is displayed on the screen, and is written into variables and stored in the database for further processing.

- Test game for coordination and concentration.

IOP Conf. Series: Materials Science and Engineering 1159 (2021) 012011 doi:10.1088/1757-899X/1159/1/012011

The game "Connect the dots without taking your hands off" is to connect the black and red squares with a line in the labyrinth in the minimum amount of time without taking your hand off. The game ends if the player loses (goes out of the screen, the line crosses itself, touches the walls of the maze) or wins.

4. Analysis of the obtained results of the ADES system

The developed information system ADES displays the results of all games in one window. It is enough to select a game from the drop-down list and the statistics of all the results of passing games will be displayed on the screen. If a sufficient number of games have been played (more than 10 playthroughs), then the system displays the trend of change in the progress rates (Figure 1).

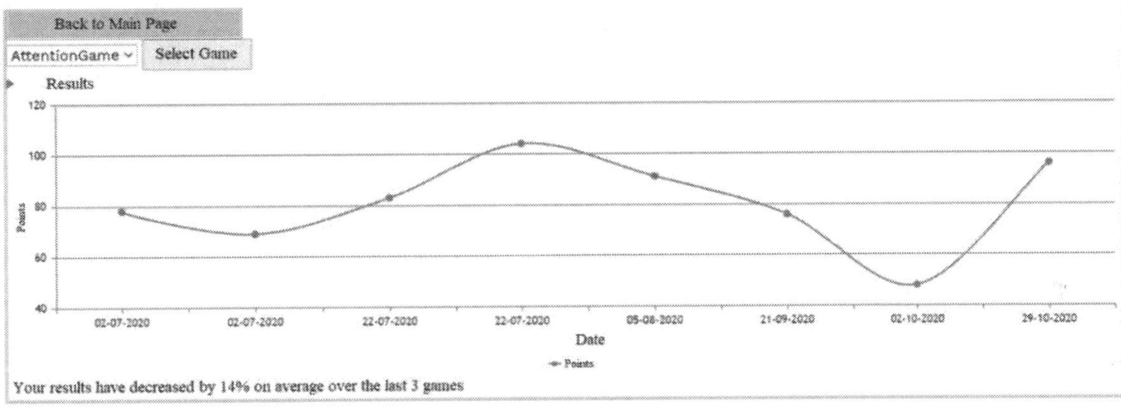

Figure 1. Results of passing the game "Switching and distribution of attention".

All game results are stored in a database, from where they can be downloaded. The uploaded data can be presented in the form of table 1.

Table 1. Fragment of test results and data processing.

Date	Time	Place	Device	State	Mood	Result		
						Correct clicks (pcs.)	Wrong clicks (pcs.)	Average response time (s)
19.08.2020	8:26	Home	Computer	Sleepy	Suppressed	33	22	1,097423
20.08.2020	10:04	Home	Computer	Vigorous	Merry	83	11	0,7130257
21.08.2020	11:01	Home	Computer	Relax	Miraculous	111	2	0,530899
22.08.2020	11:47	Home	Computer	Vigorous	Unstable	107	46	0,5493246
23.08.2020	12:55	Home	Computer	Irritated	Pensive	105	4	0,5470147
24.08.2020	14:13	Home	Computer	Vigorous	Cheerful	107	5	0,5362769
25.08.2020	16:08	Home	Computer	Tired	Cheerful	107	3	0,5435227
26.08.2020	17:10	Home	Computer	Sleepy	Suppressed	106	6	0,5392932
28.08.2020	22:48	Home	Computer	Tired	Pensive	109	6	0,5247977
20.08.2020	12:00	Home	Computer	Vigorous	Merry	104	3	0,5634636
21.08.2020	14:09	Home	Computer	Vigorous	Depressive	107	5	0,5361282
22.08.2020	15:27	Home	Computer	Vigorous	Merry	106	5	0,5394908
23.08.2020	17:15	Home	Computer	Vigorous	Smiling	105	3	0,5519719
24.08.2020	19:33	Home	Computer	Tired	Depressive	105	7	0,5329787
25.08.2020	20:45	Home	Computer	Irritated	Cheerful	103	3	0,570735
26.08.2020	21:11	Home	Computer	Vigorous	Miraculous	102	1	0,5788567
27.08.2020	21:57	Home	Computer	Worried	Unstable	100	3	0,5831272
28.08.2020	22:48	Home	Computer	Tired	Pensive	109	6	0,5247977

The results of tests using gaming technologies are processed using statistical analysis. The Wilcoxon test was used as a statistical processing method [5].

Three different pairs of emotional states of drivers were considered: “depressive” + “calm”, “depressive” + “smiling” and “smiling” + “calm”. For three pairs, the dependence of the number of correct answers, as well as the reaction time of the subject on his emotional state, was tested. Checking the dependence of the test results on the physiological state was considered in work [6].

IOP Conf. Series: Materials Science and Engineering 1159 (2021) 012011 doi:10.1088/1757-899X/1159/1/012011

Table 2 Calculation of Wilconson's criterion for the dependence of correct answers

	Name	**Value**		
	Emotional condition	Depressive	Smiling	Calm
	Condition	A	B	C
	Selection size n_i	123	135	229
	Variation series	$N_1=n_1+n_2=258$		
			$N_2=n_2+n_3=364$	
		$N_3=n_1+n_3=352$		
For N_1	**Hypothesis H_0**	The number of correct answers does not depend on whether the experiment participant is in state A or in state B		
	Hypothesis H_1	The number of correct answers depend on whether the experiment participant is in state A or in state B		
	W observed	21706		
	Significance level	0,1		
	W lower	20212		
	W upper	23207		
	Conclusion	Confirmation of the hypothesis H_0: The number of correct answers does not depend on whether the participant is in state A or in state B		
For N_2	**Hypothesis H_0**	The number of correct answers does not depend on whether the experiment participant is in state B or in state C		
	Hypothesis H_1	The number of correct answers depend on whether the experiment participant is in state B or in state C		
	W observed	16507		
	Significance level	0,1		
	W lower	14943		
	W upper	16913		
	Conclusion	Confirmation of the hypothesis H_0: The number of correct answers does not depend on whether the participant is in state B or in state C		
For N_3	**Hypothesis H_0**	The number of correct answers does not depend on whether the experiment participant is in state A or in state C		
	Hypothesis H_1	The number of correct answers depend on whether the experiment participant is in state A or in state C		
	W observed	23459		
	Significance level	0,1		
	W lower	23042		
	W upper	26233		
	Conclusion	Confirmation of the hypothesis H_0: The number of correct answers does not depend on whether the participant is in state B or in state C		

To check the dependence of the response time on the state, a similar calculation was carried out. Hypothesis H_2 was assumed - the dependence of the reaction time on the emotional state; hypothesis H_3 was chosen as a competing hypothesis - the absence of the dependence of the driver's reaction time on the emotional state. Processing the results of the experiment and testing hypotheses H_0 and H_2 in Table 3.

Table 3. Summary table with results.

Confirmation of hypothesis H_0 for a pair of states			Confirmation of hypothesis H_2 for a pair of states		
N_1	**N_2**	**N_3**	**N_1**	**N_2**	**N_3**
+	+	+	+	+	+

5. Conclusion

A methodology for testing a driver using gaming technologies to determine emotional and physiological states is presented. The presented studies have shown that the emotional state of the driver affects the reaction time and the total number of responses. On average, the ratio of the number of correct answers to all remains unchanged. These studies can be used in the training of drivers before entering a real road, as well as in studies of the influence of road situations on the condition and characteristics of the driver.

References

[1] Zhankaziev S 2017 Current trends of road-traffic infrastructure development *Transportation research procedia* T 20 pp 731–739

[2] Gorodnichev M G, Yashina M V, Polyantseva K A 2019 Modern methods of dangerous traffic monitoring *Technologies of the information society* pp 414–416

[3] Yashina M V, Dotkulova A S, Zernov A V, Korableva E V 2018 Search Optimization of Tutorial Web-Service Architecture for Cognitive Mathermatical Education *IEEE International Conference "Quality Management, Transport and Information Security, Information Technologies"* (IT&QM&IS) pp 828–831

[4] Welling L, Thomson L 2019 *Web Application Development with PHP and MySQL* (Williams)

[5] Gmurman V 2013 *Probability theory and mathematical statistics* (Yurayt-Publishing)

[6] Dotkulova A S, Yashina M V, Tatashev A G, Trofimenko Yu V 2020 Attention driver evaluation in collective traffic behavior via gaming technology *EMCTECH* (In print)

Development of a distributed mobile system «SiLOK» for the task of reconstructing a three-dimensional model of a roadbed rut

M V Yashina[1,2], I A Kuteynikov[1,2], O S Lavrov[1],

[1] Moscow Automobile and Road Construction State Technical University (MADI), Leningradsky Prospect 64, Moscow, 125319, Russia
[2] Moscow Technical University of Communications and Informatics (MTUCI) Aviamotornaya 8-a, Moscow, 111024, Russia

E-mail: ivankuteynikov09@gmail.com

Abstract. Roadbed defects are inconvenient, especially in developing countries, and can often result in vehicle damage or physical injury to passengers. By using smartphones as a source of information on roadbed conditions, drivers can be warned of approaching problem areas in real time. In addition, the location of problem areas can be saved and shared to help other drivers and road services. This paper presents a client-server mobile system "SiLOK" for reconstruction of a three-dimensional model of a roadbed rut based on a smartphone and a profilometer. The system allows the driver to register the presence of a roadbed rut, after which data about it is transmitted to the server and recorded in the database as feedback. Also, the system, based on the analysis of the route processed by other users of the system, warns the driver about approaching the roadbed rut section.

1. Introduction

The growing pace of motorization and road construction around the world is making it more and more expensive to maintain and repair existing areas. Roads are increasingly stressed and worn out faster, leading to roadbed ruts that can lead to road accidents. The problem of roadbed rut formation and its elimination has been considered one of the most important tasks for many countries of the world for a long time.

The main reasons for roadbed rut formation include wear of the top layer of the pavement as a result of the combined effect of wear and premature abnormal destruction of the asphalt concrete layer under the influence of external factors, which include, along with the impact of wheels, precipitation, temperature changes and exposure to the sun [1].

Rutting on the road leads to loss of fuel, tire wear and further damage to the vehicle. All of these problems require systematization of the collection of information about bad road conditions and direction towards analysis, driver warning and road repair.

A group of scientists from the Department of Higher Mathematics of MADI under the leadership of Professor A.P. Buslaev since the end of the 20th century has been engaged in the development of traffic flows, optimal control taking into account traffic jams and engine load modes in various road conditions, using automatic processing of data collected from mobile laboratories [2]. As a solution to the roadbed rut detection problem, a client-server system based on a smartphone running Android operating system

Content from this work may be used under the terms of the Creative Commons Attribution 3.0 licence. Any further distribution of this work must maintain attribution to the author(s) and the title of the work, journal citation and DOI.
Published under licence by IOP Publishing Ltd

and a mobile laboratory "OTROK" of the Department of Higher Mathematics "Smartphone and Laboratory Track Processing" (SiLOK) is proposed.

At the stage of collecting information, a vehicle with a smartphone with a camera installed on its frontal part moves along the road, filming a video. During the analysis phase, this data will be processed by an algorithm to detect the roadbed rut in front of the laboratory.

2. Methology

The operation of the system includes the use of computer vision algorithms for rut detection using video / photo data, as well as mathematical methods for processing data from profilometers. The use of two methods simultaneously improves the detection accuracy in cases of poor visibility or foreign objects on the road.

2.1 Recognition using computer vision methods

One of the operating modes of the three-dimensional roadbed rut model restoration system is a visual approach using video data from a smartphone, processed using computer vision algorithms without the need for machine learning algorithms.

The proposed rut detection method begins with the selection of a trapezoidal area in front of the vehicle (Fig. 1 a)) and its transformation into a black and white image (Fig. 1 b)). Binarization is carried out by comparing the brightness of each pixel with a threshold t ($0 < t < 1$). Based on the comparison results, the pixel is assigned a value of 0 (black) or 1 (white).

To clean up the image and remove noise, a median filter is applied to it (Fig. 1 c)). This filter takes the values of its neighbors for each pixel, sorts and sets this pixel to the mean value.

After all transformations to detect the rut, the found clusters of pixels are highlighted in a straight line by interpolation in brightness, which is applied to the original image (Fig. 1 d)).

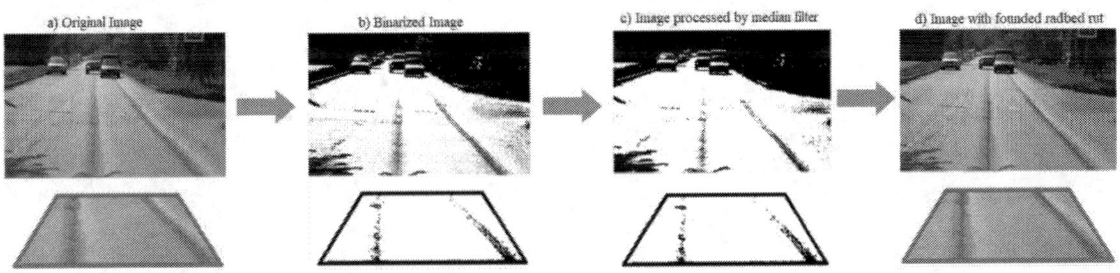

Figure 1. Image processing pipeline.

The advantages of this method include the simplicity of design and relatively low cost, depending on the price of the smartphone. The disadvantages are the requirements for illumination and weather conditions, as well as low measurement accuracy.

2.2 *Recognition using a profilometer*

The profilograph is based on laser sensors. The measuring system is used to measure the rut depth on the roadway.

A beam of light from a semiconductor laser diode, while driving, hits the roadbed surface at a frequency of 6 Hz. A detector installed in the sensor housing determines the profile of the roadbed, i.e. its deviation from the usual average level and converts the obtained value into an electrical signal, on the basis of which electronic devices can calculate the actual distance to the measuring object. The design of the laser sensor allows measurements in the range of an average height of 100 mm from its bottom edge to a distance of 300 mm. The measurement resolution of the laser sensor is 3 mm. Measurements are taken every 10 cm of the vehicle's path. All signals are transmitted digitally.

The advantages of this method are: high measurement accuracy (up to 0.1 mm) and the ability to work in bright sunlight. The disadvantages are high cost and design complexity.

3. Hardware architecture of the "SiLOK" system

The hardware basis of the system is a combination of a smartphone with a camera and a GPS module and a mobile laboratory with a profilometer and a laptop.

3.1 Mobile laboratory "OTROK"

Mobile laboratory "OTROK" ("Equipment for Telematics and Image Recognition on Wheels") (Fig 2. a)), built on the basis of the VOLSWAGEN TARNSPORTER T5 car, has combined several key properties at once. The main thing was versatility, in particular, in terms of technical training and equipment.

The OOM-MADI profilometer, installed on the mobile laboratory "OTROK" (Fig. 2. b)), was developed by the students of the Department of Higher Mathematics A. V. Tikhonov, A. N. Nigmatulin and Gorodnichev M.G. under the guidance of Doctor of Physical and Mathematical Sciences prof. A.P. Buslaev [3]. To implement the recognition function, five Dimertrix DLS-C30 laser distance sensors (rangefinders) are used, attached to a transverse rod, which is attached to the front amplifier of the mobile laboratory. Also used is the RS signal concentrator MOXA, which receives the data obtained by the rangefinders during operation and transmits them to the USB port of a laptop installed in a mobile laboratory. The laptop transfers data from the profilometer to the server.

Figure 2. a) Mobile laboratory "OTROK". b) OOM-MADI profilometer.

3.2 Terminal device for a smartphone of the "SiLOK" system

Currently, smartphones are created on the basis of System on a Chip [4]. It includes the following components (Fig. 3):

1. An application processor responsible for executing user application software using instructions from middleware and an operating system (OS).
2. A baseband processor (or modem) with its own OS components, performing radio transmission in the main frequency band and receiving audio, video and data.
3. Various user interface peripherals, including high resolution camera, an accelerometer, a gyroscope, a magnetometer, proximity and light sensors.

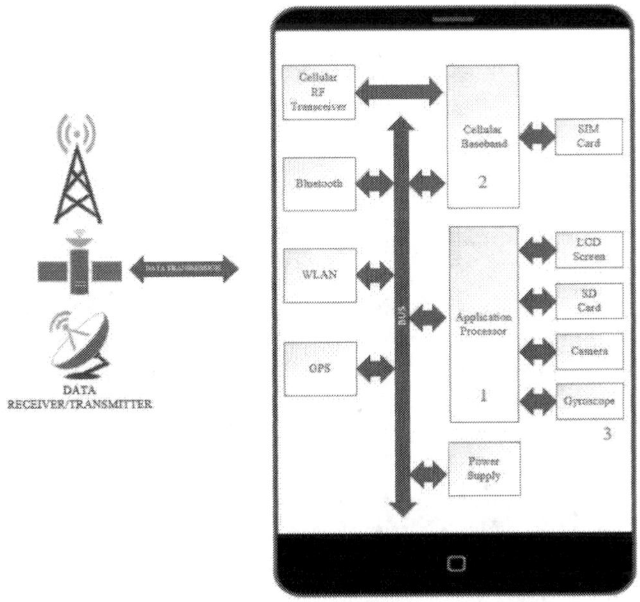

Figure 3. Smartphone Hardware Architecture.

Data from the smartphone camera along with GPS coordinates are transmitted to the server for further processing.

4. Software architecture of the "SiLOK" system

4.1 Client part

Currently, the Android operating system is the most widespread mobile operating system in the world (~ 75% of all mobile devices) [5]. It is suitable for creating applications using hardware sensors and has extensive capabilities for measuring and storing the kinematic parameters of the device and the electromagnetic characteristics of the environment. Almost all Android devices are equipped with microelectromechanical sensors, they are of two types: real (provide initial data) and virtual (provide the necessary level of abstraction between the application code and low-level device components). The client part on the smartphone is implemented on the Android platform. It is responsible for receiving the image, sending it to the application server and displaying the processing results. The JSON stream received from the server is used to display the classification results.

The client side on a smartphone has the following features:

- Ensuring the operation of the user interface - the client part includes the interface of the entire system. Using XML layouts from Android APIs, an interactive user interface is created. Using the interface, the user can access the camera and start receiving video data.
- Link to Object Extraction Components - Provides a channel for transferring the captured image to the Object Extractor component.
- Providing an interface for communication with the server - allows the client to receive messages from the server.
- Display results - responsible for displaying results from the server. Runs the JSON parser and provides the result to the user.
- Latency logging - The profiler is used to track the time spent on application components on the mobile side.

The client part on a laptop running under the Windows 10 operating system is implemented in Java. The tasks of the software product include collecting, primary processing of data received via the USB port from the profilometers, sending data to the server and receiving data, followed by visualization of a two-dimensional model and a 3D model of a road track.

Client-server communication protocol. The communication protocol is part of the client applications on both devices. The function of the protocol is to provide a standard mechanism for the exchange of data between the server and the client application. The client side uses the standard HTTP protocol to communicate with the server. For a developed application, Android provides standard APIs that must be called to implement this protocol.

The client part on a laptop running under the Windows 10 operating system is implemented in Java. The tasks of the software product include collecting, primary processing of data received via the USB port from the profilometers, sending data to the server and receiving data, followed by visualization of a two-dimensional model and a 3D model of a roadbed rut.

Client-server communication protocol is part of the client applications on both devices. The function of the protocol is to provide a standard mechanism for the exchange of data between the server and the client application. The client side uses the standard HTTP protocol to communicate with the server. For a developed application, Android provides standard APIs that must be called to implement this protocol.

4.2 Server part

The server part of the application consists of an application server, which is responsible for the implementation of machine vision algorithms for recognizing tracks from images received from a smartphone, as well as processing and interpolating data obtained from the profilometer. Before starting the algorithm for restoring the three-dimensional model of the roadbed rut, we must create a database on the database server. This database stores images, GPS coordinates and data from the profilometer. The server software includes a number of utilities for the subsequent processing of the received data, as well as from Internet scripts for automatically superimposing the received data on the maps of the Google Maps service.

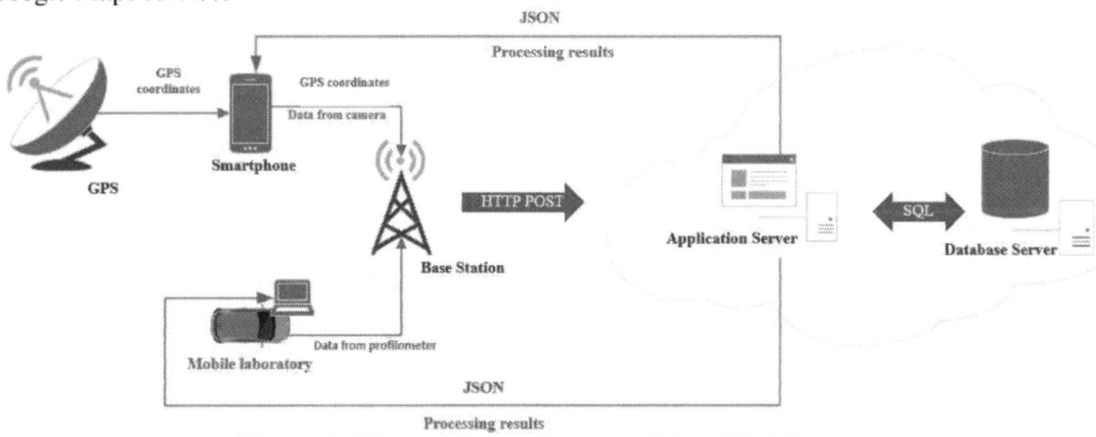

Figure 4. Client-server architecture of the «SiLOK» system.

The application server uses Java EE technology. The Jakarta Enterprise Edition platform includes libraries that provide functionality for deploying fault-tolerant, distributed, layered Java software based primarily on modular components running on a server. WildFly Application Server was selected as the server because it is a Jakarta EE certified platform for developing and deploying enterprise Java and web applications. WildFly Application Server also provides full Jakarta EE functionality as well as advanced enterprise services including clustering, caching and storage.

The server side database contains the relevant information about the image as a vector element. Depending on the type of application, the database may contain additional information about the image such as GPS or time stamp. The request from the client can also contain additional parameters to further refine the information required to access the database.

After recognition is complete, the application queries the database to retrieve the metadata for the image. The server communicates with the client via HTTP. The server communication components

IOP Conf. Series: Materials Science and Engineering 1159 (2021) 012012 doi:10.1088/1757-899X/1159/1/012012

receive information in the form of a formatted JSON file. It is then sequentially sent to the mobile client for further processing.

5. Logical scheme of the "SiLOK" software package

The developed technology is intended for automatic processing of information in real time received from a smartphone and mobile laboratory profilers about the state of the road infrastructure.

Since the camcorder shoots with a frequency of ~ 25 frames / sec. Due to the limited server performance in real time and small changes in the scene in the image, it is possible to select the image transmission interval (every 1-12 frames / sec.) and the GPS coordinates transmission interval.

In a multitasking environment, you can parallelize the running processes of subtasks and distribute them across several processors. Thus, each of the subtasks will work in parallel and independently of each other and at the same time access the same source data, including images, GPS data, data from profilometers, etc. Therefore, all records created by the system can be processed by a different class of subroutines. This conceptual feature allows for complex monitoring and information processing not only in real time, but also in laboratory conditions.

6. The results of the "SiLOK" software package

The interface of the mobile application allows to select the frame sampling rate, the transmission interval of GPS coordinates, the ability to set a new route or continue the existing one, create a route on Google Maps (Fig. 5 a)).

The application interface on a laptop allows communication with the component parts of the profiler via COM ports (Fig. 5 b)). Each of the ports associated with the laser rangefinder through the MOXA RS signal concentrator transmits data to the system in the RS-232 standard format (Fig. 5 c)). The same window is used to read / write data from the profilometer.

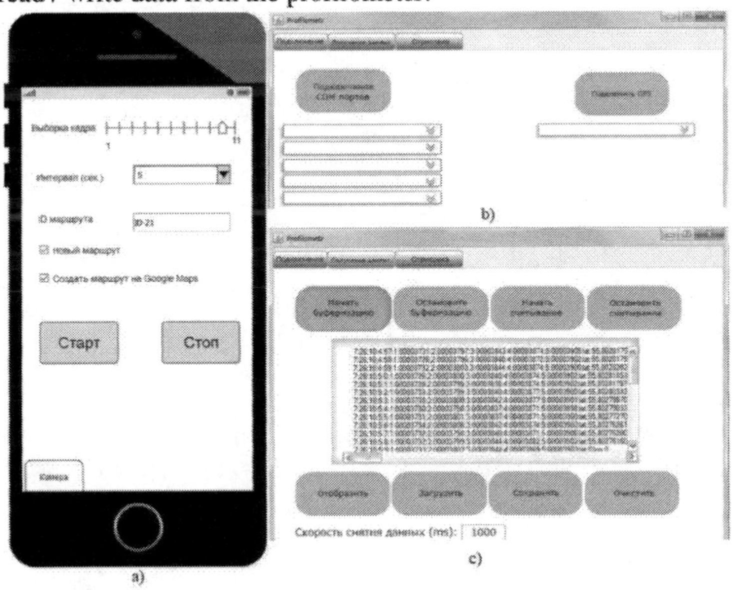

Figure 5. Interface of the "SiLOK" software package. a) Mobile application interface b) Interface for connecting to profilometers c) Receiving data from profilometers

The data from the profilometers is drawn step by step at a specified interval. Rangefinder lasers act as points on the graph. Applying linear interpolation, the points are connected by straight line segments, as a result of which a transverse profile of the road surface is drawn (Fig. 6).

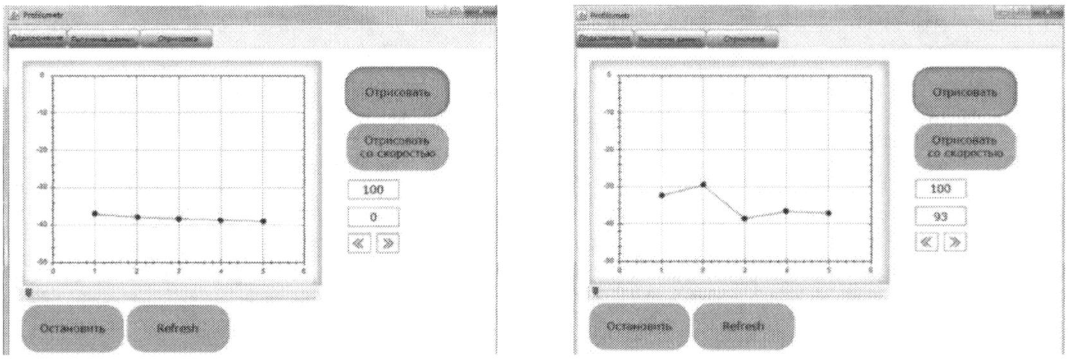

a) b)

Figure 6. a) a roadbed without a rut; b) roadbed with a rut

Based on the obtained GPS data and track detection methods, it is possible to build a route of movement [6], displaying sections with a roadbed rut on it (Fig 7.). Thus, other drivers and road services who have access to the system will be aware of the presence / absence of ruts on the route.

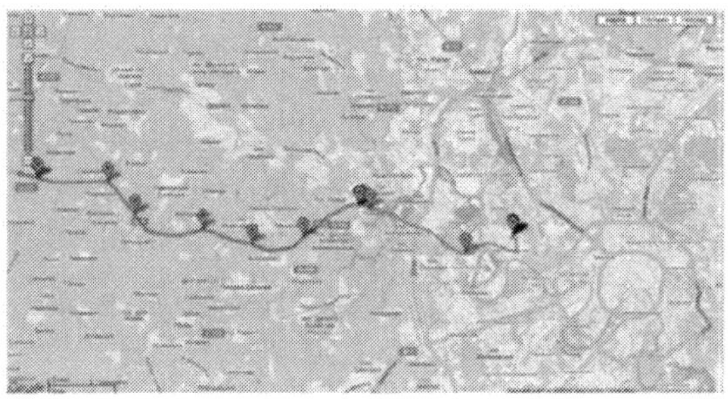

Figure 7. An example of displaying the received coordinates on the Google Maps service.

7. Conclusions

The paper presents a distributed mobile system "SiLOK" for the task of reconstructing a three-dimensional model of a roadbed rut. The principles of the system operation, applied methods and algorithms are described. The description of the client-server architecture and the primary results of the roadbed analysis are given.

As a further work, data collection for the application of machine learning algorithms is considered, which should lead to an increase in the accuracy of rut detection from a smartphone camera.

References

[1] Lugov S V and Kalenova E V 2013 The possibility of pavement wear valuation at rutting predicting *Bulletin of the Moscow Automobile and Road Construction State Technical University (MADI)* **4** 53–59

[2] Buslaev A P, Dorgan V V, Kuzmin D M, Prikhodko V M, Travkin V Yu and Yashina M V 2005 Recognition of Images and Monitoring of Road Conditions, Traffic Flows and Safety *Bulletin of the Moscow Automobile and Road Institute (State Technical University)* 4 102–109

[3] Tikhonov A V and Nigmatulin A N 2012 *Technology for automatic capture of information about the parameters of distributed dynamic systems. Mobile laboratory of intelligent monitoring OTROK: teaching aid* (Moscow: Tekhpoligraftsentr) ISBN 978-5-94385-085-1

[4] Furber S B 2000 *ARM system-on-chip architecture* (Pearson Education)

[5] The share of mobile operating systems in the world market Available at: https://gs.statcounter.com/os-market-share/mobile, last accessed 2020/10/20
[6] Abyshov R, Buslaev A, Kuprianov U and Yashina M 2011 Distributed system for maintenance monitoring of motor roads *Bulletin of the Moscow Automobile and Road Construction State Technical University (MADI)* **1** 79–85

Positioning accuracy estimation for on-board data processing at the road surface defects monitoring

M V Yashina[1,2], A G Tatashev[1,2], M Y Fomina[1]

[1] Moscow Automobile and Road Construction State Technical University (MADI), Leningradsky Prospect 64, Moscow, 125319, Russia
[2] Moscow Technical University of Communications and Informatics (MTUCI) Aviamotornaya 8-a, Moscow, 111024, Russia

E-mail: mv.yashina@madi.ru, a-tatashev@yandex.ru, fomina.m.yu@yandex.ru

Abstract. In this work, an approach has been developed to analyze data on the movement of a vehicle obtained as a result of measurements carried out using a satellite navigation system. The issues of increasing the accuracy of such measurements are of great importance in quality monitoring of road maintenance for geo-referencing of detected roadway defects. When using the proposed approach, the speed of a vehicle is considered as a stationary random process, the correlation function of which belongs to one of the specified classes. Based on the monitoring results, the value of the correlation function parameter is selected, which, in a certain sense, best matches the measurement results. The problem is also solved in a certain sense of the optimal choice of the duration of the averaging interval of the velocity values obtained as a result of monitoring, while taking into account how much, for the selected duration of the averaging interval, the sample obtained correlation function of the velocity corresponds to the correlation function from the hypothetical class for the selected parameter value.

1. Introduction

Vehicle traffic monitoring is relevant for the tasks of flow control, road safety and correct routing, as a driver assistance. One of the most problematic road conditions associated with road accidents in the Russian Federation is the poor condition of the road surface. The challenge is to improve the accuracy of GPS / Glonass vehicle speed data.

Monitoring issues on highways were considered in works [1,2]. In [1], the issues of obtaining and analyzing the results and the theory of traffic flows, including those related to the model of following the leader [3, 4], are considered. In [5], the issues of assessing and improving the accuracy of measurement results, in which the movement of vehicles is monitored using satellite navigation positioning systems, is considered.

In this article, an algorithm has been developed for analyzing changes in the speed of a vehicle on a short section of the highway using data obtained as a result of measurements carried out on the highway. The process of changing the speed on a section of the highway is characterized by a stationary random process, the mathematical expectation and correlation function of which are estimated on the basis of the measurement data carried out using a GPS sensor.

Content from this work may be used under the terms of the Creative Commons Attribution 3.0 licence. Any further distribution of this work must maintain attribution to the author(s) and the title of the work, journal citation and DOI.
Published under licence by IOP Publishing Ltd

IOP Conf. Series: Materials Science and Engineering 1159 (2021) 012013 doi:10.1088/1757-899X/1159/1/012013

2. Recognition in the theory of traffic flows (TF)

Automatic processing of video and audio information about the characteristics of TF has been developed at MADI (GTU) since the early 2000s. Technologies that perform computer processing of video recording of a multi-lane road traffic flow in laboratory conditions were tested on sections of the «Central Russia» highways adjacent to Moscow. Subsequently, the tasks of assessing the intensity, density, composition of traffic flows were implemented in real time without intermediate laboratory processing of video material. The work of recognizing TF by audio signals was also carried out, which makes practical sense in conditions of insufficient visibility. In the course of work, it turned out that the range of practical problems that can be solved by processing the signal of a video source is extremely wide. A large package of algorithms for automatic monitoring on the road, which can work in parallel, is called MonStr (Street Monitoring). This software package is based on the analysis of video and photo-stream received from video and photo cameras, which are installed in a mobile laboratory and allow research and multipurpose monitoring. Among the goals are: monitoring the condition of the road, assessing the quantitative and qualitative characteristics of the TF composition, assessing road traffic controls, assessing compliance with traffic rules. Video and photographic material obtained from the source is processed on a computer for each of the tasks. The work of the program is based on the analysis of the chromaticity matrix of the pixels of each frame of the flow.

The sources of information through which monitoring is carried out are devices classified according to the type of data acquisition: video camera, photo camera, navigation device (GPS), microphone, various sensors (light, vibration), etc. Depending on the task and the developed scheme of the algorithm, the device can be located both in the cabin of the mobile laboratory (ML) and fixed outside.

So, the fastening of the video camera is carried out on an external tripod attached to the ML body (for example, on the trunk). However, in winter, at low temperatures, fastening can be carried out inside the ML on a specially equipped tripod. The work of all devices is coordinated by a computer with an installed software package (for each subtask its own subroutine). The computer is installed in the ML compartment and carries out bi-directional operation and data collection from all connected devices, controls the recording mode of video and photographic equipment, collects data on the current coordinates received from the navigation device.

After the central computer receives data from the video camera, the frame matrix of the video stream is scanned in real time according to the mask of a given area and issues a command, depending on the task being performed, either to the video camera to switch to the recording mode, or to the camera to the photography mode. Regardless of the command issued, the received information is stored in the computer, which allows further analysis and processing of data in the laboratory. Also, the computer memorizes the navigation coordinates (latitude and longitude) and saves them to the database, which allows the binding with an accuracy of 2m.

Used devices and equipment:

- a computer of the Notebook type, or a small-sized personal computer;
- a video camera and a camera supporting the architecture of working with a computer;
- GPS navigation device;
- tripod for video and photographic equipment;
- peripherals: directional microphone; a device for measuring illumination.

IOP Conf. Series: Materials Science and Engineering 1159 (2021) 012013 doi:10.1088/1757-899X/1159/1/012013

Figure 1. Installation options for ML equipment.

3. Basics of technology for monitoring and analyzing road conditions

The control car - a mobile laboratory is moving along one of the lanes of a multi-lane road. On board of the ML there is a set "Video camera - computer". The camera operates in the observation mode, and the selected frames are processed in accordance with the current task. If, in addition, the camera is mounted on a mobile controlled tripod, the computer can control the shooting angle and optical zoom. This simulates the workings of the human eye. At the same time, the camera-computer pair, in contrast to the eye-brain ligament, can perform a large number of tasks in parallel. Let's describe the main tasks.

The problem of determining the intrinsic velocity of the ML as a function of time can be solved by means of ORZ technologies. However, even in this case, the analysis of periodic perturbations of the chromaticity of a special area of the screen (for example, pillars of illumination) can give an accuracy comparable to modern technologies.

a) b)

Figure 2. a) the control area automatically recognizes road markings; b) the section of the road marking defect is highlighted; c) the control area automatically recognizes the uniformity of the roadway and gives an estimate of its area; d) an unevenness on the roadway was revealed and its area was estimated as ~ 0.3 m^2

Similarly to the previous one, the problem of determining the *speed of vehicle moving along a multi-lane road* is solved if the speed of the ML is known. The practical significance consists in the classification of vehicles according to the speed of movement (fast, correct, slow (obstacles)) and observance of traffic rules. Note also that on the basis of the speedometer and the above-mentioned speed estimates, it is easy to determine the degree of wheel slip relative to the road surface.

Inverse problem: *checking periodic structures* on the road using information about the speed of the ML. The control of the uniformity of the placement of objects of a certain type on the road (for example, pillars of fastening of the separation barrier) is carried out. In this case, the speed of the ML does not have to be constant.

Control over the continuity of solid marking lines is implemented in several modifications. First, with stationary observation from a high point behind a large piece of a solid line (double solid line). In case of violation, the vehicle is identified and recorded in the database. Secondly, checking the quality of the markup and localizing areas where it needs to be updated. Thirdly, the analysis of the roadway for the presence of foreign components (individual objects, dirt, snow, etc.).

Dynamic clearance protection (DC) - part of the line in front of the ML, necessary for traffic safety. DC depends on the ML speed and is determined automatically. In the event of an intrusion into the DC area of any vehicle, video recording and identification of the intruder takes place. DC is necessary for the analysis of the road surface according to the described technology in conditions of saturated flows.

Integral estimation of road smoothness (flatness). By constructing the function of changing the position of a fixed point on the computer screen of the ML moving at a fixed speed, the total smoothness of the motion is determined. Then, on a section of a flat road at the same speed, the natural oscillations of the ML are determined. The results of these two functions are used to estimate the smoothness.

Determination of the geometric parameters of the road. The behavior of the projection of the ML velocity vector is analyzed. It was noted earlier that the length of the vector determines the size of the dynamic dimension. Changing the direction, in addition, allows you to evaluate other geometric parameters of the trajectory of motion: curvature, torsion.

Checking the uniformity of the road surface. Color analysis of the inviolable dynamic dimension of the ML allows to identify on the roadway such components as patches, cracks, pits, foreign objects, dirt, puddles. Each of the components has a physical nature and, therefore, can be represented as the total color components.

Determination of the area of patches. The method is based on checking the homogeneity of the road surface for the search for components - patches (5.8.) And uses formulas for transforming the perspective of the image into units of area. The parameters of the controlled area of the DC are used, such as the width, length, distance from the camera to the area and the height of the camera above the roadway.

Estimation of the volume of road surface defects. The approach is based on the estimation of the volumes of holes and potholes on the road when approximating by cylindrical bodies, and the side is separated from the base by the difference in illumination.

Determination of the secretion of the lanes. The method is based on the analysis of the color difference between two parallel lines along the movement.

Estimation of the intensity and composition of the vehicle "from the inside" in the mode of movement on one of the highway lanes. The intensity conversion is used taking into account the known velocity of the ML movement.

Constructing a mixing matrix and estimating the capacity of a complex intersection. *Building a road illumination map.* Identification of places with insufficient or poor illumination. *Road portrait or recognition technology.* A photo or video map of the road is created, which is stored in the computer's memory and serves as a reference. In the process of observation in real time, the area is recognized and compared with the standard. The goal is to determine changes in the speed of the ML with reference to the map.

4. Estimation of mathematical expectation and velocity correlation function

In this section, an algorithm has been developed for analyzing changes in the speed of a vehicle on a short section of the highway using the data obtained as a result of measurements carried out on the highway. The process of changing the speed on a section of the highway is characterized by a stationary random process, the mathematical expectation and correlation function of which are estimated on the basis of the measurement data carried out using a GPS sensor.

Suppose that as a result of measurements, n values of the vehicle speed $v_1, v_2 \dots, v_n$ were obtained. We assume that the velocity $v(t)$ over the time interval $(0, T]$, at which measurements are carried out, is a stationary random process [6]. The values of the velocity are fixed at times $i\Delta$, $i = 1, \dots, T = n\Delta$, , where Δ is the duration of the time interval after which the speed values obtained as a result of measurements are recorded. Let v_i be the speed value obtained as a result of measurement at time moment $i\Delta$, $i = 1, \dots, n$. Let m, D be the sample values of the mathematical expectation and variance of the velocity; $k(\tau)$, $\rho(\tau)$ are sample values of the correlation function and the normalized correlation function, respectively, at the value argument equal to τ . For the sample mean we have

$$m = \frac{1}{n}\sum_{i=1}^{n} v_i.$$

In order to smooth out fluctuations in the velocity values obtained as a result of measurements, we will average the velocity values over time intervals of duration $a\Delta$ (the size of the grouping interval in the terminology of the general theory of statistics [10]). The original sample $v_1, \dots, v_n$ is replaced by the sample

$$v_i(a) = \frac{1}{a}\sum_{j=(i-1)a+1}^{ia} v_j\,,\ i = 1, \dots, \left[\frac{n}{a}\right],$$

where square brackets represent the integer part of the number.

Let $k^*(l,a)$ be the value of the correlation function $k(\tau)$ estimated from the measurement results for the argument value $\tau = la\Delta$, where l is a non-negative integer, $0 \le l < [n/a]$, under the assumption that the estimation is performed with an averaging interval of duration $a\Delta$.

For the corresponding estimates, we have

$$k^*(l,a) = \frac{1}{\left[\frac{n}{a}\right] - l} \sum_{i=1}^{\left[\frac{n}{a}\right]-l} (v_i(a) - m)(v_{i+l}(a) - m)$$

(here the velocity correlation function is estimated by averaging over the products of velocity deviations from the sample mean on the intervals $((i-1)a\Delta, ia\Delta)$ and $((i+l-1)a\Delta, (i+l)a\Delta)$, with an averaging interval equal to $a\Delta$; the number of such pairs of intervals is $[n/a] - l$, since the last such pair, belonging to the entire measurement interval is the pair $(([n/a] - l - 1)a\Delta, ([n/a] - l)a\Delta)$ and $(([n/a] - 1)a\Delta, [n/a]a\Delta)$),

$$D(a) = k(0,a),$$

$$\rho^*(l,a) = \frac{k^*(l,a)}{D(a)}.$$

5. Estimation of the parameter of the velocity correlation function under the assumption that the correlation function belongs to a given class

Let us make the assumption that the normalized velocity correlation function belongs to one of the following classes:

$$\rho(\tau,\beta) = e^{-\beta\tau^2},$$

$$\rho(\tau,\beta) = e^{-\beta|\tau|},$$

$$\rho(\tau,\beta) = e^{-\beta|\tau|}(1 + \beta|\tau|).$$

For a given class and given values β and a, the mean square error of the approximation $F(\beta,a)$ is calculated by the formula

$$F(\beta,a) = \sqrt{\frac{1}{\left[\frac{n}{2a}\right]} \sum_{i=1}^{\left[\frac{n}{2a}\right]} \left(\rho^*(i,a) - \rho(ia\Delta)\right)^2} \qquad (1)$$

When choosing in practice the averaging interval of the speed values obtained as a result of measurements, the value of the error calculated by formula (1) is taken into account. If the averaging interval is too small, then it can be expected that, due to random fluctuations in the measured values of the velocity, calculated by formula (1) may be large. On the other hand, if the duration of the simulation interval is long, then information about the dynamics of the vehicle speed over the measurement time interval is lost.

6. Conclusion

Suppose that there is data on the coordinates of the points of location of the vehicle at successive equidistant points in time. For example, the coordinate values were recorded using a satellite navigation system at intervals of 1 sec. These coordinate values are used to determine the speed values of the given vehicle in intervals in seconds. Due to fluctuations in the error in determining the value of the coordinate, the obtained speed values will change abruptly. These fluctuations can be smoothed out by averaging the obtained velocity values by combining adjacent intervals of greater duration. The speed values obtained as a result of averaging more accurately approximate the real dynamics of the vehicle speed. In this work, an approach to the choice of the averaging interval has been developed, according to which the empirical correlation function of the velocity should be well approximated by the correlation function from the hypothetical class with the appropriately chosen parameter value.

References

[1] Yashina M V, Dotkulova A S, Susoev N P and Fomina M J 2019 Incommunication measurements for model validation of the follow-the leader model *2019 Systems of Signals Generating and Processing in the Field of on Board Communications* (Moscow, Russia) DOI: 10.1109/SOSG.2019.8706743

[2] Buslaev A P, Yashina M V and Volkov M M, Algorithmic and software aspects of information system implementation for road maintenance management. Theory and Engineering of Complex Systems and Dependability, pp 65-74, Proceedings of the Tenth International Conference on Dependability and Complex Systems DepCoS-RELCOMEX, June 29–July 3 2015, Brunow, Poland. DOI: 10.1007/978-3-319-19216-1 7

[3] Greenshilds D B 1934 A study in highway capacity *Proceed. Highway Research Board* vol 14 pp 448–477

[4] Treiber M and Hennecke A 2000 Congested traffic states in empirical observations and microscopic simulations *Physical Review. E 62: 1805–1824* DOI: 10.1103/PhysRevE.62.1805

[5] Haigen Min, Xin Wu, Chaoyi Cheng, Xiangmo Zhao 2019 Kinematic and dynamical vehicle model-assisted global positioning method for autonomous vehicles with low-cost GPS/camera/in-vehicle sensors *Sensors* **19** 5430 DOI:s19245430

[6] Rozanov Yu A 1982 *Introduction to the theory of stochastic processes* (Moscow: Nauka)

IOP Conf. Series: Materials Science and Engineering 1159 (2021) 012014 doi:10.1088/1757-899X/1159/1/012014

Hydraulics of the subcritical Ventury channels

A L Zuikov[1, 2]

[1] Moscow State University of Civil Engineering, 26, Jaroslavskoe Shosse, Moscow, 129337, Russia
[2] Moscow Automobile and Road Construction State Technical University, 64, Leningradsky prospect, Moscow, 125319, Russia

E-mail: zuykov54@mail.ru

Abstract. The purpose of the research is to develop and verify the method of hydraulic calculation of the Venturi channel in the flooded outflow mode. The criterion that separates the flooded and free flow regimes is the ratio of the flow depth in the throat of the Venturi channel h_m to the critical depth h_c exceeding one ($h_m/h_c > 1$). The article theoretically and experimentally substantiates a method for determining the flow rate in a flooded Venturi canal using two control cross-sections, one of which is in the upper basin of the canal, the other in its throat. The differential equation for an uneven flow in a non-prismatic channel was obtained in the form (13). The analysis showed that in a flooded Venturi channel with a subcritical flow at Froude numbers less than one (Fr <1), the denominator of equation (13) is always greater than zero; therefore, at the entrance to the channel in the region of constriction, when $db/dl < 0$, the depth is downstream decreases ($dh/dl < 0$), and in the lower diffuser, where $db/dl > 0$, the flow depth increases ($dh/dl > 0$). Thus, the curve of the flow surface has a dip in the throat of the flooded Venturi channel. Integration of differential equation (13) made it possible to obtain the function of changing the Froude numbers in the flow along the length of the flooded Venturi channel (16), and then, using relations (17), to determine the depths and flow velocities. The verification showed that the calculated data are in good agreement with the experimental ones. The deviations between calculation and experiment do not exceed ±3%. This makes it possible to recommend the outlined methods for application in engineering practice.

1. Introduction

Let us consider the method of hydraulic calculation of the Venturi flowmeter channel in a submerged water outflow (Figure 1), when the depth of the flow in the throat of the channel h_m is greater than the critical h_c. According to the current State Standard of the Russian Federation МИ 2406–97, subcritical flow regimes in water measuring channels are unacceptable. However, in practice, every year the conditions of free non-flooded outflow are violated more and more often [1]. This is directly related to global climate changes in the world with a tendency for the volume and intensity of floods and storm runoffs to increase, raising the level in front of water retention dams and in overflowing water bodies and rivers receiving extreme runoff.

The relevance of the issue is determined by the fact that if the free flow regimes in the Venturi flowmeter channel of critical depth are regulated by the State Standard of the Russian Federation МИ 2406-97, and the methods for calculating the flow in this case are well developed and have been tested by many years of practice [2 – 10], then the flooded outflow regimes are not only not regulated by the

Content from this work may be used under the terms of the Creative Commons Attribution 3.0 licence. Any further distribution of this work must maintain attribution to the author(s) and the title of the work, journal citation and DOI.
Published under licence by IOP Publishing Ltd

State Standard, but practically not investigated and methods of their calculation are absent [1 and 11]. The incorrectness of the operational accounting of extreme flows at water measuring posts is often revealed with a significant delay, when, at best, the measurement results begin to raise doubts among the service personnel, at worst, when the water retaining structures lost their stability and begin to erode with all the ensuing, including catastrophic, consequences.

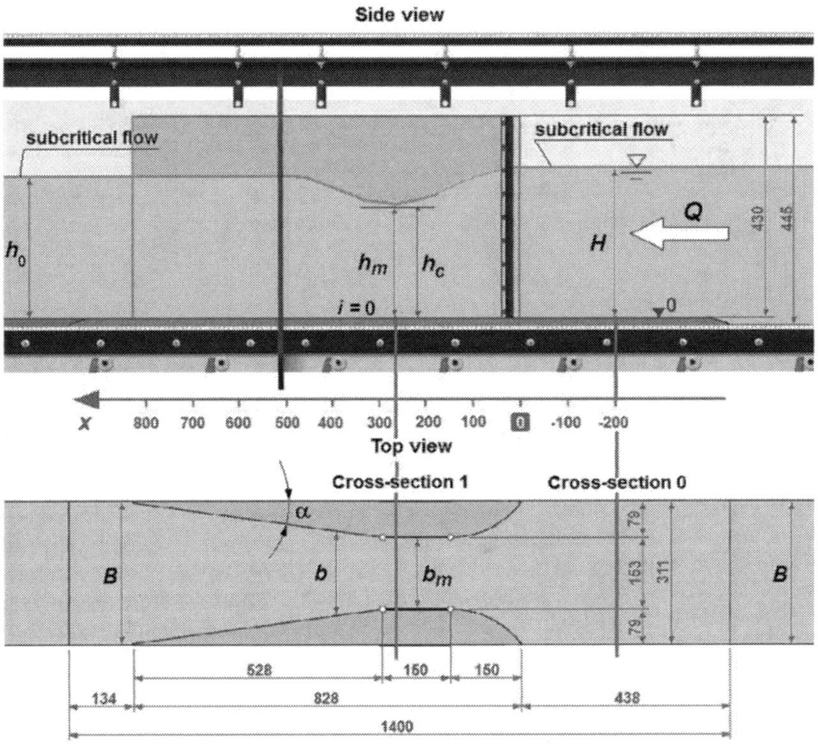

Figure 1. Model of Venturi channel in the submerged outflow mode.

The purposes of the research are to determine the criterion separating the free and flooded outflow regimes in the Venturi water-measuring channel of critical depth, to develop and verify the method of hydraulic calculation of the flow in the Venturi channel in the mode of submerged outflow.

2. Results and Discussion

2.1. Flow rate in the subcritical Venturi channel

We will assume that, due to the smooth configuration of the vertical walls in the Venturi channel, there are no vortex zones and, as a result, there are no local hydraulic losses, and, given the small length of the working section from upper pool (cross-section 0 in Figure 1) to the throat, where the flow depth h_m is minimal (section 1), we assume that the hydraulic losses along the length are also practically negligible here. Overall, this makes it possible to consider the flow-measuring Venturi flume as an ideal channel not having hydraulic losses, and to write the Bernoulli equation for cross-sections 0 and 1 in the form.

$$H + \frac{\alpha V^2}{2g} = h_m + \frac{\alpha V_m^2}{2g}, \tag{1}$$

where V and H are the mean flow velocity and depth in the upper pool at cross-section 0; V_m and h_m is the velocity and depth of the flow in the throat in cross-section 1, with h_m greater than the critical depth h_c (cf. Figure 1); α is the Coriolis (Saint – Venant) coefficient; and g is the acceleration due to gravity, $g = 9.81$ m/sec^2.

But the flow velocities at cross-sections 0 and 1 are determined through the flow rate (Q) by the equalities

$$V = \frac{Q}{BH} \quad \text{and} \quad V_m = \frac{Q}{b_m h_m}, \tag{2}$$

here B and b_m are the widths, respectively, of the upper pool and the throat of the Venturi channel.

Then Bernoulli's equation (1), taking into account equalities (2), should be rewritten as

$$H - h_m = \frac{\alpha Q^2}{2 g b_m^2 h^2}\left(1 - \frac{b_m^2 h_m^2}{B^2 H^2}\right).$$

Further denoting

$$h_m = \phi H \quad \text{and} \quad b_m = \varphi B, \tag{3}$$

as a result we get

$$Q = \varphi\phi\sqrt{\frac{1-\phi}{\alpha(1-\varphi^2\phi^2)}} B\sqrt{2g} H^{3/2}. \tag{4}$$

Formula (4) determines the flow rate of the flow following through the Venturi channel in both free and flooded outflows. To determine the flow rate through the channel, it is required to know the geometric dimensions of the channel (B and b_m) and depths in the upper pool H and in the throat of the channel h_m.

Formula (4) can be written in general form

$$Q = mB\sqrt{2g} H^{3/2}, \tag{5}$$

where

$$m = \varphi\phi\sqrt{\frac{1-\phi}{\alpha(1-\varphi^2\phi^2)}} \tag{6}$$

is a variable flow rate coefficient depending on the relations (3).

The criterion for evaluating the separation of the free and flooded outflow modes in the Venturi water measuring channel is the ratio of the depth in the throat of the channel h_m to the critical depth

$$h_c = \sqrt[3]{\frac{\alpha Q^2}{g b_m^2}}. \tag{7}$$

With a flooded outflow, this ratio should be greater than one

$$h_m / h_c > 1, \tag{8}$$

with free outflow, the ratio (8) is equal to one

$$h_m = h_c.$$

In the latter case, formulas (4), (5) and (6) are reduced to formulas for calculating the flow rates of the Venturi channels of the critical depth: (15) from paper [9] or (9) and (10) from paper [10].

2.2. Flow motion in the subcritical Venturi channel

The differential equation of steady-state smoothly varying fluid motion in a non-prismatic Venturi channel with a horizontal bottom, vertical walls and negligible hydraulic losses has the form

$$\frac{dh}{dl} = \frac{\dfrac{\alpha Q^2}{g\omega^3} \cdot \dfrac{\partial \omega}{\partial l}}{1 - \dfrac{\alpha Q^2 b}{g\omega^3}}. \tag{9}$$

here h, b and $\omega = bh$ are the depth, width and flow cross-sectional area variable along the channel length l.

Since the depth-independent partial derivative of the function of changing the cross-sectional area of the flow for a channel of variable width b and with vertical walls is equal to

$$\frac{\partial \omega}{\partial l} = h\frac{db}{dl},$$

then equation (9) can be rewritten in the form

$$\frac{dh}{dl} = \frac{\dfrac{\alpha Q^2 h}{g\omega^3}\cdot\dfrac{db}{dl}}{1-\dfrac{\alpha Q^2 b}{g\omega^3}}. \tag{10}$$

Let us now turn to an analysis of formula (10), which reflects the hydraulics of flows in non-prismatic channels with vertical walls and a horizontal bottom, to which the Venturi channels belong.

If the Froude number is written as

$$\mathrm{Fr} = \frac{V^2}{gh}, \tag{11}$$

then

$$\frac{\alpha Q^2 b}{g\omega^3} = \frac{\alpha V^2}{gh} = \alpha\mathrm{Fr}. \tag{12}$$

Let us bring equation (10) to the form

$$\frac{dh}{dl} = \frac{\alpha\mathrm{Fr}}{1-\alpha\mathrm{Fr}}\cdot\frac{h}{b}\cdot\frac{db}{dl}. \tag{13}$$

The analysis shows: at the inlet portion of the Venturi channel, where there is a calm subcritical flow with Froude numbers less than unity (Fr < 1) the denominator of Eq. (13) is greater than zero, therefore, in this area the narrowing of the channel when $db/dl < 0$, causes a decrease in the depth of the stream ($dh/dl < 0$), a the expansion of the channel ($db/dl > 0$) causes an increase in the depth ($dh/dl > 0$); in the lower diffuser of the channel, where a calm subcritical flow with Froude numbers less than one (Fr < 1) is also observed, the denominator of the Eq. (13) is also greater than zero, the depth of such a flow increases ($dh/dl <0$) with expansion of the channel ($db/dl > 0$), and a decrease in depth ($dh/dl > 0$) occurs with its narrowing ($db/dl < 0$). Thus, the curve of the free surface of the flow has a dip in the throat of the flooded Venturi channel. In this case, in the section of the maximum dip of the free surface of the flow, its depth h_m does not fall below the critical one h_c according to (7) and (8).

Let's express the derivative dh/dl in terms of the Froude number. According to equalities (11) and $\omega = bh$ we find

$$\frac{d\mathrm{Fr}}{dl} = -\frac{3Q^2}{gb^2h^4}\cdot\frac{dh}{dl} - \frac{2Q}{gh^3b^3}\cdot\frac{db}{dl} = -\frac{3\mathrm{Fr}}{h}\cdot\frac{dh}{dl} - \frac{2\mathrm{Fr}}{b}\cdot\frac{db}{dl},$$

whence

$$\frac{dh}{dl} = -\frac{h}{3\mathrm{Fr}}\cdot\frac{d\mathrm{Fr}}{dl} - \frac{2h}{3b}\cdot\frac{db}{dl}.$$

Substituting the obtained value of the dh/dl derivative into expression (13), we come to equation

$$\frac{\alpha\mathrm{Fr}-1}{\alpha\mathrm{Fr}(\alpha\mathrm{Fr}+2)}\cdot d(\alpha\mathrm{Fr}) = \frac{db}{b}. \tag{14}$$

Integrating (14), we find

$$\sqrt{\frac{(\alpha\mathrm{Fr}+2)^3}{\alpha\mathrm{Fr}}} = Cb.$$

The constant of integration C is found from the following condition. Since within the throat the flow passes through the cross-section with minimum depth h_m (cf. Figure 1), width b_m, and area $\omega = b_m h_m$, where, following (12), it's possible write

$$\frac{\alpha Q^2}{g b_m^2 h_m^3} = \alpha \mathrm{Fr}_m ,$$

then taking these conditions as the boundary ones, as a result, we find

$$C = \frac{1}{b_m}\sqrt{\frac{(\alpha \mathrm{Fr}_m + 2)^3}{\alpha \mathrm{Fr}_m}} ,$$

and finally we get

$$\sqrt{\frac{(\alpha \mathrm{Fr} + 2)^3}{\alpha \mathrm{Fr}}} = \frac{b}{b_m}\sqrt{\frac{(\alpha \mathrm{Fr}_m + 2)^3}{\alpha \mathrm{Fr}_m}} , \tag{15}$$

where the width b along the length of the Venturi channel varies from b_m to B.

With respect to the sought value $\alpha\mathrm{Fr}$, equality (15) is reduced to the cubic Cardano equation [12], which has a trigonometric solution satisfying the condition $0 < \alpha\mathrm{Fr} < 1$

$$\alpha \mathrm{Fr} = 8\frac{b}{b_m}\sqrt{\frac{(\alpha \mathrm{Fr}_m + 2)^3}{27\alpha \mathrm{Fr}_m}} \sin^3\left[\frac{1}{3}\arcsin\left(\frac{b_m}{b}\sqrt{\frac{27\alpha \mathrm{Fr}_m}{(\alpha \mathrm{Fr}_m + 2)^3}}\right)\right] . \tag{16}$$

Having determined the Froude numbers depending on the variable width b of the Venturi channel, the depths h and the flow velocity V are then calculated

$$h = \sqrt[3]{\frac{\alpha Q^2}{g b^2} \cdot \frac{1}{\alpha \mathrm{Fr}}} \quad \text{and} \quad V = \sqrt{\frac{\alpha \mathrm{Fr} g h}{\alpha}} . \tag{17}$$

Let us verification how adequately the computed formulas obtained correspond to actually observed flows. To do this, let's compare the computed values with the analogous measured characteristics of the model of Venturi channel shown in Figure 1.

2.3. Equipment, devices and instruments

Experimental studies of the model Venturi channel water gauge have been made in the Moscow State University of Civil Engineering. The following equipment, devices and instruments have been used in the studies [9 and 10]:

HM 162 scientific research hydraulic calibrating flume [13] (manufacturer: G.U.N.T. Gerätebau GmbH, Germany) with width of cross-section B = 311 mm, height 450 mm and length 15.5 m with relative roughness of the walls of the flume (hardened glass) and its bottom (stainless steel) n = 0.009;

Built-in instruments in the HM 162 flume: SHS4 80-200/40/P pump (manufacturer: Lowara S.R.L. Uniperso-nale, Italy) with maximal head 10 m, maximal delivery 150 m3/h, power 5.5 kW; Promag 10D electromagnetic flowmeter (manufacturer: Endress+Hauser Flowtec AG, France) with 0 – 150 m^3/h range of measurement, accuracy class 0.3; GSZ-100 lifting jack system for controlling the slope of the flume (manufacturer: ZIMM Maschinenelemente GmbH & Co KG, Austria) with range of variation of slope i from -0.5% to +1.75%;

HM 162.91 digital level gauge [13] (manufacturer: G.U.N.T. Gerätebau GmbH, Germany) with range of mea-surement from 0 to 455 mm correct to within 0.01 mm;

HM 162.51 model of Venturi flume (cf. Figure 1) [14] (manufacturer: G.U.N.T. Gerätebau GmbH, Germany) with height 430 mm, width and length of the gorge portion b_m = 153 mm and l = 150 mm, made of plexiglass and attached on a polyvinyl chloride plate,15 mm thick;

DLE 40 Professional digital laser rangefinder (manufacturer: Robert Bosch GmbH, Germany) with range of measurement from 0.05 to 40 m with accuracy within 0.5 mm.

All the equipment is certified consistent with the Russian Laws.

2.4. Methods and results of the experimental research

Before performing experimental studies the hydraulic flume HM 162 was set in horizontal position with a bottom slope $i = 0$ and the model of Venturi channel HM 162.51 placed in the middle part of the flume (cf. Figure 1). The digital level gauge HM 162.91 was mounted on instrument carriage and its zero adjusted relative to the floor of the model HM 162.51. The HM 162.12 specialized software package was loaded into the control computer in order to record the discharge, which was measured in course of the study by an electromagnetic flowmeter Promag 10 D.

On the control panel of the laboratory flume HM 162 operation mode or on the control computer, the flow rate was specified and the working pump SHS4 80-200/40/P of the flume was turned on.

Following stabilization of the operating mode of the flume (stabilization time 5 minutes), the flow rate values measured by the Promag 10D electromagnetic flow meter using the HM 162.12 software were written to the disk of the control computer into the generated data file; the flow measurement time was set 200 seconds with an interval between measurements of 1 second; in the process of processing, the obtained data were transferred to an Excel file, in which the average value of the flow rate (mathematical expectation) was calculated

$$Q_0 = \sum_{j=1}^{k} Q_j / k$$

and the normed standard deviation

$$\sigma = \frac{1}{Q_0} \sqrt{\frac{1}{k} \sum_{j=1}^{k} (Q_j - Q_0)^2} \,,$$

where k is the sample size, $k = 200$; Q_j is j-th element of the sample. The values of Q_0, σ and flow depth in the upstream pool H (cf. Figure 1) are written in the title lines of Table 1.

The same Table 1 shows measured digital level gauge HM 162.91 values of flow depths (or water levels) h at 11 points along the length of the model of Venturi channel HM 162.51. The distances x were determined from the input edge of the model HM 162.51 using the digital laser rangefinder DLE 40 Professional, the same meter determined the values of the width of the Venturi flume b at points x. The x-values and the corresponding b-values are recorded in the left-hand columns of Table 1.

Next, the discharge transmitted through the flume was changed with preliminarily selected step ΔQ. The total being investigated 9 regimes of transmission discharge Q_0 from 110.37 to 69.92 m^2/h. All measured values were written in units of dimensions of the measuring instruments.

Table 1. Experimental data

Q_0, m^3/h		110.37	110.37	110.37	110.37	110.37	99.95	90.14	80.15	69.92
σ		0.00245	0.00235	0.00251	0.00243	0.00224	0.00273	0.00262	0.00201	0.00226
H, mm		240.03	243.54	249.85	252.78	258.37	224.85	210.06	194.36	177.22
x, mm	b, mm	h, mm								
-100	311	233.39	248.10	246.64	248.39	252.63	227.48	211.53	190.57	176.54
0	305	233.68	244.67	250.60	255.07	261.97	224.62	212.05	189.20	176.72
100	167	198.28	211.10	225.78	221.98	234.67	187.50	179.84	164.42	152.15
220	153	177.36	191.86	209.11	205.03	219.06	168.55	159.95	146.21	131.56
300	157	186.02	198.17	204.17	213.57	219.08	177.79	165.19	153.89	137.56
400	186	214.16	217.47	221.54	233.52	242.63	198.02	186.26	174.63	154.94
500	215	211.90	226.50	229.07	230.17	241.67	207.62	192.68	173.59	158.65
600	244	222.93	229.00	231.60	243.03	249.32	204.19	196.42	182.65	163.63
700	274	213.49	222.77	234.05	242.10	245.60	207.03	187.94	175.24	166.04
800	303	222.86	227.12	229.59	245.71	248.33	203.70	194.60	181.98	159.04
900	311	224.57	227.65	241.47	236.62	250.89	203.15	192.34	179.65	159.75

The results of comparing the computational methods described in subsections 2.1 and 2.2 are summarized in Table 2 and presented graphically in Figures 2 and 3.

Table 2. Comparison of experimental and calculated data

Experimental data			Calculated values at $\varphi = b_m/B = 153/311 = 0{,}49196$					
Q_0, m³/h	H, m	h_m, m	h_c, m	$\phi = h_m/H$	m (Eq. (6))	Q (Eq. (5))	Δ
110.37	0.24003	0.17736	0.16512	0.73891		110.87	–0.00452
110.37	0.24354	0.19186	0.16512	0.78780	0.18466	110.06	0.00279
110.37	0.24985	0.20911	0.16512	0.83694	0.17396	107.74	0.02384
110.37	0.25278	0.20503	0.16512	0.81110	0.18034	113.66	–0.02982
110.37	0.25837	0.21906	0.16512	0.84785	0.17068	111.16	–0.00720
99.95	0.22485	0.16855	0.15456	0.74961	0.18929	100.09	–0.00136
90.14	0.21006	0.15995	0.14427	0.76145	0.18815	89.83	0.00343
80.15	0.19436	0.14621	0.13341	0.75226	0.18905	80.34	–0.00231
69.92	0.17722	0.13156	0.12180	0.74235	0.18986	70.25	–0.00467

According to the data in Table 2, it can be seen that the discrepancies (Δ) between the calculated (Q) and measured (Q_0) in the process of hydraulic studies, the values of flow rates were within the limit of ±3%, and the normed standard deviation is $\sigma = 0.01324$.

Figure 2 shows the graphs of changes in Froude numbers (αFr), as well as relative depths (h/H_0) and velocities ($V/\sqrt{2gH_0}$) in the flow along the length of the model Veturi channel when it is flooded in the downstream pool (h_0). The regime with a flow rate $Q_0 = 110.37$ m³/h at a tailwater level $h_0 = 218.26$ mm is considered. Hydrodynamic head here was calculated as the sum

$$H_0 = H + \frac{\alpha}{2g}\left(\frac{Q_0}{BH}\right)^2 .$$

When calculating the Froude numbers in the diffuser, hydraulic losses were taken into account.

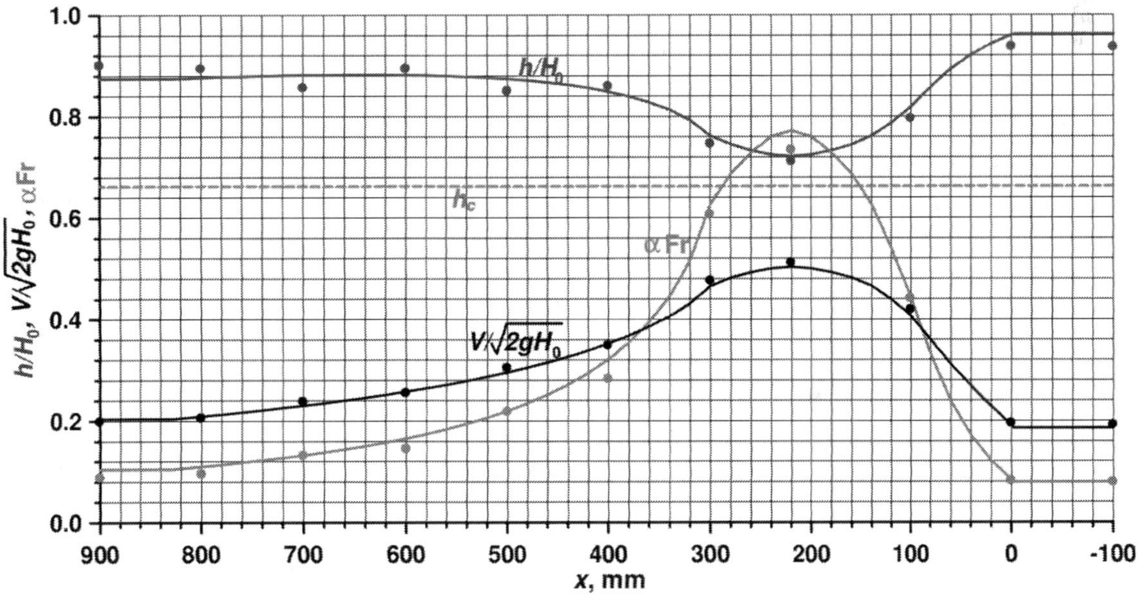

Figure 2. Functions of variation of depth h/H_0, velocity $V/\sqrt{2gH_0}$ and Froude number αFr in the flow along the length of the model subcritical Venturi channel at $Q_0 = 110.37$ m³/h.

In Figures 2 and 3, the values measured during the experiments are shown by dots, solid lines are obtained by calculation according to the methods described above.

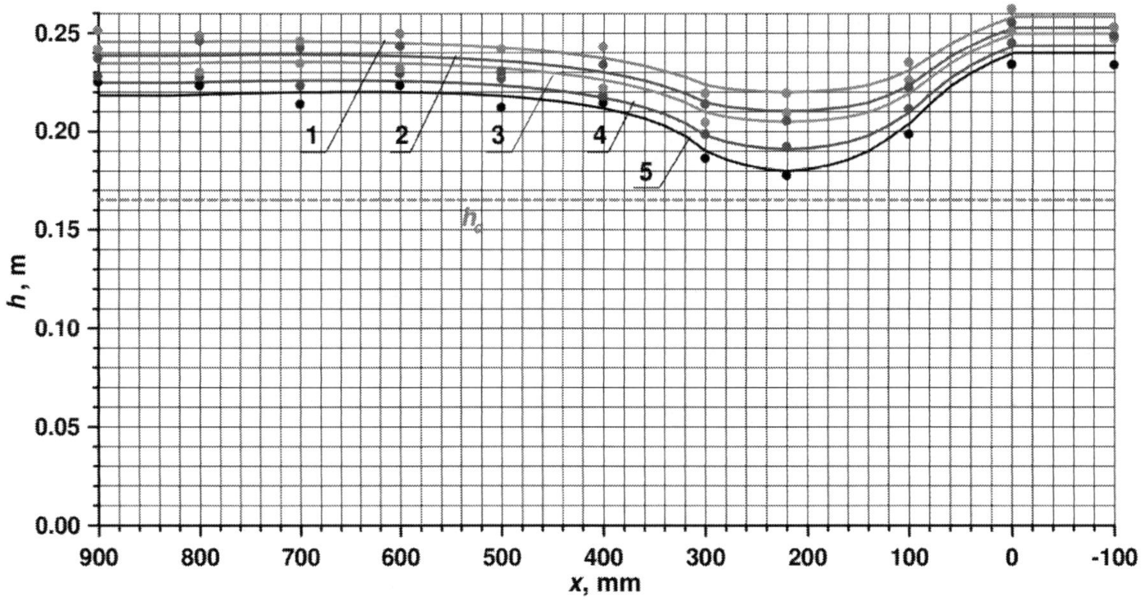

a

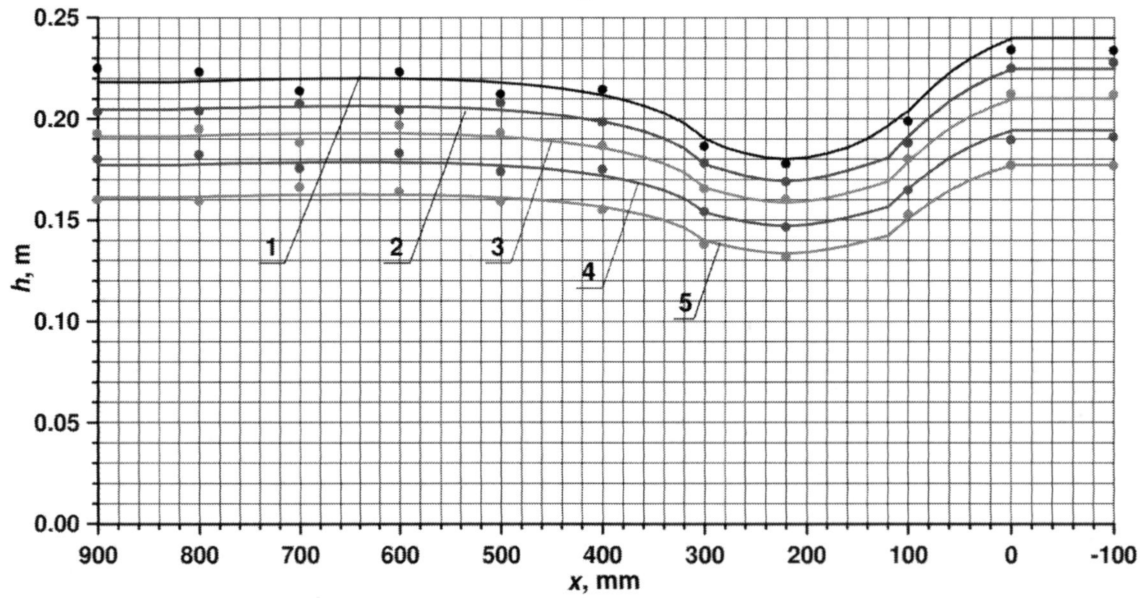

b

Figure 3. Functions of variation of the flow depth h along the length of the model Venturi channel: a – depending on the flooding in the downstream (Q_0 = 110.37 m^3/h): 1 – h_0 = 0.24561 m, 2 – h_0 = 0.23850 m, 3 – h_0 = 0.23465 m, 4 – h_0 = 0.22466 m, 5 – h_0 = 0.21826 m; b – at a variable flow rate (h_0/H_0 = 0.83 ÷ 0.87): 1 – Q_0 = 110.37 m^3/h, 2 – Q_0 = 99.95 m^3/h, 3 – Q_0 = 90.14 m^3/h, 4 – Q_0 = 80.15 m^3/h, 5 – Q_0 = 69.92 m^3/h.

In general, it can be seen that the functions of changes in Froude numbers (16), depths and flow velocities (17) along the length of the flooded Venturi channel are in good agreement with experimental data. This makes it possible to recommend the methods outlined in subsections 2.1 and

2.2 for use in practical calculations, including in the operational accounting of extreme flows at water measuring posts.

3. Conclusion

According to the State Standard of the Russian Federation МИ 2406–97, flooded modes in water measuring channels are unacceptable. In practice, the conditions of free flow are often violated due to the increase in the intensity of effluents in the context of global climate change. The incorrectness of the operational accounting of extreme flows at water measuring posts is revealed with a significant delay. The purpose of the research is to determine the criterion separating the free and flooded outflow modes in the Venturi channel, and to develop and verify the method for its hydraulic calculation in case of flooded outflows.

The criterion for separating the free and flooded outflow modes in the Venturi water-measuring channel is the ratio of the flow depth in the channel throat h_m to the critical depth hc. With flooded outflow, this ratio should be greater than one ($h_m/h_c > 1$).

The method for determining the flow rate in the flooded Venturi canal using two control cross-sections, one of which is located in the upper pool of the channel, the other in its throat, has been theoretically and experimentally substantiated.

A differential equation for an uneven flow in a non-prismatic channel is obtained in the form (13). An analysis of this equation showed that in a flooded Venturi channel with a subcritical flow at Froude numbers less than one (Fr < 1), the denominator of equation (13) is always greater than zero; therefore, at the channel inlet in the narrowing region, when $db/dl < 0$, the flow depth is down decreases downstream ($dh/dl < 0$), and in the downstream diffuser, where the channel expands ($db/dl > 0$), the flow depth increases ($dh/dl > 0$). Thus, the curve of the flow surface has a dip in the throat of the flooded Venturi channel. In this case, in the section of the maximum dip of the free surface, the flow depth h_m does not fall below the critical value h_c.

The integration of the differential equation (13) was carried out, which made it possible to obtain the function of changing the Froude numbers in the flow along the length of the flooded Venturi channel (16), and then, using (17), to determine the depths and flow velocities. When calculating the flow characteristics in the diffuser of channel, hydraulic losses must be taken into account.

The verification of the developed calculation methods showed that the data obtained by the calculation are in good agreement with the experimental ones. The deviations between calculation and experiment are less than 3%. This makes it possible to recommend the outlined methods for application in engineering practice.

References

[1] Yegorov N L, Loitsker O D and Shafranovsky M N 2016 *Water Supply and San. Techn.* **1** 66-71
[2] Castro-Orgaz O and Chanson H 2011 *Environmental Fluid Mech.* **11** 499-516
[3] Chanson H 2013 *Flow Measurement and Instrumentation* **29** 65-66
[4] Dufresne M and Vazquez J 2013 *J. Hyd. Res.* **51** 465-468
[5] Igo S W, N'wuitcha K, Palm K, Mihaescu L and Bathiébo D J 2014 *Adv. in App. Sci. Res.* **5(2)** 359-367
[6] Maghrebi M F and Ball J E 2006 *J. Hyd. Eng.* **132** 10 1044-1051
[7] Yegorov N L and Loitsker O D 2012 *Water Supply and San. Techn.* **6** 70-74
[8] Zerihun Y T 2016 *Fluids* **1(3)** 21
[9] Zuikov A L and Bakunyaeva V V 2018 *Power Technology and Engineering* **51** 6 621-627
[10] Zuikov A L, Bakunyaeva V V, Artemyeva T V and Zhazha E Yu 2018 *Magazine of Civil Engineering* **2** 76-90
[11] Howes D J, Burt C M and Sanders B F 2010 *J. Irrigation and Drainage Eng.* **136** 617-626
[12] Korn G A and Korn T M 2000 *General Publishing Company* 1-1151
[13] Linke U 2013 *HM 162. Experimental Flume 309x450 mm* 52
[14] Kellow M 2013 *HM 162.51. Venturi Flume* 59

The additive manufacturing in road construction as factor of its sustainability in conditions of climate change and hydrological catastrophe

T A Suetina[1], D D Koroteev[1,2], A I Koreneva[2], Jueru Huang[2]

[1]Moscow Automobile and Road Construction State Technical University (MADI), 64, Leningradsky prospect, Moscow, 125319, Russia
[2]Peoples Friendship University of Russia (RUDN University), Miklukho-Maklaya str.6, Moscow, 117198, Russia

E-mail: koroteev_dd@rudn.ru

Abstract. The necessity for technology incorporation has appeared in the construction sector a long time ago. A challenging area in the field of construction touches upon design, build, operation, maintenance, reconstruction and include all the steps of a process. This paper calls into the question of additive manufacturing (AM) application in the construction sector, especially road construction, describing the AM implementation. The model-driven methodology is critical to the success of manufacture. Key challenges that face construction 3D-printing are presented and discussed. The aim of this article was to show exhaustively realistic prospect of AM in road construction as factor of its sustainability, in particular in conditions of climate change and hydrological catastrophe. Furthermore, the experimental simulation was conducted to enable clean insights into the AM process. These insights could be fundamental to further experimental investigations with an eye to estimate the use of traditional already well-established construction materials. The presented opportunities elucidate the demand for on- and off-site CAD integration. The conducted experiment demonstrated all steps of AM.

1. Introduction

In the modern rapidly developing world mechanization and digitalization are universally among us. Practically every construction sector has already made the use of automation. Nowadays application of this technology has been well known for some time in construction sector as well. As opposed to almost every industry sector that successfully had involved automation [1, 2]. The fully or partly automated layer manufacturing is called additive manufacturing (AM) or 3D-printing [3]. AM could be less effective as in comparison with other sectors as a result of large size [4]. It is well known that construction is rather labor intensive, cost and highly accident industry [5, 6, 7]. Much work on the potential of AM has been carried out, however there are still some critical issues driven by few real examples. The first studies of AM implementation considered it to be a good solution in effectively management of raw materials usage [4]. Furthermore it could reduce the cost of raw materials [8]. The construction sector is expected to be improved on several main directions. The idea of providing new areas is also of particular interest [8]. The key factor for further detailed understanding and sustainable development is the understanding of the mechanical behavior of 3D-printed samples [6]. Another main

Content from this work may be used under the terms of the Creative Commons Attribution 3.0 licence. Any further distribution of this work must maintain attribution to the author(s) and the title of the work, journal citation and DOI.
Published under licence by IOP Publishing Ltd

focus could be attended to the design architectural form freedom that allows to design non-linear form without any additional formworks. The absence of formwork necessity is one of the main advantages of AM technology [1, 5, 7, 8, 9, 10, 11, 12].

This paper describes the technology of AM process drawing on the example of road construction, which is bridge with culverts (collecting pipes). Such structures are the first that can be destroyd in conditions of hydrological catastrophe, in this case, the increase of their sustainability is vital problem nowadays [13].

Most studies of AM in construction have only focused on building construction. The AM application in engineering construction and in particular in road construction has corresponding efficient approach. In the literature there are several examples of AM application in road construction. One of the bridges was consisted of metal and created by the printer called MX3D that has 6 axis robotic welding arms. It was the first additively manfactured metal bridge. The printer works on the basis of selective laser sintering (SLS) of powdered metal technology. The stell components were printed by the multinational engineering firm Arup and researchers from Imperial College London. The bridge length is 8 (10) m, width 2.5 (4) m, and located in Amsterdam, Netherlands. It was manufactured with the use of fitted robots with a welding attachment. The printing process took place off-site, as on-site construction was revealed to be too dangerous. The developed 3D-model was also used in the numerical solutions and load-testing [2, 8].

Another example is pedestrian bridge in urban park near Madrid, Spain that was finished in 2016. It was created by the Institute of Advanced Architecture of Catalonia and Enrico Dini. It is called Castilla-La Mancha 3D bridge. As stated by the authors it was the first ever 3D-printed bridge. Its span is 12 m and width 1.75 m. The bridge consists of 8 parts and was printed within the 2 months through the fused concrete powder and polypropylene reinforcement. Another interesting point is that no raw materials were used, as the fused material consisted of recycled components. The opportunity of using recycled materials proves the manufacturing technology to be ecologically friendly [2, 8].

One more bicycle bridge that was made from concrete is in Gemert, Netherlands. It was underlined that it was the first 3D-printed prestressed concrete bridge. It was manufactured by BAM Infrastructure and TU Eindhoven. The bridge length is 8 m and width 3.5 m. During the creating process 1 cm of concrete were extruded through the nozzle. The extruded concrete could maintain its form without any additional formworks after deposition. The manufacturing process was conducted partly. Then on-site all the parts were assembled together and the construction was tested [2, 14].

These examples are surprising from the view that used material was a metal that considered to be still less usable in construction. But not only structures itself could be created with the use of AM but also bridge pier sheaths that were printed and changed the old damaged ones due to scour [15]. There are some examples that show the linkage between 3D-model for AM application development and functional component topology optimisation. It proves that the obtained due to AM process 3D-model will be required for the further calculations and mechanical design of structures [16].

AM has also applications in the field of road construction. The AM application has come to be used to design, build and repair (reconstruction). In the eventuality that surrounding environment would be unfitting for the 3D-printer the structures could be printed off-site and then assemble together on-site. In Shanghai were adjusted 3D-printed sound barriers around the city highway. Their length along the S58 Shanghai-Changzhou Expresswayis 70 m. It is stated that with the use of them the highway noise would be successfully reduced. These constructions are much more effective than the traditional ones. The project was carried out by the Winsun company that is well-known for its AM application in construction sector. In December 2019 another road project was conducted in the Yunnan Province of China. The slope protection of the highway was created and tested [17, 18, 19, 20].

The paper is focused on AM application in the context of repair design, build, operation, maintenance and reconstruction with the use of CAD (computer aided design) and BIM (building information model) technologies. Digital manufacturing makes construction process more effective and can be considered as factor of improving its sustainability. Construction sector is recognized as being elusive, costly due to the use of service equipment and subjected to injury accidents. It is generally accepted that the AM application is able to resolve these conflicts [5]. This raises many questions about AM implementation but due to its technology adoption the prospect is advantageous. This paper re-examines and calls into the question of combined application of AM and BIM in construction. It has become a reality, for this reason BIM and CAD would be indispensable component. There are still numerous unexplored problems in AM technology [12, 21].

2. Materials and methods

This paper is focused on the process of manufacturing of road construction elements. The conducted experiment was used to obtain an undersized sample modelled on a probable bridge. A small sample size was chosen due to the process performance capabilities to show all steps of realistic project preparation. The model consisted of bridge in neotudor style and 3 collecting pipes, for that the special curved arch was also arranged.

Commercially available software was used. The software application used to model design and development was the Autodesk AutoCAD 2018 with the free software access for students and educators that was downloaded from https://www.autodesk.com/education/edu-software/overview. For 3D-modelling was used «3D Modeling» workspace. The model was created with milimeter measurement scale in order to be acceptable to 3D-printer measures. The device used was 3D-printer Picaso Designer X PRO. Technical specification are presented in table 1.

Table 1. Technical specification of Picaso Designer X PRO

Specification	Value
Print technology	Fused Filament Fabrication (FFF)
Extruder number	1
Nozzle number	2
Printing zone	200 x 200 x 210 mm
Printing speed	Up to 100 cm^3/h
Minimum layer thickness	10 microns (0.01 mm)
Filament diameter	1.75 ± 0.1 mm
Nozzle diameter	0.3 mm (0.2 - 0.8 mm)
Build platform	Aluminum, glass
Guide ways	Guide rail - XY Rail, steel; Z Cylindrical, steel

The printer is based on FFF-technology (fused filament fabrication) that comprises thermoplastic polymer filament melting and layer-by-layer model building up. Figure 1 shows the schematic illustration of the operating principles of 3D-printer.

The software application used to analyze the model data was PICASO 3D Polygon X software version 1.5.6 with free access. It was downloaded from https://picaso-3d.com/ru/techsupport/soft/designer-x-pro/. This programm was also utilized to create a task for 3D-printer in G-code with .plgx file extension. It was decided to joint all printed parts by dichlorethane plastic glue. The used material was 3D Printer Filament PLA by REC with diameter 1.75 mm the colour white (RAL 9003).

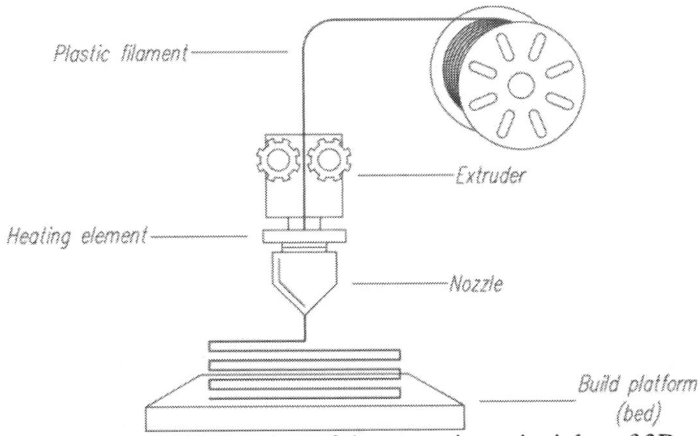

Figure 1. Schematic illustration of the operating principles of 3D-printer

Table 2. Print characteristics

Part	Number	Each size (X x Y x Z)	Print time	Plastic volume
Pipe type 1 (№1)	2	22.5 x 60 x 15 mm	4:16 h	18.8493 cm^3
Pipe type 2 (№2)	1	30 x 60 x 22.5 mm		
Bridge type 1 (№3)	2	23 x 74 x 22.5 mm	5:12 h	23.9946 cm^3
Bridge type 2 (№4)	2	67 x 65 x 37.5 mm	12:02 h	51.2646 cm^3

In the first step by creating different solid shapes and their mutual editing were received the designed bridge and collecting pipes model. The 3D-model was prepared as described above in CAD-program (figure 3a). After the full model has been done, we can then slice it in proper for 3D-printer parts. This operation was done in an attempt to reduce model size and accelerate processing and printing. The final model contained 7 parts at all: 3 collecting pipe parts and 4 composing bridge parts (figure 2). Print characteristics are presented in table 2. The sciling was pretended to be symmetrically.

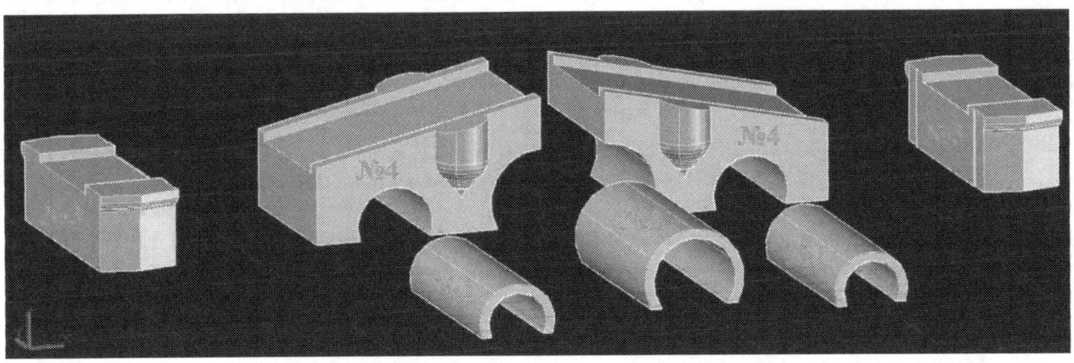

Figure 2. Model parts numeration

To determine whether any faults, holes or flipped normals the model should be carefully checked for it before the next step. This problem can be posed in terms of affordable sample size, edge thickness, engraved or embossed details. It was decided that the best method to overcome this problem were external visual examination and after programm slicing verification. The print task settings are reported in table 3.

Table 3. Print characteristics

Parameter	Value
First layer height	0.20 mm
Subsequent layer height	0.08 mm
Filling density	25%
Perimeter and filling line width	0.3 mm
Shell thickness	1 mm
Number of shell perimeters	3
Filling pattern	Octagon

As soon as this step was successfully carried out the model was adjusted to printing path. At this point the 3D-printer was adjusted to start printing. The platform and nozzle heat temperature was choosen in accordance with material. Before printing the special adhesive glue PICASO 3D was applied to increase adhesion. It was nacessary for the model not to come off the platform. The used material was selected and inserted (figure 3b). And finally the special file consisted of G-code with .plgx file extension went to 3D-printer. The G-code had fully detailed printing path dedicated to 3D-printer. And now we are ready for the next step, printing. As the printer is based on the FFF-technology the sample was created by material extrusion of molted material. The next layer cross-section was deposited before the previous had fully solidified. The procedure continued following the steps outlined above till the print head was gone throught every printing path of each layer (figure 3c,d).

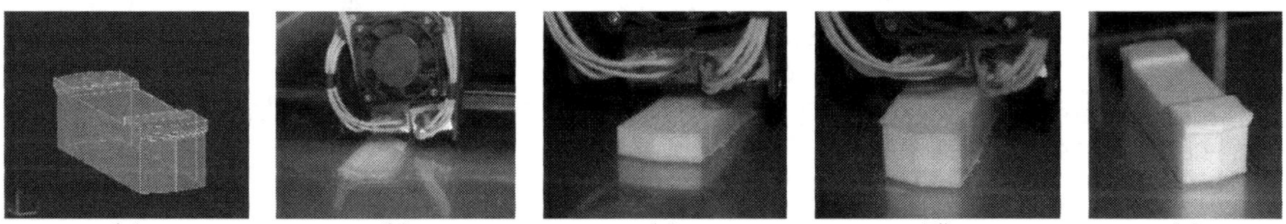

(a) (b) (c) (d) (e)

Figure 3. AM process of side part. (a) - 3D-model; (b) - first layer defining; (c) - extrusion process; (d) - higher layer printing; (e) - finished sample part.

Considerable attention must be paid when the printing process has just started. During the first layer printing the unsuitable value of heat temperature could be discovered. Therefore great care should be paid to the arranging process. After the printing was complete the printhead moved to the parking position and the build platform went lower down (figure 3e). We waited for 15-20 minutes for the glass to cool down after printing and then removed the sample from the printer.

3. Results and discussion

The AM preparing and manufacturing process is presented. Compatible construction materials are still poorly known to be used in AM but the materialtechnology development is necessary to be exercised along with other aspects such as technological processes. Our method has many attractive and practical applications concerned with AM application. During the preparation were solved many important issues of the 3D-model. We believe these problems have subsequent effect. This means that many other researches could face them.

The presented pictures showed the process how is the 3D-model step by step created (figure 3). As already mentioned above, while the model for its integrity were checking, were used both visual examination from all sides and verification in the program slicing. During the visual examination, it was necessary to take into account the printer characteristics and the principles of material extrusion.

In the program slicing and preparation for printing the main operations were controlled during the further sample construction. During the printing process, as already mentioned above, the first layer is the most important and critical, since an incorrectly selected value of build platform (bed) heat temperature can lead to poor adhesion between the ejected material and the platform itself, which will inevitably lead to the separation of the sample printed part from the build platform. For this reason the use of special adhesive glue PICASO 3D was critical.

After the model was successfully printed, it was carefully visually examined. The printed edges and surfaces are assessed as satisfactory. The subsequent solution process of the sample was carried out in accordance with the model features obtained by creating with the use of this printer. Firstly, the edges were smoothed out and sanded. Then all parts of the model were glued together with dichlorethane plastic glue (figure 4). The setting time of the model parts was taken as 30 hours for the complete bonding of the surfaces. The assembled final model was a prototype of its 3D-model with great detailed accuracy. This indicates a successful experiment and the aims and objectives fulfillment that were set at the beginning. All the difficulties that appeared during the experiment were described in this article and can be considered as a reference material for subsequent AM experiments conduction. In other words the resulting sample can be expressed as considerable progress that has been made towards the aims.

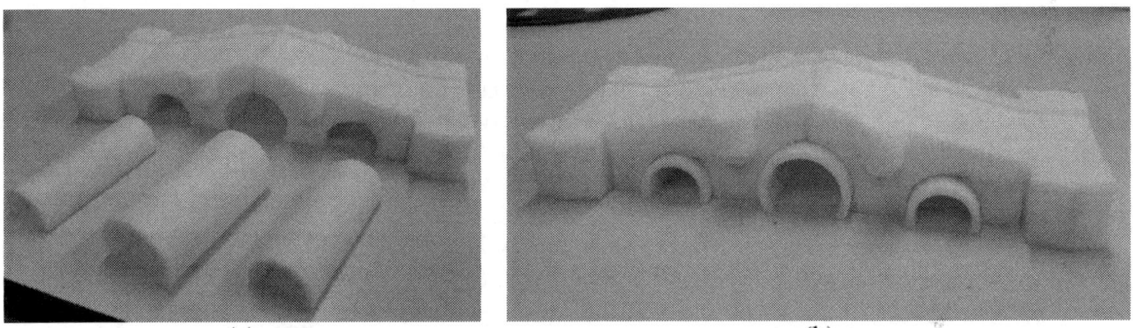

(a) (b)

Figure 4. The printed model assembly. (a) - before; (b) - after.

As stated in the Introduction, the research was conducted to reveal and show AM process. Having a model experiment enabled us to believe in AM opportunity to perform large-scale building and constructon structures, as the experiment was conducted as a scaled-down real production process. The traditional approach to create samples including molding is much more labor-intensive even on the experiment examples within the laboratory.

Our steps proceed are in the same way as [20] with the exception of the used material and atteched with it printig technology. The data preparation, design and exploring algorithms are described for pusposes of PLA plastic printing.

4. Conclusion

A striking feature of AM technology is the opportunity to extend sustainable productive development in conditions of climate change and hydrological catastrophe. As AM has become an important issue in Industrial Revolution 4.0 the technology implementation is one of the key questions. Within the next few years AM is likely to become an important component in major scientific research. Possibly by creating a more realistic model dedicated to specific surrounding environment the experimantal process could also contain calculations and mechanical analysis of behavior. The fracture mechanics is still unclear and open area for further work.

The introduction chapter touches upon some the most interesting and outstanding examples of AM implementation. This section described the liminations during past knowledge and provided an initial understanding of some processes. It was also highlighted that various alternative ways should be explored to overcome these problems and fully exploit the prospects. The second chapter described the conducted experiments and outlined the existing difficulties and potential ways to overcome them. The third chapter discussed the obtained results and the ways of interpretation. The existence of imperfection and was also mentioned in order to continue to deal with this matter. These experiments highlighted that mutuality interdependence between every manufacturing step is one of the major issues during the 3D-printing process.

The analysis did not reveal any significant data of construction materials as it should be investigated further. We are of the opinion that the validity of our model and experiment could help in AM implementation in the construction sector. It is still an emerging technology that needs standard documentation. These topics are reserved for future work to reach the targets mentioned above. And only in that case AM could access the full potential in the construction sector. Another key parameter to be in a greater detail studied is the topology optimization that plays an essential role not only in model design but also in efficient geometry and mechanical behavior of the hardened structure.

References

[1] Kazemian A, Yuan X, Meier R, Khoshnevis B 2019 Performance-Based Testing of Portland Cement Concrete for Construction-Scale 3D Printing. In 3D Concrete Printing Technology, 1st ed.; Sanjayan J G, Nazari A, Nematollahi B, Eds.; Elsevier: Amsterdam, The Netherlands, **2**, 13–35

[2] Buchanan C, Gardner L 2019 Metal 3D printing in construction: A review of methods, research, applications, opportunities and challenges *Engineering Structures*, **180**, 332-348

[3] Hambach M, Rutzen M, Volkmer D 2019 Properties of 3D-Printed Fiber-Reinforced Portland Cement Paste. In 3D Concrete Printing Technology, 1st ed.; Sanjayan J G, Nazari A, Nematollahi B, Eds.; Elsevier: Amsterdam, The Netherlands, **5**, 73–113

[4] Nam Y J, Hwang Y K, Park J W, Lim Y M 2019 Fiber-Reinforced Cementitious Composite Design with Controlled Distribution and Orientation of Fibers Using Three-Dimensional Printing Technology. In 3D Concrete Printing Technology, 1st ed.; Sanjayan J G, Nazari A, Nematollahi B, Eds.; Elsevier: Amsterdam, The Netherlands, **4**, 59–72

[5] Sanjayan J G, Nematollahi B 2019 3D Concrete Printing for Construction Applications. In 3D Concrete Printing Technology, 1st ed.; Sanjayan J G, Nazari A, Nematollahi B, Eds.; Elsevier: Amsterdam, The Netherlands, **1**, 1–11

[6] Feng P, Meng X, Chen J F, Ye L 2019 Mechanical Properties of Structures 3D-Printed With Cementitious Powders. In 3D Concrete Printing Technology, 1st ed.; Sanjayan J G, Nazari A, Nematollahi B, Eds.; Elsevier: Amsterdam, The Netherlands, **9**, 181–209

[7] Xia M, Nematollahi B, Sanjayan J G 2019 Properties of Powder-Based 3D Printed Geopolymers. In 3D Concrete Printing Technology, 1st ed.; Sanjayan J G, Nazari A, Nematollahi B, Eds.; Elsevier: Amsterdam, The Netherlands, **13**, 265–280

[8] Holt C, Edwards L, Keyte L, Moghaddam F, Townsend B 2019 Construction 3D Printing. In 3D Concrete Printing Technology, 1st ed.; Sanjayan J G, Nazari A, Nematollahi B, Eds.; Elsevier: Amsterdam, The Netherlands, **17**, 349–370

[9] Malaeb Z, AlSakka F, Hamzeh F 2019 3D Concrete Crinting: Machine Design, Mix Proportioning, and Mix Comparison Between Difference Machine Setups. In 3D Concrete Printing Technology, 1st ed.; Sanjayan J G, Nazari A, Nematollahi B, Eds.; Elsevier: Amsterdam, The Netherlands, **6**, 115–136

[10] Cao X, Li Z 2019 Factors Influencing the Mechanical Properties of Three-Dimensional Printed Products From Magnesium Potassium Phosphate Cement Material. In 3D Concrete Printing Technology, 1st ed.; Sanjayan J G, Nazari A, Nematollahi B, Eds.; Elsevier: Amsterdam, The Netherlands, **10**, 211–222

[11] Nerella V N, Mechtcherine V 2019 Studying the Printability of Fresh Concrete for Formwork-Free Concrete Onsite 3D Printing Technology (CONPrint3D). In 3D Concrete Printing

Technology, 1st ed.; Sanjayan J G, Nazari A, Nematollahi B, Eds.; Elsevier: Amsterdam, The Netherlands, **16**, 333–347

[12] Ford S, Despeisse M 2016 Additive manufacturing and sustainability: an exploratory study of the advantages and challenges *Journal of Cleaner Production*, **137**, 1573-1587

[13] Syachrani S, Jeong H S, Rai V, Chae M J, Iseley T 2010 A risk management approach to safety assessment of trenchless technologies for culvert rehabilitation *Tunneling and Underground Space Technology*, **25(6)**, 681-688

[14] Valente M, Sibai A, Sambucci M 2019 Extrusion-Based Additive Manufacturing of Concrete Products: Revolutionizing and Remodeling the Construction *Industry Journal of Composites Science*, **3(88)**

[15] Avrutis D, Nazari A, Sanjayan J G 2019 Industrial Adoption of 3D Concrete Printing in the Australian Market: Potentials and Challenges. In 3D Concrete Printing Technology, 1st ed.; Sanjayan J G, Nazari A, Nematollahi B, Eds.; Elsevier: Amsterdam, The Netherlands, **19**, 389–409

[16] Sengsri P, Kaewunruen S 2020 Additive manufacturing meta-functional composites for engineered bridge bearings: A review *Construction and Building Materials*, **262**

[17] Zhang Z, Liu Y, Liu T, Wang D 2016 Study on Teaching Experimental System for Traffic Engineering Design Base on 3D Printing *China Modern Educational Equipment*, **13**, 32-35

[18] Li Z, He F, Shi H, Qin K 2019. The application of 3D printing technology in pavement repair project is discussed *Highways*, **04**, 51-55

[19] Zhang S, Ang Y, Li M, Xin P, Du S C, Lu Z 2020 Influence of 3D Printing on Recyclability of Precast Fabricated Concrete Pavement *Slab Bulletin of the chinese ceramic society*, **08**, 2433-2440

[20] Guo B, Chen Z H, Zhang M, Zhang W B 2019 Risk Assessment Index System for Soil-Similar Slopes and Application in Highway Engineering *19th COTA International Conference of Transportation Professionals*, 799-810

[21] Lim S, Buswell R A, Le T T, Austin S A, Gibb A G, Thorpe T 2012 Developments in construction-scale additive manufacturing processes *Automation in Construction*, **21**, 262–268

Decrease in ecological damage of water throughput tubular transitions on spawning

T A Suetina[1], O N Chernykh[2] and A V Burlachenko[3]

[1]Moscow Automobile & Road Construction State Technical University (MADI), Moscow, Russia
[2] Russian State Agricultural University, Moscow, Russia
[3] RUS Mercedes Benz joint-stock company, Moscow, Russia

E-mail: t_a_souetina@mail.ru

Abstract. Intensive growth of traffic demands decrease in negative impact of water throughput constructions on ecology. Conditions of the correct installation of the tubular transitions made from steel corrugated elements (MCE) on spawning waterways which performance will allow to ensure ecological safety at movement of fish are defined. Recommendations about the organization of overcoming by fishes of difference and a velocity pressure when moving in MCE culverts are made. The obtained data on speed on a transit part of a water throughput construction are of interest at problem solving on easy access of fishes through culverts to up-stream and design of culverts. Results of pilot hydraulic studies of MCE pipes of round shape according to depths and speeds are presented on an outlet from a pipe for different operating modes, at existence and lack of a protective tray in a ground part of pipes with helical (HC) and a normal (NC) form of a corrugation. It is established that if 10% of cross sectional area of a corrugated pipe are occupied with stone dumping, then flow rate is reduced approximately by 10 - 12%. Based on results of laboratory researches and published materials, data for assessment of value of roughness coefficients of composite section of a round culvert in the form of one-point transition are provided and the targets of future researches on melioration of culverts are planned. Received results can be used at hydraulic calculations and optimization of water throughput designs in nature protection and also highway engineering.

1. Introduction

On highways the most perspective and economic is performance of water throughput constructions (such as pipes - culverts, arches, arches pipes) from metal corrugated elements (MEC) [1-3]. For an exception of hydrological and environmental disasters, safety at road and nature protection construction, it is important to inform designers and operators on operating conditions of pipes from MCE with different type of a corrugation (normal – NC and helical – HC), correctness of their hydraulic calculation and also features of interaction of MCE with an ecosystem of a water object (a spawning waterway or a stocked reservoir).

When crossing a spawning waterway, artificial water throughput constructions from MCE should answer a number of the major requirements, namely:

- provide the free admission of fish, a necessary amount of water, free migration of fishes at any stage of their development;

Content from this work may be used under the terms of the Creative Commons Attribution 3.0 licence. Any further distribution of this work must maintain attribution to the author(s) and the title of the work, journal citation and DOI.
Published under licence by IOP Publishing Ltd

- exclude considerable silting of a pipe and suspended waters;
- provide the invariable speed of a current of water and support its size according to the prevailing breed of fishes;
- provide the minimum deformation of bed of a waterway and places of spawning;
- provide maintenance of the course of a waterway or its certain site in natural or close to natural a state at minimization of perturbations in a flow;
- provide observance of legislatively established requirements to passing of fish and basic provisions of laws regarding environmental protection, protection and use of water biological resources.

On ability of the movement of fishes in a culvert influence: design of entrance and output head; size and form of pipe cross-section; length, roughness, bias and constructive registration of a benthonic part; speed, extent of filling of a pipe and biophysical parameters of fishes. Optimal conditions can provide to the migrating fish also anti-vortex devices before a tubular construction from MCE, existence of a step cascade behind it, the fish pass device on a transit part or in a diversion channel behind a culvert, etc.

It is known that pipes from MCE work in 4 main hydraulic modes: free-flow, semi-pressure head, partial and pressure head and pressure head [1-3]. As design in Russia for road pipes accept the free-flow mode which takes place at not flooded entrance to a head. At the same time the recommended restriction of filling of entrance and output section of NC should not exceed respectively 0.75 and 0.9 from height or pipe diameter. Spawning water currents at the admission of a design flowrate require a possibility of passing of fish on a pipe against the current. About 80% of water throughput pipes in the flat area of the Russian Federation have a bias 5...10% therefore it is possible to assume that work as NC and HC in the semi-pressure head and pressure head mode when crossing highways through a spawning waterway most likely economically will be ineffective. It is caused by big costs of the devices for filtration reduction, ensuring resistance of an embankment to water filtration, strengthening of the bed at the inlet and an outlet from a pipe. Therefore we will consider hydraulic aspects of passing of fishes through the road culverts from MCE working generally in the free-flow mode or partially pressure head mode (which can arise at hypothetical hydrological accident on a water object).

2. Materials and methods

Modern models of passing of fish on a culvert are one-dimensional as are constructed on a ratio of ability of swimming of fish to the average speed of a flow arising in a one-point pipe. In that case carrying out physical modeling and performance of a detailed research of distribution of flow rate in process of filling of pipe cross-section and change of depth of a flow longwise of a water throughput metal pipe from MCE with the normal corrugation which is (rather in detail investigated), and with helical, a little investigated, a corrugation is necessary. For this purpose at first it is necessary to evaluate efficiency of modern methods of calculation of roughness of the "corrugated pipe-an abrasive covering of a ground part" complex and composite "a ridged surface of a pipe - benthonic area from stone dumping", in relation to a water throughput pipe from MCE which is in Russia which is almost not explored and standardly not approved so far.

The operating experience, design and calculations of NC and HC in Russia and abroad allows to draw a number of conclusions on increase in efficiency of hydraulic work of road water throughput pipes from MCE depending on a form of a corrugation, diameter and a bias of a pipe, existence at the bottom of reinforced-concrete smooth flumes or gravel and pebble dumping. Pilot studies are conducted in laboratory of department of hydraulics of Moscow Automobile & Road Construction State Technical University (MADI) [3] on pipe model with a normal and helical corrugation diameter in nature of 1 m and 1.2 m according to a linear scale factor 5. The sizes of a normal corrugation were in NC 130X32,5 mm, in HC – 125X25 mm. The bias of pipes changed in the range from 0.01 to 0.1. The central corner φ at installation of a smooth tray made 120^0, when dumping gravel and pebble mix - $\varphi = 90^0$. After the research of influence of a design of entrance head on NC, the main part of experiments with HC has been conducted [3] with entrance to heads with a cut perpendicular to a pipe axis, and day off – in the form of a portal wall. The maximum error of measurements of a hydrostatic pressure H, depths h, an expense Q, water temperature did not exceed 4%. The program complex allowed to build pressure heads in the automatic mode, to calculate expense parameters $\theta = Q/\sqrt{g}d_p^{5/2}$ (d_p - the rated diameter of models), to determine resistance coefficients, normal depths, roughness coefficients by Manning n. At installation of a smooth tray on a bottom of pipes and values of a Reynolds number more than 350,000 for pipes with a normal corrugation, and more than 400,000 for helical corrugation pipes there came the square area that was well coordinated with natural data of Ch. Neill [4]

3. Results and Discussion

Results of laboratory researches have shown that the value of coefficient of roughness n at the free-flow and pressure head movement for NC and HC can be a miscellaneous and is considerable differ from recommended in standardizing document (SD) in Russia. It is confirmed with model and natural researches in which the value of roughness coefficient n at the free-flow movement in some cases is less, and sometimes it is more, than at pressure mode. Installation of a smooth tray on a bottom of corrugated pipes is regulated by SD for protection against abrasive destruction that allows to provide decrease in hydraulic resistance and to reduce value of coefficient of roughness n. At the same time it leads to reduction of the extent of a throat area and cross sectional area of a pipe. A tray, occupying a part of internal perimeter of a pipe (Figure 1a), creates the added resistances depending on the central corner δ and the minimum thickness of a smooth tray δ which is 5 cm in the USA, in the Russian Federation – δ = 10 cm. With change of filling of a corrugated pipe the ratio between the sizes of a smooth tray and corrugated sheet in equipotential section changes, i.e. between wetted perimeters of the smooth and corrugated parts of a pipe therefore the nature of change of roughness coefficient n depends on the sizes of a smooth tray and filling of the pipe h/d.

a b

Figure 1. Options of transitions from MCE on spawning waterways: (a) - NC with a smooth tray at the bottom: 1 – a metal pipe with a normal corrugation; 2 – packaged units of a tray; 3 – protecting cover; 4 – bolts; (b) - stone dumping in the absence of a tray in a round pipe

Lack of culverts from MCE on a spawning waterway is that tubular transitions, constraining a flow, cause it to pass through smaller cross sectional area. It leads to loss of the habitat, natural substrate and zones of a quiet current where fish could have a rest earlier. For compensation of these losses in addition to installation of a pipe with the corresponding bias as it is possible closer to a natural channel of a stream, it is necessary to bury on spawning waterways a culvert bottom below a waterway bottom (on depth not less than 2 … 3 rated diameters of a stone or 0,1...0,2 from pipe diameter). And also for preservation of the native habitat of fishes to cover a tray part of NC with the natural material steady against rated speeds of water in a pipe (a stone, gravel, coarse sand, etc.) (Figure 1a). All this provides more natural bed of substrate and increases roughness of a water throughput pipe, reducing the average speed of a flow which would be in not buried water throughput pipe of similar cross sectional area.

Forecasting of values of coefficient of roughness n for round pipes diameter d from MCE with a gravel layer at the bottom is quite difficult task what numerous researches (Strickler, testify 1923 to;

Meyer-Peter and Muller, 1948; Limerinos, 1970; Bray, 1979; Bruschin, 1985; Abt et al., 1987; Julien, 2002, etc.). These methods have been developed first of all for use in the rivers with a gravel bottom and have received widely varying results. Generally they treat model range $0.027 < n < 0.031$ when material of a surface course is linearly scaled. In practice natural values of roughness coefficient n of real melioration objects, for example, in Manitoba [5], varied from 0.035 to 0.041. It is possible to call roughness of this kind composition. For reliable assessment of roughness coefficient of a composite "a ridged surface of a pipe - benthonic area from stone dumping", it is necessary to evaluate experimentally values n in different operating modes of tubular transitions from NC and HC and to verify the existing methods of forecasting of roughness of a similar composite.

On the basis of the researches executed in MADI earlier (Figure 2) [3] it is possible to assume that at 10% the size of a gravel bed and filling of a pipe approximately 0.15 d, (that corresponds to filling of a zone of a smooth tray or gravel sketches), the value of coefficient of roughness n is expected should strive for value of coefficient of roughness for the gravel open course of a natural waterway. Thus, [1-6] recommendations provided in works are not enough for correct assessment of change of coefficient of roughness in all the range of fillings of water throughput metal pipes from NC at the different amount of gravel registration of its bottom and various modes of the movement of water in a construction.

Preliminary pilot studies have really shown reduction of flow capacity of a water throughput pipe from MCE at addition of a gravel layer that is caused by reduction of the area of a flow and increase in roughness of a composite. At the same time it has turned out that at 10% occupation of the area of NC its flow capacity is reduced by stone dumping approximately by 10 - 12% both for pipe bias $i = 0.005$, and at $i = 0.01$.

It is established that the value of coefficient n even in pipes with the uniform evenly laid gravel dumping can change for 28% only because of change of depth of a flow. Researches have shown that as in a pipe section compound – composite, the value n depends on depth of water and varies on model in the range of $0.0230 < n < 0.0256$, and in some experiences when filling from 0.4 d to 0.75 d reaches 0.029 for the gravel course, i.e. about 0.039 in nature. It is necessary to remind that for NC in the absence of additional devices in a transit part of a pipe the value of coefficient of roughness n does not depend on filling of a pipe (filling change, and respectively and the mode occurs at the free-flow mode in the range of change of filling of h/d = 0,12...0,52, and at semi-pressure mode - h/d = 0,52...0,82). Its weighted average is approximately identical and is on model $n = 0.0268$, and in real culvert $n_n = 0.035$. And at the pressure mode movement for this design the value 12% lower and is in real culvert $n_n = 0.03$. For HC without protective smooth tray at the bottom for assessment of stability of hydraulic work of water throughput transitions of round section it is expedient to accept at a pressure mode of $n_n = 0.027$, and at free-flow – $n_n = 0.03$.

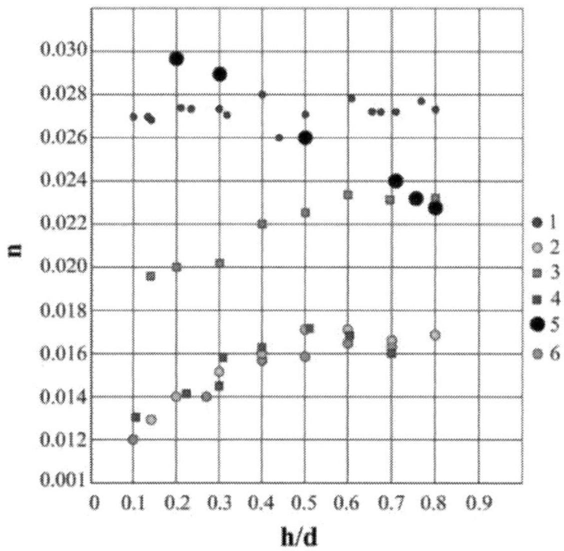

Figure 2. Dependence of coefficient of roughness across Manning *n* from change of the relative depth (*h/d*) in culverts from corrugated metal with a composite coating of a transit part at i =0.03... 0.05: 1, 3 – according to NC and HC without tray on a bottom; 2, 4 and 6 –NC and HC with a smooth tray on a bottom; 5 –NC with gravel dumping on a bottom.

It is established, for NC with a corrugation of 130x32.5 mm and a smooth tray on a bottom influence of a bias of a pipe on value *n* has an appearance

$$n = 0.02 + 0.034i \quad (1)$$

During the experiments it is received that at a free-flow mode of NC with a smooth tray on a bottom with biases $i = 0.01 \ldots 0.1$ and design filling at the inlet $h/d = 0{,}75$ value of coefficient of roughness can accept $n = 0.019$. During the work of NC with a smooth tray on a bottom in the free-flow mode with fillings at the inlet $h/d = 0.9 \ldots 1.0$ coefficient of roughness in the range of change of a bias $i = 0.01 \ldots 0.096$ are almost identical and can be accepted with a small stock equal $n = 0.021$. For HC with $d = 1.2$ m, a corrugation of 125x25 mm and a smooth tray on a bottom value of coefficient of roughness at the free-flow movement of water it is possible to accept $n = 0.0207$ (at $i = 0.03$) and $n = 0.0226$ (at $i = 0.05$), and at pressure mode– $n = 0.0234$. Results on increase in coefficient of roughness with increase in filling of a pipe from MCE with a smooth tray on a bottom qualitatively coincide with the available foreign experimental data.

Comparison of values of flow rates V_{out} at outlet from smooth and corrugated pipes shows that at the free-flow and semi-pressure head modes of their value are close only at bias $i = 0.01$ and the parameter of an expense $\theta = Q / \sqrt{g d_p^{5/2}}$ less than 0.7, and at change of a bias of pipe i from 0.03 to 0.1 in a corrugated pipe of speed it is less. At pipe bias $i = 0.03$ for the free-flow mode reduction reaches 18%, for semi-pressure head – 13%. At $i = 0.1$ at all modes and parameters of an expense reduction of speed for the free-flow and semi-pressure head modes considerably also is according to 50% and 38%, and for the first partial and pressure head mode (when the pipe bulk from entrance the head works in pressure, and at the outlet – the free-flow movement) – 30%. At the same time it is necessary to consider that existence of an protected from an erosion smooth tray on a bottom in NC can increase the flow capacity of such constructions at the free-flow and semi-pressure head modes respectively by 10% and 14 … 18% in comparison with the existing recommendations for pipes from culverts without smooth tray

(Figure 3). Thus, at pipe bias $i > 0.01$ it is established that average speed at the outlet at a corrugated pipe is less, than at smooth pipe for all modes. It allows reduce at design length of fastening of the taking-away course behind a corrugated pipe, to choose optimum and simpler type of strengthening of protection of the downstream against washout and it is more correct to define hydraulic characteristics of a flow on strengthening of an end part of tubular transition

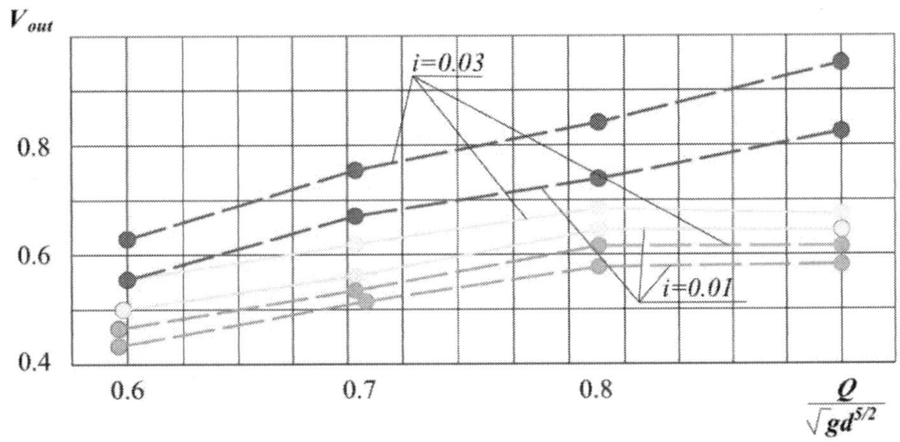

Figure 3. The schedule of dependence of $V_{out} = f(Q)$ for natural pipes diameter of $d = 1.5$ m with biases $i = 0.03$ and $i = 0.01$: - - - for the smooth pipe working in the semi-pressure head mode; - - -for NC working in the first partial and pressure head mode; - - - for a pipe from MCE with a smooth tray on a bottom.

In the hydraulic relation in modern nature protection construction and at reconstruction of road bypass tubular constructions in the course of a relining it is preferable to use a corrugated pipe instead of smooth one that is confirmed also by long-term practice of operation of corrugated pipes in Canada, France, Japan, the USA and other countries [7, 8].

Water speed at the outlet from a water throughput pipe should be close to flow rate in a spawning waterway. If speed at the outlet from a water throughput pipe exceeds maximum, then fish will not be able to overcome a current and to come into culvert, i.e. for providing a quiet outlet of fish from culvert the outlet should have whenever possible low speeds of a current of water.

To provide free passing of fish through a water throughput construction it is necessary to expel the following reasons causing negative impact of a culvert on a fish flow from MCE:
- existence of difference of water levels in a waterway and pipe;
- formation of a threshold between the river and a pipe both from tale race, and from an upper race;
- small pipe diameter;
- too big bias of a pipe;
- unacceptably high speed of a current of a water flow in a pipe;
- insufficient depth of water in a pipe;
- lack of a creek for rest of fish is lower than a culvert;
- insufficient roughness of an inner surface and bottom of a pipe.

Water throughput pipes NC on spawning waterways it is more expedient to use a round or arch design (if flow rate less than admissible washing away). The round design demands increase in diameter (from 1.5 to 7 m) in comparison with hydraulic calculation. For creation of necessary illumination in a pipe it is necessary to provide excess of a crown of a pipe over water level in 1 m and more.

Water level difference reduction (or its full elimination) and also blockings of the center of erosion lowering bottom level on the fountain site with formation of falls requires holding special constructive

IOP Conf. Series: Materials Science and Engineering 1159 (2021) 012016 doi:10.1088/1757-899X/1159/1/012016

events. So the difference of water level can be corrected by the construction of a retaining cascade by means of boulders or logs. For adult fishes each step should increase water level no more than on 25 … 30 cm. For clearing of an excess energy behind a tubular construction on a spawning waterway at heavy flow rates of water it is better to use a gravel pond.

4. Conclusions

High technical-and-economic indexes and favorable conditions for hydraulic work of corrugated pipes give the grounds to recommend them for application, at any biases in hydro amelioration and nature protection tubular constructions including with difference on spawning waterways and also at biases $i>$ 0.01 in road transitions.

Design data of water throughput constructions from MEC are appointed taking into account type of a spawning waterway, the specific list of the fishes living or coming on spawning into a small waterway, width of roadway covering. The mode of operation of such constructions is proved taking into account requirements of hydrobiontes. The dependences of coefficient of roughness based on their interrelation with a pipe bias are given. It is shown that existence of a smooth tray or the stone dumping protecting a ground part of corrugated pipes from abrasive destruction by a multiphase flow significantly changes conditions of hydraulic work: hydraulic resistances at the free-flow movement change with change of extent of filling of a pipe, as well as an flow rate and a technique of hydraulic calculation.

On the basis of preliminary results of model researches and also the published data, recommendations which will allow to eliminate migration barriers in headwaters of the small river are offered and will serve as the management for future researches of water throughput transitions from metal corrugated structures (NC, HC) on spawning waterways. Application of such culverts gives the simplest method of maintenance of migration conditions, nature, favorable for fishes, by partial deepening of a pipe is lower than a bottom of a waterway and filling of a part of pipe cross-section with gravel dumping.

Values of roughness coefficient in a composite metal corrugated water throughput pipe with gravel dumping at the bottom are for the first time received. It is established that depending on a pipe bias and the mode of course of water in the buried pipe which bottom is covered with gravel not less than for 10% the value of coefficient of roughness turns out approximately in 1.1... 1.4 times are more than values n for corrugated transitions in the absence of any tray at the bottom. It gives the chance to apply the specified values n at preliminary design stages and when calculating in the existing program systems.

It is necessary to continue researches, varying the average diameter of material sketches (D_{50}), deepening of a pipe, a corner of the sector of the fixed site of a bottom of a water throughput pipe and its size (d) in the wide range of values. For development of norms and methodical recommendations about hydraulic calculation of tubular water throughput transitions from corrugated metal structures of the increased sizes with a helical and normal corrugation, the directions for of their operation and to monitoring working in the conditions of stocked water currents it is necessary to investigate experimentally pipes which ground part is strengthened by rock filling and gabion elements.

References

[1] Altunin V I Chernikh O N and Burlachenko A V 2016 Resistance of Corrugated Metal Culvert Pipes with Elevated Abrasive Resistance *Power Technology and Engineering* **50** 4 385-390

[2] Altunin V I and Chernikh O N 2016 Resistance of a Helially Corrugated Metal Pipe Culvert // Power Technology and engineering *Power Technology and Engineering* **50** 2 July 125-129

[3] Suetina, T A Chernykh O N and Burlachenko AV 2018 Hydraulic calculation features of helically corrugated steel culverts *IOP Conference Series: Materials Science and Engineering VII International Symposium Actual Problems of Computational Simulation in Civil Engineering 1–8 July 2018 Novosibirsk Russian Federation* 456 4 5

[4] *Hydraulic design of highway culverts* 2012 US Department of Transportation Hydraulic design series 5 3d Ed Publication No. FHWA-HIF-12-026

[5] Katopodis 2005 Developing a toolkit for fish passage, ecological flow management and fish habitat works *J of Hydraulic Research* **43** 5 451-467

IOP Conf. Series: Materials Science and Engineering 1159 (2021) 012016 doi:10.1088/1757-899X/1159/1/012016

[6] *Manning's values history of research* 2010 Design data 10. American concrete pipe association 222 W Las Colinas Blvd Suite 641 Irving TX 75039
[7] ODOT *Hydraulics Manual* 2014
[8] Neill C R 2001 *Guide to Bridge Hydraulics* 2nd Ed Transportations Association of Canada

IOP Conf. Series: Materials Science and Engineering 1159 (2021) 012017 doi:10.1088/1757-899X/1159/1/012017

On the use of asphalt concrete mixes (according to the PNST)

M E Telegina[1], **G A Averchenko**[1], **I R Mukharryamov**[2], **Y E Vasiliev**[3], **E V Kotlyarskiy**[3]

[1] Peter the Great St. Petersburg Polytechnic University, St. Petersburg, Russia
[2] Saint-Petersburg State University of Architecture and Civil Engineering, Vtoraya Krasnoarmeiskaya str. 4, 190005 St. Petersburg, Russia
[3] Moscow Automobile and Road State Technical University (MADI), 64, Leningradsky Prosp., Moscow, 125319, Russia

E-mail: 79525087757@mail.ru

Abstract. The article is devoted to the preliminary national standard 184-2016 "Public roads. Asphalt concrete road mixes and asphalt concrete. Technical conditions". The purpose of the work is to identify incorrect and unsubstantiated data of the preliminary national standard in comparison with the current interstate standard 9128-2009 "Asphalt concrete road, airfield and asphalt concrete mixtures. Technical conditions". The following are ideas, some comments and suggestions that should be taken into account when developing and approving the preliminary national standard for asphalt concrete. The comparative analysis is aimed at identifying unsubstantiated technical characteristics, physical and mechanical parameters and criteria for testing asphalt concrete. The topics covered include "Grain compositions", "Road classification", "Voids", "Dust", "Sample preparation for testing". The feasibility of using the new system was evaluated. The results are presented in the form of a comparative analysis of the requirements of current standards with the requirements of preliminary standards for the general characteristics of asphalt concrete mixtures. Proposals have been developed to adjust the physical and mechanical properties of asphalt concrete in accordance with the performance indicators of St. Petersburg roads.

1. Introduction

In 2017, a number of regulatory documents relating to asphalt concrete road mixes and asphalt concrete were put into effect, which became an alternative to GOST, called preliminary national standards (PNST). PNST were developed as part of the implementation of the "Superpave" system [1, 2].

The standard applies to asphalt concrete mixes and asphalt concrete intended for the construction of structural layers of public roads and airfields. PNST 184-2016 "Public roads. Asphalt concrete road mixes and asphalt concrete. Technical conditions".

The desire to improve the quality of asphalt concrete surfaces by implementing a new design system has led the Russian community of road industry specialists to great controversy [3]. The engineers at the lab say about the structural similarity of asphalt mixtures and their physical and mechanical properties. They do not see the need to replace complete technical documentation, laboratory equipment and design technology for asphalt concrete mix. Also, in their opinion, innovations are economically inefficient, and the timing of the introduction of a new design system in

Content from this work may be used under the terms of the Creative Commons Attribution 3.0 licence. Any further distribution of this work must maintain attribution to the author(s) and the title of the work, journal citation and DOI.
Published under licence by IOP Publishing Ltd

all road construction organizations is uncertain [4, 5]. Other experts say that this design system can not be implemented "in parts", because it can lead to an uncertain result. In other words, the technical requirements of the new design system must be fully adapted to the operating conditions of Russian highways and take into account the climatic features of the terrain in which they pass. If possible, develop territorial requirements for asphalt concrete surfaces taking into account the amount of solar energy received at a given latitude, the maximum and minimum daily air temperatures [6, 7]. This will ensure that the requirements of the technical regulations of the Customs Union "Road safety" (TR CU 014/2011) are met.

Unfortunately, for a long period of consideration of the PNST, road workers were not able to come to a common opinion and start acting [8–11].

It is interesting to hear the statement of the honorary road worker M. A. Pokataev that in the 80's a similar project was adopted called "Plan for the introduction of new equipment and advanced technologies". In his opinion, the "introduction" of technologies is inefficient in all aspects, and new technologies should be stimulated by competition [12]. Many authors also consider production and technological aspects of implementing a new asphalt concrete design system [13–16].

The purpose of the work is to identify incorrect and unsubstantiated data of the preliminary national standard in comparison with the current interstate standard 9128-2009 "Asphalt concrete road, airfield and asphalt concrete mixtures. Technical conditions». The following are ideas, some comments and suggestions that should be taken into account when developing and approving the preliminary national standard for asphalt concrete.

The article is divided into sections on the following topics: "Grain compositions", "Road classification", "Voids", "Dust", "Sample preparation for testing". The feasibility of using a new asphalt concrete design system was assessed, taking into account the conditions of domestic operation of non-rigid roads. The results are presented in the form of a comparative analysis of the requirements of current standards with the requirements of preliminary standards for the general characteristics of asphalt concrete mixtures. Proposals have been developed to adjust the physical and mechanical properties of asphalt concrete in accordance with the performance indicators of St. Petersburg roads.

2. Materials and methods

2.1 PNST 184-2016. Grain composition

The traffic intensity of cargo vehicles is increasing every year, and the requirements for ABS strength in accordance with the PNST are decreasing. If in the USSR when selecting the composition of asphalt concrete mixtures of types A and B according to GOST 9128-2013 "Asphalt concrete road mixes, airfield and asphalt concrete. Technical conditions" a lot of research has been carried out, but today the changes in the new standard have no scientific justification for their application in Russia. We are talking about a mixture of A16OT, A16HE, A16OL, used in the device of the bases of road clothing under all traffic conditions, shown in table 1. Mixtures A16N, A16NN, A11NN, A11NL, which are used in the device of the lower layers of pavement, shown in table 2.

Table 1. Grain composition of mixtures for the road surface layer in accordance with table 1 PNS 184-2016

Screen size, mm	Passage through the sieve, % by weight, for types of mixtures								
	A32OT	A22O T	*A16O T*	A32O H	A22O H	*A16O H*	A32O Л	A22O Л	*A16O Л*
45.0	100	-	-	100	-	-	100	-	-
31.5	90-100	100	-	90-100	100	-	90-100	100	-
22.4	75-90	90-100	*100*	75-90	90-100	*100*	75-90	90-100	*100*
16.0	-	75-90	*90-100*	-	75-90	*90-100*	-	75-90	*90-100*
11.2	-	-	*75-90*	-	-	*75-90*	-	-	*75-90*
4.0	35-55	35-55	*35-55*	35-55	35-55	*35-55*	50-70	50-70	*50-70*
2.0	25-40	25-40	*25-40*	25-40	25-40	*25-40*	40-60	40-60	*40-60*
0.125	4-14	4-14	*4-14*	4-14	4-14	*4-16*	4-17	4-17	*4-17*
0.063	2-9	2-9	*2-9*	2-9	2-9	*2-9*	2-10	2-10	*2-10*

Table 2. Grain composition of mixtures for the road surface layer according to table 2 PNST 184-2016

Screen size, mm	Passage through the sieve, % by weight, for types of mixtures								
	A32HT	A22H T	*A16H T*	A32H H	A22H H	*A16H H*	*A11H H*	A16H Л	*A11H Л*
45.0	100	-	-	100	-	-	-	-	-
31.5	90-100	100	-	90-100	100	-	-	-	-
22.4	75-90	90-100	*100*	75-90	90-100	*100*	-	100	-
16.0	-	60-80	*90-100*	-	60-80	*90-100*	*100*	90-100	*100*
11.2	-	-	*65-80*	-	-	*60-80*	*90-100*	65-85	*90-100*
8.0	-	-	-	-	-	-	*60-80*	-	*65-85*
4.0	35-50	35-50	*35-50*	35-55	35-55	*35-55*	*40-60*	40-60	*45-65*
2.0	25-35	25-35	*25-35*	25-40	25-40	*25-40*	*30-50*	30-45	*35-55*
0.125	5-10	5-10	*5-10*	5-15	5-15	*5-15*	*5-18*	5-18	*5-18*

Based on data on the operation of Russian highways, it can be concluded that asphalt concrete mixtures containing small grain sizes are not able to meet the technical requirements for the lower layer of the road surface. Thus, it is proposed to exclude from the document asphalt concrete mixtures A16NT, A16NN, A11NN, A11NL, the feasibility of using which is not justified.

As a result, there are not so many asphalt concrete mixes that need to be leveled granulometric composition. Give the granules a certain, new border, using sieves with a cell size of 16 mm, 11.2 mm, 8 mm, 0.5 mm. It will be the guarantor of quality of asphalt concrete. Less capital-intensive changes in the part of granulometric compositions can be made in this PNST, using the standards for asphalt concrete PANK, developed in Finland. This is a document issued by the Advisory Commission on coatings (PANK GU), which presents the quality requirements used in Finland for asphalt concrete mixes and coatings, as well as raw materials for asphalt concrete. The book presents the principles of designing asphalt concrete mixes and coatings, as well as other related issues. The standards are intended for designers, contractors, developers, suppliers of raw materials, as well as for research institutions. It is worth noting that many of the researchers works are based on the analysis of data from experimental sites, which indicates the reliability of their published data [17–19].

2.2 PNST 184-2016. Classification of roads

One of the most significant parameters of road construction is the modulus of elasticity [15, 20–23]. Changes in load indicators directly affect the strength and durability of the road. The flaw in the PNST concerns the part of road classification by traffic conditions, where the calculated load is 10 tons per axle. Roads of high categories can now rely on a load of 11.5 tons per axle, which is not enough under real road operating conditions due to heavy vehicles weighing more than 26 tons. In the catalog of standard designs of non-rigid road clothes of the state company "AVTODOR", the load classes from transport are allocated, which directly depend on the number of load applications (from 3 million to 15 million or more) and the flow rate (from 20 km/h to 110 km/ h) with reliability levels of 0.88-0.98, shown in table 3.

Table 3. Assignment of the transport load class according to STO "AVTODOR" 2.25-2016. "Catalog of typical designs of flexible pavement for roads"

Total number of applications of the design load (115 kN), mln.	Speed limit			
	Slow motion*	Slow**	Standard***	High-speed****
< 3	average	average	average	average
3 – 7	high	average	average	high
7 – 15	ultra-high	ultra-high	high	ultra-high
> 15	ultra-high	ultra-high	ultra-high	ultra-high

Note:
*average flow speed < 20 km/h
**average flow speed from 20 to 70 km/h
***average flow speed > 70 km/h
****average flow speed > 110 km/h

In the PNST at a load of 10 tons and the service time of road pavement, asphalt concrete mixtures are divided into three loading conditions:

- T (for heavy conditions, if more than 3 million applications load);
- N (for normal conditions, if the load applications are from 0.3 to 3 million);
- L (for light conditions when loads are applied less than 0.3 million times).

The method for reducing loads to the calculated uniaxial 10-ton load is shown in the Appendix A of PNST. The number of applications of uniaxial loads equivalent to 100 kN (10 tons) is calculated by the formula 1:

$$N_{100} = N_i \cdot K \qquad (1)$$

where:

N_{100} - the total estimated number of applications of a single-axis ten-ton load over the service life of the road surface;

N_i - conversion factor;

K- the total calculated number of applications of the calculated uniaxial load for the service life of the road surface for a given road, determined in accordance with current regulatory and technical documents in the field of design.

The conversion factor K is calculated using the formula 2:

$$K = (Q_i / 100) \cdot 4 \qquad (2)$$

where:

Q_i- the calculated uniaxial load for a given highway, determined in accordance with the current normative and technical documents in the field of design, kN;

100 - Uniaxial load accepted in the classification according to the PNST data, kN;

4 - Conversion factor.

Comparing the accepted indicators, it becomes clear that the transport load in the PNST is much lower.

2.3 PNST 184-2016. Definitions of "Emptiness", «Dust»

In addition to the definitions of "residual porosity" and "porosity of the mineral part of asphalt concrete", the standard added the term "voids in the mineral aggregate" (PMZ). This is incorrect, since the asphalt concrete mix may contain moisture, air, vapors, etc. These substances differ in their aggregate state and density, which is very important.

Also, the purpose of introducing the following indicators is not clear: the ratio "dust-binder" (hereinafter - N/A) and "voids filled with bituminous binder" (hereinafter - PNB). The P/V ratio is actually higher than in accordance with the provisions of the PNST. The same is true for residual porosity. According to the PNST, the residual porosity data is lower than the calculated data.

According to PNST, dust is considered to be particles of mineral materials in a mixture of no more than 0.063 mm in size, although according to GOST 25100-2011 "Soils. Classification" dust has a size of less than 0.05 mm. The PNST suggests a conclusion about the effect of P/V on the binder, while if there are particles up to 0.063 mm in the asphalt concrete mixture, then these small particles have different mineralogical composition. It is very strange the ratio of N/A for PNST from 0.6 mm to 2.0 mm for all asphalt concrete mixes. This is incorrect. Let's look at the example in more detail. The data is presented in table 4.

Table 4. Example

№	Coating layer	Type of mix	Dust content, %	P/V	The boundaries of the contents of the binder, %
1	Top	A8B AHB	6 – 12	0.6	10 20
2	Top	A8B AHB	6 – 12	2.0	3 6

For example, if you arrange the top layer of the coating from a mixture of A8B or a mixture of AHB with a particle content of less than 0.063 mm in an amount of 6% to 12 % and with a ratio of P/V equal to 0.6, the boundaries of the binder content will be 10 % and 20 %, respectively. If you arrange the top layer of the coating from A8B or AHB mixture with a particle content of less than 0.063 mm in an amount of 6% to 12% and with a ratio of P/V equal to 2, the boundaries of the binder content will be 3% and 6%, respectively.

Thus, 3% of the binder content is very small, and 20%, on the contrary, a lot. It is proposed to perform calculations and derive a new ratio of P/V for each type of asphalt concrete mixture, depending on the grain composition of the mixture. The analysis of hot compositions of asphalt concrete mixtures according to GOST 9128-2013 is presented in table 5.

Table 5. The analysis of hot compositions of asphalt concrete mixtures according to GOST 9128-2013

№	Coating layer	Type of mix	P/V
1	Top	grained	1.45 – 1.55
2		coarse-grained dense	1.83 – 2.11
3		fine-grained dense	1.82 – 2.14
4		sandy	2.10 – 2.29

It can be seen that for denser mixtures, a larger amount of binder is needed to prevent leaching of the asphalt concrete pavement.

It should be noted that for the porous asphalt concrete (> 5%) proposed value filling the pores in the mineral mixture of asphalt cement are too high, and for a dense asphalt underestimated. All this makes it impossible to provide the required values of residual porosity in the asphalt concrete mix.

2.4 PNST 184-2016. Sample preparation and testing

The PNST offers the preparation and testing of samples according to the Marshall method. It should be noted that GOST 12801-67 "Mixes of asphalt concrete (hot and warm) road and airfield and asphalt concrete. Test methods", GOST 12801-71 "Mixtures of asphalt concrete road, airfield and asphalt concrete. Test methods"and GOST 12801-77 "Mixtures of asphalt concrete road, airfield and asphalt concrete. Test methods" this method is present. But in 1967-1977, this method was never implemented. Gorelyshev N. V. found that when compacting samples by ramming using the Marshall method, the crushed stone is approximately the same as when compacting samples in the laboratory (vibration compaction and compression). "The structure of asphalt concrete in a rammed sample differs from its structure in the coating, which leads to non-identical properties of the material in the sample and in the coating". The work is described in his doctoral dissertation "Research of asphalt concrete frame structure and its performance properties in road clothing" in 1978.

In addition to the above, foreign colleagues (in 1981-1982) from the University of Notingham conducted comparative studies of various compositions of bituminous materials under various test schemes: triaxial compression under cyclic load; static tests under triaxial compression; static tests under uniaxial load; tests on the Marshall test. As a result, it was found that tests for three-axis compression under cylindrical loads best reflect the actual working conditions of asphalt concrete surfaces (Fig. 1).

Static tests of samples for uniaxial compression and Marshall tests have a linear relationship between them, but the accuracy and convergence of the Marshall test is significantly inferior to tests for uniaxial compression. Thus, to obtain reliable results for Marshall, at least eight parallel samples should be tested, and three are sufficient for uniaxial compression [24, 25].

There are no innovations in the Marshall test and it is completely unjustified to switch to manufacturing samples using this method to determine the following indicators: volume density, water saturation, water resistance and the Marshall test. The number of compacting blows (50) is also unclear, while the established number is 75.

In addition, it is proposed to set the value of residual porosity for asphalt concrete used for the device of the bases and the lower layer of the coating to no more than 5%. This will improve the resistance of the layers to tensile forces, reduce cracking from below, and help avoid reflected cracks in the upper layer of the road surface.

Not justified an understatement of the brand of crushed stone for strength and lasagnette.

It is important to include in the main test indicators not only the average value of the track depth, but also the angle of inclination of the track formation curve. This is obtained immediately from the experiment. About the requirements for binders-include the current GOST 22245-90 "Road oil Bitumen viscous. Technical conditions".

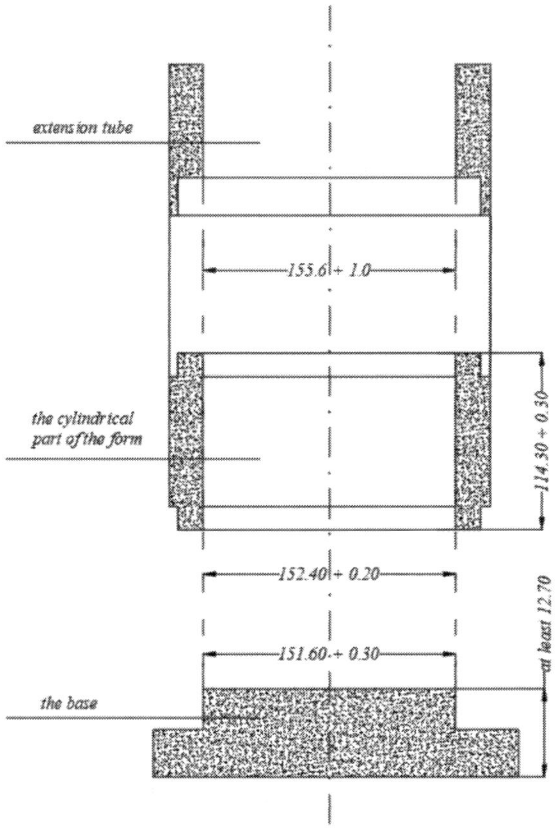

Fig. 1. Assembly form for compaction of samples

2.5 PNST 184-2016. Wear resistance in the cold season

It should be noted that in recent years, due to a sharp increase in the number of cars and the speed of their movement, one of the main indicators of the stability of coatings is their wear resistance in the cold period of the year. According to the service for the operation of the Ring road (KAD) of Saint Petersburg, the traffic intensity on high-speed lanes reaches more than 3 thousand cars per hour, and the daily rate is about 35-40 thousand cars. With an average flow speed of up to 124 km/h, and individual cars up to 220 km/h.

In this regard, it is proposed in PNST to introduce additional testing of aggregates for the top layer coating method according to the Scandinavian "Nordic-test" (EN 1097-9), as the wear resistance of the asphalt top coat on motorways and Express roads in the cold season of the actions of studded tires, especially when channeled high-speed movement becomes the main criterion of sustainability.

Approximately 300 samples of crushed stone from various rocks of different deposits were tested in the laboratory of "NCC roads". As a result, it turned out that the most wear-resistant material is crushed stone from porphyry, the next was gabbro-diabase. Wear resistance of asphalt concrete is caused by the abrasion of crushed stone, its quantity, the correctness of the selected grain composition and the binder used. According to experts on the content of the road, for one year of service, the depth of the track wear of the coating on average is about 10 mm in areas with the use of crushed mastic asphalt on gabbro - diabase and about 8 mm - when using porphyrite rubble.

3. Results and Discussion

Summarizing the results of the work, we highlight the following proposals that should be taken into account when developing and approving the PNST 184-2016 "Public roads. Asphalt concrete road mixes and asphalt concrete. Technical conditions":

1. Exclude asphalt-concrete mixes of types A16OT, A16ON, A16OL used at the device of the bases of road clothes, and A16NT, A16NN, A11NN, A11NL used at the device of the lower layers of the covering of road clothes;
2. Consider the standards for PANK asphalt concrete developed in Finland for making less capital-intensive changes in the part of granulometric compositions;
3. Consider the correct use of a transport load of 10 tons per axle instead of 11.5 tons per axle, i.e. reducing it;
4. Explain the feasibility of changing the size of dust particles from 0.05 mm to 0.063 mm;
5. Perform calculations and output a new ratio of P/V for each type of asphalt concrete mix, depending on the grain composition of the mix;
6. Note that for porous asphalt concrete (>5%), the proposed values for filling the pores in the mineral part of the mixture with bituminous binder are overstated, and for dense asphalt concrete, on the contrary, they are underestimated;
7. Set the value of the residual porosity for asphalt concrete used for the device of the bases and the lower layer of the coating, set no more than 5%;
8. Justify understating the brand of crushed stone in terms of strength and breaminess;
9. Justify the need to prepare and test samples using the Marshall method without choosing an alternative method for testing samples;
10. Add to the main test indicators not only the average value of the track depth, but also the angle of inclination of the track formation curve;
11. Introduce tests of crushed stone for the top layer of the coating according to the Scandinavian method "Nordic-test" (EN 1097-9).

4. Conclusions

1. In connection with the increase in the intensity of motor traffic demands on the quality of asphalt concrete pavements has increased significantly. Traffic safety directly depends on the quality of the road surface. Ensuring traffic safety is a key goal in the construction and operation of highways.
2. The Introduction of preliminary national standards has a huge impact on the economic, technological, production components and qualifications of road workers. Striving for high quality roads through the introduction of a new asphalt concrete design system, there is a sharp need for monetary resources, which makes it difficult for almost all road construction organizations due to their interdependence.
3. The Feasibility of using a new system for designing asphalt concrete taking into account the conditions of domestic operation of roads with non-rigid pavement is not justified today. Perhaps a more effective action would be to modernize the technical equipment of road construction services, improve the existing standards for designing asphalt concrete, and systematically introduce innovations.

References

[1] Musselman J A, Choubane B, Page G C, Upshaw P B 1998 Superpave Field Implementation: Florida's Early Experience. *Transp. Res. Rec. J. Transp. Res. Board.* **1609** 51–60
[2] Kakrasul J Superpave: How it Works and How it Can Work for Local Agencies
[3] Kiryuhin G N, Dzhumanov R B 2014 *Advantages and disadvantages of design asphalt «Superpave»*
[4] Easa S M 2020 Superpave Design Aggregate Structure Considering Uncertainty: I. Selection of Trial Blends. *J. Test. Eval.* **48** 20170682

[5] Averchenko G A, Kvitko A V, Barashev M N 2017 *Ortotropnaya kompozitnaya plita [Orthotropic composite plate deck bridge]* Utility Model Patent № RU 174705 U1E01D 19/12 (2006.01) 31
[6] Averchenko G A, Kvitko A V 2017 *Ortotropnaya kompozitnaya plita [Orthotropic composite plate deck bridge].* Utility Model Patent № RU 173490 U1, E01D 19/12 (2006.01), E01D 101/40 (2006.01) 25
[7] Radovskii B S 2007 Designing the composition of asphalt mixtures in the USA using the Superpave method *Dorozhnaia tekhnika Road Mach* 86–99
[8] Han D, Wei L, Zhang J 2016 Experimental Study on Performance of Asphalt Mixture Designed by Different Method *Procedia Engineering* 275 407–414
[9] Zhang Z Q, Tao J, Yang B, Li N L 2009 Methodology of mixing and compaction temperatures for modified asphalt mixture *Geotechnical Special Publication* **5** 34–41
[10] Fernandes F M, Pais, J C 2017 Laboratory observation of cracks in road pavements with GPR. *Constr. Build. Mater* **154** 1130–1138
[11] Hosseini F, Hossain S M K, Fu L, Johnson M, Fei Y 2015 Prediction of Pavement Surface Temperature Using Meteorological Data for Optimal Winter Operations in Parking Lots *Proceedings of the International Conference on Cold Regions Engineering* **39** 440–451
[12] Zhang H, Yu T, Huang Y 2020 Comparative analysis of HMA aggregate variability based on impacting and gyratory compaction. *Constr. Build. Mater* **242**
[13] Khasawneh M A, Alsheyab M A 2020 Effect of nominal maximum aggregate size and aggregate gradation on the surface frictional properties of hot mix asphalt mixtures. *Constr. Build. Mater* **244**
[14] Ramadan K Z, Al-Khateeb G G, Taamneh M M 2019 Mechanical properties of styrofoam-modified asphalt binders *Int. J. Pavement Res. Technol*
[15] Priyanka B A, Sarang G, Ravi Shankar A U 2019 Evaluation of Superpave mixtures for perpetual asphalt pavements *Road Mater. Pavement* 1952–1965
[16] Pereira P, Pais J 2017 Main flexible pavement and mix design methods in Europe and challenges for the development of a European method *J. Traffic Transp. Eng.* **4** 316–346
[17] Makowska M, Pellinen T 2016 Development of specifications and guidelines for hot in-place recycling in finland—outline and framework *RILEM Bookseries* **68** 851–862
[18] Makowska M, Aromaa K, Pellinen T 2017 The rheological transformation of bitumen during the recycling of repetitively aged asphalt pavement *Road Mater. Pavement Des.* **18** 50–65
[19] Changsha H S 2009 Road pavement material characterization and rehabilitation: selected papers from the GeoHunan International Conference
[20] Mohammad L N, Obulareddy S, Cooper S, Bae A 2008 Permanent deformation analysis of HMA mixtures using simple performance tests. In: Efficient Transportation and Pavement Systems: Characterization, Mechanisms, Simulation, and Modeling - Proceedings of the 4th International Gulf Conference on Roads 601–610
[21] Vallejo L E, Chik Z 2009 Fractal and laboratory analyses of the crushing and abrasion of granular materials. Geomech. Eng. 1 323–335
[22] Gao J, Wang H, Bu Y, You Z, Zhang X, Irfan M 2020 Influence of Coarse-Aggregate Angularity on Asphalt Mixture Macroperformance: Skid Resistance, High-Temperature, and Compaction Performance. *J. Mater. Civ. Eng.* **32**
[23] Association of Asphalt Paving Technologists Technical Sessions 1985 Marshall Procedures for Design and Quality Control of Asphalt Mixtures. *Asphalt Paving Technology Proceedings* **54** 265-284
[24] Ma L J, Zhang J Y, Li Z 2013 The road-test of high performance asphalt mixture with Superpave *Applied Mechanics and Materials* 1576–1579

The application of the fuzzy controller for tension system control of the scraper conveyor in the mines

Le Dinh Hieu[1,2], **Temkin I.O**[1], **Le Van Tung**[3,4]

[1] National Research Technological University "MISiS", 4, Leninsky prospect, Moscow, 119049, Russian.
[2]Hue Instrial College (HueIC), 70 Nguyen Hue street, Vinh Ninh Ward, Hue City, Thua Thien Hue province, 49100, Vietnam
[3]St. Petersburg Mining University, 2, 21 Line of Vasilyevsky Island, St. Petersburg, 199106, Russian.
[4]Quang Ninh University of Industry, 18 street, Yen Tho Ward, Quang Ninh province, 208830, Viet Nam

E-mail: levantungdktd@gmail.com, hieuhuech@gmail.com, igortemkin@yandex.ru

Abstract. The tension control of the automatic adjustment system of the scraper conveyor used for the initial tension of the chain of the scraper conveyor. A hydraulic control system was developed in accordance with the principle, and a theory of system control was created. The controller with a Fuzzy-PID controller used in the automatic tuning system was designed to limit the minimum tension point to a certain specified area. Modeling has shown that it is possible to avoid an increase in power loss due to higher tension and chain exit from the sprocket due to lower tension.

1. Introduction

As the load on the scraper conveyor becomes greater and the length becomes longer, the dynamic characteristics have an obvious influence on the performance of the drag conveyor [1, 2]. And then this will lead to instability and reduced reliability of the scraper conveyor [3-5]. Especially the dynamic chain tension among other dynamic behaviors plays an important role in improving the performance of the scraper conveyor. In this article, it is mainly considered the principle of the self-adjusting system for chain tensioning, as well as analysis methods and related control theory to ensure efficient operation of the scraper conveyor [4, 6]. The hydraulic systems were to control the tension of the scraper conveyor [8]. Moreover, the scraper conveyor was speed controlled by two PMSM motors with high stability, using the adaptive PSO-Fuzzy logic controllers [11] and the adaptive PSO-Bacterial foraging optimization speed stability control algorithm [12].

The load on the working surface of the scraper conveyor changes depending on the position of the shear (plow), and also changes the resistance to the conveyor and the tension in the chain, which leads to a change in the elastic lengthening of the conveyor chain [11]. Theoretical analysis proves that the elastic elongation of the chain of a heavy-duty 240(m) scraper conveyor is almost 0,8 m between idle and heavy-duty [13]. If the elastic elongation of the chain cannot be compensated for overtime, the tension at the point of minimum tension of the conveyor chain will decrease, even zero tension will appear. Both theories and experiments have proven that the initial tension of the scraper conveyor has

Content from this work may be used under the terms of the Creative Commons Attribution 3.0 licence. Any further distribution of this work must maintain attribution to the author(s) and the title of the work, journal citation and DOI.
Published under licence by IOP Publishing Ltd

a great influence on its running characteristics and resistance [6]. If the initial tension is too low, the operation will be unstable and the chain jump phenomenon will sometimes occur, causing the chain to get stuck and broken. In an accident, the initial voltage caused excessive resistance, resulting in high power consumption [2].

This paper presents a Fuzzy-PID controller that controls the tension of the chain conveyor under the impact of starting and stable working of the system in two cases of direct start and two-speed starting anti S type.

2. Conveyor pretension and minimum tension

The distribution of the chain tension during the operation of the scraper conveyor Figure 1 shows the dotted line is the initial, and the solid line is the tension distribution during operation, S_i is the tension at each point during stable operation, S_{0i} corresponds to the initial tension, and the drive system power is set to P_A , P_B to make $P_A = P_B = \lambda$, about the circuit voltage as direct distribution, then $S_3 - S_2 = \lambda.\left(S_1 - S_4\right)$, $W_{1h} \rightarrow W_{2h}$, $S_3 = S_4 + W_K$, $S_1 = S_2 + W_{2h}$:

$$S_4 = S_2 + \frac{\lambda.W_{2h} + W_K}{1+\lambda} \tag{1}$$

With W_{2h} : Load side-chain and resistance to movement of the load (N); W_K : No-load side chain resistance (N).

If $\lambda.W_{2h} + W_K > 0$: S_2 : point of minimum tension; $\lambda.W_{2h} + W_K < 0$, S_4 : point of minimum tension; $S_2 = S_4$: At the moment, the system has the lowest elastic elongation.

If the elastic coefficient k of the chain elasticity is constant, the elastic prolongation under pretension S_{01} is:

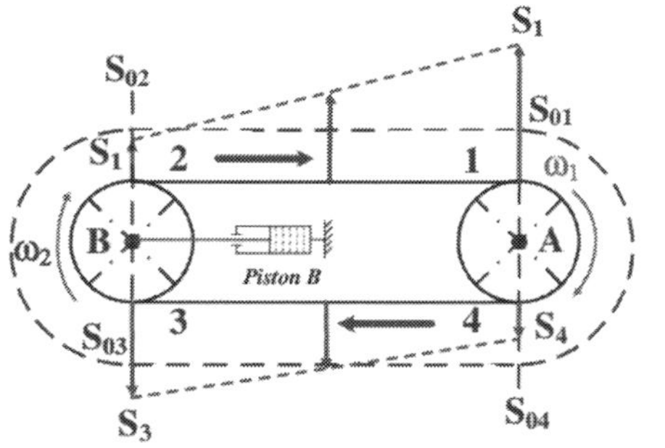

Figure 1. Tension distribution inthe scraper conveyor

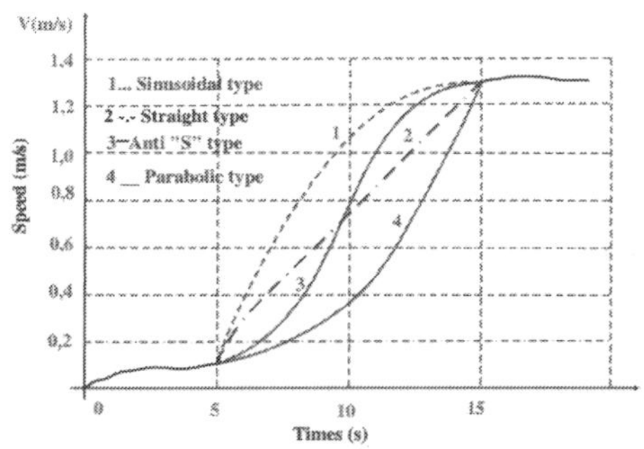

Figure 2. The speed of typical start-up [13]

$$\Delta L_0 = 2.\int_0^t \frac{S_{(t)}}{k} dt = 2.L.\frac{S_{01}}{K} \tag{2}$$

After the stable operation, the elongation is distributed along a solid line, then the elastic elongation at this time:

$$\begin{aligned} \Delta L_1 &= \frac{1}{2}\left(S_1 + S_2\right).\frac{L}{K} + \frac{1}{2}.\left(S_3 + S_4\right).\frac{L}{K} = \frac{L}{2.K}.\left(S_1 + S_2 + S_3 + S_4\right) \\ &= \left[4.S_2 + \frac{2}{1+\lambda}(\lambda.W_{2h} - W_k) + W_{2h} + W_k\right].\frac{L}{2.K} \end{aligned} \tag{3}$$

As can be seen from the above formula, if $\lambda.W_{2h} - W_k = 0$, then ΔL_1- The minimum is obtained, that the elastic elongation of the system is the smallest, which is the optimal configuration of the conveyor power.

For conveyors without automatic tension control, in order to make the tension at the point of minimum tension greater than zero during operation, the following must be done: $\Delta L_0 \geq \Delta L_1$, which the:

$$S_{01} \geq \left[4.S_2 + \frac{2}{1+\lambda}(\lambda.W_{2h} - W_k) + W_{2h} + W_k\right].\frac{1}{4} \tag{4}$$

If $\lambda = 1$ then S_2 is the minimum tension point, S_{02} must satisfy:

$$S_{01} \geq S_2 + \frac{1}{2}.W_{2h} \tag{5}$$

If $\lambda.\overline{\omega_{2h}} - \overline{\omega_k} = 0$, we have:

$$S_{01} \geq S_2 + \frac{1}{4}.(W_{2h} - W_{2h}) \tag{6}$$

Equations (5) and (6) are the initial voltage value when the engine power ratio is a specific value.

When $S_2 = [S_2]_{min}$ is substituted, the pretension is determined without an adjusting device, $[S_2]_{min}$ is the minimum allowable value for the point of minimum tension.

3. Principle of automatic pretension of the scraper conveyor

The principle of automatic adjustment of the conveyor tension is to compensate for the tension change caused by elastic elongation (or contraction) of the chain by varying the distance between the two sprockets to achieve the goal of limiting the tension to a minimum tension point within a certain range. A scraper conveyor running on a working surface can change the center distance of two sprockets only due to the hydraulic cylinder pushing the sprocket and the drive element on the housing due to the limited space. Figure 3 shows the principle of scraper conveyor tension control. Voltage signals S_A and S_B are output by the sensor detection system, S_A and S_B are calculated after analyzing and filtering the signals [8].

The actual value is then determined using the "Comparator I" signal to determine the difference between S_A, S_B and the expected delta value. The "Comparator II" signal determines if there is; whether the product exceeds the acceptable error value and outputs the trigger value. The values are given according to the following formula:

The actual value is then determined using the Comparator I signal to determine the difference between S_A, S_B and the expected delta value. The "Comparator II" signal determines if there is $\Delta S_A, \Delta S_B$; whether the product exceeds the acceptable error value and outputs the trigger value c_A, c_B. The c_A, c_B values are given according to the following formula:

$$c_A = \begin{cases} 1 & \Delta S_A < 0 \ and \ |\Delta S_A| \geq |\Delta S| \\ 0 & |\Delta S_A| \geq |\Delta S| \\ -1 & \Delta S_A > 0 \ and \ |\Delta S_A| \geq |\Delta S| \end{cases}$$
$$c_B = \begin{cases} 1 & \Delta S_B < 0 \ and \ |\Delta S_B| \geq |\Delta S| \\ 0 & |\Delta S_B| \geq |\Delta S| \\ -1 & \Delta S_A > 0 \ and \ |\Delta S_B| \geq |\Delta S| \end{cases} \tag{7}$$

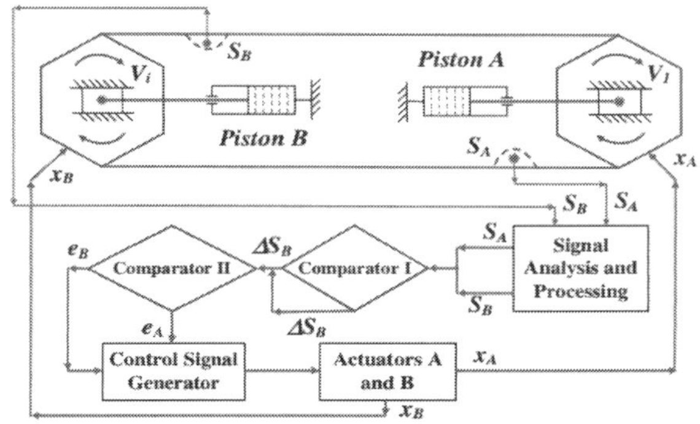

Figure 3. Schematic diagram of the auto-regulating of the scraper conveyor tension

The c_A, c_B same role, for signal generators that control hydraulic cylinder "A" and hydraulic cylinder "B", the control logic is taken as an example of c_A : if $c_A = 1$, then hydraulic cylinder "A" moves and the piston rod extends $\Delta x_A > 0$: if $\Delta x_A = 0$, then hydraulic cylinder A does not move $\Delta x_A = 0$: if $c_A = -1$, the piston rod of hydraulic cylinder A is compressed $\Delta x_A < 0$.

The hydraulic control system that implements the principle of Figure 3 is shown in Figure 4. The number "1" in the figure indicates the oil pump; the "2" is a group of bypass and safety valves; the "3" is a two-position two-way electromagnetic; the "4" solenoid proportional valves; the "5" is an accumulator that stores oil under pressure; the "6" is a 2/2 ways valve; the "7" is a 3-position 4-way valve and the "8" is a pressure switch.

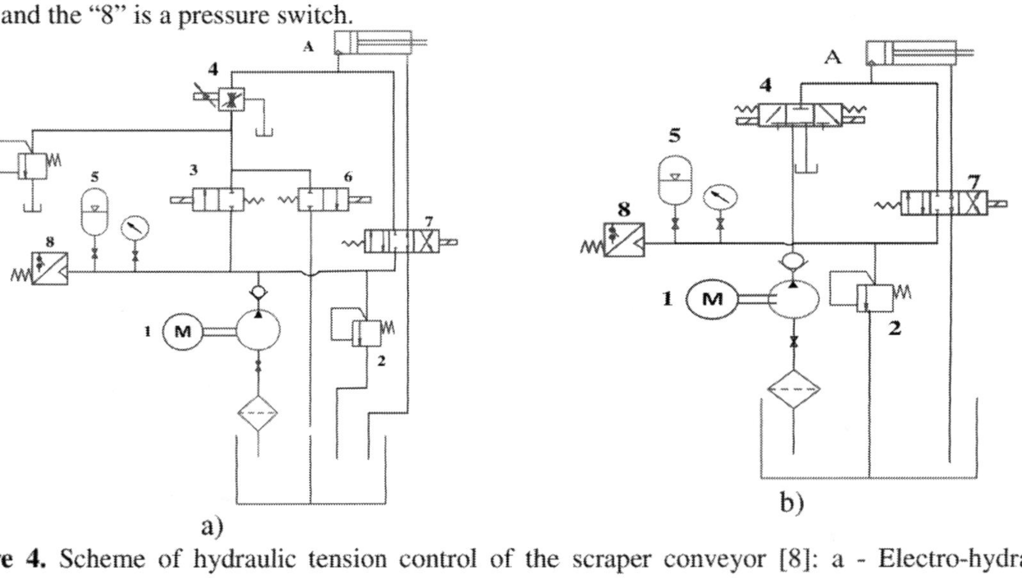

Figure 4. Scheme of hydraulic tension control of the scraper conveyor [8]: a - Electro-hydraulic proportional valve; b - Electro-hydraulic control of the servo valves.

A simplified mechanical model of a drag conveyor with an electro-hydraulic automatic tension control system is shown in Figure 4, and the corresponding dynamic differential equation:

$$[M]\left\{\ddot{X}\right\}+[C]\left\{\dot{X}\right\}+[K]\{X\}=\{\Delta F\}$$
$$\{\Delta F\}^T=\left[0,0\cdots0\cdots,\ \Delta F_A-W_{m+1}\cdot sgn(\dot{x}_{m+1}),\ \Delta F_B-W_{m+2}\cdot sgn(\dot{x}_{m+2})\right] \tag{8}$$

If the process of adjusting the tension at the initial conditions of the conveyor is investigated, the problem becomes more complex, since the force at each point of the concentrated mass of the conveyor changes over time, but if the conveyor speed stabilizes after the conveyor starts running, the study. In the case of pre-launch conditions, the problem is relatively simple. Incremental equation:

$$[M]\{\ddot{X}\}+[C]\{\dot{X}\}+[K]\{X\}=\{F\}$$
$$\{F\}^T=[F_1, W_1, W_2, W_3 \cdots, F_{j=1}, W_{j+2} \cdots W_m, F_A-W_{m+1}, F_B-W_{m+2}] \quad (9)$$

For formula (9), to take the Laplace transform, there is:

$$\{x_{(s)}\}=[H_{(s)}]^{-1}.\{\Delta F_{(s)}\}$$
$$[H_{(s)}]=[M].S^2+[C].S+[K] \quad (10)$$

The change in chain tension:

$$\Delta S_A = K.(x_{m+1}+x_m)+C.(\dot{x}m+1+\dot{x}m)$$
$$\Delta S_B = K.(x_{m+2}+x_j)+C.(\dot{x}m+2+\dot{x}j) \quad (11)$$

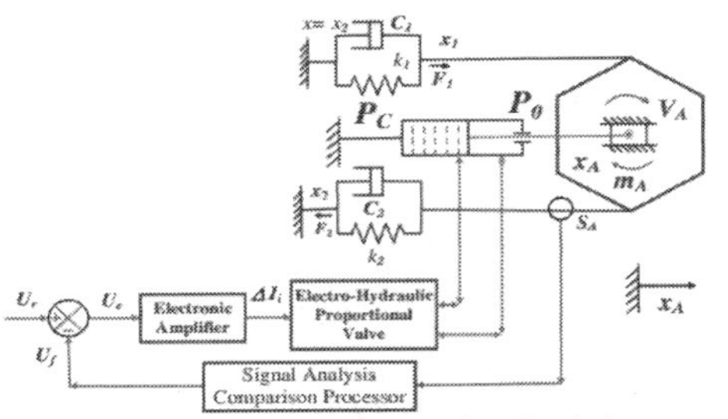

Figure 5. System Control principles

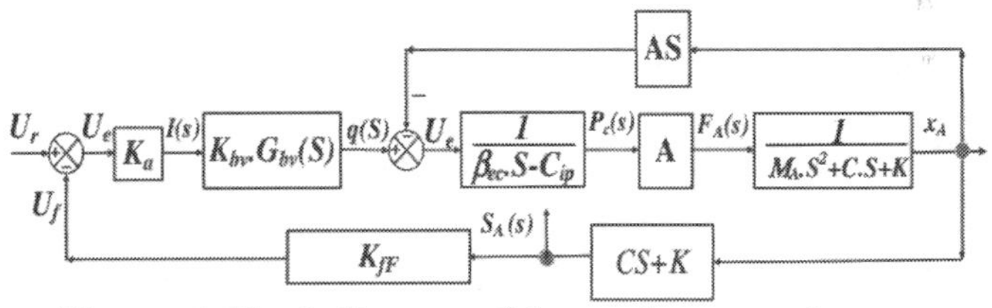

Figure 6. Block diagram of the tension control system

In order to study the basic characteristics of an even more simplified control system, a simplified closed-loop control system is shown in Figure 5. The $U_A=U_r-U_f$, U_r: control voltage signal corresponding to the desired voltage value; U_f: feedback voltage signal; $U_f=K_{fF}.S_A$: with K_{fF}: sensor gain.

$$\frac{q_{v(S)}}{\Delta I}=K_{bv}.G_{bv(S)} \quad (12)$$

4. Design Fuzzy-PID controller for scraper chain tension

Figure 5, the detection circuit is partially composed of sensors, amplifiers, filters, and operations, a comparator or the like is constructed.

On-line monitoring F_i the rate of change of output deviation voltage and deviation voltage after processing by the detection circuit, these two values are proportional to ΔF_i and $\Delta \dot{F}_i$, This part of the delay is τ_i and the control circuit is outputted by the U, U_c input value after fuzzy analysis and output control current I_{dv}, the delay for this part is $F_i \rightarrow I_{dv}$, The delay is $\tau=\tau_1+\tau_2$. The balanced equation of the cylinder:

$$P_C.A_C-P_0.A_0-P_f-F_i-F_{i+1}-m_A g.\text{sgn}(\dot{x}_A)-\eta.\dot{x}_A=m_A.\ddot{x}_A$$
$$F_i=K_i(x_A+x_i-x_{i+1})+C_i.(\dot{x}_A+\dot{x}_i-\dot{x}_{i+1}) \quad (13)$$
$$F_{i+1}=K_{i+1}(x_A+x_{i+1}-x_{i+2})+C_{i+2}.(\dot{x}_A+\dot{x}_{i+1}-\dot{x}_{i+2})$$

The pressure-flow equation of the electro-hydraulic servo valve is:

$$q_i = C_d.x_v(P_s - P_c)^{0.5}$$
$$x_v = K_d.G_{sv} \tag{14}$$

The pressure-flow equation for the cylinder is:

$$q_i = A_c.\overset{*}{x}_A + C_{ip}.(P_c - P_0) + \frac{V_c}{\beta.c}\frac{dP_c}{dt}$$
$$V_c = V_0 - A_c.x_A \tag{15}$$

In the above formulas: P_c, A_c are the oil pressure and piston area of the oil inlet chamber of the cylinder (Pa.m²); P_0, A_0 are the oil pressure and annular cavity area of the oil return chamber of the cylinder, usually $P_c \gg P_0$,Can make $P_0 \approx 0, P_f$ for the cylinder oil plug movement resistance (*N*), η- Adhesion resistance for cylinder and mass m_A (Ns/m), analysis $\eta = 0$; $m_A.g.\mu.\text{sgn}(\overset{*}{x}_A)$: Frictional resistance for mass operation (N); μ-is the mass resistance coefficient of operation; m_A- is the mass of the sprocket, tailstock, motor, coupling (or scrambler), reducer and cylinder piston (kg); P_s-System oil supply pressure (Pa); q_i- Servo valve outlet flow (m³/s); C_d - Constants related to servo valve structural parameters, fluid density; x_v the opening degree of the servo valve (m); K_d - Servo valve spool control current, provided by servo circuit; G_{sv}- Transfer function of servo valve spool displacement; C_{ip}- The internal leakage coefficient of the cylinder is related to the size and pressure of the cylinder; V_0- The initial volume of the cylinder and piping (m³); $\beta.c$ - Effective bulk modulus.

In the simulation analysis F_i and $\overset{*}{F}_i$ are used as detection signals, target value $F_{i(t)} \Rightarrow 21.000(N)$

$$\overset{*}{F}_i = K_i.(\overset{*}{x}_A + \overset{*}{x}_i - \overset{*}{x}_{i+1}) + C_i.(\overset{**}{x}_A + \overset{**}{x}_i - \overset{**}{x}_{i+1}) \tag{16}$$

In the previous section, we studied the control method of the pretension of the scraper chain and constructed the mechanical and mathematical models of the control in its simplified form. However, due to the large frictional resistance of the nose and the tail, the nonlinearity of the chain stiffness and the material. The non-uniform nature of the distribution, in fact, this control system has nonlinearity and time-varying. In order to better solve such control problems, The Fuzzy-PID control applies to this system, and now its principle is briefly described. Figure 7 is the fuzzy adjustment PID controller structure.

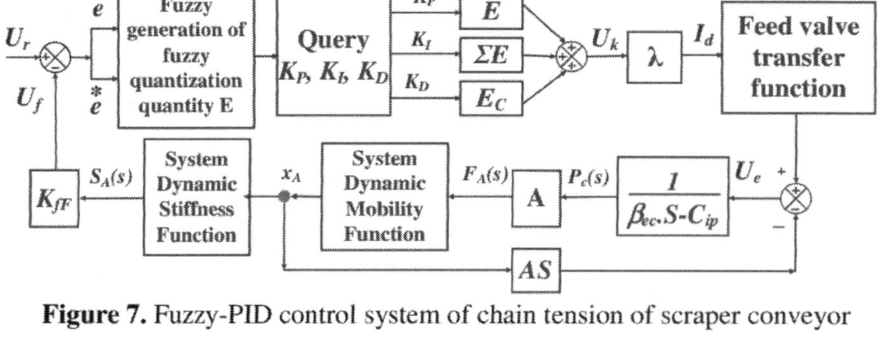

Figure 7. Fuzzy-PID control system of chain tension of scraper conveyor

Table 1. Fuzzy variable E, E_C evaluation

E/Ec	-3	-2	-1	0	1	2	3
PB	0	0	0	0	0	0.5	1
PM	0	0	0	0	0.5	1	0.5
PS	0	0	0	0.5	1	0.5	0
ZO	0	0	0.5	1	0.5	0	0
NS	0	0.5	1	0.5	0	0	0
NM	0.5	1	0.5	0	0	0	0
NB	1	0.5	0	0	0	0	0

Table 2. Fuzzy variable K_P, K_I, K_D evaluation

K_P, K_I, K_D	-3	-2	-1	0	1	2	3
PB	0	0	0	0	0.02	0.6	1
PM	0	0	0	0.02	0.5	1	0.5
PS	0	0	0.02	0.5	1	0.5	0
ZO	0	0	0.5	1	0.5	0	0
NS	0.01	0.5	1	0.5	0.02	0	0
NM	0.5	1	0.5	0.02	0	0	0
NB	1	0.5	0.2	0	0	0	0

For equations (13), (15) take the incremental equation, and turn into equations (14) into a step-by-step number to take the first-order linear term. Then we can get the linear increment equation of this equation. The Lagrangian transformation used to obtain the simplified transfer function of the system. At this time, we ignore the frictional resistance of the mass M_a, so that the transfer process of the system force is obtained, as shown in Figure 7.

$C = C_i + C_{i+1}$; $K = K_i + K_{i+1}$; K_{fF} detecting circuit gain $c^{(\tau_1+\tau_2).S} / s$ is the transfer function of the delay characteristics of the circuit system. According to the basic principles of engineering design, take $P_S = 25(MPa)$, $\overset{*}{x}_{A\max} = 0.15(m/s)$, $A_c = 0.5.\pi(0.12)^2 = 0.022608(m^2)$, then $q_{l\max} = A_c.\overset{*}{x}_{A\max} = 240(l/\min)$, the maximum flow rate of the selected electro-hydraulic servo valve is $300(l/\min)$, and the transfer function of the electro-hydraulic servo valve $G_{sv(s)}$ for:

$$G_{sv(s)} = \frac{1}{\dfrac{S^2}{\omega_{sv}^2} + \dfrac{2.\xi_{sv}}{\omega_{sv}} + 1} \tag{17}$$

In the formula: $G_{sv(s)} = 600(rad/s)$; $\xi_{sv} = 0.5$

$$K_{sv} = \frac{200(l/\min)}{400(mA)} = \frac{300.10^{-3}}{60\ .\ 40.\ 0.04} = 0.003125\ (m^3/sA)$$

$\tau_1 + \tau_2 = 80(ms)$, Other parameters are found in the manual, according to the Fuzzy-PID control scheme constructed.

Create the dynamic differential equation of the whole system and combine the equations (8) and equations (13) - (17) to obtain the dynamic response of the system.

The scraper conveyor system was a model on Matlab-Simulink R2018b software. The scraper conveyor system parameters [13-16] are described and programmed to calculate the output parameters, as shown in Figure 8 to Figure 16.

Figure 8 to Figure 10 shows the forces of chains at point 2 in Figure 1. Figure 8 shows that during the start according used to the PID controller forces of chains, the scraper conveyor tension fluctuates strongly to a peak of value 2.45.10^5(N). The oscillating force fades out after 3 cycles and stables a time of 1.25 seconds with an around the value of 2.1.10^4(N). When starting the scraper conveyor using the Fuzzy-PID controller for forces of chains. The forces of controlling the load tension are stable with decreasing fluctuation after 3 cycles, and then stable after 0.38 seconds at value 2.1.10^4(N) in Figure 9. Comparing the tension control of the scraper conveyor with the PID controller and the Fuzzy-PID controller, the results clearly show that the Fuzzy-PID controller gives much better quality in Figure 10.

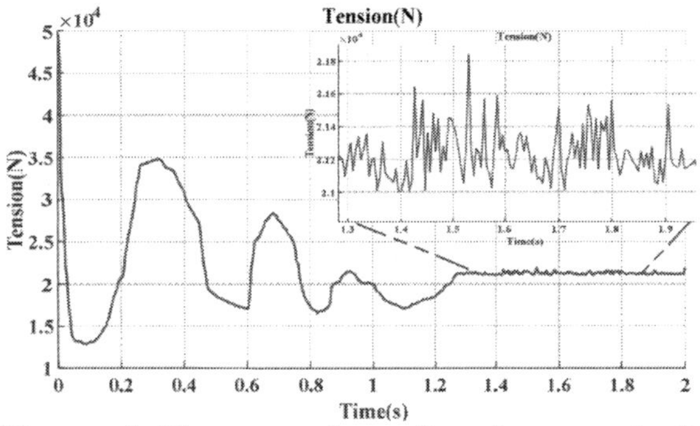

Figures 8. The comparison of tension control of the scraper conveyor with PID controller

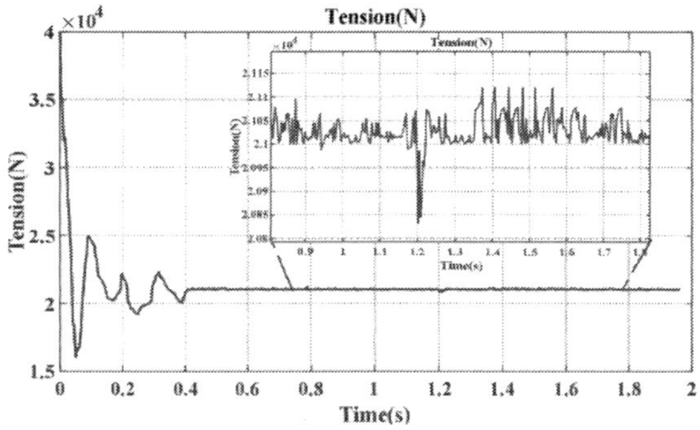

Figures 9. The comparison of tension control of the scraper conveyor with Fuzzy-PID controller

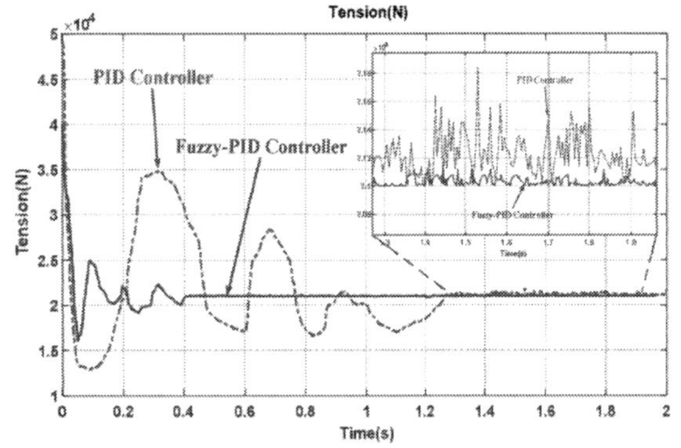

Figure 10. The comparison of tension control of the scraper conveyor with PID controller and Fuzzy-PID controller

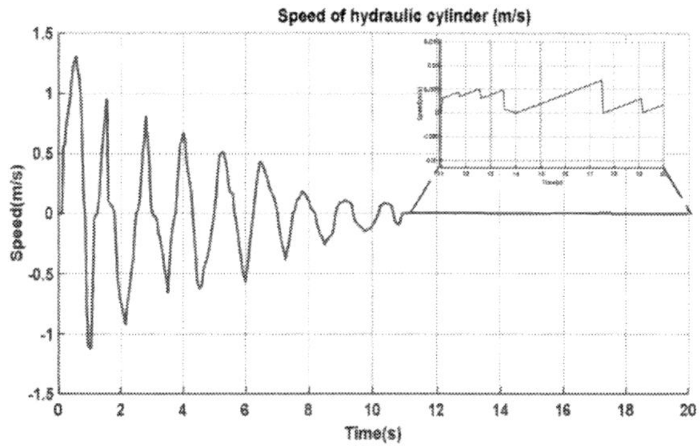

Figure 11. Flex speed of the hydraulic cylinder with PID controller

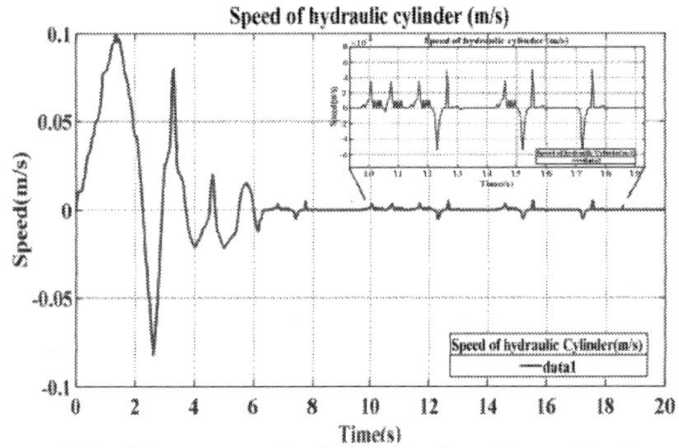

Figure 12. Flex speed of the hydraulic cylinder with Fuzzy-PID controller.

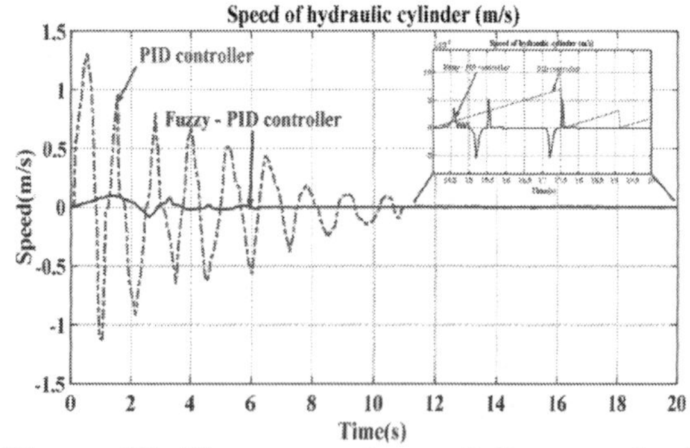

Figure 13. The comparison of flex speed of hydraulic cylinder with PID and Fuzzy-PID

IOP Conf. Series: Materials Science and Engineering 1159 (2021) 012018 doi:10.1088/1757-899X/1159/1/012018

Figure 11 shows that the start according used to the PID controller for the hydraulic cylinder, the speed of the hydraulic cylinder large fluctuations to a peak of value 1.41(m/s). The velocity of the piston gradually stops after 7 cycles and stables a time of 10.25 seconds with an around the value zero. But when starting the scraper conveyor with the Fuzzy-PID controller for forces of chains. The velocity of the piston is decreasing fluctuation after 3 cycles and a peak of value 0.11(m/s), and then stable after 5.8 seconds at an around the value zero in Figure 12. Figure 13 compares the piston velocity between the PID controller and the Fuzzy-PID controller, the results show in detail that the Fuzzy-PID controller gives better control quality and is more reliable.

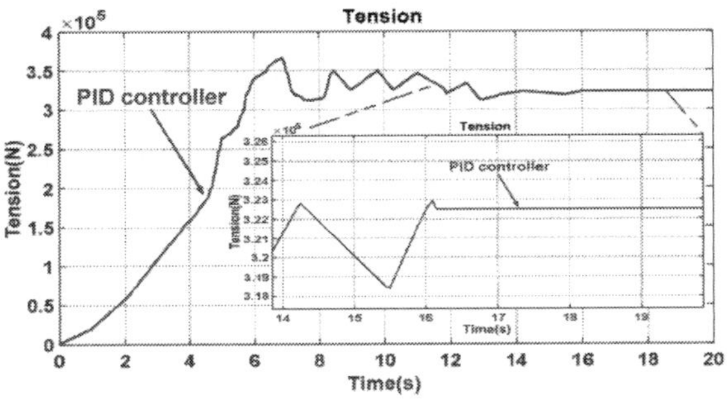

Figure 14. The process of changing the thrust of a hydraulic cylinder with PID controller

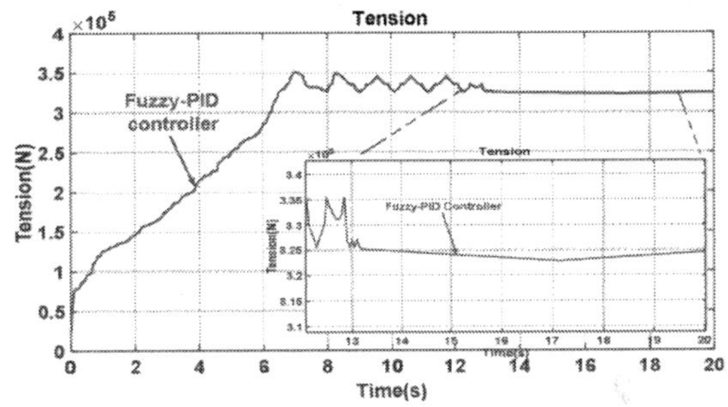

Figure 15. The process of changing the thrust of a hydraulic cylinder with Fuzzy-PID controller

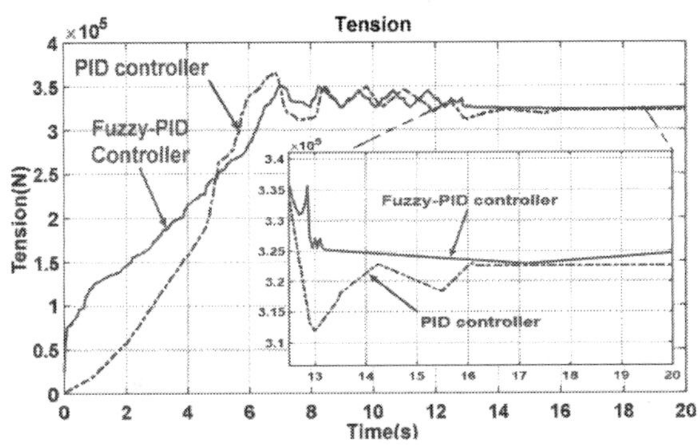

Figure 16. The comparison of changing the thrust of a hydraulic cylinder with PID controller and Fuzzy-PID controller

Figures 14 and 16 describe in detail the thrust of the hydraulic system piston acting on position B in Figure 1 to stretch the chains of scraper conveyor, respectively. Figure 16 compares the thrust of the piston between the PID controller and the Fuzzy-PID controller, the results show that thrust of the piston the Fuzzy-PID controller gives better control quality and is more reliable. It can also be seen from the figure that the system tends to be stable after $t = 12.1(s)$ at a around the value $3.24.10^5$(N). Operating conditions for starting conditions $v_{A_{max}} = 200(l / s)$ and less than the designed maximum speed. The maximum extension distance of the piston rod is $x_{A_{max}} = 0.324(m)$.

5. Conclusion

The principle of automatic adjustment of the pretension force of the scraper conveyor chain has been studied and the hydraulic control system has been developed. The system control theory is an analysis and the Fuzzy-PID controller is applied to the control system and simulation is performed. The simulation results clearly show that the Fuzzy-PID controller is capable of automatically stabilizing the tension of the chains of the scraper conveyor. The process of controlling the chain tension of the chains via the Fuzzy-PID controller, which adjusts the hydraulic valve opening, controls the scraper conveyor to operate stably, less vibration, friction between the chains, and the upper surface. In addition, the conveyor tension control system helps to reduce chain jam, gear breakage, reduce the upper surface wear of the tray with the drive chain, increase the time to change new chains and gears, and increase efficiency.

References

[1] Wang H, Zhang Q, and Xie F 2017 *IET Science, Measurement&Technology* **11(7)** 871-877
[2] Korneev S, Zotov V A and Dobronogova V 2017 *Collection of scientific papers of the Donbass State Technical University,* **2017(5)** 11-18
[3] Pingyuan X and Yandong S 2009 *Second International Conference on Intelligent Computation Technology and Automation IEEE* 33-36
[4] Pavlov V E 2018 *Bulletin of Irkutsk State Technical University* **22**(4) 135
[5] Świder J, Herbuś K and Szewerda K 2017 *MATEC Web of Conferences EDP Sciences* **94** 1-6
[6] Tkachenko A A, Osichev A V *Bulletin of NTU "KhPI"* **7** 157-162
[7] Zhetesova G S, et al 2019 *Izvestiya of higher educational institutions. Mining Journal* **2019(6)** 108-117
[8] Korneev S, V Yu Dobronogova and V Dolgikh 2016 *Collection of scientific papers of Donbass State Technical University* **3** 135-142
[9] Hu Y, et al 2018 *IOP Conference Series: Materials Science and Engineering IOP Publishing* **452(4)** 1-8
[10] Jiang S, et al 2018 *International Journal of Simulation Modelling* **17(1)** 81-91
[11] Le Dinh Hieu and Temkin I O 2018 *IEEE-ICMSC 2018-International Conference on Mechanical, System and Control Engineering, Moscow- Russian* 100-108.
[12] Le Dinh Hieu and Temkin I O 2018 *IEEE Seventh International Conference on Communications and Electronics (ICCE)-Vietnam* 196-201
[13] Le Dinh Hieu, Temkin I O, Do Thanh Lich, Agabubaev Aslan 9-9-2020 *Astrakhan State University- Russian* **2(50)** 10-21
[14] Tereshkin V M, Grishin D A, Makulov I A 2019 Journal of Mining Institute **240** 678-685
[15] Kovalchuk M S, Baburin S V 2018 *IOP Conference Series: Materials Science and Engineering IOP Publishing* **327** 1-7
[16] Zhukovskiy Y, Korolev N and Koteleva IN 2018 *IOP Conf. Series: Journal of Physics: Conf. Series* **1015(3)** 032158

Optimization of pedestrian traffic in residential areas of large cities

T V Konovalova[1], S L Nadiryan[1], M P Mironova[1], A E Litvinov[1]

[1] Kuban State Technological University, Krasnodar, 2a, Moskovskaya st., Krasnodar, 350020, Russia

E-mail: artstyleone@mail.ru

Abstract. Currently, there is a high activity in the construction of multi-storey residential buildings in large cities. This causes an increase in pedestrian flows and traffic intensity in various areas of cities, but most streets and roads are not designed for a given number of pedestrians and vehicles. In turn, this leads to a decrease in the safety of pedestrian traffic in these areas.

1. Introduction

This paper proposes a new approach to solving the problem of optimizing pedestrian traffic, the sequence of developing a model of this approach is shown in Figure 1. The proposed approach is based not on functional zoning of territories, but on the construction of the shortest connecting network of communication routes. In the proposed model, it is proposed to use stops of urban passenger transport (UPT) as points of generation of pedestrian flows. The pedestrian traffic optimization technique is based on a block diagram, which consists of eight sequential blocks. Each of the blocks performs a specific function, collection of information. The result of the whole algorithm is a traffic management project.

The first block is the collection and systematization of source information. It is impossible to get an accurate reliable result after using the technique if the initial indicators, the data were collected incorrectly. The reliability of decisions on the organization of pedestrian traffic is determined by the accuracy and reliability of the initial indicators. Therefore, the implementation of this paragraph of the methodology must be taken responsibly.[1]

The next block is the definition of reduced lengths and the construction of the shortest network of communication routes, is the development of a map with the shortest communication routes. At this stage, with the help of overhead lines, pedestrian communication routes are built and their length is determined.

When performing the third block, it is necessary to identify additional points of generation of pedestrians, these can be objects of mass attraction, stops of gas turbines. Select the links of the pedestrian network, in the work as the main links it is proposed to use the stopping points of the UPT.

The next point is to define the characteristics of pedestrian traffic: intensity, speed, density, etc. The intensity indicator plays an important role, based on the data obtained, it is concluded which of the routes can be neglected, and where, on the contrary, it is necessary to add pedestrian paths. Also, based on the characteristics of pedestrian traffic, measures are being developed to improve traffic safety associated with the use of technical means of regulation (installation of road signs, marking, etc.).

Based on the data obtained, the next section of the algorithm is performed, this is the calculation of the

Content from this work may be used under the terms of the Creative Commons Attribution 3.0 licence. Any further distribution of this work must maintain attribution to the author(s) and the title of the work, journal citation and DOI.
Published under licence by IOP Publishing Ltd

technical characteristics of the communication routes. It includes the calculation of the width of the sidewalks, the width of the pedestrian crossings, the size of the stopping areas of the UPT.

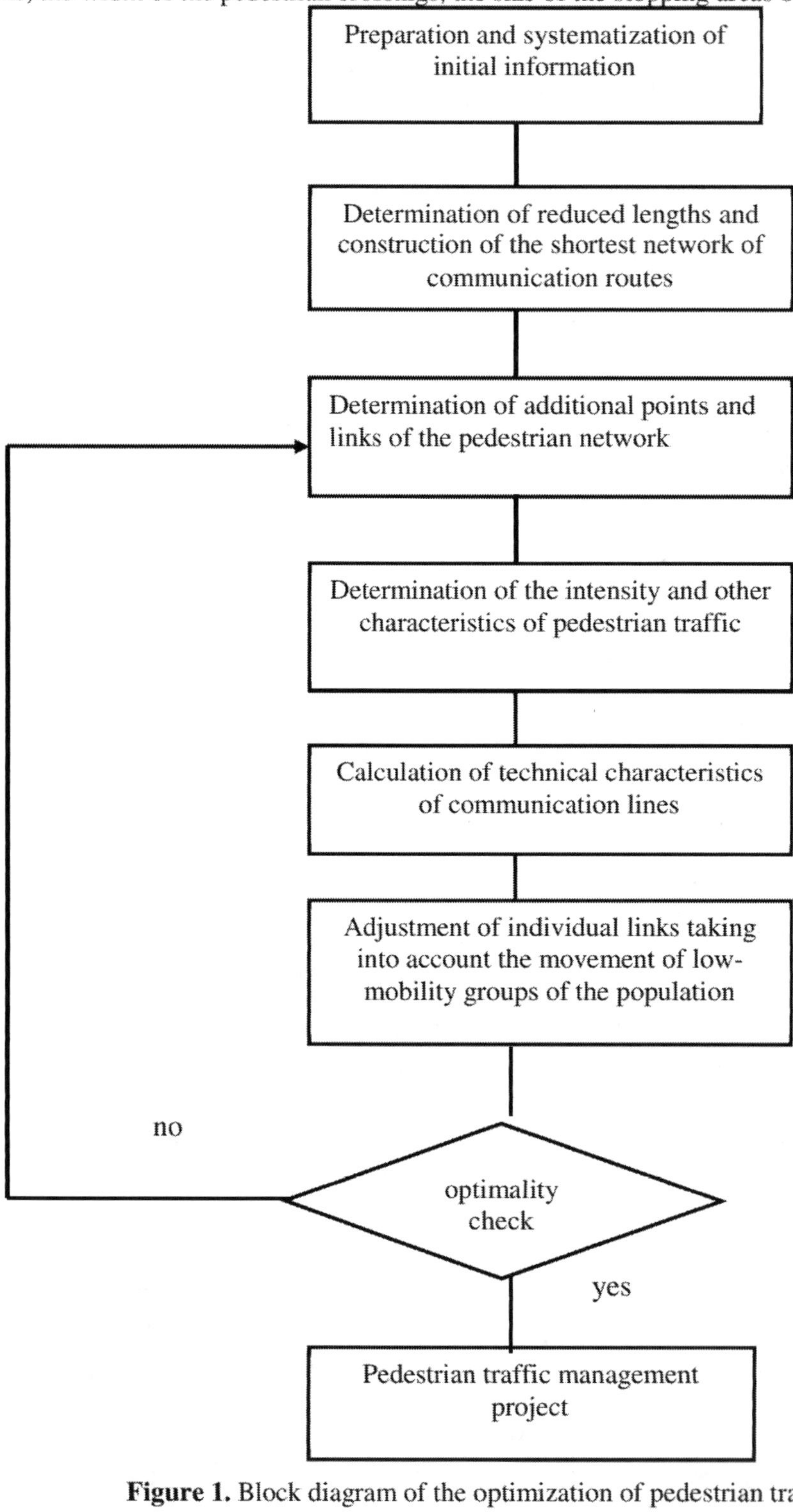

Figure 1. Block diagram of the optimization of pedestrian traffic

Currently, special attention is paid to the movement of low-mobility groups of the population (LMGP), since the UDS should be available for all categories of citizens. Therefore, in the proposed methodology there is an item called the adjustment of individual links, taking into account the movement of low-mobile groups of the population. It is necessary to conduct research on the movement of MGN, to identify areas where there is an active movement of this population group and to develop measures to create a barrier-free traffic environment.

Further, the optimality of the indicators is checked, if all indicators are normal, then a project for the organization of pedestrian traffic is developed and approved. If the data does not pass the check, it is necessary to return to the third block of the methodology and re-point everything to the standard indicators.

A model for the development of pedestrian traffic within the framework of spatial planning is being developed in accordance with the Urban Planning Code of the Russian Federation to substantiate the chosen option for the placement of infrastructure facilities based on an analysis of the modern use of the territory, possible directions of its development and projected restrictions. Similarly to these requirements, the sequence of model development "from general to specific" is determined.

To get a general idea of the planned territory of pedestrian traffic, it is possible to develop a model of its spatial organization with the allocation of enlarged zones of intensive, extensive and limited development of the territory, differing in transport accessibility.

Preparation and systematization of initial data to determine the general potential of the territory includes information on the presence of planning restrictions, natural and man-made risks of urban development and the possibilities of their reduction. A comprehensive assessment of the state and modern use of the territory involves studying the level of development of the socio-economic complex and the settlement system, assessing the possibilities for sustainable development of the territory and balanced socio-economic development, as well as identifying problem situations and priority tasks for solving them.

To analyze and assess the current state and use of the planned territory, different types of information and materials can be used: cartographic, statistical, legal, analytical about the state of the territory, its natural resource potential, including about natural conditions that affect the features of the organization of urban planning and economic activities , on the level of development of the socio-economic complex, on the restrictions on the territory.

Based on the analysis and assessment of the current state and use of the planned territory, a map is being developed showing the predominant use of the territory, zones of restrictions for organizing pedestrian traffic, incl. territories at risk of emergencies. Based on the analysis of the planning organization of the territory, a model of pedestrian traffic is being developed.

The task of urban planning studies in general [2] is to link the priority areas of socio-economic development to the territory in the form of transport infrastructure facilities based on a balanced account of environmental, economic and social factors (fundamental external conditions and restrictions) and the formation of safe, favorable conditions for human life (i.e. structuring the territory according to priority uses).[3,4]

2. Materials and methods

The implementation of these principles is ensured by the formation of a system of frameworks of the territory based on a balanced account of fundamental external conditions and restrictions on the development of the territory [5]:

- an economic framework formed by planning axes and centers: territorial production complexes, industrial centers, clusters of various profiles, industrial parks, centers of innovative development, large engineering and transport infrastructure facilities, etc .;
- a social framework formed by settlement centers (centers of administrative, organizational, economic, cultural and consumer services in the field of education, health care, social security, physical education and sports - centers of settlement systems of various ranks, formed on the basis of the most developed and conveniently located urban and rural settlements);
- an ecological framework formed by territories with a special nature conservation status (specially

protected natural areas, health resort and recreational areas, water protection zones, forest protection belts, green zones of cities, etc.).

The development model of the territory, formed by the system of economic, social and ecological frameworks, makes it possible to create favorable conditions for human life, to provide infrastructural conditions for development (including the formation of zones of advanced development of various types and "growth points"), to ensure the rational use of natural resources and conservation valuable natural complexes, and also allows predicting the rational development of various parts of the territory. Justification of the planned location of transport facilities must be carried out from the point of view of their compliance with the planned location of the economic, social or ecological frameworks and their impact on the integrated development of the territory.

With the approach described above to solving the problems of optimizing pedestrian traffic, it is possible with a high degree of reliability to determine the weight of transport links and solve the problem of placing transport infrastructure objects according to a given criterion of optimality - the minimum time spent by pedestrians on the road network.[6]

In the course of the work, overhead lines of the pedestrian route were drawn on the map of the section under consideration Figure 2.

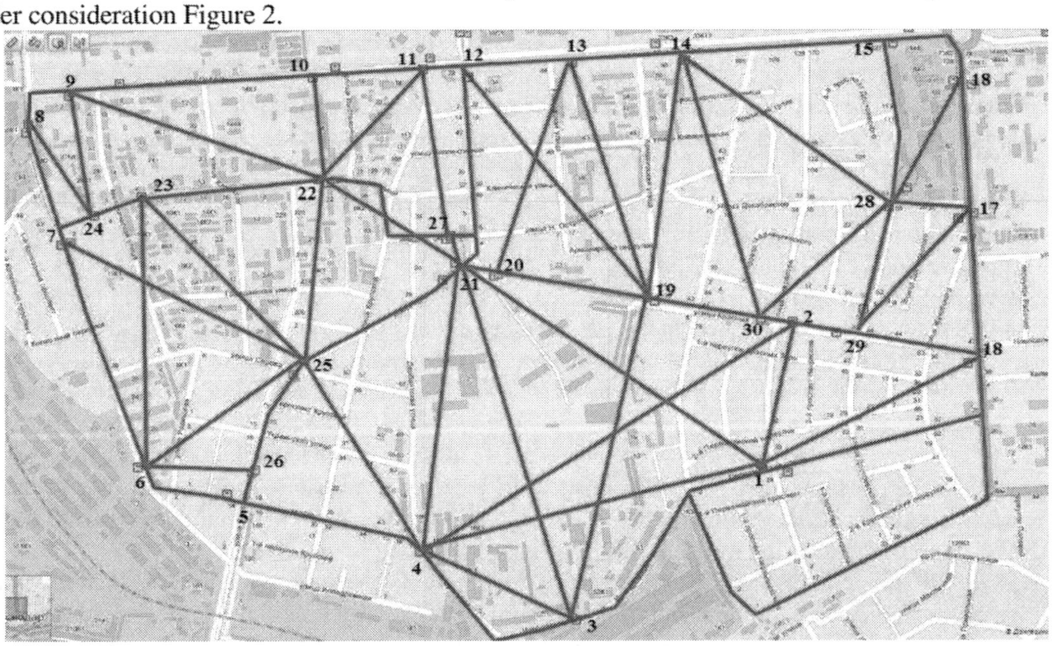

Figure 2. Matrix of pedestrian correspondence

In the figure, the numbers indicate the pedestrian generation nodes, and the red lines mark the pedestrian paths. Stops of the gas turbine are proposed as generation points. This scheme is not the final proposed structure of the communications. It is proposed to analyze the pedestrian communication routes on its basis. Add or remove some of the lines as needed.

When developing this scheme, the location of the existing residential buildings was not taken into account. It would not be rational to propose to demolish most of the development in the area under consideration to optimize pedestrian traffic. Therefore, a more detailed scheme will be presented further in the work, taking into account the existing development of the area.[7]

In the course of the work, the distance of the proposed paths of movement was measured, which is presented in Table 1. The distance was measured between the generation nodes to analyze the length of the communication paths. The table compares the existing distance between the generation points and the one proposed by the first scheme.

Table 1. Leverage between generation nodes

Shoulder	Distance along the existing UDS, m	Overhead line distance, m	Shoulder	Distance along the existing UDS, m	Overhead line distance, m
1	2	3	4	5	6
1-3	605	-	12-13	271	-
1-2	411	357	12-19	995	702
1-4	1100	811	12-21	513	-
1-18	672	579	12-27	443	-
1-21	1110	849	13-14	261	-
2-4	1480	1000	13-19	737	584
2-29	218	-	13-20	779	533
3-4	460	386	14-15	490	-
3-18	1260	-	14-19	721	587
3-19	991	792	14-28	855	596
3-21	1200	879	14-30	962	653
4-5	437	-	15-16	263	-
4-21	746	685	15-28	396	-
4-25	658	535	16-17	306	-
5-6	262	-	16-28	483	332
5-26	75	-	17-18	339	-
6-7	569	-	17-28	180	-
6-23	813	656	17-29	489	376
6-25	676	442	18-29	218	-
6-26	348	247	19-20	354	-
7-8	562	-	19-30	264	-
7-24	124	-	20-21	91	-
7-25	831	619	21-22	584	393
8-9	168	-	21-25	434	-
8-24	344	263	21-27	156	85
9-10	583	-	22-23	426	-
9-24	432	300	22-25	545	447
9-22	841	638	22-27	400	-
10-11	258	-	23-24	121	-
10-22	409	255	23-25	758	550
11-12	126	-	25-26	292	-
11-22	486	358	28-29	315	-
11-27	569	401	28-30	578	413

3. Conclusion

When studying the data obtained, reflected in table 1, the conclusions about the longest route in this area and the shortest. The table shows that most of the existing communication routes are much longer than the proposed ones. The longest distance connecting the pedestrian generation points is 1480m, and the shortest distance is 75m. On air lines the greatest is 1000m. and the smallest is 85m. The number of overhead lines is two times less than the existing number of routes. The total number of edges connecting the vertices, that is, the pedestrian generation points are 66, and 33 overhead lines.

The main goal of the work is to optimize the movement of pedestrians, to create safe traffic, taking into account the construction of the shortest driving distances. Therefore, great attention is paid to the length of the pedestrian paths.[8-10]

One of the methods used in the work is to reduce the number of paths for pedestrians, in order to streamline and direct pedestrians. This can reduce the number of accidents in the area under consideration.

References

[1] McKinsey on Sustainability & Resource Productivity 2012. *The McKinsey Quarterly* **27**
[2] Federal Road Transport Agency 2009 *Manual for elimination and prevention of road accident concentration areas during roads exploitation* ODM218.4.004 No.260-p
[3] Federal Road Agency 2009 *Recommendations regarding recording and analysis of road accidents occurring in RF roads* ODM 218.6.015 No. 853-p
[4] Konovalova T V, Kotenkova I N 2013 *Transport infrastructure* (Krasnodar: Kuban State Technological University)
[5] Sukharev O S 2015 Industrial policy and the development of transport infrastructure in Russia *National interests: priorities and security*. Vol 11 **1(286)** 2–20
[6] Pugachev I, Kulikov Y, Markelov G, Sheshera N 2016 Factor Analysis of Traffic Organization and Safety Systems *SPbOTSIC-2016* 529–535
[7] Kravchenko P, Oleshchenko E 2016 Mechanisms of Functional Properties Formation of Traffic Safety Systems *SPbOTSIC* 367–372
[8] Konovalova T, Zarovnaya L 2016 The Assessment Model for Economic Efficiency of Traffic Safety Improvements *SPbOTSIC* 311–315
[9] Kurakina E, Evtyukov S 2018 Results of studying road construction parameters condition *Architecture and Engineering* **3(1)** 29–37
[10] George R, Jana I, Joseph K 2013 International student's traffic engineering project seminar meps. *The world of transport and technological machines* **4(43)** 106–111

IOP Conf. Series: Materials Science and Engineering 1159 (2021) 012020 doi:10.1088/1757-899X/1159/1/012020

Study of the ways to increase the urban street and road network capacity

P I Pospelov[1], D S Martiakhin[2], T K Komarova[3], DM Nemchinov[4]

[1,2,3] MADI, 125319, Moscow, Leningradsky prospect, 64,
[4] TransIngProekt LLC, Tver, Universitetsky prospect, 2a

E-mail: martiakhin@mail.ru

Abstract. During the existing situation assessment on the urban road network and intersections design, it is necessary to establish the traffic capacity value of streets and its elements. When forecasting the capacity value it is necessary to turn traffic flow presented by different vehicles to uniformity expressed in the equivalent number of passenger cars by using passenger car equivalent coefficients (PCE). PCE currently used in the Russian Federation are formed on the capacity studies of the street-road network, carried out in the 70s - 80s of the last century on rural roads, where the traffic flow was mostly presented by heavy vehicles. Nowadays the major percentage of traffic flow is presented by passenger cars. At the same time, the dynamic performances of trucks and buses have changed significantly. In this regard, the question of clarifying the traffic flow composition and revising the passenger car equivalent coefficients becomes urgent. High level of streets congestion and intersections with a small heavy vehicle percentage in the traffic flow is most often observed in large cities. The studies were carried out on the the multi-lane highway in Moscow in the sections free from the intersections and junctions influence. The purpose of the study was to clarify the traffic flow composition and determine the passenger car equivalent coefficients in modern traffic conditions.

1. Introduction

The traffic load forecast in planning the development of a network and designing individual objects requires modern methods, the important components of which one is the various vehicle types reduction factor to a passenger car. The passenger car equivalent coefficients can vary depending on the drivers behavior, which in turn is influenced by the improvements in vehicle design, including active safety systems.

In this regard, the regular studies of traffic flows are required to track the dynamics of changes in parameters that affect traffic capacity.

During the highways and urban streets design, it becomes necessary to determine the capacity of traffic lanes on highways and various types of intersections.

Due to the fact that the traffic flow consists of many vehicle types, in order to determine the capacity, it is necessary to turn it into the equivalent traffic flow of passenger cars.

Nowadays in Russia, the traffic flow mostly consists of passenger cars, 2-14 tonnage trucks, minibuses, buses of small, medium and large capacity, articulated buses and 12-30 tonnage heavy trucks/

Content from this work may be used under the terms of the Creative Commons Attribution 3.0 licence. Any further distribution of this work must maintain attribution to the author(s) and the title of the work, journal citation and DOI.
Published under licence by IOP Publishing Ltd

The mixed traffic flow reduction to a homogeneous traffic flow consisting of an equivalent number of passenger cars is carried out using the passenger car equivalent coefficients.

Thus, the passenger car equivalent coefficients are a fundamental component in the highways and urban roads capacity determination.

2. The passenger car equivalent coefficients estimation methods

Most of the studies in Russia devoted to the passenger car equivalent coefficients estimation were carried out in the 70s of the last century by using several methodological approaches [3].

Method 1. It is based on the analysis of distances and headways between different vehicle types moving one after another in comparison with the movement of passenger cars.

This method compares the dynamic dimensions of the studied vehicle and a passenger car. The value of the passenger car equivalent coefficient PCE_{ij} is determined by the following ratios:

$$PCE_{ij} = \frac{d_{ij}}{d_{ii}}, \quad (1)$$

or

$$PCE_{ij} = \frac{\Delta t_{ij}}{\Delta t_{ii}}, \quad (2)$$

where d_{ij} – distance between the studied vehicles, m;
d_{ii} – distance between passenger cars, m;
Δt_{ij}– headway between the studied vehicles, s;
Δt_{ii} – headway between passenger cars, s.

Method 2. It is based on the "speed-traffic flow" analysis relationship for mixed traffic flows in comparison with a similar relationship for an equivalent traffic flow consisting of passenger cars.

Using this method, the average speed values of the mixed traffic flow and the equivalent passenger car flow were compared and the traffic at one flow speed were also analyzed. The passenger car equivalent coefficients PCE_{ij} are determined according to the following position:

$$N_{ii} = (1-p)N^{ij} + PCE_{ij}pN^{ij}, \quad (3)$$

where N_{ii} –traffic flow, consisting only of passenger cars, veh/hour;
N^{ij} – mixed traffic flow, veh/hour; p – percentage of slow moving cars.

Method 3. It is based on the capacity analysis for different traffic flow comparisons.

The traffic lane capacities are compared in homogeneous flows consisting of the considered vehicle types for the case of movement along a horizontal straightway with the traffic lane capacity of the homogeneous flow of passenger cars.

Method 4. It is based on the traffic flow density analysis for different traffic flow comparisons. The passenger car equivalent coefficients PCE_{ij} were determined from the traffic flow density ratio between the considered vehicle type flow and the passenger car flow in condition of the corresponding capacity. The density value is determined by the maximum density observed with a standing traffic flow:

$$q = 0{,}5q_{max}, \quad (4)$$

where q - traffic flow density of passenger cars corresponding to traffic capacity;
q_{max} - traffic flow density of the studied vehicle type.

Method 5. It is based on the traffic flow determination at which the maximum number of overtaking is observed.

The traffic flows at which the greatest number of overtaking of trucks by passenger cars is observed are compared. The average number of overtaking e with different traffic flow composition is determined by the following formula:

$$e = q_i(V_i - V_j)q_j, \tag{5}$$

where q_i – traffic flow density of free moving passenger cars; V_i – flow speed of free moving passenger cars; q_j – traffic flow density of slow moving cars;

V_j –flow speed of slow moving cars.

The traffic flow composition and the passenger car equivalent coefficients obtained by the methods 1-5 are presented in Table 1.

Table 1. Traffic flow composition and the passenger car equivalent coefficients obtained by the methods 1-5

Vehicle types	PCE_{ij}, values obtained by the method					Average PCE_{ij} values
	1	2	3	4	5	
Passenger cars	1,00	1,00	1,00	1,00	1,00	1,00
Motorcycles	0,75	0,70	0,68	0,40	0,72	0,65
Light trucks	1,20	1,60	1,70	1,40	1,68	1,52
Medium trucks	1,36	1,83	1,95	1,68	1,92	1,75
Heavy trucks	1,75	2,60	3,10	1,75	2,80	2,40

Nowadaysthe passenger car equivalent coefficients used in Russian Federation design manuals, are obtained on the basis of the vehicles dynamic dimensions ratio without taking into account different traffic conditions (street segments, various types of intersections, etc.) [2]. The exception is roundabounds, for the calculations of which are used passenger car equivalent coefficients, based on the gap ratio between different vehicle types when driving directly at roundabounds. In addition, over the past 20 years, a passenger car equivalent coefficients establish method has been introduced, taking into account the starting delay at signalized intersections [1].

In foreign countries the term "passenger car equivalent coefficient" was first introduced in USA road design guideline HCM 1965 (Highway capacity manual). Since that moment, foreign authors [7, 8, 13, 16, 17] have carried out a large number of studies carried out in passenger car equivalent coefficient establish for various elements of the road network.

One of many foreign approaches of the passenger car equivalent coefficients determination on the highways is a method based on the headway comparison between different vehicle types [4, 16]. In addition, this method can be applied in the passenger car equivalent coefficients determination in Russia, as it has been indicated above (method 1).

The headways are determined when a passenger car drivers move behind a passenger car drivers, and when drivers of different vehicle types follow a passenger car driver (Fig. 1).

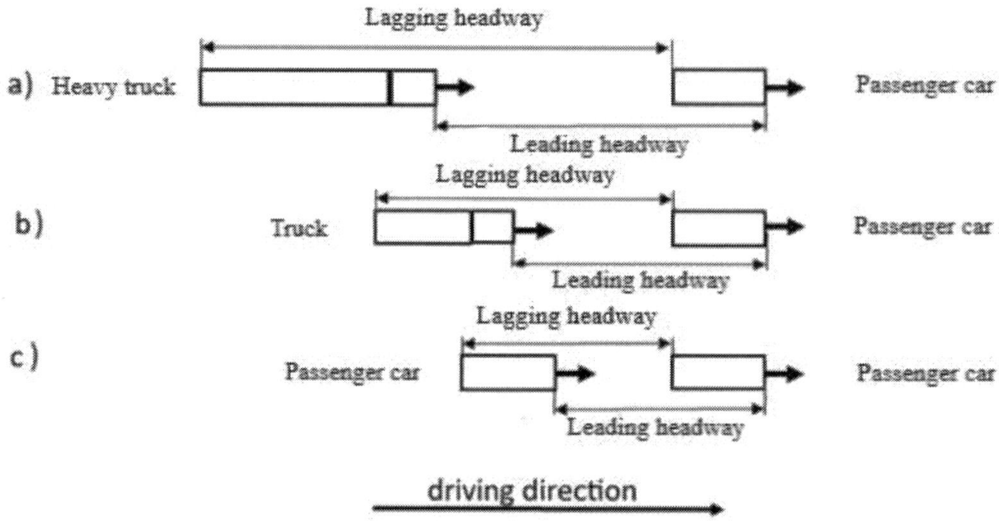

Figure 1. Headways of different vehicle types in traffic flow

In this regard, the passenger car equivalent coefficients PCE_{ij} are determined by the ratio:

$$PCE_{ij} = \frac{\Delta H_{ij}}{\Delta H_{ii}}, \quad (6)$$

where ΔH_{ij} – lagging headway of the selected vehicle type following the passenger car
ΔH_{ii} –passenger cars lagging headway.

Over the past 10-15 years, the PCE redefinition problem has been studied in India. A number of authors [9, 10, 11, 12] carried out the PCE determination researches on urban 2-lane highways, urban street segments and various types of intersections.

The fundamental task in the PCE determination is the traffic flow composition separation into vehicle types.

In Russian Federation design manuals, depending on the affiliation of a street or road to urban or rural conditions, different flow compositions and PCE are used. For urban streets are used 13 vehicle types and 14 are accepted for rural roads. In the studies devoted to the PCE determination, a number of authors [2, 3] use a simplified traffic flow composition, including 3-4 vehicle types (passenger cars, light trucks, buses, heavy trucks). The simplified traffic flow composition is also used in foreign studies [4, 7-13, 16, 17].

The American FHWA 2013 guidelines adopted a very detailed traffic flow division (13 classes, including 34 subclasses) depending on the vehicle type and its dimensions. However, Highway Capacity Manual 2010 and 2016 [14, 15] presents the average PCE values: 1.00 for passenger cars and 2.00 for all other vehicle types.

The comparison of the PCE estimation studies by foreign authors and PCE used in the Russian Federation is given in Table 2.

Table 2. Comparison of research results by foreign authors and reduction factors used in the Russian Federation

Vehicle type	PCE according to various manuals and authors researches											
	At urban and rural highways						At signalized intersections					At rounabounds (ODM 218.2.071-2016)
	SP 34.13330.2012	ODM 218.2.020-2012	SP 396.1325800.2018	Mehar, Arpan & Chandra	Roads Congress	Ahmed Anwaar&Box	Webster N	Sosin J.	Vrubel Yu.A.	Levashev A.G., Mikhailov A.Yu.	ODM218.2.020-2012	
Passenger car		1,0	1,0	1,0	1,0	1,0	1,0	1,0	1,0	1,0	1,0	1,0
Motor cycles	1,0	0,5	0,5	0,19-0,26	0,5	-	0,33	0,6	0,7	-	-	0,5
Minibus		-	1,5	-	1,5	-	-	-	-	1,093	1,1	-
Trucks with carrying capacity of:												
- from 2 t	1,3	1,1	1,5							1,179	1,2	1,5
- from 2 up to 6	1,4	1,8	2,0							1,480	1,5	-
- from 6 up to 8 t	1,6	2,1	2,5	1,34-1,58	1,5	1,37	1,75	1,6	1,4			-
- from 8 up to 14 t	1,8	2,4	3,0							1,647	1,6	-
- over 14 t	2,0	2,5	3,5									-
Heavy trucks with carrying capacity of:												
- up to 12 t	1,8	2,2					-					
- from 12 up to 20 t	2,2	2,4	4,0	1,06-1,24	4,5	1,65	-	2,8	2,3	2,231	2,2	2,3
- from 20 up to 30 t	2,7	-					-					
- over 30 T	3,2	3,3					-					
Buses with length												
- small	1,4									1,367	1,4	
- medium	2,5		2,5				2,25	1,7	2,0	-	-	2,0
- full	3,0	2,6								1,839	1,8	
articulate buses			4,0	-	3,0	-	-			2,363		3,0
Trolley buses	4,6	-	3,0				-	2,8	2,6	-	2,4	-

It is interesting to note that in mentioned studies, various traffic flow compositions and PCE determination methods were used. At the same time, foreign [5, 13] and Russian authors are similar in decision that there are a lot of factors which influence on the PCE values and it is necessary to estimate PCE for each element of the road network (street segments and highways, signalized intersections, unsignalized intersections, round abounds) taking into account the special traffic conditions (conflict points, movement directions , traffic signals, delays, drivers behaviour).

3. Data collection and analysis

The subject of research was presented by urban highways with continuous traffic, where is no influence of ramps and inflowing traffic flows on the main traffic flow.

IOP Conf. Series: Materials Science and Engineering 1159 (2021) 012020 doi:10.1088/1757-899X/1159/1/012020

This situation can be regularly observed on Moscow Ring Highway. With a high traffic density in the two right lanes, there is a flow includes heavy vehicles. Medium-tonnage vehicles move in the 2nd and 3rd lanes. Light commercial vehicles move in the 3rd and 4th lanes. The 4th and 5th lanes are occupied mainly by passenger cars.

With a high traffic density, the speed on the left lanes on the overtaking between ramps is approximately equal to the speed on the right lanes. This allows comparing the space occupied by different vehicles on the road under similar driving conditions.

During the preparation for the research, two ways of the *PCE* establish were studied.

The first method involved counting the number of cars that passed through the road cross-section with different traffic composition, determining the percentage of trucks and calculating the *PCE* values based on the ratio of the physical number of cars and the composition of the traffic flow.

The second method involved field studies by video recording the traffic stream movement of different composition in adjacent traffic lanes in a selected area. When traffic flows in adjacent lanes were at approximately the same speed, the headways between different vehicle types were fixed.

Video recording is performed by the drone located at an altitude of 50-100 m above the selected area, which allows the observation of all lanes at the same time.

The recording result is a 15 minutes video, which is a subject of further processing to obtain a database.

The data includes traffic flow, number of vehicles divided by the selected traffic composition, lagging headways between different vehicle types

The selected drone height allows determining the vehicle type in motion. When processing video materials, a road cross-section is selected, driving through which the lagging headways are established.

One of the research objectives was to define the actual traffic flow composition in Moscow. For this purpose, the traffic flow composition was adopted in accordance with the current design manuals. The obtained data analysis is based on the theory of lagging headways ratio (methods 1, 6), presented above.

During the study, it was found that some vehicle types percentage in traffic flow is rather low for reliable PCE determination in the given conditions, which confirmed the necessary to divide traffic flow in simplified composition during researches.

The results are shown in table 3 and in graphs 2 and 3.

Table 3. Results

№ n/n	Vehicle types	Vehicle types percentage in traffic flow, %	Average lagging headway values, s	PCE values
1	Passenger car	83,85	2,55	1
2	Minibus	1,01	3,52	1,38
3	Trucks with carrying capacity up to 2 tons	2,97	3,47	1,36
4	Small bus	1,71	3,16	1,24
5	Trucks with carrying capacity from 2 to 6 tons	7,49	2,58	1,02
6	Large bus	< 1,00	-	-
7	Trucks with carrying capacity more than 6 tons	1,5	3,48	1,37
8	Articulate bus	< 1,00	-	-
9	Heavy truck	1,07	-	-

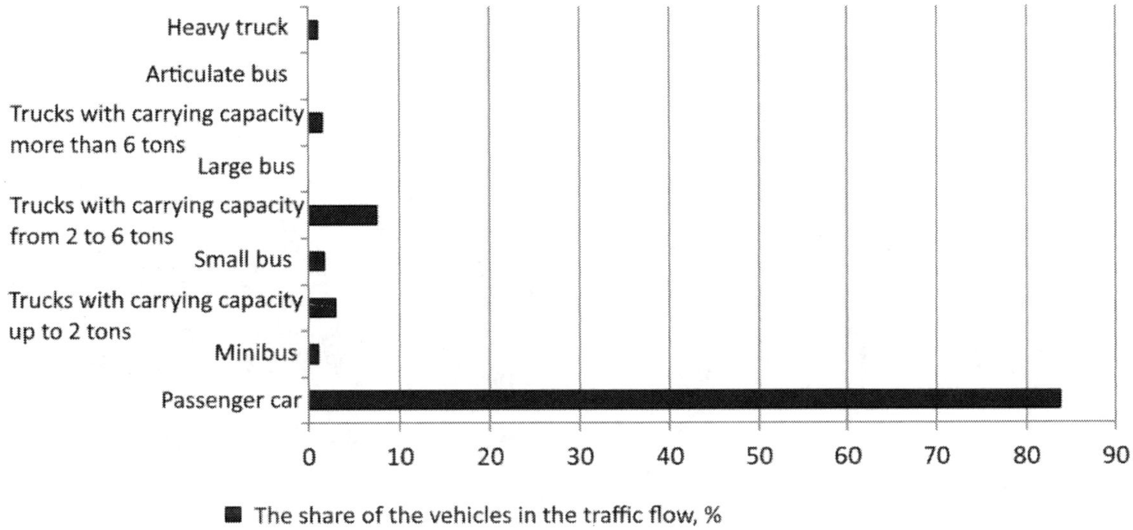

Figure 2. The Moscow Ring Higway traffic flow composition

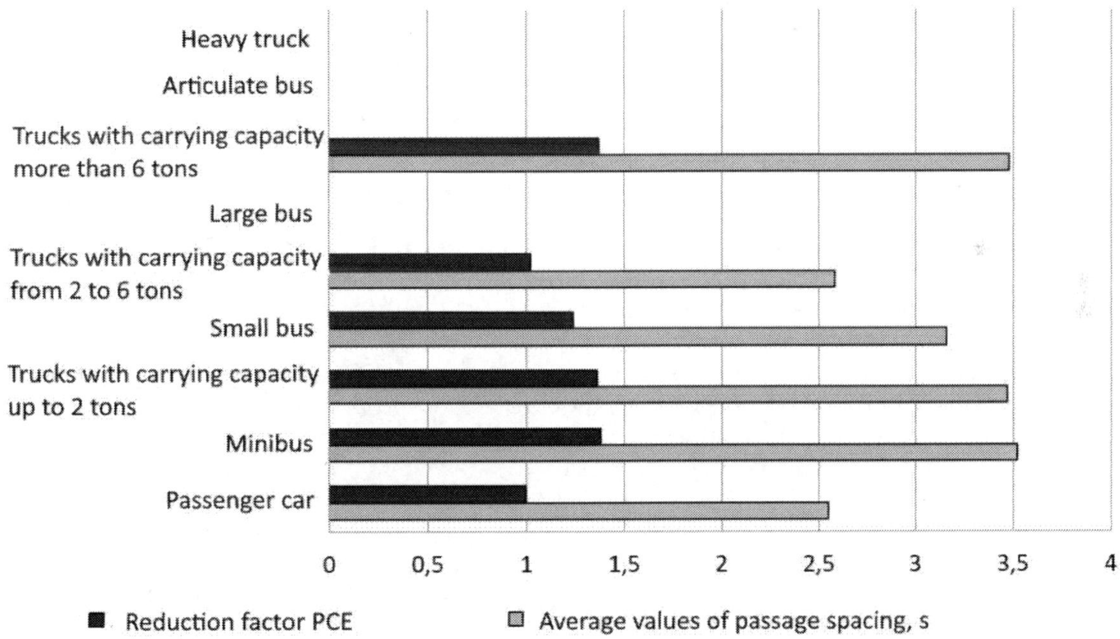

Figure 3. The Moscow Ring Highway lagging headways and PCE values

The comparison of the obtained research results with the PCE values of the Russian Federation design manuals are shown in the graph 4:

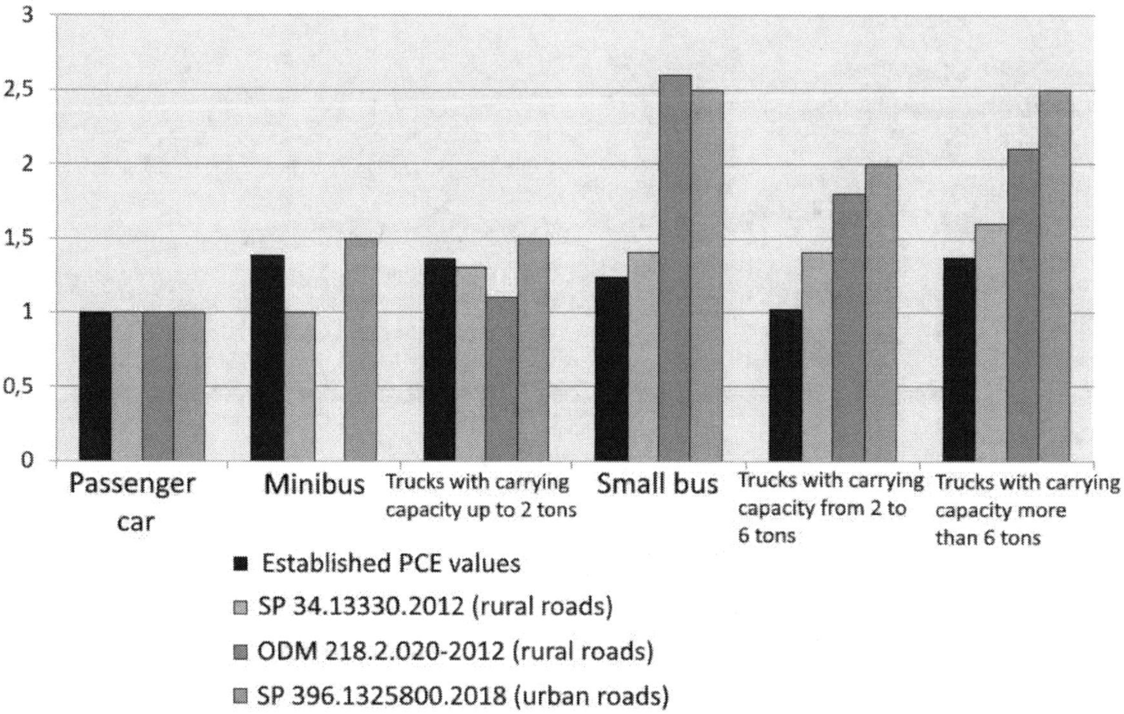

Figure 4. Obtained PCE values comparison with values in Russian Federation design manuals

4. Results and discussion

The PCE values are more dependent on the speed, traffic flow composition and road conditions. The influence of a large number of additional factors in the PCE determination results in difficulties in obtaining reliable calculated dependencies.

In the course of research, many authors accept a simplified system for the traffic flow composition: cars, trucks and buses.

The results of traffic flow the studies of the on the Moscow Ring Highway showed that the PCE values obtained in the 70s, the PCE values used at the actual design manuals and the PCE values obtained through the field observations are various.

Taking into that in urban conditions the traffic flows nature on highways and various intersection types is very different, the PCE values should be established for each situation separately.

Within the framework of the above mentioned study, the PCE values were obtained for the highway segments with continuous traffic.

Due to the fact that a huge number of factors affect the PCE values, this issue should be studied in more detail in future researches.

References

[1] Levashev A G, Mikhailov A Yu, Golovnykh I M 2007 *Design of controlled intersections: textbook* (Irkutsk: Publishing house of the Irkutsk State Technical University) 208 p

[2] Mikhailov A Yu, Golovnykh I M 2004 *Modern trends in the design and reconstruction of urban road networks* (Novosibirsk: Nauka) 267 p

[3] Silyanov V V 1977 *The theory of traffic flows in road design and traffic organization* (Moscow: Transport) 303 p

[4] Ahmed, Anwaar & Boxel, D. & Volovski, Matthew & Anastasopoulos, Panagiotis & Labi, Samuel & Sinha, Kumares 2011 Using Lagging Headways to Estimate Passenger Car

Equivalents on Basic Freeway Sections *Journal of Transportation of the Institute of Transportation Engineers* 2 5–17

[5] Elefteriadou L, Torbic D and Webster N 1997 Development of Passenger Car Equivalents for Freeways, Two-Lane Highways, and Arterials *Transportation Research Record: Journal of the Transportation Research Board* (Washington, DC: Transportation Research Board) **1572** 51–58

[6] *Highway Research Board. Special Report 87, Highway Capacity Manual.* (Washington, DC: National Research Council, Department of Traffic and Operations, Committee on Highway Capacity, 1965)

[7] Huber M J 1982 Estimation of passenger car equivalents of trucks in the traffic stream. Transportation Research Record 869. Transportation Research Board, Washington, D.C., pp 60–70

[8] Linzer E, Roess R, McShane W 1979 Effect of trucks, buses, and recreational vehicles on freeway capacity and service volume. Transportation Research Record 699. Transportation Research Board, Washington, D.C., pp 17–26

[9] Mahidadiya Ankit N and Jayesh Juremalani 2016 Global Scenario on Estimation of Passenger Car Unit: A Review

[10] Mehar, Arpan & Chandra, Satish & Senathipathi, Velmurugan 2014 *Passenger Car Units at Different Levels of Service for Capacity Analysis of Multilane Interurban Highways in India. Journal of Transportation Engineering* **140** 81–88 10.1061/(ASCE)TE.1943-5436.0000615

[11] Metkari, Mahendrakumar&Budhkar, Anuj & Maurya, Akhiles 2012 A Review of Passenger Car Equivalence for Indian Conditions *International Journal of Computer Applications*

[12] Raj, Pooja & Sivagnanasundaram, Kalaanidhi & Asaithambi, Gowri & Ravi Shankar, A. 2019 Review of Methods for Estimation of Passenger Car Unit Values of Vehicles *Transportation engineering journal of ASCE* **145** 10.1061/JTEPBS.0000234

[13] Sosin J A 1980 Delay at Intersections Controlled by Fixed Traffic Signals *Traffic Engineering and Control* Vol 21 **8/9** 407–413

[14] Transportation Research Board. Highway Capacity Manual 2010. Transportation Research Board, Washington D.C.

[15] Transportation Research Board 2016. Highway Capacity Manual 6th Edition – A Guide for Multimodal Mobility Analysis. Transportation Research Board, Washington D.C.

[16] VanBoxel D, Sinha K C, Bullock D and Mannering F L 2009 An Exploratory Study of Vehicle Class Headway Ratios as Passenger Car Equivalence Values Using Three Stage Least Squares Estimation *Presented at 89th Annual Meeting of Transportation Research Board* (Washington, DC)

[17] Webster N, Elefteriadou L 1999 A simulation study of truck passenger car equivalents (PCE) on basic freeway sections *Transp Res Part B* **33(5)** 323–336

Development of proposals for the development of the design algorithm of road pavements using geosynthetic materials

A V Korochkin[1], **V A Dumenko**[1]

[1] Moscow State Technical University, 64, Leningradskii Avenue, Moscow, 125319, Russia

E-mail: andrey_korochkin@mail.ru

Abstract. The paper deals with the manifestation of nonlinear properties of granular materials. In addition, the analysis the possibility to take into account the influence of the geogrid on the strength characteristics of the layer is made. For a numerical experiment, a pavement structure was chosen. It was designed in accordance with the calculation methodology existing in Russia and consisting of three layers of asphalt concrete, a base layer of granular incoherent materials and a base layer of sand. For a comparative analysis of the methodology, the specified structure was also designed using the algorithm that takes into account the stress-deformed state of granular particles in the bearing course layer using a mechanic-empirical model for the determination of the elastic modulus. The topic was further justified by appropriate structural calculations, taking into account the influence of a geogrid placed above the underlying layer, both in accordance with the current legal document and according to the proposed method, which is based on the change in horizontal compressive stresses in terms of the geogrid operation. The comparison of the results showed that the current method for the design of road pavements is not enough to reflect the effect of the use of a geogrid on the strength of the structure during the calculation and, in general, underestimates the requirements for the strength of pavements.

1. Introduction

According to the Decree of the Government of the Russian Federation of May 30, 2017 No. 658 [1], the increased service life of pavements was established. However, the real tools for the increase of the inter-maintenance period of structures are still at the stage of development. The problems of pavement design are primarily based on the fact that the methods of the calculation of the entire structure, as well as individual solutions, were developed in accordance with outdated requirements. It is important to note that about forty years have passed since the core of the algorithm for the calculation of the non-rigid pavement was formed. During this time, the technical capabilities of testing and design have been improved. These facts will be the advantage of the industry in order to fulfill the task.

2. Methods

Such a constructive solution as a geogrid in the base of the pavement is intended to increase the strength characteristics of the crushed stone layer, untreated with a binder and the strength of the structure as a whole.

Content from this work may be used under the terms of the Creative Commons Attribution 3.0 licence. Any further distribution of this work must maintain attribution to the author(s) and the title of the work, journal citation and DOI.
Published under licence by IOP Publishing Ltd

In order to demonstrate the effect of the geogrid, the authors calculated the pavement in accordance with Industry Road Codes 218.046-01 [2] and then the same design was recalculated taking into account the recommendations of Road industrial methodical document 218.5.002-2008 [3]. The design adopted for the calculations is shown in Figure 1. For the use in the road bed we selected a mixture S-6 (GOST 25607 [4]).

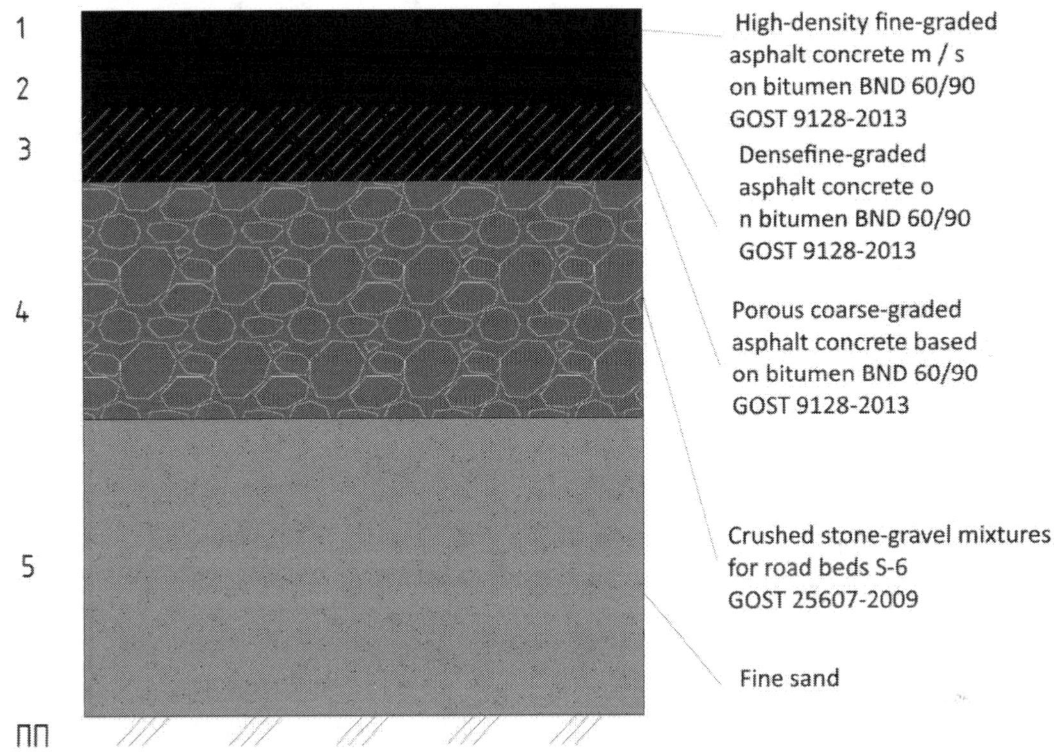

Figure 1. Analyzed structure of non-rigid road pavement (the thickness of the layers of the structure according to the design in accordance with the ODN [2]: layer No. 1 - 5 cm, layer No. 2 - 8 cm, layer No. 3 - 10 cm, layer No. 4 - 32 cm, layer No. 5 - 40 cm)

According to the results of the calculation of the effect of the geogrid, the elastic modulus of the road bed was increased by 14.6%. The thickness of the structure was reduced by 5 cm as a result of the compensation of the strength by the reinforcement of the bed.

However, it is known [5] that the obtained values are not always justified during operation, which can be explained by the inaccuracy of the method of Road industrial methodical document [3]. The study by A.E. Merzlikin [6] described some principles of work of incoherent granular materials. The physical nonlinearity of the material and the horizontal lateral thrust were especially noted. The modulus of elasticity, which is taken today in the design of pavements, is the same for the entire layer thickness for any layout of the structure and location in the horizontal direction. In fact it changes its values depending on the position of the point in question in a particular structure. The solution of the problem in its simplified version, in which there is no need to use complex processes implemented by programs using the finite element method (FEM), is a computation using a program that implements the exact solution to the theory of elasticity.

The program is developed based on the solution of A.K. Privarnikov and it calculates stresses and strains at any point in the structure.

Having received the components of the stress tensor, using an iterative approach, we apply a mathematical model of the mechanic-empirical method to calculate the elastic modulus in order to analyze nonlinear properties [7]:

$$M_r = k_1 \times p_a \times \left(\frac{\theta}{p_a}\right)^{k_2} \times \left(\frac{\tau_{oct}}{p_a} + 1\right)^{k_3} \quad (1)$$

Here k_1, k_2, k_3 – empirical nonlinearity coefficients obtained on triaxial compression units;
p_a– atmospheric pressure, MPa;
θ – volume stress, MPa;
τ_{oct} – shear stress on octahedral areas, MPa;

$$\tau_{oct} = \frac{1}{3}((\sigma_1 - \sigma_2)^2 + (\sigma_2 - \sigma_3)^2 + (\sigma_3 - \sigma_1)^2)^{1/2} \quad (2)$$

σ_1, σ_2, σ_3 – principal stresses.

The main horizontal stresses in the longitudinal and transverse directions in the axisymmetric problem statement are considered to be equal. However, the validity of such a solution, along with the fact that they do not change with the depth of the bed, must be additionally confirmed by appropriate tests. The solution of the problem by the software package can not reflect all possible operating conditions of the granular material. Therefore, the values of σ2 and σ3 were adjusted in accordance with the studies [8]. Here it is necessary to take values corresponding to the minimum part size for testing, which corresponds to the maximum particle size in the mixture.

The empirical coefficients were borrowed by the authors from the report [9] for a mixture of mineral material, which, in accordance with the requirements of GOST [4], showed the coincidence in granulometric composition with crushed stone mixture S-6. The percentage of passes is shown in Figure 2.

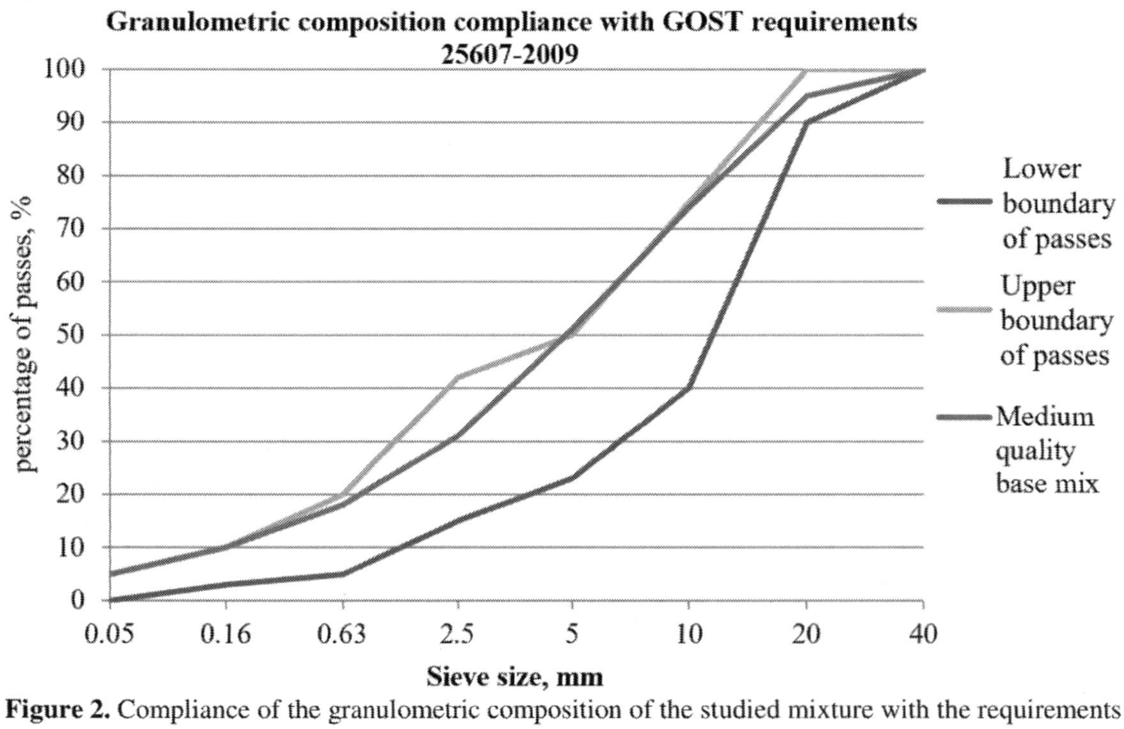

Figure 2. Compliance of the granulometric composition of the studied mixture with the requirements of GOST 25607-2009

It is known that the solution to lay the supporting bed of the base from crushed stone-sand mixtures is now quite common rather than from crushed stone. This is explained by the fact that under the action of cyclic dynamic loads due to the abrasion of grains of crushed stone, its so-called additional compaction occurs, which at the stage of operation is of a non-systemic nature. In turn, the uniform sand content reduces the amount of residual deformation of the layer. Then crushed stone really works as "stress-hardening" (a material that increases strength characteristics with increasing loads) [10].

When the values of the average weighted real modulus of elasticity of the bed are obtained, the authors of this work propose to recalculate the thickness of the layer using formula 3 [3] and iterative approach:

$$h_i = \Bigg(1.05 - \Bigg(0.71 \times \left(E_{gen}^{(i+1)} \times E_i^{-1}\right)^{1/3} \times 1.35 \times 2 \times h_i \times \left(\left(6 \times E_{gen}^{(i+1)}\right)^{-1} \times E_i\right)^{1/3} \times$$

$$D^{-1} + \frac{E_i}{E_{gen}^{(i+1)}} \times \frac{2}{\pi} \times tan^{-1}\left(D \times \left(2 \times h_i \times \left(\left(6 \times E_{gen}^{(i+1)}\right)^{-1} \times E_i\right)^{1/3}\right)^{-1}\right)\Bigg) \times E_{gen}^{i} \times E_i^{-1}\Bigg) \times$$

$$\frac{10 \times D}{1 - \left(E_{gen}^{(i+1)} \times E_i^{-1}\right)^{1/3}} \qquad (3)$$

According to the calculations by the indicated method, the increase in the required thickness of the structure relative to the results of calculations according to Industry Road Codes [2] is 15.6%.

The indicated method can be experimentally used in order to calculate the entire structure of non-rigid pavement. In foreign reports [9], we can find empirical coefficients for sands and soils for the design according to formulas 1 and 2 and using this algorithm, it is theoretically possible to calculate the strength characteristics of asphalt concrete [11].

Taking into account the influence of geosynthetics in this way, the key parameter is the horizontal stress within about 15 centimeters above the geogrid [5], which characterizes the so-called zones of increased rigidity, the appearance of which is provoked by the joint operation of the geogrid and granular material. Assuming that the grid makes the material particles act as a larger aggregate particles with the correct selection of rib size, it makes sense to use the stress values [8] for the larger part in relation to the lower block of the mixture layer.

3. Results

Thus, the thicknesses of the structural layers were calculated using four different algorithms, taking into account the fulfillment of the same strength criteria. The results are shown in Table 1.

Table 1. Results of pavement calculations with and without geogrid

№ bed	Classic calculation according to Industry Road Codes [2]	Calculation with geogrid by Road industrial methodical document [3]	Calculation taking into account special properties according to the theory of elasticity	Calculation with a geogrid according to the proposed algorithm
5	40 cm	40 cm	40 cm	40 cm
4	32 cm	27 cm	37 cm	33 cm
3	10 cm	10 cm	10 cm	10 cm
2	8 cm	8 cm	8 cm	8 cm
1	5 cm	5 cm	5 cm	5 cm

4. Discussion

The above mentioned calculation results and conclusions about the operation of the road bed are valid only for the structure analyzed in this work, because it is known that, for example, when the thickness of the pavement layers is varied or the composition of the asphalt concrete changes, the road bed will begin to resist the occurrence of structural defects in a different mode. In addition, the results of the numerical experiment show that in order to analyze the properties of the road bed, its thickness was taken as a variation parameter. In order to introduce the algorithm into general practice, the next stage in the development of the methodology can be the testing of various compositions of non-cohesive materials for the base in laboratory conditions, as well as in the composition of structures at experimental sites.

5. Conclusion

As a result of the research it is possible to make the following conclusions:

- The modulus of elasticity of granular material depends on the location of the bed in the pavement structure and on the way how the overlying beds work. In other words, the stress-deformed state directly affects the modulus of elasticity of the granular material, the change of which has a nonlinear law.
- A geogrid under a bed of granular non-cohesive material creates the necessary wedge, which increases the strength characteristics of the layer and will even work for bending within the rib strength.
- The proposed algorithm increases the design accuracy of non-rigid road pavements using geosynthetics.

Acknowledgments

The authors express their gratitude to A.E. Merzlikin, who passed away, for his contribution to the development of the design of pavements, in particular, the issue of the use of geosynthetics, as well as for his help in the development of the studied topic.

References

[1] *On the standards of financial costs and the Rules for calculating the amount of budgetary appropriations of the federal budget for major repairs, repairs and maintenance of federal highways: Resolution of the Government of the Russian Federation dated 30.05.17.* Available at: http: //docs.cntd.ru/document/436738392

[2] *Industry Road Codes 218.046-01 Design of non-rigid road pavements* (Moscow: Informavtodor, 2001)

[3] *Road industrial methodical document 218.5.002-2008 Guidelines for the use of polymer geogrids (geogrids) to strengthen the layers of road pavement made of granular materials* (Moscow: Informavtodor, 2008)

[4] *GOST 25607-2009. Crushed stone-gravel-sand mixtures for coatings and bases of highways and airfields. Specifications* Available at: http://docs.cntd.ru/document/gost-25607-2009

[5] Korochkin A V, Dumenko V A 2020 Design of pavements using geosynthetic materials *Science and technology in the road industry* **1** 6-8

[6] Merzlikin A E, Korchazhnikov Ya N 2018 *Extending the service life of non-rigid road pavements in design: trivial and non-trivial ways* **39/1** 105-116

[7] Mechanistic Empirical Pavement Design Guide (MEPDG). Software version 0.910 User's Guide: Appendix D: User's Guide-Design Software and Design Examples, 2002

[8] Sall A O 2005 *A new concept of using crushed stone materials to increase the bearing capacity of road bases* Available at: https://docplayer.ru/64871627-Ao-sall-novaya-koncepciya-primeneniya-shchebenochnyh-materialov-dlya-povysheniya-nesushchey-sposobnosti-dorozhnyh-osnovaniy.html

[9] Xiao Y, Tutemluer E 2012 *Best value granular material for road foundations* (Final report, Department of Civil & Engineering University of Illinois at Urbana-Champaign)
[10] *RESILIENT MODULUS. Shane Buchanan.* 2007 Available at: https://docviewer.yandex.ru/view/322255801/&lang=en
[11] Korochkin A V, Dumenko V A 2020 D. Christensen's model for use in calculating and designing pavements with asphalt concrete pavement (wear layers) *World of roads* **128** 38-39

IOP Conf. Series: Materials Science and Engineering 1159 (2021) 012022 doi:10.1088/1757-899X/1159/1/012022

Analysis of the properties of new forms of transitional sections of railway and highway curves

G V Velichko[1], **P I Pospelov**[2]

[1] Credo-Dialog Company LLC, 5 Malyi lane, 220014, Minsk, Belarus, E-mail:
[2] Moscow Automobile and Road Construction State Technical University, Leningrad Avenue, 64, Moscow, Russia,

E-mail: vgvkurve@gmail.com, pospelov@madi.ru

Abstract. The properties of new forms of transitional sections of curves of railways and highways are largely determined by the specific features of the design and functioning of the systems "railway line-rolling stock" and "driver-car-road". Taking them into account had a significant impact on the formalization of targets and criteria for the quality of the functioning of these systems, as well as on the mathematical means and methods of their achievement. The results obtained in this case significantly differ from the traditional ones. The geometric differences between the transitional sections of railways are in the nonlinear regularities of strictly monotonic, G4 smooth changes in the curvature of their line axis and G3 smooth cross slope, which are described by non-identical functions with variable properties. At the same time, the quality of movement, assessed at the design level of the rolling stock, is ensured by the harmonization of the length of the transition section and the variable properties of its functions with the given characteristics of other properties of the system. The differences between the transitional sections of highways consist in the inclusion of the calculated speed and the speed of its change directly into the principle of the trajectory curvature of vehicles. This ensures the safety and convenience of curvilinear movement of cars on them both at constant and variable speeds. The theoretically substantiated potential for the improvement of the quality of functioning of these systems with the proposed solutions determines the reasonability of their comprehensive testing in practice and further improvement of the theory, norms and rules for the design of transitional sections of curved rail and highways.

1. Introduction

The physical principles of the movement of vehicles on railways and roads have significant differences. Despite this, the provisions of the current paradigms are based on the general principles of the design of their routes with identical principles in the properties of their curving elements. The paradox of this situation is largely reasoned by the significant simplification of models and criteria for the assessment of the safety and efficiency of these systems on these sections of the routes. Mostly they are focused on the established mode of curvilinear movement of rolling stock and cars on a section with a constant curvature k, speed v and cross-slope of track i or the surface of the roadway.

At the same time, the main indicators of the safety and convenience of this process on circular sections of curvature of railways and highways are the non-reduced side a and the dimensionless side force coefficient $\mu = a/g$ (where g = 9.81 m / s2) directly dependent on it. The value of the non-reduced side acceleration a is calculated from the conditions of its action at the level of the top of the rail heads or the roadway surface according to the formula common for both modes of transport

Content from this work may be used under the terms of the Creative Commons Attribution 3.0 licence. Any further distribution of this work must maintain attribution to the author(s) and the title of the work, journal citation and DOI.
Published under licence by IOP Publishing Ltd

IOP Conf. Series: Materials Science and Engineering 1159 (2021) 012022 doi:10.1088/1757-899X/1159/1/012022

$$a = k \cdot v^2 + g \cdot i \qquad (1)$$

This equation allows establishing such a combination of design speed v, curvature k and side slope i, at which the balance $a \leq a_{max}$ or $\mu \leq \mu_{max}$ required for reasons of safety and convenience of movement will be provided. In order to change the mode of movement from a permanently rectilinear to a permanently curvilinear one, a transition section is arranged with a linearly varying curvature of its axis and a cross slope. Automatically it is assumed that the side acceleration acting on such transitional sections can not exceed its maximum on the circular part of the rounding. Therefore, the only controllable factor that can affect the efficiency and quality of the systems functioning on this fragment of the design of their rounding is only its length L. The provisions of the current paradigms for tracing railways and highways provide the limitation of only the minimum lengths of transition sections in terms of the maximum rate of change in the non-reduced side acceleration da/dt and along the length of the section of the curve or retraction of the elevation of the outer rail of the line over the inner one.

In some cases, they are unreasonably limited by the value of the radius of curvature R or the value of the shift of the circular curve p.

The simplicity of this design of the transition section and the corresponding "simplicity" of standardizing its parameters significantly limits the possibilities of the improvement of the quality of the systems functioning on the curves and even adjacent sections of the routes. As the speed of movement increases and the requirements for its quality increase, the negative influence of both the structures of the transition sections and simplified criteria for the assessment of their effectiveness is increasingly manifested. The limited list of their universal indicators does not take into account important features of the structure and functioning of each of these systems. For the elimination of these shortcomings, various authors proposed many alternative forms of transition sections with the laws of curvature and cross slope that are different from linear ones [1].

However, the empirical nature of the search for these solutions by the method of trial and error did not provide the formalization of the aim pursued by them and the justification of the criteria for its achievement. The positive and negative properties of these solutions were investigated using multifactorial deterministic kinematic models [2]. The obtained results allowed substantiating the goals of the improvement of the forms of transitional sections of curves of rail and highways, mathematical means and methods for their achievement.

2. Specific features of the device and functioning of the "railway line-rolling stock" system

One of the important features of railways is the construction of their line from rails with a certain curving stiffness. Therefore, polylines representing projections of their 3D models in the horizontal and vertical projection planes must provide at least the G3th order of smoothness at the junction points of the elements that make up the polyline. This requirement is observed when the values of both the mathematical functions of adjacent elements and their first, second and third derivatives change continuously at these points. The higher the order of geometric smoothness of the polyline, the more accurately it describes the model of the corresponding projection of the rail and the less it is noticeable in movement along it that the project of its model was composed of dissimilar geometric elements. The accuracy of the mathematical model of the line increases the reliability of the forecast of its functional indicators and contributes to the increase in the accuracy of its physical embodiment on the ground. The compliance with the minimum G3th order of smoothness is especially important for the so-called continuous rail. In this case, the traditional project of the line model on the rounding section with clothoid transition sections will consist of polylines with G0th and G2nd orders of smoothness that does not meet this requirement. As a result of the mechanical properties inherent in the rails, they will inevitably curve out of design. The negative consequences of this will lead to shocks and swaying of the rolling stock, which will increase the rate of breakdown of the geometry of the line, increase the wear of rails, wheels and its other elements.

The second important feature is reasoned by the fact that as a rule the elevation H of the center of mass of the rolling stock or any of its other functionally significant design points above the level of the top of the rail heads exceeds the distance S between their axes. With the inconsistent effect of the principles of the retraction of the curvature of the line and its cross slope, the kinematics of the design point deteriorates significantly. This has the most negative effect on the curvature of the horizontal projection of the trajectory of its movement, which determines the value of the centrifugal acceleration. As a result, the non-reduced side acceleration a acting at the level of the design point, will significantly deviate from the value predicted by the formula (1). The additional action of the side force on the line rails due to this acceleration will also increase in proportion to the H/S ratio.

The third important feature is reasoned by the influence on the functional indicators of the system that the curvature of the profile of the track axis k_V, as well as the vertical curvature of k_L and k_R of each of the strands of its rail have. Their values depend on the order of vertical smoothness of a symmetric or asymmetric non-linear elevation retraction, as well as on its maximum value D and the length L of the transition section, which is used in the project.

3. Studied parameters of the "railway line-rolling stock" system

The influence of the above mentioned factors on the forces of interaction between the rails and the wheels of the rolling stock are determined by the analytical dependences of the laws of physics, which are the basis of the multifactorial deterministic kinematic model shown in Figure 1 a), b) and c) fragments of the design scheme. The coordinate method for the description of the kinematics of the design point of the rolling stock M is of great importance in ensuring its adequacy. It is determined taking into account its elevation H above the level of the top of the rail heads and the geometric properties of one form or another of the transition section (Fig. 1 d)).

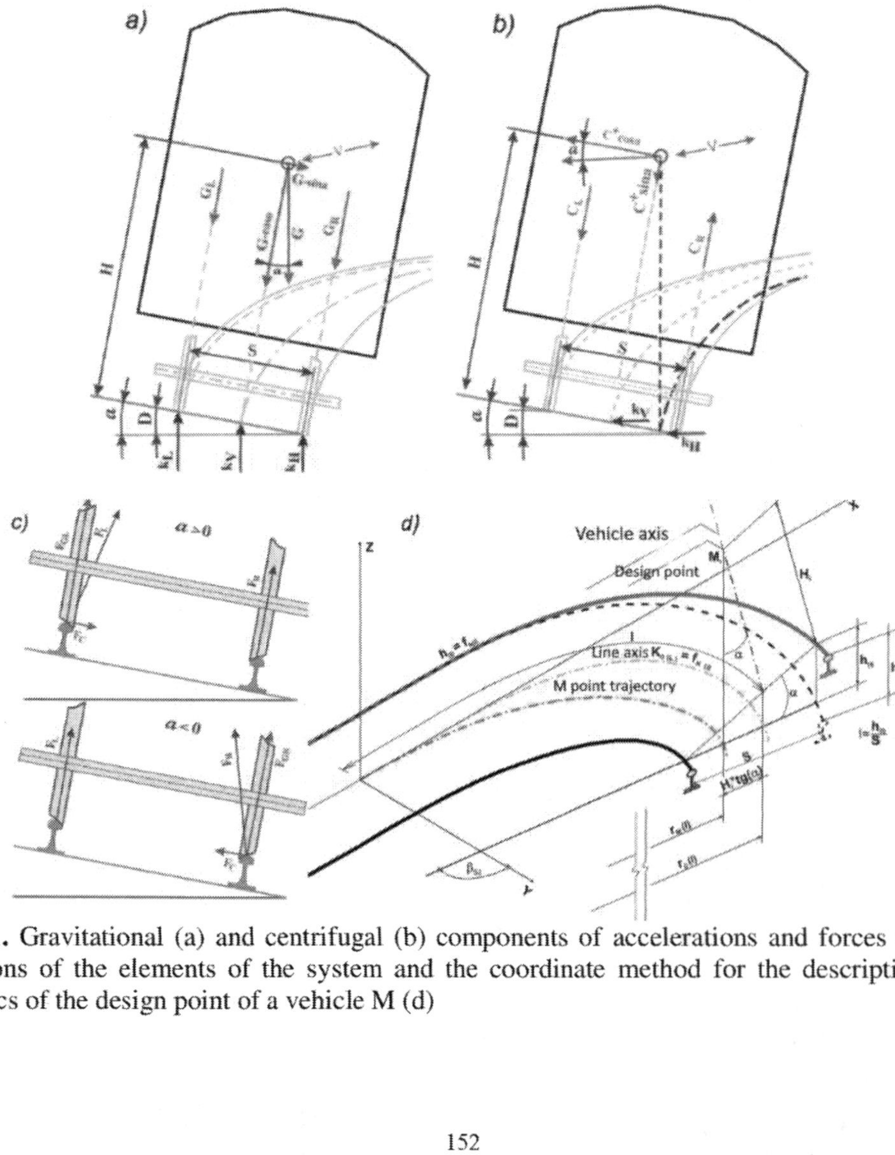

Figure 1. Gravitational (a) and centrifugal (b) components of accelerations and forces (c) of the interactions of the elements of the system and the coordinate method for the description of the kinematics of the design point of a vehicle M (d)

In accordance with this model, the coordinates $x_M(l)$ and $y_M(l)$ of the horizontal projection of the trajectory of the design point can be described by the following parametric equations:

$$x_M(l) = x_0(l) - H \cdot i(l) \cdot \sin\beta(l) \quad (2)$$

$$y_M(l) = y_0(l) + H \cdot i(l) \cdot \cos\beta(l) \quad (3)$$

where

$x_0(l)$ and $y_0(l)$ –current coordinates of the line axis, calculated in accordance with the regularity of the angle of the tangent β (l) to the line axis;

$\beta(l)$ –current angle of the tangent to the line axis, calculated in accordance with its principle and the value of the parameter U;

$i(l)$ –current cross slope calculated in accordance with its principle and the value of the parameter Z;

H – elevation of the design point M above the level of the top of the rail heads

Taking into account this model, the principle of curvature $k_H(l)$ of the horizontal projection of the trajectory of any point M located at the design level of the crew $H \geq 0$ should be described by the differential equation

$$k_H(l) = \left(\frac{dx_M}{dl}\cdot\frac{d^2y_M}{dl^2} - \frac{d^2x_M}{dl^2}\cdot\frac{dy_M}{dl}\right)\cdot\left(\left(\frac{dx_M}{dl}\right)^2 + \left(\frac{dy_M}{dl}\right)^2\right)^{-3/2} \quad (4)$$

These and other dependencies of the model allow predicting the performance of the system with one form or another of the transitional rounding sections, taking into account the factors predetermined and controlled at the design stage. Their addition with the parameter U of the principle of the tangent angle $\beta(l)$ to the line axis of its transition sections and the parameter Z of the principle of the properties of the retraction of their cross slope $i(l)$ expands the possibilities of the provision of the required quality level of the project of its model. The analysis of many different combinations of the principles of the properties of the transition sections showed that the quality of the system functioning depends on how harmoniously the values of the parameters *L*, *U* and *Z* are combined with the values of the design speed *V*, the curvature radius *R*, the design elevation of the outer rail above the inner one *D* and also the distance between the line axes *S* and the elevation H of the design point above the level of the top of rail heads.

In terms of the language of natural communication, the desired level of this quality means the comfort of passenger movement and the positive dynamics of the force interaction of the system elements. However, the criteria of the prior technical level are not suitable for an unambiguous numerical assessment of its achievement. It is mainly because of their extremely mediated and indirect connection with those indicators that can characterize the desired quality of the system functioning. Thus, for example, the so-called elevating speed of wheel lifting along the elevation of the outer rail with such indicators of the quality of the system functioning as with the non-reduced side acceleration a at the level H or with the rate of its change $\psi=da/dt$ at the same level is not always obvious and even contradictory.

Therefore, as a formalized target of the harmonization of the properties of the system, an integral indicator was proposed, by the numerical value of which it is possible to determine the degree of achievement of the required quality of its functioning [1]. The value of this indicator characterizes the variance of the oscillation amplitudes of the principle of the derivative of the change ratio of the non-reduced side acceleration acting at the calculated level H. In this case, the highest level of harmonization quality of the system properties is achieved at the minimum variance of the functional $d/dt=d^2a/dt^2$, estimated at the central segment of the transitional section between points δ_b and δ_e (Fig. 2).

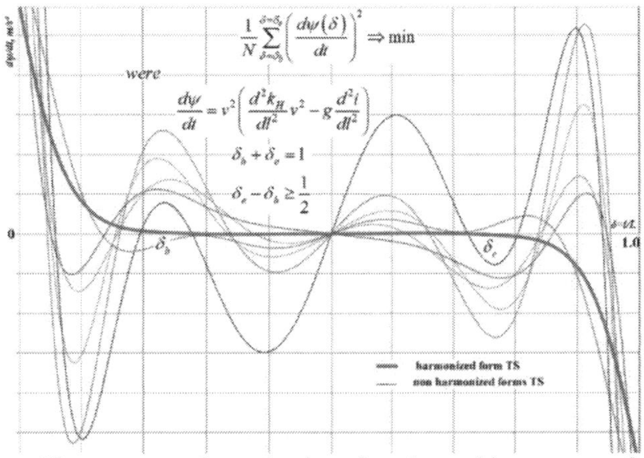

Figure 2. Sample of the shape of the diagram of the functional $d\psi/dt$ with a minimum of the variance of the oscillations amplitudes of its values in the central section δ_b - δ_e, which is the integral target of the harmonization of the properties of the system (red)

If we meet the criterion for the achievement of this integral target, each of the 4 generally recognized local quality targets of the transitional section of the railway is achieved (Fig. 3).

Target № 1
Strictly monotonic curvature k_0 of the line axis at the level of the top of the rail heads

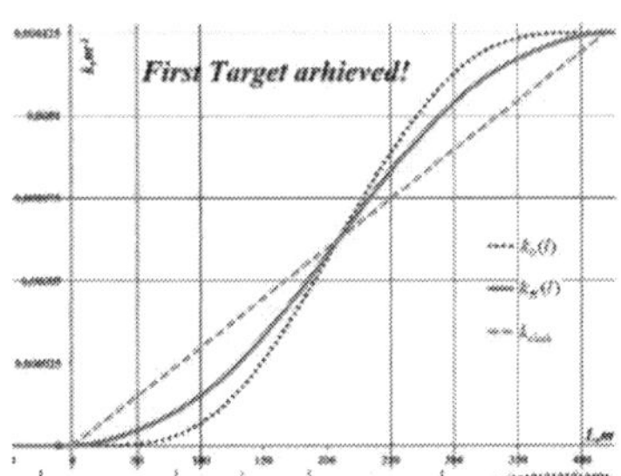

Target № 2
Quasilinear plot of NPU $a(l)$ at the design level H

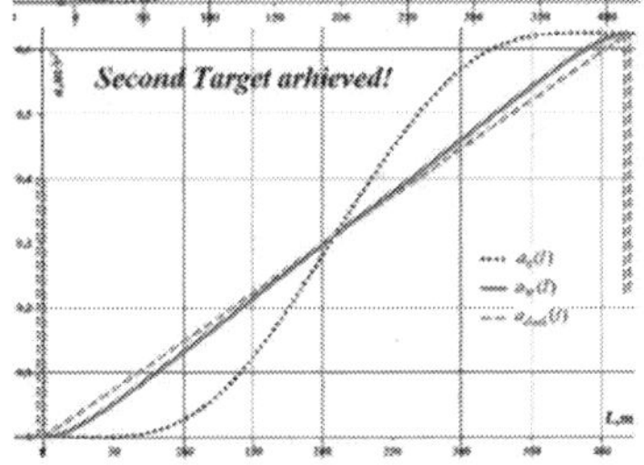

Target № 3
Continuity of change in NPU
$\Psi(l)=da/dt$
With minimal speed
Ψmax –> MIN

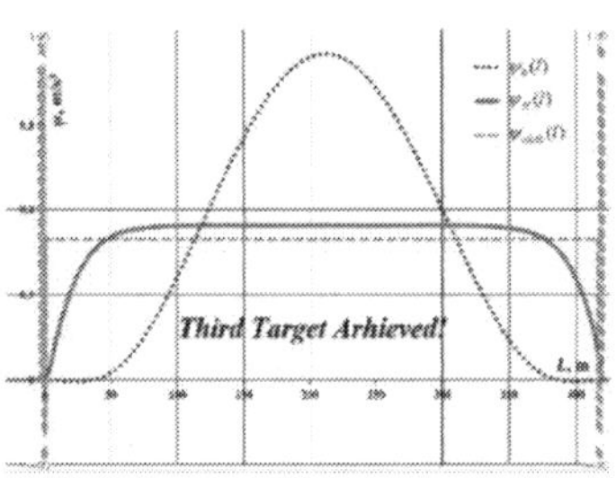

Target № 4
Strictly monotonic and smooth change in difference of reaction force of rails

$\Delta F(l)= FL - FR$

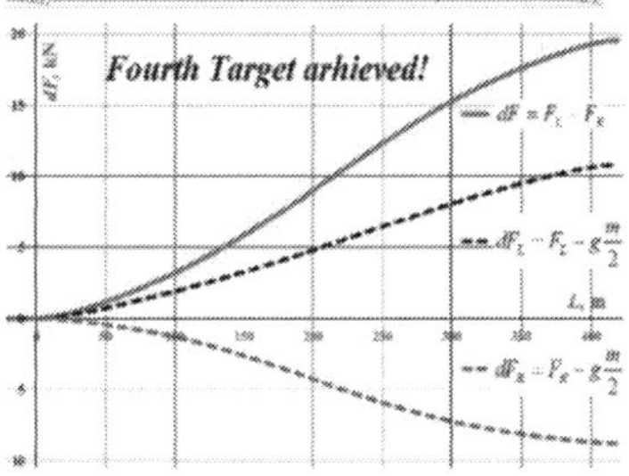

Figure 3. Local targets of the geometric and functional quality of the system, achieved by harmonizing the parameters L, U and Z at a minimum $d\psi/dt = d^2a/dt^2$

4. Comparison of the properties of new forms of transitional sections of curves of railways with properties of forms of the prior technical level

The analysis of the results of harmonization of more than a thousand forms of transitional sections confirmed the high quality of their properties, which were achieved with a minimum dispersion of the oscillation amplitudes of $d\psi/dt$ functional, which varied in the range from 10^{-8} to 10^{-5} m/s^4. Each of the forms of these transition sections corresponded to a unique variant of the combination of its predetermined properties, varying in the range of speeds V from 100 km / h to 400 km / h with such combinations of radii R and elevations D, which for each of the standard line widths of S = 1520 mm and S = 1435 mm provided the maximum values of the non-reduced side acceleration a in the range from 0.1 m/s^2 to 1.0 m/s^2 at the design levels $1800 \leq H \leq 2500$ mm.

The data selection from the results of this research at a design speed of V = 400 km / h allows comparing the lengths L of the harmonized transition sections with the lengths of a number of other forms. In accordance with the standard recommended for their use [4], their lengths are set equal to the length of the equivalent clothoid, calculated taking into account the type of used coefficient. Taking this fact into account, for the purposes of this comparison, the lengths of only basic clothoids were calculated, which ensured the speed of the wheel elevation along the elevation of the outer rail v_d = 28 mm/s. According to the results of these calculations, the graphs of the dependence of the lengths of new and clothoid forms on the calculated values of the non-reduced side acceleration a and other parameters of curve were created (Fig. 4).

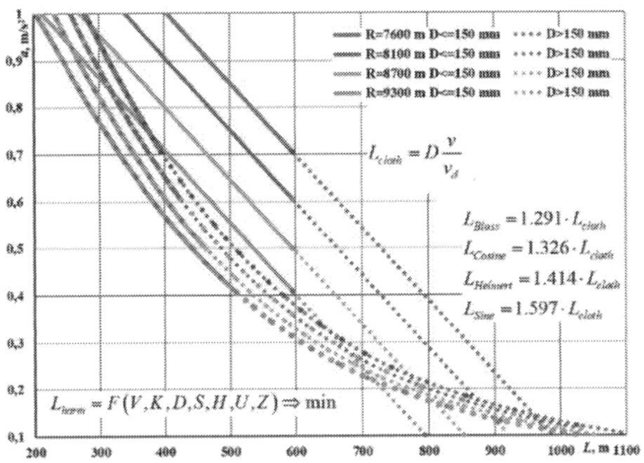

Figure 4. Graphs of the dependence of the lengths L of new and clothoid forms of the transition sections on the calculated value of the non-reduced side acceleration a and other parameters of the options for curve of the studied railways at V = 400 km / h, H = 2200 mm and S = 1520 mm

In all the research cases, the functional quality of the harmonized forms exceeded the quality of the transition sections of the forms of the prior technical level and the values of their lengths L were strictly ordered according to the values of the parameter a in the form of nonlinear graphs of implicit functions $L = f(a)$. The results of the calculation of the unique coordinates a and L of these graphs varied depending on the options for the values of other parameters H and S, as well as on the values of the parameters U and Z found at the time of the determination of each of these lengths. In contrast to this, the values of the basic lengths of the linear graphs of clothoid forms did not change, since the provisions of the current paradigm did not provide these parameters of the "railway line-rolling stock" system.

The graphs from Figure 4 show that with the nonlinear nature of the graphs of the length functions L of the harmonized forms of transitional sections, their values in the zone $0.4 \leq a \leq 0.7$ m/s^2 are 1.2 ÷ 1.5 times less than the basic lengths of the transitional sections of clothoid forms. As the smoothness of other forms of transition sections increases and the values of the corresponding coefficients increase, the difference between the lengths will increase even more. This indicates obvious problems of empirical methods of the assignment of the lengths of transition sections of traditional forms in proportion to the lengths of the basic clothoids. After all, the methods of calculating both the basic lengths of the clothoid forms of the transitional sections, or the coefficients of their proportions with the lengths of other *half-sine* forms, do not have any significant theoretical justification.

It is necessary to note that the shift of the circular curve p in the harmonized transition areas is significantly (several times!) less than in clothoid forms similar in other parameters. This creates the prerequisites for their harmonization within the existing curves with a possible increase in radii and elevations with minor shifts of their track from the original design position. Due to this, the level of operational quality of the existing transition areas can be significantly increased during the planned adjusting and tamping works.

5. Specific features of the device and functioning of the "driver-car-road" system

The safety and comfort of curved road traffic largely depends on the maintenance of a balance between the side grip coefficient φ and the side force coefficient μ provided by their pavement. Its value depends on the part of the side acceleration a that is not reduced by the curve acting on a car. It is generally accepted that this balance is observed under the condition $\varphi \geq \mu$. However, in real driving conditions, the values of the parameters φ and μ are subject to significant deviation. It is reasoned by the combined effects of many meteorological, technical, situational and behavioral factors. A radical reduction in the associated risks ensures compliance with the speed of driving curved sections of the line, which a driver must select in accordance with the current traffic conditions.

The justification of this choice and safety of its implementation largely depend on the length and regularities of the properties of the transition section of curve preceding its maximum curvature.

The most predictable and critical situation is in which a driver is forced or continues to reduce the speed of his vehicle in the transitional section of the curve. Therefore, its length L must be enough to promptly and safely reduce the initial speed v_b to the final speed v_e with a comfortable deceleration d.

In this case, the current curvature k of the transition section must be in line with the variable speed v according to the balance condition gφ≥w between the acceleration depending on the values of the friction coefficient φ and the vector sum of the side acceleration e kv2-ge partially reduced by the curve slope e with the algebraic sum of longitudinal accelerations from the slope ± gi and vehicle deceleration d (Fig. 5).

The required length of the transitional section of the curve L is determined from the condition of a uniform change in speed $v = v_b + d \cdot t$ over time t with a certain margin of time t_r, which is necessary for the driver's timely response to the beginning of the curvilinear trajectory of movement

$$L \geq 0.5 \cdot \left(v_e^2 - v_b^2\right) \cdot d^{-1} + v_b t_r \qquad (5)$$

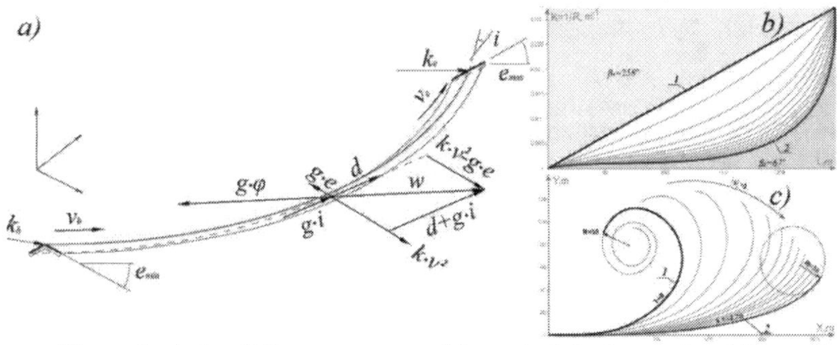

Figure 5. Model for the assessment of the safety of curvilinear movement of a car with deceleration d (a), principles of curvature k (b) and the corresponding coordinates x and y of its trajectory (c)

The principle of the curvature of its axis k is determined from the condition of uniform change in time t in the general vector of vehicle accelerations $w = w_b + j \cdot t$. Taking this into account, its principle on a horizontal section of the road should correspond to following equitation:

$$\left(\left(k_b v_b^2\right)^2 + \partial^2\right)^{\frac{1}{2}} = \left(\left(kv^2\right)^2 + \partial^2\right)^{\frac{1}{2}} + jt \qquad (6)$$

where

∂ – deceleration determining the law of curvature k, determined from the conditions of safety and comfort of movement with design deceleration d;

j – speed of uniform increase of the total acceleration vector w, determined without taking into account the superelevation e and the longitudinal slope i at the calculated value ∂;

k_b –curvature at the beginning of the transition section of curve

The indicator of movement comfort, traditionally estimated by the rate of increase of the centrifugal acceleration C, with variable curvature k and variable speed v at any point on the axis of the transition section should be calculated by the following formula:

$$C = \frac{dk}{dl} v^3 + 2 \cdot v \cdot k \cdot d \qquad (7)$$

At a variable speed v, the values of this indicator C reach their functionally significant maximum only at the end of the transition section with curvature k_e. Therefore, the value of deceleration ∂, which determines the principle of the curvature of the axis of the transition section k, should be calculated taking into account the required convenience of movement C_{max} according to the following equation:

$$j \cdot \left(1 + \partial^2 \cdot \left(k_e \cdot v_e^2\right)^{-2}\right)^{\frac{1}{2}} + 2 \cdot v_e \cdot k_e \left(d - \partial\right) - C_{\max} = 0 \qquad (8)$$

As a rule, at $C_{max} \approx 50.4/V_e$ equation (8) is achieved at $\partial \leq d$. This ensures the observance of the balance $w \leq g\varphi$ with some excess of the standard safety level of curvilinear movement with a variable speed. In Figure 6a the areas of such *a* excess near the *VGV_Curve* with a length of L = 300 m is blue. It is provided at V_b=120 km / h, V_e=60 km / h, R = 150 m, e = 60 ‰, d = -1.4 m/s^2, C = 0.4 m /s^3 and ∂ = -2.55 m/s^2. The area of shortage is red, which increases the risk of road accidents on a clothoid transition section of the same length and radius.

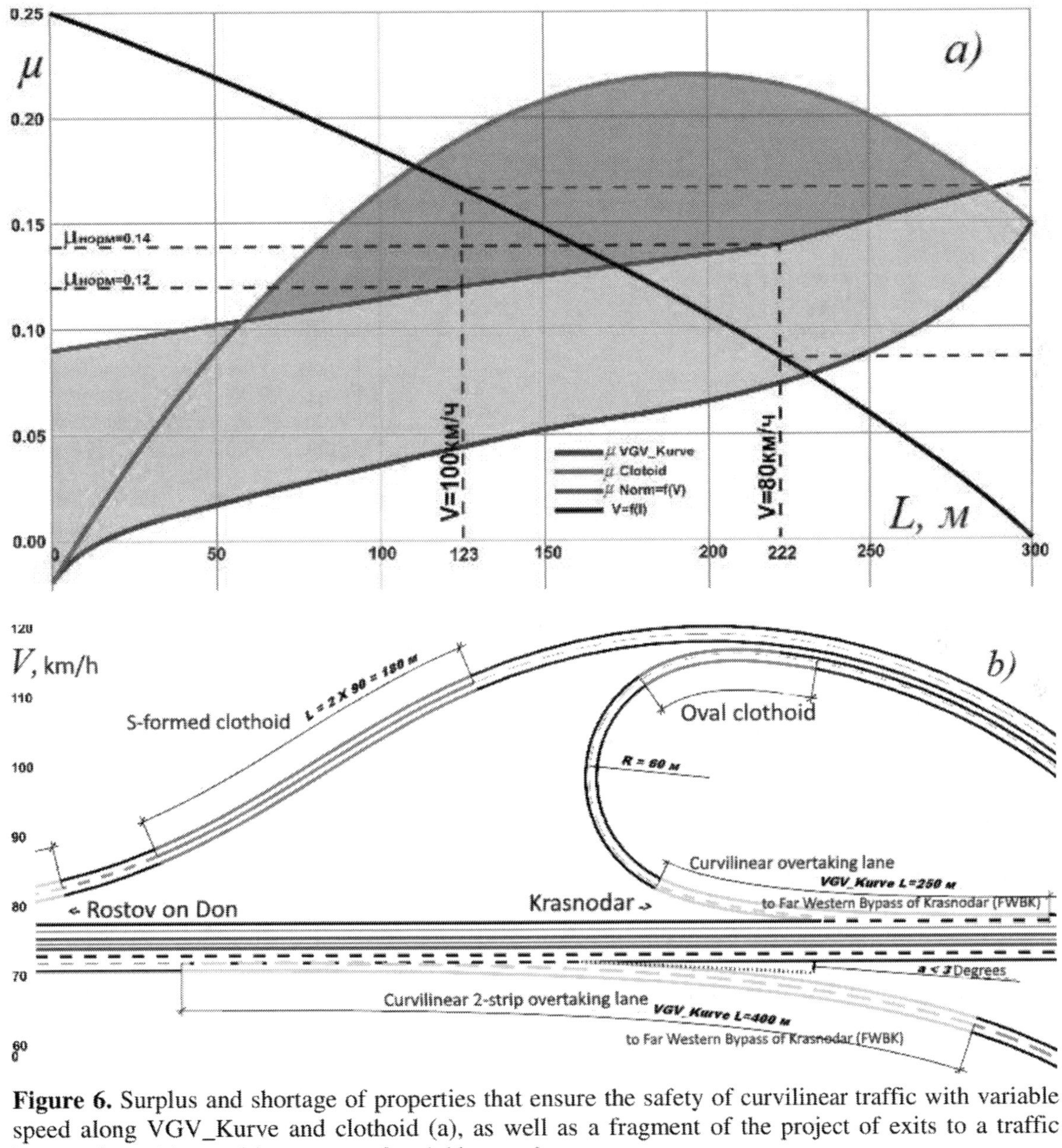

Figure 6. Surplus and shortage of properties that ensure the safety of curvilinear traffic with variable speed along VGV_Kurve and clothoid (a), as well as a fragment of the project of exits to a traffic intersection with transition curves of variable speed

6. Conclusion

The geometric and functional properties of the transitional sections of the curves of railways and highways have significant differences due to the properties of the systems that include them. The presence and essence of these differences is illustrated by the characteristics of the alternative VGV_Kurve and VGV-41H-39 forms of the transition section of curve R = 300 m, which, in traditional practice, could be implemented in the form of Clothoid with a linear elevation or superelevation (Table 1). The significance of these differences is also illustrated by the graphs of the curvature of the axes of these forms of transition sections (Fig. 7).

Table 1. Characteristics of alternative forms of the curve transition section R = 300 m

Form	Speed V (km/h)	Comfort C (m/s^3)	Length L (m)	Shift p (m)	Angle β (g, m, s)	Additional data
Clothoid	90	0.58	90	1.12	8°35`40``	–
VGV_Kurve	from 140 to 90	0.6	300	11.46	15°42`15``	d=-1.3 m/s^2, ∂= -4.22 m/s^2
VGV-41*H*-39	70	0.073	100	0.44	9°32`57``	D=150 mm, H=2400 mm, S=1520 mm a=0.342 m/s^2, Z=-0.41, U=-0.39,

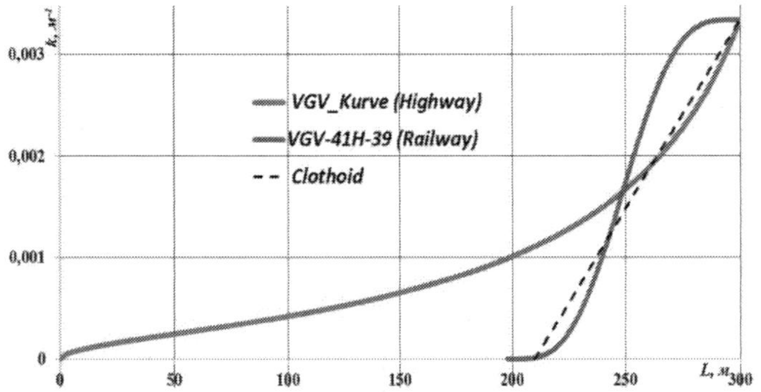

Figure 7. Differences in the principles of the curvature of the forms' axes of transitional sections alternative to the clothoid, the properties of which are consistent with the functional characteristics of the systems "railway road-rolling stock" and "driver-car-road"

The justified forecast of a significant increase in the operational quality of the functioning of these systems creates all the conditions for a comprehensive verification of the proposed solutions in practice and further improvement of the theory, current norms and rules for the design of transport infrastructure facilities with new, corresponding forms of transitional sections of their curves.

References

[1] Kufver B 1997 VTI rapport 420A *Mathematical description of railway alignments and some preliminary comparative studies* (Swedish National Road and Transport Research Institute) Available at: http://www.diva-portal.org/smash/record.jsf?pid=diva2%3A675179&dswid=4876

[2] Velichko G 2020 Quality analysis and evaluation technique of railway track + vehicle system performance at railway transition sections with various shape curves *Transport Means 2020: Proceedings of the 24th International Scientific Conference* Part II: 573-578 Available at: https://transportmeans.ktu.edu/wp-content/uploads/sites/307/2018/02/Transport-means-A4-II-dalis.pdf

[3] Velichko G 2020 Shape Harmonization of the Railway Track Transition Section & the Kinematics of Vehicle Body Design Point *Transport Means 2020: Proceedings of the 24th International Scientific Conference* Part II: 910-915 Available at: https://transportmeans.ktu.edu/wp-content/uploads/sites/307/2018/02/Transport-means-A4-II-dalis.pdf

[4] EN 13803-1:2010: Railway applications - Track-Track alignment design parameters - Track gauges 1435 mm and wider - Part 1: Plain line [Required by Directive 2008/57/EC]

Road conditions and traffic speeds

P A Lushnikov[1], N A Lushnikov[1,2], V A Kretov[2], A V Kostsov[3]

[1] ROSDORNII, 2, Smolnaya street, Moscow, 125493, Russia
[2] Russian university of transport, 9/9, Obraztsova Street, Moscow, 127994, Russia
[3] Moscow Automobile and Road Construction State Technical University, 64, Leningrad Avenue, Moscow, 125319, Russia

E-mail: lab10@mail.ru, adaoif@bk.ru, kostsov_msfs@bk.ru

Abstract. The paper shows that from the point of view of energy consumption, the most optimal is the mode of movement with a constant speed among all possible speed modes of movement due to the fact that the average value of the mechanical energy of a moving car has a minimum value when driving at a constant speed. However, in real road conditions, certain sections of road do not allow moving at the selected speed and drivers are forced to change the mode of movement of the car for various reasons. The article discusses various methods for the assessment of the unevenness of the traffic flow speed, providing the use of forward accelerations, the coefficient of variation, as well as spatial and temporal speed. The study was carried out using the experimental analytical method. The paper shows that the forward accelerations of a car moving in a traffic flow have an oscillating character. Therefore, it is difficult to determine problem areas of the road that are the reasons for the decrease in speeds. It is also difficult to determine the areas where the speeds of movement change from the linear graph of the coefficient of variation, since this graph oscillates to a greater extent than the acceleration graph. As a result of the analysis and numerical calculations, it was proposed to assess road conditions by the ratio of temporal and spatial speed, which characterizes the unevenness of the speed of movement of vehicles. It is proposed to call this ratio the coefficient of speed unevenness of the traffic flow. It is shown that this indicator is more convenient for the determination of problem sections of the road. The graphs are based on the results of experimental measurements of driving speeds on specific highways. The coefficient of unevenness of speed better reveals problem road sections, since the dependence of this coefficient on the traveled distance has a much less oscillating character, which facilitates the selection of unfavorable sections of highways, in comparison with other existing indicators. The important result of this work is the proposal to use the coefficient of speed unevenness as one of the criteria for the assessment of road conditions, for example, when diagnosing highways.

1. Introduction

The modes of movement of vehicles are largely determined by road conditions: visibility distance, roadway condition, type of intersections with other roads, etc. The speed of traffic flow is one of the key factors in the *driver-car-road system* [1-4]. As a rule, the safest driving conditions and the shortest travel time are observed at a constant vehicle speed [2, 5]. In this mode, the mechanical impact of a car on the road is also minimal, as well as energy consumption for the transportation of goods [1-4]. In this regard, there is the problem of eliminating sections on the road with uneven traffic of cars by developing measures to improve traffic conditions in such sections and equalize the speed of movement throughout the road.

Content from this work may be used under the terms of the Creative Commons Attribution 3.0 licence. Any further distribution of this work must maintain attribution to the author(s) and the title of the work, journal citation and DOI.
Published under licence by IOP Publishing Ltd

The paper discusses the use for this purpose of the coefficient of speed unevenness of movement k:

$$k = v_t / v_s$$

where

v_t – temporary average speed of a car;

v_s – spatial average speed of movement of a car **[4]**.

It is necessary to note that the authors did have the purpose to provide the comprehensive survey of existing methods for the assessment of the unevenness of movement with a detailed analysis of their advantages and disadvantages. For comparison they used only some approaches to solve the problem of leveling the speed of movement based on the use of the values of the linear speeds and accelerations, the coefficient of variation of speeds and the coefficient of speed unevenness, with the purpose of possible further use of this coefficient as an estimated indicator of the state of road conditions.

2. Problem statement

It seems reasonable to find the optimal speed of movement according to the criterion of minimum energy consumption. Let us suppose that the car moves along a straight and horizontal section of the road with a length L for time T. Then the average mechanical energy of movement E can be determined by the following expression:

$$E = \frac{\int_0^T m\dot{x}^2 dt}{2T}, \quad (1)$$

where

m – mass of a car;

$x(t)$ – movement along the studied road section, $x(0)=0$, $x(T)=L$;

t – current time.

In order to determine the minimum of E for all possible laws of motion $x(t)$, $x(0) = 0$, $x(T) = L$, *we need to solve the Euler equation*

[6]:

$$\frac{\partial L}{\partial x} - \frac{d}{dt}\frac{\partial L}{\partial \dot{x}} = 0, \quad (2)$$

where

$$L(x, \dot{x}, t) = \frac{m\dot{x}^2}{2}. \quad (3)$$

Taking into account **(3)** the equations **(2)** is:

$$m\ddot{x} = 0$$

or

$$\dot{x} = const. \quad (4)$$

That is, the average value of the mechanical energy of a moving car has a minimum value when driving at a constant speed. This means that in the studied case from the point of view of energy consumption the most optimal is the mode of movement with a constant speed among all possible high-speed modes of movement.

As it was already noted such a movement is optimal in other cases. Therefore, the issue of the determination of areas with uneven movement speeds is very relevant. In order to solve this problem, we use the concepts of temporal and spatial average speeds [4]. These speeds are defined as follows.

IOP Conf. Series: Materials Science and Engineering 1159 (2021) 012023 doi:10.1088/1757-899X/1159/1/012023

3. Research method

Let us suppose that a car moves during time T along a section of the road of length L. Let us conditionally divide this section into n equal segments with uniform traffic conditions, each of which has length h: $L = nh$. Then temporal v_t and spatial velocity v_s are determined as follows:

$$v_t = \frac{L}{T}\ ; \qquad \textbf{(5)}$$

$$v_s = \frac{\sum_{k=1}^{n} v_k}{n}\ , \qquad \textbf{(6)}$$

where

$$v_k = \frac{h}{t_k}\ ,$$

t_k–time of movement of the car along the kth segment of the considered road section

Thus, the temporary speed is determined by the ratio of the length of the road section to the time of movement along it and the spatial speed is determined as the average value of the speeds on its parts. After the transformation of the equation (5) we get:

$$v_t = \frac{L}{T} = \frac{nh}{\sum_{k=1}^{n} t_k} = \frac{nh}{\sum_{k=1}^{n} \frac{h}{v_k}} = \frac{n}{\sum_{k=1}^{n} \frac{1}{v_k}}\ . \qquad \textbf{(7)}$$

Thus, the average temporal speed (7) on the considered section of the road is the average harmonic value on its parts and the average spatial speed (6) is the arithmetic mean of the speed. Therefore, the inequation [7] is as follows:

$$v_t \le v_s\ . \qquad \textbf{(8)}$$

Moreover, the equality sign takes place only in the case when $v_k = const, k = 1, \dots, n$ i.e. in the case of movement at a constant speed. In this regard, the ratio of temporary and spatial speeds $k = v_t/v_s$ can be used as a quantitative measure of the movement unevenness. Taking into account the expression (8), the inequation $0 \le k \le 1$ takes place and $k=1$ is only on sections with a constant speed of movement.

4. Results of calculations

Let us illustrate the above mentioned case with the examples in which segments of length h = 5 m were used to determine the speed v_s.

As an example, Figure 1 shows the graphs of the spatial and temporary speeds of a car moving in a traffic flow on a section of the ring highway around St. Petersburg. It is possible to notice a sharp decrease in speed at 36 km. As a result of the study of the reasons for the decrease in speed, it was found that a road accident occurred on this section.

IOP Conf. Series: Materials Science and Engineering 1159 (2021) 012023 doi:10.1088/1757-899X/1159/1/012023

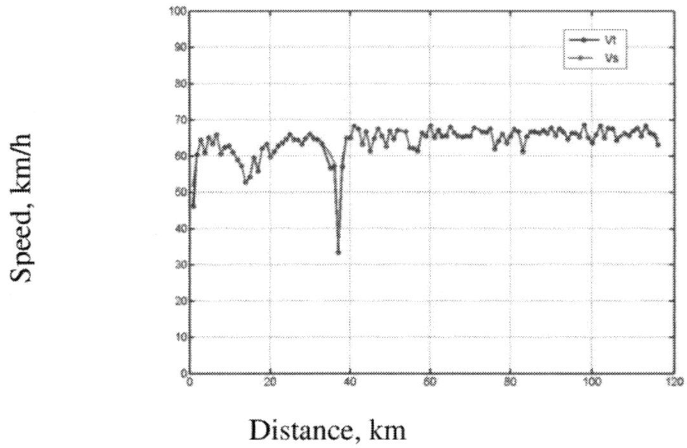

Figure 1. Dependence of spatial and temporary speeds of a car movement from the distance traveled on the ring highway section

Figures 2 and 3 show the values of indicators that can be used to assess the movement steadiness: the coefficient $k=v_t/v_s$ and the forward acceleration of a car averaged over kilometers on the same section of the ring highway. Both graphs allow highlighting an unfavorable (from the point of view of traffic steadiness) section, but the value of the forward acceleration is determined by significantly larger oscillations, which makes the determination of road sections with uneven traffic more difficult.

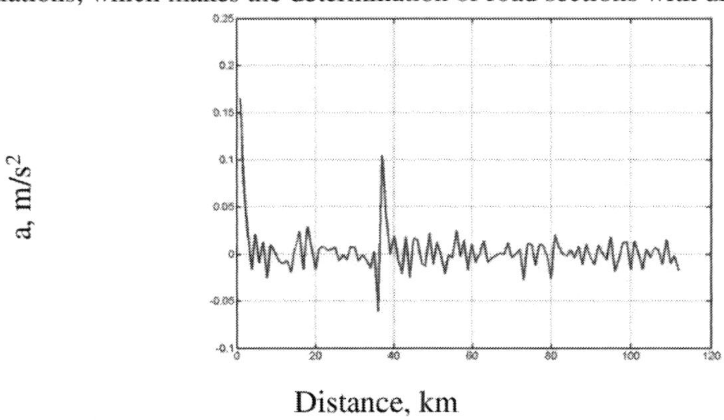

Figure 2. Dependence of the forward acceleration of a car from the distance traveled on the ring highway section

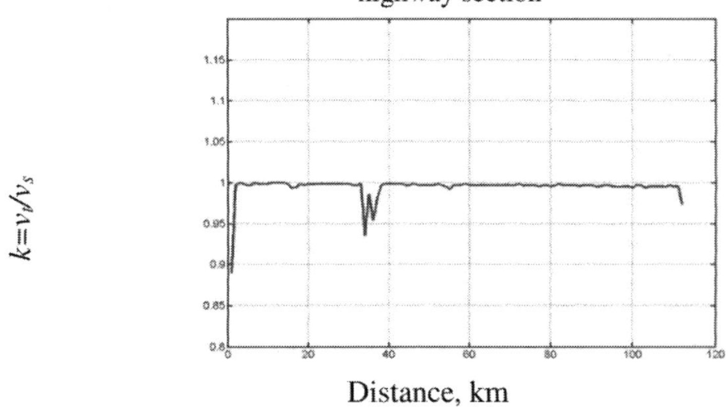

Figure 3. Dependence of coefficient $k=v_t/v_s$ from distance traveled on the section of the ring highway

Figures 4-6 show the results of speed measurements on a section of another road - the Small Moscow Ring (SMR) from the highway M-9 "Baltiya" in the direction of Zvenigorod. The values of v_t and v_s were determined after 100 m. The large value of the necessary coefficient of speed unevenness in a section with a length of 1 to 2 km (Fig. 4) is explained by the presence of the intersection with traffic light regulation.

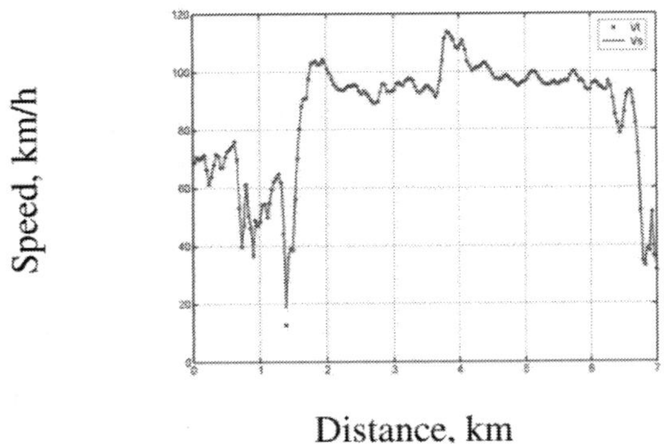

Figure 4. Dependence of spatial and temporary speeds of the movement of a car from the distance traveled on the SMR section with a step of 100 m

Figures 5 and 6 graphically show the change in the coefficient *k* and the coefficient of variation of speed depending on the path along the sections 100 m long. The dependence of the coefficient *k* on the distance traveled, in comparison with the analogous dependence for the coefficient of variation (Fig. 6), has a much less oscillating character, which makes it easier to determine the areas with uneven movement.

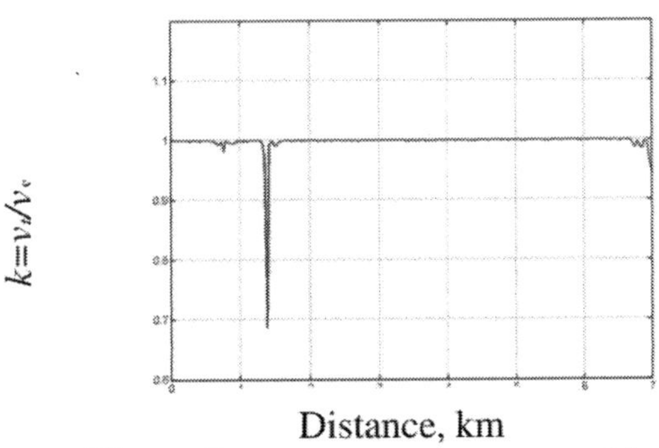

Figure 5. Dependence of coefficient $k = v_t/v_s$ from distance traveled on the section of SMR

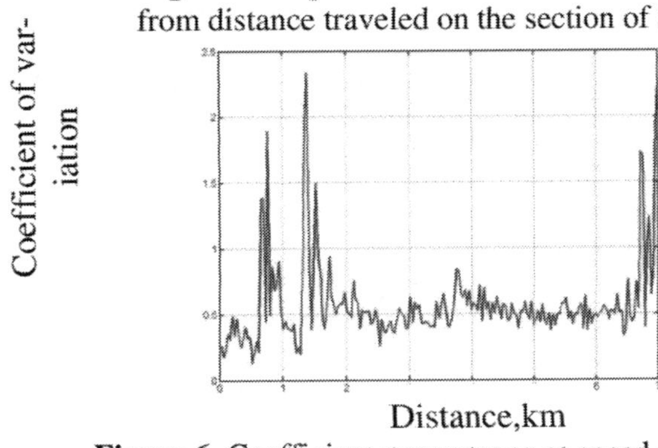

Figure 6. Coefficient dependence of speed variation from the distance traveled on SMR section

IOP Conf. Series: Materials Science and Engineering 1159 (2021) 012023 doi:10.1088/1757-899X/1159/1/012023

5. Conclusion

Thus, the research results allow drawing the following conclusions:

1. The movement of traffic flows at a constant speed is optimal from the point of view of traffic safety, minimum mechanical impact of a car on the road and energy consumption when a car is moving, as well as in other cases (for example, when a driver does not move). Therefore, the task of the determination of road sections on which traffic steadiness is not ensured is urgent.
2. In order to determine unfavorable areas with uneven movement, along with forward acceleration, linear speed of a car and coefficient of variation of speeds, the coefficient of unevenness of speed $k=v_i/v_s$ can be used, which better determined problem areas of the road, since the dependence of the coefficient k on the distance traveled has a significant less oscillating character, which makes it easier to determine unfavorable road sections, compared with other existing indicators.
3. It is proposed to use the coefficient of speed unevenness of movement as one of the criteria for the assessment of road conditions

References

[1] Silyanov V V 1977 *The theory of traffic flows in the design of roads and traffic management* (Moscow: Transport)
[2] Chvanov V V 2010 *Methods for assessing and improving road safety taking into account the working conditions of the driver* (Moscow: INFRA-M)
[3] Drew D R 1972 *Traffic flow theory and control* (New York)
[4] Haight F A 1971 *Mathematical theories of traffic flow* (New York, London, 1963, 285 p
[5] *Methodology for assessing traffic safety and transport qualities of highways* Ed. V.F. Babkov. (Moscow: Higher school)
[6] Gelfand I M, Fomin S V 1961 *Calculus of variations* (Moscow: Fizmatlit)
[7] Hardy G, Littlewood J, Polya G 1934 *Inequalities* (Cambridge Univ. Press)

IOP Conf. Series: Materials Science and Engineering 1159 (2021) 012024 doi:10.1088/1757-899X/1159/1/012024

Impact of traffic speed on the strength of highway pavements

Andrey Korochkin

Moscow State Technical University (MSTU), Leningradskii Avenue, 64, Moscow, 125319, Russian Federation

E-mail: andrey_korochkin@mail.ru

Abstract. The paper touches upon the issues related to the influence of the speed of traffic flow on the calculation of the strength of road pavements of highways. The dependence of the calculated parameters on the vehicle speed is shown. The definitions of such concepts as the design speed of movement on the road, the permitted speed of movement and the actual speed are given. The authors give the values of normative design indicators related to the speed of movement, which were determined more than 50 years ago for the speed of movement of trucks 40-60 km / h, and have remained unchanged since then. The paper also presents the results of the research on the speed of traffic flows along two highways M-2 "Crimea" and M-9 "Baltia" for 16 years, which prove that the speed limit on modern highways has changed dramatically in recent years and does not meet the requirements and design parameters. According to own research and research of many modern scientists, the author proposes to take into account the speed of 110 km / h when determining the design parameters for the calculation of road pavements and to use the dynamic coefficient equal to 1.85 (instead of 1.3) when calculating the pavement instead of standard values and take into account the time of the impact of the wheel on the coating of 0.014 s (instead of 0.1 s).

1. Problem statement

The study of the speed of movement of traffic flows on highways showed that the actual speed of movement has increased significantly in recent years. Due to this fact some of standard parameters associated with the speed limit and used in the design of road pavements no longer meet modern conditions, since they were calculated for vehicles moving at much lower speeds.

First of all, two calculated parameters require correction: the dynamic coefficient and the time of impact on the wheel of a moving vehicle on the road surface. The existing standard values of these design parameters were determined 50-30 years ago for speeds of 40-60 km / h.

Nowadays, the values of these parameters recommended by legal documents are as follows:

- Dynamic factor $F_d = 1.3$;
- Application time of the dynamic load of a vehicle car wheel to the coating t = 0.1 s

The research and analysis of scientific developments of a number of modern scientists allow recommending the following values of the design parameters for the design of pavements of highways:

- Dynamic factor F_d **= 1.8;**
- Application time of the dynamic load of a vehicle car wheel to the coating t **= 0.014 s.**

Content from this work may be used under the terms of the Creative Commons Attribution 3.0 licence. Any further distribution of this work must maintain attribution to the author(s) and the title of the work, journal citation and DOI.
Published under licence by IOP Publishing Ltd

2. Introduction

According to modern methods, the design of road pavements of both rigid and non-rigid types is carried out for static and dynamic loads. The static load is taken into account during the design of the pavement of parking lots, bus stops, cargo terminals, etc. At the same time, the pavements of the roadways of highways and city streets are design for dynamic load. Taking into account the fact that the dynamic load exceeds the static one, especially at high speeds and high traffic intensity, the strength of the pavement and its design depend on it [1].

The main design parameters for the determination of the dynamic load are the dynamic factor and the time of application of the load from the vehicle wheel to the road surface. Both of these parameters depend on the speed of traffic. That is, the speed of movement, along with the intensity of traffic, is one of the most important characteristics on which the design and construction of pavement depends [2].

It is necessary to note that for the design and operation of highways, several concepts of traffic speeds are determined. First of all, it is necessary to note that there are standard traffic speeds, that is, those that are recommended by legal documents and the actual traffic speed. Standard traffic speeds include:

- Design speed of a vehicle. The design speed depends on the road category. Depending on the design speed, geometric parameters of highways are assigned (radii of curves in the plan and longitudinal profiles, maximum longitudinal slopes, width of the traffic lane of the roadway and exits of interchanges, etc.);
- Permitted speed of a vehicle. The permitted speed is related to traffic safety and is determined by the rules of the road.

Actual traffic speed is the actual speed at which traffic is moving along the road. It is not a constant value, but depends on many factors such as: the technical condition of the road, traffic intensity, especially road conditions, means of traffic flow regulation, natural and climatic conditions, etc. In addition, the speed of movement in the traffic itself is not the same. Someone moves faster, someone slower. Therefore, such a concept as a percentage speed provision was introduced. For example, a 15% speed provision indicates the speed of slow moving cars. The 50% provision corresponds to the average speed of traffic flow of 85% provision - a speed that does not exceed the bulk of cars [3]. Thus, according to the speed of 85% provision, certain decisions are made to regulate traffic flows.

3. Methods

The study of the speed of the traffic flow, as well as the traffic intensity on the studied highways was carried out using 24-hour video recording. On the basis of which the hours of "peak" loads were determined. The studies covered all seasons: summer, autumn, winter, spring. For greater objectivity and reliability, the observations and video recording were carried out throughout the whole week, which allowed obtaining the data on traffic speeds both on weekdays and on weekends and holidays.

Thus, the video was filmed four times a year (each season), seven days a week and 24 hours a day. The determination of the speed of traffic flows was carried out in office based on the results of video recordings. This makes it possible to confidently speak about the maximum reliability and accuracy of the research results.

4. Results

4.1. Design speed of a vehicle

As it was mentioned above, during the design of motor roads, their geometric characteristics are assigned taking into account the design speed. These include: radii of curving angles, values of longitudinal slopes, width of the roadway and shoulders, etc. Design traffic speed is the maximum possible speed at which a vehicle can theoretically move on a given road. Therefore, as a rule, real road speeds are limited to values lower than the design speed.

The design speed is assigned in accordance with SP 34.13330.2012 "Highways" [4]. For roads of category IA, the design speed is 150 km / h. For motor roads of categories IB and II it is 120 km / h.

For roads of category IB and III it is 100 km / h, for IV category it is 80 km / h, V category it is 60 km / h. The design speed for streets and city roads is assigned in accordance with SP 42.1333.2011 "Urban planning" and with unregulated traffic it is 100 km / h and with regulated traffic it is 80 km / h.

4.2. Permissible speed

The permissible speed on highways is assigned in accordance with the "Traffic Regulations of the Russian Federation" (TR). According to traffic rules on highways (the highways of IA and IB categories) the permissible speed is 110 km / h, on other roads it is no more than 90 km / h. The exception is the highways of State Company Avtodor, where the permissible speed is 130 km / h. In cities, according to traffic regulations, the permissible speed does not exceed 80 km / h. Figures 1 and 2 show the road signs indicating the permissible speeds [5].

Figure 1. Road signs on highways

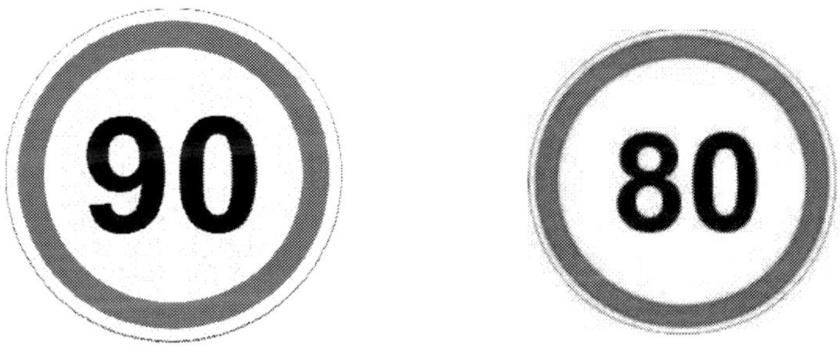

Figure 2. Road signs on highways and city streets. Actual traffic speeds

In 2003-2019 the author carried out research on a number of highways. One of the tasks of the research was to determine the actual traffic speeds on modern highways [5]. We will consider the results obtained as a result of the research on the example of sections of two roads: M-2 "Crimea" and M-9 "Baltia".

On the M-2 "Crimea" highway, the studies were carried out along traffic lines in two sections; Section 1 (six lines) km 49 + 000 - km 59 + 000; Section No. 2 (four lines) km 98 + 000 - km 104 + 000. The traffic speed on sections of these highways was of 90% provision (Table 1).

Table 1. Traffic speed on the sections of these highways of 90% provision

Year	Section №1 km/h			Section №2 km/h	
	Right line	Middle line	Left line	Right line	Left line
2003	82,0	96,5	117,5	81,5	103,2
2009	85,1	97,2	117,9	82,6	105,0
2015	87,4	98,3	118,3	88,4	106,1
2019	89,8	101,4	120,8	89,5	108,4

In addition, the maximum speeds of individual vehicles were recorded. The maximum speed of a passenger car was recorded at section No. 1 and was 184 km / h. The speed of freight car on the same section # 1 was 154 km / h.

On the highway M-9 "Baltia", a section with four lines of km 85 + 000 - km 95 + 000 was studied. The results of the study of 90% provision are presented in Table 2.

Table 2. Research results of 90% provision

Year	Km/h	
	Right line	Left line
2003	91,8	122,4
2009	93,4	123,1
2015	96,8	125,6
2019	99,1	127,5

Here the cases when the speed of cars reached 160-170 km / h, and the speed of trucks - 130-140 km / h were determined.

It is necessary to note that on the M-2 "Crimea" and M-9 "Baltia" highways the permissible speed is 110 km / h and the design speed is 120 km / h.

Usually, both in the Russian Federation and in many foreign countries, the 85% level of provision is used to determine the actual speeds. However, taking into account that in Russia the share of violations associated with exceeding the speed of traffic is very high, the author believes that for the design of road pavements, it is necessary to take 90% level of provision.

The research on traffic flow was carried out by the author using photo and video recording, as well as when driving in a traffic flow in different modes. The average speed of the traffic flow during the day was recorded. It is necessary to note that the measurements were carried out at the same sites, in the same years and at the same time as traffic intensity.

According to the results of the study, it is possible to conclude that, in general, the traffic flow is moving exceeding the permissible speed. In this case, the clause of the traffic regulations on the possibility of exceeding the speed by 19 km / h plays an important role. That is why the average speed of cars is higher than the permissible one.

4.3. Design parameters depending on traffic speed

The parameters used for the design of road pavements and related to the speed of traffic flow were mentioned above: dynamic factor and the time of interaction of the wheel with the road. Both of these parameters are normative. The problem lies in the fact that the values of these parameters were determined back in the 60s - 80s of the last century by the founders of the modern methodology N.N. Ivanov. [6], Birulya A.K. [7], Babkov V.F. [8] and others. However, at the same time, these parameters were determined for the conditions corresponding to the time when the carrying capacity of trucks was 3-10 tons and their speed did not exceed 40-60 km / h.

Nowadays on our roads, trucks and road trains with a carrying capacity of 40-60 tons carry out freight transportation, while the speed of movement, as can be seen from the materials presented above, is 90-120 km / h. That is, the conditions for the movement of traffic flows have undergone significant changes, while the calculated parameters have remained unchanged. Taking into account the above mentioned research results, the author proposes to consider the speed of traffic flow equal to 110 km / h during the design of the pavement and design parameters.

Thus, according to the current PNS 265-2018 "Design of non-rigid road pavements", put into effect by the Order of the Federal Agency for Technical Regulation and Metrology of 11.04.2018 No. 3-PNS [9], ODN 218.046-01 "Design of non-rigid road pavements" approved by the order of Rosavtodor dated 20.12.2000 No. OS-35-R [10] and "Methodological recommendations for the design of rigid road pavements (instead of VSN 197-91)", enacted by the order of the Ministry of Transport of the Russian Federation of January, 3, 2003 No. OS-1066 -p [11] road pavements should be designed for short-term repeated action of moving loads equal to 0.1 s. Exactly the same exposure time 0.1 s. VSN 46-83 was also laid, approved by the order of the USSR Ministry of Transport Construction in May 29, 1983 No. LN-550.

Now let us perform a calculation. According to PNS 265-2018, the diameter of the track of the driving wheel of a car at a standard load of 115 kN is 34 cm. The duration of the increase in the load with this diameter at different speeds will be expressed in the values presented in Table. 3.

Table 3. Duration of the load increase with a track diameter of a driving wheel of 35 cm at various speeds

Speed, km/h	110	100	60
Speed, s	0,011	0,012	0,020

As it can be seen, the load application time is hundredths of a second. At the speed of 110 km / h it is an order of magnitude less than the standard value equal to 0.1 s. That is, the standard values of the time for the application of load from the wheel to the surface are outdated and no longer meet modern requirements.

The same can be said about the dynamic factor.

5. Discussion

In the existing legal documents for the design of road pavements PNS 265-2018, ODN 218.046-01, Methodological recommendations for the design of rigid road pavements (instead of VSN 197-91), a dynamic factor equal to 1.3 is used in the design of road pavements. The same value of the dynamic factor is also reflected in the previously existing VSN 46-83 "Instructions for the design of non-rigid pavements" and VSN 197-91 "Instructions for the design of rigid pavements". This coefficient was calculated for speeds of 40-60 km / h.

The problem of the determination of the real dynamic coefficient is devoted to the research of many modern scientists - Smirnov A.V., Radovsky B.S., Suprun A.S, Konorev A.S., Woodrooffe J.H., Cebon D, etc. Thus, according to the formula proposed by B.S. Radovsky [12], the dynamic factor at a speed of 110 km/h is $F_d = 1.46$. According to the calculations of Professor A.V. Smirnov [13] at the same speed of movement $F_d = 1.78$. Konorev A.S. [14] at 110 km / h, the dynamic coefficient reaches $F_d = 2.1$. The author also conducted research in this direction [15]. The graph of the dependence of the dynamic factor on the speed of movement obtained by him is shown in Figure 3.

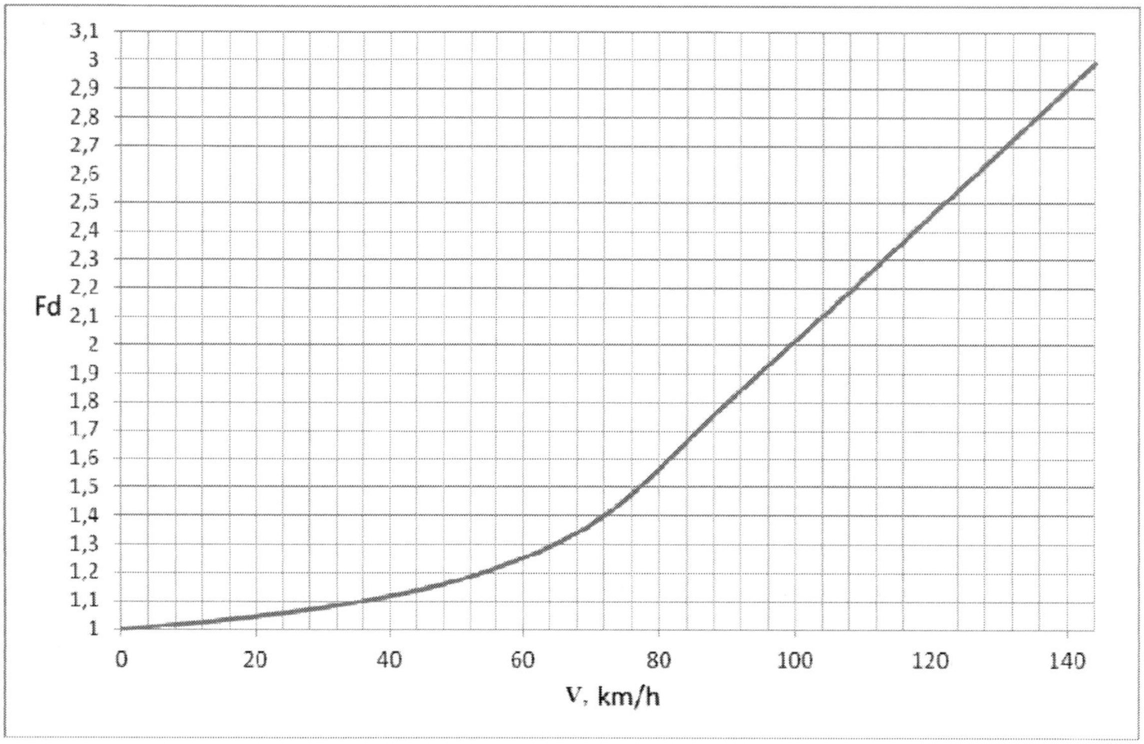

Figure 3. Dynamic factor (according to the author's research)

As it can be seen from the graph, the dynamic factor in this case for a speed of 110 km/h is equal to F_d = 2.25. The presented results are approximate, since the inertial and damping properties of the base of the pavement were not taken into account during the construction of the spatial model of the stress-deformed state. However, the obtained solution quite accurately reflects the tendency of road structure deflections to grow with the increase in the speed of traffic flows.

Having analyzed and summarized the results of various studies carried out both by the author and by other Russian and foreign scientists, the author believes that it is advisable to take the dynamic coefficient F_d = 1.85 for a speed of 110 km / h.

Thus, we found out that the calculated parameters associated with the speed of traffic flow, currently used with the existing method for the design of road pavements, are determined for road traffic conditions that are fundamentally different from those that we actually have on our roads. Consequently, as a result of their use, we get pavement designs that do not meet modern conditions, which significantly reduce the performance of pavements, their strength and durability.

6. Conclusion

The studies of traffic speeds allowed establishing that the average speed of movement of both cars and trucks on highways is growing annually and now it reaches 108.4-127.5 km/h on the extreme left lines, 89.5-112 on the extreme right lines, 0 km/h at a permisiible speed of 110 km / h (130 km/h). However, the parameters used for the design of pavements and depending on the traffic speed, such as the dynamic factor and the load time, which are currently proposed by the regulations, were calculated at a speed of 45-60 km/h. Therefore, in order to obtain more reliable results of pavement design, the author proposes to apply the design indicators calculated at a speed of 110 km / h.

In addition, during the design of road pavements the author proposes to take into account the time of application of the wheel load to the surface equal to 0.014 s, which is more consistent with the existing conditions. For the same reason, it is proposed to take the dynamic factor F_d=1.85 instead of the normative F_d=1.3.

References

[1] Korochkin A V 2018 Specificity of design and calculation of rigid road pavement with asphalt concrete pavement with asphalt concrete pavement on highways *Bulletin of the Moscow Automobile and Road Construction State Technical University (MADI)* **2** 104-109
[2] Afinogenov O P, Efimenko V N, Efimenko S V 2008 *Design and calculation of road pavements: textbook* (Kemerovo: Kuzbassvuzizdat) 371 p
[3] *ODM 218.2.020-2012 Guidelines for assessing the throughput of highways: Rosavtodor* (Moscow, 2012) 143 p
[4] *Highways. Updated edition of SNiP 2.05.02-85: SP 34.13330.2012* (Ministry of Regional Development of Russia. Moscow, 2012) 167 p
[5] *Resolution of the Government of the Russian Federation of 23.10.1993 N 1090 as amended by from 26.03.2020 "On the Rules of the Road"* (together with the "Basic Provisions for the Admission of Vehicles to Operation and the Obligations of Officials to Ensure Road Safety") Available at: https://legalacts.ru/doc/PP-RF-_1090-ot-23_10_93/.
[6] Ivanov N N 1973 *Design and calculation of non-rigid road pavements* (Moscow: Transport) 328 p
[7] Birulya A K, Mikhovich S I 1968 *Efficiency of road clothes* (Moscow: Transport) 1968 172 p
[8] Babkov V F, Andreev O V 1987 *Road design* T 1 (Moscow: Transport) 368 p
[9] *Public roads. Design of non-rigid pavements: PNST 265-2018.* (Moscow: Standartinform, 2018) 77 p
[10] *Design of non-rigid pavements: ODN 218.046-01: approved. by order of the State Service road. households (Rosavtodor) of the Ministry of Transport of the Russian Federation of 12/20/2000 N OS-35-R: entered. effective from 01.01.2001.* (Moscow: Informavtodor, 2001) 148 p
[11] *Guidelines for the design of rigid road pavements: (instead of VSN 197-91): industry road methodological document. Rosavtodor: introduction. 03.12.2003.* (Moscow, Informavtodor, 2004) 135 p
[12] Radovsky B S, Suprun A S, Kozakov I I 1989 *Designing of road pavements for the movement of heavy vehicles* (Moscow: Budivelnyk) 168 p
[13] Smirnov A V 1975 *Dynamics of road pavements of highways* (Moscow: Zap.-Sib. book publishing house) 184 p
[14] Konorev A S 2012 *Taking into account the dynamic impact of multi-axle vehicles in the calculation of road structures* (Volgograd: RSSU) 187 p
[15] Korochkin A V 2005 *Design of non-rigid road pavements: a tutorial* (Moscow: MADI) 129 p

IOP Conf. Series: Materials Science and Engineering 1159 (2021) 012025 doi:10.1088/1757-899X/1159/1/012025

Decision support system for street-road network objects repair

Yu E Vasiliev, M A Fineeva, A B Belyakov, S V Varshavsky, A A Caesar

The Federal State Funded Educational Institution of Highest Education
Moscow Automobile and Road Construction State Technical University – MADI,
Russian Federation, 125319, Moscow, Leningradskiy Prosp., 64

E-mail: vashome@yandex.ru

Abstract. The article analyzes factors influencing managerial decisions adoption on repair of road network facilities (RN). Road repair is a challenge that must take into account a number of conflicting criteria. It is shown that decision-making process is a complex mental activity that consists in choosing the most effective strategy and, at present, repair work planning should be carried out using decision support systems (DSS), that is, a class of knowledge-based systems. DSS developing necessity for repair planning of RN has been substantiated. DSS architecture is described and it is shown that DSS allows increasing knowledge base by adding new knowledge without changing the structure, as well as encapsulating new knowledge into the system, increasing intellectual resource used in decision-making process. The parameters are determined, on the basis of which rules in DSS are formulated, which are necessary for making decisions on determining scope, timing, cost and sequence of repair and restoration work. Production rules examples are given. The general scheme of algorithm for planning repair and restoration works at RNF, as well as algorithm scheme for complex indicator calculating of transport and operational state - ensuring movement design speed dare described. Planning process formalized description and control actions development has been completed. It is shown that intelligent decision-making is an iterative, multi-level procedure, within which decisions table is analyzed to assess technical condition of RN facilities based on regulatory and technical documentation analysis. Situation description is consistently expanded by taking into account the quality of RN road marking, traffic intensity, road construction materials volumes, repair and restoration work cost, traffic management, seasonality and weather conditions, etc. (the number of taken into account parameters and levels of their description is not limited). The advantages of proposed approach to DSS implementation are described. It is shown that proposed DSS version provides situational adaptive intelligent control with ability to issue justified recommendations for carrying out repairs. The prospects for work development in this direction were considered and conclusion was made that management digitalization of repair activities, the digital transformation of management model and business processes in the field of RN repair will increase profits, increase operational efficiency, reduce repair costs and improve its quality.

1. Introduction

High cost, insufficient efficiency and management decisions validity, uncertainty of operating conditions and weather conditions - these are only some of the factors influencing decision-making on planning repairs at road network objects (Fig. 1).

Content from this work may be used under the terms of the Creative Commons Attribution 3.0 licence. Any further distribution of this work must maintain attribution to the author(s) and the title of the work, journal citation and DOI.
Published under licence by IOP Publishing Ltd

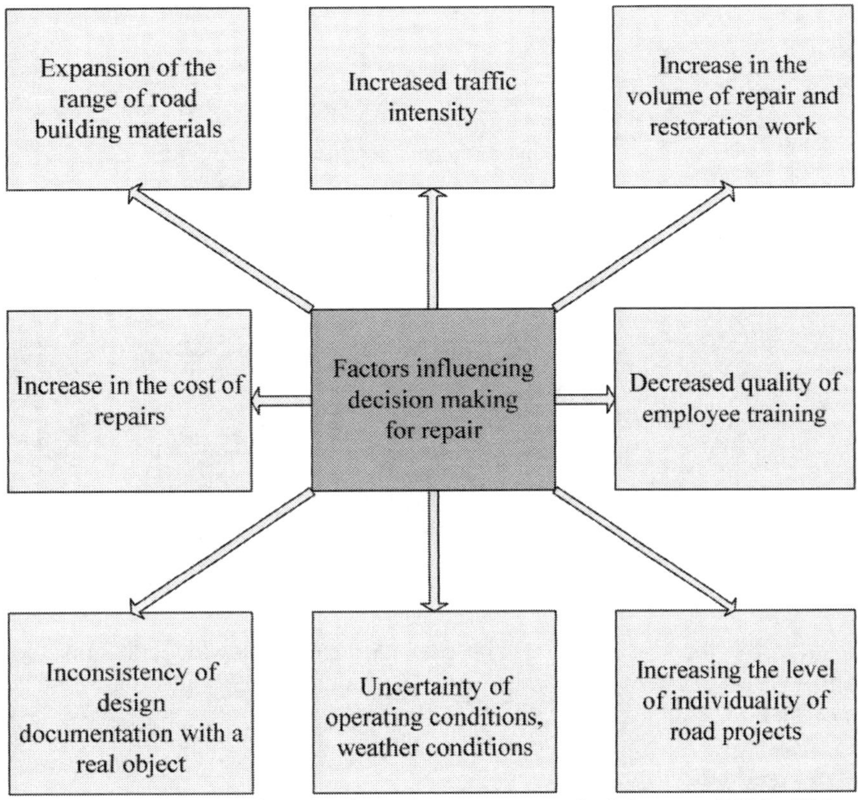

Figure 1. Factors affecting the adoption of management decisions on RN facilities repair

After repair, the roads must be in operation for 12-15 years. However, service life until the next repair is 3-5 years, since during the repair the technology may be violated, low-quality materials are used, or wrong repair method is chosen [5, 9].

It should be admitted that at present, financial resources amount allocated to carry out repair and restoration work at RN is insufficient, as well as the existing tools to support rational decisions choice [2,6]. With its use, it is not possible to determine the volume, timing and sequence of repair work, to investigate reasons for plans non-fulfillment and to assess work performed quality. In the course of work planning, there is a subjective factor, and employees competence is far from perfect. Maintenance works are characterized by large amounts of financing, a large number of subcontractors, various regulatory documents presence, a wide range of road construction, and production processes technological complexity. Due to these circumstances, it is necessary to clearly plan, organize the work and evaluate their quality. At the same time, well-grounded decisions adoption that ensure repair work implementation at RN facilities in the shortest possible time, for minimum money and with maximum quality is a multi-criteria optimization problem, the effective solution of which is not possible without attracting modern information technologies and DSS. Another circumstance that significantly complicates the decision-making procedures is the lack of subject area digital model, which should take into account all variety of different information about the object.

Structure of developed DSS [2,15,17] includes the following main components (Fig. 2):

IOP Conf. Series: Materials Science and Engineering 1159 (2021) 012025 doi:10.1088/1757-899X/1159/1/012025

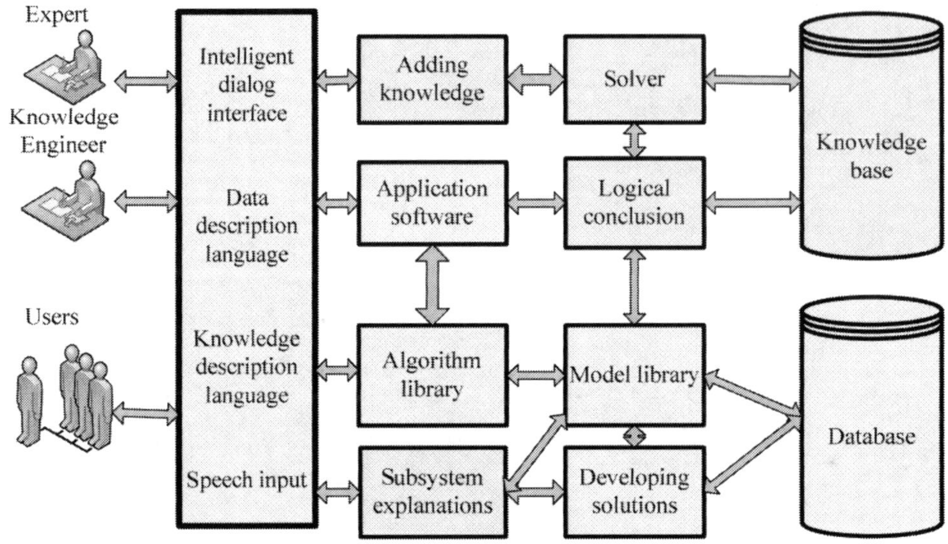

Figure 2. DSS architecture for road network facilities repair

DSS allows you to increase knowledge base by adding new knowledge without structure changing, as well as to encapsulate new knowledge into the system, increasing intellectual resource used in decision-making process.

2. Models and methods

Intelligent decision making is an iterative, multilevel procedure. First, a decision table is analyzed for assessing RN elements technical state, created on the basis of regulatory and technical documentation analysis. The situation is described in knowledge description language (KDL). Then situation description is expanded by taking into account the quality of RN roadway markings, traffic intensity, road building materials volumes, repair work cost, traffic management, seasonality and weather conditions, etc. The number of considered parameters and levels of their description in the system is not limited. Thus, the system provides situational adaptive intelligent control with ability to issue reasonable recommendations for carrying out repair and restoration work (Fig. 3).

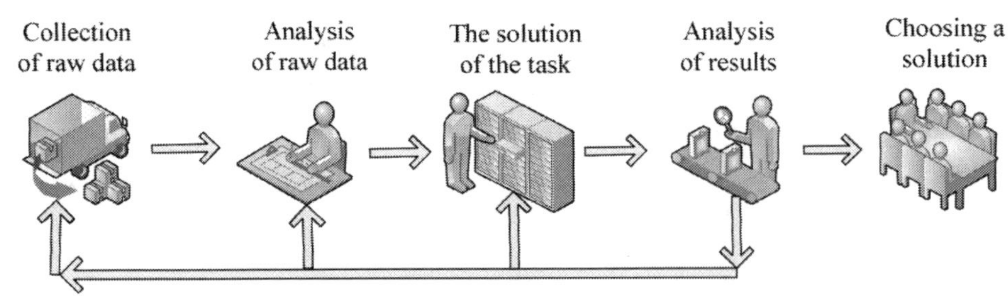

Figure 3. Iterative procedure for finding and choosing a solution

Road operational state is characterized by indicators vector (list of defects) $X = \{X_i\}, i = 1, \ldots, 10$. The X vector elements are measured by a certified mobile road laboratory [3,4] (Table 1).

IOP Conf. Series: Materials Science and Engineering 1159 (2021) 012025 doi:10.1088/1757-899X/1159/1/012025

Table 1. Classifier of detected roadway objects

№	List of objects and their designation		Object characteristics
1.	Separate crack	SC	Arbitrary location and length on roadway
2.	Crack mesh	CM	Multiple cracks of arbitrary location and length on roadway
3.	Pothole	PH	Recess on carriageway with ragged edges
4.	Repair card	RC	A patch on roadway, laid during the previous repair
5.	Spike	SP	Joining old and new road surfaces
6.	Expansion joint	EJ	
7.	Artificial unevenness from asphalt concrete	AA	Speed bumps made of asphalt concrete
8.	Artificial unevenness of collapsible type	AC	Speed bumps, assembled from standard elements
9.	Manhole hatch	MH	City utilities well cover
10.	Water intake grate (82x40)	WI	Large size storm drain element
11.	Water intake grate (30x30)	WG	Small storm drain element
12.	Carpet	CR	Gas service inspection hatch
13.	Core	CE	Circular cut for testing pavement materials
14.	Road markings	RM	Marking applied to roadway (sign, solid line, dotted line, pedestrian crossing, bus stop, double line, stop line, vehicle parking)
15.	Outside of roadway	OR	Helper class that contains areas in which you do not need to detect objects
16.	Background	BG	Auxiliary category that contains empty or undetectable objects

Based on examination results, a decision is made on damage severity and RN elements state assessment (Table 2).

IOP Conf. Series: Materials Science and Engineering 1159 (2021) 012025 doi:10.1088/1757-899X/1159/1/012025

Table 2. Technical condition evaluation of RN elements

Damage types		I Normal (Y1)	II Satisfactory (Y2)	III Unsatisfactory (Y3)	IV Emergency (Y4)
1	2	3	4	5	6
Strength	X1 – Separate crack (SC)	Distance over 40m.	Distance 6-40m.	Distance 3-6m.	Distance up to 3m.
	X2 – Crack mesh (CM)			Area up to $10m^2$	Area over $10m^2$
	X3 - Pothole (PH)		Single potholes at distance more than 20m.	Single potholes at distance more than 10-20m.	Single potholes at distance up to 10m.
Objects	X4 – Repair card (RC) X5 - Spike (SP) X6 – Expansion joint (EJ) Artificial unevenness from asphalt concrete (AA) Artificial unevenness of collapsible type (AC) Manhole hatch (MH) Water intake grate (82x40) (WI) Water intake grate (30x30) (WG) Carpet (CR) Core (CE)				
Road markings (RM)	X7 - Marking line deviation		Up to 2 cm in length and 0.5 cm. in width	Up to 5 cm in length and 1 cm. in width.	More than 5 cm in length and 1 cm. in width.
	X8 - Marking height	-	No more than 4 mm. Reflectors no more than 10 mm	No more than 6 mm. Reflectors no more than 15mm.	More than 6 mm. Reflectors more than 15 mm
	X9 - Wear and tear of thermoplastic markings (at the control section 50 m)		Up to 15%, and paint markings wear up to 35% of its area.	Up to 25%, and paint markings wear up to 50% of its area.	Exceeds 25%, and paint markings wear up to 50% of its area.
	X10 - Not demarcated		Linear size of the old marking track is up to 2 cm.	Traces are poorly visible or do not match the newly applied markup	Traces are clearly visible or do not match the newly applied markup

Based on this table, control actions matrix is formed to assess RN elements technical state (matrix $\boldsymbol{R}$) with elements $\boldsymbol{R_{ij}}, \boldsymbol{i = 1, \dots, 10, j = 1, \dots, 4}$.

These parameters serve as the basis for rules formulation in decision support system for determining the volume, timing, cost and sequence of repair and restoration work and allow formulating requirements for measurement parameters for mobile laboratory [1,7,15].

The following rules can be cited as examples of such rules

1. IF track has a depth more than 40 mm, THEN mill the top layer and lay a new one, ELSE carry out patching.
2. IF roadbed has collapsed, THEN restore pavement in places where roadbed is repaired, reinforce the shoulders, ELSE to carry out major repairs.
3. IF pavement width does not correspond to standards, THEN make expansion of roadbed, ELSE carry out major repairs.
4. IF coating does not comply with standards, THEN restore coating, ELSE to carry out repairs using bitumen-mineral materials.
5. IF pavement on reinforcement lanes and shoulders does not meet standards, THEN restore pavement, ELSE take measures to maintain roadbed.
6. IF there is surface damage, THEN it is necessary to protect asphalt concrete pavements from surface damage with sealing impregnating materials, ELSE apply a new mixture or put repair cards [8].
7. IF adhesion properties do not correspond to standards, then it is necessary to restore coating adhesion properties by means of protective layers, wear layers or surface treatment, ELSE to replace coating.

In the Russian Federation, basis for road repair is its emergency condition. This condition is preceded by a road maintenance phase. It is at this life cycle stage that defects appearing on coating are eliminated by patching, in which coating defects have an absolutely awkward irregular shape and there are a lot of them. Determining volumes of even a single pit accurately enough is a very difficult task.

When evaluating RN object, state vector elements $\boldsymbol{Y} = \{\boldsymbol{Y_j}\}$ can take four values: $\boldsymbol{Y_1}$ - normal, $\boldsymbol{Y_2}$ - satisfactory, $\boldsymbol{Y_3}$ - unsatisfactory, $\boldsymbol{Y_4}$ - emergency. Each state of vector element $\boldsymbol{X} = \{\boldsymbol{X_i}\}$ (list of defects) must correspond matrix element $\boldsymbol{R_{ij}}, \boldsymbol{i = 1, \dots, 10, j = 1, \dots, 4}$. Based on elements analysis $\boldsymbol{R_{ij}}$ DSS makes one or another control action (decision) $\boldsymbol{U_k}, \boldsymbol{k = 1, \dots, m}$. (Fig. 4).

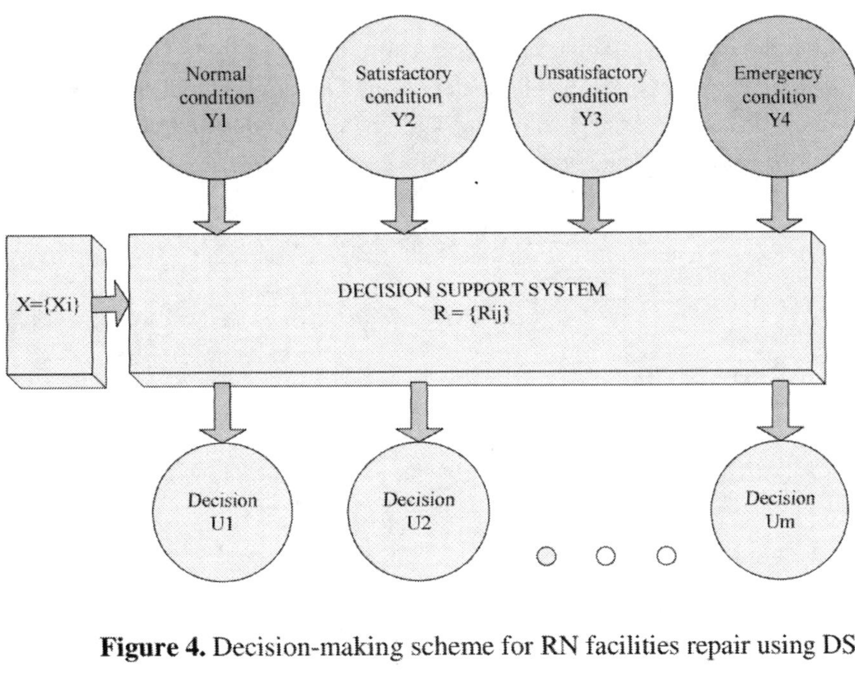

Figure 4. Decision-making scheme for RN facilities repair using DSS

The main task of decision maker is not to implement labor-intensive calculations of scheduling and forecasting the situation, but to solve weakly formalized problems. These tasks include forming goals tasks, creating conditions for goal achieving $\boldsymbol{C}$, as well as alternative solutions selection $\boldsymbol{U} = \{\boldsymbol{U_k}\}, \boldsymbol{k} = \boldsymbol{1}, \ldots, \boldsymbol{m}$. The role of decision makers is especially great at requirements analyzing stages for the n-th RN facility using regulatory framework, determining type of repair and restoration work, and also including the site in title list.

From the management theory point of view, almost any production activity, including repair work planning at RN object, is a controlled process, the purpose of which is to achieve a given goal:

$$y_{nl} = F_{nd_{v-1}}\left[t; C_n; y_{n1}, y_{n2}, \ldots, y_{nl_n}; x_{nr}; r_n; u_{Ynd_{v-1}}\right], \qquad (l = 1, .., L_n) \tag{1}$$

where t – current time

C_n – given target of n-th process;

$Y_{nd_{(v-1)}}(t) = \left[y_{n1}(t), y_{n2}(t), \ldots, y_{nl_n}(t)\right]$ - preceding states values vector;

x_n - parameters of n-th process that specify processes characteristic properties;

r_n - dedicated resource for n-th process implementation;

$u_{Ynd_{(v-1)}}$ - control action in a local time interval $d_{(v-1)}, (v = 1, \ldots, M_n)$.

As an initial option, a baseline plan for repair and restoration work is drawn up with an appointed start date $d_{(v-1)} = d_0$:

$$P_{Ynd_{(v-1)}}(t) = \left\{C_{pn}; P'_{Ynd_{(v-1)}}(t); x_{pn}; r_{pn}, \chi_n; K_n\right\}, \tag{2}$$

where C_{pn} – planned goal $P_{Ynd_{(v-1)}}(t)$. It is different from the original goal C_n.

$P'_{Ynd_{(v-1)}}$ - detailed repair program for the moment $t_h = t_{d0}$;

x_{pn} - n-th process parameters;

r_{pn} - planned process resources;

χ_n - process development restrictions;

K_n – some aggregated criterion of planned and real processes compliance.

Preparation phase actions can be represented by functionality [20]

$$\Theta: \theta\left[Y_{nd_0}(t), t\right] \rightarrow P_{Ynd_0}(t), \tag{3}$$

which defines base plan $P_{Ynd_0}(t)$, corresponding to process.

Implementation phase involves some process transformation $Y_{nd_{(v-1)}}(t)$ в $Y_{nd}(t)$ due to control actions occurrence $u_{Y_{nd}}$ to corresponding parameters x_n and resources r_n followed by process initiation $Y_{nd_{(v-1)}}(t)$ at moments $t_{d_{(v-1)}}, (v = 1, \ldots, M_n)$. The result of controlled process is control actions formation u_{Ynd}. Transition from one state to another is accompanied by actions sequence execution (management activities) m_n, tied to time intervals:

$$P_{mnd_{(v-1)}} = P_{mn}\left[m_n; r_{vn}; T; t_d^k; d_{(v-1)}\right] \tag{4}$$

where m_n - managed process identifiers;

r_{vn}- n-th process management resource ($r_{vn} \subset r_n$);

T - calendar periods;

t_d^k - scheduled date d for m_n;

$d_{(v-1)}$ – last date of plan revision.

At the moment $t_h = t_{d0}$ the control action (CA) u_{nd} is formed, taking into account restrictions χ_n and criterion K_n, included in general plan $P_{nd_0}(t)$:

$$\Psi_m: \psi_m\left[Y_{nl}, P_{mnd_{(v-1)}j}, P_{Ynd_{(v-1)}}, d_v\right] \rightarrow u_{nd}. \tag{5}$$

Thus, control action leads to plan adjustment:

$$\Lambda_p: \lambda_{pd_v}\left[P_{Ynd_{(v-1)}}(t), u_{pnd_v}\right] \rightarrow P_{Ynd_v}(t). \tag{6}$$

In the course of planning, the decision maker makes constant adjustments to the baseline plan. All managed activities $Y_{nd_{(v-1)}}(t)(n = 1, \ldots, N)$ are under his control, as well as schedule drawing up stages $P_{Ynd_{(v-1)}}(t),. (n = 1, \ldots, N)$. Master repair plan $P_m = \cup_n P_{mnd_{(v-1)}}$ is managed action plans consolidation [12]. Planned events manager $P_{mt^k_{d_v}}$ is a schedule of scheduled events for date d and is generated based on projection $P_{m\tau}$ for a given date d_v. In addition, the task is to reconcile interrelated activities whole set $d_v \times P_{m\tau} \rightarrow P_{m\tau d_v}$ by time $H: \eta[P_{m\tau d_v}; t^k_{d_\tau}] \rightarrow P_{mt^k_{d_v}}$.

In the course of planning, decision-maker evaluates all sub-processes characteristics [18] and forms an aggregated assessment of state. Develops control actions and makes the necessary adjustments to repair and restoration work baseline plan [18]. That is why efficiency and very possibility of DSS using during planning depends on decision maker qualifications. The same high requirements are imposed on experts qualifications involved in DSS development [14].

3. Results

Planning repair and restoration works process at RN consists in sequential implementation of the following steps [20] (Fig. 5):

1. Drawing up a defective statement based on RN diagnostics results using a mobile road laboratory [2,3]. Diagnostics results report contains objective and accurate information about the current state of RN facilities and is the basis for making decisions on title list formation.
2. Requirements analysis for RN i-th section using regulatory framework. Required design speed of traffic, road throughput, traffic safety, ability to pass road trains with a permitted mass and axle load are considered as requirements. It is their non-observance that leads to need for repair [10].
3. Determination of safety factors for design speed and transport and operational indicator KP_n according to MADI method [8].
4. In the case of a limited amount of funding, transport and operational indicator is adjusted downward.
5. Choice of the type of repair and restoration work is made: maintenance, repair, overhaul [9]. Based on sums values analysis R_{ij} by matrix rows R.
6. The calculation of repair and restoration work annual volume is carried out taking into account indicators complex improvement K_{pI_j} [8].
7. In case of improvement in transport and operational indicator value, this object is included in title list.
8. These actions are repeated for all RN objects, as a result of which an annual plan for carrying out repair and restoration work at RN is formed.

Each defect has its own description in automated road database (ARD). For description, a different set of attributes is used, determined by defect type (Table 3), and information on them is also entered into ARD. ARD records information on RN diagnostics results, data on traffic intensity and road traffic accidents, as well as information on repair measures taken.

Work planning according to movement estimated speed criterion (MADI method) [8] is carried out according to diagnostics results by determining estimated speed safety coefficients K_{pcij}, where *i*- road section, and *j*- number coefficient. Calculated coefficients are compared either with the specified values of road transport and operational state indicator KP_n (when assessing road technical level indicators), or with its maximum permissible values (when assessing road operational level indicators). In limited funding, this indicator is considered on terms of normative up to the limit values: $KP_{tp} = (0.5 - 1.0)KP_n$. If coefficients calculated values are $K_{pcij} < KP_n$, then measures for RN section maintenance or repair are used.

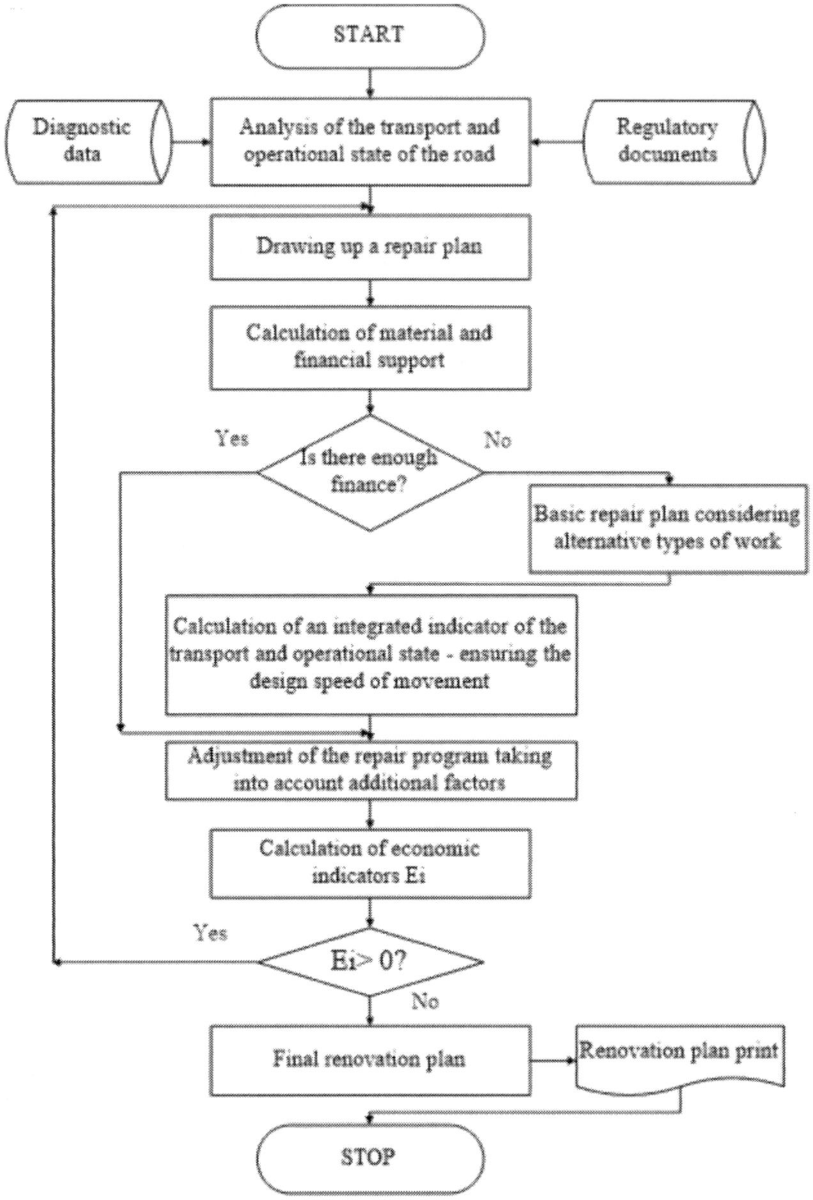

Figure 5. General scheme of planning algorithm for repair and restoration work on RN

In process of planning road repair work, it is taken into account that repair work performed must be carried out in comprehensive manner [8]. By increasing pavement strength, pavement flatness is also improved at the same time due to leveling layers laying. That is, as a result of this work, the influence of $\boldsymbol{K_{pc6}}$ and $\boldsymbol{K_{pc7}}$ will be eliminated. Simultaneously with influence elimination of individual jointly acting factors $\boldsymbol{K_{pcij}}$ on the *i*-th characteristic road section during road repair work, indicators can be partially improved, which is important for effectiveness assessing of assigned road repair work. The final choice of work types and volumes is carried out depending on selected financing conditions, according to transport effect size, based on increase in the complex indicator of transport and operational state of road before and after repair.

Transport effect is calculated from difference in transportation cost before and after repair at traffic flow average speed, calculated taking into account estimated speed provision coefficient $\boldsymbol{K_{pcij}}$ [8]. With sufficient funding, the transport effect value on goods and passengers transportation is taken as a criterion for work priority. With limited resources, there is a need for rational distribution of annually allocated funds for road repaired sections. Repair work type and sequence is determined according to criterion that takes into account differences between repair work performed in terms of overhaul life. First of all, they correct those road parameters that contribute to the greatest reduction in transport costs per unit of funds invested in road section repair or reconstruction, avoiding additional costs due to road underrepair.

Finally, option with maximum value of actual integrated transport and operational indicator $\boldsymbol{KP_{f(after)}}$ for considered road sections or road network as a whole is selected. Based on calculations results, roads title list being repaired or reconstructed, provided with allocated amount of financing, is compiled, the implementation of which gives the greatest transport effect to road users (Fig. 6).

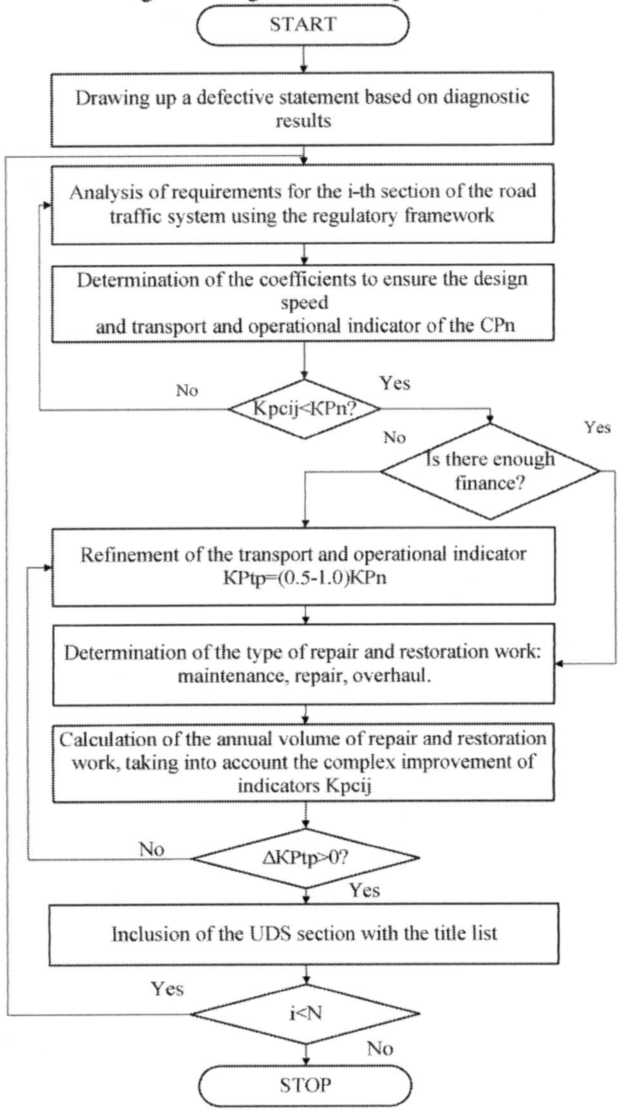

Figure 6. Algorithm diagram for calculating complex indicator of transport and operational state - ensuring design speed of movement

IOP Conf. Series: Materials Science and Engineering 1159 (2021) 012025 doi:10.1088/1757-899X/1159/1/012025

4. Results discussion

RN repairing is a complex task, in solving which it is necessary to take into account a number of conflicting criteria. In this regard, decision-making process on repair measures rational organization, their linking in order implementation and in time. A competently drawn up plan for carrying out repair work should offer the best option for organizing traffic to eliminate traffic jams and congestion on roads. When organizing repairs, not all criteria are purely economic (downtime in traffic jams, transport and environmental safety), so it is not possible to form a single criterion. As criteria, one can consider repair work cost, necessary resources, repair work time, and their quality. It is most expedient to use stepwise optimization method, when the best value is searched for according to one criterion with fixed values of all other control variables.

Decision-making process is a complex mental activity, consisting in choosing the most effective strategy and, at present, is carried out using DSS, that is, a class of knowledge-based systems [11]. Intelligent decision making is an iterative, multilevel procedure. First, a decision table is analyzed for assessing RN elements technical state, created on analysis basis of regulatory and technical documentation. The situation at KDL is described. Then situation description is expanded by taking into account RN road marking quality, traffic intensity, road construction materials volumes, cost of repair and restoration work, traffic management, seasonality and weather conditions, etc. The number of considered parameters and levels of their description is not limited. Thus, the DSS proposed version provides situational adaptive intelligent control with ability to issue reasonable recommendations for repair work.

System allows you to simply build up knowledge base by adding new knowledge without changing structure and to encapsulate new knowledge into the system, increasing intellectual resource used in decision-making process.

The main advantages of proposed DSS are:

1. Original KDL presence, namely, the language of declarative knowledge representation, close to natural language. This language is a universal tool for working with knowledge. Descriptions on KDL are in proposals list form related to any object. The list of possible sentences and their composition are specified in domain model.
2. Subject area objects are mapped according to semantic rules, into the subject area model, which ensures their semantically correct representation, independent of applied algorithms.
3. Knowledge base (KB) does not require additional information interpretation, which allows you to work with information directly using an intelligent dialog interface. This leads to the fact that any changes made in domain model are immediately available to users.
4. Graphic elements library presence provides management of KB content and allows you to create user interface for solving specialized tasks.
5. Standard solutions library availability [14] - domain model fragments that can be used in various applied problems. Fragments of created domain model for one task can be successfully used when creating a model for another task. This makes it easy to create new applied solutions from existing solutions fragments.
6. Availability of associated procedures library [16] - KB applied logic is built as a set of procedures associated with KB specific entities. The associated procedures are called automatically when accessing domain model and procedures results are also entered into the model. This ensures that procedures are isolated from each other and that you can easily change logic of any procedure without having to adjust other procedures. Associated procedures can be written in various programming languages.
7. Availability of domain model constructor - allows you to create and change domain model in accordance with new requirements for solving a problem during KB operation entire time.
8. Developed universal Internet interface - a web client that provides work taking into account web technologies specifics.

5. Conclusion

Created DSS, built in accordance with OMSD methodology (Object Model for System Design) [13], is a universal tool for creating rational plans for carrying out repair work at RN facilities. Its development will make it possible to apply artificial intelligence methods in promising field of decision-making for repair RN facilities, based on diagnostic results obtained by mobile road laboratories.

DSS introduction in road repair planning field will increase technological efficiency of repair activities, analyze defects using the knowledge base, make an informed choice of road construction materials and technologies, and manage repair work. Digitalization of repair activities management, digital transformation of management model and business processes in the field of RN repairs will increase profits, increase operational efficiency, reduce repair costs and improve their quality.

References

[1] Badalyan N P, Lyublinsky M S, Chashchin E A, Shemanaeva L I 2018 Hardware complex for information support system for making standard management decisions *Bulletin of the National Polytechnic University of Armenia. Electrical engineering, power engineering* 2018 **1** 32-45

[2] Boyarkina O O, Shkalikova A A 2016 Intelligent decision support systems *Modern research and innovation* **12** Available at: http://web.snauka.ru/issues/2016/12/75361

[3] Vasiliev Yu E, Belyakov A B 2008 Automobile road scanner "ADS-MADI" *Science and technology in the road industry* **2** 10-11

[4] Vasiliev Yu E, 2008 The future of diagnostics belongs to mobile laboratories *Science and technology in the road industry* **1** 3-7

[5] GOST R 50597-93 "Highways and streets. Requirements for operational state permissible under the road safety conditions of"

[6] Danilevich S B, Kolesnikov S S 2008 On the choice of control results reliability indicators *Legal and Applied Metrology* 2008 **2** 48-51

[7] Krylov M A, Kurganov K I, Chashchin E A Hardware and software complex for monitoring, modeling and predicting road surface condition and wear *Internet magazine "Transport structures"* 2018 **1** Available at: https://t-s.today/PDF/04SATS118.pdf DOI: 10.15862 / 04SATS118

[8] ODN 218.0.006-2002. Sectoral road rules for diagnostics and assessment of highways state (instead of VSN 6-90). Basic provisions. Date of issue 03.10. 2002.

[9] Highways acceptance diagnostics *DorTekhProekt+* Available at: http://dtproekt.ru/index.php/14-types-of-work

[10] Order of the Federal Highway Agency of June 27, 2002 N OS-568-r "On the implementation of" Methodological recommendations for asphalt concrete restoration of pavements and road bases by cold regeneration means"

[11] Skobelev P O 2013 Intelligent resource management systems in real time: development principles, industrial implementations experience and development prospects *Supplement to theoretical and applied sci-tech journal "Information Technologies"* **1** 1-32

[12] Baker K R, Trietsch D 2013 Principles of sequencing and scheduling (John Wiley & Sons)

[13] Booch G, Maksimchuk R, Engle M, Young B, Conallen J, Houston K. 2007 *Object-Oriented Analysis and Design with Applications* 3rd Edition (Addison-Wesley Professional) ISBN-13: 978-0-201-89551-3

[14] Burkov V N et al. 2015 Introduction to theory of control in organizations (CRC Press)

[15] Lioublinskiy M S, Shemonayeva L I, Kotov V V, Chashchin Ye A 2017 Information Support for Making Standard Management Decision with the Help of Hardware-Software SystemAdvances in Economics, Business and Management Research Vol 38 410-414

[16] Labadie N, Prins C, Prodhon C 2016 Metaheuristics for Vehicle Routing Problems (John Wiley & Sons)

[17] Leitao P, Colombo, A W, Kamouskos S 2016 Industrial automation based on cyber-physical systems technologies: Prototype implementations and challenges *Computers in Industry* Vol 81 11-25
[18] Ovacik I M, Uzsoy R 2012 Decomposition methods for complex factory scheduling problems (Springer Science & Business Media)
[19] Pinedo M 2016 Scheduling. Theory, Algorithms, and Systems (Springer)
[20] Radhika V Mahalle, Priyanka V Narkhede, Priya A Lokhande, Reetu M Mundane 2018 Application for Smart Bus Monitoring *International Journal of Innovations in Engineering and Science*, Vol 6 3 92-96. e-ISSN: 2456-3463

Development of a database for information support of virtual digital analysis in design and testing of road infrastructure elements

I V Demyanushko[1], L F Samigullin[1], N E Surkova [2]

[1,3]Department of Construction Mechanics, Moscow Automobile and Road State Technical University (MADI), 64, Leningradsky pr., Moscow, 125319, Russia
[2]Department of Automated Control Systems, Moscow Automobile and Road State Technical University (MADI), 64, Leningradsky pr., Moscow, 125319, Russia

E-mail: demj_ir@mail.ru

Abstract. Ensuring traffic safety on the roads and reducing the number of deaths due to road accidents is a priority task both in Russia and in the global community. Assessment of safety in the event of vehicle crashes, oncoming lane exits, collisions with road barriers and other elements of the road infrastructure (ERI) requires the use of a wide range of modern and innovative information technologies - special environments for automated design of Digital Twins of road infrastructure elements, virtual tests with collision modeling and obtaining results on different input parameters of the simulation, modern automated testing of the road infrastructure elements. All these practical works, research and development are carried out at MADI together with scientific team LLC M&PQ to improve road safety. The article deals with the prospects of development of these works, a natural transition to another level of research. Based on the intellectual technologies in the field of road construction, work has begun on creating a special information environment, the so-called "Virtual Digital Proving Ground", and, among other things, to provide it, a database (DB) of computer-aided design and virtual testing of digital models ERI, which will improve the efficiency and quality of work. The use of the ERI database will allow to implement the automated control of the developed ERI digital models, to implement tools for designing new innovative structures, to predict the behavior of accident participants, to develop reasonable standards in the field of road barriers of the future, to carry out examination by requests of various organizations.

1. Introduction

Russia's economy today faces the most complex task of transition to a fundamentally new technology of "digital economy" in terms of its scale and content, the fundamental feature of which is the unification of material and digital (virtual) worlds. Solving this global problem requires the development of fundamentally new business processes, automation and intellectualization in all sectors of the economy, leading to a systematic approach to solving the problems of the economy as a whole. A fundamentally different approach to the development of the country's transport system requires priority development - the industry faces the challenge of developing, implementing and operating a single intelligent transport system that will provide domestic companies with effective and safe interaction to create products and services that are competitive on the world market.

Content from this work may be used under the terms of the Creative Commons Attribution 3.0 licence. Any further distribution of this work must maintain attribution to the author(s) and the title of the work, journal citation and DOI.
Published under licence by IOP Publishing Ltd

The implementation and functioning of a single intelligent transport system includes the solution of many tasks, we will focus on the most important ones.

"Road digitalization" requires the development of special information support that will allow to store data for the description of the location, size, construction and technical characteristics of roads, their structural parts, elements of engineering and artificial structures, as well as their changes during the life cycle of the road. Such direction can be conventionally called GIS-modeling (GIS - geoinformation system - a system for collection, storage, analysis and graphical visualization of spatial (geographical) data and related information about the necessary objects) [1].

The second direction, which has been widely developed in recent years, is the evolution of computer-aided design (CAD) systems into the concept of information modeling of capital construction objects (Building Information Model - BIM) [2]. BIM technology is understood as such an approach to the life cycle of the modeling object, at which the information model gathers and stores all the necessary design, technological, economic and other necessary information about the components of the interrelated elements of the modeling object. When using BIM, information about the model allows one to automatically create drawings and reports to perform analysis of the project, manage the operation of the object, etc., providing unlimited opportunities for management decisions [3].

Both directions require the development of special information support for solving the problem of storing large volumes of both structured and unstructured data in one way or another. Just as the very concept of a single transport system includes a set of components interconnected with each other, but which can be considered and studied separately, so the information support is more effectively developed for individual system components with the possibility of further integration into a single information space of a single intellectual transport system [4].

Since 2010, scientific and practical work has been carried out in the Moscow Automobile and Road State Technical University (MADI) by the staff of the Department of Construction Mechanics and later, by a small innovative enterprise (LLC M&PQ) - on the problems of road safety regarding elements of road infrastructure (ERI), which the safety in case of traffic violations depends on. These are various types of road safety barriers, lighting supports, signs, advertising supports, crossings, acoustic panels, anti-tank devices etc. In a relatively short period of time, the MADI team, together with LLC M&Q, developed a methodology for virtual computer analysis of crash situations and crash-tests when vehicles collide with road infrastructure elements, as well as the digital models of wire rope safety barriers, concrete safety barriers, frontal barriers, mobile frontal barriers and other types of road infrastructure elements. More than one thousand virtual tests (crash-tests) of specific ERI objects were carried out on orders of industry enterprises. Methods and scenarios of virtual testing were developed in accordance with the relevant regulatory documents [5]. Virtual testing is now an increasingly important part of the road safety system. The application of virtual testing in assessing compliance with safety requirements, such as road barriers, has now become part of EU, US and EurAsEC road barrier standards. Our accumulated experience makes it necessary to switch to new innovative information technologies in the field of road safety through the automation of ERI design and testing. Such technology should be the development of a special intellectual environment - the information system (IS) of the Virtual Digital Proving Ground for crash testing against vehicle restraint system. This article deals with the development of a database (DB), which should become the basis for the information support of the "Virtual Digital Proving Ground for crash testing against vehicle restraint system".

2. Models and methods of information support development for the "Virtual Digital Proving Ground".

To develop the structure of the database it is necessary to perform the following actions:

- to set design requirements,
- to analyze and formalize these requirements in the form of development tasks;
- to build an information model of the process of performing these tasks in reality.

Tasks, which the developed database should provide with all necessary data:

- development and analysis of new digital ERI models (DM),

- modeling of new digital models of road infrastructure elements with the ability to vary the elements of design and materials;
- comparative analysis of virtual and field test results;
- change of test conditions due to variation of parameters;
- prompt control over compliance of virtual tests performed with regulatory documents and developed methods;
- verification and validation of digital models and test results.

To build an information model of the process, we will call it the "Virtual Digital Proving Ground", we will use the Business Process Management Notation (BPMN) method, which has proved to be a perfect method for improving operational activity and visualizing the process in conjunction with explicit business rules and operational and technical requirements. Figure 1 shows the resulting model.

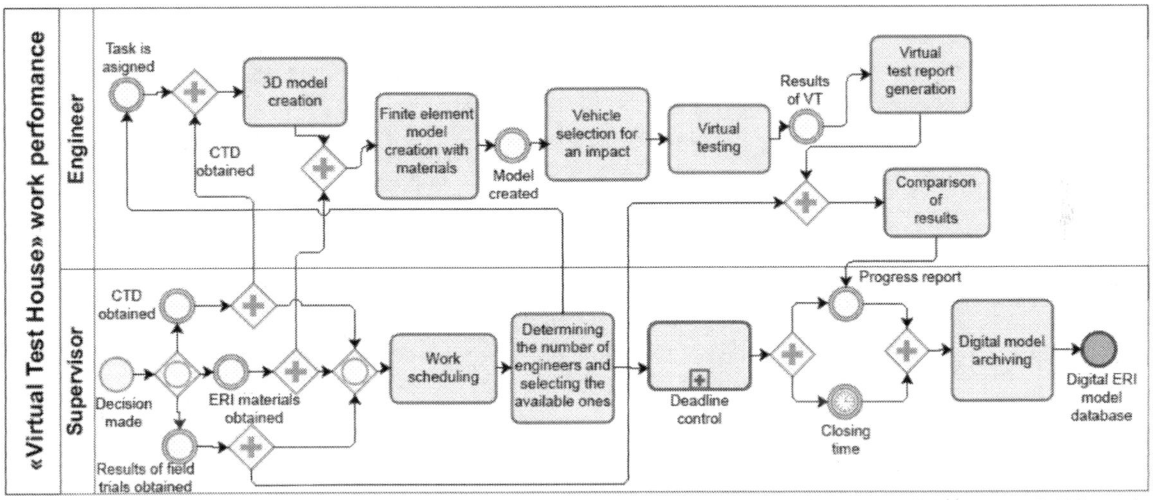

Figure 1. Information model of the main process of virtual testing with the help of the created "Virtual Digital Proving Ground"

Creation and analysis of the obtained model allows to reveal all necessary data to support the process and to form the database structure, which should store the description of information objects (Table 1). In the course of database design there was a problem of organizing storage of some data. When describing entities, it is necessary to define domains of each attribute. For the majority of attributes it is simple enough to do it, but a part of entities have attributes which should store design and drawing documentation of different types - construction and technical documentation, 3D-model in the accepted code, specialized files with extension .k. Examples of such data are shown in Figure 2.

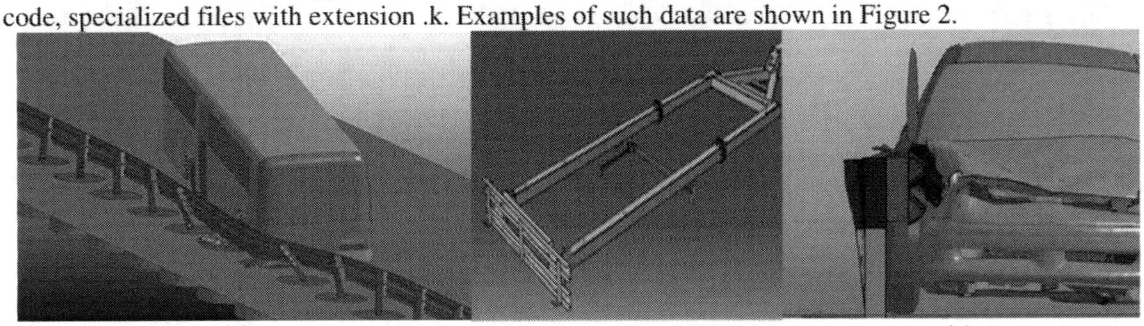

(a) (b) (c)

Figure 2. Examples of data fragments for database storage: (a) - temporary fragments (or video) of the bus hitting the road barrier - file with the extension.k; (b) - 3D model in code, e.g. Solid Works, mobile frontal barrier (MFB); (c) - Fragments of a movie (video - file) showing a car hitting a road barrier during virtual tests - file with the extension .k

IOP Conf. Series: Materials Science and Engineering 1159 (2021) 012026 doi:10.1088/1757-899X/1159/1/012026

Table 1. Description of entities

Entity Identifier	Description
Staff	Engineers and managers at different levels, other employees (under contract)
Contract	Contract with customer; Order from a counterparty; Internal task from laboratory manager to engineer; Supplement to the contract
Customers	Customer companies Counterparties Executors (as a customer)
Elements	Element (part) of a road barrier or other element of road infrastructure, for which there is a digital model
Barriers	Road design of any class, consists of the elements provided by the customer
Vehicle	A vehicle that is used as a collision vehicle in tests
Road	These roads (surface material, climatic conditions) in test
Materials	Materials of the elements according to the customer's standard and/or laboratory tests data
Tests	Virtual or field test data (list by test laboratory maintenance station)
Application	External document from the customer, on the basis of which the contract (laboratory form) is created
ERI	Elements of road infrastructure, except for barriers
Method	Methods of virtual testing (laboratory maintenance station)
Parameters	Calculation parameters (list of boundary conditions for the laboratory maintenance station)
Protocol	Laboratory document (form) for the performed tests of one contract and one barrier or ERI
NormDoc	Normative documents, according to which virtual tests are conducted (GOSTs, industry-specific road guidance documents etc.).

Other entities have attributes where scans of important legal or technical documents, which are files with the extension .pdf, should be stored.

It is necessary to make a decision how to organize the storage of such files. DBMS SQL Server 2019 was selected as the target DBMS. In this DBMS it is more convenient and efficient to store files with the .pdf extension in a table with an attribute, the domain of which is varbinary (max) or image, but with the addition of an attribute that will store the pdf file name.

Organization of file storage with other extensions requires choosing one of the methods. There are the following ways to solve this problem:

1. Save such files in a certain file system location on the database server and a link to it - in the corresponding entity attribute (domain type - text or hyperlink). The approach under consideration has many flaws:

- files are not deleted when deleting the corresponding database record (link integrity is not supported);
- a problem occurs while trying to update a file;
- violation of synchronization between the database and the file system when rolling back a transaction (transaction integrity is not supported);
- when backing up and restoring information in the database, asynchronization with the file system may occur;
- files are not subject to access restrictions imposed by the database.

2. Store files directly in the database, in a field with a domain of BLOB type (image, varbinary(MAX)). With this way of storing files, the following happens:

- database volume increase,
- database fragmentation,
- lower data processing performance.

There are special procedures for solving this problem in different DBMS. Their use will solve the problem of reducing the speed of downloading a file from the database, but does not solve the problem of increasing the database volume.

3. SQL Server 2019 DBMS features for storing large volume files - FileTable based on FileStream:

- easy handling of file names
- two-way link integrity
- guaranteed transactional integrity.

To make a decision, it is necessary to define the following parameters and their priorities in the interaction of users with the database of such attributes:

- file volume - 500 kB-3.5 mB on average
- need to work with the file and make corrections in the creation environment
- importance of file download speed.

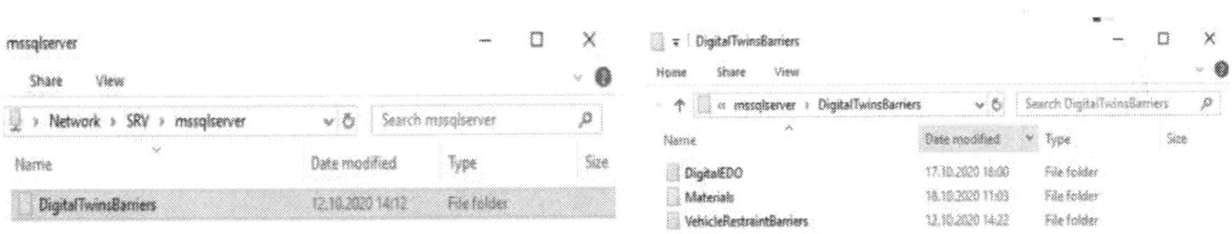

Figure 3. View of the file table directory on the server

Files of technical drawings, 3D-models, specialized files with extension .k are rather advantageously stored in file tables, thus creating a storage of digital models, managed both from the database side and from the application side. An example of a directory of such file tables created is shown in Figure 3.

After the decision to store large volume files in file tables and perform all stages of database design, the database was implemented, a part of its structure (mainly related to the road barriers) is shown in Figure 4 [6].

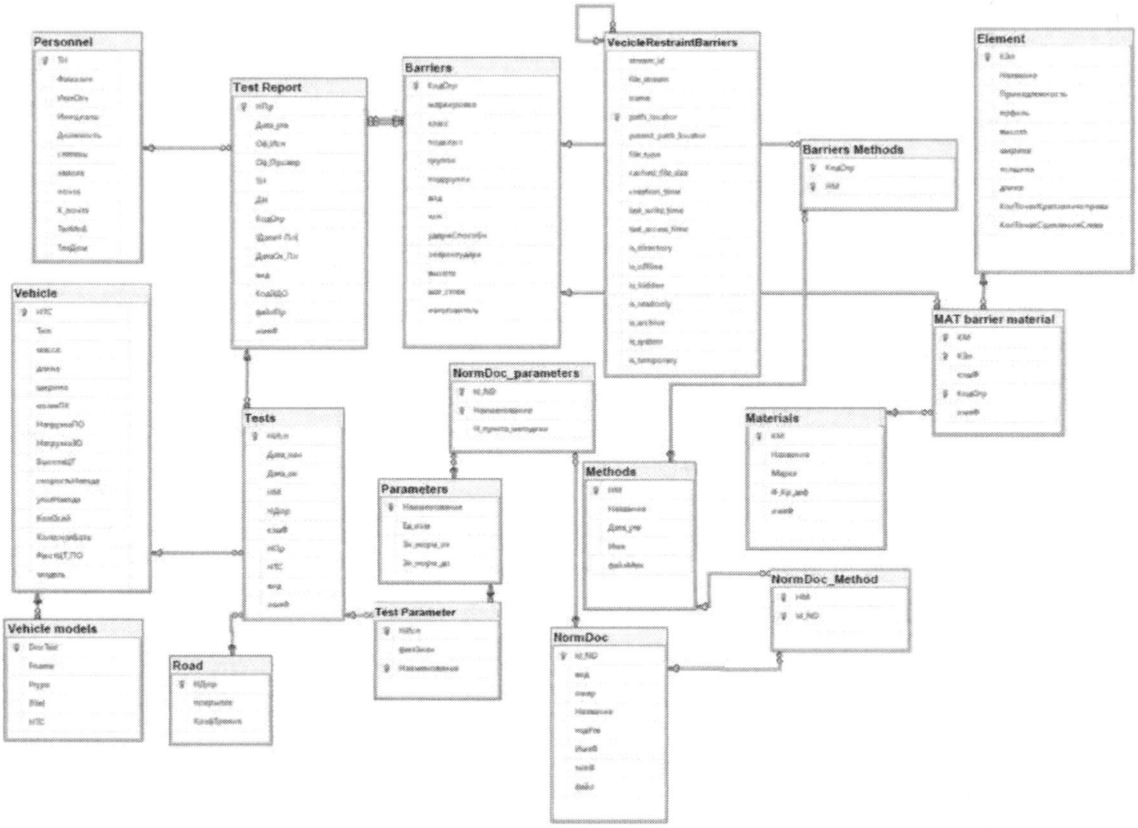

Figure 4. Structure of the "Holding side road barriers" database. (Vehicle Restraint Barriers (VRB))

3. Findings and results

The implemented database is the information basis for the creation of the "Virtual Digital Proving Ground" and will ensure the automation of all data handling processes. The advantages of working in such an intellectual environment are obvious: increasing the speed of search and processing of already existing 3D models, improving the performance and quality of new models created, conducting virtual tests, creating all necessary technical and report-related documentation and conducting validation analysis of the virtual results obtained.

The use of the "Virtual Digital Proving Ground" will increase the quantity and quality of the validation of digital models of road infrastructure elements, which gives us reason to consider the developed models as highly adaptive digital twins of ERI (ERI DT). The term "Digital Twins" is justified by a mandatory comparative analysis of the calculated parameters obtained during each virtual test with standard criteria. Necessity of such analysis of the received estimations of ERI digital twins is connected with the increased responsibility for the carried out ERI certification, which then provide safety of constructions installed on roads [7].

The important task, where the use of "Virtual Digital Proving Ground" possibilities plays a crucial role, is the task of reconstruction of road accidents. Digital simulation of possible collision situations of a car (initial speed, trajectory) on a road barrier and the variant analysis allow to reconstruct process of accident accurately enough and to define the reasons, a share of participation of a road barrier (or any element of a road arrangement), its elements and materials in absorption of energy of blow, indexes of damage for colliding cars and, what is not less important, for cars which have got on a way of its movement in a counter lane.

Thus, the "Virtual Digital Proving Ground" will also allow one to create a **digital twin of the accident** (DT) with a blow on the road barrier on the basis of complete information about the accident [8].

Development of experimental research methods for validation of virtual digital models aimed at damping studies in structural elements, construction of refined mathematical models of materials and structural elements, full automation of modeling processes, design and testing of EDI models allow us to consider the software product "Virtual Digital Proving Ground" as a BIM technology and to use it in the future as an integrated part of BIM-model of a single transport system, which will solve the most important problem of developing road safety systems with the help of modern innovative technologies.

4. Financial support

The work is largely facilitated by financing under the Development - Digital Technologies Program with support from the Foundation for Assistance to Small Innovative Enterprises in Science and Technology (Foundation for Assistance to Innovations) (№ 47GRCTS10-D5/56182), the Federal Road Transport Agency (Rosavtodor) of the Russian Federation supports the creation and processing of regulatory documents on ERI safety.

References

[1] Trofimenko Yu V, Yakubovich A N, Yakubovich I A et al. 2020. Modeling of Influence of Climate Change Character on the Territory of the Cryolithozone on the Value of Risks for the Road Network, International Journal of Online and Biomedical Engineering, Vol 16 **7** 65–74 https://doi.org/10.3991/ijoe.v16i07.14557

[2] Ostroukh A, Pomazanov A 2014 Realtime Development and Testing of Distributed Data Processing System for Industrial Company. Middle East Journal of Scientific Research, Vol 20 **12** 2184–2193 DOI: 10.5829/idosi.mejsr.2014.20.12.21106

[3] Akimov D, Krug P, Ostroukh A et al. 2016 The Simulation Model of Autonomous Truck Caravan Movement in Terms of an Extreme and Non-Stationary Environment. International Journal of Applied Engineering Research, Vol **11** 9 6435–6440

[4] Akimov D, Krug P, Ostroukh A et al. 2017 Development of an Automobile Robot System Model Based on Soft Computing in an Unsteady Environment. ARPN Journal of Engineering and Applied Sciences, Vol 12 **11** 3433–3439

[5] Demyanushko I V, Skobelev K D, Chernova A P et al. 2019 Otsenka kompetentnosti i kvalifikatsii laboratorii, provodiashchikh ispytaniia elementov dorozhnogo obustroistva [Assessment of Competence and Qualification of Laboratories Conducting Tests of Road Infrastructure Elements]. *Dorogi. Innovatsii v stroitel'stve* **82** 82–86

[6] Surkova N E, Ostroukh A V 2014 Metodologiia strukturnogo proektirovaniia informatsionnykh sistem [Methodology of Structural Design of Information Systems] *(Krasnoyarsk. Publishing House Science and Innovation Center)* 144 p

[7] Wright L, Davidson S 2020 How to Tell the Difference Between a Model and a Digital Twin. *Advanced Modeling and Simulation in Engineering Sciences* Vol 7 **1** https://doi.org/10.1186/s40323-020-00147-4

[8] Demyanushko I V, Tavshavadze B T, Mikheev P S et al. 2020 Povyshenie bezopasnosti dorozhnogo dvizheniia pri provedenii remontnykh rabot [Improving Road Safety During Repair Work] *Mir dorog* **131** 102–107

IOP Conf. Series: Materials Science and Engineering 1159 (2021) 012027 doi:10.1088/1757-899X/1159/1/012027

The deliberate results distortion problem in the road construction materials laboratory and its disposal

Y E Vasiliev[1], **U O Menkina**[1], **V I Kochnev**[1], **I Y Sarychev**[1]

[1]The Federal State Funded Educational Institution of Highest Education Moscow Automobile and Road Construction State Technical University – MADI, 64, Leningradskiy Prosp, Moscow, 125319, Russia

E-mail: Yu.vasilev@madi.ru

Abstract. Article considers the issue of asphalt concrete mixtures quality control necessity and analyses the road construction materials laboratory activity with detection description of an unacceptable random or a large systematic (shift) errors. In the process of producing asphalt concrete mixtures, there are many different factors that can influence their quality. In particular, such as: an inevitable change in raw materials properties, which is associated, among other things, with mineral components heterogeneity, for example, in grain size composition, moisture. There may be deviations in materials dosage, temperature variations and various other indicators, it is also necessary to consider subjective factor on performers part. Economic conditions for laboratory tests results distorting include: interest lack in maintaining information reliability and further strict control. There may also be a certain incentive for employee that determines appropriate behavior. Such incentives can both interest employee in providing reliable data, and can direct his actions towards a conscious distortion of information provided. At the present stage of laboratory testing level development of building materials, an important role in making management decisions is played by reliability and objectivity of information received. However, in road construction, laboratory test results distortion is quite large and plays an important role in industry. To increase results reliability of testing materials, recommendations are given for strengthening internal control in organizations, as well as external control by customer organization and / or regulatory authorities to ensure test results reliability.

1. Introduction

Nowadays information objectivity and reliability play an important role in making management decisions. Quality control system in industry needs significant adjustments, since there are real problems associated with this factor.

Work organization of any laboratory is an ergatic system, in which much depends on performer behavior, which is the main key link in such a system. Proceeding from this, “human factor” plays a significant role, since the performer is responsible for quality and timeliness of work, which determines image and profit of organization [1].

In construction sector, situation cannot always be under control, this is due to delay in obtaining test results, which is an objective reason for decrease in their reliability. This situation applies to many building materials, which is associated with time allotted for routine tests, which, as a rule, significantly exceeds the time required to establish quality parameters and decide on material use.

Content from this work may be used under the terms of the Creative Commons Attribution 3.0 licence. Any further distribution of this work must maintain attribution to the author(s) and the title of the work, journal citation and DOI.
Published under licence by IOP Publishing Ltd

There is also impossibility of using negative feedback, which provides, if necessary, correction implementation [2].

2. Problem statement

There are mistakes that are made by contractor during laboratory tests, they can be unintentional and deliberate. The latter are most often due to corresponding motivation aimed at deliberate action, which is associated with data distortion of laboratory workbook. Private conversations with laboratory staff showed that their leaders outlined criteria in which tests were to be carried out, including under dismissal threat. Consequently, there is a fairly definite motivation on the part of such organizations leaders.

At the same time, employees realize that it is impossible to correct situation due to untimely incoming information. Uselessness in tests carried out pushes ordinary employees of organizations, for example, to actually refuse in some cases from relevant tests actual conduct, thereby turning them exclusively into a virtual project.

To carry out bitumen incoming quality control in accordance with current regulatory documents, it takes up to 4 hours, according to GOST 22245-90 "Viscous road oil bitumen. Technical conditions". But, there are different situations, such as bitumen binder delivery to an asphalt concrete plant (ACP) in accordance with supply contract, in which there is a clause related to plant's obligations to discharge bitumen within one hour (in some cases, two hours) in order to release bitumen truck. Moreover, it is not uncommon for bitumen to bypass intermediate tanks and go into technological process. Thus, ACP first drains bitumen, releases bitumen truck, then after a few hours plant's laboratory [3] has opportunity to assess bitumen binder quality.

3. Research questions

Reasoning In all cases, bitumen properties assessment is carried out with a delay in relation to its appearance in technological process at ACP, and if deviations are recorded, it is not possible to make corrections.

If test results are negative, then substandard bitumen cannot be used for asphalt concrete mixture preparation and there should not be such bitumen at ACP. For this, data are somewhat "edited" in laboratory's logbook, which in turn does not entail any costs for enterprise, except for laboratory worker "some inconvenience", which is expressed in struggle with professional conscience. But as a rule, the company's management motivates its employees.

At present, after appearance of such a concept as bitumen grade according to PG (in accordance with GOST R 58400.1-2019), situation has worsened due to longer time for determining this indicator. This time is 72 hours of determination. Therefore, it can be concluded that theingress of substandard bitumen on the ACP and its becoming a component of theasphaltconcrete mixture is inevitable.
for them.

4. Purpose of Study

Consider and analyze results deliberate distortion problem in road building materials laboratory and options for their elimination.

5. Research methods

The article uses the following methods: analysis, synthesis, generalization and classification of obtained information, interviews of employees, comparison and observation.

The real way out in these conditions is to take control bitumen samples and carefully store them at customer, so that in conflict situation event, you can extract these samples, test them and find out the cause. Despite this, there is a problem: in the Soviet standards for testing bitumen, there were mandatory indications of method convergence and reproducibility, which made it possible to objectively assess differences in the results obtained, including between results obtained in laboratory and results obtained during testing control sample, but now in editions of the standards GOST 33142-

2014 "Public automobile roads. Viscous oil road bitumens. Method for determining softening point. Method "Ring and Ball"" and GOST 11506-73" Petroleum bitumens. Method for determining softening point of ring and ball", approved in recent years, this section has been transformed into a phrase that does not allow drawing conclusions:

Measurements accuracy is ensured by:

- compliance with requirements of this standard;
- carrying out a periodic assessment of measuring instruments metrological characteristics;
- carrying out periodic certification of equipment.

The person conducting measurements must be familiar with requirements of this standard."

The same situation arises in case of asphalt concrete, when quality control existing methods of asphalt concrete mixtures in accordance with the current standard make it possible to obtain quality control result not earlier than 12 hours from moment the mixture was shipped to consumer, which fundamentally suppresses possibility of operational control and timely correction of composition in event of necessity.

In accordance with this, by time information on asphalt concrete mixture quality is received, the latter has already been laid into structure and it is no longer possible to fix anything without pavement disassembling.

As a result, currently carried out quality control of asphalt concrete mixtures is often carried out formally, and data that are entered in laboratory's logbook cannot always be considered objective.

It is possible to reveal fact of asphalt concrete mixture properties indicators deviation from standard values only in sampling cores (cuttings) process from road surface at facility, but probability of this is negligible when assessing cores (cuttings) indicators in three places per 7000m^2. At the same time, results will be influenced by laying and compaction technology [4]. In addition, it is not always possible to say with certainty from which batch asphalt mixture was laid at a given coring location. In this regard, it can also be noted that, in fact, only average density and water saturation are subject to assessment, since all other indicators are determined on re-molded samples, and re-molding process can also significantly affect the test results.

Control carried out in this case allows only to record substandard material presence.

6. Research findings

Thus, examples considered indicate that there is no way to quickly assess material quality, as a result, a possible correction of data obtained occurs.

It is recommended that internal controls be strengthened within organization to improve material test results reliability. But due to the fact that questions related to test results possible distortion during their recording in laboratory's working log, as a rule, are motivated by organization management, which leads to doubt in ensuring results reliability of monitoring these studies.

In this regard, it is possible to ensure test results reliability through external control implementation by customer organization and / or regulatory authorities. But in this case, facts revealing likelihood of test results distortion is unlikely, since it is practically impossible to prove this fact.

Intentional mistakes made by performer are committed, since it is practically impossible to prove the fact of test result distortion when entering it into laboratory log. Contractor is confident in his own impunity, given that laboratory tests results, as a rule, are fully characterized by dialectics laws, including that "you cannot enter the same water twice" or, accordingly, you cannot get the same results.In any case, it is often not possible to double-check test result, especially when it comes to destructive tests (strength, frost resistance, etc.). In addition, there is innocence presumption principle, which also protects performer to some extent.

In this case, material or other benefits can be considered as motivation, in which "human factor" is triggered due to performer deliberate actions.

7. Conclusion

Laboratory that controls road building materials quality must provide a reliable representation of products quality, which is necessary for making appropriate organizational and technical decisions.

Effectiveness of any construction organization, one way or another, depends on provision and use of reliable information. However, this condition is not always met due to various circumstances, for example, delay in obtaining test results and problems elimination for enterprises. Therefore, inaccurate information may arise at information processing and data transformation stage.

Information received quality can significantly affect the management decision-making process.

As practice shows, vast majority of deliberate falsification cases occur at the stage of compiling primary documents, collecting and registering data.

Counterfeiting main goals include overstating raw materials quality indicators and finished products in order to embellish organization's activities results, including concealing the fact of using or releasing substandard products. In this case, the main way of results distorting is associated with entering any "corrected" results into laboratory's work log.

Reliable information distortion is carried out by performer under organizational reasons influence.

At many asphalt concrete plants, where experienced laboratory assistants work, true results are saved and analyzed accordingly [4].

So, in many laboratories of asphalt concrete plants, they use accelerated methods for assessing physical and mechanical indicators. In this case, various techniques or some of their modifications are used, for example, recommendations for accelerated determination of average density, water saturation, and swelling of asphalt concrete, and others [5]. Due to fact that these methods do not have an official status, obtained results using them are purely indicative. However, in a number of laboratories, data obtained during accelerated tests are recorded in workbook as acceptance certificates. In some laboratories, in addition to accelerated ones, tests are additionally carried out according to a standard method. In this case, received data is recorded, as a rule, in different journals, respectively, in "black" and "white".

It should be noted that in Russian legislation (clause 2 of Article 18.19 of the Code of Administrative Offenses of the Russian Federation), liability is provided only for inaccurate test results provision of products subject to mandatory confirmation of conformity, which in no way applies to asphalt mixtures production, subject to exclusively voluntary confirmation of conformity. Responsibility for product quality control results distortion is not provided at all. This in turn gives rise to impunity in this matter.

In connection with the above, in order to ensure asphalt concrete durability, as well as optimal funds spending, a different system of road building materials and works quality control is needed.

Quality control at the stage of their preparation at an asphalt concrete plant plays a significant role in ensuring asphalt concrete mixtures and asphalt concrete quality. Deviations from material regulatory properties, admitted at the production stage, cannot be eliminated at construction and installation work stage. Timely operational information about produced asphalt concrete mixture parameters provides in most cases a real opportunity to avoid road surfaces premature destruction.

Thus, operability based on the current regulatory documents does not have ability to quickly and influence one of the main means of asphalt concrete pavements production - asphalt concrete mixture quality [6].

Rapid methods of determining asphalt concrete physical and mechanical characteristics are required to promptly prevent rejects.

Such methods were proposed in the middle of last century, but were not implemented in current regulatory and technical documents development.

Official reason for this is correlation lack between standard and accelerated test methods.

However, here the question arises: any test method is, to one degree or another, conditional. Including asphalt concrete strength values, determined according to standardized method, are in no way taken into account in road structure calculation, and in this regard, to assert that asphalt concrete ultimate strength at 20 or 50 ° C, determined according to standardized method, is "true" value hardly qualified.

At the same time, it should be noted that value obtained is "significantly delayed" by time when this information is really needed, when, if necessary, it would be possible to intervene in production process and correct something or prevent substandard products appearance.

In this regard, it would be more logical to apply methodology described in Information Letter No. 91 of the SoyuzdorNII "Accelerated Test Methods of Road-Building Materials" in 1957, in which it was said that accelerated test methods could be applied provided that central laboratory a preliminary check of asphalt concrete physical and mechanical properties was carried out in accordance with technical conditions (non-accelerated methods) requirements. At the same time, it was indicated that indicators compliance with accelerated methods of determination and conventional ones should be checked.

Currently, there is a methodology for finding correspondence between different test methods (GOST R ISO 5725-1-2002 - GOST R ISO 5725-6-2002. Accuracy (correctness and precision) of methods and measurement results), including when testing heterogeneous materials, a typical representative of which is an asphalt concrete mixture.

It should be noted that accelerated method should be universal and should not depend on asphalt concrete composition, raw materials quality, temperature and test method.

Express method for determining properties must meet the following requirements:

- result obtaining in the shortest possible time (within 1-2 hours) after asphalt concrete mixture production;
- ensuring required test results accuracy and stability (reproducibility and repeatability in relation to standard method);
- possibility of implementing proposed methods on the basis of equipment available in laboratory.

For an objective method assessment, it is necessary to have information about processes that occur in asphalt concrete at the early stages of formation, including during cooling.

In connection with the above, it is possible to outline the main ways to ensure test results reliability. These are events such as:

- Accelerated test methods introduction;
- control samples selection by customer;
- participation in interlaboratory comparison tests;
- building control charts.

To remedy this situation, it is necessary to massively develop and implement accelerated test methods that could provide necessary information prompt receipt. These express methods are designed to help improve products quality [7]. So, in MADI, in the Shared Use Center, an express method for assessing bituminous binder properties has been developed, which provides an opportunity to quickly assess product quality.

However, in many laboratories, employees "introduce" their own test methods, which allow obtaining necessary information about properties parameters of asphalt concrete mixtures in the shortest possible time, which provides a real opportunity, if necessary, to make any adjustments to production process. In some laboratories, freshly formed samples are subjected to accelerated cooling in cold water or a refrigerator, or they organize samples testing while they are still hot.

In any case, performers pursue a good goal associated with obtaining operational information about properties parameters of asphalt concrete and adjusting them if necessary. However, question arises about feasibility of "inventing" its own test methods, while in the Belarus Republic back in 2008 the state standard of Belarus STB 1536-2008 "Asphalt concrete road, airfield and asphalt concrete mixtures. Express - Test Methods" was approved, which regulates accelerated testing possibility of asphalt concrete mixtures.

The methods regulated by this standard are not indisputable, may need to be adjusted accordingly, but they exist and have an official status. In this regard, apparently, it would be advisable to use them, instead of using "home-grown" methods.

When conducting tests, parallel samples should also be taken, which should be kept at the Customer and, in case of any conflict situation, carry out tests. This approach disciplines performer and encourages him not to follow deliberate distortion path of test results. Thus, control sample availability at the Customer provides a real opportunity to ensure test results reliability, for example: when, in case of tests without warning at a construction site, products quality significantly increased [8].

GOST 5725 states that for a number of materials there is no such thing as a "standard", and this fully applies to road construction materials (GOST R ISO 5725-1-2002 - GOST R ISO 5725-6-2002). In this regard, it is possible to assess asphalt concrete quality only in process of carrying out interlaboratory comparison tests. All laboratories are strongly encouraged to participate in interlaboratory comparison programs. Without participation in such programs, laboratory cannot regularly check correctness of its work. This, along with regular reproducibility evaluations, is absolutely essential to ensure high quality results. One of the simplest methods of such control is to compare arithmetic mean and standard deviation obtained in this laboratory with corresponding indicators of other ISI participants using the same instruments and test methods [9]

Control cards are currently widely used without fail in medicine, mechanical engineering, pharmaceuticals and other industries: almost everywhere, except in the construction industry. Control maps, for all their simplicity of construction, are extremely essential. First of all, because of its visibility. If the laboratory has a work log that records the results of the relevant tests from day to day, it is often not possible to make an analysis. All that is possible is to highlight abnormal values which should be already marked by the performer. It is impossible to identify any regularities in the change of the indicator values, such as systematic error, drift, shift or random error. Built control card makes possible to assess the quality of this analytical series based on detection rules of unacceptable random and beginning large systematic (shift) errors. [10].

References

[1] Pavlov V V 1975 *Beginnings of the ergatic systems theory* (Kiev, NaukovaDumka)

[2] Dosko S I, Kirenkov V V, Mikitenko V G 2017 Using the duality principle in solving inverse problems of evaluating test results *Bulletin of the Bryansk State Technical University* Vol 2 **55** 185–190

[3] Vasiliev Yu E 2011 Bitumen quality. How is it determined? *Science and Technology in the Road Industry* Vol 2 27–28

[4] Vasiliev Yu E, Shlyafer V L 2007 Ways to improve the properties assessment of asphalt concrete mixtures *Highways* 6 24–26

[5] Yaromko V N, Zhaylovich I L 2005 On some quality improving aspects of asphalt concrete pavements *Constructionandrealestate* 25

[6] Ramos A L, Kotlyarsky E V, Vasilev Y E, Popov V A, Sarichev I Y 2019 A theory of quantifying the effect of jet blasts from aircrafts on temperature of the surface course of asphalt-concrete airfield pavement *Dilemascontemporaneos-educacionpolitica y valores, Special Issue* Vol 6 1–26

[7] Menkina U O, Zumbadze B G, Rasskazova A A, Adou Jean-ClaudeAlexi, Vasiliev Yu E 2019 *Rapid method for properties assessing of bitumen Innovative materials and technologies in transport construction and road facilities* (Moscow: LLC Tekhpoligrafcenter) Vol 3 5–10

[8] Lermit R 2007 Concrete technology problems (Moscow, LKI PublishingHouse)

[9] Vasiliev Yu E, Kotlyarsky E V, Kochnev V I et al 2020 Interlaboratory collaborative evaluation experiments for testing petroleum road bitumen and bitumen asphalt (Moscow, Tekhpoligrafcenter LLC)

[10] Cooper U G 1998 Basics of laboratory research quality control. Quality control manual. Retrieved from: https://www.qcnet.com/Portals/89/PDFs/Q-1109_RU.pdf

IOP Conf. Series: Materials Science and Engineering 1159 (2021) 012028 doi:10.1088/1757-899X/1159/1/012028

Computing information and control complex for control of road construction machines

O I Maksimychev[1], K N Mezentsev[1], A N Yakubovich[1], A V Libenko[1], V M Guryanov[1]

[1] Department of Automated Control Systems, Moscow Automobile and Road Construction State Technical University (MADI), 64, Leningradsky Prospect, , 125319, Moscow

E-mail: Maksimychev@asu.madi.ru

Abstract. The article considers a model of the digging process carried out during the operation of a road-building machine, identifies the controlled and control parameters of the model. Based on the results of numerical simulation, it is shown that a rigid control algorithm is capable of providing effective results only in cases where most of the operating parameters of the machine do not go beyond the optimal area. To improve the efficiency of control and expand the permissible range of parameters of the state of the machine, it is proposed to use control based on fuzzy logic. The structure of synthesis of algorithms for fuzzy control of road-building machines is given. For the purpose of practical implementation of a fuzzy control algorithm, a basic model has been built, which includes 10 linguistic rules, on the basis of which chains of actions are formed during the operation of a machine. When implementing a fuzzy control model for the main parameters of the state of the machine, it is proposed to use triangular membership functions. The fuzzy inference surface for the fuzzy control algorithm is obtained and visualized, which makes it possible to efficiently search for the optimal trajectory of the control parameter function. An agent-based traffic model is considered, which describes the movement of road-building vehicles on a site and uses active and proactive algorithms. The results of numerical modeling confirm the feasibility of considering a variety of road-building machines at the site of joint work execution as a multi-agent system with elements of self-organization (CDPS-system). The maximum effect of self-organization is observed when the distance between the machines is changed, when they are uniformly located on the site. The expediency of combining local systems of automatic control of road-building machines into a single information and control complex is substantiated, a list of information and control tasks to be solved by the complex is formulated, and its structural diagram is synthesized.

1. Introduction

On-board computer systems of road-building machines provide the necessary information and functional interaction of technological processes. Robotic systems create conditions for the transition to fully intelligent (smart) digital production of construction work and the emergence of new technologies [1]. Digital models of technological processes allow making planning decisions in an automatic mode. The introduction of information management systems (IMS) accelerates the timing of road construction works, this becomes possible through the use of digital twins of construction objects and technological machines [2].

Content from this work may be used under the terms of the Creative Commons Attribution 3.0 licence. Any further distribution of this work must maintain attribution to the author(s) and the title of the work, journal citation and DOI.
Published under licence by IOP Publishing Ltd

The production of linear earthworks sets the task of organizing and developing optimal solutions for the coordinated management of all machines located at the construction site. Such control is possible only with a comprehensive consideration of all parameters of machine control at the time of execution of technological processes. Research of control models [3] and analysis of structures of automatic control systems (ACS) of machines [4] show the possibility of coherent control of technological operations, in this case it becomes possible to increase the productivity of earth-moving operations only by creating new control algorithms [5, 6] without making changes in the mechatronic organization of earth-moving machines for various purposes. The changes affect only the software and hardware control complex at the local level of machines, thereby increasing the efficiency of technological processes.

Possibilities of effective control of an autonomously functioning machine to a decisive extent depend on the full consideration of all physical laws that determine both its movement and interaction with the environment in the process of performing work functions. The design stage is the first stage at which the optimal relationship between mechanical, electronic and software components of the machine is formed, modeling and verification of both the physical model and the control logic are performed - before the actual implementation is made [7]. Already in the design process, the efficiency and reliability of control at the stage of future operation (necessarily taking into account numerous operational uncertainties) should be considered as one of the main goals [8].

A large number of works have been devoted to automatic control systems for the main elements of mobile machines, including the development of the necessary physical models. The control circuit of the steering wheel electro-hydraulic system, carried out with the joint use of the voltage servo controller and the noise rejection controller, is shown in [9]; the simulation results confirmed the effectiveness of the proposed method, taking into account the fast response and control accuracy. The construction and verification of a dynamic model of clutch control was performed in [10]; As a result, it was possible to satisfactorily solve the main control problems characteristic of this functional element: the presence of complex nonlinear dynamics of the interaction of a fluid and mechanical systems, a time-varying load, a short duration of time allotted for the development of a control action, and lack of information about the duration of the upcoming movement of the object. Optimization of automatic transmission control is considered in [11]: a dynamic model of the transmission system is built, a linear-quadratic regulator is used to achieve the optimal clutch pressure, software verification of two possible states is performed (in the absence of load without a slope and full load with a maximum slope). The application of impedance control to the active suspension system is shown in [12], automatic control was implemented using two internal loops, which were power control of the drive using reverse linearization and a fuzzy control loop according to the impedance rules.

The basic means of automatic control of the operation of road construction machines are PID controllers. A promising direction of their use is data control (or modelless control). As a result, it is often possible to improve the closed-loop behavior of linear, and especially non-linear systems, using very simple control structures. As a result, the unknown or poorly known dynamics of external influences on the object is compensated by control actions, and random disturbances are discarded without any training or online identification [13]. Systems with delays are of particular interest due to the increased difficulty of automatic control. In [14], using the Hermite-Beeler theorem, a complete set of stabilizing parameters of the PID controller is determined for both stable and unstable systems with an open control loop; in the second case, a necessary and sufficient delay condition for the existence of stabilizing controllers is determined. It was revealed that the stabilizing set in the space of the integral and derivative gain is a trapezoid, triangle or quadrangle. Based on the analysis of the stability of control based on PID controllers, it was shown in [15] that, in the parameter space, the stability regions in the general case are convex polyhedra.

The issues of developing and optimizing the trajectory of movement, both for single machines and for their combination, are traditionally given great attention. A simple automatic trajectory tracking system for a single unmanned vehicle moving over rough terrain is shown in [16]. The successful use of an agent with a hybrid behavioral architecture for autonomous vehicle navigation is considered in [17], where more specific actions associated with the performed technological operation are determined from a set of basic motion actions (forward, turn, and so on). Planning the optimal path is essential for the efficient operation of autonomous machines. A heuristic algorithm based on the principles of cuckoo search is shown in [18]; as a result of the algorithm execution, a smooth optimal trajectory is formed. Comparison of two algorithms for trajectory planning in real time, providing an optimal trajectory with avoiding static obstacles, was performed in [19]. The first of the compared algorithms corresponds to the method of an artificial potential field, modified with the help of dummy loads; the second algorithm uses Voronoi diagrams and includes Delaunay triangulation to avoid routes on which traffic blocking is possible. Both of the considered algorithms are quite efficient. Saving computational resources is discussed in [20], where it is shown that navigation can be provided using a relatively small number of if-then rules using the Tagaki-Sugeno fuzzy inference method and fuzzy sets represented by normalized triangular functions.

Automatic control systems have hierarchical levels of control, in which three are usually distinguished: strategic, tactical and executive [21]. In modern electromechatronic systems at the strategic level, control subsystems with elements of expert systems and artificial intelligence are often used, and at the lower levels - fuzzy and neurosubsystems [22].

2. Models and algorithms

2.1. Digging process model

Bulldozer and scraper are considered as the main types of earth-moving vehicles (EMV), as machines with similar principles and control structures at the local level. The process of digging with a bulldozer on a crawler track can be represented by the equation:

$$\frac{dV_A}{dt} = \frac{1}{m(P_T - P_R)}, \tag{1}$$

where P_T is the total driving force of the bulldozer; P_R is total resistance to movement; t is time; V_A is the actual speed of the bulldozer when digging; m is the reduced average mass of the bulldozer.

It is customary to evaluate the driving force of the propellers by the amount of their slip:

$$P_{T,i} = f(\delta_i), \tag{2}$$

where δ_i is the slip coefficient of the i-th propeller.

The slip coefficient evaluates the positive slip of the propellers and is determined by the expression:

$$\delta_i = \frac{V_{T,i} - V_{A,i}}{V_{T,i}}. \tag{3}$$

The theoretical speed of the propeller at a constant rolling radius of the tracked course is a linear function of the angular velocity of the propeller:

$$V_{T,i} = r_i \omega_{A,i}, \tag{4}$$

herewith

$$\frac{d\omega_{A,i}}{dt} = \frac{M_{A,i} - P_{T,i} r_i}{J_i}, \tag{5}$$

where $\omega_{A,i}$, is the angular speed of the engine; $M_{A,i}$ is torque; J_i is reduced moment of inertia; $P_{T,i}$ is driving force of the i-th mover; r_i is rolling radius.

Torque value:

$$M_A = \frac{M_e \eta_M i_M}{K_A}, \tag{6}$$

where η_M is the efficiency of the engine transmission; i_m is the total gear ratio.

The analytical expression of the engine torque Me(ω) when solving the problems of machine dynamics for the regulatory characteristic has the form:

$$M_e = M_N \left(\omega_x - \frac{\omega_e}{\omega_x} - \omega_N \right), \tag{7}$$

where M_N is the rated torque of the engine; ω_x is the maximum angular speed of the engine shaft at idle; ω_N is nominal angular speed of the motor shaft. Current value of angular velocity in the range from ω_x to ω_N:

$$\omega_e = i_M \omega_{A,i} \,. \tag{8}$$

2.2. *Algorithm for hard control of the digging process*

The structure of the automatic control system is shown in Fig. 1. The executive bodies are usually electro-hydraulic servo drives with fixed displacement pumps. This determines the relay characteristics of the control unit and the entire device as a whole. The automatic control device maintains the value of the controlled variable X_p at a given level X_0 by changing the position of the working body of road-building machines h. This is done as follows. When the value X_1, proportional to X_p, deviates from its specified value X_0, an error signal сигнал appears. Depending on the magnitude and sign of the error signal, the control unit moves the working body of the road-building machines through the servo drive of the executive body and changes the digging depth towards decreasing the error.

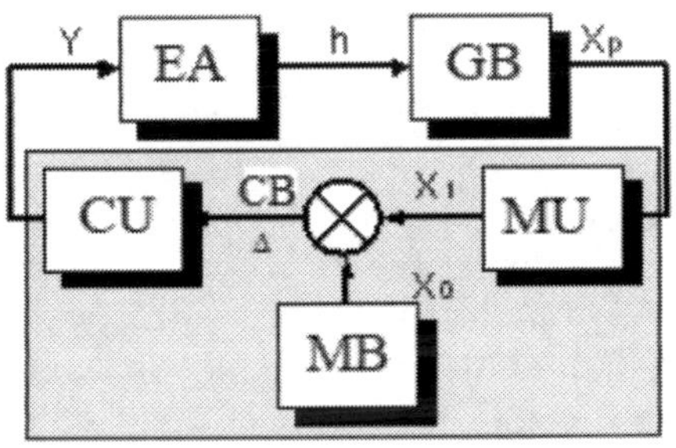

Fig. 1. Functional diagram of the ACS by the process of digging the soil. MU - measuring unit; CB - comparing block; MB - master block; CU - control unit;. X is the controlled value; h - control value (digging depth); Y - output signal of the control device, under the action of which the executive body changes the value of the digging depth.

The controlled value of the contour that maintains the maximum tractive power is the slip coefficient of the most skidding propulsion unit, which is determined by expression (3) through the theoretical and actual speed of the machine. The setpoint (setpoint) in this circuit is equal to the optimal value of the slip coefficient $\delta \approx 10\%$, at which the maximum traction force and traction power are realized. The controlled value of the contour that maintains the set value of the free power of the engine is the angular velocity of its crankshaft ω_e. The value of the setpoint of the angular speed of the engine ω_0 is determined by the operator during tuning, which achieves the equality of the implementation of the free power of the diesel engine and the maximum tractive power for specific soil conditions and parameters of the propellers.

The algorithm allows to develop control actions, while analyzing the current value of the angular speed of the engine ω_e, the theoretical speed for each propeller $V_{T,1}$and the actual speed of the machine $V_{T,2}$. The optimal system of automatic control of the bulldozer digging process is double-circuit. One circuit: it maintains the given free power of the diesel engine during the digging process, by automatically changing the depth of cutting the soil as a function of the angular speed of the diesel engine shaft. The other, with excessive slipping of any of the propellers, forcibly pushes the dump.

2.3. *Fuzzy algorithm of digging process control*

Fuzzy logic control model for EMV is shown in Fig. 2. The control principle consists in the implementation of fuzzy closure of the system, i.e. introduction at the tactical level of feedbacks on controlled parameters, as well as the introduction of a fuzzy setting of the control law. The system under consideration is not rigidly controlled. The task of control is the sequential implementation of some evolutions by the working body of the bulldozer, leading to the achievement of the optimal mode, for example, a steady motion with the necessary characteristics for cutting the soil; in other words, the goal is reflex control.

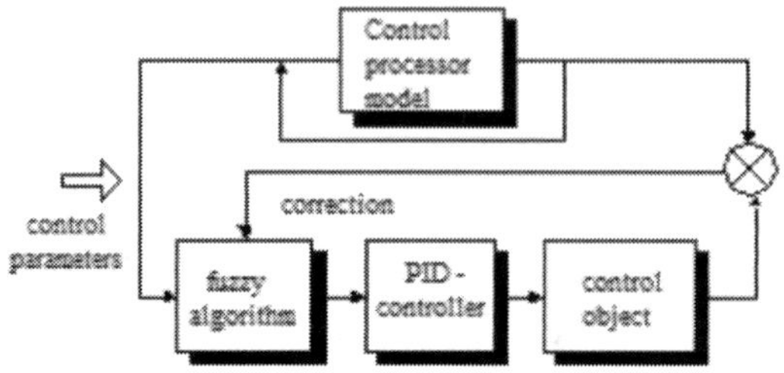

Fig. 2. The structure of synthesis of algorithms for fuzzy control of technological processes of EMV.

The main function assigned to the fuzzy controller is the formation of corrective corrections to the coefficients of the PID controller, depending on the current coordinates of the system. In this case, a PID controller with a corrective fuzzy controller is a non-linear system.

2.4. Agent model of traffic of road construction vehicles

The totality of EMV at a road construction site is considered as an agent-based model. Taken together, many machines form an agent-based system such as CDPS (Cooperative Distributed Problem Solving). In this case, individual agents interact in a single system based on the general technological plan of movement and the design depth of soil removal. Modeling motion within the framework of the accepted CDPS paradigm allows us to identify patterns that must be taken into account when developing an autopilot algorithm. To formally represent an individual agent in a multi-agent system, two sets are distinguished: a set of states of the working environment $E=\{e_1,e_2 \ldots e_n\}$ and a set of agent actions $A_C=\{a_1,a_2 \ldots a_m\}$. Three sets of agent movements are taken into account:

- R is the set of all possible displacements (E and AC);
- RAC - set of all movements that ended with the action - soil removal;
- RE - the set of all possible movements that led to a change in the agent's environment.

Changing the environment of the agent for one turn of the agent, the condition for completing his work and the description of the environment as a tuple are written as:

$$\left.\begin{array}{l} t=(e_0,a_0,\ldots,e_n,a_n) \\ If\ \ t(e_0,a_0,\ldots,e_n,a_n)=\otimes\ \ then\ \text{end} \\ Env=(E,e_0,t) \end{array}\right\}. \quad (9)$$

In a more compact form, the behavior of the scraper agent is described by the following function:

$$Ag: R^E \Rightarrow A_C. \quad (10)$$

Within the framework of the model under consideration, two algorithms for the functioning of the agent can be distinguished: reactive and proactive. Reactive behavior requires the agent to have a subsystem of perception of the environment (analysis of the state of the soil, the presence of obstacles, the presence of a number of other agents) and perception of the subsystem of action (removal of soil, bypassing obstacles, making a maneuver to avoid collisions with other agents). Modeling the behavior of a reactive agent is performed using the functions:

$$\left.\begin{array}{l} action:\langle information\rangle \rightarrow A_C \\ \langle information\rangle = E \Rightarrow action: E \rightarrow A_C \end{array}\right\}. \quad (11)$$

Proactive behavior requires an agent to process the information received from its environment to make a decision on further actions and is described by the following relationships:

$$\left.\begin{array}{ll} action: I \rightarrow A_C, & next: I \times Per \rightarrow I \\ see(e) = \text{Perception}, & next(i_0, see(e)) \\ action:\left(next(i_0, see(e))\right), & action: I \rightarrow A_C \end{array}\right\}, \quad (12)$$

where i_0 is the agent's internal state, I is the set of internal states.

To study the agent's behavior, we will use a safe movement model based on the Nagel – Schreckenberg movement model [23]. The agent inspects the environment in the direction of its movement in a cone of a given radius and angle. If there are other agents (road-building machines) in the cone of view, then it is necessary to identify the nearest agent and if the distance to it is less than the critical one, then braking is performed, otherwise a further increase in the scraper speed is permissible.

3. Results of numerical modeling and structural synthesis of the control system

The main dependences obtained using the model of the digging process (1) - (8) are shown in Fig. 3. These dependencies underlie the control carried out according to a rigid algorithm and, as will be shown below, significantly limit its efficiency.

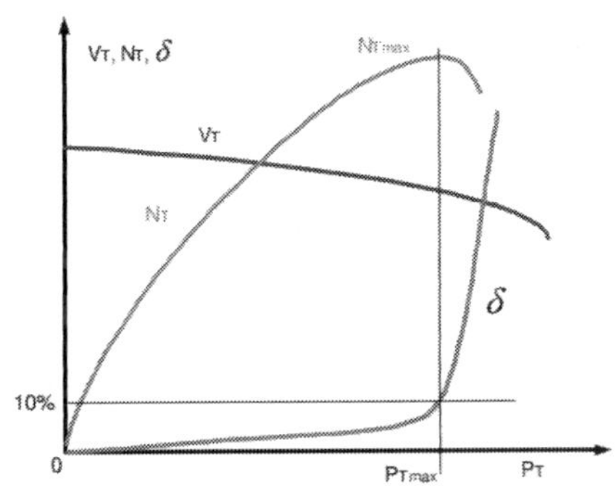

Fig. 3. Dependence of traction power N_T, actual speed v_T and slip coefficient δ as a function of driving force P_T.

The base of fuzzy linguistic rules, developed as part of the implementation of the fuzzy control algorithm, describes the necessary sequence of actions, in which the values and directions of movements of the working body are determined. The basic model consists of ten linguistic rules governing the choice of control commands. Control parameters are reduced to three terms. The proposed rules are based on the analysis of the results of interviewing management experts, based on the linguistic descriptions of the management rules, chains of action sequences are formed:

1. IF "n is PS" AND "a is Z" THEN "h is NB" (engine speed is increased, the speed is unchanged, a signal should be given for the maximum lowering of the working body);
2. IF "n is ZP" AND "but there is NS" THEN "h is NS" (the engine speed is slightly higher than the set ones, and it drops, a signal should be given for a slight lowering of the working body);
3. IF "n is Z" AND "but is Z" THEN "h is Z" (engine speed is set and unchanged, do not send a signal to the working body);
4. IF "n is NS" AND "but there is NS" THEN "h is ZP" (the engine speed is lower than the specified one, and it drops, then a signal should be given for a small rise of the working body);
5. IF "n is NM" AND "but there is NS" THEN "h is PS" (the engine speed is minimum and the frequency drops, a signal should be given to raise the working body);

6. IF "n is NB" AND "but is NS" THEN "h is PB" (the engine speed is less than the minimum allowable and the frequency drops, then a signal should be given for the maximum lift of the working body);
7. IF "n is NB" AND "but there is PS" THEN "h is PM" (the engine speed is less than the minimum permissible, and the frequency is increasing, a signal should be given for a large lift of the working body);
8. IF "n is NM" AND "but there is PS" THEN "h is ZP" (the engine speed is minimal, and the frequency is increasing, a signal should be given for a small lift of the working body);
9. IF "n is Z" AND "but there is PS" THEN "h is ZP" (the engine speed is set and the frequency increases, a signal should be given for a small rise of the working body);
10. IF "n is ZP" AND "but there is PS" THEN "h is Z" (the engine speed is slightly higher than the specified one, and the frequency increases, stop the movement of the working body).

The fuzzification procedure is performed for the following input variables:

- "Engine speed": a set of values T1 = {less than the minimum allowable; minimum; below specified; given; slightly higher than specified; elevated};
- "Actual speed": many values: T2 = {decreases; is equal to zero; growing};
- "Change of position": a set of values T3 = {maximum lowering; slight lowering; Stop; slight rise; climb; big rise; maximum lift}.

The membership functions for the control used in the fuzzy control algorithm have a triangular shape. The numerical implementation of the fuzzy control algorithm was carried out within the framework of one system. The system for designing fuzzy control systems MatLab (Fuzzy Logic Toolbox) is integrated into the developed structure. The resulting fuzzy inference surface is shown in Fig. 4.

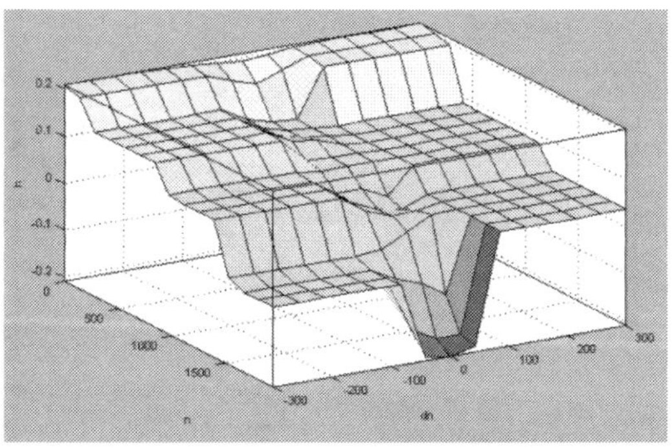

Fig. 4. Visualization of the fuzzy inference surface for the fuzzy control algorithm:
n - engine speed; dn / dt = a is rate of change of engine speed; h is the amount of movement of the working body;

The numerical implementation of the agent-based motion model was carried out using the NetLogo software environment for multi-agent modeling [24]. When building an agent-based NetLogo model, the movement of agents took place in a two-dimensional virtual space. The simulation was carried out in a discrete mode of operation of a model processor. The agents worked in a closed cycle: removing soil at the site, leaving the site, entering the site, etc. The work continued until the soil sample was reached to the specified depth. It was assumed that the agent moves with a speed, the nature of the change of which obeys a uniform distribution law, while taking into account the distance to the driving agent ahead. Upon reaching a critical distance, the agent performed braking. This behavior of agents led to different variants of their positioning on the track, depending on the value of the critical technological distance. The scraper speed was maintained in the range $0 \leq v \leq v_{max}$ where v_{max} is the maximum speed allowed on the road section during operation. In addition, the size of the distance to the agent in front was supposed to ensure that there was no collision. The simulation results are shown in Fig. 5.

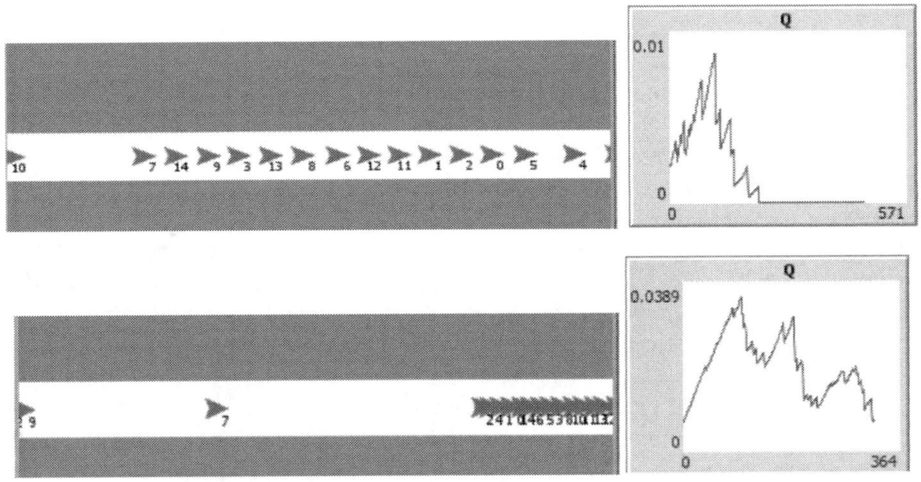

Fig: 5. Optimal and critical travel distances

Based on the proposed control model, it becomes possible to synthesize a self-organizing structure that unites at the information-hardware level any number of construction machines of various technological purposes and thereby allows synchronizing their interaction at a construction site. Such a complex interaction is possible only with the use of methods and technologies that will make it possible to divide the entire data stream into two streams: the first is the macro-level, where an expert assessment of the situation as a whole is carried out, and the second is the performance of an accurate calculation or synthesis of the target control function after the expert assessment is made. The hardware implementation of such systems can be implemented using structured hierarchical computing systems, which are based on local control systems for technological machines. Figure 6 shows a scraper control complex, a distinctive feature of which is a distributed monitoring and control system for its aggregate-functional purpose.

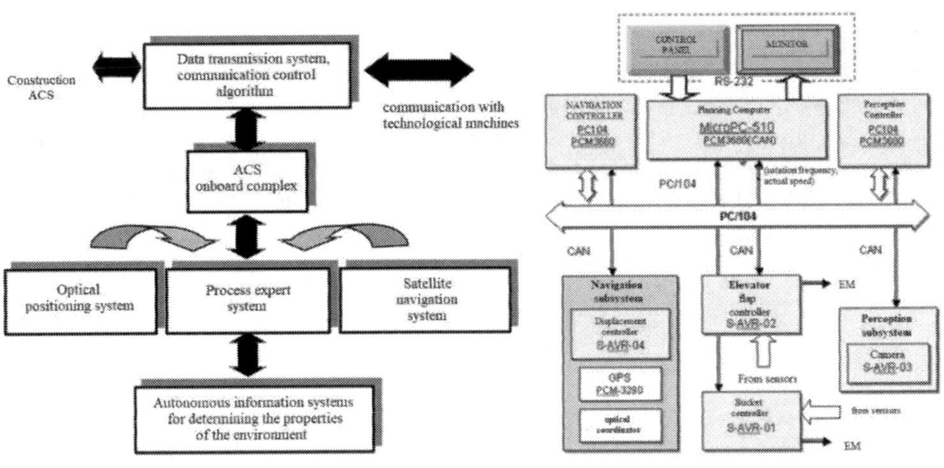

Fig. 6. Block diagram of the system and scraper control controller.

4. Discussion and conclusions

In Fig. 3 it can be seen that in the process of digging the soil, the maximum productivity is achieved when the maximum tractive power $N_{T,max}$. The $N_T=f(P_T)$ function is extreme and reaches its maximum at maximum driving force stability, which is achieved at $\delta = 10\%$ for tracked propellers. Fig. 3. it can be seen that when digging, as the resistance forces increase, the driving force grows and, despite the decrease in the actual speed, the tractive power increases. With an optimal slip ratio, power is maximized. In this case, the engine should not be overloaded, for which it is necessary to stabilize its angular speed of rotation at the level corresponding to the implementation of the maximum driving force P_T, max. In general, the analysis of single-loop automatic control systems that implement a rigid algorithm for controlling the digging process shows that such systems can only give an effect if the other parameters do not leave the optimal zone.

So, when stabilizing the angular speed of the engine of the machine, a large margin of adhesion mass is required to ensure the absence of any noticeable slipping of the propellers. When driving by slipping the propellers, the installed engine power should be overstated and not used when digging. To eliminate these disadvantages, it is proposed to control the digging process along two circuits: according to the maximum free engine power and according to the maximum tractive power. One circuit in both cases stabilizes the engine power at a predetermined maximum level, and the other traction power or slip coefficient. There is no technical implementation of these principles, which makes it impossible to assess the structure of the system and the quality of control over the digging process.

The mathematical apparatus of fuzzy logic is most appropriate for use in control control and analysis of large arrays of heterogeneous information. Visualization of control parameters (Fig. 4) makes it possible to efficiently search for the optimal trajectory of the control parameters function. In this case, the accuracy of the data obtained depends on the number of ACS training operations performed and the accuracy of fuzzy control rules. In general, the use of fuzzy control methods makes it possible to obtain a high-quality transient process without using cumbersome computational procedures typical of the classical control method.

The results of modeling the interconnected work of agents allow us to conclude that road-building machines in the technological process of road construction can be considered as a multi-agent system with elements of self-organization. The self-organization effect is observed when the distance between the scrapers changes (Fig. 5). The optimal choice of distance allows you to evenly position the equipment on the site when performing road works. The simulated algorithms for the movement of agents can be the basis for the construction of an adaptive mechanism for the autopilot of road construction equipment.

Automation of technological processes of earth-moving and other types of machines and equipment, carried out at the level of individual types of machines, leads to an increase in their productivity, but does not allow increasing the efficiency of work in the composition, a set of machines. Accordingly, the unification of the local automatic control systems of the EMV into a single information and control complex is an objective necessity. For this, it is effective to use hardware and software computing systems built on the components of expert systems, and fuzzy decision-making logic. At the same time, the control of the flow of cars based on the agent-based motion model allows you to control and manage technological interaction. The principles and technologies shown in this work for constructing information and control systems based on robotic control principles and implementing control algorithms based on fuzzy logic open up new opportunities for construction production and support the life cycle of roads.

References

[1] Maksimychev O I, Vas'kovskij A M 2012 New directions in automation of road construction technologies *Bulletin of the Moscow automobile and road State technical University* (Moscow: MADI) **3(30)** pp 53a-57 (in Russian)

[2] Maksimychev O I, Bojkov V N 2019 Support for the life cycle of road construction projects in the digital economy paradigm *CAD and GIS of highways* **1(12)**, pp 10-15 (in Russian)

[3] Maksimychev O I, Ostroukh A V, Pastukhov D A, Nuruev Y E-O, Karelina M Y, Zhankaziev S V 2016 Automated control system of road construction works *International Journal of Applied Engineering Research* **11(9)** pp 6441-6446

[4] Maksimychev O I, Sergeeva D A 2017 Tools for information and control systems for road construction automation support *Automation and control in technical systems* **2(23)** p 13 (in Russian)
[5] Maksimychev O I, Vinogradov V A, Gur'yanov V M, Hakmana Vidzhesinghe T N 2015 Technologies for monitoring and designing algorithms for managing earthmoving vehicles *In the world of scientific discovery* **6(66)** pp 40-51 (in Russian)
[6] Maksimychev O I, Ivakhnenko A M, Ostroukh A V, Ephimenko D B, Zbavitel P Y, Karelina M Y 2016 Technology of monitoring and control algorithm design for earth-moving machine *International Journal of Applied Engineering Research* 11(9) pp 6430-6434
[7] Josef Adolfsson, Amos Ng, Petter Olofsgård, Philip Moore, Junsheng Pu, Chi-Biu Wong 2002 Design and simulation of component-based manufacturing machine systems *Mechatronics* **12(9-10)** pp 1239-1258 doi: https://doi.org/10.1016/S0957-4158(02)00027-2
[8] Andreas Wurm, Dieter Bestle 2015 Robust design optimization for improving automotive shift quality *Optimization and Engineering* **17** pp 421–436 doi: https://doi.org/10.1007/s11081-015-9290-1
[9] Chengqiang Yin, Qun Sun, Jian Wu, Chengqiang Liu, Jie Gao 2018 Development of Electrohydraulic Steering Control System for Tractor Automatic Navigation *Journal of Electrical and Computer Engineering* **ID 5617253** p 7 doi: https://doi.org/10.1155/2018/5617253
[10] Xingyong Song, Zongxuan Sun 2011 Pressure-Based Clutch Control for Automotive Transmissions Using a Sliding-Mode Controller *IEEE/ASME Transactions on Mechatronics* **17(3)** pp 534-546 doi: https://doi.org/10.1109/TMECH.2011.2106507
[11] Heng Zhang, Xinxin Zhao, Jianning Sun 2020 Optimal Clutch Pressure Control in Shifting Process of Automatic Transmission for Heavy-Duty Mining Trucks *Mathematical Problems in Engineering* **ID 8618759** p 9 doi: https://doi.org/10.1155/2020/8618759
[12] Mohammad Mehdi Fateh, Seyed SinaAlavi 2009 Impedance control of an active suspension system *Mechatronics* **19(1)** pp 134-140 doi: https://doi.org/10.1016/j.mechatronics.2008.05.005
[13] JorgeVillagra, BlasVinagre, InésTejado 2012 Data-driven fractional PID control: application to DC motors in flexible joints *IFAC Proceedings Volumes* **45(3)** pp 709-714. doi: https://doi.org/10.3182/20120328-3-IT-3014.00120
[14] Silva G J, Datta A, Bhattacharyya S P 2002 New results on the synthesis of PID controllers *IEEE Transactions on Automatic Control* **47(2)** pp 241-252 doi: https://doi.org/10.1109/9.983352
[15] Ackermann J, Kaesbauer D 2003 Stable polyhedra in parameter space *Automatica* **39(5)** pp 937-943 doi: https://doi.org/10.1016/S0005-1098(03)00034-7
[16] Mehmet Bodura, Ehsan Kianib, Hasan Hacışevki 2012 Double look-ahead reference point control for autonomous agricultural vehicles *Biosystems Engineering* **113(2)** pp 173-186 doi: https://doi.org/10.1016/j.biosystemseng.2012.07.005
[17] García-Pérez L, García-Alegre M C, Ribeiro A, Guinea D 2008 An agent of behaviour architecture for unmanned control of a farming vehicle *Computers and Electronics in Agriculture* **60(1)** 39-48 doi: https://doi.org/10.1016/j.compag.2007.06.004
[18] Prases K. Mohanty, Dayal R. Parhi 2016 Optimal path planning for a mobile robot using cuckoo search algorithm *Journal of Experimental & Theoretical Artificial Intelligence* **28(1-2)** pp 35-52 doi: https://doi.org/10.1080/0952813X.2014.971442
[19] Gómez E J, Martínez Santa F M, Sarmiento F H M 2013 A comparative study of geometric path planning methods for a mobile robot: Potential field and voronoi diagrams *2013 II International Congress of Engineering Mechatronics and Automation (CIIMA)* (Bogota), pp 1-6 doi: https://doi.org/10.1109/CIIMA.2013.6682776
[20] Iluminada Baturonea, Andrés Gersnoviez, Ángel Barriga 2014 Neuro-fuzzy techniques to optimize an FPGA embedded controller for robot navigation *Applied Soft Computing* **21** pp 95-106 doi: https://doi.org/10.1016/j.asoc.2014.03.001
[21] Yakubovich A N, Yakubovich I A, Trofimenko Yu V, Shashina E V 2019 Intelligent Management System of the Automobile Road's Technical and Operational Condition in the Cryolithozone

Proceedings Systems of Signals Generating and Processing in the Field of on Board Communications SOSG **2019'03** doi: https://doi.org/10.1109/SOSG.2019.8706742

[22] Orlov S V, Maksimychev O I 2017 Information system for planning the use of automation of technological processes in road construction *Automation and control in technical systems* **2(23)** pp 7 (in Russian).

[23] Arnaud Banos 2015 *Agent-based Spatial Simulation with NetLogo 2. Advanced Concepts* (London: ISTE Press – Elsevier)

[24] Mezencev K N 2015 *Multi-agent modeling in the NetLogo environment* (Saint-Petersburg: LAN Publishing House) (in Russian)

Testing of cement-concrete road surfaces for wear resistance

V V Ushakov [1], **Yu E Vasiliev** [1], **S M Dmitriev** [1], **G G Dyakov** [1], **O Agaryshev** [2]

[1] Moscow Automobile and Road Construction State Technical University (MADI), Moscow, Russian Federation
[2] LLC "LafargeHolcim Rus", Moscow, Russian Federation

E-mail: madi-ushakov@mail.ru

Abstract. The article presents the results of a study of the wear resistance of cement concrete coatings of highways. The tests were carried out on the passenger module of the universal complex for testing the road surface and car tires KUIDM-2 "Carousel", built on the territory of the MADI Educational and Scientific Center (MADI ESC). On a circular stand, the real conditions of interaction of a studded tire of a car wheel with a cement concrete surface were modeled. Road pavements were arranged on the track of the universal ring complex in the form of segments - concrete slabs 100 x 120 cm in size and 10 cm thick. The concrete slabs were subjected to repeated abrasion by spiked rubber. Research has made it possible to establish the influence of the nature of crushed stone, its strength and size of fractions on the wear resistance of cement concrete pavements.

1. Introduction

The problem of rut formation on road surfaces is characteristic not only of the highways of the Russian Federation, but also of other countries of the world [1]. In contrast to non-rigid road pavements, rutting on cement concrete pavements is not a consequence of plastic and permanent deformations, but a consequence of wear and abrasion of the surface layer of the pavement, mainly in winter under the influence of studded rubber. This leads not only to a decrease in the turnaround time, but also to a deterioration in road safety [2, 3]. During the period of atmospheric precipitation, water accumulates in the track, which leads to a decrease in the adhesion qualities of the road surface and the appearance of the aquaplaning effect when vehicles are moving at high speeds. The study of the effect of the composition of concrete, the strength of crushed stone and the size of its fractions, as well as construction technology on the abrasion of cement-concrete coatings under the action of studded rubber, is an urgent task.

2. Statement of the problem

The main advantages of cement-concrete pavements of highways in comparison with asphalt-concrete ones are their high strength and durability, which directly depend on the quality of cement, strength of coarse aggregate and its fractions, operating conditions, as well as construction technology [4]. The appearance of a track on cement concrete pavements leads to high economic costs in the repair of highways. Within the framework of the study, the task was set to conduct comparative tests for abrasion by studded rubber of cement-concrete pavements with various concrete compositions. Research allows

Content from this work may be used under the terms of the Creative Commons Attribution 3.0 licence. Any further distribution of this work must maintain attribution to the author(s) and the title of the work, journal citation and DOI.
Published under licence by IOP Publishing Ltd

to establish the influence of the nature of crushed stone, its strength and size of fractions on the wear resistance of cement concrete pavements.

3. Research methods and materials

To determine the effect of studded tires on the cement concrete pavement, tests were carried out on the passenger module of the universal complex for testing road pavement and automobile tires KUIDM-2 "Carousel", built on the territory of the MADI Educational and Scientific Center (MADI ESC). The universal complex KUIDM-2 "Carousel" (Fig. 1) is designed for comparative tests of road surfaces and automobile tires in conditions as close as possible to operational ones [5]. In terms of tactical and technical characteristics, it surpasses all known domestic and foreign analogues. Provides the ability to optimize the parameters of materials for road surfaces and car tires, is currently the only system in the Russian Federation that provides research of the "car-road" system in the conditions of real interaction of elements.

Fig. 1. Circular stand KUIDM-2 "Carousel" (MADI)

At the circular stand "Carousel", the real conditions of interaction of a studded tire with a cement concrete surface were modeled. Cement-concrete pavements were arranged on the track of the passenger module of the universal ring complex in the form of segments - concrete slabs 100 x 120 cm in size and 10 cm thick. The concrete slabs were subjected to repeated abrasive action of spiked rubber.

Table 1. Parameters of the passenger module of the universal ring complex

№	Parameters	Units	Values
1	Road load	kg	550
2	Tire pressure	atm.	2,4
3	Linear wheel speed	km/h	85-100
4	Wheel trajectory radius	m	16
5	Track gauge on concrete slab	mm	210
	Wheel parameters:		
6	Radius	inch	16
7	Profile width	mm	205
8	Profile width to height ratio	%	65
9	Number of thorns per 1 running meter	psc.	90

IOP Conf. Series: Materials Science and Engineering 1159 (2021) 012029 doi:10.1088/1757-899X/1159/1/012029

When preparing concrete mixtures for the construction of cement-concrete pavements, the following materials were used:

1. Cement PC 500-D0-N in accordance with GOST 10178-85;

Cement CEM I 42.5N in accordance with GOST 31108-2016.

2. Microsilica MKU-95 (bulk density 300-400 kg / m^3).

3. Rubble (GOST 8267-93):

- rubble from porphyrite fraction 5-15 mm;
- rubble from gabbro-diabase, fractions 5-10 mm and 10-20 mm;
- rubble from gabbro-diabase after electrohydrodynamic crushing of 4-8 mm fraction;
- rubble from granite of fraction 5-10 mm and 10-20 mm.

3. Natural coarse sand of I class (washed) (GOST 8736-2014).

4. Chemical additives:

- air-entraining additive Micro Air 125;
- «curing-TAL» set retarder.

Table 2. Type of crushed stone and the size of its fractions in the preparation of concrete mixes

Concrete composition number	Rubble type	Rubble fraction size, mm
№1	gabbro-diabase	4 – 8
№2	gabbro-diabase	5-10 и 10-20
№3	granite	5-10 и 10-20
№4	gabbro-diabase	5-10 и 10-20
№5	granite	5-10
№6	porphyrite	5-15
№7	porphyrite	5-10
№8	porphyrite	5-15
№9	granite	5-10

On slabs of concrete of compositions No. 3 and No. 9, the surface of freshly laid concrete was treated with a brush with a plastic bristle. Such a technological operation is often performed in the construction of cement-concrete pavements in order to increase the coefficient of adhesion of the pavement to the car wheel [6]. The surface of the cement concrete pavement of composition No. 8 was cleaned from the mortar part with a metal brush using the "Wash beton" technology to create a rough surface.

Every 25 thousand passes of the wheel with studded rubber, one track was taken to measure the track depth with a three-meter rail with a wedge gage (Fig. 2). The maximum number of wheel passes with studded tires was 250 thousand.

Fig. 2. Measurement of track depth with a wedge gage

4. Research results

The studies performed allowed us to establish that the highest wear resistance was demonstrated by concrete coatings of compositions No. 8 and No. 6, made using porphyrite of fraction 5-15 mm as a coarse aggregate. After 250 thousand passes of a wheel with studded rubber, the average track depth on cement-concrete pavements was 4.67 and 5.03 mm, respectively. In addition, it should be noted that the application of the "Wash beton" technology, "exposure" of a large aggregate on the surface of the road surface, made it possible to slow down the abrasion of concrete under the action of studded rubber, when the mortar part was removed from the surface of the coating. Low wear resistance was shown by coatings made of concrete of compositions No. 1 and No. 5, using crushed stone of fine fractions from gabbro-diabase fractions 4-8 mm and granite fractions 5-10 mm. The average track depth on the considered cement-concrete pavements was 10.50 and 8.17 mm, respectively (Fig. 3).

During the first 25 thousand passes of the wheel with studded rubber, all cement-concrete coatings, with the exception of the coating made of concrete of composition No. 8, had the highest abrasion resistance. During this period, abrasion of the mortar part formed on the surface of the cement-concrete pavement occurs during its construction. The highest abrasion after the first 25 thousand passes of the wheel was shown by the coatings made of concrete of compositions No. 3 and No. 9, made with the use of crushed granite. On these coatings, the surface of freshly laid concrete was brushed with plastic bristles. Later on, wear on these concrete pavements slowed down.

The test results showed that the strength of crushed stone and the size of its fractions have a significant effect on the wear of cement-concrete pavements under the action of studded rubber. The use of a durable filler (porphyrites) and its optimal fractions can reduce the abrasion of cement-concrete pavements under the action of studded rubber by 1.5-2 times.

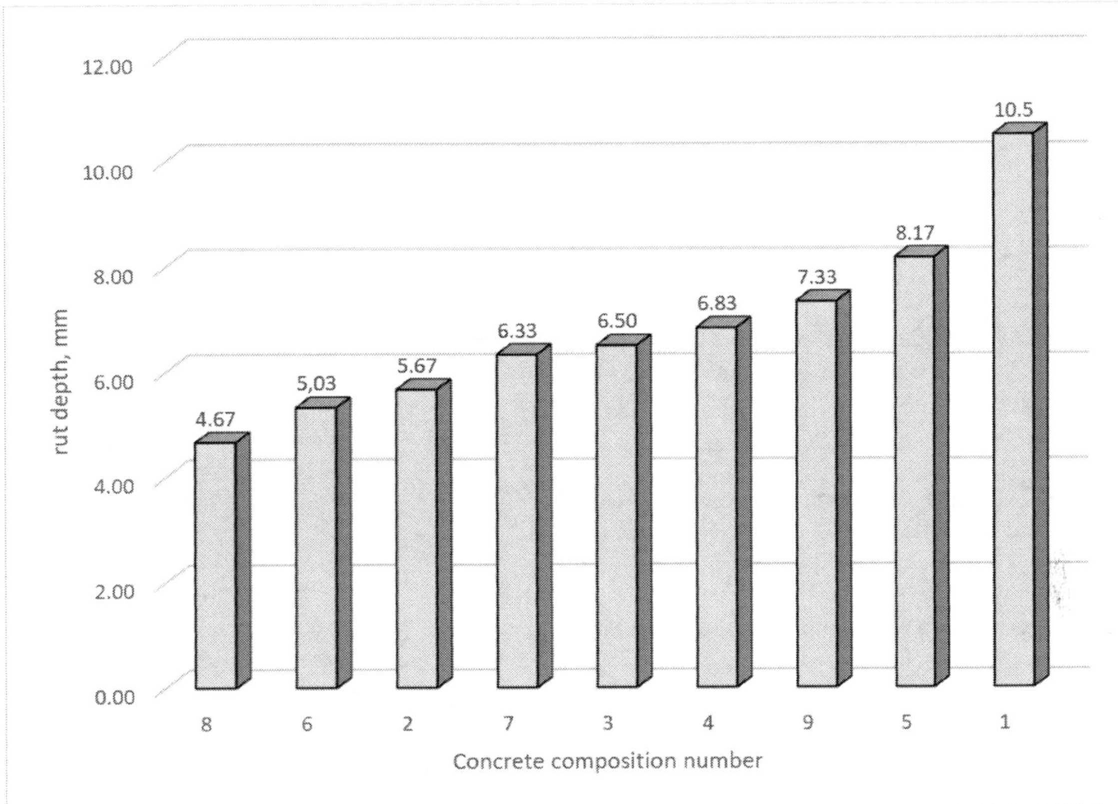

Fig. 3. The depth of the track on cement-concrete pavements after 250 thousand passes of a car with studded rubber in one track.

Fig. 4. Track on the concrete pavement of the ring track.

5. Conclusion

As a result of the studies performed, it was established:

1) the size of the coarse aggregate fractions in concrete affects the wear of cement-concrete coatings. With an increase in the size of crushed stone fractions, the wear resistance of concrete increases;

2) the wear of cement-concrete pavements under the influence of spiked rubber is influenced by the strength of crushed stone. The use of a durable filler (porphyrites) and its optimal fractions in the preparation of concrete allows to reduce the abrasion of cement-concrete coatings under the action of studded rubber by 1.5-2 times. Construction of two-layer cement-concrete pavements will provide a significant reduction in the consumption of expensive crushed stone, using it only in the upper layer of the pavement;

3) removal of the weak mortar part from the surface of the coating using the "Wash beton" technology, which allows to slow down the wear of the cement-concrete coating under the action of studded rubber.

References

[1] Li X, Zhang K, Bahadori A and Muhunthan B 2020 *Journal of Materials in Civil Engineering* **32(3)** 04020023
[2] Ushakov V V and Dyakov G G 2014 *Sci. and Engr. for highways* **67(1)** 30-32
[3] Vasiliev Yu E, Ivachev A V and Bratishchev I S 2014 Investigation of the stability of road building materials to wear rutting in conditions close to operational *Naukovedenie* **5(24)** 11TVN514
[4] Ushakov V, Jelagin D, Dyakov G, Dmitriev S and Goryachev M 2020 *Int. Conf. on Digital Solutions for Automobile Industry (DS-ART 2019)* (Cholpon-Ata) **832(1)** (IOP Conference Series: Materials Science and Engineering) 0120372019
[5] Prikhodko V M and Vasiliev Yu E 2014 *Promyshlennoe i grazhdanskoe stroitel'stvo* **12** 37-40
[6] Korochkin A V 2019 *Stroitel'nye materialy (Construction materials)* **7** 21-27

IOP Conf. Series: Materials Science and Engineering 1159 (2021) 012030 doi:10.1088/1757-899X/1159/1/012030

Numerical modeling of multilayer asphalt-concrete roads based on the finite element method

I V Demyanushko[1], V M Stain[1], A V Stain[2]

[1] Department of Construction Material Technology, Moscow Automobile and Road Construction State Technical University (MADI), Leningradskiy prospect, 64, Moscow, 125319, Russia
[2] LLC "MIP Research Institute of Mechanics and Quality Problems", Leningradskiy prospect, 64, Moscow, 125319, Russia

E-mail: demj-ir@mail.ru, vamis@yandex.ru

Abstract. The task of creating progressive technologies for the construction of highways and high-speed railways require the development of new methods of computational and design analysis based on computer technologies and modern methods of mechanics, considering the complex, physically and geometrically nonlinear nature of the problems being solved, the peculiarities of the operation of multilayer structures and the cyclical nature of transport loads. Studies on the application of the finite element method (FEM), carried out at the Department Construction material technology, MADI, have shown the effectiveness of this method for calculating road surfaces and high-speed railways, considering the real work of bases, structural features of longitudinal and transverse seams, etc. These studies made it possible to identify such features of the behavior of road structures that could not be described by existing engineering techniques. The presented finite element models for solving various problems of analyzing the behavior of pavement structures have shown the prospects of using the developed approaches for digital modeling of multilayer roads with asphalt concrete pavement.

1. Introduction

During the operation of road and airfield pavements, cracks form in them, which, spreading both in the horizontal direction and deep into the pavement, over time can lead to fragmentation of the pavement and the impossibility of its further use. Considering the annually increasing intensity of traffic flows and an increase in the weight of vehicles, the process of destruction of coatings can be significantly accelerated, and the real terms of durability of coatings will significantly differ from those laid down in the project. In this regard, it becomes necessary to develop modern methods for calculating and designing road and airfield pavements, considering the latest advances in computer technologies and methods of fracture mechanics. The analysis is based on the development of finite element adequate spatial models for analysis. Our works on the application of numerical methods of nonlinear spatial mechanics to solving the problem of modeling the behavior of multilayer road surfaces are presented in [1–4]. In this paper, we consider some new approaches to the construction and verification of the adequacy of some digital models of road structures, which make it possible to take a certain step in the development of methods for virtual digital analysis of the work of multilayer road pavements.

Content from this work may be used under the terms of the Creative Commons Attribution 3.0 licence. Any further distribution of this work must maintain attribution to the author(s) and the title of the work, journal citation and DOI.
Published under licence by IOP Publishing Ltd

2. Finite-element model of a laboratory installation of a 4-point bend of a beam

In Russian Federation and abroad, laboratory installations of 4-point bending of rectangular beams are most often used to study the fatigue life of asphalt concrete (Fig. 1).

Tests for 4-point bending are most consistent with the operating conditions of asphalt concrete in the structures of real road surfaces and therefore allow you to evaluate the fatigue life of various asphalt concrete mixtures in the road surface.

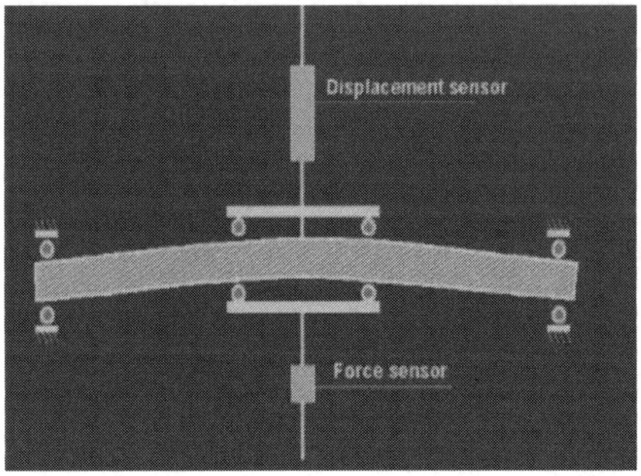

Figure 1. Design diagram of the installation for fatigue testing of a girder asphalt concrete sample for 4-point bending

Fatigue life testing of asphalt concrete mixtures requires the use of complex expensive equipment and a significant investment of time. Therefore, replacing at least some of these tests with virtual tests using finite element models is an urgent task for a researcher.

In our work, to assess the durability of an asphalt concrete beam specimen at 4-point bending, we used methods of fracture mechanics. The finite element model (FE model) of the experimental setup is shown in Figure 2.

Figure 3 shows the simulated vertical displacement distribution in a finite element model of an asphalt concrete specimen at a load frequency of 10 hertz and a crack length of 31.5 mm.

The effect of load frequency on the durability of an asphalt concrete specimen is illustrated by the graph in Figure 4.

In [5], based on the results of field experiments, it was shown that there is a relationship between the speed of a heavy vehicle (with an axle load of 100 kN) and the vibration frequency of the asphalt concrete pavement

$$f = 0.277 v^{0.944} \quad (1)$$

where f is the vibration frequency of the asphalt concrete pavement;

v is vehicle speed.

In the same work, based on the results of laboratory tests of asphalt concrete samples for 4-point bending, a relationship was proposed between the frequency of application of a load to an asphalt concrete sample and the speed of a heavy vehicle in the form

$$f\,[\mathrm{Hz}] \sim 0.5 v\,[\tfrac{\mathrm{km}}{\mathrm{h}}] \quad (2)$$

where f is the frequency of the load applied to the asphalt concrete sample;

v is vehicle speed.

It follows that, using formulas (1 - 2), and the results of calculations using a finite element model of an experimental setup for testing on a 4-point bend, it is possible to judge the influence of the speeds of a heavy vehicle on the durability of the asphalt concrete road surface.

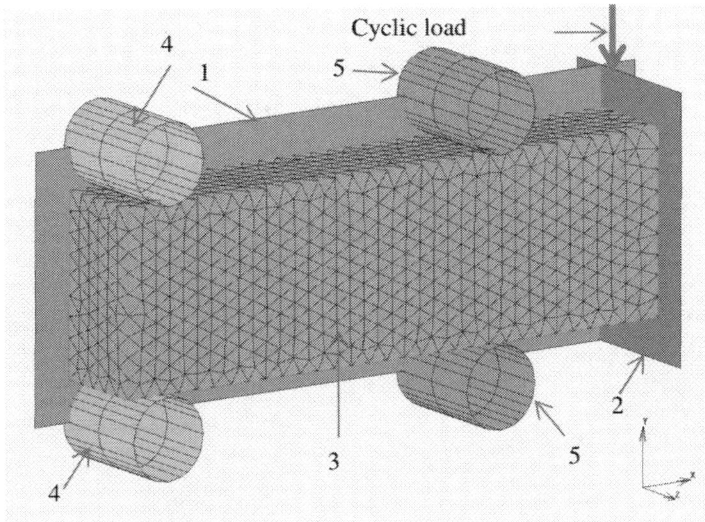

Figure 2. Finite element model of the experimental setup: 1 - longitudinal plane of symmetry of the experimental setup; 2 - transverse plane of symmetry of the experimental setup; 3 - asphalt concrete sample; 4 - stationary steel cylindrical supports; 5 - steel cylinders transmitting a cyclic load to an asphalt concrete sample

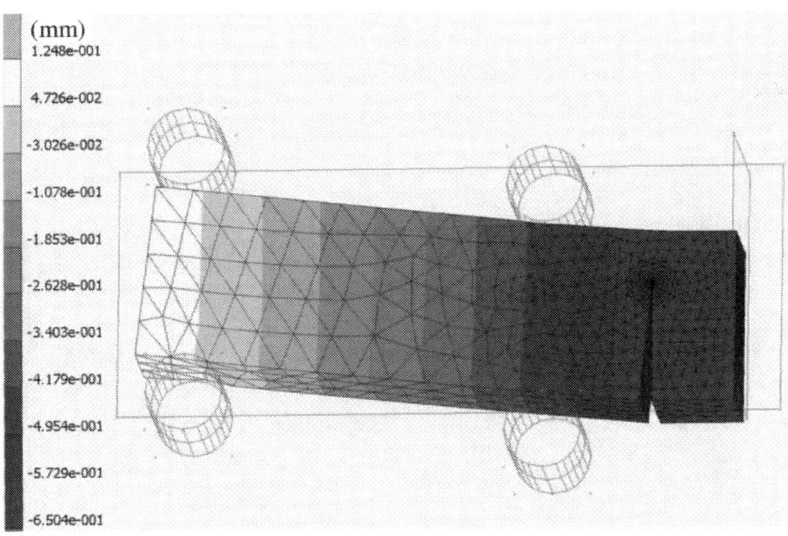

Figure 3. Distribution of displacements in the finite element model at a load frequency of 10 Hz and a crack length of 31.5 mm

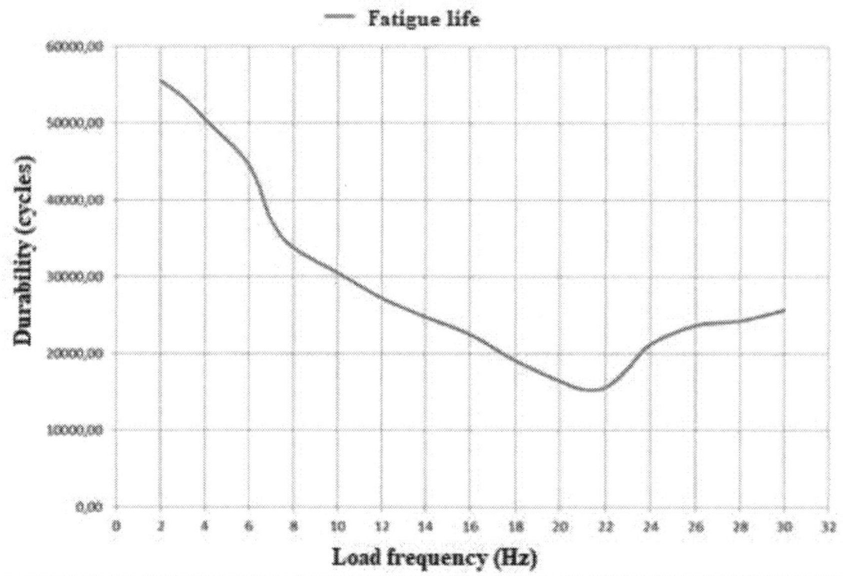

Figure 4. Graph of changes in the durability of an asphalt concrete sample depending on the frequency of load application

It follows from the graph in Figure 4 that an increase in the speed of heavy vehicles above 42 km/h should lead to an increase in the fatigue life of asphalt concrete pavements.

3. FE model of temperature interaction of AC coating with the environment

To create a FE model of the thermal interaction of the road surface with the environment, the results published in [6] were used. In this work, a one-dimensional heat exchange of an asphalt concrete pavement with the environment was investigated for three days. The solution obtained by the finite difference method was compared with the experimental results.

In our work, the same problem was solved by the finite element method. A finite element model of the road surface in a one-dimensional setting is shown in Figure 5.

The coating, in accordance with Figure 6, is exposed to solar and atmospheric radiation. The heated coating radiates heat to the atmosphere, is cooled by wind currents and exchanges heat with the underlying coating layers and the substrate. All of these effects are modeled using appropriate initial and boundary conditions.

The results are illustrated by the graphs shown in Figure 7. As can be seen from the comparison of these graphs, the results obtained in this work by the finite element method (FEM) are in good agreement with both the experimental results and the results obtained by the finite difference method (FDM).

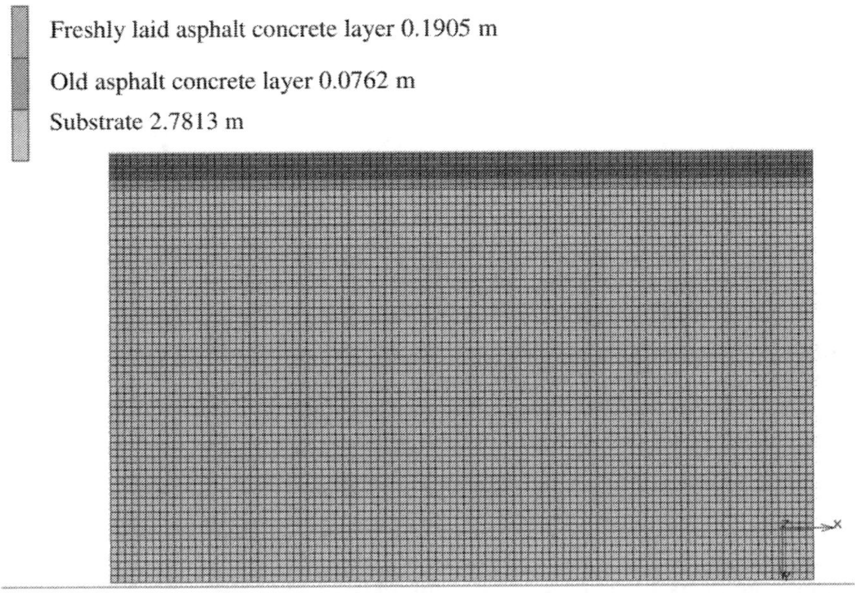

Figure 5. One-dimensional finite element model of asphalt concrete pavement for calculating heat transfer between the pavement and the environment

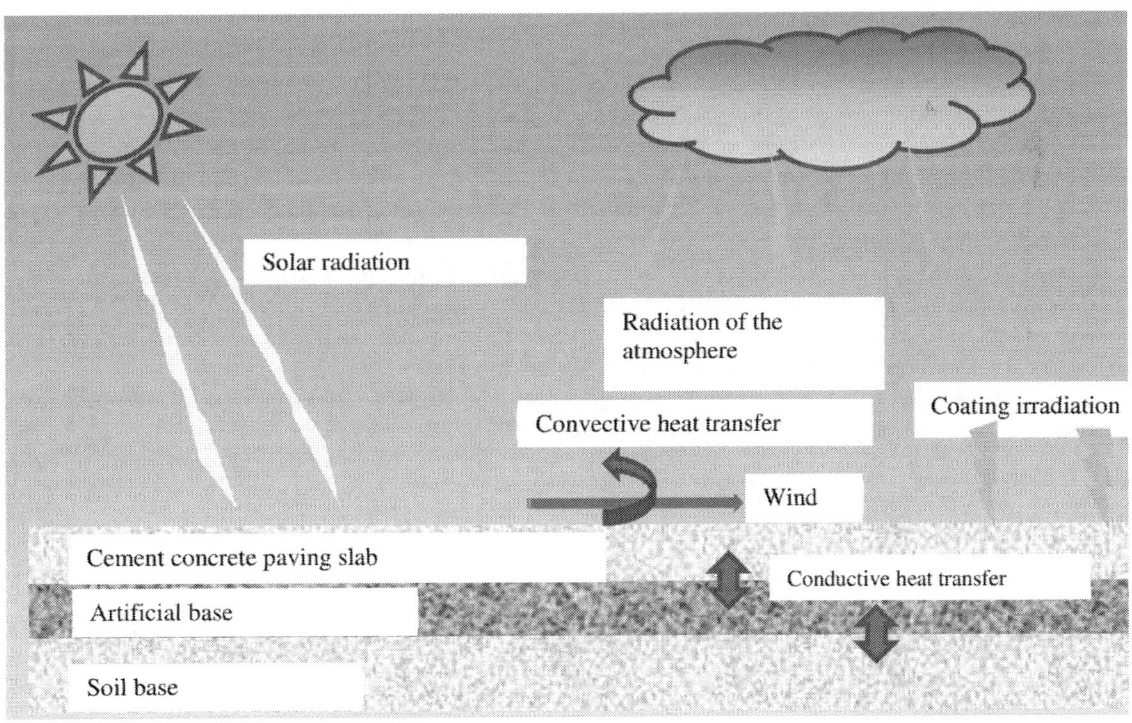

Figure 6. Diagram of the components of the thermal balance of the road surface with the environment

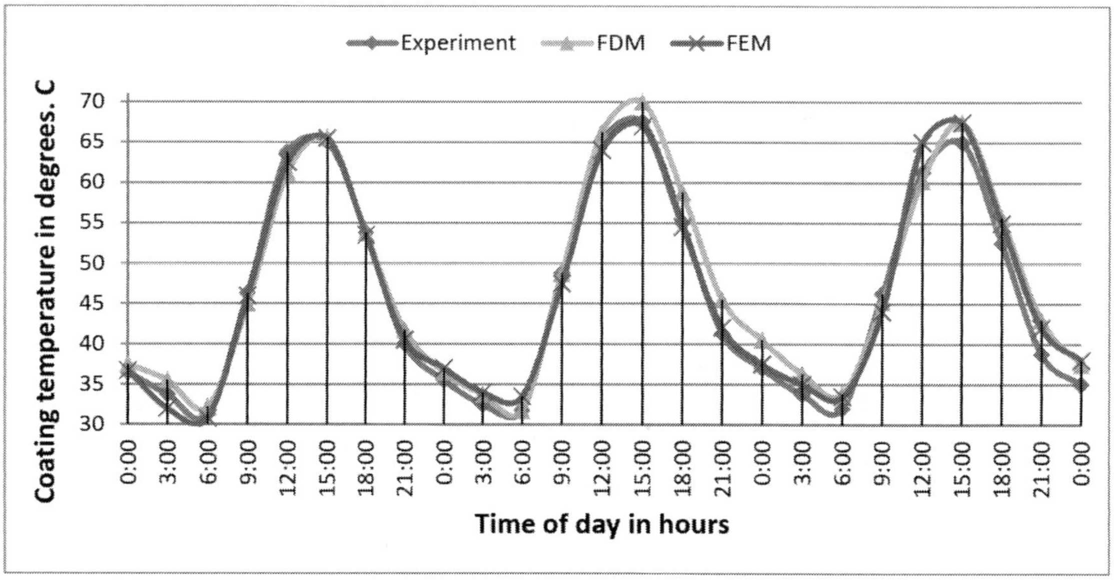

Figure 7. Graphs of temperature changes at a depth of 12.7 mm from the coating surface for three days, obtained experimentally, by the finite difference method (FDM) and the finite element method (FEM)

The proposed finite element model can be used to study the thermal interaction of the road surface with the environment in a flat and spatial setting.

4. Influence of the thickness of the asphalt concrete layer on the fatigue life of the road surface and the study of rut formation

The axisymmetric finite element model of the AC coverage gives the results of calculating stresses and strains close to the results obtained using the spatial finite element model, but, at the same time, requires significantly fewer computing resources. Therefore, to assess the effect of the thickness of the layer of asphalt concrete on the fatigue life of the road surface, it was the axisymmetric finite element model that was chosen (Figure 8). It is used to analyze the durability of AC coatings using fracture mechanics methods.

The load from a car wheel in this model was transferred to the coating through a circular contact area with an area of 0.02 m^2. The contact pressure was taken to be 0.83 MPa. The calculated scheme of the load on the asphalt concrete pavement from the vehicle (trailer) is shown in Figure 9a.

At a trailer speed of 80 km/h, the load versus time can be represented as shown in Figure 9b. It was assumed that the repeated passage of the trailer took place 30 seconds after the previous one. Base layers were considered as linearly elastic bodies. The material of the AC layer was assumed to be viscoelastic.To determine the fatigue life of the AC coating, a crack propagation calculation method was used, known as the Virtual Crack Closure Technique.

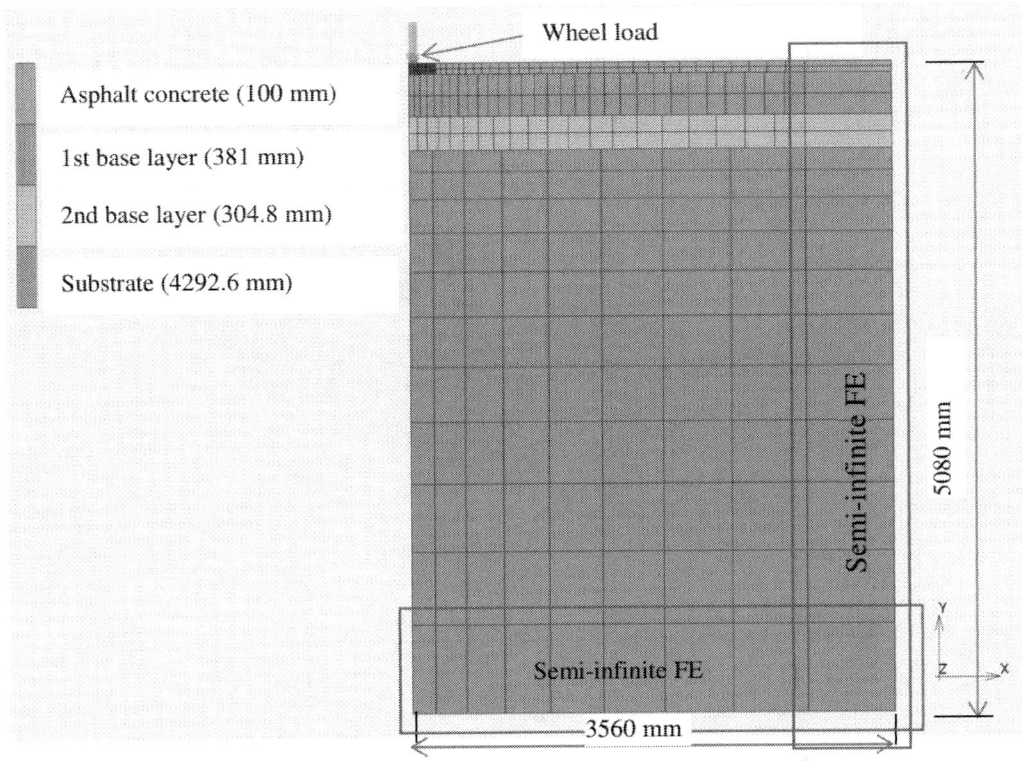

Figure 8. Axisymmetric finite element model of asphalt concrete pavement created in the Marc® and Mentat® finite element package

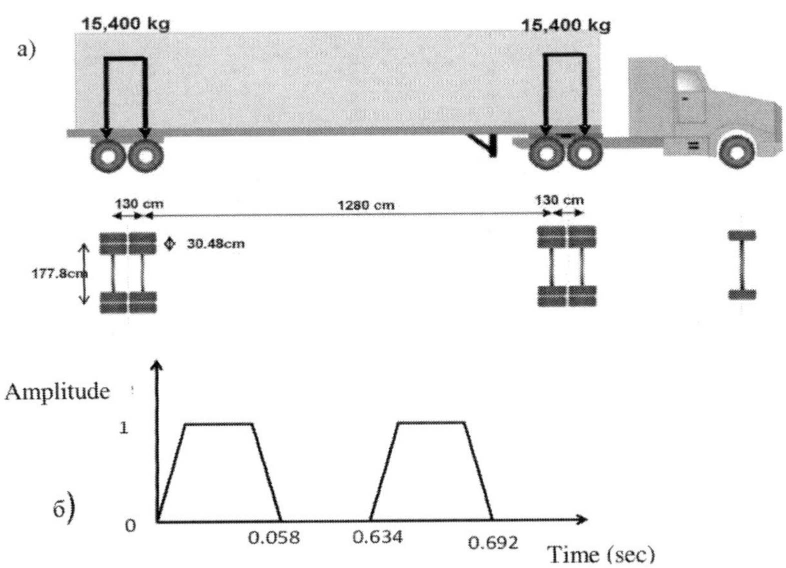

Figure 9. Design diagram of the load on the asphalt concrete pavement from the vehicle (trailer): a) - load diagram; b) - for a trailer speed of 80 km / h

As a result of the numerical experiment, it was found that for relatively thin AC coatings 5 and 10 cm thick, fatigue cracks appear on the lower edge of the AC layer and develop from bottom to top. At the same time, for relatively thick AC coatings with a thickness of 15 and 20 cm, fatigue cracks appear on the upper edge of the AC layer and develop from top to bottom.

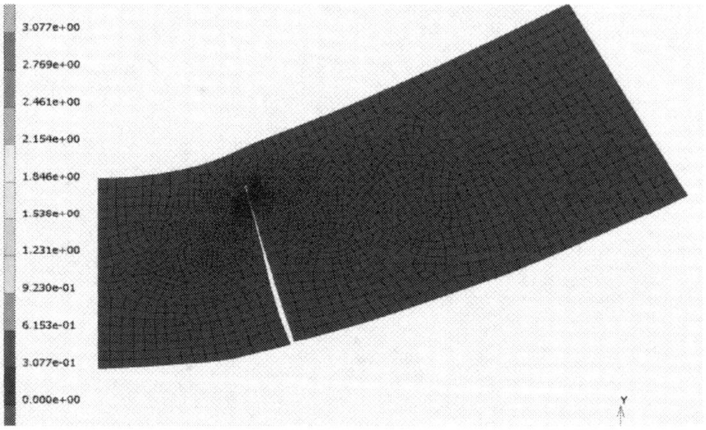

Figure 10. Crack position corresponding to 1.10491×10^7 loading cycle of AC coating

These results are consistent with the results of studies by other authors.

As an example, illustrating the results of calculations of the fatigue life of AC pavements under the action of a transport load, the results obtained for a road surface with an AC layer thickness of 10 cm are presented.

Figure 10 shows the crack position corresponding to the 11.10491×10^7 loading cycle of the AC coating. Deformations for clarity of figures are increased by 1000 times. The results of studying the dependence of the durability of the AC coating on the thickness of the AC layer are presented in the form of a graph shown in Figure 11.

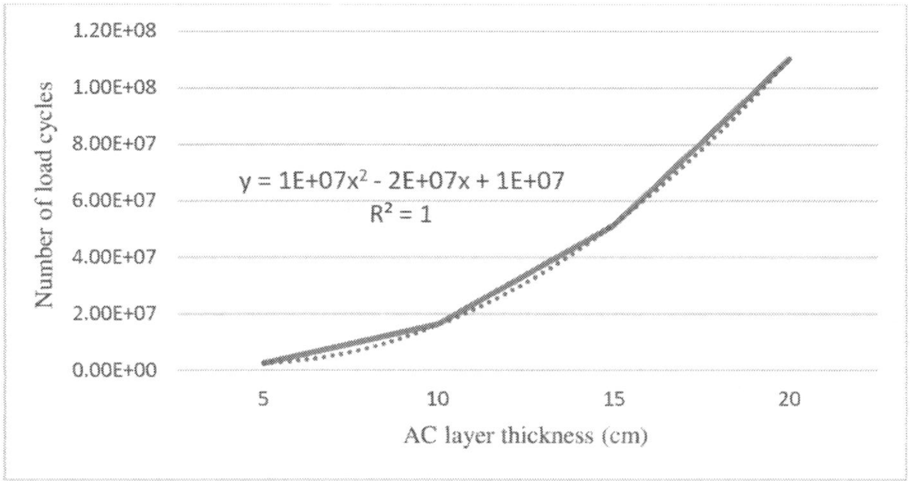

Figure 11. Graph of the dependence of the durability of the AC coating on the thickness of the AC layer

It follows from this graph that the durability of the AC coating increases with an increase in the thickness of the AC layer, and the relationship between the thickness of the AC layer and the durability of the coating can be approximated by a polynomial of the 2nd degree (red curve)

$$Y(x) = 10^7 x^2 - 2 \times 10^7 x + 10^7 \quad (3)$$

with an approximation reliability $R^2 = 1$.

Within the framework of the axisymmetric task (Figure 8), the issue of the appearance of a rut in the asphalt concrete pavement under the action of a cyclic load that changes over time was also considered in accordance with Figure 9.

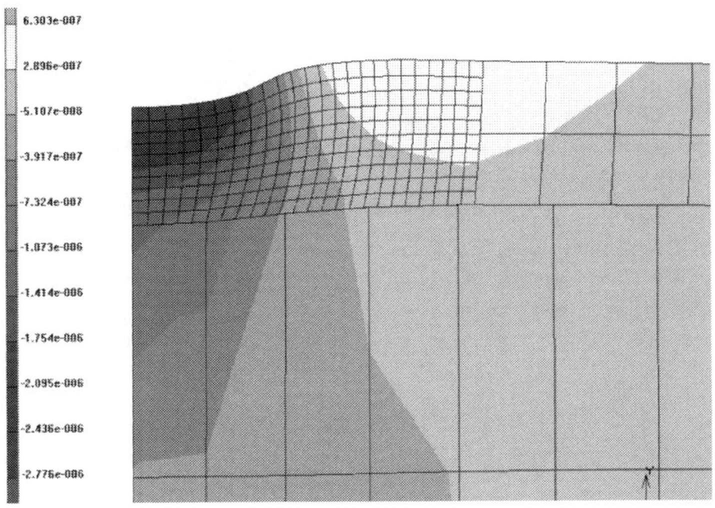

Figure 12. Fragment of the FE model, reflecting the distribution of the residual vertical displacements of the surface of the asphalt concrete pavement after 1000 load cycles (the displacements are increased by 10000 times for the clarity of experiment)

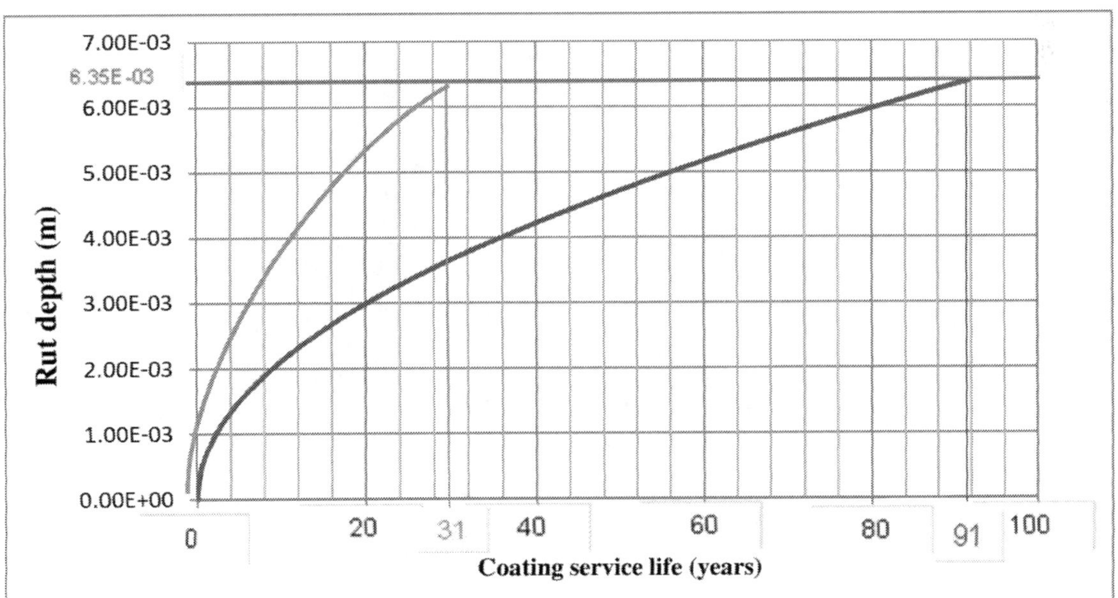

Figure 13. Comparison of service life up to the formation of the maximum track depth of 6.35 mm, obtained by the mechanical-empirical method (MEPDG) (green curve) and using the FE package MARC-MENTAT (red curve)

IOP Conf. Series: Materials Science and Engineering 1159 (2021) 012030 doi:10.1088/1757-899X/1159/1/012030

Figure 12 shows a fragment of the FE model showing the distribution of the residual vertical displacements of the surface of the asphalt concrete pavement after 1000 load cycles.

The results reflecting the increase in the track depth in the asphalt pavement depending on its service life are shown in Figure 13.

The maximum track depth is considered to be 6.35 mm (horizontal blue line).

The red curve practically coincides with a similar curve constructed in [6] based on the results of calculations in the ABAQUS finite element package. A significant discrepancy between the results found by the mechanical empirical method (MEPDG) and the finite element method, in [7], is explained by the lack of consideration in the finite element model of the viscous plastic properties of asphalt concrete and the mechanisms of structural destruction of the material.

5. A model for studying the effect of the asphalt concrete bearing layer on the deformations of the ballastless topside of the railway track

In modern operating conditions of railway tracks with an increase in the speed of rolling stock, axle load, an increase in the mass and length of trains, one of the main problems is the deformability of the track. The use of a ballastless structure - a ballastless track superstructure (BTSP) - is one of the options for increasing track stability.

The results of long-term monitoring show that the load-bearing layer of asphalt concrete in the BVSP structure reduces deformation of the subgrade, which has a beneficial effect on reducing the cost of the life cycle of railway equipment.

The Moscow-Kazan-Yekaterinburg high-speed rail structure was taken as the basis for creating a FE model of the high-speed rail with BTSP and a carrier layer of asphalt concrete, which, in turn, is based on the ballast-free structure of the superstructure of the CRTS III (China Railway Track System III) track. developed in China.

The task of creating a digital model for studying the operation of the asphalt concrete layer of BTSP was proposed by N.V. Bystrov.

The spatial model of the ballastless rail track structure created in the licensed FE complex Ls-Dyna is shown in Figure 14.

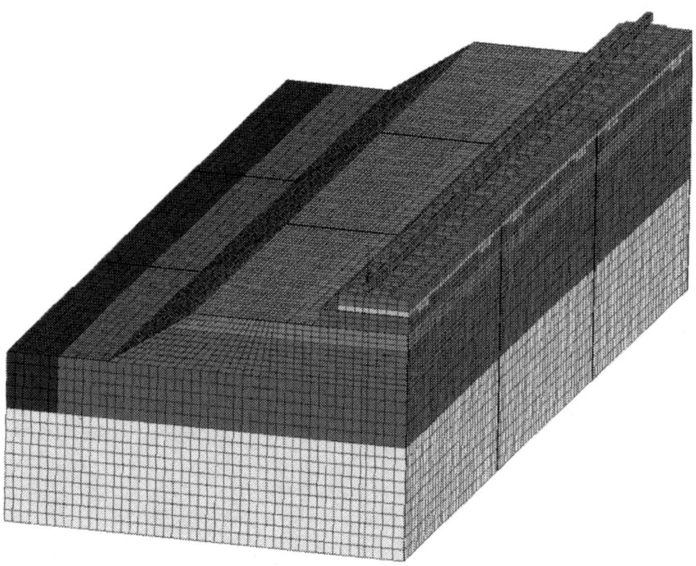

Figure 14. Spatial image of the FE model of a ballastless railway track

The calculation results are presented in Figures 15 and 16.

The vertical displacements of the base of the BTSP with an asphalt concrete layer thickness of 30 cm at a time instant of 0.071999 seconds are shown in Figure 15. Graphs showing how the maximum deflections of the foundation slab varied over time depending on the asphalt concrete layer thickness are shown in Figure 16.

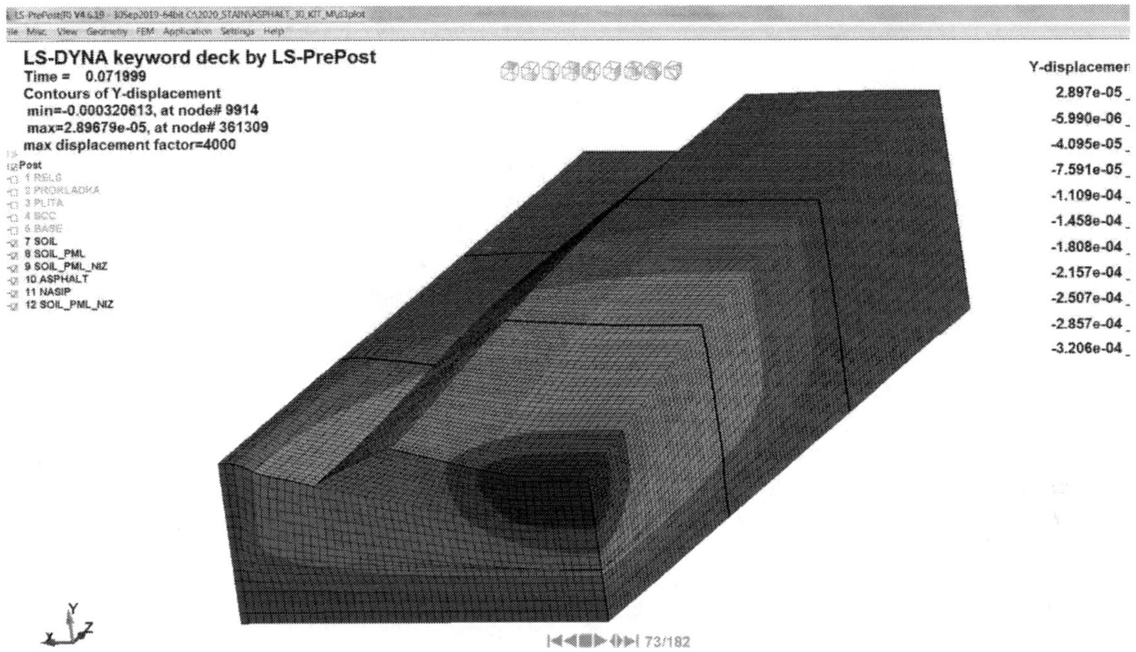

Figure 15. Vertical displacements of the BTSP base with a layer of asphalt concrete of 30 cm at a time of 0.071999 sec. The movements are enlarged 5000 times for the clarity of experiment

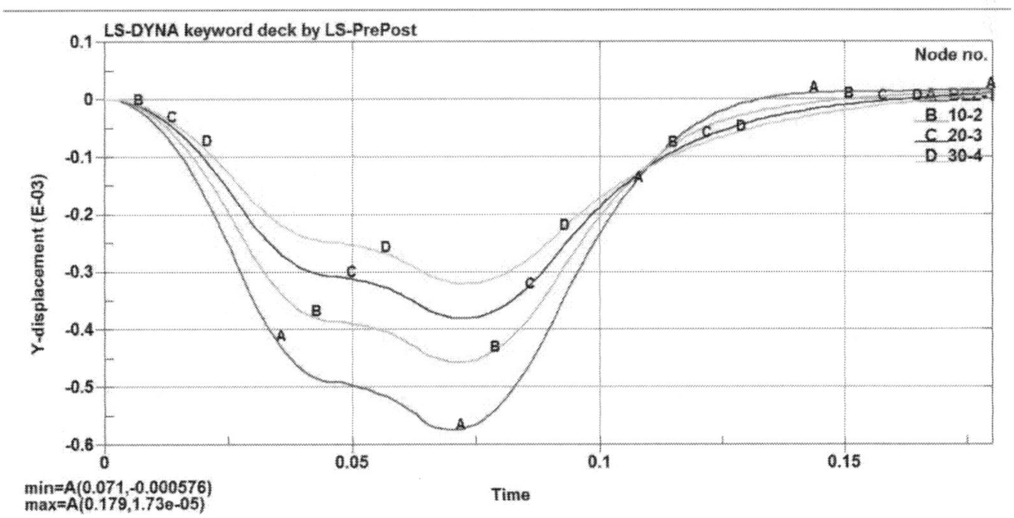

Figure 16. Graphs of the vertical displacements of the foundation slab versus time for the thickness of the asphalt concrete slab: A-0 cm; B-10 cm, C-20 cm, D-30 cm.

In accordance with work [8], it is assumed that the maximum displacement for the lower edge of the foundation slab is 0.5 mm, then the maximum displacement equal to 0.576 mm (Figure 16), obtained for the construction of a railway track without a layer of asphalt concrete, exceeds the permissible by 15.2%.

IOP Conf. Series: Materials Science and Engineering 1159 (2021) 012030 doi:10.1088/1757-899X/1159/1/012030

However, even with a 10 cm thick layer of asphalt concrete in the railway track structure, the maximum displacement of the bottom edge of the foundation slab becomes 0.458 mm, which is 8.4% less than the permissible displacement. With the thickness of the asphalt concrete layer equal to 30 cm, the maximum displacement of the lower edge of the foundation slab becomes 0.321 mm, which is 35.8% less than the permissible displacement.

6. Conclusion

The task of creating progressive technologies for road construction requires the development of new methods of computational and design analysis based on computer technologies and modern methods of mechanics, considering the complex, physically and geometrically nonlinear nature of the problems being solved, the peculiarities of the operation of multilayer road structures and the cyclical nature of transport loads. At present, the methods of fracture mechanics are widely used abroad in the study of the stress-strain state of road structures. In this case, any impacts, both transport and climatic, are considered.

The results obtained by various researchers have not yet been reflected in foreign regulatory documents, but judging by the intensity of the development of this direction in recent years, the norms for the calculation and design of road structures based on the use of fracture mechanics methods may appear within the next five years.

Previous studies and the latest results presented in this work on the application of the finite element method (FEM) for modeling the loading and destruction of structures of multilayer pavements, performed at the Department of Construction Mechanics, MADI, have shown the effectiveness of this method, including for digital modeling of roads with asphalt concrete pavement.

References

[1] Demyanushko I V, Stain V M, Stain A V 2013 *Transport construction* **8** 14–17
[2] Demyanushko I V, Stain V M, Zubov N V 2018 *Bulletin of the Moscow Automobile and Road Construction State Technical University* (MADI) **2** 32–37
[3] Demyanushko I V, Stain V M, Stain A V 2020 *Road Power* **97** 56–59
[4] ODM 218.3.051-2015 154
[5] Mollenhauer K, Wistuba M, Rabe R 2009 *2nd Workshop on four-point bending University of Minho* 16
[6] Kamil E, Joby D, Jay G and Patrick E 2008 *Portland Cement Association Final Report* 139
[7] Kim Y R, Ban H, and Im S 2010 MATC *Final Report No. 223*, (Lincoln, Nebraska
[8] Jie Z, Xianhua C, Qinghong F, Gang X, and Degou C 2019 *Appl. Sci.* **9** 19

IOP Conf. Series: Materials Science and Engineering 1159 (2021) 012031 doi:10.1088/1757-899X/1159/1/012031

Development and scientific justification a set of the high-speed road construction machines with hardware-percent integration and intellectual interactions algorithm of various technical systems

G V Kustarev[1], A V Ushkov[1]

[1] Moscow Automobile and Road Construction State Technical University, 64, Leningradsky prosp., Moscow, 125319, Russia

E-mail: Ushkov.AV@yandex.ru

Abstract. The description of complex technical systems with the implementation of hardware and software integration through the introduction of additional controls is presented. Recommendations for increasing the service life of the road surface and preventing premature repair of asphalt concrete surfaces are presented. A review of possible difficulties in implementing new technologies in road construction companies is performed. The project of modernization of new equipment is described, considering the research conducted in the field of finding alternative solutions to eliminate the "human" factor during road construction works. The interaction of the components of the complex for high-speed construction of asphalt concrete pavements is considered. The work of machines is described from the energy point of view, which is based on various interaction processes. The main components of the system structure and principles are considered. Multi-factor processes of road construction equipment operation are explained. The structure of algorithms for implementing an integrated machine management system is presented. The structure of hardware interaction through mathematical modeling to direct control of machine functionality in real time is described. Asphalt paving is represented by a multi-factor process, where one of the significant factors of the quality of the laid asphalt concrete is the road construction equipment used, which provides the required quality parameters of the road surface. The article describes a method for creating complex structural elements that allow creating artificial intelligence systems in the future.

1. Introduction

When building and repairing road surfaces, contractors are increasingly faced with the problem of introducing new technologies and machines into the road construction process [1]. A group of scientists from the Department of Road-building machines MADI jointly with the plant "BETSEMA" was designed and developed high-performance range of machines (figure 1), consisting of haul trucks with bottom discharge loader asphalt mix, paver and band sealing equipment. In the resulting full-scale samples, it is possible to configure various parameters and operating modes of each machine. After the introduction of these machines into operation, the task appeared not only to manage each machine in the complex, but also the entire complex. This task is determined by several factors that affect the choice of equipment suitable for road construction. The first factor is economic feasibility. Not every organization can afford to purchase new equipment for work, and even more so if we are not talking about one car,

Content from this work may be used under the terms of the Creative Commons Attribution 3.0 licence. Any further distribution of this work must maintain attribution to the author(s) and the title of the work, journal citation and DOI.
Published under licence by IOP Publishing Ltd

IOP Conf. Series: Materials Science and Engineering 1159 (2021) 012031 doi:10.1088/1757-899X/1159/1/012031

but about a whole complex of cars. For small developing companies, there is a solution – to purchase such equipment on lease, but then every minute of equipment downtime will cost "a pretty penny", and to coordinate all processes so that there are no downtime is a very difficult task.

Figure 1. Complex of machines for high-speed technology of laying asphalt concrete surfaces.

But even if a road construction company is ready to accept such conditions, there is a second important factor-the competence of its employees when working with new equipment using new technology. After all, even when using the old classic technology of laying asphalt concrete pavements, when considering the work of the road construction team, it is almost always possible to talk about violations of technology, and violation of technology leads to loss of quality of the roadbed [2], which in turn leads to decrease in the service life of the road surface and premature repairs. If we consider the new technology of high-speed construction of asphalt concrete pavements, then the nuances and undesirable problems associated with the" human " factor become much greater. And if we consider the fact that the equipment used in this technology is new and not familiar to ordinary employees of a contractor, then thoughts about introducing any innovations disappear by themselves.

Based on the results of a large survey of road construction organizations, the main problems and shortcomings of new equipment and technology that most concern organizations were identified [3]. And based on the received material, it was decided to create a project for the modernization of new equipment in accordance with the requests. There was a goal-to improve road construction machines used in high-speed technology for the construction of asphalt concrete pavements, so as to eliminate as much as possible the possible negative impact of humans on the process of creating a road surface. Since it is necessary to minimize human involvement to solve the task, it is possible to implement a project of such complexity without changing the processes of construction technology only using a hardware and software integration complex and an intelligent algorithm for interaction of various technical systems.

2. Description of hardware and software integration

To understand the implementation processes in technology, we need to understand the overall goals of integration:

- reducing the cost of equipment operation (achieved by reducing the time for synchronization processes);
- increase the speed of performing the main processes of machine operation in a joint mode;
- improving the quality of the road surface;
- increase the safety of personnel (by minimizing participation in the processes of paving).

You can get closer to achieving common goals by looking at the structure and principles of the system in detail.

For each of the machines used in the technology, there is a so-called external (technological) environment, which includes equipment, equipment and objects of work.

When considering each machine of the complex, the main components can be identified. The first component is mechanical, a device that has a working organ as the final link in the chain. The second component is electronic, which includes power converters, executive motors, and drive units. And the third component is a computer, a device that is responsible for control, or through which the device or machine is controlled.

The first component is designed to convert the movements of the component elements into the movement of the working body and may include brakes, clutches, gears, mechanical links, and so on.

The second component includes a number of microelectronic devices, converters and electronics of measuring circuits, all sensors and sensor systems responsible for collecting data on the actual state of the external and internal environment.

The third component is used to control the process of mechanical movement in real time with the ability to process information from sensors. Coordination of control of movements of working bodies depending on external conditions is carried out, the organization of data exchange with peripheral devices is carried out.

Since the operation of the machine from an energy point of view is based on various pneumatic, hydraulic, or combined processes, it is necessary to add appropriate sensors and transducers to the system. Each element of the chain must be connected not only by hardware, but also by software, in order to exchange data in real time.

3. The intelligent algorithm of the interaction of technical systems

Turning to the interaction of technical systems, I would like to note the fact that even without the introduction of an integrated control system, machines can perform work in a rational range solely by setting the appropriate parameters by the machine operator [4]. Setting any operating mode by the operator of the machine may cause the machines to be out of sync with each other, since any change in the external environment, any unpredictable factor entails a change in the operating mode parameters of each machine in the complex. Since the laying of asphalt concrete pavements is a multi-factor process, it is necessary that road construction equipment both meets the parameters of variability of external conditions, and adequately responds to such changes (figure 2), adjusting the operating modes in accordance with the established parameters of the external environment. In addition to the external environment is crucial to the interaction of complex machines because the machines move each other by performing the operation between the machines must be linked [5,6] to the first machine could send information from sensors on the next machine in the processing chain, and they, in turn, was able to set modes and parameters of work in accordance with the incoming information.

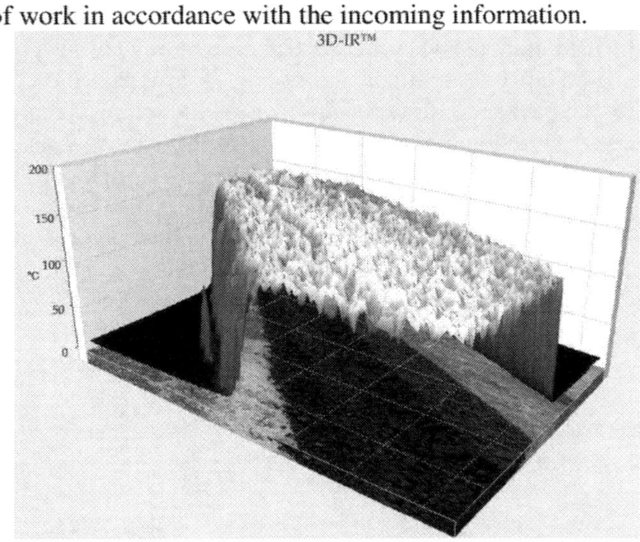

Figure 2. Influence of external factors on the cooling of the asphalt mix in the roller.

It is advisable to create the implementation of such algorithms based on several integration points (figure 3), thus forming an intelligent network of interactions between the machines of the complex [8]. At the same time, the essence of the created automatic systems is the ability to make a choice from a variety of variations and make optimal decisions based on previously obtained experience and rational analysis of external and internal environmental influences without the direct participation of the operator.

4. Theoretical background to the creation of the "smart machine" system

Any software component strives to become an element of artificial intelligence in the future. When creating complex structural elements, various approaches to the possibility of creating artificial intelligence systems may arise:

1. Logical.

This approach is based on Boolean algebra. The most interesting technique in this approach is the concept of fuzzy logic. A distinctive feature of the method is additional conditions, for example, veracity can take alternative intermediate values in addition to "Yes/no"(1/0) - the values "I don't know" (0.5), "rather Yes than no" (0.75), "rather no than Yes" (0.25).

2. Structural.

A special feature of this method is attempts to create artificial intelligence by modeling the structure and operation of the human brain.

3. Evolutionary.

Attempts to create artificial intelligence systems by creating a basic model and setting parameters and rules for changes in development (evolution).

4. Simulation.

This system uses the concept of "black box". This is a device that is a software module or random data set, where any information about the internal content is missing, but the input and output data are known. For example, an object whose behavioral characteristics are simulated is essentially a "black box", and it doesn't matter what processes take place inside, the main thing is that the model behaves exactly the same in similar simulated situations.

The most common and promising system is Logical, which involves creating algorithms for artificial neural networks and fuzzy logic.

5. Conclusion

In modern conditions of development of computer technologies, it is important to maintain a balance between the interaction of equipment and computer technologies that allow you to create complexes that provide an unprecedented quality of road construction work to repair and create new, durable asphalt concrete surfaces. Software packages allow for optimal interaction between machines within the same complex for high-speed construction of asphalt concrete surfaces, as well as between machines and the external environment, which has a significant impact on the operation modes of equipment. This allows you to increase the productivity and overall efficiency of the entire complex, which in turn will lead to an improvement in the quality of the resulting road surface, reduce the time to create such a section of road and improve performance.

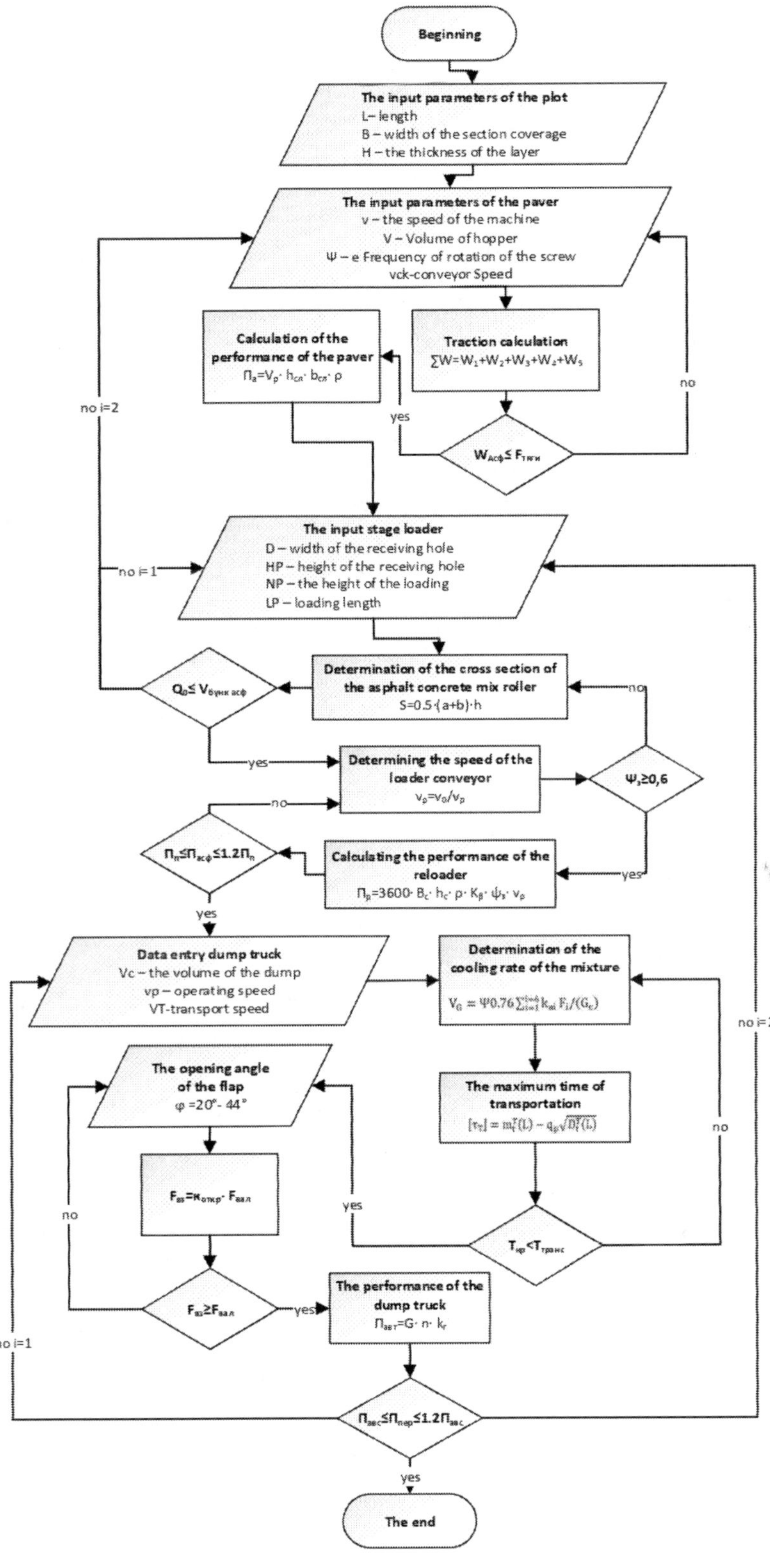

Figure 3. Algorithm for calculating the main parameters of the complex for high-speed construction of asphalt pavements

References

[1] Ushkov A V, Kustarev G V, Sobchenko E V 2014 Complex of machines for innovative technology of high-speed construction of asphalt concrete coatings *Transport business of Russia* **1** 188-191

[2] Ushkov A V 2013 *The Possibility of using foreign technologies in the construction of asphalt roads in Russian conditions* (71st scientific-methodical and scientific-research conference MADI, international seminar "problems of improvement of the design of building, road, municipal and airport vehicles") (Moscow: Moscow State University Press) **1** pp 95-98

[3] Ushkov A V 2014 Analysis of modern technologies and systems of machines for creating asphalt concrete coatings *Innovations and investments* **3** 241-243

[4] Shestopalov K K Morozov R V 2013 The parameters of the dump truck with bottom loading and reloader of asphalt-concrete mixture **7** 28-31

[5] Seliverstov N D 2016 Study of systems for the transportation and handling of asphalt mixtures road *Repair. Recovery. Modernization* **10** 27-30

[6] Kustarev G V, Ushkov A V, Shestopalov K K 2012 *Scientific justification for the design and creation of a dump truck with bottom unloading and an asphalt mixture reloader: research report scientific hands* (Moscow: Moscow State University Press) p 341

[7] Vorobyov V A, Suvorov D N, Kotlyarsky E V 2009 *Computer modeling in the automation of asphalt concrete mix production. Book 2. Practical developments* (Moscow: Publishing house of the Russian engineering Academy) p 732

[8] Balovnev V I 1994 Modeling of processes of interaction with the environment of working bodies of road construction machines: textbook. manual 2 *Mashinostroenie* p 432

Prospects for the use of composite and polymer materials in transport construction

V S Nadezhdin[1], A S Gruzdev[1], O V Titov[1]

[1] Department of construction mechanics, Moscow Automobile and Road State Technical University (MADI), 64, Leningradsky ave., Moscow, 125319, Russia

E-mail: shashki18@gmail.com

Abtract. The article discusses the prospects for the use of road construction elements made of composite and polymer materials. Currently, the main materials used in the production of road infrastructure elements are metal, concrete and plastic. The use of modern polymeric materials will eliminate the disadvantages of the materials used with the same advantages. The study considers the possibility of replacing steel energy-absorbing elements of frontal road barriers with energy-absorbing elements, which are made of a polymer material based on ultra-high molecular weight polyethylene (UHMWPE). The comparison was carried out on the basis of digital modeling of a vehicle collision with a frontal road barrier according to the energy absorption criterion using the LS-Dyna nonlinear dynamics software package. The developed digital models are validated by bench and field tests. Based on the results of the virtual tests, the dimensions of the energy-absorbing elements of the polymer structure of the frontal road fence were selected, which in their properties is not inferior to the used metal analogue. The results obtained prove the prospects of using modern composite and polymer materials in road safety systems.

1. Introduction.

Currently, metal, concrete, reinforced concrete and plastics are used for the production of road safety systems. At the same time, the overwhelming majority of barrier and cable retaining road fences are made of steel, parapet and curb fences are made of concrete or reinforced concrete, and dividers, signal posts and water barriers are mainly made of plastics (Figure 1) [1].

The disadvantages of metal fences include high cost, expensive repairs, rather complicated installation (specialized equipment is required), as well as corrosion, especially in aggressive environments.

Concrete and reinforced concrete fences have a high holding capacity when vehicles run over them, which does not allow vehicles to drive into the oncoming lane, however, installation and replacement of concrete and reinforced concrete barrier-type fences is carried out only with the help of special equipment. A significant drawback is the fact that in a collision with a vehicle, concrete and reinforced concrete fences practically do not deform and do not protect a person from injury and damage.

One of the most important advantages of plastic elements of road infrastructure (ERI) is the low cost. Another advantage of plastic products is the ability to manufacture products of almost any shape, size and color. Plastic structures are often placed on a temporary basis to mark repairs and hazardous areas. The disadvantage of plastic is the inability to provide a high holding capacity due to its low mechanical

Content from this work may be used under the terms of the Creative Commons Attribution 3.0 licence. Any further distribution of this work must maintain attribution to the author(s) and the title of the work, journal citation and DOI.
Published under licence by IOP Publishing Ltd

strength, therefore, when a collision occurs, the vehicle is able to leave the roadway or drive into the oncoming lane. Plastic products are also very sensitive to low temperatures and it is quite easy to leave mechanical damage on such structures (scratches, cuts, tears etc.).

Figure 1. Element of road infrastructure

Currently, the volume of use of composite and polymer materials for the production of ERI is growing every year. Improving road safety criteria requires the use of new materials with predetermined properties in structures. Materials with ultra-high strength, hardness, heat resistance, corrosion resistance, chemical inertness, UV resistance and other properties and their combination are required.

The main advantages of composite and polymer materials over traditional ones (metals, concrete, reinforced concrete, plastic) are:

1) a unique combination of properties uncharacteristic for other materials (strength, deformation, impact, elasticity, temperature, rheological, adhesive, electrical, frictional, heat-conducting etc.);

2) the ability to control the properties of composite and polymer materials by changing the properties of the matrix and binders or the use of special additives;

3) preservation, when changing the structure and properties of the low density of the material, and, consequently, the weight of the entire structure, as well as the possibility of almost 100% recycling.

One of the main disadvantages of composite and polymer materials is their high cost at the stage of technology creation, production of experimental products and its certification. In mass production, the use of advanced composite and polymer materials will reduce the weight of ERI structures and their cost without reducing consumer characteristics.

The growing range of materials used in the field of ERI production requires the development of testing both samples of materials (in order to obtain strength and mechanical characteristics) and parts of ERI structures (to obtain the given and consumer characteristics). A correctly developed and applied test method will allow validating the digital model at a high level, and, as a result, will lead to a verified virtual experiment.

2. Materials and equipment.

This study compares the metal structure of the frontal road fence (Figure 2) and the proposed structure made of polymer material. [2]. UHMWPE was considered as the material of the energy absorbing element. Determination of mechanical and strength characteristics was carried out using a Tira Test

2300 testing machine with a maximum machine force of 100 kN and a maximum loading speed of 1000 mm/min. Tensile loading was carried out at different loading rates (20-500 mm /min). As a result of the experiment, the value of the elongation of the sample was recorded, as well as the value of the force of resistance to elongation of the sample. According to the test results, mechanical and strength characteristics were determined, which is reflected in [3]. The rest of the characteristics were determined from studies [4, 5].

Figure 2. Frontal road metal fence

Having determined the main mechanical characteristics of UHMWPE, it was decided to develop a fence structure with an energy-absorbing element in the form of a cylinder (Figure 3).

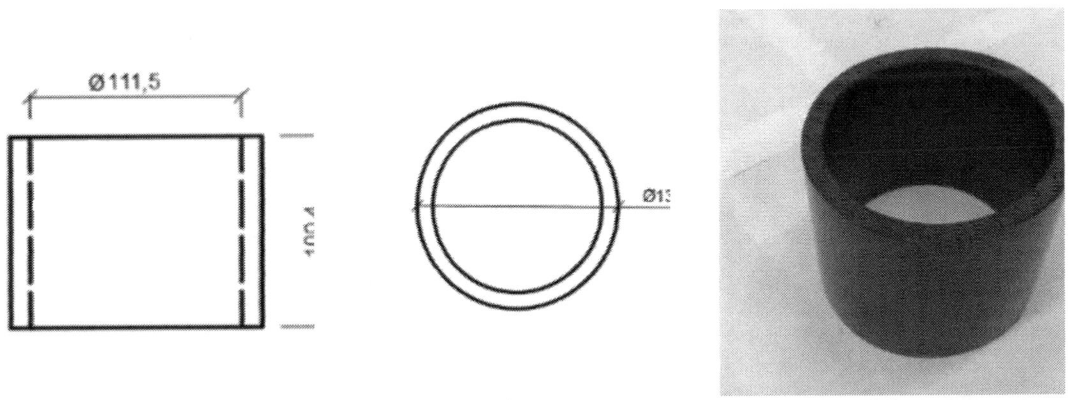

(a) (b) (c)

Figure 3. Sample of energy-absorbing UHMWPE element: (a) – sketch, frontal view; (b) – sketch, top view; (c) – full-scale sample

To determine the magnitude of the energy absorption of such an element, compression tests were carried out. The tests were carried out on a Tira Test 2300 testing machine with a maximum machine force of 100 kN and a maximum loading speed of 1000 mm/min. The test scheme is shown in Figure 4.

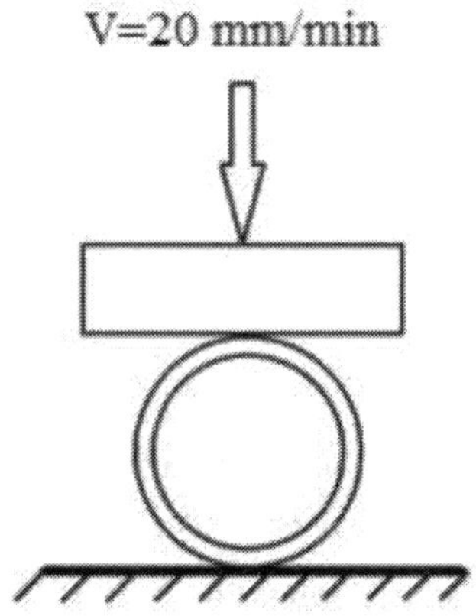

Figure 4. Compression test scheme for a sample of an energy-absorbing UHMWPE element

As a result of a full-scale experiment, a diagram of dependence $P=f(u)$ of the effort on the movement of the active traverse of the testing machine was built (Figure 5). According to the results of experimental data on the diagram, the value of energy absorption was $U_{exp} = 0{,}326$ kJ. The sample after testing is shown in Figure 6.

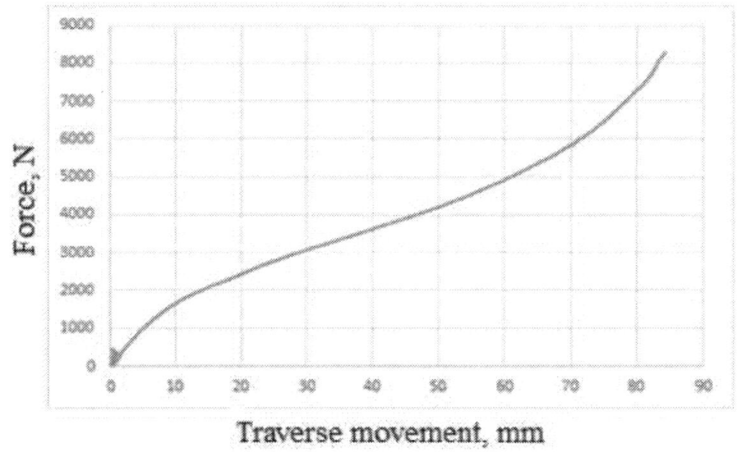

Figure 5. The graph of the dependence of the resistance to compression on the traverse movement

Figure 6. Sample of energy absorbing element at the moment of full compression

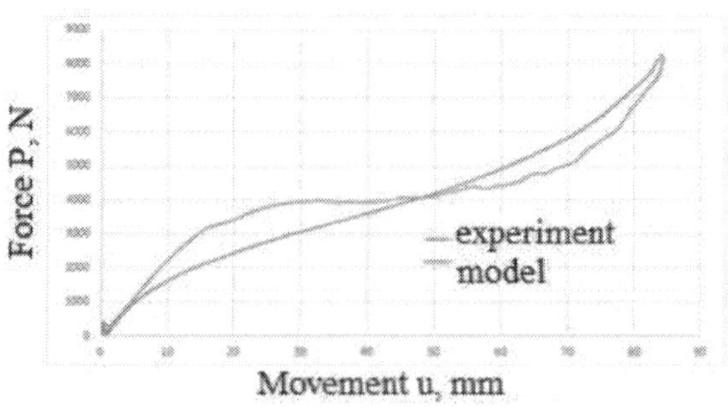

Figure 7. Graph of dependence of forces on displacements of a compressive element

Analyzing the appearance of the sample after testing, it was noted that after three days the residual deformations were less than 10%, from which it can be concluded that with such a design of the energy-absorbing UHMWPE element can be reused after significant deformations, in contrast to steel energy-absorbing elements.

3. Model and methods.

Virtual tests were carried out on the basis of the finite element method (FEM) using a multipurpose software package Ls-Dyna (Licence SIC Research Institute of mechanical and quality problems №2034, under contract №12/12/16 from 12 December 2016) [6-9]. As a result of the virtual experiment, a graph *P=f(u)* of the dependence of the force on the movement of the compression plate was built, which was compared with an identical graph obtained in the course of a full-scale experiment (Figure 7). The compression process of a sample of an energy-absorbing UHMWPE element is shown in Figure 8.

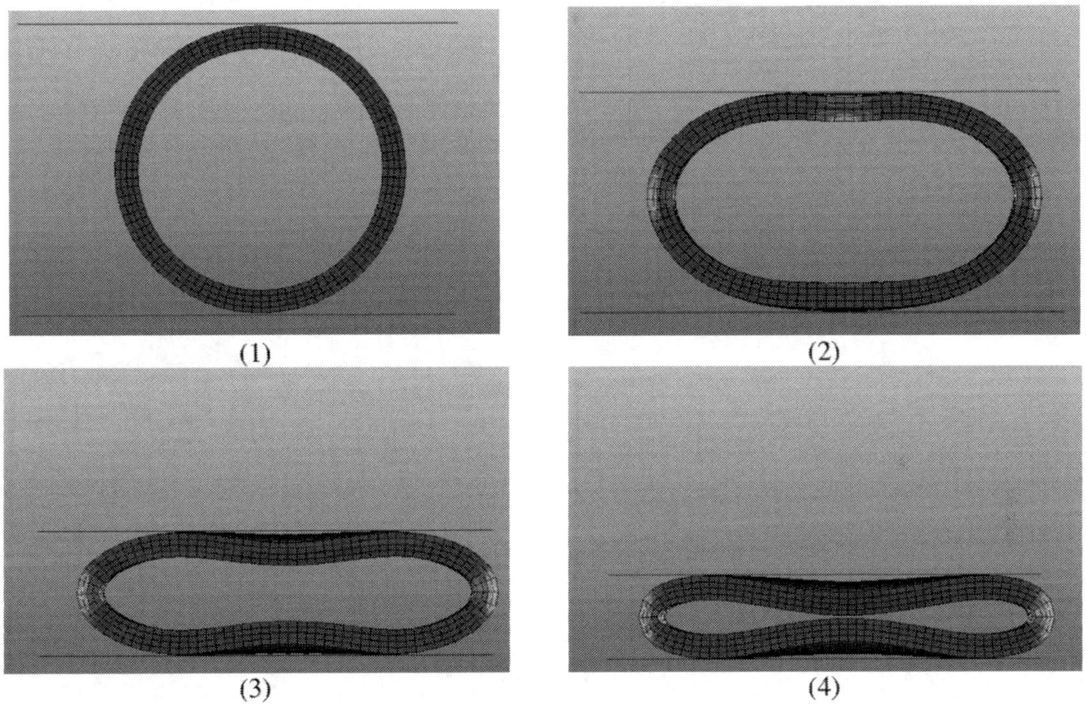

Figure 8. The process of compression of the FE model of a sample of an energy-absorbing element made of UHMWPE

According to the results of the virtual experiment, the energy absorption of the UHMWPE sample was $U_{FE} = 0{,}335\ kJ$.

Let's consider one of the possible design options for the sections of energy-absorbing elements in the form of UHMWPE cylinders (Figure 9).

Figure 9. Possible UHMWPE cylinder assembly

Figure 10. Model of steel energy absorbing elements (honeycomb)

To compare the energy absorbing capacity, a model of energy absorbing elements in the form of metal honeycomb sections was also built (Figure 10).

To determine the value of the energy absorbing capacity, a virtual test model for the compression of the energy absorbing elements in the assembly was built. The test scheme is shown in Figure 11. The loading process was realized by means of an absolutely rigid plate, moving at a speed of 20 mm /min.

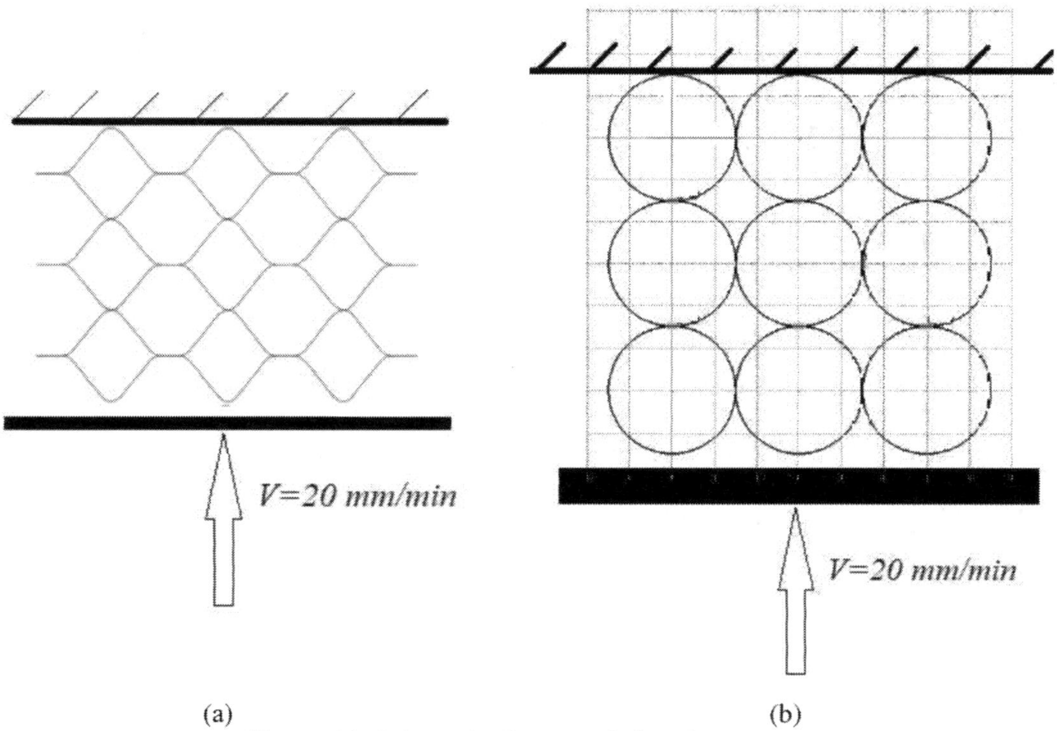

Figure 11. Schematic diagram of virtual tests.
(a) – metal honeycomb, (b) – polymer pipes

4. Results

According to the results of virtual tests, the values of energy absorption capacity were obtained. Energy absorption diagrams are shown in Figure 12.

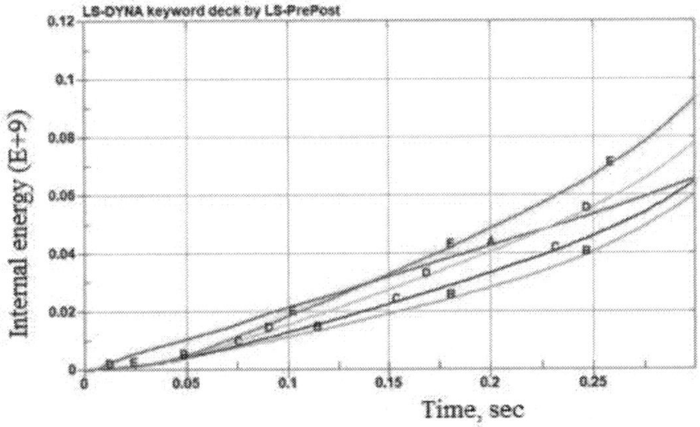

Figure 12. Energy absorption diagrams

Table 1 shows the range of values of energy absorption by UHMWPE sections with an outer diameter of 90 mm during loading, and the mass of the structure. For comparison, Table 2 shows similar indicators for steel elements.

Table 1. Energy absorption and weight of UHMWPE structure

Parameter	Thickness 8mm	Thickness 9mm	Thickness 10mm	Thickness 11mm
Energy absorption	6,0*10^7 J	6,42*10^7 J	7,73*10^7 J	9,24*10^7 J
Weight	37,99 kg	42,49 kg	46,92 kg	51,32 kg

Table 2. Energy absorption and weight of the steel structure

Parameter	Weight	Energy absorption
Honeycomb made of steel with a wall thickness of 2 mm	64,72 kg	6,52*10^7 J

5. Conclusions.

Based on the results obtained, it can be preliminarily concluded that UHMWPE can be used as a material for energy-absorbing structural elements. With an outer diameter of 90mm and thicknesses from 8mm to 11mm, the closest energy absorption to steel is achieved.

6. Acknowledgements. The authors are grateful to their supervisor, Doctor of Technical Sciences, Professor I.V. Demyanushko for guidance, comments and advice in carrying out this research work.

7. References

[1] American Association of State Highway and Transportation Officials. Roadside Design Guide, 4th Edition 2011. American Association of State Highway and Transportation Officials (Washington D.C., 2011)

[2] Bligh Roger, Nauman Sheikh, Wanda Menges, and Rebecca Haug 2005 *Portable Traffic Barrier for Maintenance Operations* Report FHWA/TX-05/0-4692-1 (Texas Department of Transportation, Austin)

[3] Demiyanushko I V, Nadezhdin V S, Gruzdev A S, Titov O V 2020 Influence of loading rate on strength characteristics of ultra high molecular weight polyethylene (UHMWPE) *IOP Conference Series: Materials Science and Engineering* **747(1)** 012016
[4] Mohseni Taromsari S, Salari M, Bagheri R, Faghihi Sani M A 2019 Optimizing tribological, tensile & in-vitro biofunctional properties of UHMWPE based nanocomposites with simultaneous incorporation of graphene nanoplatelets (GNP) & hydroxyapatite (HAp) via a facile approach for biomedical applications *Composites Part B: Engineering* **175** 107181
[5] Xiao S, Lanceron C, Wang P, Soulat D, Gao H 2019 Mechanical and thermal behaviors of ultra-high molecular weight polyethylene triaxial braids: the influence of structural parameters *Textile Research Journal* **89(16)** 3362-3373
[6] LS-DYNA® Keyword User's Manual. Livermore Software. Technology Corporation, Livermore, CA, 2016
[7] Demiyanushko I, Karpov I, Tavshavadze B 2019 Applications Fem Analysis in Researches Reliability and Dynamics of Road Safety Barriers at Collisions With Cars *Mechanisms and Machine Science* **73** 1651-1658
[8] Büyük M, Atahan A O, Kurucuoğlu K 2018 Impact Performance Evaluation of a Crash Cushion Design Using Finite Element Simulation and Full-Scale Crash Testing *Safety* **4** 48 https://doi.org/10.3390/safety4040048
[9] Wright L, Davidson S 2020 How to tell the difference between a model and a digital twin *Adv. Model. and Simul. in Eng. Sci.* **7:13** https://doi.org/10.1186/s40323-020-00147-4

Comparative evaluation of the performance of different asphalt concrete road pavements on reinforced concrete road structures

N N Arous[1], M G Goryachev[1]

[1] Moscow Automobile and Road Construction State Technical University (MADI) 64, Leningradski Avenue, Moscow, 125319, Russia

E-Mail: 92nawararous@gmail.com, Rudeger@rambler.ru

Abstract. Comparative design of structures allows at an early stage of project development to assess their behavior and predict service life. Approximate numerical methods, for example, the finite element method, are widely introduced to simulate the operation of pavements on road structures. The variety of modern asphalt concrete poses the problem of finding the most effective design solutions based on the results of comparing the performance indicators of road pavements. The article presents the results of modeling by the finite element method of the work of road pavements, consisting of a two-layer asphalt-concrete package, on a reinforced concrete slab of bridge. Various options for combinations of asphalt concrete are considered. The well-known American method of the Asphalt Institute was used as a methodological basis. For the criterion of the limiting state, the indicator of accumulated fatigue cracks in the amount of 10% of the wheelpath area is taken. Preliminary recommendations are given on the field of application of the considered road pavements on reinforced concrete bridges in the conditions of the Syrian Arab Republic.

1. Introduction

The traditional methods of designing pavements on highways, based on solving the problem of a multilayer system "pavement (surface, base, sub-base) on subgrade", cannot be applied to the design of pavements of a bridge deck. In global practical experience, most often, road pavements of bridges are assigned without calculation, taking into account the accumulated experience based on the results of monitoring the condition and statistics on the actual service life of pavements under various operating conditions. However, in case of a lack of statistical material, effective designs of road pavements can be proposed based on modeling their operation in specified modes.

Asphalt pavements on bridge deck are much thinner than pavement on subgrade. Its thickness is about 8-12 cm. because of the different stiffness of stresses and strains, they must be considered separately for both concrete and orthopedic surfaces.

The most common types of damage and defects to pavements on bridge structures are cracks, rutting, corrosion, various forms of permanent deformations, and potholes. Corrosion of the surface course is damage that is largely due to the Corrosion and grinding resistance of coarse aggregate(crushed stone). Shear defects (formation of waves , corrugation, transverse unevenness, patches) can be minimized with the correct designation and determination of the binder grade for asphalt concrete and it's ratio.

Content from this work may be used under the terms of the Creative Commons Attribution 3.0 licence. Any further distribution of this work must maintain attribution to the author(s) and the title of the work, journal citation and DOI.
Published under licence by IOP Publishing Ltd

IOP Conf. Series: Materials Science and Engineering 1159 (2021) 012033 doi:10.1088/1757-899X/1159/1/012033

Potholes are a "neglected" defect that usually occurs with inadequate current maintenance of the bridge deck.

Therefore, predicting the growth of fatigue cracks in asphalt concrete layers will be most indicative for assessing the serviceability by the strength of road pavements of bridges. It is appropriate and expedient to perform such express-forecasting of resistance to the impact of multiple traffic loads of road pavements using computer modeling using the finite element method [1, 8].

2. Method of investigation

The fatigue life of an asphalt pavement may be defined as the magnitude of traffic, expressed as the number of equivalent standard axles (ESAs) (single axle, dual wheels carrying 80 kN load) that it can withstand before distress in terms of cracking of a certain percentage of the wheelpath area is observed. The tolerable cracking level may vary depending on design traffic loading but generally ranges from 10 to 45%.

The fatigue life of pavements is determined based on the state of stresses and strains of the structure and material properties obtained from laboratory and experimental results.The most common methods for calculating fatigue life are the Asphalt Institut method, the AASHTO 2004 method, the Shell method, the University of Nottingham method and the Center for Road Research and transport in Belgium method.The listed methods have certain similarities. However, the most widely used method is the Asphalt Institute. The Asphalt Institute model (Asphalt Institute 1981) is also based on strain and stiffness but also takes into account the influence of effective binder volume (V_b) and air voids (V_a) on fatigue life [7, 11, 12].

The Asphalt Institute model is given by the following equation [14]:

$$N_f = 18.4 \times C \times \left(6.167 \times 10^{-5} \times \varepsilon_{ha}^{-3.291} \times |E|^{-0.854}\right) \quad (1)$$

where N_f – Number of repetitions to fatigue cracking;

ε_{ha} – Tensile strain at the bottom of the asphalt layers;

E – Stiffness modulus of the material, MPa;

C – Adjustment coefficient.

$$C = 10^M \quad (2)$$

$$M = 4.84 \times [V_b/(V_b + V_a) - 0.69] \quad (3)$$

where V_b – Effective binder content (%);

V_a – Air voids (%).

Modern pavement designs on bridge decks must include at least two layers of asphalt concrete. At the same time, various types of asphalt concrete are used , in both the upper and lower layers. Since the upper layer is designed to be as a wearing course, it is rarely constructed and laid with a thickness of more than 5 cm, often limited to only 4 cm thick.The lower layer of the pavement, which is responsible for the overall structural stability, is assigned, as a base course , with a thickness of at least 4 cm.

As part of a numerical experiment, the simulation of the operation of six different road pavements was performed, consisting of various combinations of varieties of asphalt concrete, arranged on reinforced concrete slabs of bridge deck. For all the options under consideration, the thickness of the upper coating layer was taken equal to 4 cm, and the lower coating layer 4 cm (Figure 1).

In (Figure 1), the following abbreviations and acronyms are adopted: Dense AC rous asphalt concrete based on bitumen grade 60/90; SMA PBB 60 - Stone-mastic asphalt concrete based on polymer-bitumen binder PBB 60 ; Mastic AC - Mastic asphalt concrete based on bitumen-rubber composite binder BRCB [3, 5, 6, 10, 16].

The properties and design characteristics of asphalt concrete strongly depend on their temperature. Under real operating conditions, the temperature of asphalt concrete changes significantly. Taking this condition into account greatly increases the complexity of the experiment.

Taking into account that for the conditions of the Syrian Arab Republic the average annual temperature is about 15 ... 25 °C, it was decided to set the design characteristics of asphalt concrete corresponding to a temperature of 20 °C.

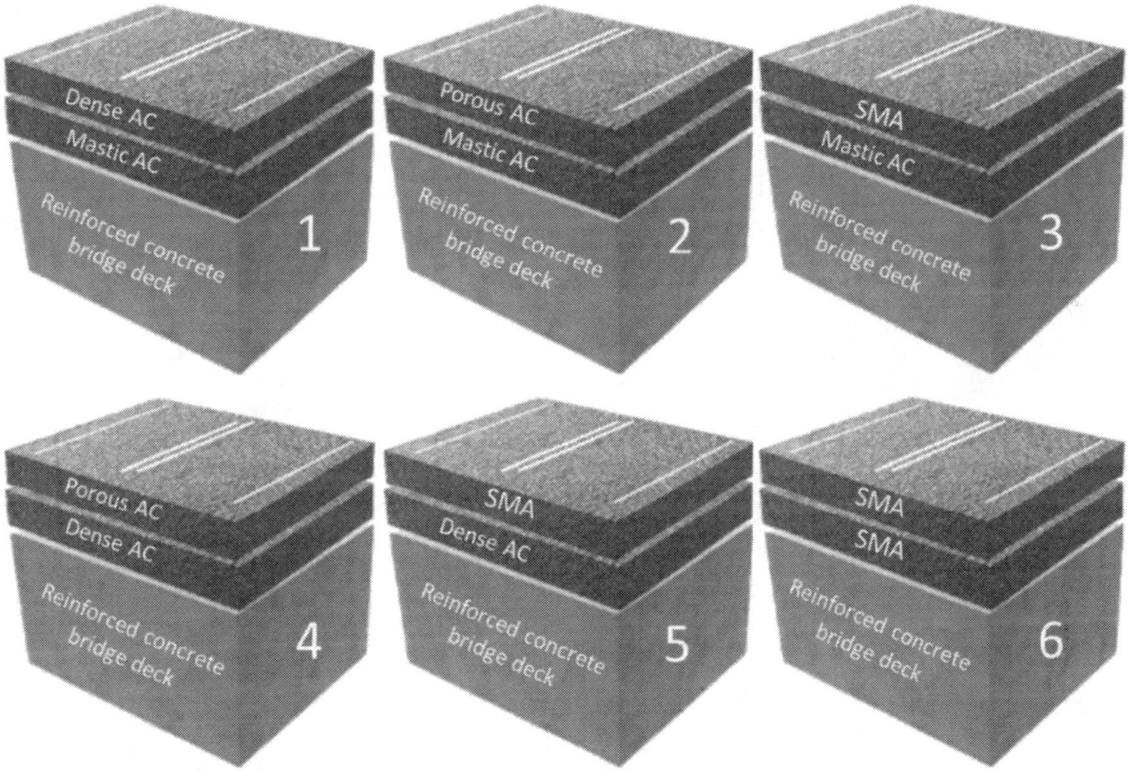

Figure 1. The typical layouts pavement layers on the bridge decks.

The table 1. shows the physical-mechanical properties of the asphalt concrete of the investigated and studied pavements [2, 9, 13, 15].

Table 1. Physical-mechanical properties of the asphalt concrete, adopted for experimental modelling.

Asphalt concrete type	Elastic modulus (E) at a temperature of +20 ° C, MPa	Poisson's ratio	$V_b(\%)$	$V_a(\%)$
Dense AC	1800	0.35	5.5	4.0
Porous AC	1200	0.35	4.0	6.0
SMA PBB 60	2600	0.35	6.5	3.0
Mastic AC	2900	0.35	7.0	2.0

IOP Conf. Series: Materials Science and Engineering 1159 (2021) 012033 doi:10.1088/1757-899X/1159/1/012033

3. Research results

Finite Element Simulation (FEM) was performed using EverStressFE software. The deformations that occur in the layers of road pavements on reinforced concrete bridges were calculated.

The thickness of the reinforced concrete slab is taken to be 20 cm, and the calculated wheel load was 40 KN, the contact pressure of the wheel on the coating was 690 kPa (Figure 2).

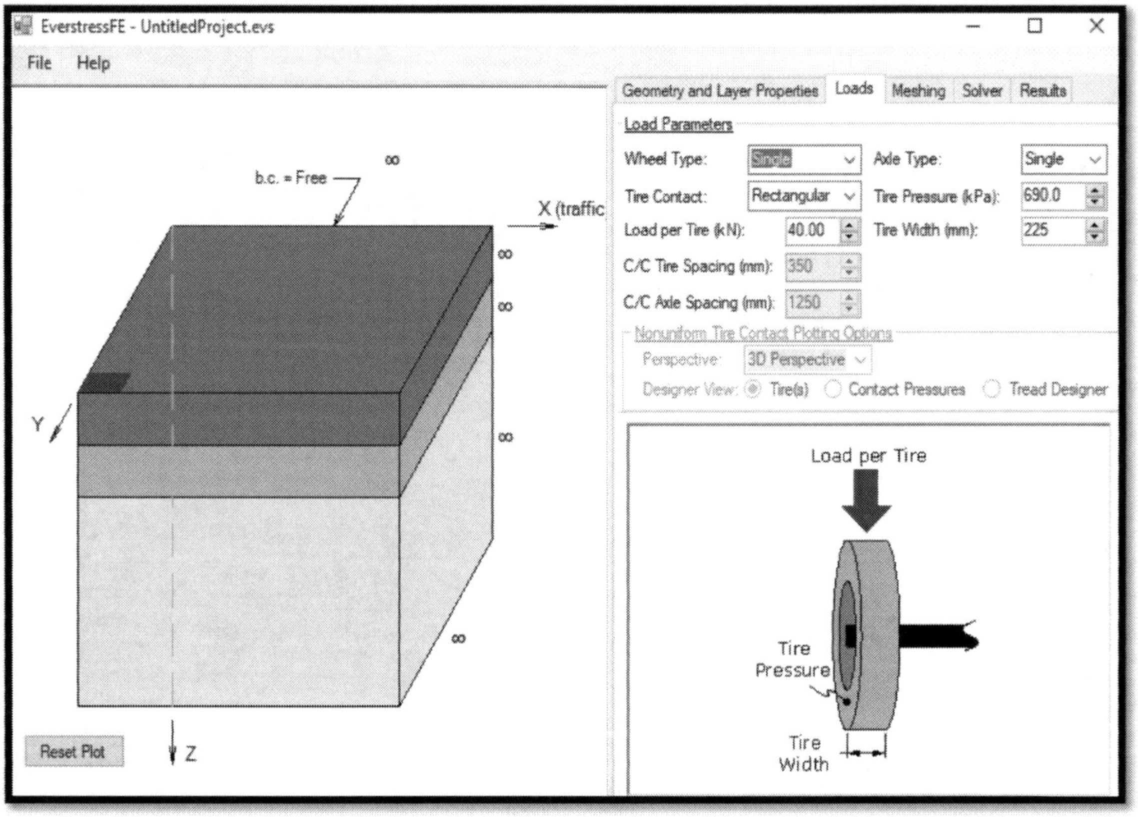

Figure 2. Interface of the wheel load design parameters input window.

On the basis of computer finite element method (FEM) calculations and review of the literature the strains which occur in the different bridge pavements placed on reinforced concrete bridges decks were shown (Figure 3).

The performance of pavements is often assessed and judged by the number of applications of the design load before the beginning of the limit state according to the criterion under consideration [4].

Depending on the selected pavement structure, the discrepancy in performance according on the criterion of fatigue cracking was quite important. Thus, the variance of designs **No. 2** and **No. 3** exceeded 5.5 times (Figure 4).

The results obtained showed that the best performance according to the criterion of fatigue cracking is possessed by pavements with a top layer of (SMA) stone-mastic asphalt concrete on a polymer-bitumen binder, so these design solutions can be recommended for bridges on roads with the highest traffic density.

On bridges located on roads with a load corresponding to III and IV categories, we can recommend the more economical option **No. 1**. In the case of a low-traffic local road network, the design of option **No.2** can be applied. Such recommendations can be taken as a basis for the development of catalogs of typical road pavements on reinforced concrete bridges.

IOP Conf. Series: Materials Science and Engineering 1159 (2021) 012033 doi:10.1088/1757-899X/1159/1/012033

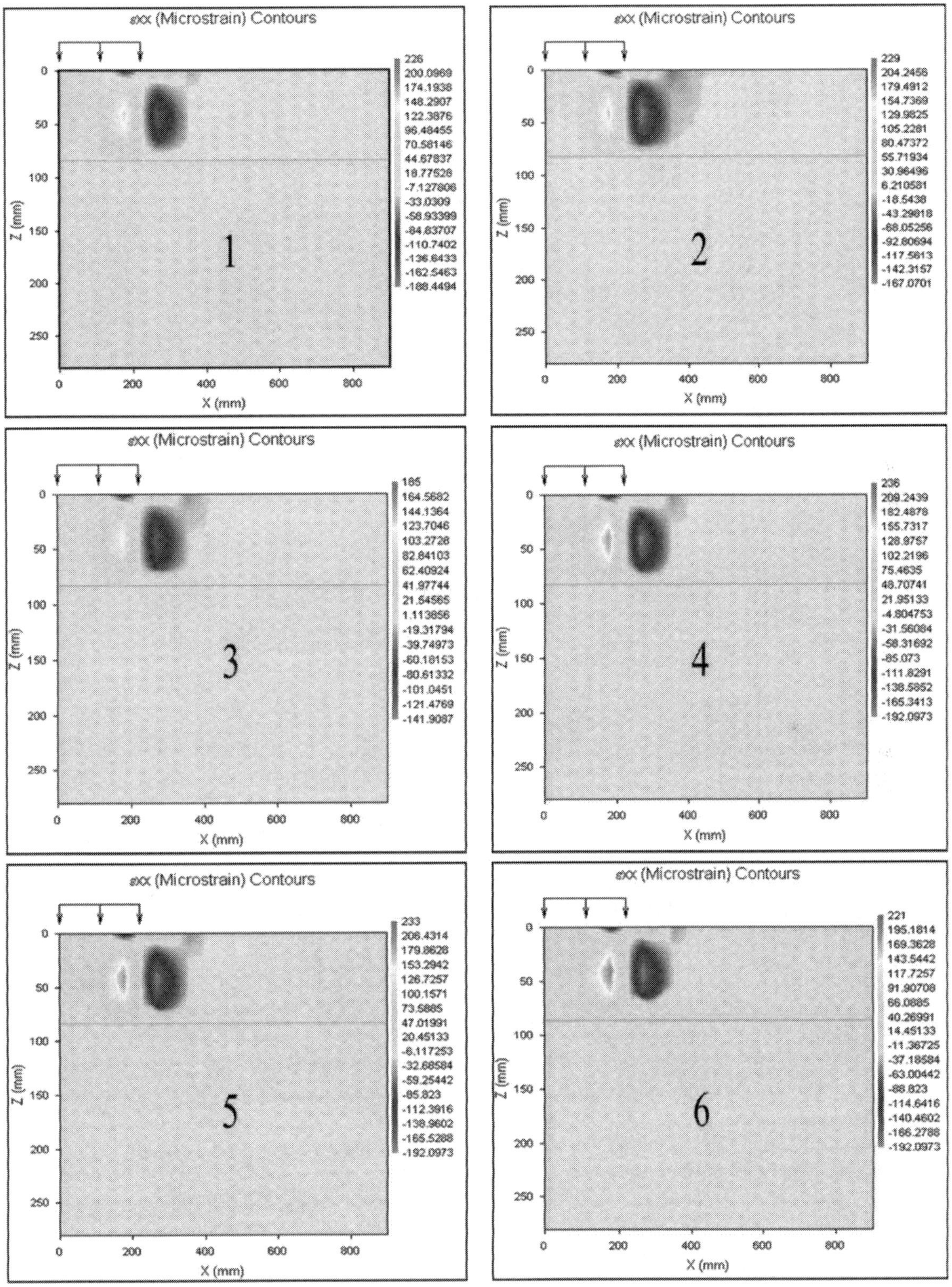

Figure 3. Graphical results of the performance of asphalt bridge pavements from a numerical experiment.

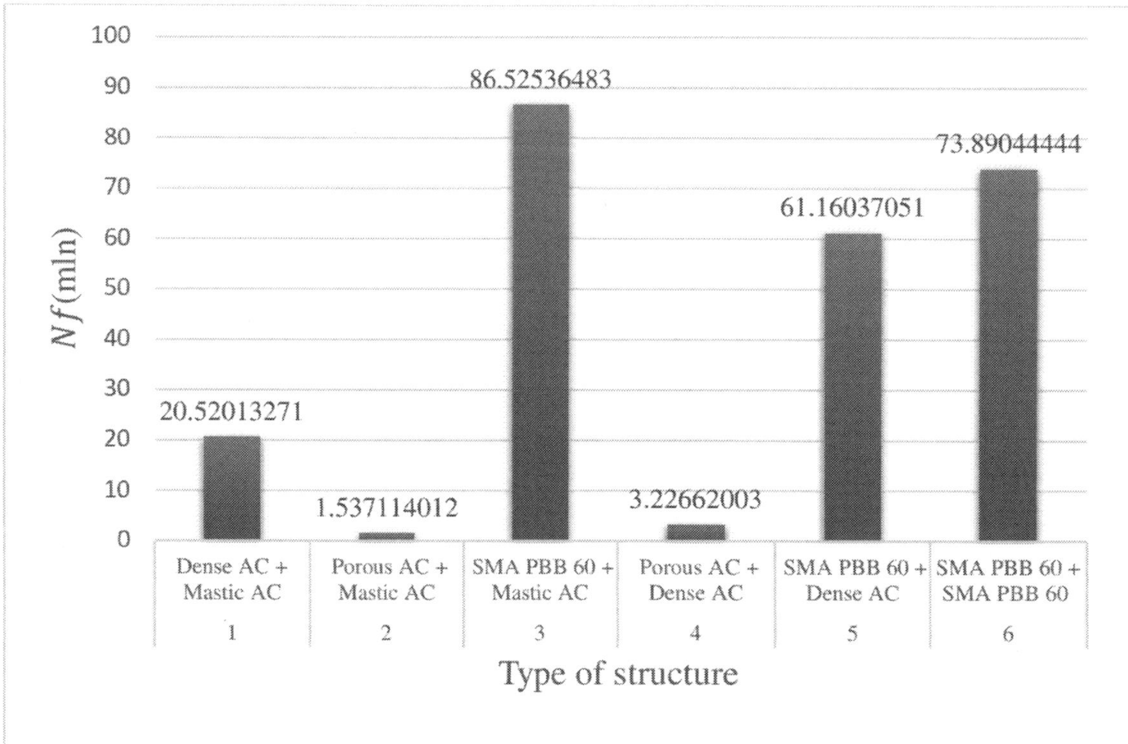

Figure 4. Fatigue life of different asphalt bridge pavements.

4. Conclusion

To evaluate the performance of two-layer asphalt bridge pavements on a rigid reinforced concrete base, the finite element method was applied. The Asphalt Institute algorithm developed by American specialists was used. It has been established that pavements with layers of (SMA) stone-mastic asphalt concrete on polymer-bitumen binder have the highest performance, expressed in the number of applications of the design load, according to the criterion of fatigue cracking. The results obtained will contribute to the development of standard design solutions for bridge structures in the Syrian Arab Republic. In the absence of a clear and reliable design method for designing paving layers over bridges. The method studied is considered the basis for a comprehensive study to develop a method for designing asphalt pavement layers over bridges.

References

[1] Caliendo C 2012 Stresses and strains prediction model of asphalt pavements on concrete bridges. *International Journal of Civil Engineering Research* **3(3)** 223-239

[2] DND MO-004/2004. *Recommendations for the selection of asphalt concrete on bitumen rubber composite binders for top coating coats and wear layers*

[3] European Asphalt Pavement Association (EAPA) 2013 Asphalt pavements on bridge decks *EAPA Position paper* (Brussels, Belgium)

[4] Goryachev M G 2020 Increasing the reliability of highways. *Textbook Manual*. MADI. M-158 c

[5] GOST 31015-2002 Bituminous stone mastic mixtures and stone mastic asphalt. Specifications

[6] GOST 9128-2013 Asphaltic concrete and polimer asphaltic concrete mixtures, asphaltic concrete and polimer asphaltic concrete for roads and aerodromes. Specifications

[7] Judycki J 2014 Analysis and design of flexible and semirigid pavement structures. *wkl.* Warsaw

[8] Kim T W, Baek J, Lee H J and Lee S Y 2014 Effect of pavement design parameters on the behaviour of orthotropic steel bridge deck pavements under traffic loading. *International Journal of Pavement Engineering* **15(5)** 471-482
[9] Kostin V I 2009 *Crushed stone-mastic asphalt concrete for road surfaces* (Nizhny Novgorod)
[10] Krauss P D, Lawler J S and Steiner K A 2009 Guidelines for selection of bridge deck overlays, sealers and treatments *NCHRP Project* 20-07
[11] Olidis C and Hein D 2004 Guide for the mechanistic-empirical design of new and rehabilitated pavement structures materials characterization: Is your agency ready In *2004 Annual Conference of the Transportation Association of Canada*
[12] Pilat J and Piotr R 2010 Asphalt pavements: an academic textbook. *Communication and Communications Publishing House*
[13] PNST 265-2018 Automobile roads of general use. Flexible pavement design
[14] Pokorski P, Radziszewski P and Sarnowski M 2016 Fatigue life of asphalt pavements on bridge decks. *Procedia Engineering*, *153*, 556-562
[15] Pouget S, Sauzeat C, Di Benedetto H and Olard F 2012 Modeling of viscous bituminous wearing course materials on orthotropic steel deck. *Materials and structures* **45(7)** 1115-1125
[16] STO NOSTROY 2.29.113-2013 Pavement arrangement on bridges and artificial structures. Moscow 2014

Ready-mixed concrete recycling system

O I Maksimychev[1], A V Ostroukh[1], N A Filippova[1], C B Pronin[1], A V Libenko[1]

[1] Moscow Automobile and Road State Technical University (MADI), 64, Leningradsky prospect, Moscow, 125319, Russia

E-mail: maksimychev@yandex.ru, ostroukh@mail.ru

Abstract. This article describes the technological diagram and equipment required for establishing a Ready-Mixed Concrete Recycling System, which allows to divide concrete into separate components, such as cement milk, crushed stone and sand. The waste from demolished concrete structures can be reused to reduce the environmental burden during their life cycle and ensure economic use of the construction resources. The recycled aggregate concrete is a sustainable concrete waste which in the long term can replace the demand for natural aggregate, a process which would, in turn, lead to its preservation. Most concrete industries have been observed to be reluctant in the production of recycled aggregate concrete and utilization in its maximum potential. Industries are yet to embrace it not only due to its uncertain material performance but also due to its unexplored manufacturing plant operations which are yet to be established. The system is using wireless communication technologies and data exchange between sensors and data acquisition devices. The raw materials obtained by the showcased recycling process could be reused in subsequent production cycles of ready-mixed concrete preparation. The following process could help reduce the growing amount of industrial waste generated with constant infrastructure improvements, which could be beneficial both from an environmental and economical standpoint.

1. Introduction

Concrete production, like any other production cycle, is not complete without waste generation. The growing problem of maintaining a clean environment and the reuse of any industrial waste determines the problem of reusing sludge and concrete that remains on the walls and in the cavities of concrete pumps and concrete mixers during their operation. In the intervals between cycles, the structures are washed and the resulting composition is removed. Before the advent of such technologies called recycling, these wastes were usually discharged into the environment. Currently, there is a possibility of their reuse, which positively affects both the state of nature and the consumption of components for the production of concrete.

2. Materials and Methods

Modern engineering technologies completely solve the problem of waste-free production of ready-mixed concrete [1, 2]. Processing complexes allow using the obtained raw materials in subsequent production cycles. The main task of the recycling complex is the separation of the mixture into separate components, such as cement milk, crushed stone and sand (Fig. 1).

Content from this work may be used under the terms of the Creative Commons Attribution 3.0 licence. Any further distribution of this work must maintain attribution to the author(s) and the title of the work, journal citation and DOI.
Published under licence by IOP Publishing Ltd

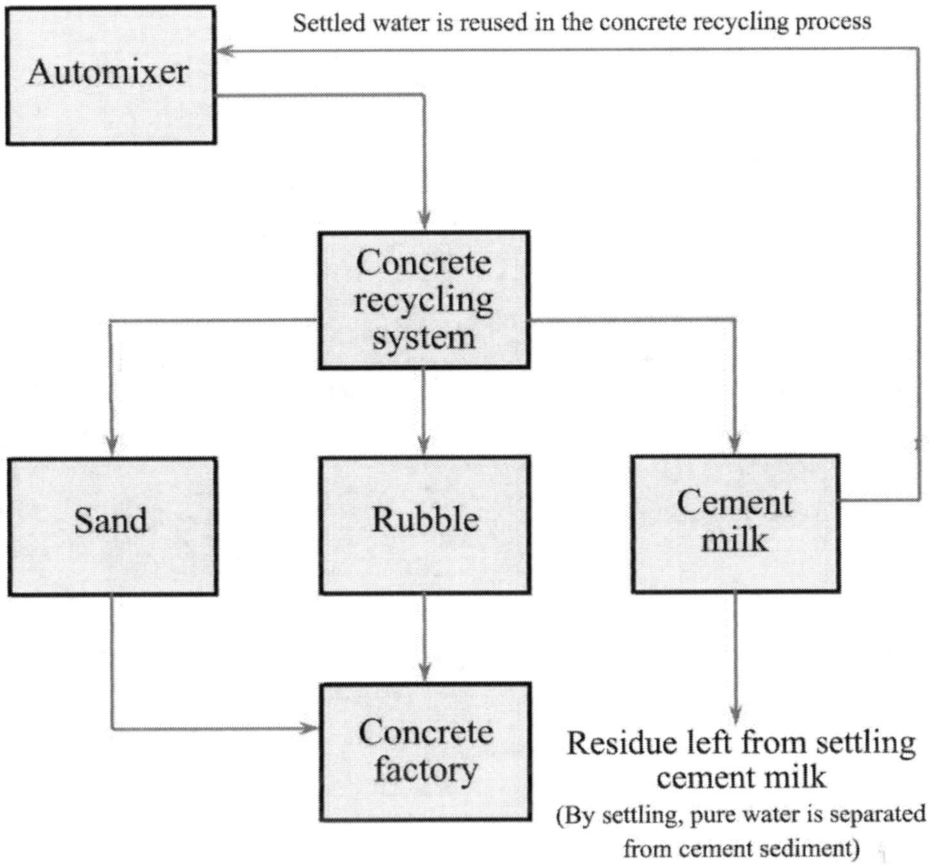

Figure 1. Recycling scheme for ready-mixed concrete

The transformation of waste is carried out directly at the factories, the reuse of slag allows you to save on the manufacture of new volumes, and there is no useless mass left.

A complete recycling system consists of the following main processes:

- screening and separating inert materials system;
- recycling waste materials, concrete (actually recycling);
- reuse of water.

Each of these three components can be used autonomously, but in a single complex it is a complete system for thorough transformation and use of recyclable materials.

3. Results and Discussions

The work of the system of units is carried out in three operating modes: two, operating on automation, and manual. The first of the self-acting modes performs the task of flushing waste from a concrete mixer, and the second one - converting large volumes of concrete waste [3, 4]. The essence of the system consists in a kind of splitting of concrete into its components, which are sorted, removed into different containers and sent back for production.

On a significant scale, the percentage of solid residues is greater, therefore, to separate them into sand and crushed stone, a system of sifting and separation of inert materials is used [5, 6]. The plant operates as follows: the raw material is loaded into the receiver, after which the raw material enters the sieve. Then the raw material is sieved and divided into fractions: the sand is taken out to one side, and crushed stone comes out on the other (Fig. 2).

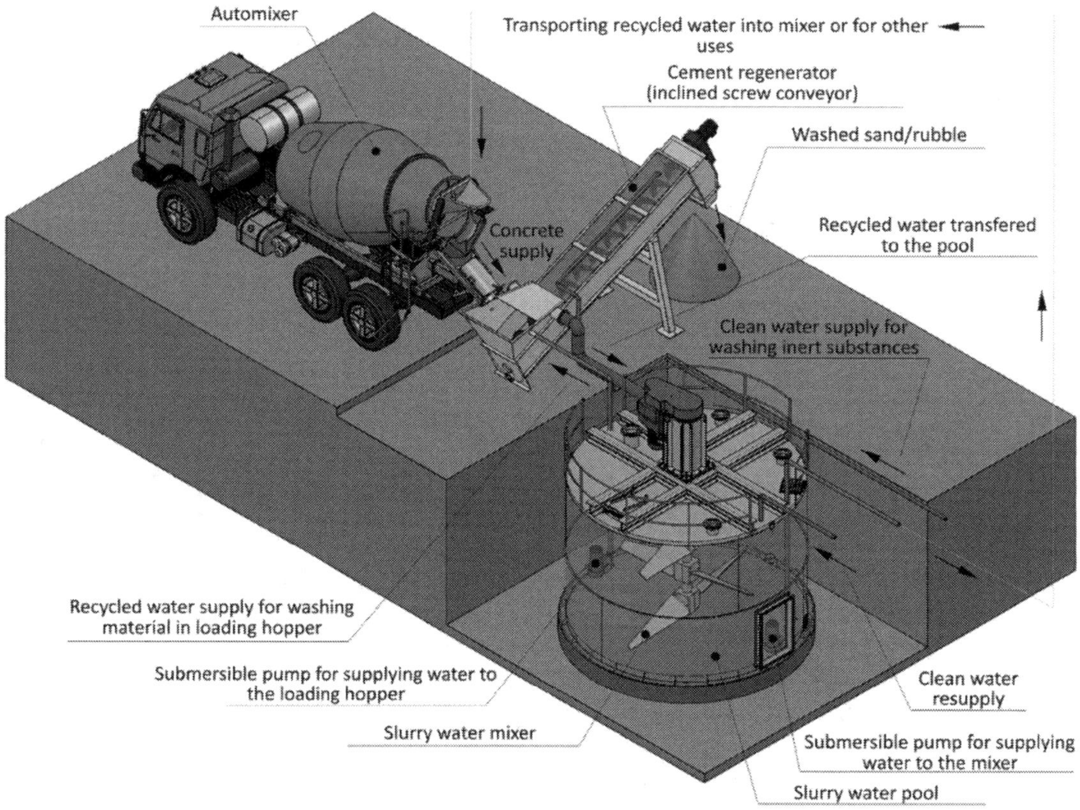

Figure 2. Equipment for the ready-mix concrete recycling system

The beginning of the full cycle of the process - concrete recycling - is the moment when the spray can of a concrete mixer rests on the auto-start lever. At the same time, a sound signal is given, at the same time the operation of the electric motor of the sludge mixer, as well as the pump supplying water to the mixer, is turned on. When the pump is turned off, the water supply stops, and the alarm goes off. The driver can prolong the action of the mixer, and, if necessary, manually continue to supply water using the control panel.

A special filter passes the slurry water to separate the remaining large residues of raw materials, after which the water returns to the storage pool. The so-called garbage is discharged outward, reaching along the screw rod to the unloading passage. Drainage and purity of water is regulated manually or using an electronic panel.

All elements of the system (pump, electric motor) will continue to work for some time to completely recycle the remaining material. Then the agitator passes from the active phase to the stage of periodic stirring. This system provides for the operation of each of the functional units for half an hour.

In the case of the set automatic mode or for a long period of inactivity, prevent the formation of thick in the storage pool. For this purpose, the control system autonomously starts the agitator from time to time. In addition, it itself controls the constant boundary of the water: if it drops to the critical line, the pump starts to supply clean water - this is the ability to maintain a certain density of the sludge water, and its mixing is a suitable state for reuse.

Since the recyclable materials are processed in high quality, there is little waste sludge at the outlet. The proportion of the converted material is small and does not affect the quality factor of the material being produced. Therefore, the processed product is suitable for use in production with small volumes.

4. Conclusion

Recycling technology can significantly reduce the amount of waste generated and makes it possible to create a closed production cycle. Even pure water used to separate fractions is in a state of constant circulation and is used more than once.

Acknowledgments

The research was funded by a grant from the Russian Science Foundation for the event "Research by world-class scientific laboratories in the framework of the implementation of the priorities of scientific and technological development of the Russian Federation" of the President's program of research projects implemented by leading scientists, including young scientists.

References

[1] Ostroukh A V, Nedoseko I V., Pudovkin A.N., Nuruev Y.E. 2015. Development of the Automated Control System for Concrete Plant with Two Units Concrete Mixing. International Journal of Applied Engineering Research **10(17)** 37792-37798

[2] Ostroukh A, Maksimychev O, Nikolaev A, Kolbasin A, Nedoseko I 2016 Development of automation of the drying and milling unit for the mineral powders plant *ARPN Journal of Engineering and Applied Sciences* **11(9)** 5717-5721

[3] Anastasiou E, Papachristoforou M, Anesiadis D, Zafeiridis K, Tsardaka E-C 2018 Investigation of the Use of Recycled Concrete Aggregates Originating from a Single Ready-Mix Concrete Plant *Applied Sciences* **8(11)** 2149 https://doi.org/10.3390/app8112149

[4] Vieira L D B P, de Figueiredo A D, Moriggi T and John V M 2019 Waste generation from the production of ready-mixed concrete *Waste Management* 94 146-152

[5] Dosho Y 2020 Contributions of Concrete Recycling Technology toward Sustainable Development: Standards and Technologies Related to Recycled Aggregate Concrete in Japan *Journal of Materials and Engineering Structures "JMES"* **7(4)** 525-549

[6] Xuan D, Poon C S and Zheng W 2018 Management and sustainable utilization of processing wastes from ready-mixed concrete plants in construction: A review *Resources, Conservation and Recycling* **136** 238-247

IOP Conf. Series: Materials Science and Engineering 1159 (2021) 012035 doi:10.1088/1757-899X/1159/1/012035

Railroad cement warehouse digital system

Y E Vasiliev[1], A V Ostroukh[1], C B Pronin[1], I Y Sarychev[1], I.E. Kotlarsky[1]

[1] Moscow Automobile and Road State Technical University (MADI), 64, Leningradsky prospect, Moscow, 125319, Russia

E-mail: vashome@yandex.ru, ostroukh@mail.ru

Abstract. In the article the concept of Railroad Cement Warehouse Digital System (RCWDS) with two silos. The article briefly analyzes the ways of unloading hopper, as well as basic equipment for mechanical and pneumatic conveying of cement hopper in the cement silos. The system is a set of hardware maintenance, information, mathematical and software for management of technological objects of railroad warehouse of cement. The system allows you to track trends in cement consumption and accumulation and manage its stocks rationally. The system also allows remote calibration of all silos, further reducing costs and staff time on site. The system is using technological equipment and industrial controllers that allow monitoring silo filling, the occurrence of excess pressure and reduce the risk of dust release into the atmosphere. The system is scalable and can be further modified to control a large number of silos. In accordance with the proposed concept, RCWDS provides the optimum level of automation in the collection and processing of information for generating control signals and transmit them without loss and distortion to the actuators in order to achieve the most efficient warehouse operation cement.

1. Introduction

The current situation in the concrete market is such that cement is often a scarce commodity, and concrete manufacturers prefer to stock large quantities at once to ensure the smooth running of the concrete plant.

In many countries, it is common to transport cement by rail in hopper wagons. In this case, the actual issue is the unloading of cement from under the hopper and its supply to the cement truck. The proximity to the concrete plant of the railway tracks makes it possible to install automatic equipment for receiving cement from railway hopper cars, and at the same time loading the service silos or cement warehouse.

Construction of new and reconstruction of existing long-term storage warehouses and cement terminals with automated control systems [1, 2, 3] is currently an urgent task and is unlikely to cease to be such in the foreseeable future.

2. Materials and Methods

2.1. Methods for unloading cement from hopper cars into storage silos

There are various ways to unload cement from hopper cars, each with their own advantages and disadvantages.

Content from this work may be used under the terms of the Creative Commons Attribution 3.0 licence. Any further distribution of this work must maintain attribution to the author(s) and the title of the work, journal citation and DOI.
Published under licence by IOP Publishing Ltd

2.1.1. *Unloading hopper wagons using sliding auger conveyors*

This is a mechanical method, which involves the mechanical extraction of cement from under the car using an auger, and its subsequent transportation to silos, also by a vertical screw conveyor (Fig. 1).

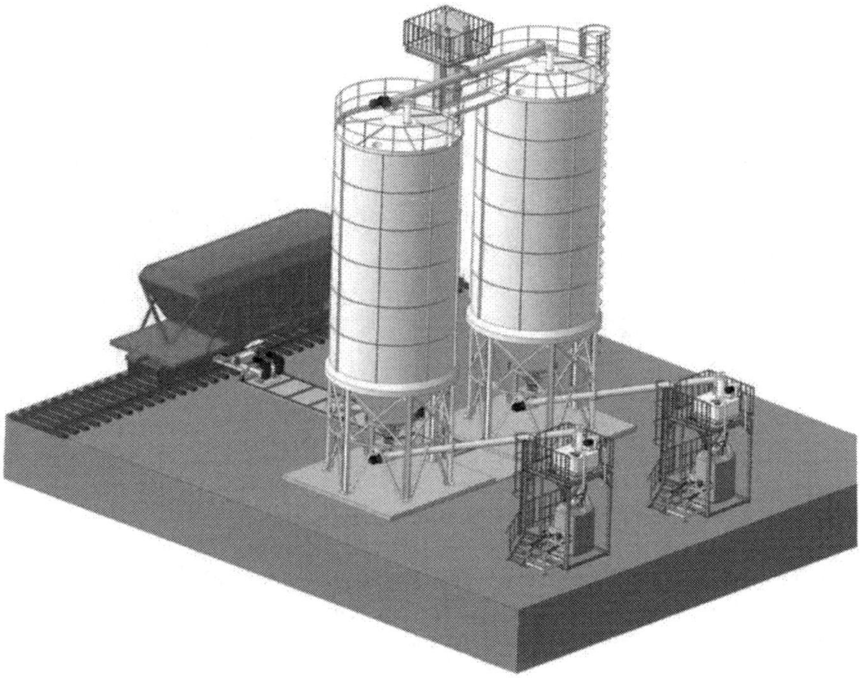

Figure 1. Unloading hopper cars using rolling auger conveyors

The horizontal auger fits into the space between the sleepers. To do this, it is necessary to place there a protective box with a section of 400-450 mm, in which the screw feeder is laid. The unloading of cement into the inlet of the auger is carried out through a magnetic grate, which catches metal inclusions present in the transported cement, and extension boards, the upper frame of which is attracted to the neck of the car, thereby ensuring minimal spreading of cement. The auger feeder feeds material into a vertical auger conveyor which transports the material to one of two silos. The choice of the silo for loading is carried out by means of the overhead mechanism.

If necessary, the auger feeder and conveyor can be dismantled and transported to another site.

This technical solution does not allow for high performance of unloading bulk material, because the height of the space between the neck of the car and the rolling auger is minimal. But with the correct location of the system, it does not require coordination of its placement and minimizes construction work. This unloading system also has restrictions on material intake, which means that not all hopper car models can be efficiently unloaded. If you use high-quality components, then the advantages of this system include low maintenance costs, and, as a result, a quick return on investment.

2.1.2. *Unloading hopper wagons using a pit*

Using the under-rail space for unloading the car (Fig. 2). can provide high productivity of the process, use various models of hoppers, and makes it possible to place pneumatic chamber pumps in the pit for pumping material into service silos.

Figure 2. Unloading hopper wagons using a pit

This option is convenient if the pit already exists and there is project documentation approved by JSC Russian Railways (JSC Russian Railways), otherwise it will take considerable time and money to agree on the project. In addition, in many regions there are difficulties with the level of groundwater, which makes it impossible to build a full-fledged pit.

At the location of the car, a shelter from wind and precipitation is being constructed, which makes it possible to effectively use the dedusting system of the unloading process. To prevent spreading of material under the car along the railway tracks, the equipment set includes easily mounted, front boards.

To ensure the normal operation of the equipment in the pit, the package includes electric heaters, which automatically maintain the required temperature. Taking into account that the temperature in the pit rises during operation of the equipment, the heaters are started before starting work, or in between short stops.

2.1.3. Unloading of hopper cars with partial use of the under-rail space.

With this option, transportation from under the car is carried out mechanically, as an option - by a screw feeder (Fig. 3).

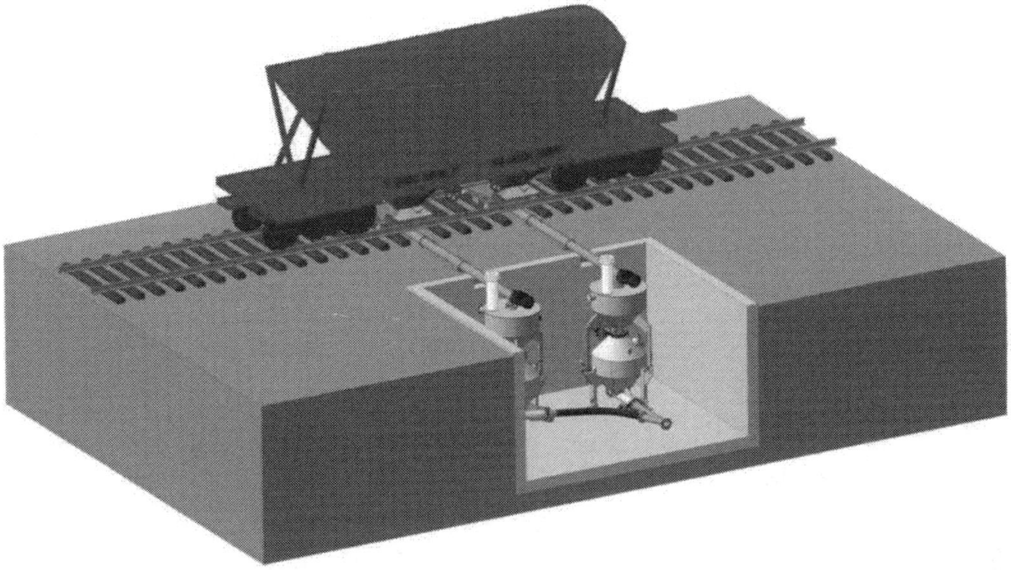

Figure 3. Unloading of hopper cars with partial use of the under-rail space

The performance of such an arrangement may be less than the option with a pit, because the space between the neck of the car and the receiving auger is not large, so the productivity will depend on the material flow and the operation of the auger. In addition, this unloading technology is not suitable for every hopper model, because the unloading throat of the car is significantly larger than the intake throat of the auger, which will not allow efficient unloading of material. Therefore, it is very important to know what type of carriages will be supplied with material.

A significant advantage of this option is the shallow depth of the auger, which makes it possible to place the system in areas where there is high groundwater. This option greatly facilitates the coordination of the project with Russian Railways, but does not exclude it.

2.2. Automation object

The object of automation is a railroad terminal for transshipment of cement from hopper cars to cement trucks (Fig. 4) [4, 5].

Figure 4. Automation object

From an economic point of view, the organization of a cement warehouse on the basis of several large-volume cement silos (for example, 300 or 500 tons) is optimal.

The main requirements for its functionality of a cement terminal are:

- time for unloading one wagon less than one hour (including wagon cleaning);
- minimization of energy consumption during cement transportation;
- prevention of over-grading of cement;
- reduction of dusting in the area of unloading wagons;
- prevention of its caking during long-term storage;
- measurement of the current level of cement;
- controlled shipment of cement to the consumer;
- reliability of the terminal equipment.

It is proposed to use this method of unloading hopper cars for the automated control system for technological processes of the railroad cement warehouse. The system consists of a receiving hopper located in a pit under the railway tracks, where cement is unloaded from hopper cars, a screw conveyor, through which cement from the receiving hopper enters a pneumatic chamber pump, from which, under the pressure of compressed air, cement is fed through pipes to silos (see. fig. 4).

The implementation of the automation project is carried out in stages. At the first stage, a system is automated, which consists of a receiving device and two silos for handling cement. At the next stages, it is possible to increase the number of silos by a multiple of two (4, 6, 8, etc.).

2.3. *Pneumatic chamber pumps*

Considering the real quality of the cement supplied for transshipment and the non-guaranteed "cleanliness" of the wagons used for its transportation, along with the requirement for the warehouse equipment to operate with maximum reliability, it was decided to use pneumatic chamber pumps as cement transportation.

The pneumatic chamber pump is designed for pneumatic transportation of bulk materials with a fraction of not more than 2 mm. The product is transported in a cyclic mode by means of compressed air from an external compressor, at a pressure of no more than 6 bar. Pneumatic chamber pumps are successfully used for unloading cement from railway hopper cars with further supply to storage silos.

According to their design, the following types of pneumatic chamber pumps are distinguished:

- pneumatic chamber pumps with manual control;
- pneumatic chamber pumps with an automatic control system;
- pneumatic chamber pumps with an automatic control system and a shut-off valve from pneumatic transport.

A manually operated pneumatic chamber pump is optimal for minor and / or intermittent product handling operations. Depending on the volume, they can be used, for example, as part of a big bag unloading station or as part of an aspiration unit and providing the collected dust 1-3 times a day.

Pneumatic chamber pumps with an automatic control system ensure efficient operation at short and medium distances - up to 80 m. It is possible to feed material at a distance of up to 500 m with a significant loss of productivity.

The disadvantages of these models of pumps from some manufacturers is the irrational air flow rate at long delivery distances and the inability to combine a group of pumps into one pneumatic transport line. The unloading of the hopper car with the supply of cement to the silos can be done with only one pump, which significantly increases the unloading time.

Pneumatic chamber pumps with an automatic control system and a shut-off valve from pneumatic transport are currently the most modern mechanisms for pneumatic conveying and can be used at significant feed distances (over 500 m). These pumps work perfectly in pneumatic conveying systems not only in single mode, but also in pair mode (Fig. 5), which increases productivity by 1.5-2 times.

Figure 5. Twin pneumatic chamber pumps with automatic control system and shut-off valve

Unloading of the hopper car can be carried out by two pumps with cement supply through one pneumatic transport line to the group of silos.

To achieve the maximum energy efficiency of the cement pumping process, high-tech pneumatic chamber pumps with an automatic control system and a shut-off valve were selected, which provide flexible control of the technological process at each phase.

3. Results and Discussions

The RCWDS of the cement warehouse is designed to control the control of the equipment of cement warehouses of various volumes and with a receiving device for unloading hopper cars in automatic, semi-automatic and manual remote modes of operation. An example of a mnemonic diagram of an operator station, developed using a modern SCADA system, is shown in Figure 6 [6, 7].

IOP Conf. Series: Materials Science and Engineering 1159 (2021) 012035 doi:10.1088/1757-899X/1159/1/012035

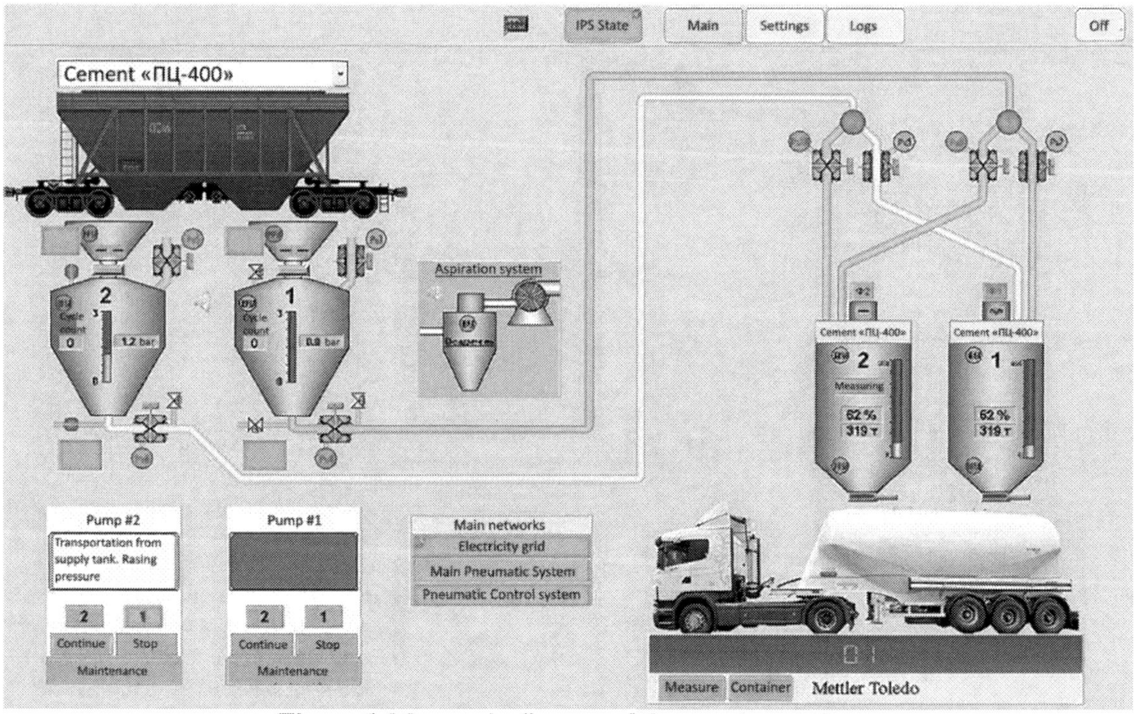

Figure 6. Mnemonic diagram of the operator station

The typical technological equipment of the cement warehouse includes:

- sensors of filling level of storage tanks and bunkers-consumers (threshold and tracking);
- filters for cleaning outgoing air with an automatic cleaning system;
- controlled switches of cement flow;
- pneumatic chamber pumps (see Fig. 5) complete with valves, gates, pressure and level sensors;
- fans and filters of the intake system aspiration;
- auxiliary pressure gauges for monitoring air pressure in the control and pumping pneumatic networks;
- consoles for loading consumer bunkers.

The RCWDS of the cement warehouse performs the following functions:

- constant monitoring of the filling levels of the storage tanks (Fig. 7) and consumer bins;
- automatic execution of cement pumping processes.

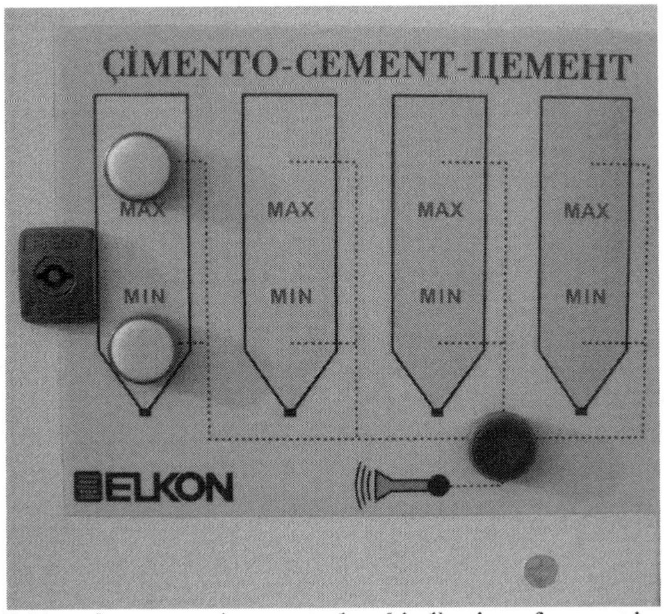

Figure 7. Cabinet for automation, control and indication of cement levels in silos

This provides:

- selection of sources and receivers as directed by the operator or automatically, taking into account the cement grades and filling levels;
- installation of mechanisms for switching the flow to the required positions (route preparation) with the inclusion of pre-start alarm bells;
- starting the operation of pumps in a cyclic mode and monitoring their operation until the end of pumping;
- monitoring the state of the line and all necessary conditions during pumping, automatic shutdown of the process and issuing messages to the operator in case of deviations;
- termination of the pumping process at the command of the operator, or upon filling the receiver tank, or emptying the source tank;
- purging of the used sections of the route to release material;
- recording of the pumping operations (work log) with fixing the time, cement grade and filling levels at the beginning and at the end of pumping.

The system provides:

- possibility of simultaneous synchronous operation of several pumps on one outlet line to increase productivity;
- simultaneous independent execution of several pumping processes on non-intersecting sections of the route.

4. Conclusion

By combining an efficient way of unloading hopper cars, high-tech pneumatic chamber pumps, modern technological solutions, effective and fully functional automation systems, it is possible to achieve high efficiency of cement handling processes, good timing and high reliable operation of the entire technological complex on the types of cement existing in the region.

Acknowledgments

The research was funded by a grant from the Russian Science Foundation for the event "Research by world-class scientific laboratories in the framework of the implementation of the priorities of scientific and technological development of the Russian Federation" of the President's program of research projects implemented by leading scientists, including young scientists.

References

[1] Jing S and Meng Q 2009 Research and design of remote real-time supervisory system based on OPC in cement enterprise *IEEE International Conference on Automation and Logistics*, 761-765 doi: 10.1109/ICAL.2009.5262822

[2] Ostroukh A V, Nedoseko I V, Surkova N E and Bulatov B G 2016 Automated Control System for the Milling Unit of Mineral Powders Plant *International Journal of Applied Engineering Research* **11(4)** 2625-2628

[3] Ostroukh A, Maksimychev O, Nikolaev A, Kolbasin A and Nedoseko I 2016 Development of automation of the drying and milling unit for the mineral powders plant *ARPN Journal of Engineering and Applied Sciences* **11(9)** 5717-5721

[4] Hettiarachchi B D, Dilum Bandara H M N and Samarasekera N A 2018 Automated Multi-Plant Scheduling of Ready-Mixed Concrete Trucks *IEEE International Conference on Service Operations and Logistics, and Informatics (SOLI)* (Singapore) pp 43-48 doi: 10.1109/SOLI.2018.8476729

[5] Syahputra R H, Komarudin K and Destyanto A R 2018 Optimization Model of Ready-Mix Concrete Delivery Route and Schedule: A Case in Indonesia RMC Industry *3rd International Conference on Computational Intelligence and Applications (ICCIA)* (Hong Kong) pp 21-25 doi: 10.1109/ICCIA.2018.00012

[6] Teixeira M A, Salman T, Zolanvari M, Jain R, Meskin N, Samaka M. 2018 SCADA System Testbed for Cybersecurity Research Using Machine Learning Approach *Future Internet* **10(8):76** https://doi.org/10.3390/fi10080076

[7] Aghenta L O and Iqbal M T 2019 Development of an IoT Based Open Source SCADA System for PV System Monitoring *IEEE Canadian Conference of Electrical and Computer Engineering (CCECE)* (Edmonton, AB, Canada, 2019) pp 1-4 doi: 10.1109/CCECE.2019.8861827

Capacity of merging areas of multi-lane roads with two-lane entrance ramps

A V Kostsov[1], **A M Tanatova**[1]

[1] Moscow State Automobile and Road Technical University, 64, Leningradsky prospect, Moscow, 125319, Russian Federation

E-mail: kostsov_msfs@bk.ru

Abstract. The discrepancy between transport demand and supply in the largest cities of the Russian Federation determines the need to create urban high-speed roads on their territory. Currently, there is no data on the capacity of sections of multi-lane road junctions with two-lane exit ramps, taking into account traffic conditions in large and major cities of Russia. The article presents the results of studies of the throughput of such sections of interfaces. The research was carried out using simulation methods after calibration of the model taking into account traffic conditions on the roads of the largest cities of the Russian Federation. The obtained research results are recommended for a reasonable assignment of the type of conjugation of two-lane exits with multi-lane highways on the territory of large and largest cities in Russia.

1. Introduction

Sections for merging travel paths of highways and urban high-speed roads are considered to be one of the main sources of increased accident number and congestion on highways (Hounsell and McDonald, 1992; Evans et al., 2001, J. Lu, P. Liu, B. Behzadi, 2007, J.A. Barria, S. Thajchayapong, 2011, M.W. Levin, S.D. Boyles, 2016). Despite a significant number of works devoted to this topic, there has not been developed until now a unified approach to determining the throughput capacity of highway merging sections. The concept of "confluence" of traffic flows is understood by most researchers as an elementary section consisting of two traffic lanes, within which the confluence of two traffic flows (the basic and the secondary) takes place. When multi-lane ramps are merged with the basic segment of a highway, a complex confluence section is formed consisting of several elementary sections for the confluence of traffic flows (Fig. 1), within which cars perform changeover maneuvers related both to the need for a mandatory lane change due to changing road conditions (MLC maneuver), and for reasons related to the desire of a car driver to move in a certain lane (DLC maneuver).

Content from this work may be used under the terms of the Creative Commons Attribution 3.0 licence. Any further distribution of this work must maintain attribution to the author(s) and the title of the work, journal citation and DOI.
Published under licence by IOP Publishing Ltd

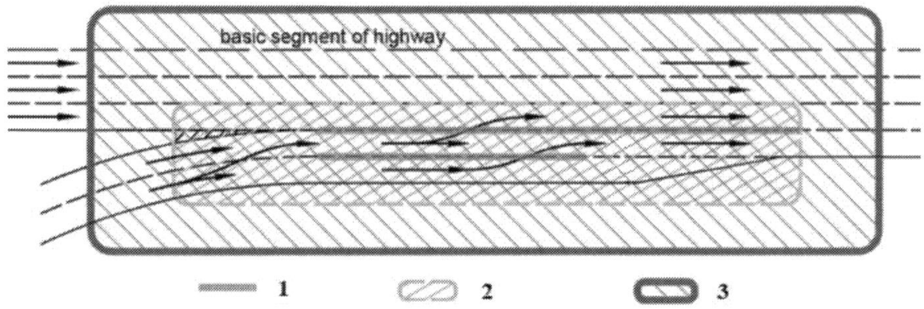

Figure 1. Scheme for determining the throughput capacity of a travel path merging section: 1 - an elementary merging section, 2 - a complex merging section, 3 - a traffic flow merging section.

The variety of schemes representing the mutual arrangement of elementary sections for merging traffic flows and the characteristics of the drivers' behaviour (national peculiarities of legislative regulation, the mentality of drivers, the composition of the traffic flow) does not allow to develop until now uniform recommendations for determining the throughput capacity of traffic flow merging sections, which is noted in a number of works devoted to studies of the throughput capacity of traffic flow merging sections (J. Shen, W. Li, F. Qiu, S. Zheng, 2015). This paper presents the results of studies devoted to the throughput capacity of traffic flow merging sections of highways with two-lane ramps, taking into account traffic conditions on the territory of the Russian Federation.

2. Method of studies devoted to the throughput capacity of merging sections of multi-lane roads with two-lane ramps

Traditionally, the Poisson distribution and the exponential distribution related with it inversely have been most widely used to determine the throughput capacity of traffic flow merging sections (Gross, D. and Harris, C, 1998). The Poisson distribution is most valid if a certain event constantly occurs with a sufficiently high probability; therefore, this distribution corresponds to the nature of the movement of congested traffic flows along road sections with heavy traffic (in conditions of continuous traffic, without traffic control devices). With the distribution of the car following distances according to Poisson's law, the distribution of the time intervals between the cars will be continuous exponential. Then we can write the probability density function for occurrence of intervals in the flow (1):

$$P = \lambda e^{-\lambda x} \tag{1}$$

Where $\lambda -$ distribution function parameter

The exponential distribution does not give satisfactory results for convergence of empirical data and theoretical values in the conditions of car traffic "in columns", which can be observed when cars move along two-lane highways in a mode close to the throughput capacity mode. The exponential distribution shows satisfactory convergence only when driving with some margin of capacity along two-lane roads, or along multi-lane roads where overtaking is possible.

Assuming multiple entries of secondary direction cars into the basic traffic flow with a sufficient time interval to complete such a maneuver, the ratio (2) could be written:

$$N = M * \frac{e^{-\frac{M*\Delta T_{\text{rp}}}{3600}}}{1-e^{-\frac{M*\Delta T_{\text{rp}}}{3600}}} \tag{2}$$

Where

$M -$ Traffic intensity along the lane adjacent to the confluence area, reduced number of cars / hour;

N_c — Throughput capacity of the secondary movement direction, reduced number of cars / hour;

ΔT_{rp} – Time interval used in the interaction of traffic flows, depending on road conditions, s;

This method of determining the throughput capacity of the confluence sections gives good convergence with the results of observations in the study of elementary sections for traffic flow confluence on single-lane speed transition ramps (for example, Drew, 1967; Darzentas, 1981).

Throughput capacity investigation for more complex merging sections, for example, merging sections of two-lane ramps, is associated with the study of confluence sections consisting of several elementary sections. The close relative position of the elementary merging sections determines their significant mutual influence. Therefore, there is not enough data on the throughput capacity of elementary merging sections to assess the throughput capacity of complex confluence sections; it is necessary to simulate their joint work, taking into account the performance of DLC maneuvers by cars.

Taking into account the limited sampling capabilities concerning sections for merging two-lane ramps with highways and high-speed township roads in the territory of the largest cities of Russia, studies of the throughput capacity of sections for merging travel paths were carried out using simulation methods based on the single-car movement models developed by Wiedemann (1974) and lane change modeldeveloped by Sparmann (1978), allowing to take into account the joint work of elementary merging sections (hereinafter SW-model).

Before carrying out a numerical experiment using the simulation method, the SW-model was calibrated taking into account the traffic conditions on multi-lane roads in cities of the Russian Federation. The model was calibrated on the basis of comparing the throughput capacity values of the merging sections obtained from the calculation using the SW-model (Fig. 2, SW-model) and the throughput capacity values of the merging sections obtained by the calculation according to formula (2) using the data on the distribution of boundary following distances (ΔT_{rp}) for cars on elementary merging sections (Fig. 3), which were obtained by the authors within the intersections at different levels of the roads in the largest cities of the Russian Federation: Moscow, St. Petersburg, Yekaterinburg and Novosibirsk in 2018 (Kostsov, 2018). The length of the speed transitional lane (the section of merging traffic flows) when carrying out calibration calculations is assumed to be 260m. A good convergence of the results obtained by the calculation method according to (2) and using the SW model was obtained due to the results of calibration calculations.

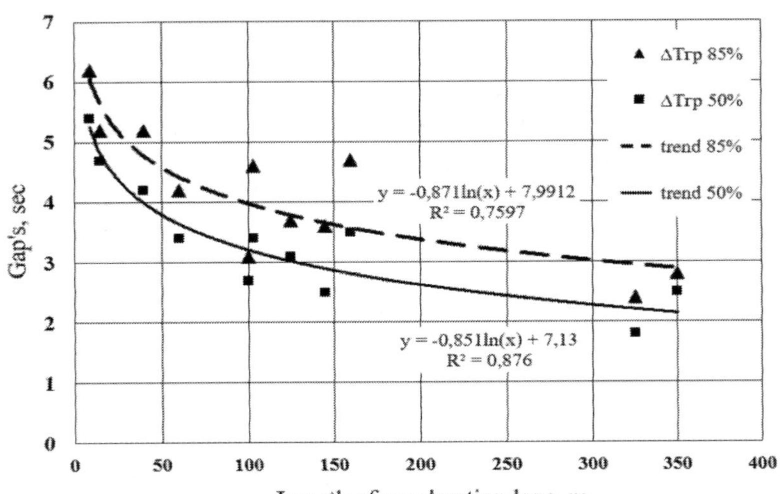

Figure 2. Traffic gaps of cars at the acceptance rate of 50% and 85% in confluence zones of traffic flows.

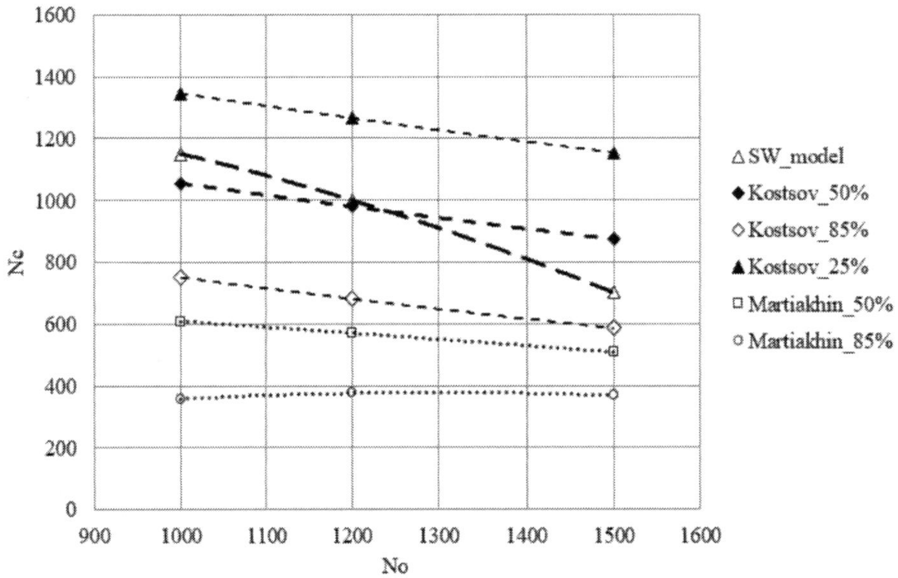

Figure 3. The throughput capacity of the traffic flow merging sections: No - traffic intensity of the basic traffic flow, Nc - intensity in the secondary traffic direction

3. Investigation of the capacity of the sections of mergings of multi-lane roads with two-lane ramps by the method of simulation

In order to study the throughput capacity of the sections for merging multi-lane roads with two-lane ramps, a classification of traffic flow confluence sections was developed, which is shown in Fig. 4. The type of speed transition lanes according to a parallel scheme is adopted. The lengths of the ramp sections of the transitional speed lanes, and the full-width section lengths are taken in accordance with the recommended values specified in the regulatory documents in force in the Russian Federation (SP 34.13330.2012). The values of the main geometrical parameters adopted in the research are given in Table 1.

Table 1. Values of the main geometric parameters

L, L_1, m	L_0, L_{01}, m	L_2, m	Slope, %
180	80	60	0

A numerical experiment was carried out for each merging section type for four-lane, six-lane and eight-lane highways within the range of possible traffic intensities on the base segment (N_0).

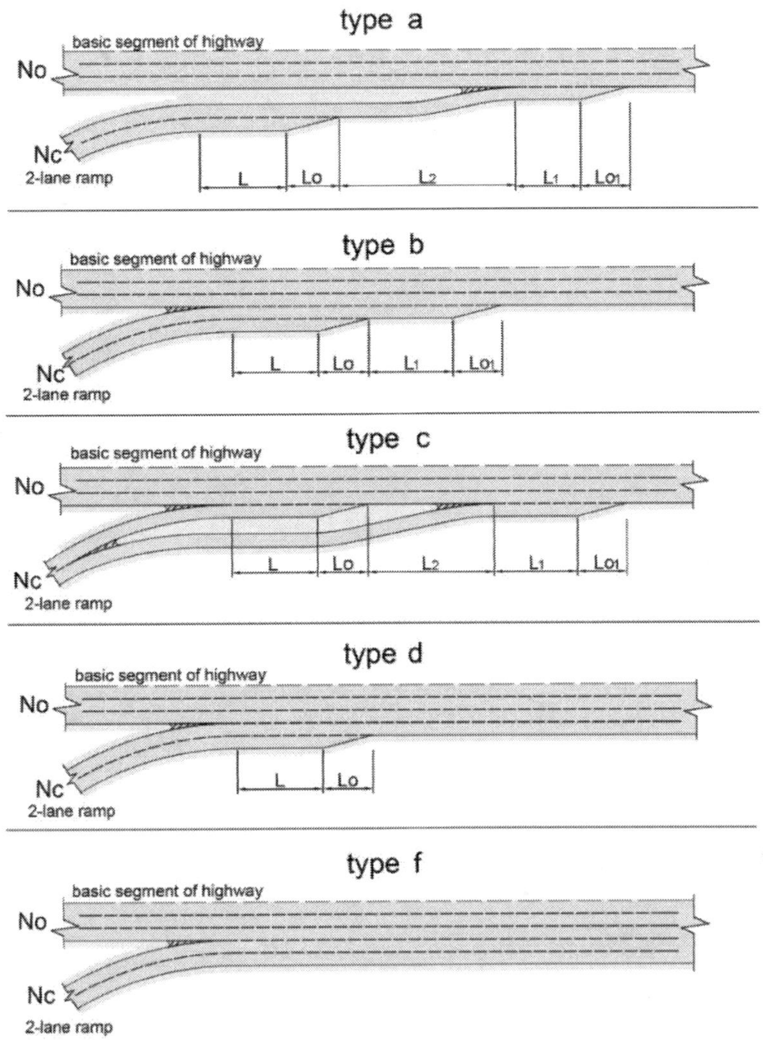

Figure 4. Types of travel path merging sections of two-lane ramps

4. Results and discussion

Statistical processing of the experimental results obtained by means of simulation modelling made it possible to obtain data on the throughput capacity of two-lane ramps depending on the car movement intensity in the basic direction and the type of merging section for two-lane ramps upon traffic conditions on the territory of the Russian Federation. The throughput capacity of the merging section made according to type (a) does not exceed 2200 reduced number of cars / hour. Arrangement of two-lane merging by type (a) is recommended in narrow conditions of construction and reconstruction, when it is impossible to place other types of two-lane ramp merging with a highway.

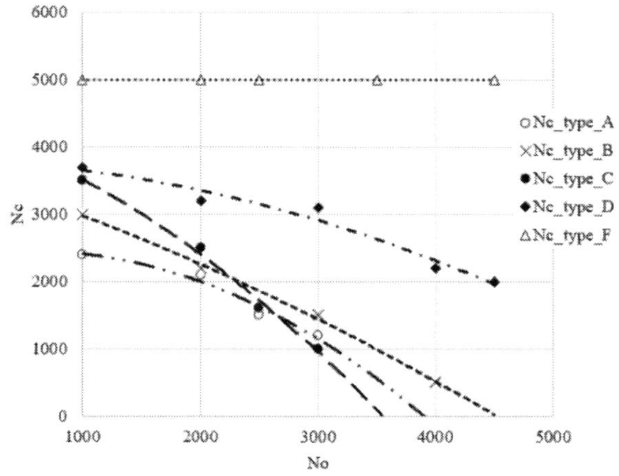

(a)

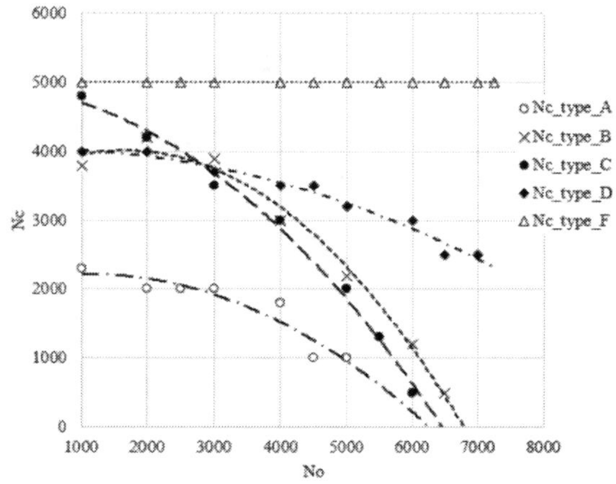

(b)

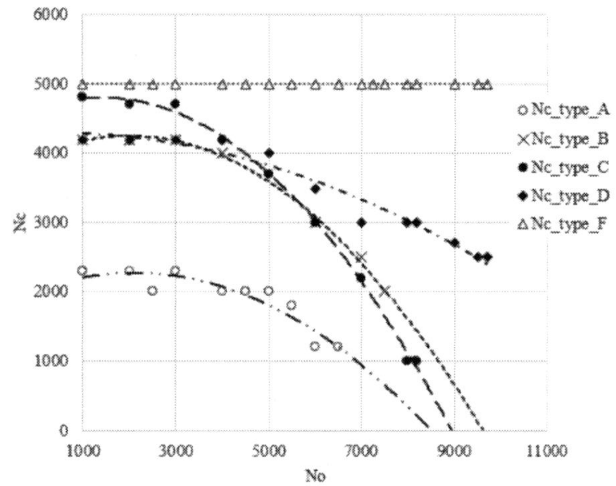

(c)

Figure 5. Simulation results - curves for the dependence of the throughput capacity of two-lane ramps on the traffic intensity in basic direction: (a) - for 4-lane roads, (b) - for 6-lane roads, (c) - for 8-lane roads.

The arrangement of adjoining roads according to types (b) and (c) allows the merging of traffic flows without increasing the number of lanes in the basic direction and is recommended for use when the ratio of traffic intensities does not exceed those indicated in Fig. 5. When the ratio of traffic intensities does not allow the arrangement of adjoining roads of type (b) and (c), it is necessary to use adjoining roads of type (d). If the intensity ratio does not allow the type (d) merging to arrange, it is recommended to use the type (f) merging with the arrangement of two additional lanes in the main direction.

5. Resume

Based on the studies carried out, the dependences of the throughput capacity of two-lane ramps on the traffic intensity in the basic direction were established, which are shown in Fig. 5. The research results can be extended to the mergings of two-lane ramps in large cities and metropolises of Russia.

6. Conclusion

The results of studies for two-lane ramps presented in the paper can be used to the throughput capacity calculate the capacity of multi-lane highways at the merging ramps of two-lane ramps.

References

[1] Levin M W, Boyles S D 2016 A multiclass cell transmission model for shared human and

autonomous vehicle roads *Transportation Research Part C: Emerging Technologies* Vol 62 103–116

[2] Lu J, Liu P, Behzadi B 2007 Safety Evaluation of Freeway Exit Ramp *Proceedings of the TRB Annual Meeting CD-ROM* **07–1293**

[3] Barria J A, Thajchayapong S 2011 Detection and classification of traffic anomalies using microscopic traffic variables *IEEE Transactions on Intelligent Transportation Systems* Vol 12 **3** 695–704

[4] Drew D R 1967 Gap acceptance characteristics for ramp-freeway surveillance and control *Highway Res Rec* **157** 108–134

[5] Darzentas J 1981 Gap acceptance: myth and reality *Proc 8th Int Symp on Transportation and Traffic Theory* (eds. V. F. Hurdle, E. Hauer and G. N. Stewart) Toronto pp 175–192

[6] Hounsell N and McDonald M 1992 *An Investigation of Flow Breakdown and Merge Capacity on Motorways* Contractor Report 338, Transport Research Laboratory, UK

[7] Evans J, Elefteriadou L and Gautam N 2001 Probability of breakdown at freeway merges using Markov chains *Transp Res B* **35** 237–254

[8] Kostsov A V 2018 Research of critical gaps of traffic flows on freeway interchange *5th International Conference on Road and Rail Infrastructure* (Zadar, Croatia)

[9] Sparmann U 1978 Spurwechselvorgange auf zweispurigen *BAB-Richtungsfahrbahnen, Straßenbau und Straßenverkehrstechnik* University of Karlsruhe, Heft 263 (in German)

[10] Fransson E 2008 *Driving behavior modeling and evaluation of merging control strategies - A microscopic simulation study on Sirat Expressway Department of Science and Technology* (Linköping University, Sweden)

[11] Martyahin D 2008 *Povyshenie propusknoj sposobnosti pri proektirovanii s"ezdov gorodskih transportnyh razvyazok* [Increasing capacity in the design of the urban freeway interchanges] (Moscow) (in Russian)

[12] SP 34.13330.2012 *Avtomobilnyie dorogi* [Russian Set of Rules SP 34.13330.2012 Automobile roads] Moscow, 2012 (in Russian)

[13] Izmenenie №1 k SP 34.13330.2012 *Avtomobilnyie dorogi* [Amendment No. 1 to Russian Set of Rules SP 34.13330.2012 Automobile roads]. Moscow , 2017 (in Russian)

[14] Gross D and Harris C 1998 *Fundamentals of Queuing Theory* (3rd Edition, John Wiley, Chichester)

[15] Shen J, Li W, Qiu F, Zheng S 2015 Capacity of Freeway Merge Areas with Different On-Ramp Traffic Flow *Promet – Traffic&Transportation* Vol 27 **3** 227–235

IOP Conf. Series: Materials Science and Engineering 1159 (2021) 012037 doi:10.1088/1757-899X/1159/1/012037

Mathematical model of an intelligent decision support system for road quality management

E M Abbasov[1] N N Teodorovich[2], N P Sidorova[2] and T S Abbasova[2]

[1] Department «Electrical engineering and electrical equipment», Moscow Automobile and Highway State Technical University (MADI), 64, Leningradsky ave. Moscow, 125319, Russia

[2] Department of Information Technologies and Control Systems, The State-Funded Educational Institution of Higher Education Moscow region University of Technology (TU), 42, Gagarin St., Korolev, Moscow region, 141070, Russia

E-mail: abbasova_univer@mail.ru

Abstract: The experience of operation of decision support systems is analyzed, it is shown that the existing principles of creating these systems do not take into account the failure and change in the performance of individual service element, which depend on the current load on the system. The definition of an adaptive decision support system as an intelligent system that works in an automated mode and is able to function in a changing environment, its internal structure, knowledge base, regulatory framework for management, support tools, system users is given. The tasks of construction a mathematical model of an adaptive decision support system and determining its objective function are set. A distributed decision support system, which receives information containing data from various input sources was studied. For construction a mathematical model of the system under investigation, the procedures for selecting mathematical models and methods of queuing theory were used. To assess the quality and effectiveness of an adaptive decision support system, simulation modeling is proposed.

1. Introduction

High quality decision-making in the analysis of complex problems, characterized by the processing of large volumes of information on the quality of highways, is difficult to achieve without the involvement of computer technology only on the basis of subjective expert assessment of analysts and specialists.

Operating experience of automated decision support systems (DSS) showed [1 ... 3] that the existing principles for their creation do not take into account the change in the speed of action of individual service elements (EO) as the load on the system increases. The situation with the speed of DSSS can be significantly improved if the system is adapted to adapt to changing conditions of the external and internal environment in order to optimize functioning [4].

2. Method

Usually the creation of automated DSS, which uses artificial intelligence methods, is associated with solving many problems. These include information modeling, intelligent design, the generation of

Content from this work may be used under the terms of the Creative Commons Attribution 3.0 licence. Any further distribution of this work must maintain attribution to the author(s) and the title of the work, journal citation and DOI.
Published under licence by IOP Publishing Ltd

formal models, using methods for representing knowledge. Of no less importance is the development of software with computer support, the improvement of the architecture of the information and communication network to support decision-making, and many other tasks.

The quality of functioning of intelligent DSS significantly increases with the introduction of additional modules that allow taking into account the output, the change in the speed of individual service elements depending on the load on the system. In fact, we are talking about the transformation of DSS into an adaptive system.

In DSS, requests are used as information processing tools. The servicing element (SE) of the requests is the server from which data is collected (on which the multidimensional data warehouse multidimensional data warehouse is located). This data is used in the preparation of reports on the results of collection and analysis of system data. If the resources or performance of the server is insufficient, then the management data store can be installed on another computer / computers (second and third SE). To take into account the factors characterizing adaptive DSS, the following enlarged groups of instrumental computer facilities can be distinguished:

– adaptive knowledge base for the rapid accumulation of information about changing conditions of the situation;

– adaptive software and hardware support for choosing the best calculation plan when making a decision;

– an adaptive interface that can adapt to the individual user features and current requirements for the solution.

Selection among the existing models, further development and effective use of the mathematical model of the DSS is carried out in accordance with the chosen optimization criterion that takes into account the operating conditions of the system and the tools with which it is implemented. In [5] various procedures for selecting and evaluating models are described.

Usually the procedure governing the choice of a mathematical model is carried out by means of the formation of a preference function. It is necessary to take into account that in DSS the important tools are the requests for data processing.

The preference function has the form of (linear or nonlinear) convolution of the form [5]:

$$u_j = \sum_i k_{j,i} k_{j,i}\,{}_{,i=\overline{1...n}}\ , \tag{1}$$

where u_j – the value of the preference function when choosing the j-th model, r_i – (estimate) the quality of the i-th criterion, k_i – the weight of the i-th criterion.

When choosing the j-th model, the modified convolution can be estimated as follows:

$$u'_j = \left(\frac{r_1k_1+r_2k_2}{r_3k_3}\right) + r_4k_4 + r_5k_5 + r_6k_6, \tag{2}$$

where u_j – modified convolution when choosing the j-th model, r_1 – performance (the number of requests serviced per unit of time), r_2 – optimization criterion, r_3 – costs (energy, production, etc.), r_4 – rate of DSS adaptation, r_5 – an indicator of the accuracy of processing requests and visualizing the results of processing requests, r_6 – time costs indicator for DSS adaptation, k_i – weight of the i-th optimization criterion.

On the basis of relations (1) and (2), an algorithm for selecting mathematical models is constructed. To minimize the development costs of these models, it is advisable to apply linear programming methods. From the point of view of the result, models should be described by similar characteristics:

$$F = \sum_{j=0}^{n} c_j x_j \rightarrow min \tag{3}$$

$$\sum_{j=0}^{n} h_{i,j}x_j \le b_i, \qquad d_j \le x_j \le D_j, \qquad i = \overline{1 \dots m}, j = \overline{1 \dots n}$$

where $F = f(x_j)$ – objective function, $g_i(x_j) \le b_i$ – confines, $d_j \le x_j \le D_j$ – border conditions, x_j – group of j-type models, $h_{i,j}$ – Required resource of the i-th type type when developing j-th model, b_i – limitations of the i-th resourse (value), c_j – cost of developing the j-th group of models.

In determining the objective function, the (total) cost of model development, the time of solving the problem (scenario), and the indicators of financial, time or personnel (professional) resources are often taken into account [6]. Obviously, the lower cost of developing a model and the shorter time of solving the problem do not always lead to effective results.

It is advisable to take the number of requests processed per unit time as the objective function of adaptive DSS, which is a measure of the efficiency of using an information resource. DSS with SE is presented as a discrete event queuing system. The number of requests processed per unit time tends to the maximum and depends on the following variables: the probability of using EO at different loads, the coefficient of use of EO at different loads, the probability of refusal to service EO, the probability of several service elements working simultaneously, and the average number of requests in the system.

Restrictions on the modes of operation of EA are associated with a restriction on the number of functioning and backup EA, serviced requests and denials of service. More details are considered in [7]. This choice is due to the fact that SE and operations with them are typical functions realized by discrete processes, and computer systems and communication channels for transferring the results of computation in infocommunication interaction with the support of decision making and remote control can be represented as networks of single- and multichannel queuing systems (QS) with failures and various maintenance disciplines [8].

It is assumed that the investigated DSS is distributed, and it receives information containing data from various input sources that is written to databases (data warehouses).

Under such initial assumptions, the problem to be solved is reduced to finding the maximum efficiency of using the information resource of adaptive DSS in the conditions of its normal functioning. A means of achieving this goal is the development of algorithms, methods and means of solving the problems of integration and data quality of various suppliers of information resources.

Based on the selected objective function, a simulation model of adaptive DSS can be constructed. In the process of conducting theoretical and experimental studies, this model allows us to determine the change in the parameters of the change in the number of service elements depending on the intensity of a given flow. The efficiency of using an information resource is increased by reducing the cost of maintaining the data of the EA and reducing the proportion of lost requests due to large queues.

3. Mathematical model

The formula for calculating the probability of the initial state of the system is obtained by setting a limit on the maximum possible number of SE in DSS

$$p_0 = \left\{1 + \sum_{S=1}^{E_{max}} \left(\frac{\psi^{S-1}}{(S-1)!}\right)^K \left(\frac{\Psi}{S}\right) \left[\frac{(\Psi/S)^K - 1}{\Psi/S - 1}\right]\right\}^{-1}, \tag{4}$$

where S – the number of SE; E_{max} – the maximum number of SE elements; Ψ – load on the system; K – the number of system states, which depends on the length of the queue.

The obtained result p_0 corresponds to the state of the system when there is no service request in it, i.e. the system is completely idle.

We determine the probability of the presence of at least one request in the system

$$p = 1 - p_0. \tag{5}$$

We calculate the probability of several S (in the presented example, one, two, and all three) SE

$$P_S = p_0 \left(\frac{\psi^{S-1}}{(S-1)!}\right)^K \left(\frac{\Psi}{S}\right) \left[\frac{\left(\frac{\Psi}{S}\right)^K - 1}{\frac{\Psi}{S} - 1}\right]^{-1}. \tag{6}$$

The values of the SE utilization coefficients were obtained with the addition of a restriction on the maximum possible number of SE [7]

$$a_n = \sum_{S=n}^{Emax} P_S = p_0 \sum_{S=n}^{Emax} \left(\frac{\psi^{S-1}}{(S-1)!}\right)^K \left(\frac{\Psi}{S}\right)\left[\frac{(\Psi/S)^K - 1}{\Psi/S - 1}\right] = 1 - p_0 - \sum_{S=1}^{n-1} P_S. \quad (7)$$

where n – the number of requests.

Requests are refused only if two conditions are met at the same time: all available EOs are occupied and all places in the queues for EO are occupied. The probability of this event [9]

$$p_r = \frac{\psi^{max \cdot K}}{(max!)^K} p_0. \quad (8)$$

4. Results of calculations and simulation

The effectiveness of introducing the adaptability property into DSS was checked by comparing the results of its functioning with two other common systems. As such, two systems of the limiting type are chosen

- M/M/1/100 system based on one SE (1), in which the incoming flow of requirements is Poisson's (M), the distribution of time for servicing obeys the exponential law (M), an almost infinite queue is allowed (100)
- the M/M/3 system is built on the basis of three SE and assumes the complete absence of queues.

Table 1 presents the results of calculations of the probability of the presence of at least one query in the system. In graphical form, these results are shown in Figure 1. In addition to the calculations, simulation was carried out in the GPSS environment, which showed similar results.

Table 1. Probability values for the presence of at least one request in the system.

System in question	System load, Ψ						
	0	0,5	1	1,5	2	2,5	3
M/*M*/3		0,334	0,500	0,600	0,666	0,716	0,750
Adaptive DSS	0	0,469	0,752	0,877	0,932	0,961	0,974
M/*M*/1/100		0,503	0,996		1		

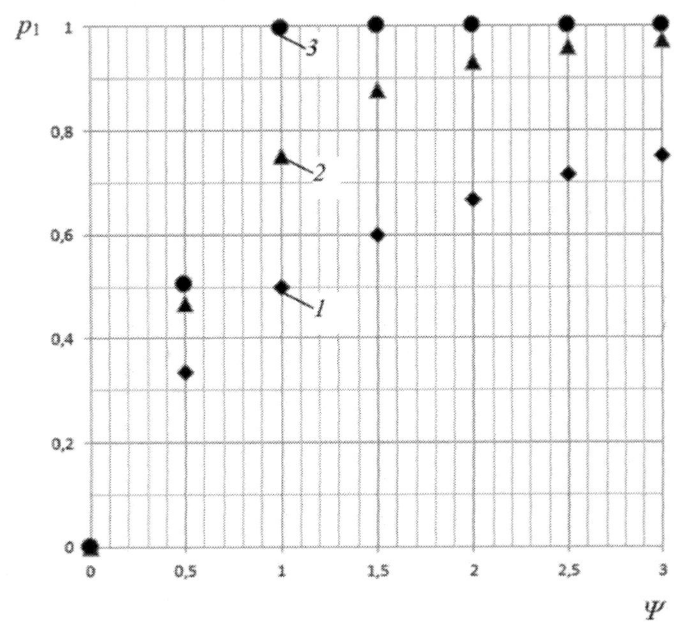

Figure 1. Dependencies of the probability of using the first element of service at different loads for decision support systems: *1* – *M*/*M*/3; *2* – adaptive; *3* – *M*/*M*/1/100.

Obviously, adaptive DSS (points 2) occupies an intermediate position between a system without *M*/*M*/3 queues (points 1) and a system with an almost infinite *M*/*M*/1/100 queue (points 3), confirming the conclusions of theoretical calculations. The system operates in the optimum mode, preventing neither excessive downtime in operation (as in the first case), nor operation in emergency mode (as in the case of the *M*/*M*/1/100 system).

By determining the percentage difference between the two numbers, it is estimated that on average, an adaptive system provides 28% more time for processing requests than a system without queues, and 11% less downtime than a system with an infinite queue.

You need to know the percentage of denial of service when all EO and all the places in the queue for them are occupied. The results of the calculation of the probability of denial of service according to the formula (8) are presented in table 2.

Table 2. Denial of service probability values

System in question	**System load, Ψ**										
	0	**0,5**	**1**	**1,5**	**2**	**2,5**	**3**	**3,5**	**4**	**4,5**	**5**
M/*M*/3		0,013	0,061	0,137	0,210	0,280	0,348	0,402	0,449	0,49	0,529
Adaptive DSS		0		0,009	0,037	0,090	0,160	0,230	0,305	0,363	0,418
M/*M*/1/100	0			0			0,012	0,146	0,252	0,337	0,399

The dependencies of the probability of denial of service at different load on the DSS are presented in Figure 2. It is obvious that the adaptive system (points 2) provides a lower probability of denial of service than a system without queues M / M / 3 (points 1) but greater than a system with almost endless queue of M / M / 1/100 (points 3).

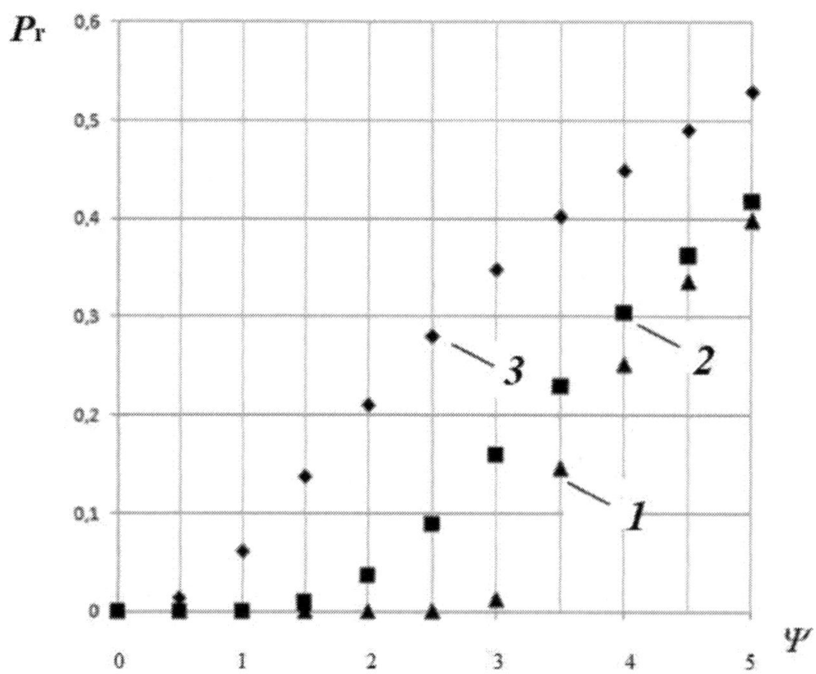

Figure 2. Dependence of the probability of a denial of service failure on various load on the system: 1 – M/M/3; 2 – adaptive; 3 – M/M/1/100.

The implementation of tools and management mechanisms of DSS in a single information and analytical complex that includes information-modeling systems provides expansion of functionality, additional analytical support for DSS and performance of information processing.

5. Conclusions

1. Introduction to the DSS adaptation options successfully solves the problem of maximizing the number of requests processed per unit time.
2. The adaptive system on average provides a 28 percent increase in employment by processing requests compared to a queue-free system, combined with an 11% reduction in downtime compared to a system that allows an infinite queue.
3. The probability of failure of the adaptive system compared to the system without queues on average does not exceed 5.8%.

References

[1] Kirikov I A, Kolesnikov A V, Listopad S V, Soldatov S A 2015 Heterogeneous intelligent computer decision support systems: models of coordination and coherence *Systems and means of informatics* **vol 25 2** pp 96–110
[2] Egorov S L 2010 Development of the architecture of data warehouse in the process of designing a decision support system *News of higher educational institutions. The Ural region* **4** pp 18–21
[3] Volkov V U, Volkova V V. The tools of cognitive modeling in decision support systems. *Herald of the International Academy of System Studies. Informatics, ecology, economics*, 2016, vol 18, №1, pp 104–108
[4] Tkachuk E O. 2002 Requirements for adaptive decision support systems *Herald of TRTU* **2(25)** pp 248–251
[5] Ageeva A F 2018 Review of modern decision support systems created using the agent-based approach *Electronic information systems* **4(19)** pp 29–46
[6] Ovsyannikov E K, Mozgirev B T. 2004 The choice of mathematical models for decision support systems *Proceedings of SPIIRAS* **vol 1 2** pp 36–46
[7] Artyushenko V M Akimkina E E 2018 Simulation model of adaptive decision support system *Herald of Computer and Information Technologies* **2** pp 46–54
[8] Kleinrock L 2013 Queuing Theory (Moscow: Book on demand) 429 p
[9] Akimkina E E 2017 Optimization of data processing in decision support systems with service elements *Bulletin of the Voronezh State University. Series: System Analysis and Information Technology* **2** pp 90–97

IOP Conf. Series: Materials Science and Engineering 1159 (2021) 012038 doi:10.1088/1757-899X/1159/1/012038

Application of digital technologies at road monitoring

A A Fotiadi[1], E V Zhustareva[1], V I Bochkarev[1], S A Gnezdilova[2] and E D Partsevsky[1]

[1]Moscow Automobile and road construction state technical university (MADI), Moscow, Russia
[2]Belgorod State Technological University named after V.G. Shukhov, Belgorod, Russia

E-mail: fotiadi@gmail.com

Abstract. Largely, the condition of the road network determines the economic and social development of any country. Annual monitoring of roads and analysis of their condition allow identifying in advance the sections where repairs are needed. In order to develop a correct and effective repair strategy, it is necessary to be able to analyze large volumes of road monitoring data. Therefore, the development and introduction of specialized software into practice are required for gathering, storing, processing and analyzing a great amount of data. The paper examines current specialized software developed by the specialists of the Department "Construction and Maintenance of Roads" of MADI (software products: ProfORG and Vsteets2) to process road monitoring data. This software makes it possible to record, store, process and analyze information using the DVS-3 road mobile laboratory. The article presents the research results of changes in the longitudinal and transverse roughness of the pavement surfaces performed using the ProfORG and Vsteets2 software.

1. Introduction

The economic and social progress of any country is considerably determined by the condition of roads. The sections of roads requiring rehabilitation may be revealed ahead of time by annual monitoring, that makes it feasible to conduct necessary works for maintaining roads in optimal conditions of use. Obtaining reliable information on the condition of the road pavement surface and its correct processing and analysis are prerequisites for developing an effective repair strategy [1].

Recently, road mobile laboratories, which are equipped with devices for digitizing measurement information, have been used to monitor the condition of roads. Within the annual monitoring, the databases on the transport and operational condition of highways are created. Therefore, there is a need to develop and introduce specialized software into practice with the aim to record, store, process as well as to analyze a large volume of the data obtained.

Such technologies allow identifying regularities of changes in the major transport and operational characteristics, developing forecasting models, and determining the strategy of repair works as well.

2. Problem statement

Many European countries use the monitoring of road pavement surface. In 2010, Denmark, Iceland, Norway and Sweden implemented a joint project NordFoU - Pavement Performance Models (PPM)

Content from this work may be used under the terms of the Creative Commons Attribution 3.0 licence. Any further distribution of this work must maintain attribution to the author(s) and the title of the work, journal citation and DOI.
Published under licence by IOP Publishing Ltd

[2]. Within the framework of the project, various technical and operational characteristics of flexible road pavements were investigated. Researches resulted in

- developing a pavement performance model describing changing in the condition of flexible pavement structures;
- defining calibration coefficients of the model with regard to natural and climatic conditions in research areas;
- developing software application for prognostication of pavement performance at a network level.

Particular attention is paid to road monitoring in the United States. In this country, the instruction on the design of road pavements based on the surveys of the condition of road pavements for many years was developed. The instruction includes the calculations on predicting the major damages to road pavements and changes in transport and operational characteristics [3].

In Russia, roads are also monitored annually. For 10 years, the specialists of MADI have been performing the annual monitoring on sections of the R-22 "Kaspiy" federal road going through the Moscow, Ryazan, Tambov, Voronezh and Volgograd regions. The total length of the monitoring sections was about 775 km. The survey sections had different pavement surface design. Traffic density, depending on the location of the sections, ranged from 3,500 to 18,000 car/day. During monitoring, measurements of longitudinal and transverse roughness of road pavement surface were conducted for each lane, and local damages were recorded using video filming. Works were performed by means of the DVS-4-IK road mobile laboratory.

The survey during several years gave the possibility to develop the system for the control of the road pavement condition aiming at practical application on the R-22 "Kaspiy" road. Also the recommendations on improving the complex evaluation method of the highway maintenance quality in terms of the International Roughness Index 4] were given. Moreover, a remote method for forecasting damages of the cement-concrete pavement was offered [5]. At present, research works continue.

3. Methods and materials

The DVS-4-IK mobile laboratory of the MADI department "Construction and Maintenance of Roads" was used for monitoring of road pavement surface. The laboratory in question is a car equipped with a set of measuring equipment for obtaining, storing and processing data concerning the condition of the road pavement surface and facilities on the road. Figure 1 shows a general view of the DVS-4-IK mobile laboratory.

Figure 1. General view of the DVS-4-IK mobile laboratory.

To obtain and collect data the laboratory is equipped with the following systems:

- video filming;
- georeferencing;
- measuring the microprofile of the carriageway in the longitudinal direction;
- measuring the microprofile of the carriageway in the transverse direction.

To record, store and process the obtained data the laboratory is equipped with the following set of equipment:

- computer;
- monitor;
- data entry means;
- electronic unit for data collection;
- software.

Special ProfORG and Vsteets2 software for researches was developed by the engineers Osipov I.A. and Vaschenko A.V. (MADI).

The obtained data of the longitudinal and transverse roughness of the pavement surface and the continuous video recording of the road condition were processed and systematized in the office. The results of measurements of longitudinal and lateral roughness of road pavement surface of selected road sections of each year inspection could be displayed on the monitor using the ProfORG software. This program made it possible to view video at any point of the road. Thus, a visual identification of the reasons of inadequate values of the longitudinal and transverse roughness of the pavement surface was possible. Figure 2 and 3 show the fragments of the ProfORG program.

To determine the number of local damages on the road pavement as well as to evaluate them, the Vsteets2 program was used, which allowed defining the geometric dimensions of damages in terms of the extent (length of cracks on the pavement surface) or the area (in the case of a pothole). Figure 4 demonstrates a fragment of Vsteets2 program.

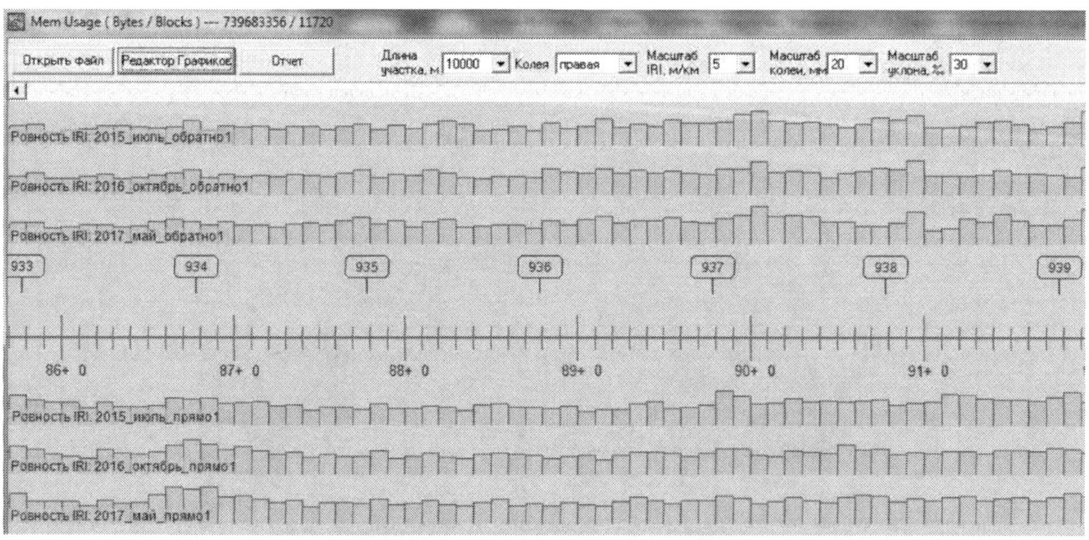

Figure 2. Data on longitudinal roughness of the pavement surface in the ProfORG software: green columns are the values of the IRI along the traffic lane in the forward (at the bottom) and reverse (at the top) directions during survey periods - May, October, July; red scale is a distance in kilometers; blue rectangles indicate kilometers.

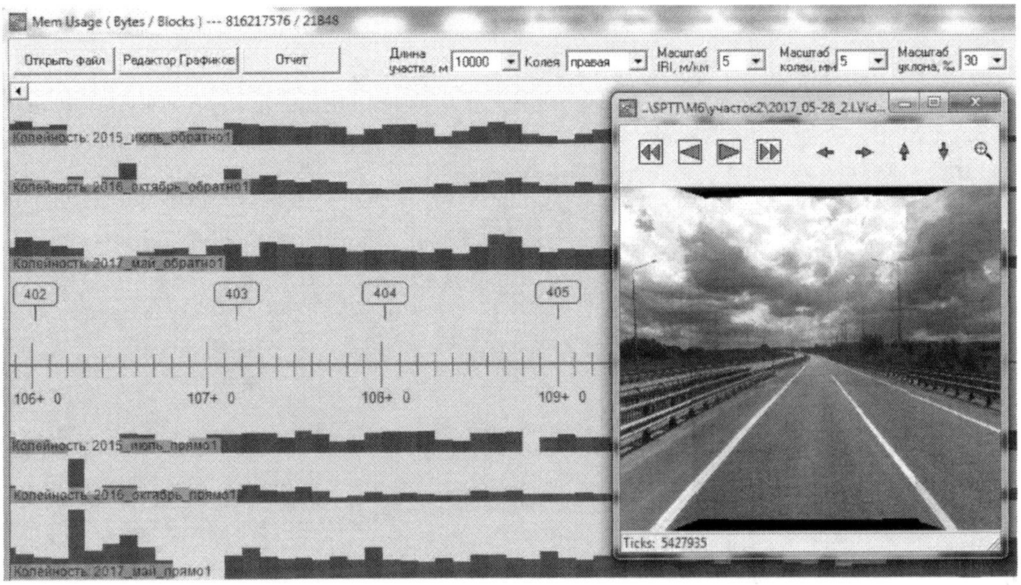

Figure 3. Data on the transverse roughness of the road pavement surface and video image of the measurement section in the ProfORG software: blue columns are the values of the rut depth along the traffic lanes in the forward (at the bottom) and reverse (at the top) directions during the survey periods - May, October, July; red scale is a distance in kilometers; blue rectangles indicate kilometers.

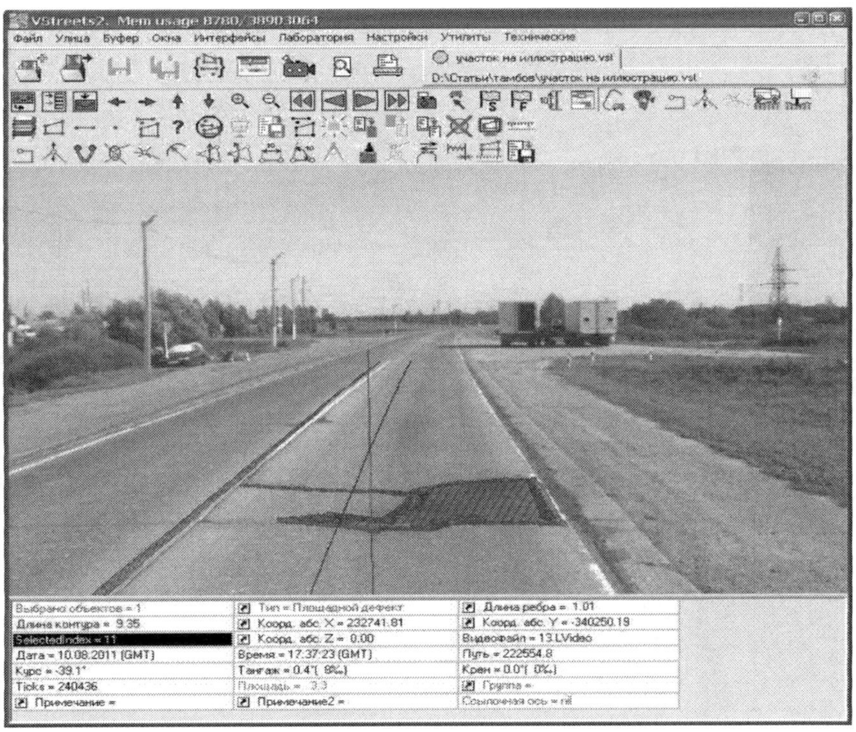

Figure 4. Determination of the area of a local damage of road pavement surface in Vsteets2 software: damage area is highlighted in red.

IOP Conf. Series: Materials Science and Engineering 1159 (2021) 012038 doi:10.1088/1757-899X/1159/1/012038

4. Results and discussions

Based on the annual monitoring of the condition of the pavement surface of the R-22 "Kaspiy" federal road the intense increase in the values of indicators (dynamics of change) of the longitudinal and transverse roughness at different density of traffic was established. Figure 5 shows the results of measurement of the longitudinal roughness after 6 years of inspection, and Figure 6 indicates the results of measurement of transverse roughness after 5 years of inspection.

An annual increase in the International Roughness Index ΔIRI depending on the initial value of IRI_0 and traffic density is given in Table 1. The values of the annual increase in the rut depth for asphalt-concrete and cement-concrete pavements, depending on the traffic density, are presented in Tables 2 and 3.

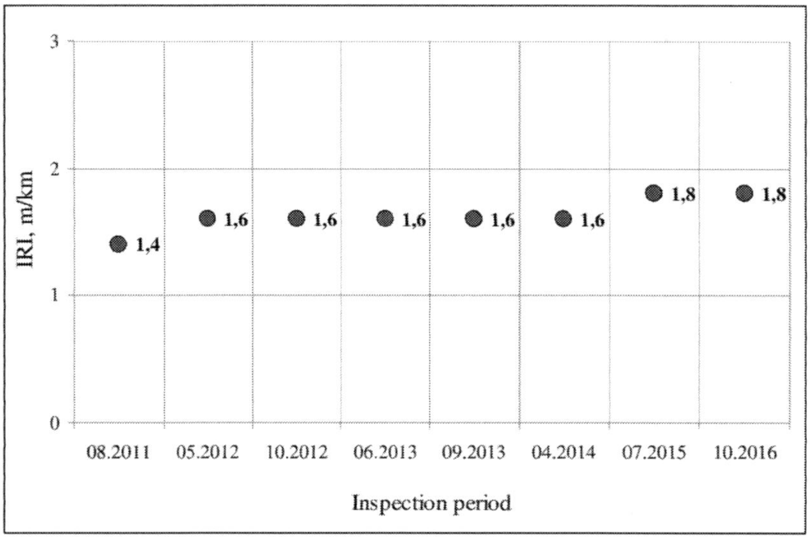

Figure 5. Values of the IRI on the section 198+500 – 202+350 km of the R22 "Kaspiy" federal road.

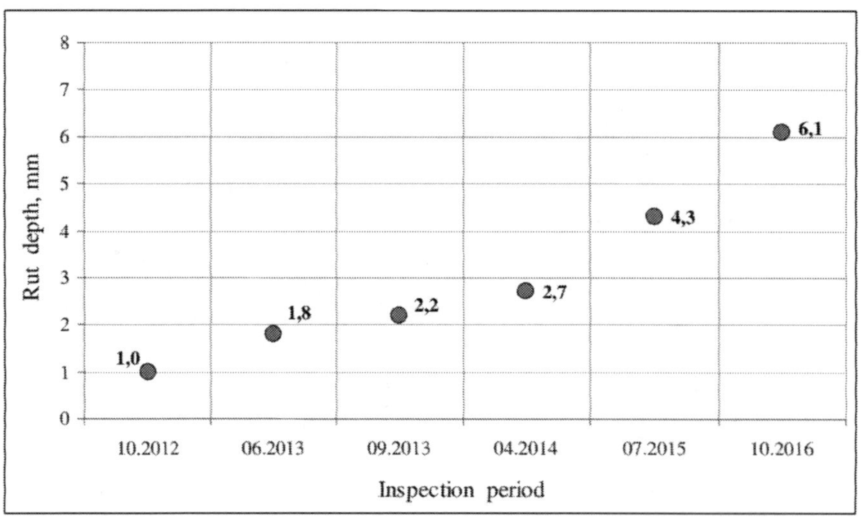

Figure 6. Values of a rut depth on the section 250+000 – 260+000 km of the R22 "Kaspiy" federal road.

Table 1. Annual increase in the International Roughness Index

Traffic density in one lane, car/day	Annual increase in the International Roughness Index ΔIRI at IRI_0 value, m/km		
	< 1.5	1.5 – 2.2	≥ 2.2
<2 500	0.10	0.10	0.15
2 500 – 5 000	0.10	0.15	0.20
5 000 – 10 000	0.10	0.20	0.20
10000 – 20 000	0.15	0.20	0.25
> 20 000	0.15	0.25	0.25

Table 2. Annual increase in a rut depth for asphalt-concrete pavement

Traffic density in one lane, car/day	Annual increase in a rut depth ΔK, mm
< 1 000	1.00
1 000 – 2 500	1.25
2 500 – 5 000	2.00
5 000 – 10 000	2.50
10000 – 20 000	3.00
> 20 000	6.00

Table 3. Annual increase in a rut depth for cement-concrete pavement (data for the first two years of operation)

Traffic density in one lane, car/day	Annual increase in a rut depth ΔK, mm
< 2 500	0.50
2 500 – 5 000	1.00
5 000 – 10 000	1.25
10000 – 20 000	1.50
> 20 000	2.50

5. Conclusion

The use of digital technologies when monitoring roads enables quickly obtaining and processing a large amount of data on the transport and operational condition of roads, systematizing and storing this information as well.

The database, resulted from the annual monitoring of the road condition, permits to resolve various scientific and practical problems:

- to establish dependencies between individual technical and operational characteristics of roads;
- to simulate mathematically the change in the technical and operational characteristics of roads over time;
- to specify the methods of calculating and evaluating the condition of road pavement surface;
- to schedule repair works efficiently and in proper time [6].

The results of survey show that an intensive increase in the value of the IRI depends to a greater extent on the initial value of IRI_0 than on the traffic density.

It has been established that the transverse roughness decrease occurred more intensively than longitudinal one, and depended mainly on the traffic density.

References

[1] Uglova E V, Saenko S S 2016 Review of tools for the road structure state management *Russian journal of transport engineering* Vol 3 **1** Available at: http://t-s.today/PDF/02TS116.pdf (in Russian)

[2] https://www.researchgate.net/publication/282867684_NordFoU_PPM2_-_Validation_of_Performance_Models
[3] http://onlinepubs.trb.org/onlinepubs/archive/mepdg/2appendices_OO.pdf
[4] Zhustareva E V, Bochkarev V I 2020 The complex method of estimation of highway maintenance quality taking into account the International Roughness Index *IOP Conf. Series: Materials Science and Engineering* **832** 01203
[5] Fotiadi A A, Gnezdilova S A, Strekha IS 2020 Remote Method for Predicting Damage to Cement Concrete Pavements *Lecture Notes in Civil Engineering* **Vol 95** pp 333–339
[6] Radovsky B, Merzlikin A 1998 Definition of the regulatory terms of service of transportation facilities in the design *Science and technology in the road industry* **3** 9–12

Methodology and examples for assessing the environmental performance of certain road traffic management measures

S V Shelmakov[1], T U Grigoreva[1]

[1] Moscow automobile and road state technical university (MADI), 64, Leningradsky Prosp., Moscow, 125319, Russia

E-mail: shelwood@yandex.ru

Abstract. The article describes the methodology for assessing the impact of road traffic management (RTM) measures on the environment and public health. As an approbation of the methodology, such measures as ban on the movement of heavy trucks, ban on the movement of cars worse than the third ecological class and organizing the movement of non-motorized transport (NMT) are considered. Assessment of these measures impact on the environment was to provide an assessment of the pollutant emissions by the transport flow, the maximum and annual average ground-level concentrations of pollutants at the adjoining territory, the traffic flow noise characteristics (TFNC) and the equivalent sound levels at the adjoining territory. These estimates formed the basis for calculating public health risk indicators. Studies have shown that the greatest efficiency can be provided by measures to organize the movement of NMT and ban on the low environmental classes' vehicles movement. It is also shown, that in addition to the traffic flows characteristics, specific features of the territory under consideration also have a significant impact on the results.

1. Introduction

According to [1], documentation on road traffic management (RTM) should be developed taking into account environmental safety and reducing the vehicles negative impact on the environment.

RTM measures are quite diverse and require a specific approach to their formalization in order to assess the potential environmental impact of their implementation. Many small activities, such as installing pedestrian traffic lights or artificial bumps, are a priori insignificant in environmental terms. Decisions on their implementation made mainly based on their effectiveness assessment in ensuring the safety of road users. But such measures as, for example, the organization of calm traffic zones, zones with a ban on entry of cars low environmental classes, or the creation of a citywide bicycle infrastructure, undoubtedly require justification of environmental efficiency.

It seems that decision-making on the implementation of most local RTM measures should be based on examples of best engineering practice: the measures that have proven effective in terms of improving road safety and/or in terms of improving the environmental situation in domestic and foreign practice can be implementing without any additional justification. On the contrary, deviation from the best engineering solutions should be justified, and only in some extraordinary cases. Examples of best engineering practice should be applied in the form of albums of typical solutions for organization of road traffic.

Decisions on the implementation of more complex RTM measures (not included in the albums of standard solutions) should be made using the methodology described below.

Content from this work may be used under the terms of the Creative Commons Attribution 3.0 licence. Any further distribution of this work must maintain attribution to the author(s) and the title of the work, journal citation and DOI.
Published under licence by IOP Publishing Ltd

IOP Conf. Series: Materials Science and Engineering 1159 (2021) 012039 doi:10.1088/1757-899X/1159/1/012039

2. Object of research

This article considers a kilometer section of a 4-lane city street with an estimated capacity (in one direction) of 3000 vph, the actual traffic intensity during peak hours is 2200 vph (i.e., the traffic situation can be described as pre-traffic), the average flow velocity is 20 km/h, and the flow density is 65 v/km. Traffic composition: passenger cars – 85%, trucks – 10%, road trains – 1%, buses – 4%.

As examples, we considered four measures for RTM:

1. Removing congestion (this option is the basic one);
2. Removing congestion and narrowing the roadway to 2 lanes (for example, with the organization of non-motorized transport (NMT) – bicycles, scooters, segways, etc. – along the free lanes);
3. Removing congestion and ban on movement on this street section of the cargo vehicles with a gross weight more than 3.5 tons;
4. Removing congestion and ban on movement on this street section of the vehicles worse than the third ecological class.

The forecast scenario for a **first (base) measures**: the estimated street capacity will not change and will be 3000 vph, the projected traffic during peak hours going to grow up to 2800 vph, the average flow velocity is 60 km/h, the flow density will increase slightly and will be 25 v/km. The flow composition will not change: cars – 85%, trucks – 10%, road trains – 1%, buses – 4%.

The forecast scenario for a **second measures**: the estimated street capacity reduced to 1500 vph (organization for two-lane bike paths in one direction a capacity of about 2600 bph, so the total street capacity (in passenger traffic) in comparison with the first variant) will increase from approximately (when the passenger car interior filling ratio is equal to 1.6.) 4800 pph to 5000 pph, the projected traffic during peak hours will be 1400 vph, the average flow velocity is 60 km/h, the flow density will be 25 v/km. The composition of the flow will not change: cars – 85%, trucks – 10%, road trains – 1%, buses – 4%.

The forecast scenario for a **third measures**: the estimated street capacity will not change and will be 3000 vph, the forecast traffic intensity during peak hours will increase slightly (due to a denser flow without long trucks) and will be 2850 vph, the average flow velocity will not change and will be 60 km/h, the flow density will also increase slightly and will be 30 v/km. The flow composition will change: cars – 93%, trucks (gross weight less than 3.5 tons) – 3%, road trains – 0%, buses – 4%.

The forecast scenario for a **fourth measures** will be formulated as follows: the estimated street capacity will not change and will be 3000 vph, the forecast traffic intensity during peak hours will not change and will be 2800 vph, the average flow velocity will not change and will be 60 km/h, the flow density will not change and will be 25 v/km. The composition of the flow will also not change, but its structure will change according to environmental classes: only Euro-3 and higher cars with proportional distribution will remain.

For the examples of RTM measures under consideration, we will set the necessary characteristics of transport flows in a simplified version. The composition of traffic in all scenarios we assume the following:

- passenger vehicles – all "middle" class with a petrol engine with a uniform distribution on environmental classes (Euro 0,1 – 20%; Euro-2 – 20%; Euro-3 – 20%; Euro 4 – 20%; Euro 5 – 20%);
- cargo vehicles – small class (gross weight up to 3.5 t) – 80% (all petrol); large class (gross weight over 3.5 t) – 20% (all diesel) with a uniform distribution of environmental classes (Euro-0,1 – 20%; Euro-2 – 20%; Euro-3 – 20%; Euro-4 – 20%; Euro-5 – 20%);
- road trains – all diesel with a uniform distribution by environmental class (Euro-0,1 – 20%; Euro-2 – 20%; Euro-3 – 20%; Euro-4 – 20%; Euro-5 – 20%);
- buses – small class (gross weight up to 3.5 t) – 70% (all petrol); large class (gross weight over 3.5 t) – 30% (all diesel) with a uniform distribution of environmental classes (Euro-0,1 – 20%; Euro-2 – 20%; Euro-3 – 20%; Euro-4 – 20%; Euro-5 – 20%).

3. Description of the method

The methodology for assessing the environmental and public health impact of the RTM measures implementation is as follows:

- a base and one or more forecast scenarios for the implementation of RTM measures are formulated;
- based on the expert assumptions or transport modeling, the impact of the considered RTM measures on the traffic flows characteristics for each scenario is determined;
- based on the methodology for estimating pollutant emissions by motor transport, the maximum single (with an averaging period of 20...30 minutes) pollutant emissions on the studied street sections are determined for each scenario;
- based on the methodology of noise estimation, the traffic flow noise characteristic (TFNC) is determined for each scenario;
- based on the calculating methodology of pollutant dispersion in the atmosphere, the ground-level maximum single (i.e., characterizing the most unfavorable scattering conditions) concentrations are calculated at the selected points for each scenario;
- based on the characteristics of non-stationary traffic flow and meteorological parameters, long-period (average annual) pollutant concentrations are calculated at the selected points for each scenario;
- based on the methodology of sound propagation in the environment, maximum and equivalent sound levels are calculated (and, if more accurate estimates are needed, sound pressure levels) at selected points for each scenario;
- based on methods for assessing the public health risk from air pollution and noise factors, appropriate risk indicators are calculated for those population groups that are negatively affected by the traffic flows under consideration, for each scenario;
- based on cost-of-living and health methodologies, economic damage is calculated for each scenario;
- based on a comparison of the base scenario with the forecast(s), the economic effect of the considered measures implementation under the RTM is estimated;
- the information obtained is used when making a decision of the considered RTM measure implementation.

This is the most complete sequence of actions. In some cases, it is not necessary to reach economic estimates of the damage in absolute terms (in money), but rather enough to obtain relative estimates of the RTM measures effectiveness (as a percentage). It is sufficient to estimate changes in the levels of primary impact on the environment, i.e. calculate the change in pollutant emissions and TFNC forecast scenario relative to the base. In other words, there is no need to carry out the most labor-intensive and highly error-prone calculations of pollutant scattering and sound propagation, as well as all subsequent stages. The possibility of such a "truncation" of a calculations sequence is to the linear relationship of all estimated parameters with the magnitude of primary impacts, i.e. pollutant emissions and TFNC.

In this article, this methodology was used to assess the environmental performance of 4 road traffic management measures before the stage of calculating ground-level concentrations of pollutants and noise levels. The assessment of risk and environmental and economic damage for the examples under consideration will be discussed in the next article.

4. Results and Discussion

4.1. Determination of the maximum single pollutant emissions on a street section

The traffic characteristics for each scenario serve as input data for assessment methodologies of pollutant emissions from vehicles. The goal is to estimate the maximum single pollutant emissions (g/s) from vehicles on the studied street section.

The considered examples were calculated in the COPERT 5.3 program, which implements the EMEP/EEA methodology [2]. The results are shown in Fig. 1 ... 4.

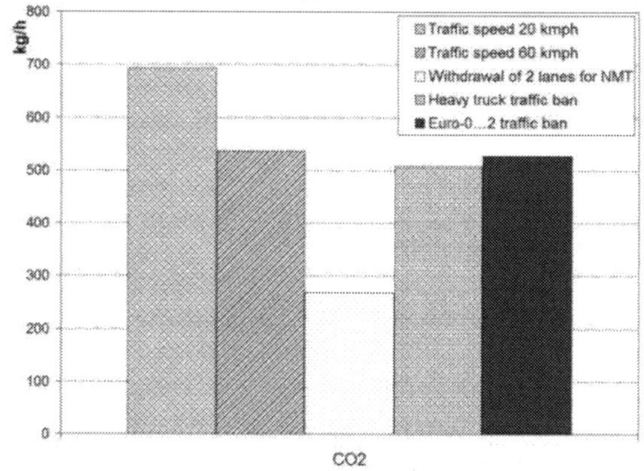

Figure 1. CO_2 emissions for the considered scenarios of the RTM measures

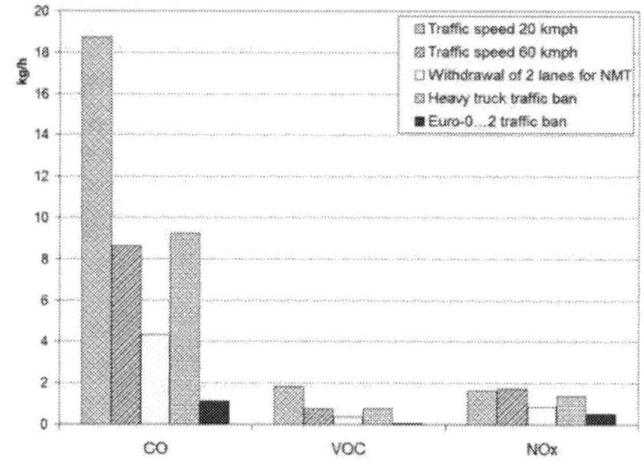

Figure 2. CO, VOC and NOx emissions for the considered scenarios of the RTM measures

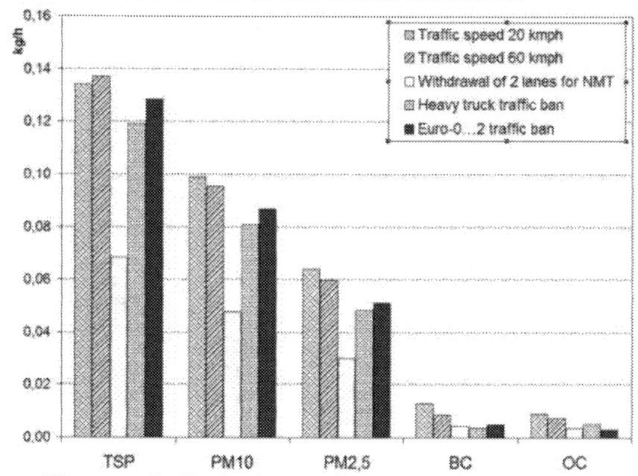

Figure 3. Emissions of various fractions of dispersed particles for the considered scenarios of the RTM measures

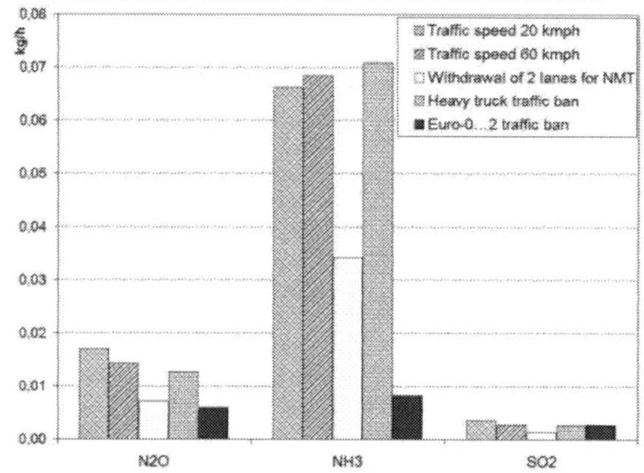

Figure 4. Emissions of N_2O, NH_3 and SO_2 for the considered scenarios of the RTM measures

From these Figures, it is clear that the best effectiveness for all pollutants is provided by the scenario with the 2 lane removal for the organization of NMT traffic: the reduction of pollutant emissions is proportional to the decrease in traffic intensity.

In second place by effectiveness is the ban on the old "dirty" cars movement. In this case, the effect is most pronounced for normed pollutants (CO, VOC, NO_x) removed from exhaust gases by their purification systems, and less pronounced for pollutants, the emission of which is correlated with fuel consumption (CO_2, SO_2) or is caused not only by fuel combustion processes, but also by tire and brake wear (dispersed particles of various fractions).

Measures related to changes in traffic mode and traffic flow composition (classical RTM measures) are less effective and the effect is differentiated by various levels.

4.2. Determination of the traffic flow noise characteristics (TFNC)

TFNC calculations for these examples were conducted in accordance with [3]. The results are shown in Fig. 5. The range of possible values shown as error lines is due to possible combinations of factors other than the default values. A decrease in the L_{Aeq} is possible if there is a low-noise road surface (crushed-mastic asphalt concrete), and an increase in the L_{Aeq} – in the case of a noisy road surface (rough surface treatment) in combination with a 10% and longitudinal slope.

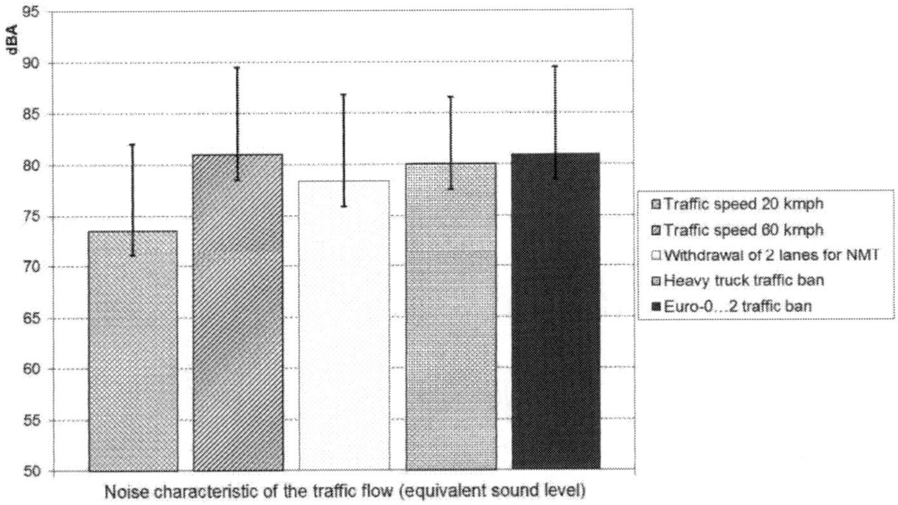

Figure 5. Noise characteristics of the transport flow (TFNC) for the considered RTM measures

As follows from the examples considered, the change in the TFNC was most affected by a decrease in the velocity and intensity of traffic flow. The change in the flow composition was less effective, since the share of heavy-duty cargo vehicles in the considered examples is initially small. The change in the flow ecological structure did not change the TFNC, since the methodology [3] is not sensitive to this factor.

4.3. Evaluation of the ground-level maximum single pollutant concentrations

The calculation of the pollutant dispersion in the atmosphere was performed according to [4].

To calculate the pollutant dispersion, a number of additional input data must be set. Urban building was not accounted for, so as not to introduce additional uncertainty in the results. The meteorological parameters are chosen to characterize typical conditions in the Central European part of Russia and South-Western Siberia (shown in Figure 7 as error lines).

The highest concentration levels (in fractions of MPC) are observed for nitrogen dioxide. The concentration reaches its maximum values directly above the roadway, rapidly decreasing with distance (Fig. 6). The non-linear nature of this relationship demonstrates the importance of urban planning solutions for the mutual location of the highway and adjacent buildings. One of the most effective measures of protection against the road traffic negative impact is protection by distance.

Ban on movement for trucks can reduce the area of excessive air pollution at 19 m, the removal of one lane for introduction NMT traffic allows to reduce the area of excessive air pollution at 42 m, the ban on movement vehicles up to the ecological class Euro-3 allows to exclude the formation of excessive air pollution zone.

To assess the impact of pollutants on the territory adjacent to the road, average concentrations were calculated in the 500-meter zone adjacent to the roadway (Fig. 7).

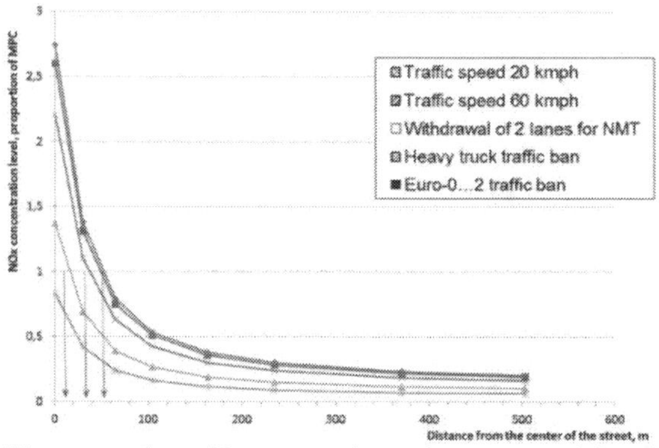

Figure 6. Changes in the ground-level concentration of nitrogen dioxide depending on the distance from the road for the considered RTM measures

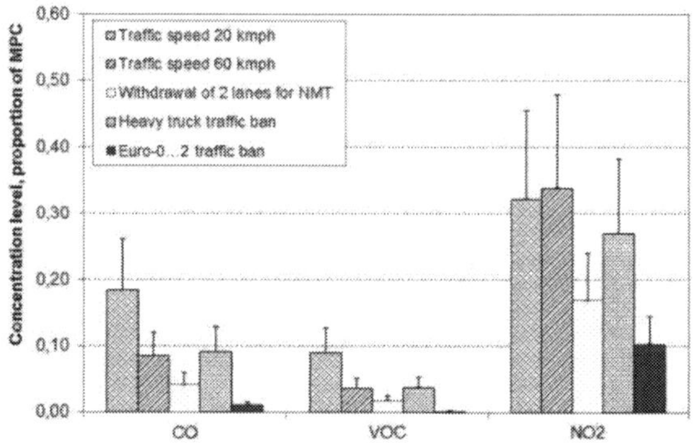

Figure 7. Average pollutant concentrations in the 500-meter zone adjacent to the roadway, for the examples considered

Comparing the obtained results of ground-level concentrations with the results of emissions shown in Fig. 1 ... 4, it is necessary to note the relative hazard influence of a particular pollutant set by the MPCMR values. Thus, although nitrogen dioxide is released in smaller quantities, it creates a more dangerous concentration, since its MPC is much lower (harder) than that of the other pollutants shown in the figure. As for the effect on the pollutant concentration of the actual road traffic management measures under consideration, the analogy with the effect of these measures on the emissions of certain pollutants discussed above should be noted.

4.4. Estimation of the maximum and equivalent sound levels

For the examples under consideration, calculations of noise propagation in the adjacent territory are performed according to [5]. Urban buildings and the underlying surface were not taken into account, so as not to introduce additional uncertainty in the results. The results are shown in Figure 8.

This figures shows that the dependence of the equivalent noise level as a function of the distance from the road is non-linear. This indicator, as well as in the case of ground-level concentrations, can be effectively suppressed by increasing the distance from the road to the protected objects. Thus, proper planning of roads (streets) and adjacent buildings can significantly reduce the acoustic impact of traffic on the environment and public health.

Banning the movement of trucks will reduce the excess noise zone by only 2 m, removing one lane of the roadway for the NMT traffic organization will reduce the excess noise zone by 10 m, both by reducing the TFNC and by shifting the noise source position at 3.5 m (lane width) to the axis of the road.

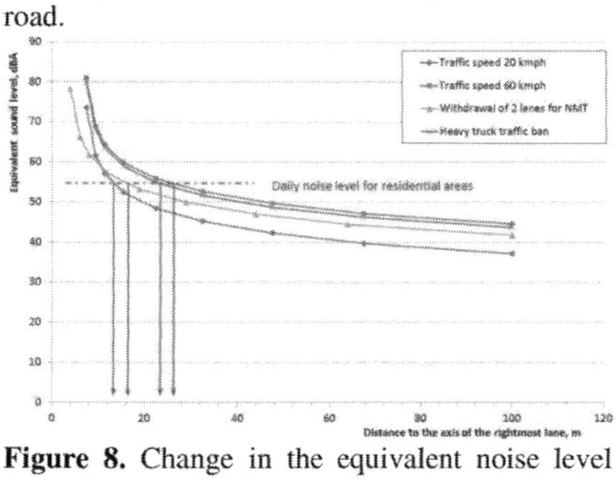

Figure 8. Change in the equivalent noise level depending on the distance from the road for the considered RTM measures

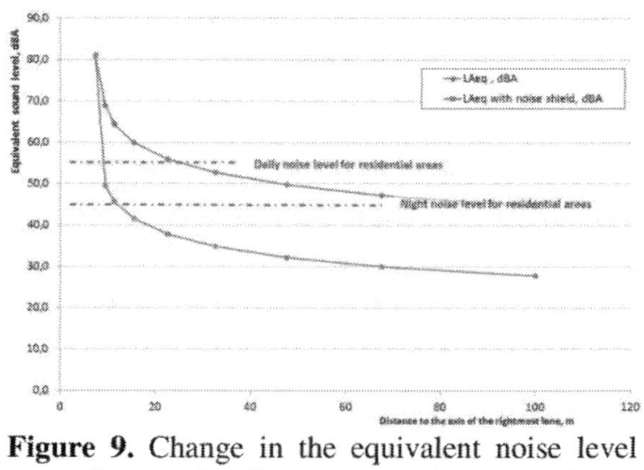

Figure 9. Change in the equivalent noise level depending on the distance from the road, when a noise shield was installed

To compare the effectiveness of the considered RTM measures on acoustic comfort in the territory adjacent to the street, calculations of the equivalent noise level were made, taking into account the installation of a 4 m high noise shield along the edge of the roadway. The results for the noisiest base scenario (velocity 60 km/h without congestion) are shown in Figure 9. This figure shows that the installation of a 4-meter noise shield allows to completely eliminate the zone of excess noise, both during the day and at night. This example shows the importance of applying engineering measures to protect against noise propagation, along with RTM measures.

5. Conclusion

Based on the analysis of domestic and foreign experience, the structure and content of the methodology for assessing the environmental and socio-economic effectiveness of road management measures are determined.

Studies have shown that the greatest effectiveness can be provided by measures to organize the NMT traffic and ban the traffic of low environmental classes vehicles. It is also shown that in addition to the characteristics of traffic flows, specific features of the territory under consideration also have a significant impact on the results.

References

[1] Federal Law No. 443-FZ of 29.12.2017 (as amended on 15.04.2019) "On the organization of road traffic in the Russian Federation and on amendments to certain legislative acts of the Russian Federation".

[2] EMEP/EEA emission inventory guidelines 2009, 2013, 2016. Available at: http://www.eea.europa.eu (accessed 20.10.2020).

[3] SP 276.1325800.2016 Set of rules. The building and grounds. Design guidelines protection from the noise of the traffic.

[4] Methodology for calculating the dispersion of emissions of harmful (polluting) substances in the atmospheric air. (Order of the Ministry of natural resources of Russia dated 06.06.2017 No. 273).

[5] SP 51.13330.2011 Noise protection.

IOP Conf. Series: Materials Science and Engineering 1159 (2021) 012040 doi:10.1088/1757-899X/1159/1/012040

Assessment of the road transport infrastructure facility functionality loss risk resulting from climate change

A N Yakubovich[1], Y V Trofimenko[1], I A Yakubovich[2] and E V Shashina[1]

[1]Technosphere Safety Chair, Moscow Automobile and Road Construction State Technical University (MADI), Leningradsky avenue, 64, Moscow, Russia, 125319
[2]Operation and Servicing of Motor Vehicles Chair, Moscow Automobile and Road Construction State Technical University (MADI), Leningradsky avenue, 64, Moscow, Russia, 125319

E-mail: Shelvlad@gmail.com

Abstract. The cause of a road bridge functionality loss risk in the conditions of climate changes in the Arctic is a decrease in its load-bearing capacity as a result of permafrost soil degradation. Modeling of the soil temperature regime of the bridge piling in Yakutsk for 2 years was performed and the minimum load-bearing capacity of the bridge piling was determined for this period. It was assumed that the average annual air temperature increased by 2 degrees Celsius in combination with an increase in the contrast between summer and winter temperatures. The obtained value of the climate risk level (438 units) allows to continue operating the facility without implementing engineering and technical measures aimed at stabilizing the soil temperature regime.

1. Introduction

Climate change significantly increases the risk of dangerous phenomena and events occurrence in all spheres of human activity. The main types of climate risks include [1]: an increase in the severity of emergencies, an increase in the frequency of their occurrence, the spread of these phenomena to new territories where they were not previously observed, as well as a significant acceleration of slow natural processes, which, in particular, include the degradation (thawing) of permafrost. Taking into account the significant distribution of permafrost and cryolitic soils on the territory of Russia (60-65% of its total area), a systematic decrease in the load-bearing capacity of thawing soils in the cryolithozone is particularly dangerous for the road network and transport infrastructure.

Forecasting climate risks for engineering facilities requires the most reliable estimates for the climate parameters values. The main one (but not the only) is the air temperature in the surface layer. Taking into account the natural variability for individual years, climate change can manifest itself in relation to either its individual parameter (as a change in the average value, variance, frequency of occurrence, or probability density), or several parameters simultaneously (multivariate analysis) [2]. In [3], a methodology for recording the values of climate parameters that change over time on the territory of the road network by means of an intelligent information and measurement system is shown.

Methodological approaches to climate forecasting are very diverse. Individual climate parameters forecasting is shown in [4], where the combined use of global and regional models of atmospheric circulation makes it possible to obtain highly accurate data on future atmospheric precipitation in the territory. Forecasting the intensity of solar radiation using the random subspace method [5] allows obtaining a correlation coefficient of 0.98 between the expected and subsequently observed values of

Content from this work may be used under the terms of the Creative Commons Attribution 3.0 licence. Any further distribution of this work must maintain attribution to the author(s) and the title of the work, journal citation and DOI.
Published under licence by IOP Publishing Ltd

this parameter. The use of a neural network for forecasting future air temperature based on its past values is shown in [6]; the average error is 0.907 °C. Multidimensional climate models used for Central and Eastern European regions showed an expected warming of up to +3 °C by 2050, and up to +5 °C by 2100 [7].

Modeling of climate systems on the territory of the cryolithozone has additional features. Accounting for heat transfer in frozen soils (including permafrost soil thawing), as a necessary condition for effective modeling of changes in the atmosphere climate parameters, is justified in [8]. The high sensitivity of discontinuous permafrost to long-term warming is shown in [9]; the expected rate of its degradation increases on steep slopes facing the South. The rate of increase in air temperature in the Arctic is expected to be 1.5-2 times higher than in the rest of the world [10]; the increase in permafrost temperature by 2026 will be +2 °C. Forecast estimates of the road network settlement due to permafrost thawing at its base range from 1 to 3 cm (with scenario assumptions of air warming by +2 °C), while clay soils with high humidity are the most vulnerable [11].

A quantitative assessment of a facility operational suitability measure reducing (which occurs as a result of certain changes that are unfavourable for the facility operation) is the amount of risk. For complex facilities, the risk is determined by taking into account the probability of malfunction of individual elements (individual risk) and the joint influence of all elements on the facility functioning as a whole (system risk) [12]. The technology of risk assessment using expert analysis of the road bridges state (based on the classification of identified defects for each facility under consideration) is shown in [13].

The reasons for the increase in accidents and reduced safety of buildings and structures built on permafrost are systematically considered in [14]. The most dangerous is the melting of permafrost soils of high humidity, for which a number of methods for calculating strength and settlement are not applicable, which are effectively used for soils with low humidity [15]. To quantify the risk to a transport infrastructure facility in the cryolithozone, it is mandatory to take into account the ice content in the soil; as a result, it is possible to forecast climate risks for the roadbed with a correlation coefficient for predicted and subsequently observed damage levels of about 0.74 [16]. Using the response surface to speed up the computational procedure allows obtaining forecasted estimates of climate risks for large areas. In [17], the backbone road network on the territory of the Magadan region (within the boundaries of the 300x400 km section) is considered; the forecasted risk of +2 °C warming ranged from 78 points on a 1000-point scale (dry sandy soils on the continental territories of the region) to 360 points (aqueous clay soils in the coastal zone of the sea of Okhotsk).

Reducing climate risks to an acceptable level for newly constructed facilities is achieved by enlarging the bearing part of the foundations by an amount determined by the solution of the probabilistic and economic problem [18]. For existing facilities, risk reduction is possible due to engineering and technical measures that counteract the impact of climate change and maintain the temperature regime of permafrost soil within the design conditions.

2. Research methods

The quantitative assessment of climate risk R includes the following steps:

- identification of the climatic conditions taken into account during the facility designing and determination of the climate parameters values separated by the modeling interval Δt for at least one year (the base climate represented as the **C0** data array);
- formulation of scenario assumptions about the changed climate and determination of the corresponding climate parameters combined in the **C** array;
- recording of the soil conditions **G** (type of soil, humidity, permafrost layer temperature at the depth of seasonal temperature fluctuations, etc.);
- performing simulations of heat transfer in the soil during the period reflected in the CO and C arrays, with determination of the soil temperature dynamics (represented by the **D0** and **D** arrays – for the base and modified climate, respectively);

- determining the values of the facility condition criterion indicator K (reflecting the load-bearing capacity of the soil) throughout the entire modeling period, identifying its minimum values in the conditions of the basic and modified climate;
- based on the decrease value in K in the changed climate (compared to the base climate), the facility functionality indicator U in the new climate conditions is determined, as well as the associated risk R.

In this paper, the climate risk in relation to the road bridge pile foundation is assessed that occurs when the average annual air temperature increases by a value of ΔT=2 °C in combination with an increase in climate contrast ΔT^{+}=2 °C in the absence of risk reducing engineering and technical measures. Contrast refers to an additional increase in temperatures during the summer period (the duration of which was determined by the condition T≥0 °C), with a simultaneous decrease in temperatures during the rest of the year; the value of this decrease was assigned so that the average annual air temperature as a whole increased by 2 °C.

The cross-section of the pile is 40x40 cm, the depth of immersion is 6 m. The location of the bridge, which determines the climatic conditions of its operation, is Yakutsk. Indicators of the base climate state, determined by the results of long-term instrumental observations, are shown in table 1.

Table 1. Indicators of the basic climate of Yakutsk.

The average year's month		The average air temperature, °C			The average wind speed v, m/s
Ordinal number	Number of days	Maximum T_{max}	Minimum T_{min}	Average T_{mid}	
1	31	–37.7	–46.6	–42.6	1.1
2	28	–30.4	–41.7	–35.9	1.1
3	31	–14.4	–30.1	–22.2	1.7
4	30	–0.2	–14.6	–7.2	2.6
5	31	12.0	–0.5	5.8	3.1
6	30	21.7	8.1	15.4	2.9
7	31	25.2	11.7	18.7	2.6
8	31	21.7	8.6	14.9	2.5
9	30	12.1	0.7	6.2	2.4
10	31	–3.6	–12.4	–8.0	2.2
11	30	–24.1	–32.9	–28.3	1.6
12	31	–35.3	–43.5	–39.5	1.1

The air temperature increase by the value of ΔT for the forecast climate state was assumed to be the same for all days of the year. The contrast increase suggested that the temperature during the period of positive temperatures was increased by ΔT^{+}, and during the period of negative temperatures – decreased by ΔT^{-}; the value of this decrease was determined from the condition that the average annual air temperature remained unchanged. Both parameters of the changed climate were represented as random variables $\Delta\tilde{T}$ and $\Delta\tilde{T}^{+}$, distributed according to the normal law, with average values of +2 °C. The variability of the normal distribution of these parameters was determined from the condition of temperature deviation from the average value by the value ε=±2°C no more than 1 time in 50 years, which corresponded to the probability p±2=0.01 (figure 1, a).

For a $\Delta\tilde{T}$ normal distribution where $\Delta\tilde{T}$=2 and $p_{\varepsilon}=2p_{\pm2}=0.02$:

$$\int_0^4 \frac{1}{\sigma\sqrt{2\pi}} e^{-\frac{(\Delta T-2)^2}{2\sigma^2}} d(\Delta T) = 0{,}98 \quad (1)$$

Solving equation (1) with respect to the standard deviation, we get σ=0.86. Thus, the normal distribution law of a random variable $\Delta\tilde{T}$ is characterized by the parameters $\Delta\bar{T} = +2°C$ and $\sigma_{\Delta T}$=0,86 °C:

$$f(\Delta T)=\frac{1}{0{,}86\sqrt{2\pi}}e^{-\frac{(\Delta T-2)^2}{2\cdot 0{,}86^2}} \quad (2)$$

For the three implementations used, the probabilities of their occurrence are (figure 1, b):

$$p_1=\int_{-\infty}^{1,5} f(\Delta T)d(\Delta T)=0{,}2805,$$
$$p_2=\int_{1,5}^{2,5} f(\Delta T)d(\Delta T)=0{,}439,$$
$$p_3=\int_{2,5}^{+\infty} f(\Delta T)d(\Delta T)=0{,}2805. \quad (3)$$

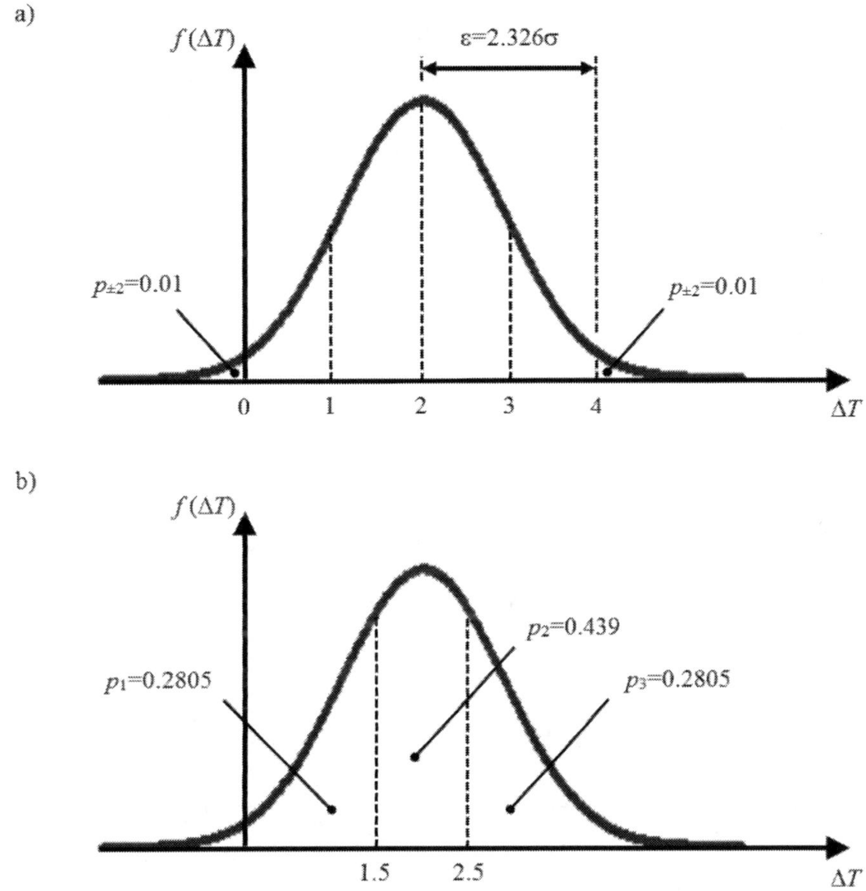

Figure 1. Accounting for the climate change probabilistic nature:
a) – determination of the standard deviation $\Delta\tilde{T}$ for the normal distribution law,
b) – determination of the $\Delta\tilde{T}$ realizations probability.

3. The results of climate risk modeling

The computation process parameters were adopted as follows:

- the design soil elements dimensions when modeling the dynamics of the temperature regime were $\delta_x=\delta_y=0.1$ m;
- simulation quantum (the time interval during which the simulated parameters were considered unchanged) was $\Delta t=900$ s;
- the simulation period duration was two full years, which corresponded to $t_{max}=63{,}072{,}000$ s, or $N_{\Delta t}=70{,}080$ simulation quanta.

Three implementations of interrelated random variables $\Delta\tilde{T}$ and $\Delta\tilde{T}^{+}$ were considered, combined as $\tilde{X}_1\left(\Delta\tilde{T},\Delta\tilde{T}^{+}\right)=\{+1°\text{C},+1°\text{C}\}$, $\tilde{X}_2=\{+2°\text{C},+2°\text{C}\}$ and $\tilde{X}_3=\{+3°\text{C},+3°\text{C}\}$. For each of the three combinations, the changed climate temperature indicators were determined (table 2). The average monthly wind speeds values for all three variants of climate change were taken from table 1.

Table 2. Temperature indicators for three climate change variants.

Month i	ΔT=+1 °C, ΔT^{+}=+1 °C			ΔT=+2 °C, ΔT^{+}=+2 °C			ΔT=+3 °C, ΔT^{+}=+3 °C		
	T_{max}	T_{min}	T_{mid}	T_{max}	T_{min}	T_{mid}	T_{max}	T_{min}	T_{mid}
1	–37.4	–45.8	–42.0	–37.3	–45.0	–41.4	–37.3	–44.3	–41.0
2	–30.1	–40.9	–35.3	–30.0	–40.1	–34.7	–30.0	–39.4	–34.3
3	–14.1	–29.3	–21.6	–14.0	–28.5	–21.0	–14.0	–27.8	–20.6
4	1.8	–13.8	–6.6	3.8	–13.0	–6.0	5.8	–12.3	–5.6
5	14.0	1.5	7.8	16.0	3.5	9.8	18.0	5.5	11.8
6	23.7	10.1	17.4	25.7	12.1	19.4	27.7	14.1	21.4
7	27.2	13.7	20.7	29.2	15.7	22.7	31.2	17.7	24.7
8	23.7	10.6	16.9	5.7	12.6	18.9	27.7	14.6	20.9
9	14.1	2.7	8.2	16.1	4.7	10.2	18.1	6.7	12.2
10	–3.3	–11.6	–7.4	–3.2	–10.8	–6.8	–3.2	–10.1	–6.4
11	–23.8	–32.1	–27.7	–23.7	–31.3	–27.1	–23.7	–30.6	–26.7
12	–35.0	–42.7	–38.9	–34.9	–41.9	–38.3	–34.9	–41.2	–37.9

To determine air temperatures and wind speeds at random times, interpolation was used, the base points for which were the values of T_{max}, T_{min}, T_{mid} and v, assigned to the midpoints of the corresponding months. As a result, based on the data in table 2, for each simulation quantum $0 \leq N \leq N_{\Delta t}$, T_{max}, T_{min}, T_{mid}, and v were determined. Than the air temperature T_a corresponding to this time quantum was determined.

Soil temperature dynamics modeling was performed for three variants of climate changes, as well as for the basic state of the climate. For each soil mass temperature state, the value of the criterion indicator K was calculated, which reflected the maximum possible load on an individual pile. Based on the analysis of changes in K values during the modeling period, the minimum values for each of the four climate states were determined (table 3).

Table 3. Minimum values of the criterion indicator K.

Climate state	Criterion indicator K	Day	Functionality indicator U
Base climate	2,602	640	1,000
$\tilde{X}_1$	2,461	650	0,765
$\tilde{X}_2$	2,331	650	0,548
$\tilde{X}_3$	2,231	650	0,382

For each variant of climate change, the functionality indicator values U were determined, corresponding to the worst, during the average year, criterion indicator K. The dependence $U(K)$ was assumed to be linear on the section $0<U(K)<1$. The value of the criterion indicator for the basic climate state K_0=2.602 was considered a point on the border of the transition area; the used reserve value was Δ=1.3. Accordingly, the other extreme point of the transition area was characterized by the value $K_0/1.3$=2.002 (figure 2).

For all three variants of climate change, the point corresponding to the functionality indicator is located on the transition section of the graph shown in Fig. 2. Values of the functionality indicator determined by linear interpolation between points $U(2{,}002)=0$ and $U(2{,}602)=1$ are shown in the last column of table 3.

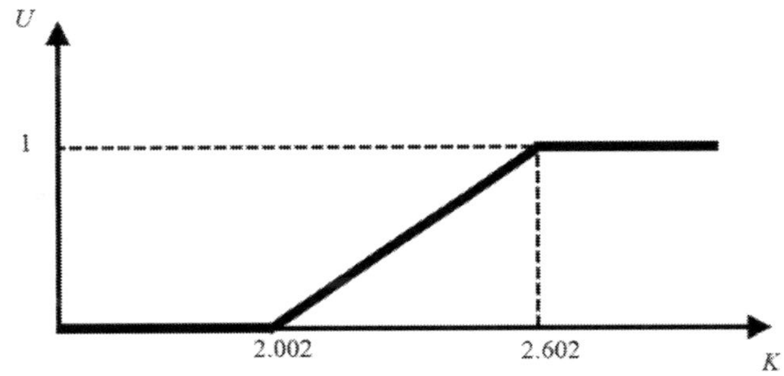

Figure 2. Dependence of the functionality indicator U on the criterion indicator K

Climate risk R was calculated as the weighted average value of negative consequences (due to reduced facility functionality) for all variants of climate change (table 3), determined taking into account the probabilities of their occurrence (3) and normalized to the range [0; 1000]:

$$R = 1000\sum_{i=1}^{3} p_i\left(1 - U_i\right) = 438 \quad (4)$$

4. Discussion and conclusions

When interpreting a quantitative risk assessment, the following should be considered. The minimum possible risk (within the range used [0; 1000]) R_{min}=0 occurs in climatic operating conditions that fully correspond to the facility design. Under these conditions, the safety margin for the base bearing capacity is maximum, based on the rich experience of operating similar facilities, there is no reason to doubt its safe operation. The maximum possible risk R_{max}=1000 means that this safety margin is completely depleted. At the same time, even the value R=1000 cannot be interpreted as an immediate destruction of the object due to insufficient load-bearing capacity of its base. Physical destruction in this case is possible if a number of operational factors are combined unfavorably (for example, when several types of loads, each of which has its own safety margin, simultaneously reach their calculated values), which is very unlikely. However, the safety of a facility when there is no safety margin of the base bearing capacity at R=1000 obviously cannot be considered sufficient, and the continuation of its operation in its current state should be immediately stopped – despite the continued integrity of the facility shape.

With an allowance for the abovementioned, the climate risk R=438 should be considered average. The operation of the facility in the changed climatic conditions may continue, but in the medium term it is necessary to develop and implement engineering and technical measures aimed at risk reducing. These measures should be aimed on adjusting the soil temperature regime and bringing it to the design characteristics corresponding to the base climate conditions.

It should also be noted that the scenario assumptions used in the simulation about the changed climate (warming by +2 °C) are very cautious. As it was shown in this article earlier, the expected warming in Europe in a number of scenarios is up to +3 °C by 2050, and the increase in air temperatures in the Arctic is expected to be 1.5-2 times more significant. Accordingly, in relation to the facility under consideration, it is necessary to systematically monitor the climatic parameters on the territory of its operation. If a higher rate of warming is recorded than was taken into account in determining the climate risk in this paper, it is necessary to re-evaluate the risk in accordance with the new scenario assumptions about the climate.

References

[1] Glasser R 2020 *International Journal of Disaster Risk Science* 11(2) pp 152–154
[2] Qunqun L, Shiquan W and Bin G 2016 *Discrete Dynamics in Nature and Society*
[3] Yakubovich A N, Yakubovich I A, Trofimenko Yu V and Shashina E V 2019 *Proceedings Systems of Signals Generating and Processing in the Field of on Board Communications, SOSG 2019'03*

[4] Donghyuk K, Kyoung J L, Chun H J, Jichul R, Jae E Y, Seong J K, Dong S K and Younghun J 2014 *Advances in Meteorology*
[5] Yao D and He J 2019 *Mathematical Problems in Engineering*
[6] Zao Z and Yuan D 2020 *Mathematical Problems in Engineering*
[7] Belda M, Skalák P, Farda A, Halenka T, Déqué M, Csima G, Bartholy J, Torma C, Boroneant C, Caian M and Spiridonov V 2015 *Advances in Meteorology*
[8] Ki-Hong M and Wen-Yih S 2015 *Advances in Meteorology*
[9] Bonnaventure P P and Lewkowicz A G 2011 *Climatic Change* 105(1-2) pp 293–312
[10] Kaplan J O and New M 2006 *Climatic Change* 79(3-4) pp 213–24
[11] Yakubovich A, Mayorov S, Pyatkin D and Yakubovich I 2020 *Advances in Intelligent Systems and Computing* 1116 (Switzerland: Springer Nature) pp 924-933
[12] Hochrainer-Stigler S, Pflug G, Dieckmann U, Rovenskaya E, Thurner S, Poledna S, Boza G, Linnerooth-Bayer J and Brännström Å 2018 *International Journal of Disaster Risk Science* 9(4) pp 561–567
[13] Mansour D, Moustafa I, Khalil A, Mahdi H 2019 *Ain Shams Engineering Journal* 10(4) pp 695-704
[14] Kronik Y A 2017 *Soil Mechanics and Foundation Engineering* 54(3) pp 198–201
[15] Yao X, Qi J and Wu W 2012 *Acta Geotechnica* 7(3) pp 193–202
[16] Zhang C, Zhang H, Zhang M and Gong Q 2019 *Mathematical Problems in Engineering*
[17] Yakubovich A and Yakubovich I 2020 *Advances in Intelligent Systems and Computing* 1258 pp 486-495
[18] Khrustalev L N and Emel'yanova L V 2013 *Soil Mechanics and Foundation Engineering* 50(1) pp 19–23

IOP Conf. Series: Materials Science and Engineering 1159 (2021) 012041 doi:10.1088/1757-899X/1159/1/012041

Distributed generation in small-scale power generation. Search for energy-efficient solutions by automation of complex power systems

T V Golubchik[1], **A S Kulikov**[1], **D Y Taratutin**[1]

[1] Moscow Automobile And Road Construction State Technical University, Leningradsky Ave, 64, Moscow, 125319

E-mail: tvgolubchik@mail.ru

Abstract. The Paper presents data on the control of a generating set from an external source in parallel operation with the network. An experimental study was carried out on the example of a real generator set that provides a production facility with electricity. The economic feasibility of the simultaneous use of several sources of electricity is considered. The advantages and disadvantages of usage a generator set for parallel operation with the mains are considered. Considered solutions for the automation of complex systems in the small power industry.

1. Introduction

Scientific and technological progress, centralization of industry facilities, increase of population density in cities driven by an increase of development density are the factors responsible for an increased demand for electric and thermal power. Over the past 30 to 40 years, electricity consumption in developed areas has increased tenfold while the main burden can be attributed to the already densely populated areas and technology parks. The method of modern separate economic management prioritizes the infrastructure and transport accessibility of the facilities, which inevitably leads to centralization of loads, and increased energy consumption.

Given other positive aspects of load centralization, there is a number of negative factors, including the duration of the design of the power supply system and the construction of network infrastructure facilities. The reconstruction of electric mains takes place over the decades, not to mention the limitation of output due to the insufficient capacity of generating, nuclear, hydro and thermal power plants. In the old areas, we have to work with networks designed and constructed 40-50 years ago, and ultimately not designed for the increased loads. Therefore, we have a serious shortage of electricity both in industrial and residential areas. This factor severely constrains manufacture, economic and technological growth of economic facilities in the modern world. The cost of connection to electricity mains is growing exponentially. The denser the area is developed, the more complex and costly are the measures to connect a facility to a power source located on its territory, and the more electricity is needed.

2. Isolated power supply and network management

The From the economic point of view, the self-contained power supply consists of three fundamental components: the cost of initial capital investments, the cost of fuel, and the cost of maintenance and operation. Summing up the three values, comparing them to the cost of connection to the external mains,

Content from this work may be used under the terms of the Creative Commons Attribution 3.0 licence. Any further distribution of this work must maintain attribution to the author(s) and the title of the work, journal citation and DOI.
Published under licence by IOP Publishing Ltd

IOP Conf. Series: Materials Science and Engineering 1159 (2021) 012041 doi:10.1088/1757-899X/1159/1/012041

and the cost of kW/h of electricity we get an answer to the question of the expediency of construction of an entire power center. Often, the cost of self-contained power supply is higher, so the issue changes to the plane of off-line operation and reduction of energy consumption, or relocation of production to a more resource-friendly area.

Fortunately, the experience of using hybrid sources, branched self-contained power systems, automation systems in power centers make it possible for us to approach the solution of the issue comprehensively: combine heterogeneous power systems with a network, use diesel and gas power plants to remove peak loads, use solar panels, energy storage devices, etc.

It stands to reason that such approach requires a complex systematic approach, the energy map of a facilityhas to be developed first considering the availability of certain resources (limit of connection to the power supply network, gas pipeline, wind potential, prospect of installation of solar panels, etc.). When creating facilities' energy map, allowance is made for daily, weekly, monthly, and seasonal maximum and minimum loads, nature of loads, stability and continuity of consumers. Based on the above, the required amount of electricity, maximum and average load values, as well as the category of electrical receivers have to be calculated.

3. Issues in operation of a network of different sources

Power quality parameters directly depend on the stability of the generating sources. No such problem occurs during operation from such network. When using the equipment of self-contained power centers, the designer often tries to use the same-type equipment that is a part of the power complex, with the same-type device of the control gear, comparable to the values of internal resistance of sources. This is due to the fact that when the load changes, especially its reactive component, each source reacts differently and compensates for changes in the network. As a negative factor, self-induced oscillations and, in general, instability of the system are possible.

The next issue is the need to develop a complex process automation system to divide the load between the generating sources. Such system should centrally distribute the load between the generating systems, considering their load, the required level of redundancy and the correction factors (e.g., the daily coefficient for a grid source or solar panels, etc.). Such system is quite an expensive and elaborate automated complex, but its use can provide the economic effect of the entire combined system.

Another obstacle in network operations – the demands of regional network companies of strict control over electricity export to the network (some internal network companies allow export, but under strict control and according to the agreed specifications). Once again, this factor makes it necessary to apply a centralized power center control system with a load control for each of its elements and a protection system against export, considering a number of factors (unloading of consumers, emergency shut-off, reactive power of asynchronous motors, etc.).

In summary, the combination of self-contained power centers and network sources is an effective complex solution for the delivery of electricity to the consumer in conditions of an increased demand, which is a feasible task these days, yet requiring an integrated approach to automation of the process and administration of power generation.

4. Implementation of a parallel network management solution and power distribution

As an example, let us examine a small power system implementing the principle of power distribution and control between a stand-alone source and a network.

Technical specifications of the power system:

- Power supply of consumers by 1 and 2 categories of reliability,
- Input 1, transformer 10/0.4 kV, 1000 kVA, model TMG-21.
- Input 2, transformer 10/0.4 kV, 1000 kVA, model TMG-21,

• Combined bus section (2 sections), islanding is not carried out, parallel operation of transformers is not provided,

• Input 3 from the gas-reciprocating power plant (hereinafter referred to as the GRPP) designed through the circuit breaker for consumers of the 1st bus section, parallel to input 1,

• TN-C-S five-wire system.

The electrical diagram is shown in Fig. 1

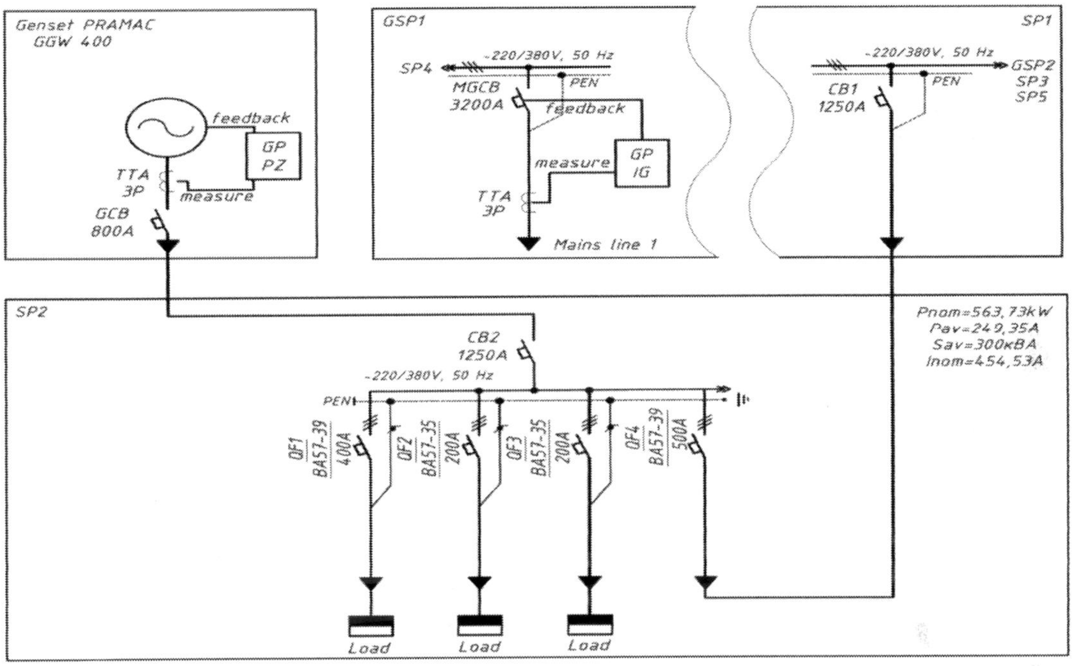

Figure 1. Electrical diagram.

In this project, the Customer is not limited by the capacity of the network connection, but due to the individual tariff and the high cost of 1 kW/h, the calculation was based on the choice of a gas power plant for parallel operation with Input No. 1 for its maximum unloading. By means of a calculation, the maximum power value of the GRPP was obtained based on the limits of gas connection, i.e. 400 kW of active power in the maximum mode, as well as the schedule of average daily loads in summer and winter seasons, which is shown in Fig. 2.

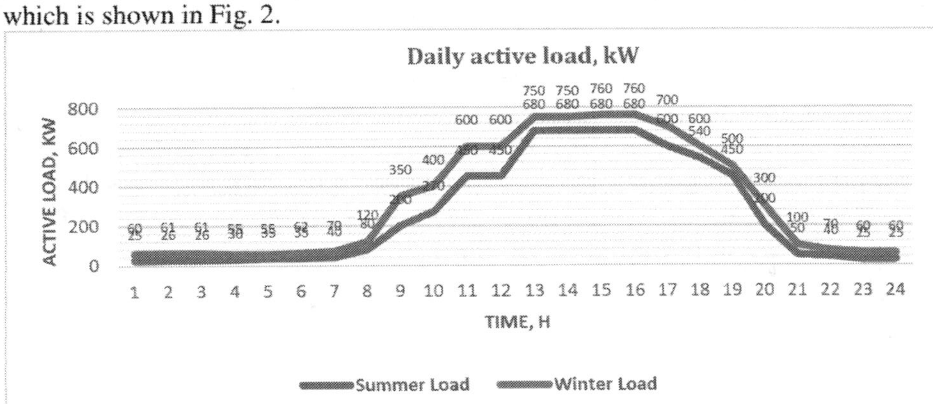

Figure 2. Average daily graph load.

For this system to operate in parallel with the existing network, it was necessary to arrange an automation system performing the following functions:

• Control of power consumption from the network.

• Automatic start and stoppage of the GRPP when the set load is reached. This is necessary for efficient use of the power plant's life and operation in an effective power range.

• Control over the distribution of load between the GRPP and the GRPP load priority network in an effective range of 60-80%.

• Protection against export to the network by adjusting the load capacity of the GRPP with its unloading while reducing the load of the consumer.

• System dispatch for monitoring of running processes and protection of equipment in the event of malfunctions.

In the arrangement of parallel work with the network, organizational measures are taken, in particular, the development of technical specifications for the distribution of power and the amount of exports to the network (including for transitional processes), the design of the power supply system, the design of the automation system and the development of a set-up and testing program.

The automation block diagram is shown in Fig. 2. The PRAMAC model GGW400G was used in the implementation of this GRPP system. The GENERAC GRPP controller, PowerZone model was used as automation equipment. To monitor the controller of power system (network and power distribution, as well as the logic of operation of the whole automation system) ComAp, model IM-NT-BB IV5 was used. In addition, the KLEA 320P power quality analyzer, manufactured by Klemsan, was used This device made it possible to obtain parameters of active, reactive and full power with high accuracy and speed, which was a necessary prerequisite for reducing the duration of transition processes in operation of the GRPP.

To control the algorithm of system operation, a discrete and analogue PLC logic was applied, cascade PID (proportional-integral differential) control of a controller secondary to GRPP was also used for power distribution, as well as for the engine speed control (fast governor) and alternator excitation control (slow governor).

The Main Control Unit was written using the original IM-NT-BB controller tools (ComAp) in the PLC Free Programming Unit, which was a non-trivial task and required long-term testing, debugging and correction of the installation's operational remarks.

The protection against export to the network is carried out through the discharge of the load by reducing the engine speed and at high unloading – disconnection of the network switch and switching to island mode – operation from the GRPP. 120 kW power reserve setting to enable starting of heavy starting motors according to the load chart and technical specifications.

The percentage of exports to the network is insignificant and is due only to technological transitions at load discharge, and is approx.. 0.01% for active and 0.2% for reactive power, the power factor in the range from 0.77 to 0.97 is stable with the control accuracy of 0.1 subject to constant load.

This automated system has the 3rd degree of automation according to GOST 50783-95 and is also integrated into SCADA (Supervisor Control and Data Acquisition) system for monitoring operational parameters. . Technical record-keeping of electric energy and gas consumption was arranged. Later, integration into the High Level Control System for monitoring and diagnostics is carried out by installing the APM (operator's automated working station) and developing the remote monitoring site based on the Wonderware software.

The automation block diagram is shown in Fig. 3.

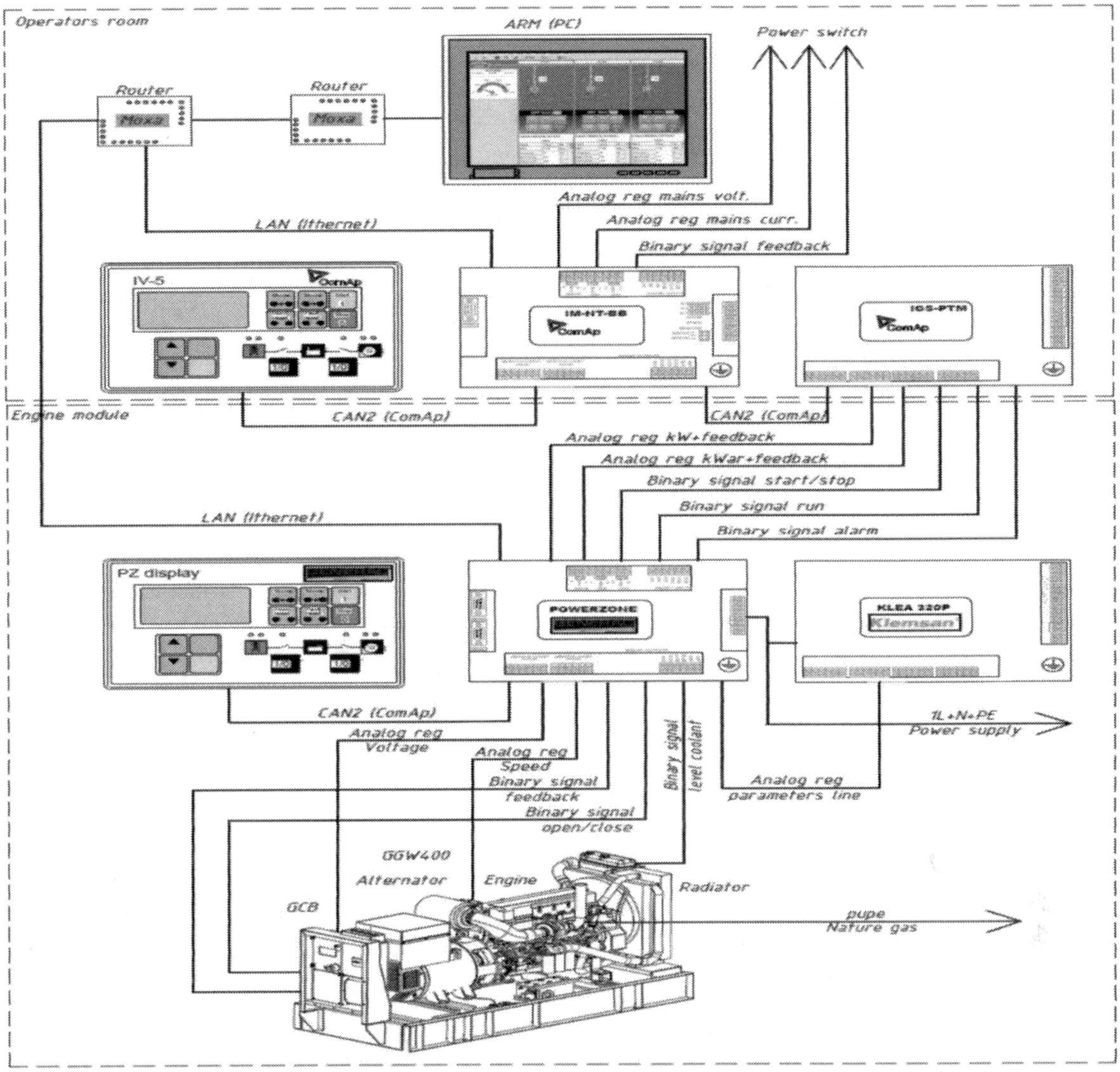

Figure 3. Automation block diagram.

From a technical point of view, the following activities were carried out:
Initially, the active/reactive power distribution has to be stabilized between the generating units and the network.

In this case, all load changes (active and reactive) compensate the gas power plant, being the main source of power, see Fig 4.

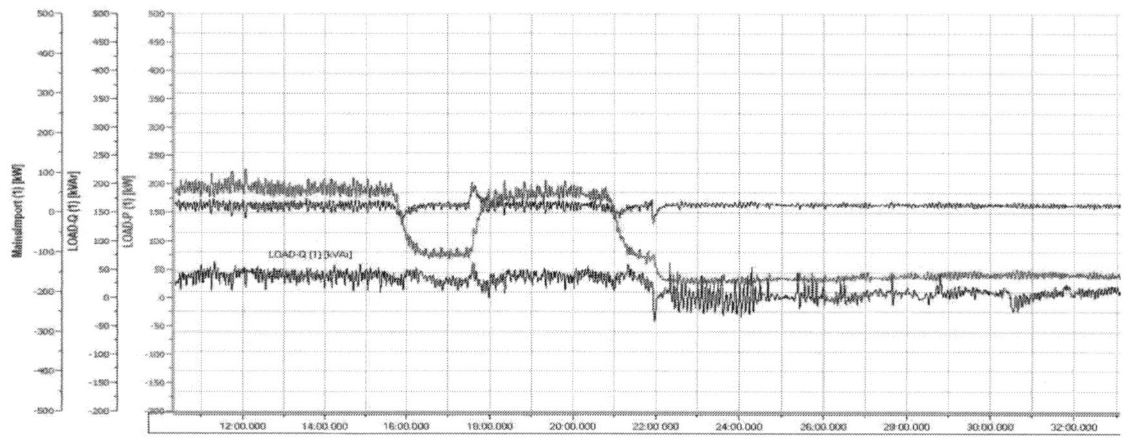

Figure 4. Load distribution. MainImport, kW –Import of load from main grid, LOAD Q, kVAR – power of the network source, LOAD P, kW – power produced by the GRPP.

Fig. 4 shows the reset and ascention of active power received by the GRPP, which are caused by a change in the consumer's load while the network is maintained at a minimum stable level. In this case, condition is also met to maintain the generating capacity power reserve at the gas-reciprocating power plant. Otherwise, it would be possible to switch the network switch off. Here, we can also see some fluctuations in reactive power, they are related to the operation of the asynchronous motors in the network, as well as their launch.

The second important point in setting the power distribution is to provide acceptable power-ascension/drop properties when the equipment is introduced into the network. This is to avoid any loss of electricity quality, emergency shut-down of the equipment, and to ensure that the system is operating smoothly in general.

Fig. 5 shows a graph of a power ascension with a high overshooting value.

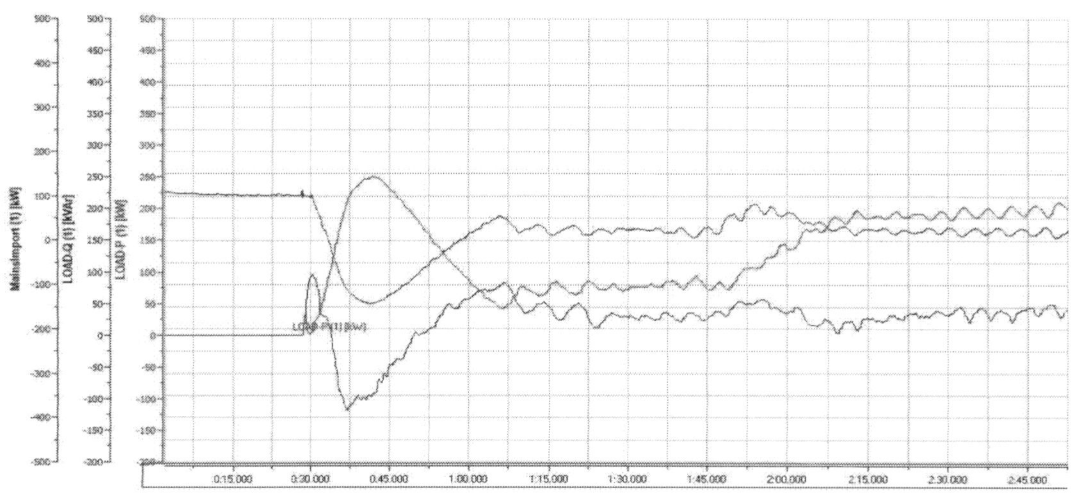

Figure 5. Power ascension and overshooting. MainImport, kW –Import of load from main grid, LOAD Q, kVAR – power of the network source, LOAD P, kW – power produced by the GRPP.

This graph shows that the regulator needs additional setting. Also, the relationship between the network and the generator in transition processes is clearly visible, as well as the need for accurate adjustment of the PID settings of the GRPP controller – the harmonization of the controller factors and the feedback parameters of the adjustable value. Fig. 5 shows continuous fluctuations in the active power, which means that the governor of the GRPP engine has to be adjusted.

System settings during debugging were adjusted to acceptable values, overshooting in the process of ascension/droping of load of approx.. 5-7%, without auto-oscillations and static controller error.

The third important point is to maintain stable and even operation of the power system under varying and constant loads.

This factor is implied by the previous one and is an indicator of the stability of the system as a whole in a continuous-duty mode. To do this, the system is tested for 10-16 hours and point corrections are made to the factors of PID controller.

The fourth point of adjustment is the arrangement and logic of on/off switching on the mains switch. In this case, this entirely depends on the need to maintain the power reserve and is controlled by the power distribution controller. In our case, the impact on the power switch has been completely reduced to the functions of protection against export of electricity to the network. As may be required, a system is created with the switch disconnected and connected with reverse synchronization on command from the managing controller in case of a lack of power in the system.

The fifth point is to protect against export of electricity to the network. This factor is the result of arrangement of the energy system as a whole and the comprehensive sustainable work of the previous four points. In particular, it depends on the duration of transient processes at unloading/loading of the GRPP by the consumer, the accuracy of setting of the cascade of power distribution controllers and GRPP management taken together, the stability of maintaining the set parameters by the speed controller and the generator's excitation controller. them.

5. Analysis, solutions, and conclusions

It is possible to use the example of this system to analyze the technical solution as a whole, as well as for a follow-up revision of the concept of distributed generation.

Advantages:

- Possibility of partial replacement of electricity consumption from the network with electricity produced by self-contained sources.
- Quality of electricity produced in accordance with GOST 33105-2014.
- Provision of 3rd degree of automation according to GOST 50783-95.
- Automatic power distribution between generating units based on current load, and rapid power change as network load changes.
- Flexibility of the system in terms of equipment redundancy, repair, provision of high category for electrical receivers by using a stand-alone generating unit.

Disadvantages:

- Complexity of setting and debugging of the system.
- Cost of the solution, even though the system is simple (3 sources, of which 2 are mains supply sources), it requires significant development and implementation costs, testing on a case-by-case basis.
- Duration of operation in the test mode as a consequence of the previous item.
- The power distribution principle is not implemented, considering the operation of the equipment in the energy-efficient range. This solution is discrete and oriented on generation from the main source (in our case - GRPP), and unloading of the latter.
- No feedback for consumers – unloading/loading takes place without feedback to the system, which complicates management and imposes high requirements to the system's responsiveness to changing loads.

From all of the above it follows that for the complex automation of such energy solutions, and especially in case of operation from several generating sources, new technical solutions for power distribution are needed, and, if possible, the departure from complex customized software systems to reliable and proven industrial solutions, in particular, due to the inability to perform long-term testing of the power system at the implementation site. In particular, an effective solution would be to use a controller taking lead in arrangement of power distribution according to a given and pre-configured algorithm, having I/O for feedback with equipment, a terminal block for frequency/reactive unloading of the network, with a built-in function of controlling the loading of generating units in a given range and switching it on/off if necessary, with a set power reserve.

This device would improve the energy efficiency of distributed generation solutions, reduce the cost of implementation for such projects, move from experimental systems to industrial ones and sustainable solutions, and would provide a pronounced economic effect.

References

[1] Andreev E B, Popadko V E 2005 Software tools for process control systems in the oil and gas industry (Moscow: "Oil and Gas")

[2] Denisenko V S 2006 PID Regulators: Principles of Building and Modification *Modern Automation Technologies* **4**

[3] Golubchik T V, Yutt V E, Taratutin D Y 2019 Automated Control System for Parallel Operation of Five Generator Sets in the Electric Power System *Institute of Electrical and Electronics Engineers* 2019 International Science and Technology Conference "EastConf".

[4] Korkov S A 2011 Calculations of electrical consumption in the energy survey of an industrial enterprise: A training and methodical manual (Izhevsk: "Udmurt University" publishing house)

[5] Serdtseeva A V 2016 Development of automated systems of control of technological processes *The Herald of UlGTU*

[6] Trutnev S Y, Marchenko A A 2014 Analysis and investigation of parallel operation of the synchronous generator and static converter *Bulletin of the Kamchatka State Technical University* (Petropavlovsk-Kamchatsky)

Processing of expert information for anti-pollution systems research

A M Reznik[1] A V Pogodin[2] and N V Logacheva[2]

[1] Department «Electrical engineering and electrical equipment», Moscow Automobile and Highway State Technical University (MADI), 64, Leningradsky ave. Moscow, 125319, Russia
[2] Department of Information Technologies and Control Systems, The State-Funded Educational Institution of Higher Education Moscow region University of Technology (TU), 42, Gagarin St., Korolev, Moscow region, 141070, Russia

E-mail: pogodin@bk.ru

Abstract: The introduction in Russia of the Euro-5 standard and the tightening of rules on emissions of harmful substances has resulted in an overdue improvement in the methods and algorithms of calculation and design of toxicity decreasing systems to enhance the functioning of devices of these systems. Despite widespread methods of simulation and computer modelling for the study and testing of the proposed organizational and technical measures, it is relevant to assess the increase of exhaust recirculation motor units with the help of experimental studies. The purpose of the study is to evaluate emissions of harmful substances of the car exhaust and suggest an experimental device based on the apparatus and methodology for analysis of carbonyl compounds and volatile organic compounds. In assessing of the effectiveness of reducing the toxicity, significant attention is paid to methods of experimental studies. As an indicator of the effectiveness in reducing the toxicity of motor units, the volume of emissions of harmful substances was selected. The choice of the components of the experimental system for the automated processing of the environmental inspection results of motor vehicles was substantiated. These components are: software tools that implement the conduction of the test by a cold start of the engine and chassis dynamometer settings, as well as monitoring the progress of the experiment; interactive systems to transmit information between the posts of ecological police, private persons and legal entities, companies for the maintenance of vehicles.

1. Introduction

The introduction in Russia of the Euro-5 standard and the tightening of rules on emissions of harmful substances has resulted in an overdue improvement in the methods and algorithms of calculation and design of toxicity decreasing systems to enhance the functioning of devices of these systems. Despite widespread methods of simulation and computer modeling for the study and testing of the proposed organizational and technical measures, it is relevant to assess the increase of exhaust recirculation motor units with the help of experimental studies [1].

When testing standards for emissions of hazardous substances, the connection to a unified automated information system of technical inspection of motor vehicles (motor units) [2] shall be

Content from this work may be used under the terms of the Creative Commons Attribution 3.0 licence. Any further distribution of this work must maintain attribution to the author(s) and the title of the work, journal citation and DOI.
Published under licence by IOP Publishing Ltd

provided for increasing the responsibility of technical inspectors for the more competent issuance of a diagnostic card or an international certificate of technical inspection of the vehicle (motor unit). However, the rules for inspection, when connected to a single information system remain unchanged.

The standards and methods of quantification of motor vehicle emissions and their danger to the environment and the biosphere are extremely difficult to implement in practice. European research methods are characterized by costly test processes using sophisticated equipment, and many measurement procedures are also difficult to implement in practice [3].

2. Method

For the purpose of studying the composition and quantity of the exhaust gases of a vehicle, a quantitative specific emission of each individual component of the composition of the exhaust gases based on engine load for one kilometer, taking into account regional environmental standards of background contamination of the surface layer of the atmosphere is usually determined [4]. In connection with the proposals for the replacement of vehicle tax for the environmental tax it is advisable to introduce environmental police posts which will be combined with inspection stations. For measurement of the quantitative composition of the exhaust gases of the vehicle on the ecological police posts it is advisable to use the instrument complex:

- Chassis dynamometer with a fan to blow air onto the vehicle during the test - This is necessary to simulate the air flow that cools the car.
- Classic electronic impactor with Teflon filters, which use accurate scales for weighing (the impactor is a device for fractional precipitation of aerosol particles).
- Portable pump for sampling the exhaust gases during the test.
- Special containers for samples of exhaust gases.
- Apparatus for high-performance liquid chromatography.
- Apparatus for gas chromatography of mass-spectrometry.
- Display on which the driver (environmental police officer) views the current parameters taken from the vehicle (momentum, work of the throttle, lambda-control, and so forth);
- Data collection system programmed for self-regulation of cold start tests on New European Driving Cycle.
- Personal computer (PC) with the necessary software to work with this equipment for ensuring the control of environmental standards of vehicles.

To study the composition and quantity of the exhaust gases it was proposed to identify the most harmful factors of the exhaust gases of the vehicle. The most toxic components of exhaust gases of gasoline engines are: carbon monoxide (CO), nitrogen oxides (NOx) and hydrocarbons (SnHm).

Carbon monoxide is produced during the combustion of fuel-air mixtures with a certain lack of oxygen, and also due to the dissociation of carbon dioxide that occurs at high temperatures.

The reason for the formation of hydrocarbons is heterogeneity of fuel-air mixture in the combustion chamber as well as uneven temperatures and pressures in various parts of the chamber. In some areas of the combustion chamber fuel is not consumed since there is a breakage of a chain reaction of hydrocarbon oxidation. Nitrogen oxides in the exhaust gases are formed due to the reversible oxidation of nitrogen with oxygen under high temperatures and pressure in the combustion chamber. The Increasing of the maximum temperature of the work cycle and an excess of oxygen are the main factors for the formation of nitrogen oxides. PC software is used to communicate all of the hardware installation, obtaining the experimental values from this hardware, the conduction of cold start test of the engine and setting the chassis dynamometer, as well as monitoring the progress of the experiment.

The ecological cleanliness of exhaust must lie in the construction of domestic aggregates and motor vehicles at the design stage. Furthermore, the operating characteristics of toxicity should remain stable [5].

This research method can be applied to domestic cars at the development stage, and also for monitoring the environmental performance of exhaust (especially considering the fact that the vehicles with environmental class lower than the Euro 3 are prohibited from entering the Third Ring Road).

Multi-Automated Information System (MIS) for the processing of the results of environmental inspection of vehicles that is produced after appropriate examination. The MIS will include expert subsystem of diagnostics and forecasting. Expertise includes experimental measurement on the stand by the environmental police officers through specialized interfaces and a single database, ensuring the integrity and availability of personal data during their processing. The results of the examination and information about how the test of assessing motor unit exhaust was conducted must be available to employees of service stations to collect information for better understanding of what criteria the exhaust gases have been exceeded. This is necessary for a more accurate and rapid determination of what is not working properly, because some errors cannot be recorded in the driving unit of the vehicle.

3. The implementation of environmental control at the Centre of traffic management (CTM) and use of Automated Data Processing Center (ADPC)

The realization of ecological monitoring of vehicles was proposed with the assistance of CTM, the structure of which is shown in Fig. 1. Special positions for measuring the environmental performance of cars and trucks must be set up. Some systems of CTM are already well-developed, for example, automated traffic management system (ATMS). A hardware-software complex information system is used for the automated processing of the results of environmental inspection of motor vehicles during the implementation phase. Organizational and technical procedures for the sustainable functioning of the complex have yet to be developed [6].

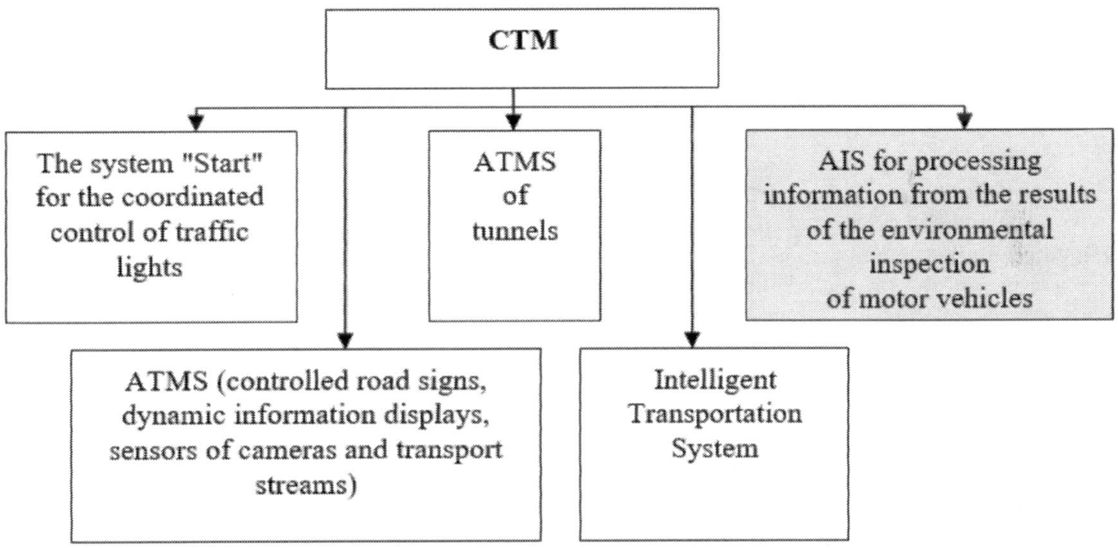

Figure 1. Structure of the CTM.

Functions of AIS that are developed for the processing of the results of motor vehicles environmental:

- Reducing the number of vehicles with poor environmental performance;
- Analysis of carbonyl compounds and volatile organic compounds;
- Analysis of particulate emissions;
- Informing drivers about the requirements for the environmental performance of vehicles.

Expert subsystem will use the Automated Data Processing Center (ADPC) to gain access to databases of traffic safety inspection. For information on vehicle registration and information about the driver, as well as for checking from where the car was sent to the environmental emissions test centre. For example, vehicles with an ecological class lower than Euro 2 are forbidden within the

MKAD ring road, and vehicles with an ecological class lower than Euro 3 are forbidden within the TTR (third transport ring). In the event of a car being stopped within the TTR and sent to the test centre, it became clear in the test that its environmental class was below the permissible level in this area. The test results are then sent to the Department of Traffic Safety Inspectorate, in which the car is registered, for reconciliation with the traffic officer. The traffic officer, according to the test results, will write the decision of an administrative offence. It is necessary to develop a resource (knowledge base), where information about a particular car and standards for exhaust will be available and regulated by the manufacturer, the test results and information about the factors of external and internal environment during the test (room temperature, pressure, humidity, amount of fuel), as well as all information about how the test was conducted.

This is necessary to ensure that in cases where the vehicle does not pass the test for ecological compatibility of exhaust on various parameters, or passes, but its ecological class is lower than that stated by the manufacturer (say the passport vehicle complies with environmental class Euro-6, and the test showed that the ecological class of the car matches only Euro-4), specialized service stations (SSS) can access this information for a better understanding of which particular node in the car may not be working properly, as some errors are not recorded in the control unit of the car, for example the swirl flaps in the intake manifold on the Volvo XC70 with the motor D4 are not always recorded in the control unit and can adversely affect the ecological compatibility of the exhaust, and the driver will be unaware of this.

The opportunity for responsible citizens, who want their car to be environmentally-friendly, to register for a test over the Internet should exist, and partitions must be created for private and legal entities if a transport company wants to test a fleet of vehicles to avoid problems that may arise in the event of violations of environmental regulations. It is essential that companies that conduct technical inspections have access to information about tests for environmental friendliness so that they have enough information to decide on the car's technical condition – a diagnostic card, which is required for obtaining insurance on Vehicles from 2019.

4. Conclusions

An experimental device for evaluating the exhaust of the car is suggested, based on the apparatus and methodology for analysis of carbonyl compounds and volatile organic compounds. The device is implemented in practice. Compared regulated and unregulated emissions, including particulate matter from both gasoline and gasohol types of fuel.

Additional analysis of fuel economy, along with the possibility of predicting the health impact when burning the fuel with low methanol content, as an alternative was substantiated. Also, the choice of the components of the ES for the automated processing of the environmental inspection results of motor vehicles was substantiated. These components are: software tools that implement the test by a cold start of the engine and chassis dynamometer settings, as well as monitoring the progress of the experiment; interactive systems to transmit information between the posts of ecological police, private persons and legal entities, companies for the maintenance of vehicles.

References

[1] Zhao H, Ge Y S, Hao C X, Han X K, Fu M L, Yu L X, et al 2010 Carbonyl compound emissions from passenger cars fueled with methanol/gasoline blends *Science of the Total Environment* **vol 408 (17)** 3607–3613

[2] Liang B, Ge Y S, Tan J W, Han X K, Gao L P, Hao L J, et al 2013 Comparison of PM emissions from a gasoline direct injected (GDI) vehicle and a port fuel injected (PFI) vehicle measured by electrical low pressure impactor (ELPI) with two fuels: gasoline and M15 methanol gasoline *Journal of Aerosol Science* **vol 57** 22–31

[3] Dai P P, Ge Y S, Lin Y M, Su S, Liang B 2013 Investigation on characteristics of exhaust and evaporative emissions from passenger cars fueled with gasoline/methanol blends **vol 113** 10–16

[4] Wang X, Ge Y, Liu L, Peng Z, Hao L, Yin H, Ding Y, Wang J 2015 Evaluation on toxic reduction and fuel economy of a gasoline direct injection- (GDI-) powered passenger car fueled with methanol–gasoline blends with various substitution ratios *Applied Energy* **vol 157** 134–143
[5] Heywood J B. 1988 Internal combustion engine fundamentals (New York: Mc-Graw Hill) 328 p
[6] Artyushenko V M, Abbasova T S. Increasing noise immunity of electric communication channels in high-speed telecommunication systems 2014 *Biosciences Biotechnology Research Asia* **11** 277–279

The development of standards for road restraint systems

I V Demiyanushko[1], B T Tavshavadze[1], I A Karpov[1], S S Petrosyan[2]

[1]Department of Structural Mechanics, Moscow Automobile and Road Construction State Technical University (MADI), 64, Leningradsky Prospect, Moscow, 125319, Russia
[2] M&PQ LLC, 64, Leningradsky Prospect, Moscow, 125319, Russia

E-mail: t_beka@mail.ru

Abstract. The article discusses requirements for road restraint systems, gives recommendations for their classification, formulates the requirements for layout of cameras for recording tests, proposes a method for determining the values of the dynamic deflection and the working width of the road barriers. This work proposes a classification of road barriers; gives the estimates of the influence of tolerances in weight and collision speed of a vehicle in the field tests of road barriers on the values of dynamic deflection and the working width; introduces the concept of category for the dynamic deflection and working width of a barrier; supplements the testing methods for different types of road barriers.

1. Introduction

As the construction of public highways intensifies and the speed limit and traffic intensity increase, the road safety problem also becomes increasingly more important. The greatest number of road traffic accidents (RTA) with serious consequences are related to vehicles crossing to a wrong lane or leaving the carriageway. The most effective means of improving road safety is by installing road restraint systems. However, most of the currently available national standards were developed on the basis of studies carried out in 2005-2012. The emergency of the novel tools, i.e., virtual testing [1,2,3], made possible studying the mechanics of the barriers in more detail and formulate clearer and unambiguous requirements for road barriers. Therefore, it is necessary to revise and update all existing regulatory documents and standards.

2. Materials and main results

All requirements for road barriers can be divided into the following main categories: classification, technical requirements, testing methods, application rules. The rules for the use of road barriers are given in the updated GOST R 52289-2019 "Automobile roads for general use. Technical Facilities for Traffic Management. Rules of application of traffic signs, markings, traffic lights, guardrails and delineators".

The draft of updated GOST 33127 "Automobile roads of general use. Road restraint systems. Classification" [5] specifies categories of road barriers according to their functional purpose, groups and subgroups according to the functional purpose of road retaining lateral barriers. By the principle of operation, road retaining lateral barriers are subdivided into types: rail-type; parapet; cable; combined, and other types of structures.

Content from this work may be used under the terms of the Creative Commons Attribution 3.0 licence. Any further distribution of this work must maintain attribution to the author(s) and the title of the work, journal citation and DOI.
Published under licence by IOP Publishing Ltd

GOST 33127-2014 divides lateral restraint barriers into the following subclasses: non-deformable and deformable. Studies have shown that the classification of lateral restraint barriers into non-deformable and deformable was established back in the 1970's, when such fences as "curved beam" (figure 1) or rail-type barriers with a reinforced concrete post (figure 2) were used [6]. These constructions of road barriers are outdated and do not meet the restraint requirements of modern metal rail-type barriers (figure 3). Thus, the currently used types of lateral restraint barriers can be either deformable (rail, cable, combined) or non-deformable (parapet). Based on the above, the draft GOST 33127 lacks the classification of lateral restraint barriers into subclasses.

Figure 1. Curved beam barrier

Figure 2. Barrier fence with a reinforced concrete post

Figure 3. Modern road rail-type barrier design

Particular attention was paid to the classification of road frontal barriers, installed in places of traffic flow separation. Studies have shown the need to classify road frontal barriers by types: parallel (with parallel lateral edges in the top-view), non-parallel (in the form of the asymmetrical trapezoid in the top-view) and symmetric (in the form of a symmetric trapezoid in the top-view) (figure 4). This classification will allow the installation of road frontal barriers with the required configuration, contributing to their adequate perception by drivers in specific road conditions.

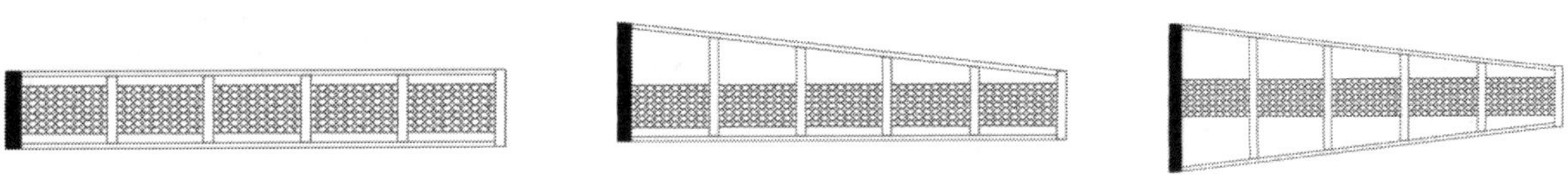

(a) *(b)* *(c)*

Figure 4. Frontal barrier types: (a) parallel, (b) non-parallel, (c) symmetric

Figure 5 shows the general recommended classification scheme for road barriers.

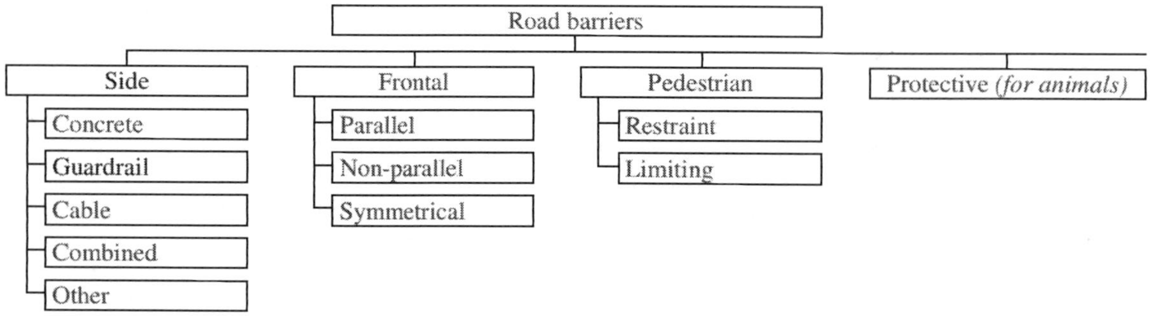

Figure 5. Classification chart of road barriers

Draft GOST 33128 "Automobile roads of general use. Road restraint systems. Technical requirements" [7] gives the following main consumer characteristics of road lateral barriers: restraint capacity, dynamic deflection, and the working width of the barrier. These parameters are determined by full-scale field tests, where a vehicle with a maximum weight and a given speed collides with a road barrier (figure 6) [7].

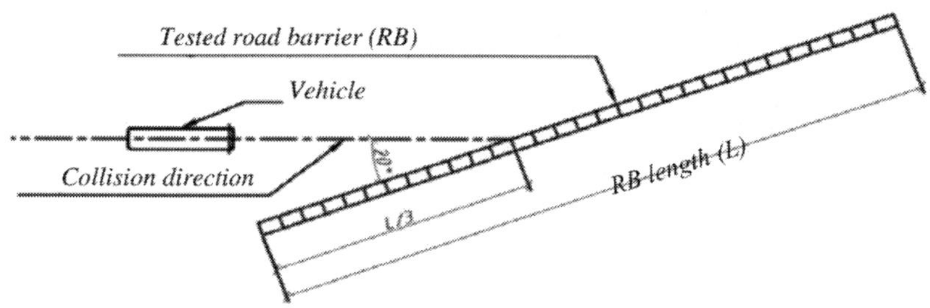

Figure 6. Road lateral barrier testing scheme

However, full-scale tests of the same structure of the barrier gave different values of the dynamic deflection and the working width of the barrier. To determine the reason for the obtained variations, the analysis of the field-testing method was done [7]. The analysis has shown that the test results differed because of the tolerances in the mass of the testing vehicle, ranging from 0.4 to 0.7 t, depending on the total mass of the testing vehicle (from 12.0 to 22.0 t), with a collision speed tolerance of ± 5 % [8]. Impact energy value for restraint capacity U4, determined according to GOST 33129-2014 [8] are given in Table 1.

Table 1. Impact energy values as a function of weight tolerances and vehicle collision speed tolerances

Restraint capacity	Required impact energy, kJ	Total mass of the bus, t	Collision angle, °	Collision speed, km/h	Actual impact energy
U4	300	15.5	20	68.01	323.51
U4	300	15.0	20	67.0	303.88
U4	300	14.5	20	65.99	285.01

The analysis has shown that the difference in the values of the actual impact energy reaches up to 10%. In order to determine the effect of tolerances of maximal mass and collision speed on the values of dynamic deflection and working width of the barrier, the side barriers were virtually tested using the LS-DYNA software package. The model was validated by comparing the results of virtual testing with the results of field tests carried out at Federal State Unitary Enterprise "NAMI", State Research Centre of Motorized Vehicles, Spare Parts and Accessories (figure 7). The reliability of the model for consumer characteristics of the barrier exceeded 90%.

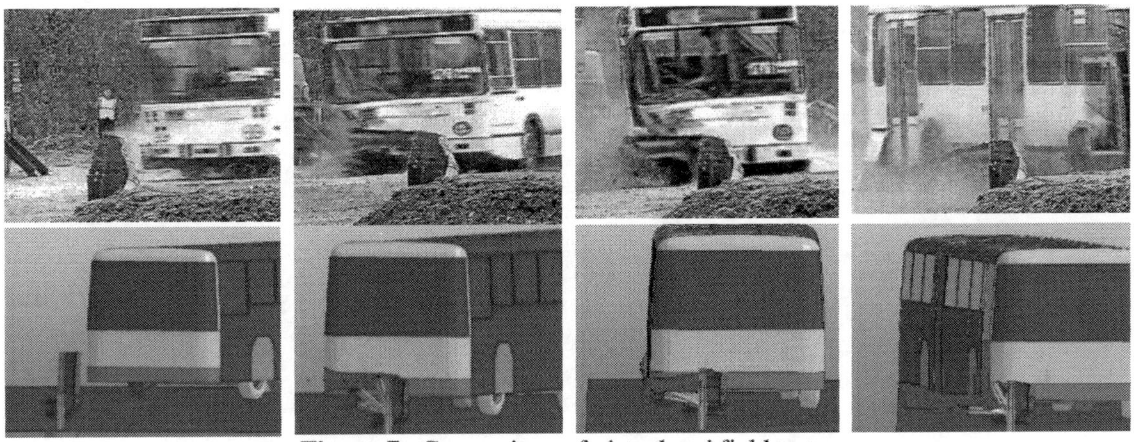

Figure 7. Comparison of virtual and field tests

After a validated model was obtained, the values of the mass of the bus, the speed of the bus collision, and the impact place were varied for the analysis. The results of virtual testing are given in Table 2.

Table 2. Virtual testing results for lateral rail-type road barrier

Test No. / Parameters of the barrier	Impact between the posts				Impact on the post			
	1	2	3	4	1	2	3	4
Vehicle weight, t	15	15.5	14.5	13	15	15.5	14.5	13
Collision speed, km/h	67	68	66	70	67	68	66	70
Impact energy, kJ	304	324	285	288	304	324	285	288
Dynamic deflection, m	0.99	1.06	0.93	0.85	0.96	1.01	0.88	0.86
Working width, m	1.34	1.51	1.48	1.42	1.18	1.40	1.42	1.49
Injury severity index	0.15	0.15	0.15	0.21	0.15	0.16	0.18	0.16
Number of deformed posts, pcs.	7	8	6	8	7	7	5	7
Number of deformed beams, pcs.	5	6	4	6	5	5	4	6
Lane exit angle, deg	5	8	7	7	6	8	7	7
Tilt angle, deg	11	18	11	25	9	13	10	21

The studies have shown that the values of the dynamic deflection and the working width of the barrier can differ by up to 0.2 m. These characteristics of the barrier are also influenced by the impact section aimed at the barrier.

The obtained results were verified by analyzing more than 50 full-scale field test reports and about 170 virtual test reports. The analysis has shown that the field tests are carried out under different weather conditions, and therefore it was decided to study the effect of the tire-road surface traction coefficient (φ) on the dynamic deflection of the barrier and the heading angle of the testing vehicle. Figure 8 gives an example of the influence of the tire-road surface traction coefficient on the deflection of a cable barrier with a restraint capacity of U4 (300 kJ) obtained in virtual testing.

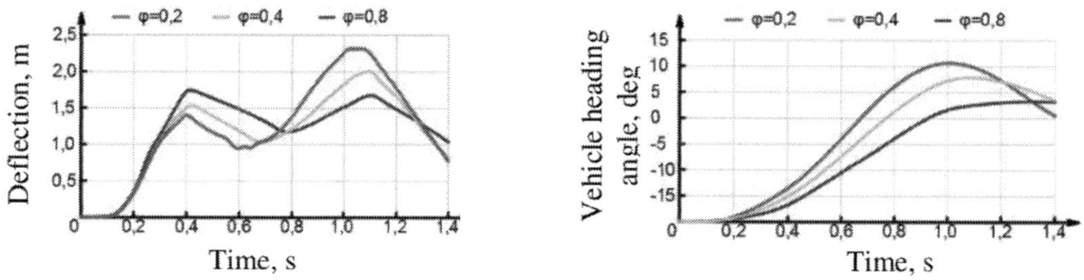

Figure 8. The influence of the tire-road surface traction coefficient on the deflection of a cable barrier with post pitch 2.0 m

The test results show that the tire-road surface traction coefficient significantly affects both the deflection of the barrier and the trajectory of the vehicle during the collision. Therefore, the tire-road surface traction coefficient shall be regulated in the standards for road barriers testing.

The analysis of the protocols of full-scale field tests also revealed that many of them do not contain values of the dynamic barrier deflection and only present the residual deflection values. The dynamic deflection of the barrier shall be determined based on video recording of the testing vehicle collision with the tested road barrier. It should be noted that the accuracy of its measurement largely depends on the competence of the testing laboratory, as well as the correct placement of video cameras. The dynamic deflection of a road barrier is one of the main characteristics of the barrier and under no circumstances can it be considered equal to the residual deflection (except for parapet barriers). Studies carried out at MiPK LLC have shown that the difference in the value of the dynamic deflection of the rail-type barrier with the residual deflection is about 20% (figutr 9), and for cable barriers the difference exceeds 100% (figure 8) [9].

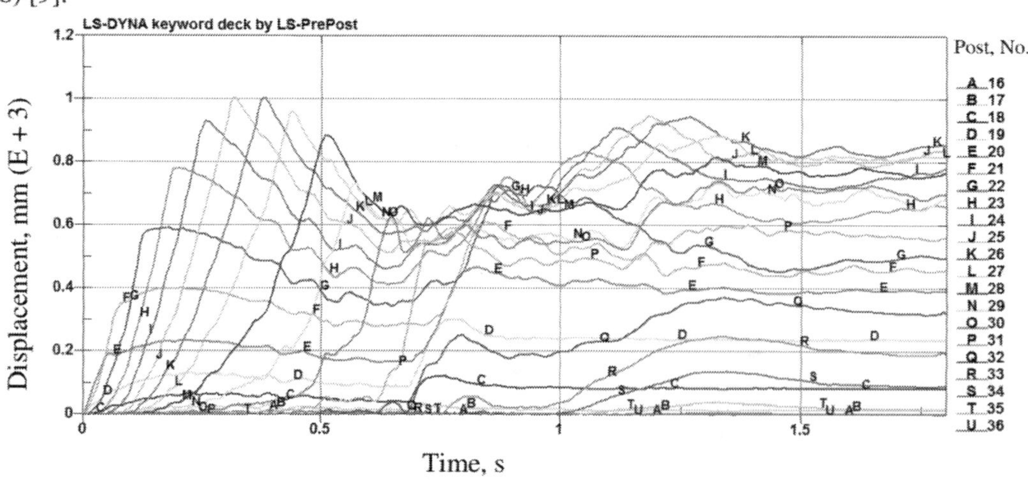

Figure 9. Dynamic deflection changes of the rail-type barrier during collision

Thus, to date, a methodological framework for solving the following tasks has been formed:

1) Develop of a method for determining the dynamic deflection and the working width of the barrier that minimized the difference in the test results of the same barrier structure;

2) Introduce requirements for the location of the impact point of the vehicle on the road barrier.

3) Introduce requirements for the tire-road surface traction coefficient

Solving these tasks requires in first place increasing the accuracy of measurement of the dynamic deflection of the barrier. For this, it is suggested to record the collision of the testing vehicle with the road barrier using high-speed video cameras. The location of the cameras is shown in figure 10.

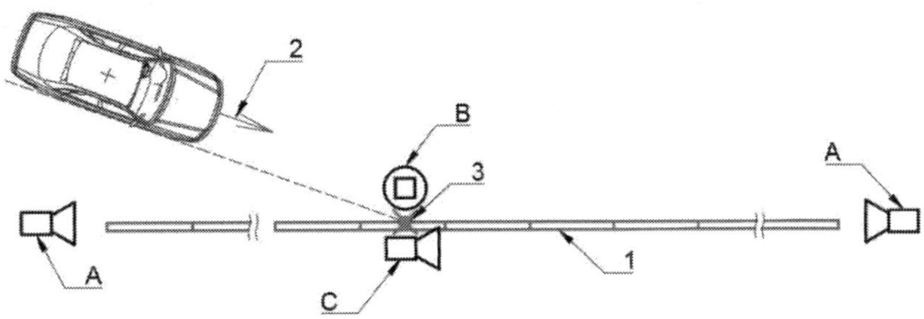

A, B, C - high-speed video cameras; 1 - tested lateral restraint barrier; 2 - movement direction of the test vehicle; 3 - impact point

Figure 10. Recommended location of high-speed video cameras during full-scale tests of lateral barriers

High-speed video cameras A should be located at the beginning and end of the barrier. High-speed video camera B is installed above the impact point and perpendicular to the horizontal surface of the test site. It should record the top view of the impact point, covering the area 6.0 m before and 6.0 m after the impact point. High-speed video camera C should be installed above the barrier at an angle to the horizontal surface of the test site and record the impact (collision) of the test vehicle on the road barrier.

To account for the difference in impact energy due to the tolerances in the mass of the vehicle and the speed of the collision, introducing an impact energy coefficient (K) is recommended; the coefficient is determined by dividing the design impact energy E_D by the actual impact energy E_A:

$$K = \sqrt{\frac{E_D}{E_A}}$$

It is recommended to determine the final value of the dynamic deflection of the barrier (normalized dynamic deflection, *Dn*), given in the design documentation, as the product of the dynamic deflection D_M measured in field tests and the impact energy coefficient:

$$D_n = D_i * K$$

A normalized working width of the barrier can be determined in a similar manner. However, even in this case, the distortion of the video records can lead to measurement error in the range of 0.1 to 0.2 m. Therefore, it is recommended to introduce a category of dynamic deflection and the working width of the barrier, determined by the Tables 2 and 3. It should be noted that a similar approach is implemented in the European standard EN 1317-2 [10].

Table 3. Dynamic barrier deflection categories

Dynamic barrier deflection category	Normalized dynamic barrier deflection, D_N
D1	$D_N \leq 0.4$
D2	0.4 <Dn ≤ 0.7
D3	0.7 <Dn ≤ 1.0
D4	1.0 <Dn ≤ 1.2
D5	1.2 <Дn ≤ 1.5
D6	1.5 <Dn ≤ 1.8
D7	1.8 <Dn ≤ 2.1
D8	2.1 <Dn ≤ 2.5
D9	2.5 <Dn ≤ 3.0
D10	D_N >3.0

Table 4. Barrier working width categories

Barrier working width category	Normalized working width of the barrier, R_N
D1	$D_N \leq 0.4$
D2	0.4 <Dn ≤ 0.7
D3	0.7 <Dn ≤ 1.0
D4	1.0 <Dn ≤ 1.2
D5	1.2 <Дn ≤ 1.5
D6	1.5 <Dn ≤ 1.8
D7	1.8 <Dn ≤ 2.1
D8	2.1 <Dn ≤ 2.5
D9	2.5 <Dn ≤ 3.0
D10	D_N >3.0

To solve the 2nd task, additional virtual tests were performed. The results of the study allowed identifying the most critical cases of a vehicle - barrier collision [9]. Thus, in field tests, the impact of a vehicle should be tested:

a) In the middle of the block for parapet barriers;

b) In the middle between adjacent posts for rail-type barriers;

c) At a distance not exceeding 1/2 of the distance between adjacent posts, but no more than 0.75 m for cable barrier posts; if there is an interlacing of cables in the structure of the cable barrier with respect to the post, the barrier should be installed in a manner that during testing the lower cable on the first post from the point of impact of the testing vehicle was located on the outside of the barrier (Figure 11).

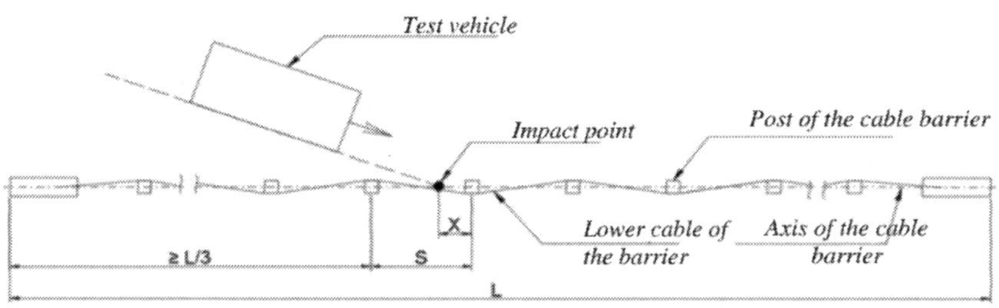

L is the total length of the cable barrier; S is the pitch of the posts; X is the distance from the barrier post to the point of impact of the vehicle on the cable barrier

Figure 11. Cable barrier test scheme

The tire-road surface traction coefficient at the test section must be at least 0.3 [11]; this value shall be measured immediately before testing and indicated in the full-scale test report.

The data obtained in the research generally allow making alterations and addenda to the drafts of GOST 33128 and GOST 33129.

3. Conclusions

Based on the results of the research, this work proposes a classification of road barriers; gives the estimates of the influence of tolerances in weight and collision speed of a vehicle in the field tests of road barriers on the values of dynamic deflection and the working width; introduces the concept of category for the dynamic deflection and working width of a barrier; supplements the testing methods for different types of road barriers. This information is suitable for updating GOST 33127, GOST 33128, and GOST 33129.

Acknowledgements

The study was carried out by M&PQ LLC on behalf of the Federal Road Agency (ROSAVTODOR) under the state contract No. 31/1-2019 dated December 04, 2019, "Updating GOST 33127 "Automobile roads of general use. Road restraint systems. Classification", GOST 33128 "Automobile roads of general use. Road restraint systems. Technical requirements", GOST 33129 "Automobile roads of general use. Road restraint systems. Methods of testing"

References

[1] Demiyanushko I, Karpov I 2019 Applications Fem Analysis in Researches Reliability and Dynamics of Road Safety Barriers at Collisions with Cars *Advances in Mechanism and Machine Science. Proceedings of the 15th IFToMM World Congress on Mechanism and Machine Science. Mechanisms and Machine Science* Vol 73 pp 1651-1658

[2] Demiyanushko I, Karpov I, Tavshavadze B 2017 Developments of non-linear dynamics FEM simulation of the impact performance of road safety barriers with use of experimental validation of models *Mathematical and Numerical Aspects of Dynamical System Analysis 14th International Conference "Dynamical systems theory and application"* (Lodz, December 11-14) pp 165-174

[3] Michal K, Jana M 2018 FEA simulation and probability approach of the road barrier crash tests *Proceedings of the 12th fib International PhD Symposium in Civil Engineering 2018* (Faculty of Civil Engineering of the Czech Technical University in Prague; Czech Republic) pp 493-501

[4] GOST P 52289 *Traffic control devices. Rules of application of traffic signs, markings, traffic lights, guardrails and delineators* Available at: http://www.gostinfo.ru/catalog/gostlist

[5] GOST 33127-2014 *Automobile roads of general use. Road restraint systems. Methods of control* Available at: http://www.gostinfo.ru/catalog/gostlist

[6] Shesterikov V 2013 Improving the regulatory framework and structures of road barriers *Car roads* **3** 52-57

[7] GOST 33128-2014 *Automobile roads of general use. Road restraint systems. Technical requirements* Available at: http://www.gostinfo.ru/catalog/gostlist.

[8] GOST 33129-2014 *Automobile roads of general use. Road restraint systems. Methods of control* Available at: http://www.gostinfo.ru/catalog/gostlist

[9] BS EN 1317 – 2. *Road restraint systems: Part 2: Performance classes, impact test acceptance criteria and test methods for safety barriers including vehicle parapets* (London: BSI Standards Publication, 2010)

[10] Tavshavadze B T 2019 D*evelopment and justification of the methodology for calculations, testing and certification of road restraint barriers of the rail type: dissertation for the doctor candidate degree* (Moscow)

[11] GOST R 50597-2017 *Automobile roads and streets. The requirements to the level of maintenance satisfied the traffic safety. Methods of control* Available at: http://www.gostinfo.ru/catalog/gostlist.

IOP Conf. Series: Materials Science and Engineering 1159 (2021) 012044 doi:10.1088/1757-899X/1159/1/012044

Methodology and results of experimental studies of precision weighing systems of refueling complexes of upper stages and spacecraft of space rockets

A N Sova[1], V M Shulga[1], O A Valyaev[1], S A Zolotoy[1], A A Loginov[1]

[1]Department of transport installations, Federal state budgetary educational institution of higher education Moscow automobile and road state technical University (MADI), 64, Leningradsky prospect, Moscow,125319, Russia

E-mail: a.sova@madi.ru, slsova@mail.ru

Abstract. This article presents the main provisions of the new methodology, scientific and methodological approach and the results of experimental studies of precision weighing systems of refueling complexes of upper stages and spacecraft of space rockets. The article presents the following results of experimental studies of precision weighing systems: experimental substantiation of the provisions and conclusions of the metrological model of the refueling complex; experimental confirmation of the metrological characteristics of weighing devices of platform scales and platform two-channel scales in terms of weighing the delivered doses; experimental determination of the actual dose delivery error according to the indications of platform scales and two-channel platform scales with the supply of propellant components from transport and refueling containers; experimental testing of PC software in order to identify possible errors, as well as to make possible adjustments to the software for more accurate tracking of the main refueling operations; selection of optimal values of low flow rate for a more accurate cutoff of the dose and practicing the operator's actions at the stage of the cutoff operation; determination of cutoff errors based on statistical processing of experimental data. The article also presents in detail the experimental unique scientific installation created by the authors, identifies the directions of its practical application in testing new weighing systems, in research work and in the educational process.

1. Introduction

Experimental studies of precision weighing systems (WS) with the supply of propellant components directly from the transport and refueling container (TRC) using bench equipment simulating refueling complexes (RC), play a significant role in solving an important problem situation of increasing the accuracy of refueling complexes of upper stages and space vehicles (SV) space rockets (SR) [1-4]. These studies were carried out in the joint research laboratory of the Department of Transport installations of MADI and ZAO ELVES (Closed Joint Stock Company) under the scientific supervision of the leading researcher of the MADI research department, Candidate of Technical Sciences, Senior Researcher V.M. Shulga. When conducting experimental studies of precision weighing systems, the following goals were set: experimental substantiation of the provisions and conclusions of the metrological model of refueling complexes; experimental confirmation of the metrological characteristics of the weighing devices (WD) of the platform weighing device (PWD) and weighing device platform two-channel

Content from this work may be used under the terms of the Creative Commons Attribution 3.0 licence. Any further distribution of this work must maintain attribution to the author(s) and the title of the work, journal citation and DOI.
Published under licence by IOP Publishing Ltd

IOP Conf. Series: Materials Science and Engineering 1159 (2021) 012044 doi:10.1088/1757-899X/1159/1/012044

(WDPTC) in terms of weighing the doses issued; experimental determination of the actual dose delivery error according to the indications of the PWD and WDPTC with the supply of RFC from the TRC; conducting experimental testing of the PC software in order to identify possible errors, as well as to make possible adjustments to the software for more accurate tracking of the main refueling operations; selection of optimal values of low flow rate for a more accurate cutoff of the dose and practicing the operator's actions at the stage of the cutoff operation; determination of cutoff errors based on statistical processing of experimental data.

Experimental studies were carried out by issuing control doses of water from a filling tank installed on the weighing module of the weighing system into a control container installed on a control balance. To carry out the experiment, we assembled experimental installations, the description of which is given in this article in Section 2 "Materials and Methods". Refueling complexes with a volume of 100 and 250 liters are used with PWD-500, with WDPTC-1000 - 250 liters, with WDPTC-2500 - 999 liters, with WDPTC-5000 - 2000 liters. For the dose of each nominal value, as a rule, 10 experimental points were obtained. Doses were dispensed at a constant pressure of 0.3 ± 0.001 MPa. In the experiments, a correction was introduced for the mass of displaced air in the filling tank and for the mass of displaced air from the control tank (CT). The actual mass of the i-th dose was calculated based on the readings of the control weights. When dispensing each i-th dose, the dose cut-off error was recorded, which is equal to the readings of the Icon board. The processing of the experimental results was carried out according to the method described in this article in Section 2 "Materials and Methods".

2. Materials and methods

Experimental installations and stands were created on the basis of the hydro laboratory. The stand includes main and auxiliary equipment. The main equipment includes: a pneumatic system, a water supply and drainage system, a set of high-precision control scales, a set of filling tanks, a set of control tanks (CT), gauges of excess pressure, temperature, humidity and atmospheric pressure, gas-hydraulic dispensing shields, sets of connecting hoses, weights of class M1 with a nominal value of 20 kg, a set of weights of 1 g - 1 kg and 1 kg - 10 kg class M1, weights of class M1 with a nominal value of 500 kg.

Ancillary equipment includes: a beam crane with a lifting capacity of 2000 kg, a hydraulic trolley with a lifting capacity of 1.5 t, a hydraulic stacker with a lifting capacity of 1.5 t, a universal foundation with embedded parts designed for the installation of stationary weighing devices with Max up to 15,000 kg. The composition of the stand allows testing the methods and technologies of refueling in the dose range from 0.4 to 2500 kg. The description of the stand is given in [4, 9]. A mobile set of special bench equipment for research in the range from 40 to 10,000 kg has also been created. Experimental installations are simplified models of refueling complexes. In this case, the following rule should be adhered to: the maximum load of the check balance should be 2 ÷ 5 times less than the maximum load of the working weighing devices. This rule reflects the real conditions of refueling, under which the mass of the dispensed doses is 2 ÷ 10 times less than the mass of the filled with the refueling component. In all experiments, the total mass of the filling container with water should be equal to the gross mass of the refueling complex. If necessary, to fulfill this condition, the required number of weights is additionally placed on the platform of the weighing device. In all experiments, the devices used as check weights are adjusted immediately before work is carried out, and the adjustment is carried out in the range of delivery of control doses. The result is a high actual weighing accuracy on the checkweigher. Actual error values, as a rule, do not exceed the following values: ±1 g for VM24001 scales, ±5 g for special control scales (SKS) SKS-100, ±10 g for SKS-400, ±10 g for PWD-500, ±20 g for WDPTC-1000. Figure 1 shows a pneumohydraulic diagram of an experimental setup with a WDPTC-2500 working balance. This weighing device is intended for refueling complete with fuel filling station V = 999 l in the dose range from 200 to 1200 kg. The filling tanks are filled from the preparation tank E2 using the P1 pump. The required pressure in the system is generated and maintained using a pneumatic system. If the lines are spilled, the water is drained into the drain tank (DT). Dose cutoff is performed by valves V1 (high flow rate) and V2 (low flow rate, which is provided by the Throttle 1 nozzle).

IOP Conf. Series: Materials Science and Engineering 1159 (2021) 012044 doi:10.1088/1757-899X/1159/1/012044

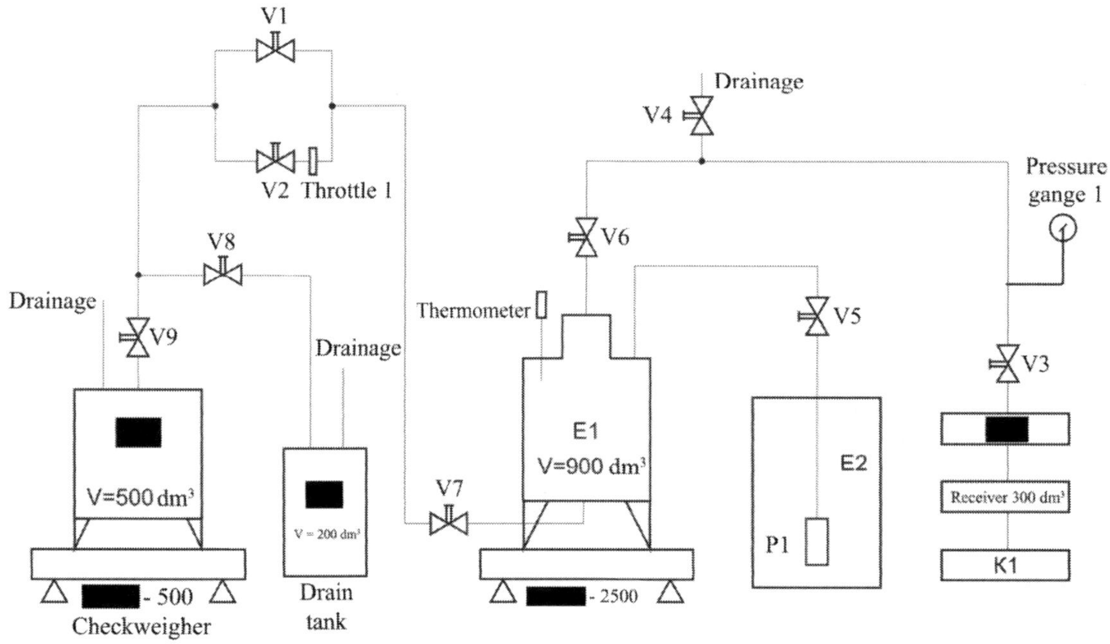

Figure 1. Diagram of an experimental setup with WDPTC-2500

A general view of the experimental setup is shown in Figure 2. The figure shows the WDPTC platform with a filling tank located on it; control rack based on a 19-inch cabinet; two mobile digital displays made in explosion-proof design; dispensing shield, which includes, along with other valves, cut-off valves and a nozzle. Figure 3 shows the 46" video wall panel included with the unit, which mirrors the PC monitor image and allows the refueling supervisor and commissioners to monitor the refueling process.

Figure 2. General view of the experimental setup with WDPTC-2500

Figure 3. General view of the video wall panel

The panel displays information obtained during the development of the technology of filling the upper stage (dose of the oxidizer weighing 292.00 kg). The plant circuits similar in structure are used in experiments with the use of VD PWD-500 and WDPTC-1000. The difference lies in the volume of the filling and control tanks, the maximum load and the type of working and control weights. Table 1 shows the data for the equipment used in experiments with PWD-500 and WDPTC.

Figure 4 shows a variant of the installation with operating scales PWD-500 and a filling tank with V = 250 liters.

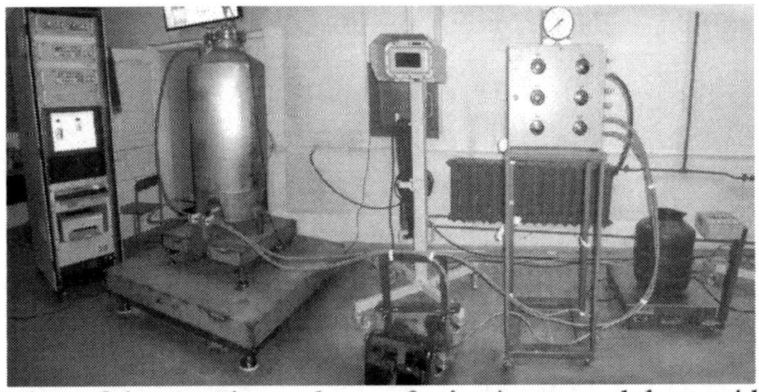

Figure 4. General view of the experimental setup for issuing control doses with the simultaneous use of weighing devices PWD-500 (main channel) and WDPTC-1000 (backup channel)

The installation, assembled in this version, was used to test the technology of work in case of emergency situations, therefore PWD-500 was installed on the WDPTC-1000 platform. A SKS-100 balance and a control container with V = 50 liters were used as control weights. The installation with WDPTC-5000 has a somewhat more complex scheme (see Figure 5). During research, filling tanks E1.1 and E1.2 are simultaneously filled with water from tanks E2.1 and E2.2 using pumps P1 and P2.A general view of the experimental setup (without check weights) is shown in Figure 6. The figure shows the WDPTC platform with two refueling tanks located on it; a control stand, a mobile digital display, a dispensing board, which, along with other valves, includes shut-off valves and a nozzle.

Figure 7 shows check scales WDPTC-1000 and PWD-500 with control tanks with V = 1000 l and V = 500 l. This installation allows you to issue control doses of water weighing up to 1300 kg. Experimental equipment for the study of the metrological characteristics of the complex of refueling facilities for the RB and spacecraft of the ILV requires separate consideration. For these purposes, a mobile set of special bench equipment was created in the following composition: special control scales SKS-500 (Max = 500 kg); special control scales SKS-10 (Max = 10 t); capacity V = 1 m^3; capacity V = 10 m^3; pump with a maximum capacity of 16 m^3 / h.

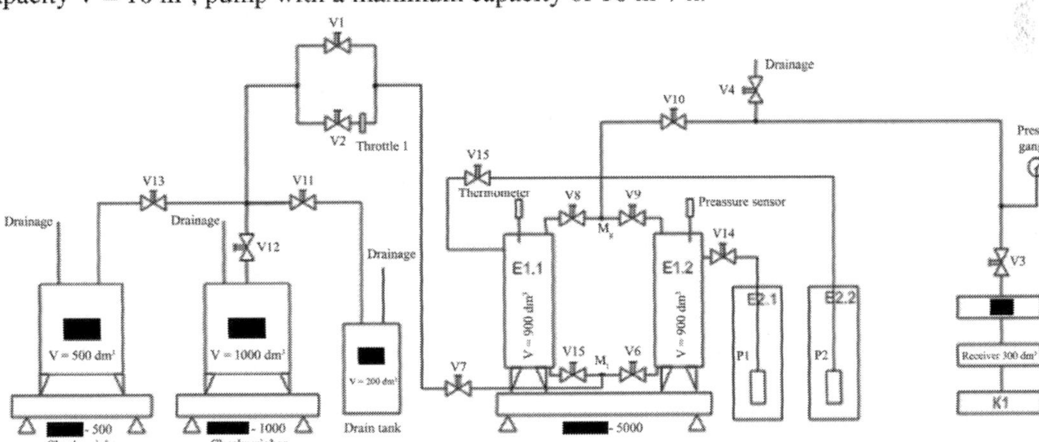

Figure 5. Diagram of an experimental setup with WDPTC-5000

Before testing, the SKS-500 balance was adjusted, and their actual error did not exceed 10 g in the entire weighing range from 0 to 500 kg. SKS-500 complete with a capacity of 1 m^3 made it possible to weigh doses in the range of 40 - 400 kg. For research in the range from 500 to 10,000 kg, special control scales SKS-10 with a maximum load of 10,000 kg were specially developed [5].

Table 1. Composition of experimental equipment

Working scales	PWD-500	WDPTC-1000	WDPTC-2500	WDPTC-5000
Control scales	VM24001 SKS-100 SKS-400	SKS-100 PWD-500	SKS-100 PWD-500 WDPTC-1000	PWD-500 WDPTC-1000
Filling volume containers, liters	250	250 900	900	1800
Control volume capacity, liters	50 200	50 200 500	100 500 1000	500 1000

Figure 6. Installation for experimental testing of the technology of filling products with WDPTC-5000

Figure 7. Check scales WDPTC-1000 and PWD-500

The scales used load cells of the SB4C4MI7.5 type with a nominal load of 5000 kg. To complete the scales in terms of electronic equipment, an ADC-M unit and a TVM weight terminal were used. Based on the tests carried out for the purpose of type approval, the VSK-10 were registered with the Federal Information Fund for Ensuring the Uniformity of Measurements. The limits of permissible error of the balance are given in Table 2. The use of this equipment makes it possible to weigh the control doses of water dispensed by the filling complex in the range from 40 to 10000 kg. A series of three test dips showed that the actual error of the scales in the range of 10-10000 kg when loaded lies in the range from minus 0.4 kg to 0.0 kg, and when unloaded - from minus 0.4 kg to plus 0.3 kg. The set with the VSM-10 includes a plastic tank with V = 10 m^3 and a submersible pump, the capacity of which in the real scheme was 6.5 m^3 / hour, which made it possible to pump 10 tons of water from the control tank to the STRC complex in 1.5 hours.

Table 2. Limits of permissible error of scales

Load range	*mpe*, kg
20 to 500 kg	±1,0
500 to 2000 kg	±2,0
2000 to 10000 kg	±3,0

The main task of research on experimental facilities is to determine the actual error in the dose delivery using the developed weighing equipment. Filling of spacecraft and booster blocks can be carried out: 1) at a constant pressure of the displacing gas in the refueling complex (filling tank); 2) with a constant gas mass in the refueling complex (the gas pipeline is closed). In the overwhelming majority of cases, refueling is performed at constant gas pressure. Option 2 is used for refueling small doses of propellant components into the tanks of small spacecraft or propulsion system (PS) for stabilization, orientation and launching (SOL) of the upper stage. On bench equipment, both refueling options can be tested. Let's consider the work according to the first option. The mass of the dispensed dose and the working excess pressure are set. The water temperature is measured. Corrections for the mass of the displaced air in the filling tank and the correction for the mass of displaced air from the control tank are calculated. The operator selects the number of the main channel, according to the indications of which the dose will be delivered, and enters it into the PC memory. During dispensing, the monitor screen displays indications of the main and reserve channels, and on digital displays in the «filling room» only the indications of the main channel are displayed, through which the dose is cut off. Let us denote by the letter I the readings of the main channel, and: Iinit - initial readings, Itek - current readings during dispensing, Ifin - final readings after the end of dispensing.

Consistently in accordance with the methodology substantiated and developed by V.M. Shulga, refueling operations are performed: 1) equipment test; 2) input of parameters (in this operation, the operator enters all the parameters of the refueling, in particular, MGD, MAIR.TRC and preliminaries - a part of the dose given out at a high flow rate); 3) filling of highways; 4) the passage of the mains (this operation is carried out until the complete disappearance of air bubbles in the transparent section of the issuing main); 5) control exposure; 6) delivery of a dose.

In parallel with the solution of this main task, the development of software is carried out, which is ultimately intended for the implementation of product filling technologies.On the bench equipment, the technology of weight dosing is being tested when refueling small-sized spacecraft. The minimum dose for these devices starts at 0.4 kg. For refueling small-sized devices, modular weighing devices (weighing device with dosing (WDD), weighing device with dosing and control container (WDDCC)) can be used, as well as mobile special control scales of the SKS type.The equipment is also used in the development of multi-portion weight dosing methods.The experimental development of work technologies in emergency situations deserves a separate discussion. These situations are also simulated on bench equipment, and the technology of work aimed at the successful completion of refueling is experimentally worked out.

3. The experimental studies of precision weighing systems of refuelling complexes of upper stages and spacecraft of space rockets

When solving the problem of increasing the accuracy of weighing rocket fuel components at the stage of experimental studies of precision weight measuring systems of refueling complexes of upper stages and spacecraft of space rockets, it is necessary to determine the main metrological characteristics of stationary weight measuring systems (SWMS). Weighing devices (WD) with a vertical capacity of V = 4.2 m3 are designed for filling the upper stage tanks at filling stations. The maximum load of the devices is limited to 3000 kg and 5500 kg for various types of aircraft WD. These loads correspond to the mass of fuel and oxidizer, which can be used to fill filling tanks with a volume of 4.2 m3 at a filling factor of 0.9. The absolute error of the WD did not exceed +0.1 kg ÷ -0.2 kg in the load range from 0 to 3000 kg during loading and unloading, while a slight nonlinearity of the readings was observed. A similar character of nonlinearity is also present in the characteristics of another type of aircraft WD with an absolute error of +0.1 kg ÷ -0.3 kg in the load range from 0 to 5500 kg during loading and unloading. In experimental studies, doses were dispensed using the mixed dosing method. The experimental research methodology was as follows: 500 kg of a component (fuel and oxidizer, respectively, for certain types of aircraft, were poured into the filling tank [9]. The platforms of the devices were loaded up to loads of 3000 and 5500 kg relative to the mass of the empty container. The main part of the dose was «given out» by removing the weights from the platform. The remaining 10 kg were dispensed as a component into a drain container installed on a check balance with Max = 250 kg, the error of which at loads of 10 ÷ 50 kg is ± 0.05 kg Doses of the component were given out with a constant mass of helium in the filling tank and with a closed drain line in the drain tank. Therefore, the actual mass of the delivered dose was equal to the sum of the mass of the weights and the readings of the control weights for the i-th dose. During experiments with fuel, it was found that there is an incomplete filling of the lines. This meant that there were stagnant zones in the highway system, due to the fact that the highways from the filling tank were built into an essential highway system. The introduction of the «vacuuming of mains» operation into the refueling technology made it possible to ensure complete filling of the mains. In the oxidizer filling system, the lines were filled completely without evacuation. Since two doses of the component from one filling container are sequentially filled into the upper stages tanks of space rockets, two doses of the component were also dispensed sequentially during the tests. Three dispenses of two doses for fuel and oxidizer were carried out. The results of issuing control doses are presented that the absolute error in the delivery of amylin doses lies in the range from 0.0 kg to minus 0.4 kg, and the dose cutoff error by pneumatic valves lies in the range from minus 0.1 to plus 0.1 kg.

In general, the experimental values of the dose delivery error are much less than the permissible error of refueling the of upper stages tanks, equal to ± 6 kg for a dose of 2000 kg, and less than the passport error of the aircraft WD in a special mode, equal to ± 2 kg. In relative units, the experimental values of the dose delivery error are within the range from 0.00% to + 0.02%.

In experiments on the «dose delivery» of a dose of 1000 kg using the aircraft VD, similar results are observed: the experimental error (0.0 ± -0.2 kg) is less than the permissible error of refueling (± 3 kg) and the WDDF passport error in special mode (± 1 kg). In relative units, the experimental values of the dose delivery error also lie in the range from 0.00% to minus 0.02%.

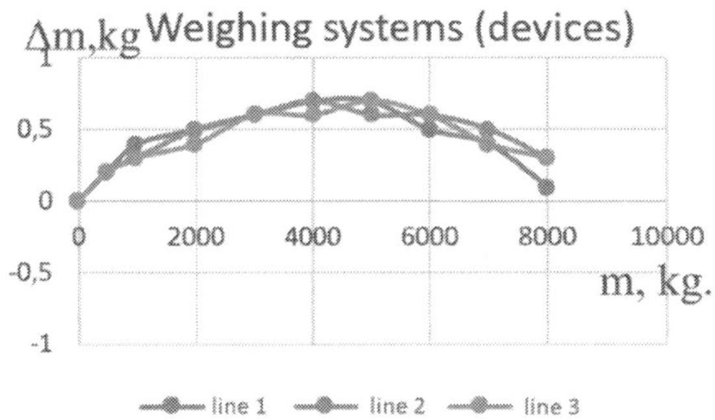

Figure 8. Deviation of readings WS (WD) from the nominal value of the mass of the weights

Experimental studies WS (WD) with a horizontal capacity V = 10 м3, designed for refueling upper stages tanks with naphthyl at the filling and neutralization station (FNS), the load range of which is limited by the capacity and is equal to 7000 kg., based on the dispensing doses weighing 6090 kg by the combined method: 6080 kg were "dispensed" by removing weights of class M1 from the platform, and the remaining 10 kg - with water into a control container installed on a SKS-400 control balance. Three days before the start of the active stage of experimental research on the platform WS (WD) 2000 kg weights were installed, which simulated the mass of the container. Then, 3 hours before the start of the dose of the propellant component, weights with a mass of 9000 kg were installed on the platform, simulating the intake of the component into the container. A container V = 120 l with water was also installed on the platform. The dose was cut off using a pneumatic valve [8]. Water was dispensed at a constant mass of air in the capacity; a correction was introduced for the mass of air displaced from the control capacity.

Averaging the error over 9 doses, starting with the second, gives a value of minus 0.54 kg, which is a systematic error. The dosing flow rate was 0.1 kg/sec. The standard deviation at a confidence level of 0.95 is equal to $S_{0,95} = \pm 0{,}1$ kg characterizes the random error due to the following components: random

IOP Conf. Series: Materials Science and Engineering 1159 (2021) 012044 doi:10.1088/1757-899X/1159/1/012044

error WS (WD) (dynamic and electromagnetic noise); random error due to valve operation. Thus, the absolute error in delivering a dose with a mass of 6090 kg, calculated as the sum of the systematic error and the standard deviation at a confidence level of 0.95, is equal to ±0.64 kg, or ±0.011% in relative units, which is in good agreement with the data shown in Figure 8.

4. Conclusion

In conclusion, it should be noted, that the developed and put into operation stationary weighing devices have an actual error, which is an order of magnitude less than the passport values of the permissible error limit. This gives grounds for increasing the accuracy of filling by approving the type of a single sample (or a series of WS) with higher metrological characteristics. Thus, the results of studies of stationary weighing devices developed and put into operation can serve as a basis for increasing the accuracy of weighing and dose delivery by implementing a scientific and methodological approach in accordance with [10] based on type approval of single samples WS (WD) (or a batch of aircraft WD), in operation, with higher metrological characteristics in the case of confirmation of these characteristics by experimental studies.

Designed, manufactured and tested in the course of scientific research and the above bench equipment formed the basis of a research laboratory created on the basis of the Department of Transport installations of MADI as part of the formation of a ground work for interaction of MADI with strategic industrial partners - AO M.V. Khrunichev (Joint Stock Company), AO Moven and AO TSENKI within the framework of the creation of a scientific and educational world level in terms of experimental development of equipment and scientific and methodological support of an innovative scientific and production center for research, development and technological work, projects and programs. In the research laboratory of the Department of Transport installations of MADI, along with the research described above, laboratory work of students and research of applicants, graduate students and doctoral students of MADI are also carried out.

References

[1] Denisov O E, Zolin A A, Ivanov M F, Kobyzev S B, Matveeva O P, Khlybov V F, Chugunkov V V 2016 *Launch equipment and technological processes of fuel preparation and rocket refueling: textbook. manual* (Moscow: Publishing house of MSTU named by N E Bauman)

[2] Vorobiev E V, Denisov O E and Kuznetsov V I 2014 *Design of special purpose vehicles: a tutorial* (Moscow: MADI)

[3] *Technological objects of ground infrastructure of rocket and space technology (engineering manual)* ed. I V Barmina (Moscow: Polygraph RPK, 2012)

[4] *Investigation of the actual metrological characteristics of precision weighing systems and the results of substantiation and development of recommendations to improve the accuracy of filling spacecraft tanks and upper stages of space rockets with propellant components: research report (concluding)* ed. V M Shulga (Moscow: MADI, 2019)

[5] Special control scales SKS. Specifications: TU 4274-034-45081993-13 (Moscow: Closed Joint Stock Company "Elves", 2013)

[6] GOST R 8.736-2011 *State system for ensuring the uniformity of measurements. Multiple direct measurements. Methods for processing measurement results.* Basic provisions (Moscow: Standartinform Publishing House, 2013)

[7] Burdun G D, Markov B N 1984 *Fundamentals of Metrology* (Moscow: Standards Publishing House)

[8] Shut-off valves. Specifications: P 96261–00.025 M1 ТУ (SPb., ZNS 3T, 2009)

[9] Shulga V M, Nazarov V N, Borisov V G, Lebedev A G, Runov V V 2014 *Determination of the mass of doses of propellant components using WDDF-5 and WDDF-10 devices: measurement technique* (recommendation No. FR.1.28.2014.19022) (Moscow: Publishing house of Rosstandart)

[10] *Federal Law of the Russian Federation* dated 26.06.08 No. 102-F3.

IOP Conf. Series: Materials Science and Engineering 1159 (2021) 012045 doi:10.1088/1757-899X/1159/1/012045

Production of modifying additives for highways

A D Kolosov, V V Kondrat'ev, R V Kononenko and E A Guseva

Irkutsk National Research Technical University, Lermontov Street, 83, Irkutsk, 664074, Russia

E-mail: kvv@istu.edu

Abstract. The article presents a description and operating parameters of the technological process for the production of modifying additives for concrete and asphalt concrete based on exhaust gas dust in the production of silicon by operations. Described is a foam product, which is a raw material for obtaining fullerene-like carbon, a modifying additive MD2, and a chamber product for spherical silicon dioxide, a modifying additive MD1. The regulated parameters have been worked out, the regulation limits and the norms of expenses are given. The development of a functional diagram is based on the analysis of the technological process, as an object of automation, in which the main control and controlled influences are distinguished, which must be regulated and maintained within specified limits, in order to obtain the final product with the required quality characteristics. The implementation of this scheme is presented in this work. It combines a flow diagram and technical means measuring, controlling and regulating these flows, in order to ultimately obtain products of the desired quality.

1. Introduction

At present, the involvement of waste in various industries [1-3] and concern for the ecology and health of the population [4-5] are gaining in popularity.

Dust from the gas cleaning system of silicon production [6-9], containing nanoparticles of silica and carbon, with the help of a hydraulic removal system through pipelines, goes to the sludge field of CJSC KREMNIY.

It seems rational to consider in parallel two options for taking material for processing for modifying additives for concrete and asphalt concrete.

According to the first, the material is taken directly from the line of the wet removal system through the receiving hopper (RH) and is sent directly to the flotation process.

The second option is associated with the release of the sludge field. The content of the sludge field is a dark gray suspension with a hygroscopic moisture content of more than 85%. At the site in front of the sludge field, work was organized to take a moistened product using chamber pumps and pump it into truck tanks.

Further, the moistened nanodispersed dust enters the technological operation for the preparation of sludge. The final product is nanodispersed dust, which enters the receiving hopper (RH) - at the initial stage of the scheme for the production of modifying additives for concrete and asphalt concrete.

As a result of processing nanodispersed dust, 2 modifiers are obtained. One contains carbon nanotubes, fundamental research and application in various fields of which are described in [10], the second - silicon dioxide, the use of which in concrete has been widely studied [11-15]. The use of

Content from this work may be used under the terms of the Creative Commons Attribution 3.0 licence. Any further distribution of this work must maintain attribution to the author(s) and the title of the work, journal citation and DOI.
Published under licence by IOP Publishing Ltd

these additives in asphalt concrete and road concretes contributes to an increase in their reliability, resistance to growing loads, climatic extremes, and humidity.

2. Description and operating parameters of the technological process for the production of modifying additives for concrete and asphalt concrete based on exhaust gas dust in the production of silicon by operations

General description of the technological process.

From the RH through the feeder auger (FA1), the rotation speed of which is set by the operator remotely, the initial product goes for mixing with water into the repulpator (RP). The water supply to the RP is carried out from the water tank WT1, which is a closed-type tank.

The obtained homogeneous pulp from the RP enters the flotation machine (FM). At the same time, flotation reagents (a mixture of kerosene and pine oil) and compressed air are supplied to the FM from the capacity of flotation reagents (CF) to carry out the flotation process. Here the pulp is divided into two products - foam and chamber.

The foam product is a raw material for obtaining fullerene-like carbon - modifying additive MD2.

The chamber product for spherical silicon dioxide - modifying additive MD1.

The chamber product is fed through the conveyor to the drum vacuum filter (F1), as a result of which the hygroscopic moisture content of the pulp at the outlet is 6-8%, the filtrate is returned to the WT1water tank.

Further, the pre-dried slurry enters the rotary calcining bake (B1), where the temperature is maintained at 600-660 °C. At the exit from the calcining furnace, the hygroscopic moisture content of the product is 2-3%.

This product is spherical silicon dioxide, a modifying additive MD1, which is fed to the filling line (FL1), adapted for filling the marketable product in big bags or in paper bags.

The frothy product of the flotation process enters the reactor (R) for washing with hydrofluoric acid, which comes from the acid capacity (AC1) in order to clean the MD2 concentrate from impurities SiO_2.

The enriched foam product enters the drum vacuum filter for drying. As a result of the operation, we obtain a paste-like concentrate, which is sent to the foam product washing tank (WT) for conditioning. From the water capacity (WC2), the flushing liquid through the dosing device enters the WT washing tank. Further, the conditioned pulp is sent by a screw feeder to the coagulator (C) for thickening the particles of the solid phase. The coagulant is supplied to the coagulation process through a dispenser from the coagulant tank (CT).

The amount of supplied pulp and coagulant, as well as the output of thickened pulp and sediment, is controlled within the limits indicated in Table 1.

The thickened slurry is fed to a vacuum filter by means of a pump, where it is dehydrated, the resulting paste-like concentrate is sent to drying, which results in a commercial product, fullerene-like carbon. The finished product is fed to the packaging complex for filling in big bags or paper bags.

The regulation limits and flow rates are given in Table 1.

2.1. Technological control standards

The development of a functional diagram is based on the analysis of the technological process, as an object of automation, in which the main control and controlled influences are distinguished, which must be regulated and maintained within specified limits, in order to obtain the final product with the required quality characteristics. The implementation of this scheme is presented in Table 1. It combines a flow pattern and technical means for measuring, controlling and regulating these flows, in order to ultimately obtain products of the desired quality.

Table 1. Technological control of parameters.

No	Parameter name	Position	Measurement limit	Instrument functions				
				Indications	Registration	Regulation	Control	Signaling
1	Furnace top temperature 1	TT 1a	600-650 °C					
2	Furnace bottom temperature 1	TT 2a	640-660 °C					
3	Furnace top temperature 2	TT 3a	350-400 °C					
4	Furnace bottom temperature 2	TT 4a	470-410 °C					
5	Air flow supplied to the cell	UIR 5a	10-15 Nl/m^3 of pulp	+	+			
6	Air line pressure	UIR 5a	0.2-0.4 MPa	+	+			
7	Water consumption from the network in WC1	FIRT 6a	0.2-1.3 m^3/h	+	+			
8	Water consumption from WC1 to RP	FIRT 7a	1.2-1.4 m^3/h	+	+			
9	Pulp consumption from RP to FM	FIRT 8a	1.4-1.5 m^3/h	+	+			
10	Consumption of flotation reagent from CF to FM	FIRT 9a	0.002-0.003 m^3/h	+	+			
11	Consumption of acid-washed foam product from R to F2	FIRT 10a	0.16-0.21 m^3/h	+	+			
12	Hydrofluoric acid consumption from AC to R	FIRT 11a	0.12-0.14 m^3/h	+	+			
13	Water consumption from WC2 to WT	FIRT 12a	0.9-1.1 m^3/h	+	+			
14	Consumption of washed foam product from WT to C	FIRT 13a	0.95-1.15 m^3/h	+	+			
15	Coalescer consumption from CT to C	FIRT 14a	0.003-0.004 m^3/h	+	+			
16	Water consumption from the network in WC2	FIRT 15a	0.2-1 m^3/h	+	+			
17	Raw material level in RH	LITR 16a	20-80 %	+	+			
18	Water level in WC1	LIT 17a	70-85 %	+				
19	Pulp level in RP	LITR 18a	60-85 %	+	+			
20	Flotation reagent level in CF	LITR 19a	25-95 %	+	+			
21	Pulp level in FM	LITR 20a	99 %	+	+			

22	The level of the foam product washed by the acid in the reactor	LITR 21a	40-70 %	+	+			
23	Level in the foam product rinsing tank	LITR 22a	60-85 %	+	+			
24	Coagulator level in CT	LIT 23a	25-95 %	+				
25	Hydrofluoric acid level in AC	LIT 24a	50-75 %	+				
26	Water level in WC2	LIT 25a	70-85 %	+				
27	Pulp level in C	LITR 26a	70-85 %	+	+			
28	Top level in F1	LA 27a					+	
29	Lower level in F1	LA 27б					+	
30	Upper level in F2	LA 28a					+	
31	Lower level in F2	LA 28б					+	
32	Upper level in F3	LA 29a					+	
33	Lower level in F3	LA 29б					+	
34	pH of hydrofluoric acid to AC	QIT 30б	3-5 pH	+				
35	Dust concentration SiO_2 in the air of the working area	QITA 31a	0.5 mg/m^3	+				+
36	Concentration of vapors of hydrofluoric acid in the air of the working area	QITA 32a	0.5 mg/m^3	+				+
37	Pulp density at the outlet of the repulpator	DIRT 33a	1060-1120 kg/m^3	+	+			
38	Density of the foam product at the outlet of the reactor	DIRT 34a	900-950 kg/m^3	+	+			
39	Density of the slurry at the outlet of the washing tank	DIRT 35a	950-980 kg/m^3	+	+			
40	Moisture content of SiO_2 after B1	MIRT 36б	2-3 %	+	+			
41	Humidity of CNT after B2	MIRT 16б	2-3 %	+	+			

2.2. Temperature measurement

To measure the temperature due to high temperatures in the furnaces (up to 750 °C), the technologist must turn on temperature meters, record the readings in the log, make sure that the readings are within the permissible values according to Table 1, if the maximum permissible values are exceeded, turn off the installation and inform the management.

2.3. Measuring pressure and air flow rate into the cell

The Rosemount-3051 SMV is designed to measure the flow rate and pressure of air supply to the cell.

When measuring the flow rate, standard orifice devices in accordance with GOST 8.586-2005 are used as primary transducers. Measurements are carried out in accordance with GOST 8.586-2005.

The technologist must turn on (if necessary) metering devices for pressure and air flow, record the readings in the log, make sure that the readings are within the permissible values according to Table 1, if the maximum permissible values are exceeded, turn off the installation and inform the management.

2.4. Flow measurement of liquids and slurries

To measure the flow rate of liquid and slurry, a Rosemount 8700 device is used. To measure the flow of liquids and slurries, the technologist must turn on (if necessary) metering devices, record the readings in a log, make sure that the readings are within the permissible values according to Table 1, if limit values turn off the unit and inform the management.

2.5. Level measurement

Metran-150 hydrostatic pressure (level) sensors are designed to measure the level in water and fluor tanks. To measure the level in the receiving hopper, a non-contact radar level gauge Rosemount 5600 is intended. To measure the level in a flotation machine, a guided wave level gauge Rosemount 5300 is intended.

To measure the level, the technologist must turn on (if necessary) metering devices, record the readings in the log, make sure that the readings are within the permissible values according to Table 1, if the maximum permissible values are exceeded, turn off the installation and inform the management.

2.6. Level alarm

The Rosemount 2100 Vibrating Level Switches are designed to alarm the level when the drum filter bucket is overfilled.

To signal a level overflow, the technologist must turn on (if necessary) the metering device, in case of an alarm, turn off the installation and inform the management.

2.7. PH measurement

A Yokogawa PH450G device is designed to measure ph.

To measure ph, the technologist must turn on (if necessary) the metering device, record the readings in the log, make sure that the readings are within the permissible values according to Table 1, if the maximum permissible values are exceeded, turn off the installation and inform the management.

2.8. Concentration measurement

To analyze the air in the working area for the concentration of silicon dioxide dust and hydrofluoric acid vapors, and to signal if their permissible values are exceeded, a universal stationary gas analyzer GANK-4C (R) is intended, using the appropriate chemical cassettes.

To measure the density, a device manufactured by LLC "Piezoelectric" Density meter 804 was selected.

The moisture meter M-Sens 2 was chosen to measure the moisture content of bulk media.

To measure concentrations, the technologist must turn on (if necessary) the metering device, record the readings in the log, make sure that the readings are within the permissible values according to Table 1, if the maximum permissible values are exceeded, turn off the installation and inform the management.

3. Conclusion

The paper presents the results of the development of a resource-saving technology for the production of modifiers for asphalt concrete from metallurgical waste. The regulated parameters have been worked out, the regulation limits and the norms of expenses are given. A functional diagram has been developed based on the analysis of the technological process as an automation object, in which the main control and controlled influences are distinguished, which must be regulated and maintained within the specified limits, in order to obtain the final product with the required quality characteristics.

One of the main directions of further improvement of this technology is to reduce the cost of the resulting products, automate production, as well as expand the scope of application of the obtained modifiers in the construction industry.

References

[1] Lobanov D A, Sheshukov O Y, Egiazaryan D K, Nekrasov I V, Ovchinnikova L A 2020 EAF and LF slag co-processing: Study for wasteless utilization *AIP Conference Proceedings* **2313** 050019

[2] Sheshukov O, Mikheenkov M, Vedmid' L, Nekrasov I, Egiazaryan D 2020 Mechanism of ion-diffusion solid-phase reduction of iron oxides of technogenic origin in the presence of the liquid phase and without it *Metals* **10(12) 1564** 1–12

[3] Nekrasov I V, Sheshukov O Y, Metelkin A A, Sivtsov A V and Tsymbalist M M 2016 Slag conditions in electrosmelting: A review *Steel in Translation* **46(6)** 435-442

[4] Konyuhov V Yu, Gladkih A M, Galyautdinov I I and Kiseleva T Yu 2019 Power industry of future is renewable sources *IOP Conference Series: Earth and Environmental Science* **378** 012047

[5] Gladkih A M, V Yu Konyuhov, Galyautdinov I I and Shchadova E I 2019 Green building as a tool of energy saving *IOP Conference Series: Earth and Environmental Science* **350** 012032

[6] Sivtsov A V, Yolkin K S, Kashlev I M and Karlina A I 2020 Processes in the Charge and Hearth Zones of Furnace Working Spaces and Problems in Controlling the Batch Dosing Mode during the Smelting of Industrial Silicon and High-Silicon Ferroalloys *Metallurgist* **64(5-6)** 396-403

[7] Sivtsov A V, Elkin K S, Pankov V A and Karlina A I 2021 Specific Features of the Electric Mode of the Technological Process of Smelting of Commercial Silicon *Metallugrist* **64** 923-930

[8] Yolkin K S, Yolkin D K, Nemarov A A, Sysoev I A and Karlina A I 2018 Conduct of reduction smelting of metallic silicon: Theory and practice *IOP Conference Series: Materials Science and Engineering* **411** 012029

[9] Yolkin K S, Yolkin D K, Kolosov A D, Ivanov N A, Shtayger M G 2018 Technologies, which allow to reduce an impact of metal silicon production on the environment *IOP Conference Series: Materials Science and Engineering* **411(1)** 012028

[10] Ponomarev A, Steshenko D and Rassokhin A 2018 Development of technology for production of fire-resistant nanocomposite constructional rebar and structural elements based on it *MATEC Web of Conferences* **245** 04001

[11] Svintsov A P, Galishnikova V V and Stashevskaya N A 2020 Dataset on the effect of nano-modified additives of concrete mixes technological properties for winter concreting *Data in Brief* **31** 105756

[12] Kondratiev V V, Karlina A I, Guseva E A, Konstantinova M V, Gorovoy V O 2018 Structure of Enriched Ultradisperse Wastes of Silicon Production and Concretes Modified by them *IOP Conference Series: Materials Science and Engineering* **463(4)** 042064

[13] Kondratiev V V, Karlina A I, Guseva E A, Konstantinova M V, Kleshnin A A 2018 Processing and Application of Ultra disperse Wastes of Silicon Production in Construction *IOP Conference Series: Materials Science and Engineering* **463(3)** 032068

[14] Petrushenko I K 2018 DFT Calculations of Hydrogen Adsorption inside Single-Walled Carbon Nanotubes *Advances in Materials Science and Engineering* **2018** 9876015

[15] Petrushenko I K and Petrushenko K B 2019 Physical adsorption of hydrogen molecules on single-walled carbon nanotubes and carbon-boron-nitrogen heteronanotubes: A comparative DFT study *Vacuum* **167** 280-286

Nanostructures based on carbon and silicon dioxide to improve the properties of building and structural materials

V V Kondrat'ev, M V Konstantinova, R V Kononenko and A D Kolosov

Irkutsk National Research Technical University, Lermontov Street, 83, Irkutsk, 664074, Russia

E-mail: kvv@istu.edu

Abstract. The article presents hardware and technological solutions for obtaining a product containing silicon dioxide from nanodispersed dust of gas cleaning of silicon production, intended to improve the properties of building and structural materials. To obtain nanostructures of carbon and silicon dioxide of the required quality, allowing to improve the properties of building and structural materials, a complex resource-saving technology of their high-tech production was developed. The necessary characteristics of nanodispersed modifying additives have been determined. A technological scheme has been developed; technological operations of thermal vortex enrichment are described. The material balance of thermal vortex enrichment is given.

1. Introduction
Development of technologies at present are energy and resource conservation [1-3], green technologies [4]. Development of technologies allowing to improve the quality of associated mineral raw materials without prejudice to the main technological process of silicon production [5-8] makes it possible to obtain nanodispersed dust (ND) containing carbon in the form of carbon nanotubes (CNTs), fullerene and silicon dioxide. The modifying effect of CNTs [9-10] and silicon dioxide [11-14] is becoming increasingly popular in various industries, including construction. One of the promising directions for the development of the use of these nanoparticles is road construction.

2. Developed product
This product was developed as part of a project to create a high-tech production of nanostructures based on carbon and silicon dioxide.

2.1. Characteristic of output product
The output product described in this article is a powdery concentrate of nanostructures containing silicon dioxide.

Content from this work may be used under the terms of the Creative Commons Attribution 3.0 licence. Any further distribution of this work must maintain attribution to the author(s) and the title of the work, journal citation and DOI.
Published under licence by IOP Publishing Ltd

Table 1. Characteristics of the output product containing silicon dioxide.

Indicator name	Indicator values	Method of measurement
Mass fraction, %:		
SiO_2	not less than 98	EN 196-2
CaO	no more than 0.3	EN 451-1
SO_3	-	EN 196-2
K_2O	no more than 0.3	EN 196-2
Na_2O	no more than 0.1	EN 196-2
Fe_2O_3	no more than 0.1	GOST 2642.5-97
Al_2O_3	no more than 0.3	GOST 2642.4-86
MgO	no more than 0.2	GOST 2642.8-97
P_2O_5	-	GOST 2642.10-86
Cl	-	EN 196-2
H_2O	no more than 0.3	GOST 2642.1-86
SiC	-	GOST 26564.1-85
C_{free}	-	GOST 2642.15-97
pH	7.5±0.5	GOST 2642
Mass fraction of losses on ignition at 950 °C,%	no more than 0.8	EN 196-2
Specific surface (according to BET), m^2/g	not less than 16	ISO 9277
Bulk density, kg/m^3	to 360	GOST R 54246-2010
Granulometric composition of the packaged product (after long-term storage and self-coagulation processes), %	not less than 63.5	
Individual nanoparticles:		GOST R 8.755-2011
Agglomerates: - small, more than 1 micron;	no more than 30.0	
- medium, more than 10 microns;	no more than 5.0	
- large, more than 45 microns	no more than 1.5	

2.2. Technological scheme of thermal vortex enrichment

The general process flow diagram is shown in Figure 1.

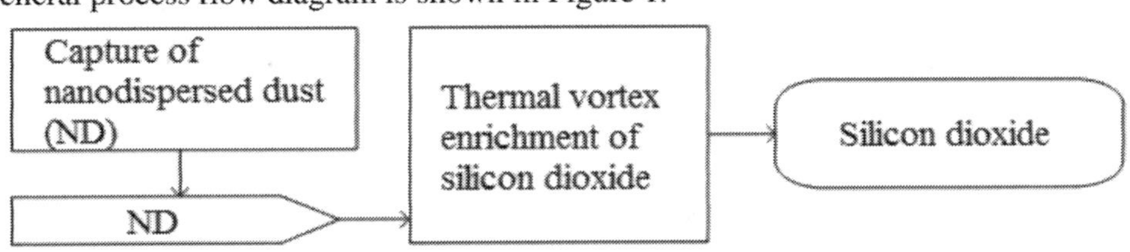

Figure 1. Technological scheme of thermal vortex enrichment.

2.3. Characteristic of raw materials

Nanodispersed dust for performing thermal vortex enrichment is a product of collection in the gas cleaning system of silicon production.

Table 2. Quantitative composition of nanodispersed dust.

Parameter	Content, %
Dust chemical composition	
- SiO_2, not less than	84.3
- Fe_2O_3, no more than	0.6
- Al_2O_3, no more than	1.0
- CaO, no more than	2.1
- C, no more than	11.5

To perform thermal vortex enrichment, materials with the characteristics presented in Table 3 are used.

Table 3. Characteristics of the materials used.

No P/p.	Name of raw materials, materials, intermediate products	GOST, OST, STP, TU, regulations or methods for the preparation of raw materials
1	Natural gas	GOST 5542-2014

2.4. Description of the thermal vortex enrichment

The process is a thermal vortex enrichment of nanodispersed dust. The dust is fed tangentially to the bottom of the furnace. Burning gas is also tangentially fed through the second inlet pipe. Due to tangential input and natural convection, an upward vortex with a temperature of 700-850 °C is formed in the working chamber of the furnace. At a given temperature, the carbon contained in the initial nanodispersed dust is oxidized to carbon dioxide. Decarbonized nanodispersed dust together with carbon dioxide is taken from the upper part of the furnace. Decarburized nanodispersed dust is a product of MD 1. A description of the implementation of thermal vortex enrichment is given in Table 4.

Table 4. Description of the implementation of thermal eddy enrichment.

No P/p.	Action	Result
1	Check the serviceability of the equipment, the integrity of the connections and branch pipes	Thermal vortex enrichment equipment is ready for operation
2	Switch on the supply of burning gas to the working chamber of the furnace	The supply of burning gas to the working chamber of the furnace is switched on
3	Wait until the working chamber warms up to operating temperature, determined with a thermometer	Working chamber is ready for thermal enrichment of nanodispersed dust
4	Switch on the supply of nanodispersed dust to the working chamber of the furnace	The supply of nanodispersed dust to the working chamber of the furnace will turn on
5	Wait until the end of thermal enrichment of the required amount of nanodispersed dust, determined visually by the amount of the output product	The process of thermal enrichment of nanodispersed dust is completed
6	Turn off the gas supply	Thermal vortex enrichment completed

2.5. Standardized technological parameters

Standardized technological parameters of thermal vortex enrichment are shown in Table 5. Control of parameters is carried out using the measuring equipment of the experimental stand. The layout of the technological equipment is shown in Figure 2.

Table 5. Parameters of the thermal vortex enrichment process.

No P/p.	Parameter	Working value	Limit values
1	Temperature, °C	700 – 850	650 – 900

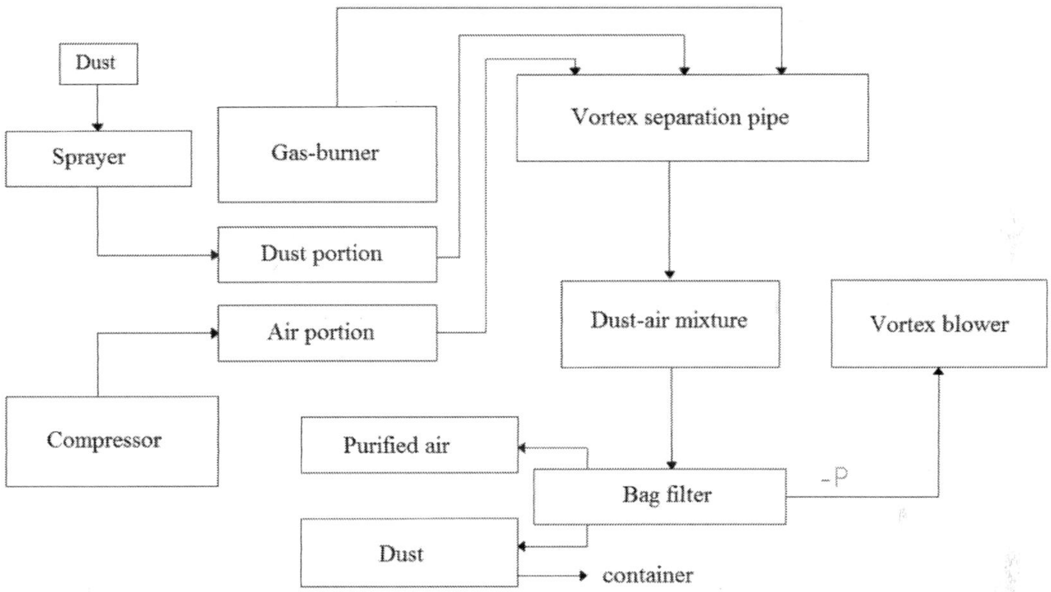

Figure 2. Layout of technological equipment.

2.6. Material balance

The material balance of thermal vortex enrichment is shown in Table 6.

Table 6. Material balance of thermal vortex enrichment.

Parish		Consumption	
Nanodispersed dust, kg	1	Concentrate MD1, kg	0.85
Natural gas, kg	1	Carbon dioxide*, kg	1.15

* Carbon dioxide consumption is based on pure carbon.

The technological process, which is the implementation of thermal vortex enrichment, does not produce waste that requires neutralization.

3. Conclusion

The developed technology of thermal vortex enrichment of dust from gas cleaning of silicon production allows not only to improve the ecological situation, to increase the profitability of the main production, but also to obtain innovative nanostructures for modifying road materials, concrete and asphalt concrete. The requirements for the quality of the produced modifying nanoparticles and the quality of the feedstock and materials used for their production from the waste gases of an ore-thermal furnace for silicon production have been established.

References

[1] Bulatov Y, Kryukov A, Suslov K, Shamarova N 2019 Ensuring postemergency modes stability in power supply systems equipped with distributed generation plants *Proceedings of the 10th International Scientific Symposium on Electrical Power Engineering, ELEKTROENERGETIKA 2019* p 38-42

[2] Lombardi P, Arendarski B, Suslov K, Shamarova N, Sokolnikova P, Pantaleo A M, Komarnicki P 2018 A Net-Zero Energy System Solution for Russian Rural Communities *E3S Web of Conferences* **69** 01013

[3] Lobanov D A, Sheshukov O Y, Egiazaryan D K, Nekrasov I V, Ovchinnikova L A 2020 EAF and LF slag co-processing: Study for wasteless utilization *AIP Conference Proceedings* **2313** 050019

[4] Gladkih A M, V Yu Konyuhov, Galyautdinov I I and Shchadova E I 2019 Green building as a tool of energy saving *IOP Conference Series: Earth and Environmental Science* **350** 012032

[5] Sivtsov A V, Yolkin K S, Kashlev I M and Karlina A I 2020 Processes in the Charge and Hearth Zones of Furnace Working Spaces and Problems in Controlling the Batch Dosing Mode during the Smelting of Industrial Silicon and High-Silicon Ferroalloys *Metallurgist* **64(5-6)** 396-403

[6] Sivtsov A V, Elkin K S, Pankov V A and Karlina A I 2021 Specific Features of the Electric Mode of the Technological Process of Smelting of Commercial Silicon *Metallugrist* **64** 923-930

[7] Yolkin K S, Yolkin D K, Nemarov A A, Sysoev I A and Karlina A I 2018 Conduct of reduction smelting of metallic silicon: Theory and practice *IOP Conference Series: Materials Science and Engineering* **411** 012029

[8] Yolkin K S, Yolkin D K, Kolosov A D, Ivanov N A, Shtayger M G 2018 Technologies, which allow to reduce an impact of metal silicon production on the environment *IOP Conference Series: Materials Science and Engineering* **411(1)** 012028

[9] Petrushenko I K 2018 DFT Calculations of Hydrogen Adsorption inside Single-Walled Carbon Nanotubes *Advances in Materials Science and Engineering* **2018** 9876015

[10] Petrushenko I K and Petrushenko K B 2019 Physical adsorption of hydrogen molecules on single-walled carbon nanotubes and carbon-boron-nitrogen heteronanotubes: A comparative DFT study *Vacuum* **167** 280-286

[11] Ponomarev A, Steshenko D and Rassokhin A 2018 Development of technology for production of fire-resistant nanocomposite constructional rebar and structural elements based on it *MATEC Web of Conferences* **245** 04001

[12] Svintsov A P, Galishnikova V V and Stashevskaya N A 2020 Dataset on the effect of nano-modified additives of concrete mixes technological properties for winter concreting *Data in Brief* **31** 105756

[13] Kondratiev V V, Karlina A I, Guseva E A, Konstantinova M V, Gorovoy V O 2018 Structure of Enriched Ultradisperse Wastes of Silicon Production and Concretes Modified by them *IOP Conference Series: Materials Science and Engineering* **463(4)** 042064

[14] Kondratiev V V, Karlina A I, Guseva E A, Konstantinova M V, Kleshnin A A 2018 Processing and Application of Ultra disperse Wastes of Silicon Production in Construction *IOP Conference Series: Materials Science and Engineering* **463(3)** 032068

Technology for the production of modifying additives for concrete and asphalt concrete

A S Sukhanov, A D Kolosov, V V Kondrat'ev and R V Kononenko

Irkutsk National Research Technical University, Lermontov Street, 83, Irkutsk, 664074, Russia

E-mail: kvv@istu.edu

Abstract. The article presents the results of the development of modifying additives for concretes and asphalt concrete, consisting of spherical SiO2 and fulleron, extracted from silicon production waste. The use of these additives can reduce the destruction of asphalt concrete pavements, increase water and frost resistance. The proposed technology is energy- and resource-saving, converting metallurgical production wastes into the class of "associated mineral raw materials" and improving the environmental situation in regions with the presence of metallurgical industries. The paper presents studies of the feedstock and characteristics of waste gases of silicon production, the characteristics of the main technological equipment for the production of modifiers, the instrumental - technological scheme of the production process, the chemical composition and quality indicators of the resulting spherical silicon dioxide. The order and relationship of the processes that make up the technology of the associated additives based on production of silicon have been determined. The economic effect of the use of these modifiers in asphalt concrete pavements is achieved by increasing the durability and increasing the turnaround time.

1. Introduction

One of the most common types of road surfaces is asphalt concrete. In recent years, the speed and traffic density of vehicles has increased. This leads to the destruction of asphalt concrete pavements and, as a result, to the need to make repairs earlier than the standard terms. One of the main reasons for the destruction of road surfaces is the insufficient water and frost resistance of asphalt concrete. However, one of the most effective ways is the use of modifying additives. In modified asphalt concrete, special compositions are used as a binder, including petroleum bitumen and a modifier. Modifiers are additives that change the physical and mechanical properties and structure of asphalt concrete in the right direction.

One of the variants of such modifiers is silicon dioxide [1-5], the other is carbon nanotubes and fullerene-like carbon [6-7]. Both of these components are contained in the dust of the gas purification plant of silicon production [8-11]. There are solutions for the automation of the technological process for the production of such modifiers. It is also worth noting the high importance of creating modifying additives specifically from production waste in order to improve the environmental situation [12].

Content from this work may be used under the terms of the Creative Commons Attribution 3.0 licence. Any further distribution of this work must maintain attribution to the author(s) and the title of the work, journal citation and DOI.
Published under licence by IOP Publishing Ltd

2. Characteristics of raw materials, basic technological materials and equipment

The raw material for the associated production of modifying additives for concrete and asphalt concrete is dust from the gas cleaning system, waste gases from the ore-thermal furnace. The characteristics of waste gases from the production of silicon in an ore-thermal furnace are shown in Table 1.

Table 1. Characteristics of gases removed from ore-thermal furnaces.

P/p No.	Parameter name	Value
1	Power of ore thermal furnaces (RTP), MVA	16.5
2	Volumes of gases removed from the furnace umbrella, nm^3/h	200 000
3	Temperature of gases removed from the furnace umbrella, °C	230-260 peak to 350
4	Volumes of gases removed from the taphole of the furnace, nm^3/h	70 000
5	Temperature of gases removed from the taphole of the furnace, °C	60-80
6	Dust content of gases removed from the taphole of the furnace, g/nm^3	0.697
7	Concentration of pollutants in gases at the outlet of the furnace compartment, g nm ^{3}:	
	- of dust SiO_2	3.0
	- SO_2	0.042
	- NO_x	0.04
	- CO	0.122
8	Chemical composition of dust, %:	
	- SiO_2	84.3
	- Fe_2O_3	0.6
	- Al_2O_3	1.0
	- CaO	2.1
	- C_{free}	12.0
9	Dispersed composition of dust, %:	
	- less than 30 microns	73
	- more than 30 microns	27
10	Physical properties of dust:	
	- bulk density, g/dm^3	200
	- angle of repose, degrees	
	static	72
	dynamic	52
	- abrasiveness	yes
	- explosion and fire hazard	no
	- wettability	97
	- specific electrical resistance (resistivity), Ohm*m	□10^{11}

2.1. *Characteristics of the main equipment*

The equipment used in the associated production of modifying additives for concrete and asphalt concrete based on the dust of the gas cleaning system is shown in Table 2.

Table 2. The equipment used in the associated production of modifying additives for concrete and asphalt concrete based on the dust of the gas cleaning system in the production of silicon.

Pos.	Name	Destination
RH	Receiving hopper	Feeding the initial dust by the auger into the RP
WC1	Water capacity	Water storage and supply to RP
RP	Repulpator	Mixing of initial dust from PH with water from WC1
FM	Flotation machine	Flotation of nanotubes and deposition of SiO_2 nanoparticles
CF	Capacity of flotation reagents	Storage and supply of flotation reagents in FM
F1	Drum vacuum filter	Primary drying
B1	Rotary calcining bake 1	Calcining the chamber product from F1
R	Reactor	Washing of foam product from FM with hydrofluoric acid from AC
AC	Acid capacity	Storage and supply of hydrofluoric acid in R
WC2	Water capacity	Storage and supply of water to the power supply unit
F2	Drum vacuum filter	Separation of acid and foam after R
WT	Foam product washing tank	Washing the dried foam product
C	Coagulator	Thickening of solid phase
CT	Coagulant tank	Coagulant supply in C
F3	Drum vacuum filter	Drying of carbon product
B2	Bake	Calcining the carbon-containing product
FL1, FL2	Filling lines	Packaging products

2.2. Hardware-technological scheme

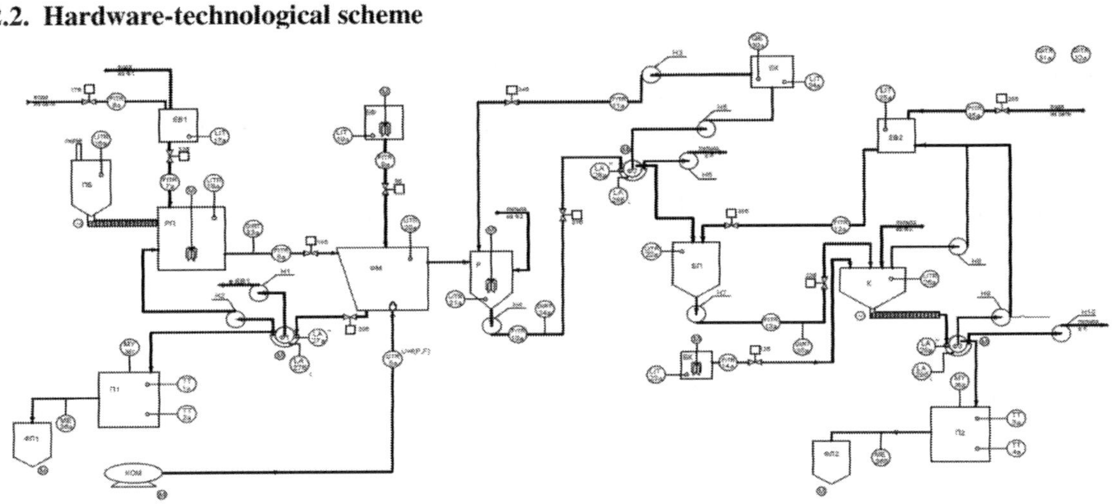

Figure 1. Technological scheme for the production of modifying additives for concrete and asphalt concrete based on dust from waste gases in the production of silicon.

IOP Conf. Series: Materials Science and Engineering 1159 (2021) 012047 doi:10.1088/1757-899X/1159/1/012047

Table 3. Chemical composition and quality indicators of the nodular silicon dioxide of the modifying additive.

Indicator name	Indicator values	Method of measurement
Mass fraction,%:		
SiO_2	not less than 98	EN 196-2
CaO	no more than 0.3	EN 451-1
SO_3	-	EN 196-2
K_2O	no more than 0.3	EN 196-2
Na_2O	no more than 0.1	EN 196-2
Fe_2O_3	no more than 0.1	GOST 2642.5-97
Al_2O_3	no more than 0.3	GOST 2642.4-86
MgO	no more than 0.2	GOST 2642.8-97
P_2O_5	-	GOST 2642.10-86
Cl	-	EN 196-2
H_2O	no more than 0.3	GOST 2642.1-86
SiC	-	GOST 26564.1-85
C_{free}	-	GOST 2642.15-97
pH	7.5±0.5	GOST 2642
Mass fraction of losses on ignition at 950°C, %	no more than 0.8	EN 196-2
Surface area (by BET), m^2/g	not less than 16	ISO 9277
Bulk density, kg/m^3	to 360	GOST R 54246-2010

The technological process must comply with the requirement:

- POT RM 016-2001 "Interindustry rules on labor protection (safety rules) when working in electrical installations."

Measures to ensure industrial fire safety and working conditions must comply with the legislative norms and requirements of the Russian Federation.

The equipment used must ensure the safety of work during operation and repair, comply with the requirements.

3. Conclusion

Developed technology:

- describes the technology for the production of modifying additives for concrete and asphalt concrete based on dust from waste gases in the production of silicon;
- establishes requirements for the quality of raw materials and materials used for the production of modifying additives from the waste gases of the ore-thermal furnace for silicon production;
- determines the order and interrelation of the processes that make up the technology of the associated production of modifying additives based on carbon in the production of silicon.

The developed technology contains provisions, the implementation of which is mandatory for all technological personnel of the workshop for the production of modifying additives for concrete and asphalt concrete. Additives produced using this technology can act as modifiers for concrete and asphalt concrete of highways.

References

[1] Ponomarev A, Steshenko D and Rassokhin A 2018 Development of technology for production of fire-resistant nanocomposite constructional rebar and structural elements based on it *MATEC Web of Conferences* **245** 04001

[2] Svintsov A P, Galishnikova V V and Stashevskaya N A 2020 Dataset on the effect of nano-modified additives of concrete mixes technological properties for winter concreting *Data in Brief* **31** 105756

[3] Galishnikova V V, Abdo Sh and Fawzy A M 2020 Influence of silica fume on the pervious concrete with different levels of recycled aggregates *Magazine of Civil Engineering* **93** 71-82

[4] Kondratiev V V, Karlina A I, Guseva E A, Konstantinova M V, Gorovoy V O 2018 Structure of Enriched Ultradisperse Wastes of Silicon Production and Concretes Modified by them *IOP Conference Series: Materials Science and Engineering* **463(4)** 042064

[5] Kondratiev V V, Karlina A I, Guseva E A, Konstantinova M V, Kleshnin A A 2018 Processing and Application of Ultra disperse Wastes of Silicon Production in Construction *IOP Conference Series: Materials Science and Engineering* **463(3)** 032068

[6] Petrushenko I K 2018 DFT Calculations of Hydrogen Adsorption inside Single-Walled Carbon Nanotubes *Advances in Materials Science and Engineering* **2018** 9876015

[7] Petrushenko I K and Petrushenko K B 2019 Physical adsorption of hydrogen molecules on single-walled carbon nanotubes and carbon-boron-nitrogen heteronanotubes: A comparative DFT study *Vacuum* **167** 280-286

[8] Sivtsov A V, Yolkin K S, Kashlev I M and Karlina A I 2020 Processes in the Charge and Hearth Zones of Furnace Working Spaces and Problems in Controlling the Batch Dosing Mode during the Smelting of Industrial Silicon and High-Silicon Ferroalloys *Metallurgist* **64(5-6)** 396-403

[9] Sivtsov A V, Elkin K S, Pankov V A and Karlina A I 2021 Specific Features of the Electric Mode of the Technological Process of Smelting of Commercial Silicon *Metallugrist* **64** 923-930

[10] Yolkin K S, Yolkin D K, Nemarov A A, Sysoev I A and Karlina A I 2018 Conduct of reduction smelting of metallic silicon: Theory and practice *IOP Conference Series: Materials Science and Engineering* **411** 012029

[11] Yolkin K S, Yolkin D K, Kolosov A D, Ivanov N A, Shtayger M G 2018 Technologies, which allow to reduce an impact of metal silicon production on the environment *IOP Conference Series: Materials Science and Engineering* **411(1)** 012028

[12] Lobanov D A, Sheshukov O Y, Egiazaryan D K, Nekrasov I V, Ovchinnikova L A 2020 EAF and LF slag co-processing: Study for wasteless utilization *AIP Conference Proceedings* **2313** 050019

Development of an algorithm for the introduction of innovative technologies in the construction of roads

I V Dudina[1], Y A Skorobogatova[2], O V Arkhipkin[3], and A S Bovkun[4]

[1] Bratsk State University, 40, Makarenko, Bratsk, 665709, Russian Federation,
[2] Baikal State University, 11 Lenin St., Irkutsk, 664003, Russian Federation,
[3] Irkutsk State University,1, Karl Marx St., Irkutsk, 664003, Russian Federation,
[4] Irkutsk National Research Technical University, 83, Lermontov str., Irkutsk, 664074, Russian Federation

E-mail: Bovas87@yandex.ru

Abstract: This article provides an analysis of the current state of the road industry in Russia. There are factors that increase the demand for innovative road construction technologies. The authors have developed an algorithm for the introduction of innovative technologies in road construction. The algorithm developed by the authors will allow road construction enterprises to make informed decisions in choosing the directions of their development. It also enables an enterprise to more clearly identify its market positions and take measures to improve them, solve problems and increase the efficiency of its activities.

1. Introduction

Roads are a public good because anyone can use them. The quality of state and municipal management is assessed by the condition of the roads. The development of the economy of the region and the country directly depends on the development of the road network in a given area. Federal, regional and municipal authorities exercise control over this area of activity and its financing [1].

At the same time, it should be especially noted that the motorization of the population has grown significantly over the past 10-15 years. In 2007–2019 the number of own cars increased 1.6 times per 1,000 people. The number of public service buses has grown 2.4 times per 100 thousand people. Trucking industry grew 1.3 times (in 2019, the share of this transport in the total freight turnover reached almost 5%). All this increases the load on the country's road network. The study showed a decrease in bus passenger traffic. It is mainly due to the accelerated development of air transport, which is used by the population for the farthest movements across the country. The number of road traffic accidents, which are accompanied by injuries and deaths, remains quite high (although in 2007-2019 it was possible to reduce the value of this indicator by 2 times. This level is still noticeably higher than in most developed countries of the world) [2].

2. Theory

Currently, our country ranks among the ten countries with the worst roads. Each Russian region is funded differently. Therefore, it should be noted that the quality of the road depends on the well-being of the region. So, for example, Moscow and St. Petersburg have the most well-groomed roads in contrast to other parts of Russia [3]. For example, the annual volume of the road construction market

Content from this work may be used under the terms of the Creative Commons Attribution 3.0 licence. Any further distribution of this work must maintain attribution to the author(s) and the title of the work, journal citation and DOI.
Published under licence by IOP Publishing Ltd

of St. Petersburg reaches 15 billion rubles. The northern capital spends an average of 3-4 billion rubles on major repairs, 6-7 billion rubles are spent on patching roads, and the rest of the money goes to construction [4].

The Russian leadership pays close attention to this problem. To improve the road situation in the country, the Federal Target Program "Development of the Transport System (2010-2020)" was approved. The total mileage of roads planned for reconstruction is 19,800 km. [5, p.39]

Table 1 shows the dynamics of the construction and repair of highways in Russia and the volume of financial support for these works over the past 3 years.

Table 1. Dynamics of construction and repair of highways in Russia and the volume of financial support for these works [6]

Road budget line	year		
	2017	2018	2019
Repaired public roads with hard surface, thousand km, total	19,20	19,29	23,47
federal significance	8,84	8,44	7,63
regional or inter-municipal significance	10,36	10,86	15,84
The total amount of financial support for the implementation of projects (programs) of the State Program of the Russian Federation "Development of the transport system", billion rubles.	921,3	954,9	1509,0
Total amount of financing for the national project "Safe and High-Quality Roads", billion rubles.			589,5
Total funding	430.2	575.9	548.1

Russia is a unique country that has individual climatic and territorial characteristics, seasonal freezing of soil, different population density and traffic intensity. All this requires the use of innovative developments of a different nature [6].

The following factors explain the need for innovative road construction technologies:

- a constant increase in the number of cars, an increase in the share of passenger cars with significant dynamic characteristics and trucks with increased axle loads. This is necessary to improve the consumer properties of the road and its quality [7];
- significant motorization and mobility of the population, the intensity of traffic flows. These indicators can significantly increase the level of road congestion and reduce traffic congestion that occurs in large cities. This is necessary in order to use improved traffic management systems and to develop highways and bypass roads that meet international standards;
- the high cost of all road-building materials and modern high-performance equipment at the same time with an increase in the repair and overhaul of roads. It will improve the pricing methods of road construction;
- the use of innovative ideas and projects for the construction and maintenance of highways. This contributes to the improvement of the quality characteristics of the roads and the quality of road works.;
- substantial repair and construction of district roads to connect settlements and road networks. This requires the study and use of economically viable technologies and materials, the preparation of new standards for the design, construction and operation of such roads.;
- basic consideration of traffic safety and environmental regulations in road construction.

It is obvious that the lack of innovative scientific activity will not give the harmonious development of all significant industries [8, 9].

3. Statement of the problem

Based on the analysis of the innovative activity of road construction in the Russian Federation, we present a number of problems:

- the main consumers of innovative products do not consider the application of innovations as one of the priority tasks of their practice;
- these properties have a direct impact on the construction cost of roads and the required amount of costs for their maintenance during operation;
- low length of reconstructed, under construction and repaired roads;
- Russian annual costs for the construction of one kilometer of roads are higher than foreign ones.

The main task is to develop an algorithm for the introduction of innovative technologies in road construction.

4. Results

As Russian Deputy Prime Minister Marat Khusnullin noted, in 2020 the volume of road infrastructure construction grew by 14% despite the coronavirus pandemic [10]. In the spring of 2020, this was made possible by the decision to continue road construction despite the epidemic. In 2020, Russian regions received almost 255 billion rubles for the construction and repair of roads as a part of the Safe and High-Quality Highway National Project. The mechanism of temporarily free funds allowed us to achieve such high rates. With the help of it, the regions received an additional 100 billion rubles for the repair and construction of roads.

In the short term, we can say that the coronavirus has positively influenced the current state and development of the road construction industry due to additional financial investments in this industry. However, in the long term, funding for the road construction market will be reduced. In 2020, the country's GDP, according to the Bank of Russia, will decrease by 4-5%. The fall in GDP caused by:

- final household consumption fell due to lower real incomes and restrictions imposed to contain the spread of coronavirus [11].
- the decline in investment activity of companies was reflected in the fall in investment in fixed assets.
- decline in industrial production.
- decrease in gross fixed capital formation.
- increase in the unemployment rate, etc.

It can be predicted that due to the fall in the country's GDP, the Government will not provide additional funding to the road industry. In the future, the growth rate of the road industry can range from 1% to 10% per year.

Road construction companies need to focus on certain aspects of their business to gain a competitive advantage in the market under restrictive measures [12,13].

The use of innovations can significantly change the situation and the rationale for decisions, because it affects both the quality characteristics and the cost of roads. From these positions, the following types of innovations can be distinguished:

- aimed at improving quality at constant costs;
- aimed at reducing costs with constant quality indicators;
- aimed at improving quality and reducing costs.

Among the listed types of innovations, the most attractive are innovations aimed at improving quality and reducing costs in road construction [14,15].

Considering that the introduction of innovative technologies should be systematic, we have developed an algorithm for the introduction of innovative technologies in the construction of roads (Figure 1).

The algorithm developed by the authors will allow road construction enterprises to make informed decisions in choosing the directions of their development and enable the enterprise to more clearly identify its market positions, take targeted measures to improve them and solve the problems of increasing the efficiency of its activities.

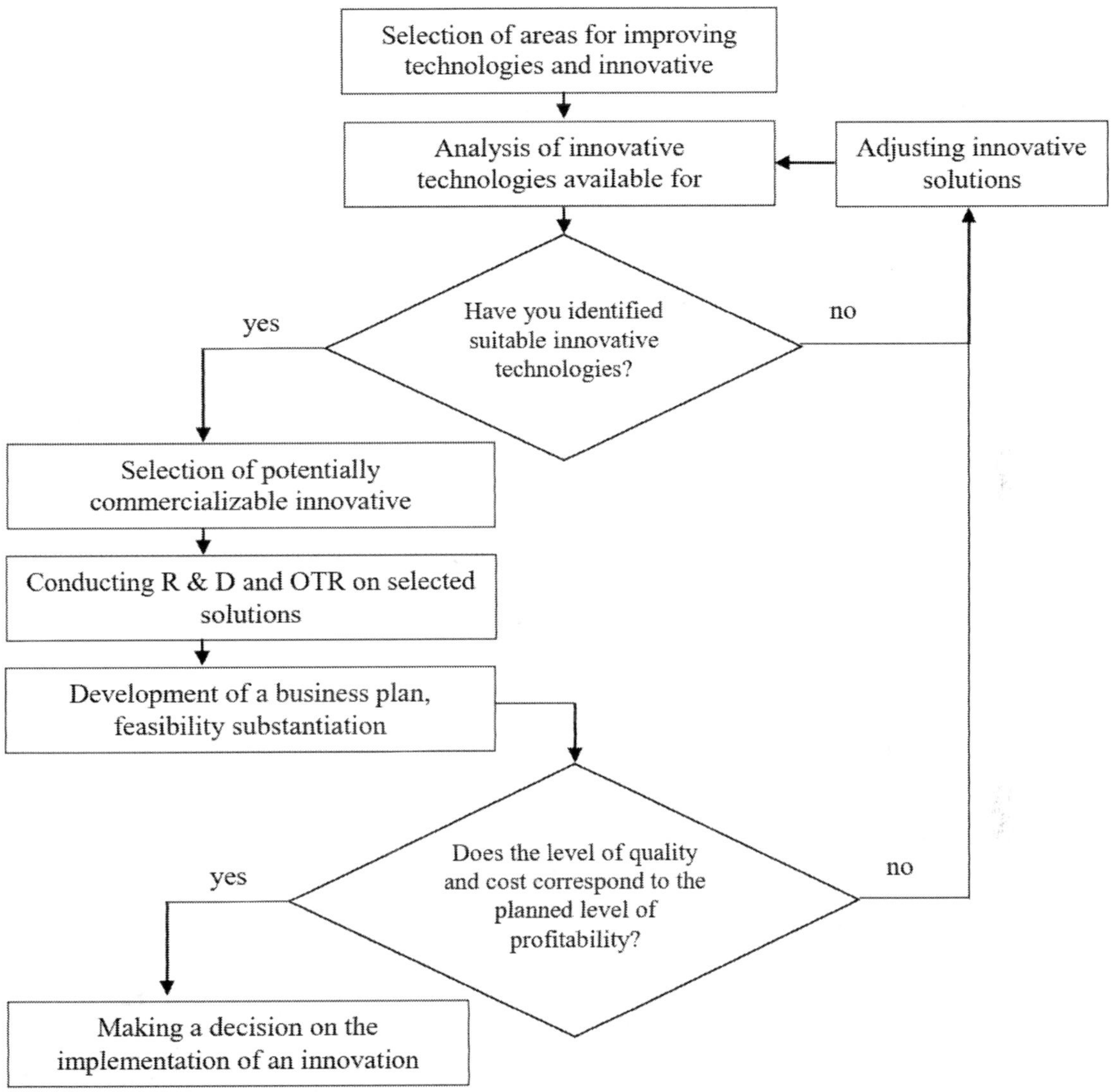

Figure 1. Algorithm for the introduction of innovative technologies in the construction of roads

5. Conclusion

Thus, it should be emphasized that Russian road construction companies have a rather low level of competitiveness [16]. Every year, foreign companies are leaders in the road industry because they are interested in introducing innovative technologies. The introduction of innovative technologies allows the enterprise to increase its competitiveness.

Each business entity must constantly improve and introduce innovative technologies. The introduction of innovative technologies is necessary for any enterprise, regardless of its size, profit or age. Innovation is the engine of progress.

As a result, an enterprise that implements innovative technologies expects:

- improvement of transport and operational indicators of roads;
- longer service life;

- improving road safety;
- best environmental conditions;
- cost reduction;
- the use of effective means of engineering equipment and road construction, as well as modern information technologies and communication systems.

References

[1] Getmanova I A, Zavarzina T S 2018 State and trends of development of road construction business in Russia. *Collection of articles of the II International Scientific and Practical Conference "Modern directions of scientific research". Publishing house: Limited Liability Company "Center for Scientific Research and Consulting" (Samara)* 54-58

[2] Blagodir A L, Grebenkina I A, Grebenkina S A 2020 Connected territories of the Russian Federation due to the creation of intelligent transport systems *Legislation* **1** 77–85

[3] Rogov S 2013 Overview of the traffic market *The risk granter* **9**

[4] Ignatieva A V, Pakhomov A V 2020 Implementation of national projects at the municipal level: first results and ways to improve efficiency. *Municipal property: economics, law, management* **1** 11–17

[5] Roy O M 2015 Road construction as an object of public administration. *National priorities of Russia* **4 (18)** 36-41

[6] Voroshilov N V 2020 Main results of the implementation of the national project "Safe and high-quality highways" in 2019. *Municipality: economics and management* **3 (32)** 124-139

[7] Grasmik L V and Molodan I V 2014 The problem of competitiveness of road construction industry enterprises *Actual problems of aviation and cosmonautics* **10 (2)** 300-301

[8] Khalturin R A 2011 State and experience of road network construction in Russia and abroad. *Economic Sciences* **1 (74)** 223-226

[9] Letalov A 2016 The situation in the field of road construction and mechanical engineering *E Vesti* Available at: http://www.e-vesti.ru/ru/road-development-2016/

[10] More roads were built in 2020 than a year earlier. 2020 *Rossiyskaya gazeta* Available at: https://rg.ru/2020/12/25/v-2020-godu-postroeno-bolshe-dorog-chem-godom-ranee.html

[11] Danilenko M I 2020 Assessment of the competitiveness of the enterprise *Bulletin of the Academy of Knowledge* **4 (39)**

[12] Kormakov L F 2019 Modernization of road transport infrastructure – the starting stage of the integrated development of rural areas *Economy of agricultural and processing enterprises* **9** 75–80

[13] Svetnik T V 2014Transformation of the construction industry into an effective locomotive of economic growth *Bulletin of Irkutsk state technical University [Socio-economic and social Sciences]* **12 (95)** 336-342

[14] Bykov D N, Khomkalov G V 2018 Construction materials market in competitive conditions *Izvestiya vuzov. Investment. Construction. Realty* 2018 **3** (26)

[15] Samarukha A V 2018 Topical Trends of Improving Economy of Regions and Municipal Entities in Siberia *Baikal Research Journal* vol 9 **3**

[16] Svetnik T V 2017 Systemic nature of the hoodwinked investors' problem in Russia *Baikal Research Journal* 2017 vol 8 **4** DOI: 10.17150/2411-6262.2017.8(4).26

Mixture substantiation of the powder composition for bearing units seats restoring of rolling bearings

E Yu Lyakhov, V A Zorin

Moscow Automobile and Road Construction State Technical University (MADI), 64, Leningradsky Prospect, Moscow, 125319, Russian Federation.

E-mail: zam-upr@gmail.ru, madi-dm@list.ru

Abstract. The article substantiates optimal and most promising composition of a powdery thermoplastic composite polymer material used to restore worn-out rolling bearing units seats with combination of electrostatic and vibro-vortex coating methods. A laboratory installation for applying and forming coatings by a chamber electrostatic method is presented, with the help of which coatings were applied to steel 45 and AL-4 aluminum samples, after which they were melted in an oven at a temperature of 220 °C for 10 minutes. Based on the studies carried out, the optimal composition of the powdery polymer composition was determined: epoxy oligomer - 32 ... 40 %, powdered glass fiber - 9.8 ... 11.6 %, the rest - polyamide 12. At the same time, the coatings obtained from this composition on aluminum alloys substrates have insufficient adhesive strength. Therefore, the search for ways to increase it is relevant, and the solution of this issue will create a reliable technology for restoring rolling bearings seats.

1. Introduction

The operation of cars and road-building machines is the longest period of product's life cycle, realizable capabilities completeness of machines during operation is determined by adopted maintenance and repair system, technological equipment used, operating materials quality, qualifications and motivation of the service personnel, and adopted technological processes of restoring operability.

Bearing units state of the rolling bearings has a significant impact on automotive and road construction equipment durability and performance, operating modes of which are characterized by specific operating conditions [1].

Experience of using automobile and road-building equipment, numerous studies carried out in the field of durability and various operational properties studying of bearing assemblies show that wear or linear dimensions and geometric shape inconsistency of the seats with regulatory requirements lead to coordinated arrangement violation of all parts of mechanisms. As a result, the static and dynamic loads on the mechanisms parts increase, wear intensity of rolling bearings seats and elements increases. A great influence on accuracy of mechanisms parts arrangement during their operation is exerted by the one-sided nature of the loads taken by rolling bearings, which leads to a relative body parts walls deformation due to changes in residual linear stresses.

Optimal polymer composition choice for bearing assemblies restoration should be based on the theoretical foundations of viscoelastic bodies destruction processes under the mechanical loads action [2].

Literature data analysis shows that powdery thermoplastic polymers and compositions based on them are the most suitable for restoring bearing seats, and a directed change in polymers properties during their processing by composition mixture optimizing, intensifying the adhesive joints formation is one of

Content from this work may be used under the terms of the Creative Commons Attribution 3.0 licence. Any further distribution of this work must maintain attribution to the author(s) and the title of the work, journal citation and DOI.
Published under licence by IOP Publishing Ltd

IOP Conf. Series: Materials Science and Engineering 1159 (2021) 012049 doi:10.1088/1757-899X/1159/1/012049

the most expedient ways of creating coatings with high and stable physical and mechanical properties.

2. Problem formulation

Electrostatic is the most promising method for applying powders to the bearing seats repairable surfaces. It is based on the ability of polymer powders to acquire an electrical charge when placed in a high voltage electrostatic field and then transferred to the part to be coated. The method is characterized by high productivity, possibility for complex mechanization and automation of the process, ease of adjusting coating thickness, allows you to cover products made of heterogeneous materials. With a combination of electrostatic and vibro-vortex (chamber) methods, coatings with a thickness of 1 ... 2 mm can be obtained [3].

However, the existing mixture compositions for chamber electrostatic method do not meet modern repair production requirements for coatings used to restore rolling bearings bore holes, and existing equipment does not allow obtaining uniform coatings over the hole entire surface. It is recommended to use polyamide coatings to restore parts that perceive vibrational loads and work in aggressive environments [4].

Powders based on polyamide 12, a polymer material with a complex of valuable physical, mechanical, technological and operational properties, are promising.

The coatings obtained from P-12 have satisfactory wear resistance, a wide range of operating temperatures (-20 ... + 100 °C), and absorb shock loads well [5]. However, particles of polyamide powders when heated poorly retain an electric charge due to a decrease in electrical resistance, which complicates obtaining coatings process in an electric field. Polyamide coatings have significant shrinkage and high internal stresses, and their adhesion strength is insufficient [3]. These features limit the possibility of their use as coatings for restoring bearing assemblies seats.

Nowadays, the industry has mastered various epoxy powder polymers production, which are used mainly to create protective and decorative coatings. These powders, in contrast to polyamide ones, are charged faster in an electric field and more fully concentrate the charge. The coatings obtained from them have good adhesion strength and hardness [6], but it is impossible to use these coatings in their pure form to restore the mounting holes, in view of their high fragility and low wear resistance.

It is possible to assume that polyamide coatings properties can be purposefully changed by introducing epoxy oligomers into them. This should help, on the one hand, improve composition powder particles electrodeposition, adhesion strength increase, on the other hand, the polyamide should give the coatings increased impact strength and flexibility [4]. The hardness, wear resistance and dimensional stability of coatings can be increased by filling the composition with crushed glass fiber [7].

Based on available data analysis, it can be concluded that polymer composition creation based on polyamide P-12 and an epoxy oligomer using crushed fiberglass as a filler will make it possible to obtain coatings on worn-out holes of bearing assemblies with necessary physical and mechanical properties.

Bearing assemblies operating conditions and repair companies experience show that coatings used to restore the seats of rolling bearing housing parts must provide the following indicators:

- have high wear resistance during assembly, operation and disassembly;
- have high impact strength (not less than 4.6 N·m);
- have a high ability to electrodeposition (not less than 150 g / m^2).

Thus, the aim of the study is to develop a powdery polymer composition for obtaining coatings on bearing assemblies bore by electrostatic chamber spraying method.

3. Materials and methods

The research was carried out in two stages. At the first stage, composition components were chosen and dependence of the physical and mechanical properties on components percentage was studied. At the second stage, the search for the optimal mixture composition was carried out.

Filler effect on composition quality is the least predictable. Therefore, studies were carried out for five levels of filler content.

As criteria for mixture optimizing of the composition, limiting powder electrodeposition, coating impact strength, and repaired bore holes wear resistance during mounting and dismounting of bearing assemblies are taken.

Powder composition was prepared by dry mixing in a ball mill for 1 hour at a drum rotation speed of 80 ... 100m^{-1}, followed by sieving through a vibrating sieve with a mesh № 0.2.

Composition mixture: polyamide 12, epoxy oligomer P-EP-219 and crushed glass fiber.

The composition was applied by a chamber electrostatic method (Fig. 1), followed by reflow in a thermal furnace at a temperature of 220 °C for 10 min.

The adhesion strength was determined for coatings applied to steel 45 and AL-4 specimens according to the procedure shown in Fig. 2, impact strength - for coatings applied to flat samples. Fittings wear resistance was assessed by the number of assembly-disassembly cycles of mates with 20 μm initial interference until the zero position of the tested mates was reached, i.e., until a constant bore size is achieved.

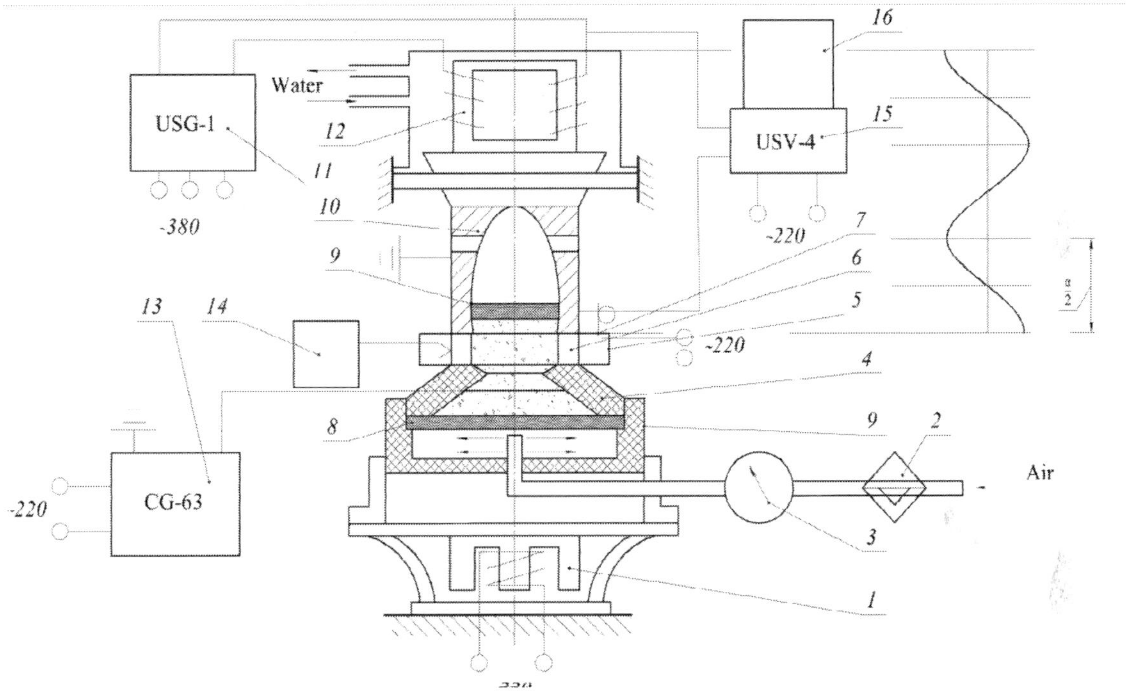

Figure. 1. Laboratory installation scheme for applying and forming coatings by a chamber electrostatic method: 1 - electromagnetic vibrator; 2 - filter drier; 3 - manometer; 4 - fluidized bed bath; 5 - heater; 6 - part to be coated; 7 - polymer coating; 8 - a ring with an electrode grid; 9 - porous partition; 10 - waveguide; 11 - ultrasonic generator; 12 - magnetostrictive transducer; 13 - cascade generator; 14 - thermostat; 15 - vibrometer; 16 - oscilloscope.

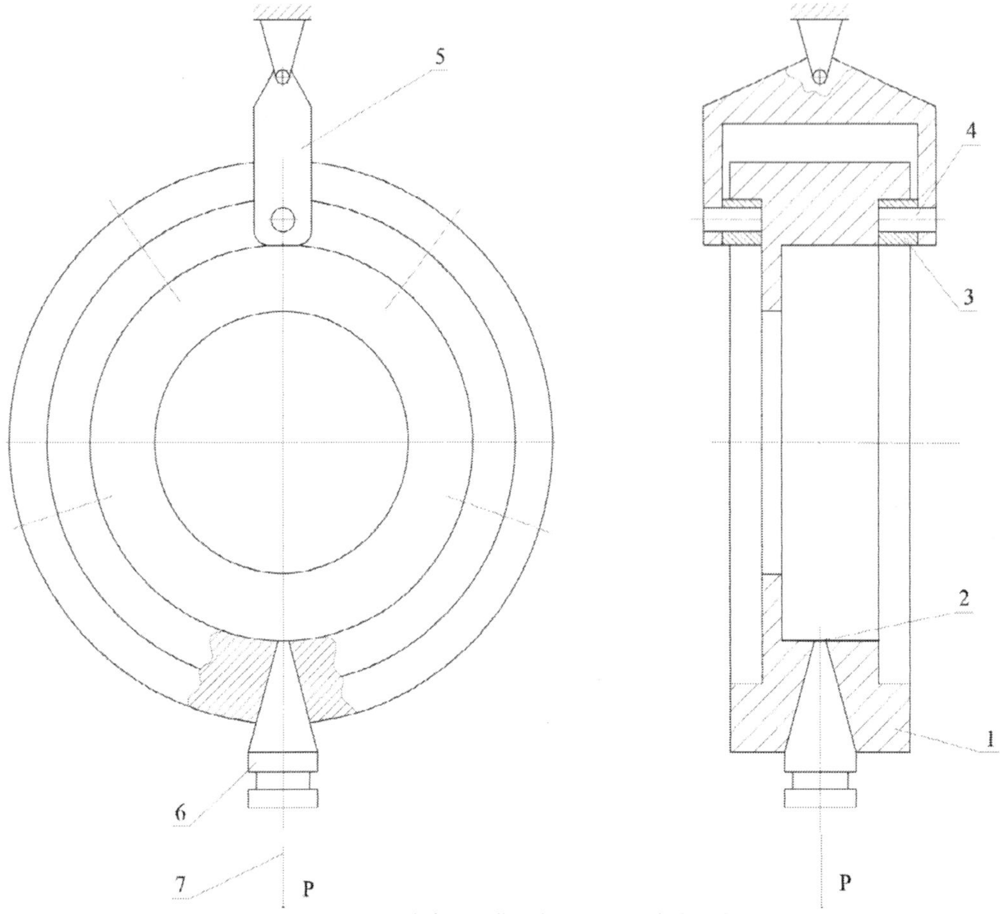

Figure. 2. Scheme of device for determining adhesive strength by the pins method: 1 - sample; 2 - PKM; 3 - roller; 4 - roller axis; 5 - self-centering mechanism; 6 - pin; 7 - breaking force;

4. Results and discussion

As a result of experimental data analysis, the dependence of the physical and mechanical properties on percentage of constituent compositions was obtained:

- with an increase in the amount of P-EP-219, powder adhesion strength and electrodeposition increase, other indicators decrease;
- with an increase in the amount of finely dispersed glass fiber, wear resistance increases during assembly, operation and disassembly;
- mixture composition affects the impact strength ambiguously.

The search for optimal composition mixture showed that maximum value for assembly and disassembly cycles of joints with an initial interference of 20 μm until reaching the zero position is 8, and impact strength did not decrease below 4.6 N·m. Thus, the optimal composition mixture is: epoxy oligomer - 32 ... 40%, powdered glass fiber - 9.8 ... 11.6%, polyamide 12 - the rest.

Physical and mechanical properties of the composition for constituents limiting content conditions are shown in Table 1.

Table 1. Physical and mechanical properties of powders and coatings obtained from the recommended composition

composition mixture			physical and mechanical properties				
polyamide 12	P-EP-219	fiberglass	EDP[a], g/m^2	n[b]	i[c], N·m	adhesion strength, MPa	
						steel 45	AL4
66.4	32	11.6	162	8	4.77	29.2	18.16
58.4	40	11.6	174	8	4.6	29.35	18.15
68.2	32	98	162	8	4.9	28.86	20.16
60.2	40	9.8	174	8	4.73	30.85	21.07

[a] Powder electrodeposition, g/m ^{2};

[b] Number of assembly-disassembly cycles of joints with an initial tension of 20 microns until reaching zero position;

[c] Impact strength N·m;

5. Conclusion

Thus, the developed composition ensures coatings production with high wear resistance (n = 8), good powder electrodeposition of the (EDP 162-174 g/m^2) and high impact strength (y = 4.6-4.9 N·m). These indicators make it possible to use the composition for restoring worn-out rolling bearing seats. However, it should be noted that the coatings obtained from this composition on aluminum alloy substrates have insufficient adhesive strength. Therefore, the search for ways to increase it is relevant, and the solution of this issue will create a reliable technology for rolling bearings seats restoring.

References

[1] Zorin V A, Lyakhov E Yu 2018 Durability analysis of bearing seats restored with polymer materials *Interstroymech-2018* 337–342

[2] Lyakhov E Yu 2020 Modeling of surfaces fatigue fracture processes of automobiles bearing units restored with polymer materials *Herald of the Moscow Automobile and Road Construction State Technical University (MADI)* **3** 10–12

[3] Yakovlev A D, Dzor V F and Kaplan V I 1979 *Powder polymeric materials and coatings based on them* (Leningrad: Chemistry)

[4] Bazhenov S L 2014 *Mechanics and technology of composite materials: Scientific edition* (Dolgoprudny: Intellect)

[5] Kolyasko I V 1980 *Research and development of technology for restoring parts agricultural machinery with polyamide coatings: Ph.D. thesis in Engineering Science* (Moscow)

[6] Dudchak V P 2003 Theoretical prerequisites for adhesive strength study of polymer composite coatings with a base *Electro material processing* Vol 39 **1** 27–30

[7] Baurova N I and Zorin V A 2018 *Polymer composite materials application in machinery* (Moscow: INFRA-M)

The concept of functioning of the on-board quality control system for anti-icing pavement treatment

K P Mandrovskiy[1], Y S Sadovnikova[2]

[1] Department of Road Construction Machines, Moscow Automobile and Road Construction State Technical University (MADI), 64, Leningradskiy prospect, Moscow, 125319, Russia
[2] Department of Airports, Engineering Geology and Geotechnics, Moscow Automobile and Road Construction State Technical University (MADI), 64, Leningradskiy prospect, Moscow, 125319, Russia

E-mail: jana.sadovnikova@yandex.ru

Abstract. The article discusses the concept of an electronic system for ensuring the quality of anti-icing processing of sidewalks and other elements of road construction. The object of research is the quality of particle distribution of anti-icing material. In particular, such quality indicators as the width of the processing band, the density of distribution of the material over the coating and the degree of lateral deformation of the processing zone are considered. The subject of the research is the general principle of the control system for the numerical values of the distribution quality indicators. Such a system is based on mathematical models of the movement of individual particles of a material along a flat rotating disk and in air. The original mathematical basis was developed by the authors of the article, taking into account the analytical equations of motion of a particle on a rotating horizontal disk obtained earlier by other researchers and using Newton's second law. The article describes the algorithm for the functioning of the distribution quality control system, which takes into account geometric, regime and meteorological parameters. The system is designed to work with a symmetrical way of feeding material from the tray to the disc. In addition, the physicochemical properties of the particles of anti-icing material and numerical quality criteria are taken into account, such as the required width of the processing strip, the rate of density of distribution (consumption) of the material and the maximum permissible degree of lateral deformation of the processing zone. The block diagram of the system is presented. As a result of the operation of the proposed system, it becomes possible to change the values of the operating parameters of the equipment in the process of material distribution, providing the required values of quality indicators. Thus, the negative impact on the ecological environment of the city will be reduced and the economic efficiency of anti-icing processing of sidewalks will be increased.

1. Introduction

The existing methods of quality control of anti-icing treatment of pavements can be conditionally divided into three groups: control of quality indicators at the stage of certification tests of the machine for the distribution of anti-icing materials; in the process of distributing anti-icing material by the machine; based on the results of a technical and economic analysis of the effectiveness of anti-icing measures. At the stage of certification tests, the width of the distribution of anti-icing material, the step of adjusting the width of the distribution of the material, the uniformity, density and asymmetry of the distribution of the material are controlled.

Content from this work may be used under the terms of the Creative Commons Attribution 3.0 licence. Any further distribution of this work must maintain attribution to the author(s) and the title of the work, journal citation and DOI.
Published under licence by IOP Publishing Ltd

IOP Conf. Series: Materials Science and Engineering 1159 (2021) 012050 doi:10.1088/1757-899X/1159/1/012050

With the direct implementation of operations on anti-icing processing, the following is monitored: the modernity of the beginning of the distribution of anti-icing material, compliance with the prescribed deadlines for the elimination of winter slipperiness; uniform distribution of material within the carriageway; actual distribution density of the material. The most common way to control the quality indicators of anti-icing processing is onboard electronic control systems. Such systems are able to control the width, direction of distribution and the actual flow rate (distribution density) of the deicing material. In doing so, they take into account the air temperature, the speed of the machine and the operating mode of the distribution equipment. Evaluation and control of the uniformity of material distribution is not performed by modern on-board systems.

The technique of technical and economic analysis of the efficiency of machines for the distribution of materials evaluates the quality indirectly. It uses two indicators: actual economic efficiency based on specific statistics and surveys, and potential (theoretical) economic efficiency.

2. Problem Statement

It is fair to say that the described quality control methods are used for road and airfield surfaces. The quality of the distribution of anti-icing material on the surfaces of sidewalks, footpaths and other elements of road construction is assessed visually. Often the material is distributed manually without observing the norms and special control of the degree of uniformity. Such processing causes increased difficulties for the movement of pedestrians and poses an environmental hazard to humans, animals, plants, soil and microorganisms. Uneven, lumpy distribution of the material indicates a low efficiency of its use and leads to an increase in the cost of anti-icing treatment. This indicates the relevance of developing a new approach to the issue of ensuring the quality of the distribution of anti-icing material on sidewalks.

In this article, a distribution scheme is proposed for use, in which material is fed from a hopper to a disk through a tray with an annular cross-sectional shape (symmetric method). With this method of feeding, particles of material leave the disk simultaneously from each sector bounded by the blades of the disk, at the same speeds. In the conventional method, the material is fed asymmetrically relative to the center of the disc and this creates the preconditions for the subsequent uneven distribution of particles on the disc and in the air. The symmetrical method can be rationally implemented using a disk located at some distance from the machine. Then all the advantages of this method will be maximized. For the proposed equipment and feeding method, it is required to develop an on-board quality control system for the distribution of anti-icing material particles. Such a system should ensure control of the width of the processing strip, the uniformity of material distribution and compliance with the rate of material consumption, taking into account not only the operating and geometric parameters of the equipment, but also meteorological parameters. These include wind speed and direction.

3. Materials and methods

The functioning of the quality control system for anti-icing treatment of sidewalks should be based on mathematical models of the movement of material particles along a flat rotating disk and in the air. The works [1-3] are devoted to the description of the motion of solid particles on a disk using inhomogeneous differential equations of the second order. Modeling the motion of particles in air using Newton's equations is presented in detail in [4, 5]. The results of these studies make it possible to calculate, for given initial conditions, the flight range of particles in the air (width of the processing strip) and to determine, in the presence of wind, the degree of deviation of the processing strip width from the required initial one (which is set in the absence of wind).

The distribution can be considered qualitative if the required processing bandwidth is observed, the consumption rate (distribution density) of the material per 1 m^2 of pavement and the degree of lateral deformation of the processing area does not exceed the specified maximum value.

The actual density q_f of the material distribution should be equal to the normalized q_n. Considering the symmetrical way of feeding the material onto the disc, the actual density (flow rate) is defined as:

IOP Conf. Series: Materials Science and Engineering 1159 (2021) 012050 doi:10.1088/1757-899X/1159/1/012050

$$q_f = \left(\rho\left(\pi h_l(r_2^2 - r_1^2)\right)\right) / S_f,$$

where ρ – the density of anti-icing material, kg/m^3;
S_f – the actual area of the processing strip, m^2;
r_1 – the inner radius of the annular tray, m;
r_2 – the outer radius of the annular tray, m;
h_l – the disc blade height, m.

The width of the processing lane is set taking into account the maximum width of the sidewalk. The degree of lateral deformation is an important quality indicator, since it allows you to assess how much the width of the processing strip has changed when exposed to wind and quickly make a decision to minimize this impact. It is obvious that an increase in the degree of lateral deformation leads to an increase in the number of machine passes, and, consequently, to an increase in the processing time of the pavement.

The purpose of the on-board control system is to provide a numerical assessment of quality indicators in the process of anti-icing processing and to offer the operator solutions to eliminate the negative impact of the external environment on the processing area. These solutions include options for varying the operating parameters of the distribution equipment, namely, the rotational speed of the disk, the height of the disk above the pavement, and the speed of the machine. In this case, the system takes into account the values of the geometric parameters of the equipment, such as the outer and inner radii of the circular section of the chute, the volume of the hopper with material, the diameter of the disc, the length and height of the disc blades, the angle of inclination of the disc to the axis of rotation, the angle of inclination of the blades to the radial position. The physicochemical properties and particle size distribution of the anti-icing material are also taken into account: density, mass, volume and radius of a particle.

The on-board control system (Figure 1) should contain a control unit with a microcontroller and an associated sensor unit that monitor the speed of the machine and the value of the material consumption, as well as a control panel with a display. The values of the operating parameters are entered into the control unit database. In addition, the calculated dependences of the width of the processing strip and the actual consumption of material on the operating parameters, such as the speed of the machine, the height of the disc above the coating and the rotational speed of the spreading disc, are introduced. Additionally, the sensor unit must be equipped with a sensor for measuring the speed of the machine (speed sensor), a manometer (pressure gauge), a sensor for measuring the flow (supply) of material to the disk (flow meter), sensors for measuring the speed and direction of the wind (anemometer), and also for measuring linear (linear displacement transducer – LDT, Figure 1) and angular displacements of the disc, namely, the height of the disc above the pavement, the angle of inclination of the disc to the axis of rotation (angular displacement transducer – ADT1, Figure 1) and the rotational speed of the disc (ADT2, Figure 1).

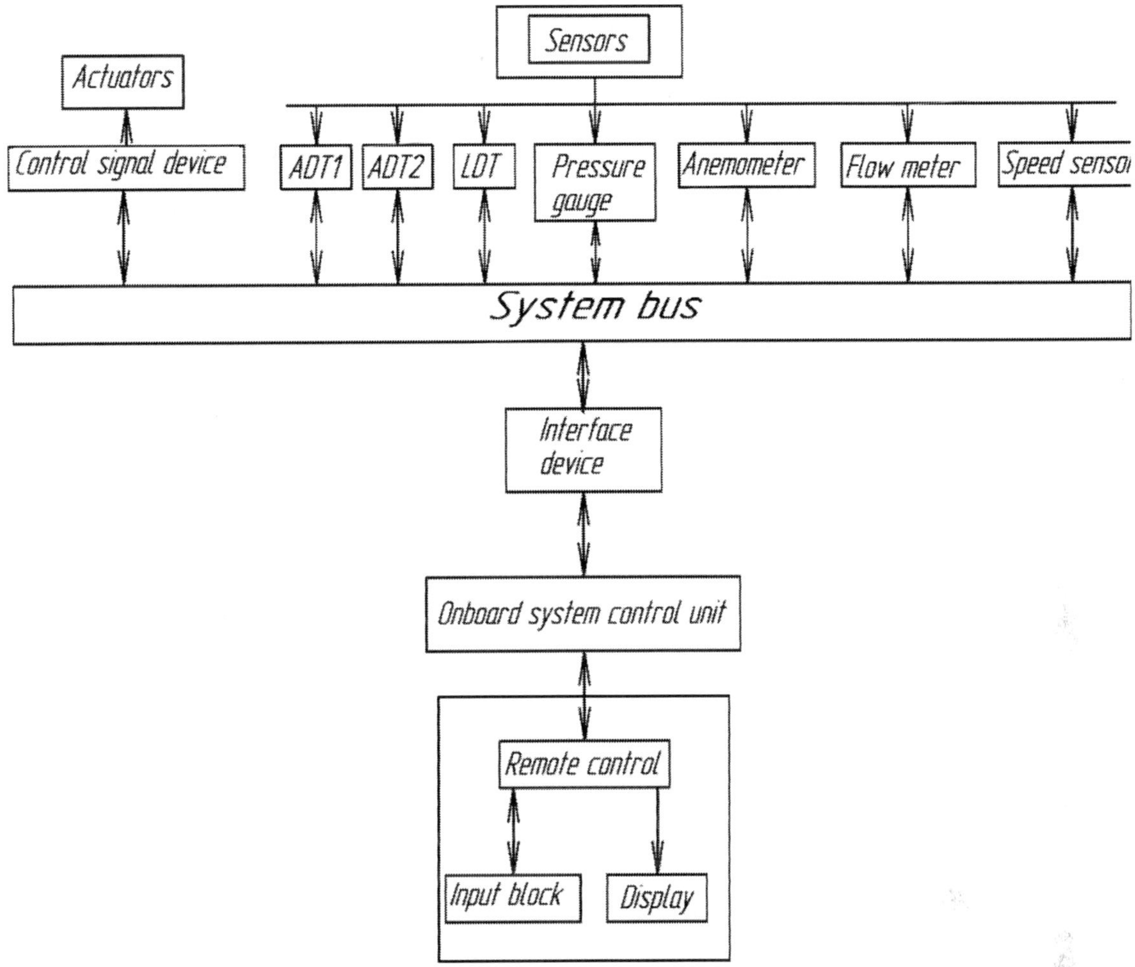

Figure 1. Structural scheme of the system

In addition, when forming the information base in the control unit, the values of the geometric parameters of the distribution equipment are taken into account: the diameter of the distribution disc, the length, height and number of disc blades, the outer and inner radii of the circular section of the chute, and the volume of the hopper with material. The control unit is entered with the current values of meteorological parameters, that is, the actual values of the direction and speed of the wind, as well as a database on the physical and chemical properties and particle size distribution of the anti-icing material. In the memory of the microcontroller, information is entered with a description of the mathematical models of the movement of material particles through the distribution equipment and in the air, as well as the calculated dependences of the width of the processing strip, the degree of lateral deformation of the processing zone and the actual consumption (distribution density) of the material on geometric, operating and meteorological parameters. Numerical criteria of the distribution quality are also introduced - the normalized distribution density of the material under ideal conditions and the permissible deviation of the width of the processing strip under real distribution conditions from the width in the absence of wind. Directly in the process of material distribution, the width of the processing strip, the actual distribution density of the material per 1 m^2 of pavement is calculated and the degree of lateral deformation of the processing zone is numerically determined. If the required values of the above quality indicators are not achieved, as well as if the degree of lateral deformation of the processing zone exceeds the permissible maximum value, then the system selects the operating parameters of the equipment using databases with mathematical models and numerical criteria for the distribution quality.

The display shows possible options for varying certain operating parameters to improve quality indicators. The specific choice is made by the operator using the initial data input unit.

4. Discussion

It is necessary to clearly explain the essence of the concept of functioning of the quality control system for pavement processing. To do this, consider how, with the initial parameters set, individual quality indicators can be controlled.

Let the initial geometric parameters be the following data (Table 1).

Table 1. Initial geometric parameters of disk equipment

Parameter	Value
Length of disc blade	175 cm
Height of disc blade	5 cm
Angle of disc blade	45°
Outer radius of tray	10 cm
Inner radius of tray	4 cm
Diameter of disk	180 cm
Angle of inclination of the disk to the axis of rotation	90°

A solid anti-icing material with a density of 1150 kg/m^3, with an average particle diameter of 0.25 cm and a required distribution density of 90 g/m^2 was selected for distribution. The required width of the processing strip is 0.52 m, the maximum allowable degree of lateral deformation of the zone is set equal to 10% of 0.52 m.

Meteorological conditions are described by the following parameters: wind speed 8 m/s, north-east wind, steady direction, moderate, the speed vector is directed at an angle of 10 degrees. to an axis perpendicular to the direction of travel of the machine. The air temperature is -5 ^{0}C, the air density is 1.3163 kg/m^3, the dynamic viscosity of the air is 0.000017 Pa·s.

The initial operating parameters of the equipment are described in Table 2.

Table 2. Initial operating parameters of the equipment

Parameter	Value
Disc height above cover	400 cm
Disk rotation speed	390 rpm
Machine speed	20 km/h

After entering the initial data, the width of the processing strip is calculated, which in the absence of wind is 0.52 m.

Taking into account the effect of wind on the processing area, the calculated actual value of the processing strip width is 0.4609 m. The degree of lateral deformation of the processing area is 11.37%.

It can be concluded that the required width of the processing strip has not been reached and when the maximum permissible degree of lateral deformation of the processing zone is set at 10%, the quality of anti-icing processing has not been achieved.

Reducing the negative consequences of deformation of the processing zone can be achieved if the height of the disc above the coating is changed to 0.5 m. Then, the width of the processing strip will be 0.52 m

Another option can be proposed, namely, increasing the value of the disk rotation speed to 7 s^{-1}. Then the strip width will be 0.53 m, therefore, a complete elimination of the degree of deformation of the treatment zone is achieved.

The system, based on the calculated data, provides the operator with the opportunity to choose one or another strategy for controlling the parameters of distribution equipment. A control panel with a display is used to display and enter information.

It should be noted that such systems for equipment distributing material along sidewalks using a symmetric feeding method have not been described previously, therefore, the development of the concept needs further development and research.

5. Conclusion

The concept of the on-board anti-icing quality control system proposed in the article is a new approach to solving the problem of eliminating the uneven distribution of anti-icing material on sidewalks and other elements of road construction. It is based on the implementation of a symmetrical method of feeding material from a tray to the surface of the disc and involves placing the disc on a remote rod. The quality control system assumes taking into account the meteorological parameters of the external environment, which have a significant impact on the shape of the processing zone, since it is not characterized by large dimensions. Such a system does not include expensive equipment and can be used for winter maintenance of sidewalks, bridges, footpaths, providing an economical use of materials and reducing the negative environmental impact on the environment and the human body.

References

[1] Hofstee J W and Huisman W 1990 Handling and spreading of fertilizers: part 1, physical properties of fertilizer in relation to particle motion *Journal of Agricultural Engineering Research* **47(4)** 213–234

[2] Liedekerke P Van, Tijskens E, Dintwa E, Anthonis J and Ramon H 2006 A discrete element model for simulation of a spinning disc fertilizer spreader I. Single particle simulations *Powder Technology* **170** issue **2** 71–85

[3] Villette S, Cointault F, Piron E and Chopinet B 2005 Centrifugal Spreading: an Analytical Model for the Motion of Fertiliser Particles on a Spinning Disc *Biosystems Engineering* **92** issue **2** 157–164

[4] Mandrovskiy K P and Sadovnikova Y S 2018 Characteristics of the droplet motion of a liquid antifreeze reagent *Magazine of Civil Engineerin* **03** 14–26

[5] Mandrovskiy K P and Sadovnikova Y S 2019 Modeling the Uniform Treatment of Coatings with an Anti-Icing Liquid Reagent *Mathematical Models and Computer Simulations* **11(5)** 842–849

IOP Conf. Series: Materials Science and Engineering 1159 (2021) 012051 doi:10.1088/1757-899X/1159/1/012051

Changing the spatial structure of the rocker arm of the tipping mechanism in a garbage truck

N M Andryukhov[1], **S A Pavlov**[1], **K D Mikhailov**[1]

[1.] Moscow automobile and road construction state technical university, (MADI), 64 Leningradsky Prospekt, Moscow, 125319, Russia

E-mail: Andryukhov_nm@mail.ru

Abstract. Building a model of spatial structure of the rocker arm and its design optimization are related to the problem of excess weight of the rocker arm in the garbage tipping system. The simulation was aimed at producing a structure where stresses are minimized. The body of the rocker arm model was determined using the Equivalent Static Load Method (ESL). Based on the simulation, the design of the rocker arm was developed. The weight of the improved rocker arm was reduced by 10.8 kg while maintaining the level of stress produced by loads.

1. Introduction

The container tipping system is an important part of the garbage truck design (Figure 1). Its main function is to pick up the waste container and rotate it so that the waste is dumped into the hopper of the truck. The system consists of support frame 1, folding hydraulic cylinder 2, rocker arm 3, pull rod 4 and support stand 5, as shown in Figure 2. When the piston extends, the folding hydraulic cylinder sets the waste container with support stand in motion rotating it counterclockwise and tips the container over into the hopper.

The initial position is achieved by retracting the piston of the folding hydraulic cylinder. In this case, it is difficult to determine the operating conditions of the rocker arm, which is affected by the system of forces, since it is a part of a four-link mechanism. The currently used rocker arm is bulky and heavy, which reduces the overall load capacity of the garbage truck.

The increasing volume of waste collection led to the idea of optimizing the design of the rocker arm and the spatial arrangement of its elements. In the study [1], optimization was carried out using the finite element method in the ANSYS software environment. Another research [2] presents a static analysis of the forces applied to the rocker arm under maximum load conditions. The finite element method is also used to analyze the strength of the rocker arm in the study [3]. Papers [4,5] analyze dynamic loads in container tipping systems. Paper [6] is focused on validating the design changes to small waste collection truck made in the stress analysis environment.

The aforementioned works focus on the analysis of one operating position of the rocker arm simulated as an elastic body, which is quite different from the actual process of tipping a container.

For solving the above issue and data preprocessing, the Hyperworks environment was used. The complex stress state of the rocker arm in real-life conditions and the optimal load were addressed using the Motionsolve module, and the optimization of the spatial structure was carried out using Optistruct.

The article discusses a fully loaded rocker arm, and the analysis of the design optimization is carried out using a dynamic model, where the rocker arm is considered as an elastic body with a hinged pull rod, which is more consistent with the real-life operation.

Content from this work may be used under the terms of the Creative Commons Attribution 3.0 licence. Any further distribution of this work must maintain attribution to the author(s) and the title of the work, journal citation and DOI.
Published under licence by IOP Publishing Ltd

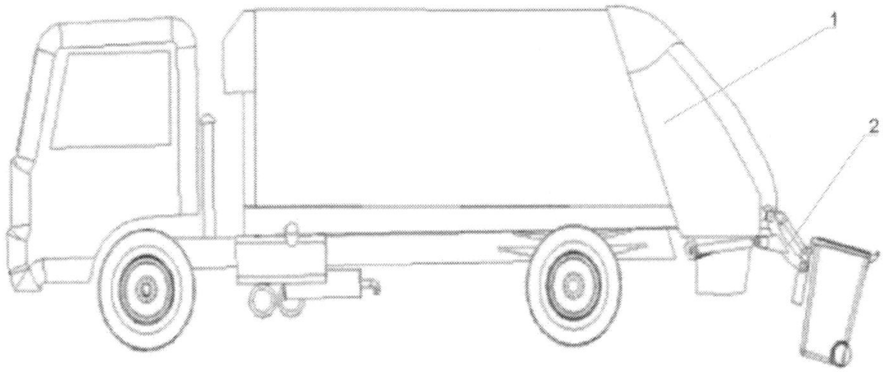

Figure 1. Garbage truck with garbage compaction system: 1. Garbage truck bin, 2. Container tipping system

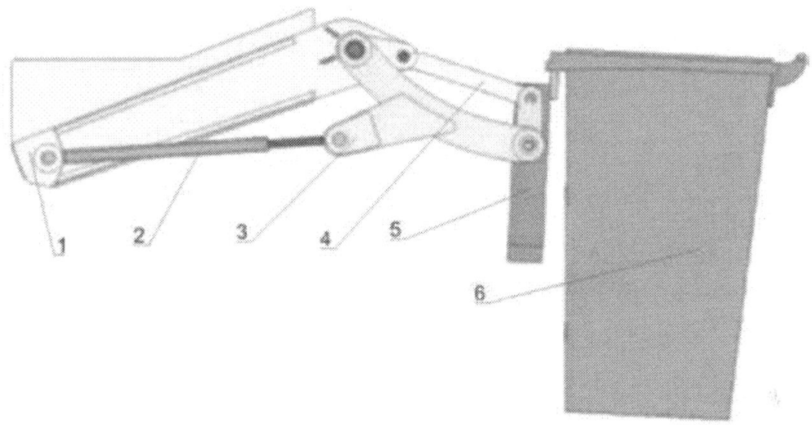

Figure 2. Container tipping system: 1. Support frame 2. Hydraulic cylinder, 3. Rocker arm, 4. Pull rod, 5. Support leg, 6. Waste container.

2. The theory of elastic body and topological optimization

A complex mechanical system, consisting of several objects connected by hinges, is called multi-link. It can be described as a rigid system of elastic bodies joined by rigid links. In such case, the deformation of the bodies should be addressed in the simulation.

A flexible body is associated not only with the general motion of the system, but also has its own elastic deformation. It can be described as a set of nodes of a finite element model, with the deformation as a linear superposition of the shape in this model. The nodes in the local coordinate system will perform translational and rotational motion relative to a slight displacement of the model coordinate system.

In this article, the rocker arm is considered as an elastic body and other components of the tipping system as rigid bodies.

The main idea of optimization is to transform the structure to reach optimum distribution of material. Optimization methods chosen are variable thickness method, homogenization method and relative density method.

The mathematical optimization model with the variable thickness method is simple, but the optimized object can only be applied to two-dimensional space.

The homogenization method is a physical description method and is widely used. The basic idea is to introduce the microstructure shown in Figure 3 into the topological model. The area of the microstructure is defined in the expression $\Omega_s = \int(1-ab)d\Omega$, the density function of the element is $\rho = (1-ab)\rho_s$, where $0 \leq a \leq 1$, $0 \leq b \leq 1$, Ω is the computational domain, Ω_s- object domain, ρ_s - material density, a and b - design parameters.

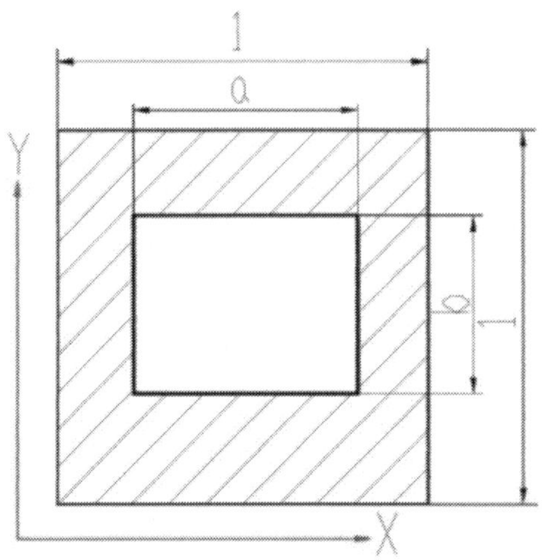

Figure 3. Elementary microstructure

The relative density method is a common topological optimization method, and its main idea is to introduce not an elementary microstructure, but a hypothetical material, the relative density of which varies from 0 to l.

The material model can be described by a system of equations:

$$\rho(x) = x_e \rho_0 \quad (1)$$

$$E(x) = x_e^{\rho} E_0 \quad (2)$$

where: ρ_0 and ρ – initial and final density of the topological structure; E_0 and E – initial and final elasticity of the topological structure of homogeneous objects, respectively; x_e – the relative density of the hypothetical material.

The relative density of the hypothetical material is used as a design variable, so optimization translates into optimal material distribution in the rocker arm design. The algorithm of the topological optimization process is shown in Figure 4.

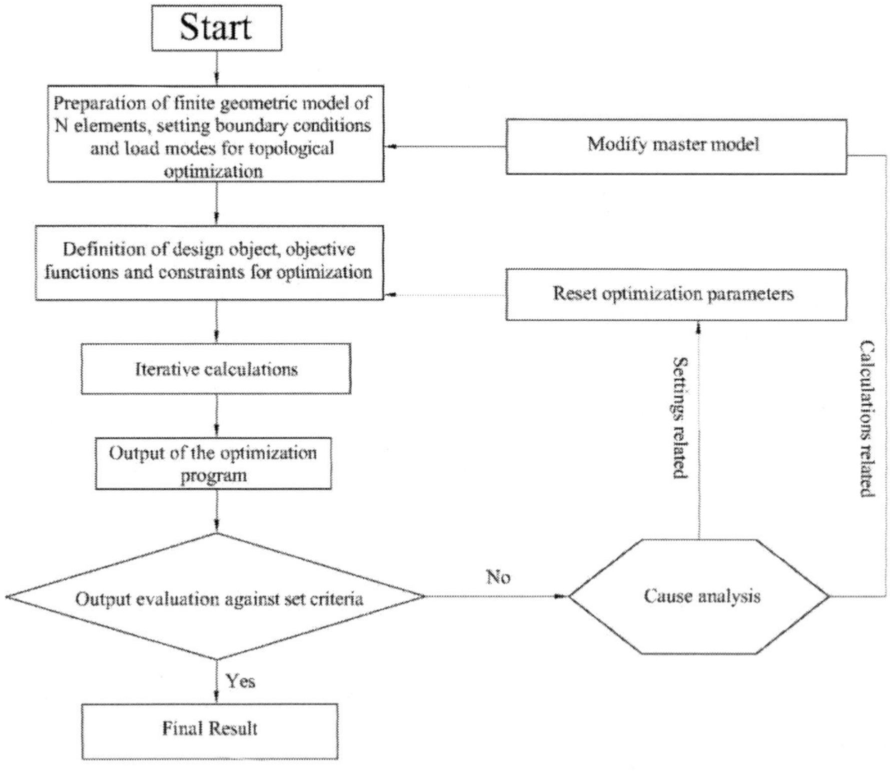

Figure 4. Algorithm of Typical Topological Optimization Process

3. Building of the master model of the rocker arm and the model of topological optimization

In the original design the rocker arm is welded from steel plates, properties shown in Table 1.

Table 1. Properties of steel plates

Elastic modulus (MPa)	Poisson's ratio	Density (t mm-^{3})	Tensile strength (MPa)	Permissible stress (MPa)
210,000	0.28	7.9e-09	235	156

Following the analysis and calculation, the maximum stress of the rocker arm has been determined at 90.3 MPa, which is much less than the permissible value, as shown in Figure 5. The weight of the rocker arm reached 24.5 kg, which is relatively heavy.

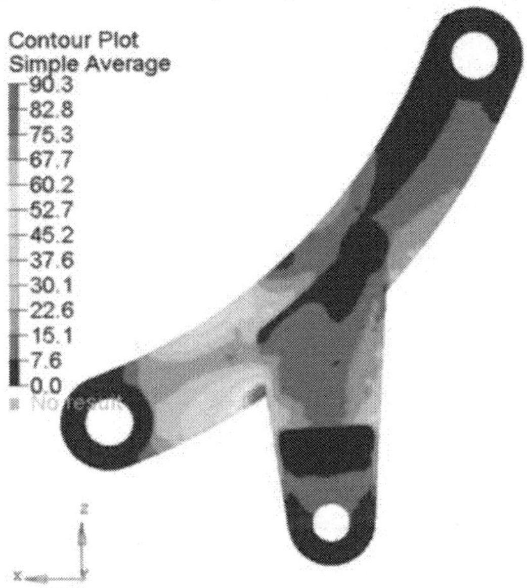

Figure 5. Stress level of the original structure

The main space of original rocker is filled, and the three pivot points are set as non-project areas. (fig. 6.)

Figure 6. Topological model of the rocker arm (non-project areas shown in blue color)

The objective function and the constraints must be defined prior to the topology optimization. The objective function is the minimum number of design areas and elements, while the constraint is imposed at the permissible stress level of 156 MPa.

4. Results of topological optimization of the rocker arm

The optimization calculations were performed in three iterations, the results for second and third iterations shown in Figure 7, 8.

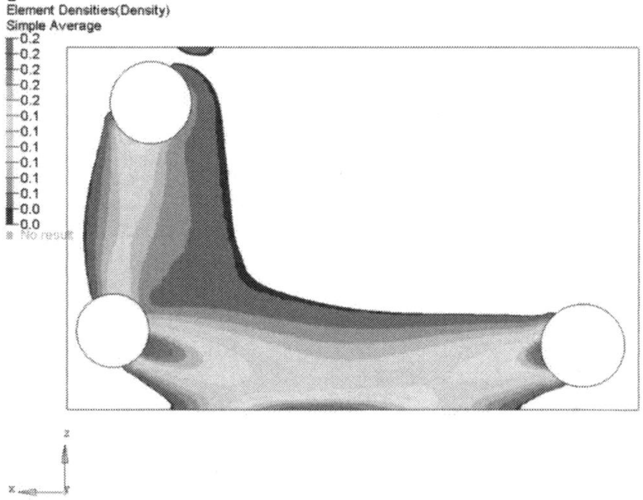

Figure 7. The second iteration of topological optimization

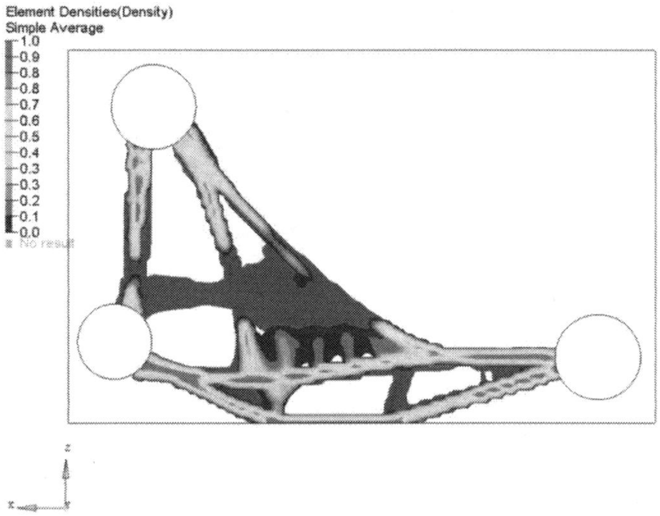

Figure 8. The third iteration of topological optimization

After considering design and operational factors, the structure and design of the rocker arm took on the following form (Fig. 9).

Figure 9. Optimized rocker arm

The level of stresses in the optimized rocker has been imitated (Fig. 10).

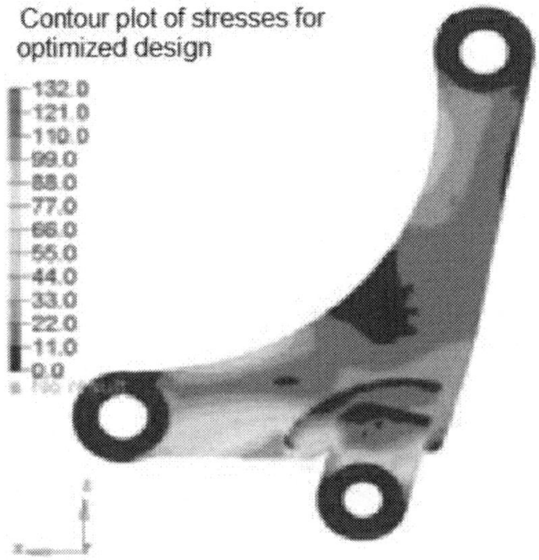

Figure 10. Stress level in the optimized rocker arm design

Optimization of the design of the rocker arm of the tipping system in a garbage truck and modelling the spatial arrangement of the material allowed to maintain maximum stress levels below the permissible level. Its value for the optimized design is about 132 MPa.

Optimization of material distribution allowed to reduce the weight of the rocker arm to 13.7 kg, which is 44% less than the weight of the original design. Thus, the main optimization goal, which was weight reduction of the existing rocker arm, has been achieved.

References

[1] Balovnev V I and Danilov R G 2019 Development of garbage trucks in Russia *Construction and road machines* **11** 3–11

[2] Kargapol'tseva N P 2019 Removal and disposal of MSW: garbage trucks for solving urgent problems *Municipal solid waste* **11(161)** 24–25

[3] Mironov S E and Safyannikov N M 2008 Construction of technologically invariant topology sketches based on the description of the topology in specific design codes *Izvestia SPBGETU LETI* **2** 19–25

[4] Kiryan V I, Dvoretsky V I and Malgin M G 2012 Calculation of local stresses in the zones of welded joints of large spatial structures *Automatic welding* **4 (708)** 3–7

[5] Suknev S V 2001 Taking into account the scale effect in strength calculations of structural elements with stress concentrators *Factory laboratory. Diagnostics of materials* **67** 57–6

[6] Vitus M Tabie, Yesuenyeagbe A K Fiagbe 2014 Weight optimization of A lift-Tipping mechanism for small solid waste collection truck *International journal of scientific & technology research* **3(11)** 8–12

[7] Andruchov N M, Kulik K M, Pysnina D Yu 2020 Garbage trucks with replaceable containers *Education. Transport. Innovations. Construction* (Electronic Materials vol 3) ed Fedosov B B (Omk: SibADI) chapter 1 pp 4–8

Assessment of the deformation and strength properties of parts manufactured using additive technologies, depending on the connection method and operating conditions

I S Nefelov[1], **N I Baurova**[1]

[1] Department of Manufacture and repair of automobiles and road machines, Moscow Automobile and road construction state technical university, 64, Leningradskiy avenue, Moscow, 125319, Russia

E-mail: nbaurova@mail.ru

Abstract. The manufacture of dimensional parts on 3D printers with a small working area for printing leads to the need to separate the model into parts, print them separately and join them after the end of printing. The paper considers various ways of joining plastic products obtained by 3D printing methods. Samples were made, parts of which were connected in the process of research in the three considered ways. The tests were carried out in accordance with the requirements of GOST 11262-2014 "Plastics. Tensile tests". The evaluation of the deformation and strength characteristics of the joints at different temperatures (+ 20°C, -30°C, - 50°C) was carried out. It has been established that negative temperatures do not have a negative effect on the deformation and strength characteristics of the joints of parts made of ABS plastic. The results of tensile tests are presented. The use of the technology proposed by the author for joining plastic parts at enterprises engaged in the manufacture and repair of dimensional products will allow expanding the range of products without significant changes in the material and technical base.

1. Introduction

Additive technologies have a huge potential for the production of products in various industries [1-2]. However, their distribution in the repair industry is relatively small and is mainly determined by the technologies that were developed in the second half of the twentieth century for the restoration of metal products. Taking into account the technical leap in the development of 3D printing, attention should be paid not only to the use of these technologies in manufacturing, but also in the repair of equipment [3-4].

Almost any road-building vehicle, automobile and other equipment has plastic products in its design. Moreover, a significant part of these parts is made of ABS plastic [5]. During operation, various kinds of defects may arise, associated with both natural wear and tear and emergency impacts. The most serious are those damages in which there is separation and loss of structural elements.

Restoration of plastic parts by traditional methods is carried out by gluing (used if the broken structural element is preserved), surfacing with filler rods approximately similar in configuration to the lost element, or molding with special compounds [6-7]. But all these methods do not allow you to accurately recreate the required geometry of the part.

The restoration method proposed by the authors consists in making an exact copy of the missing element on a 3D printer and connecting it to the part. The purpose of this work is to determine the

Content from this work may be used under the terms of the Creative Commons Attribution 3.0 licence. Any further distribution of this work must maintain attribution to the author(s) and the title of the work, journal citation and DOI.
Published under licence by IOP Publishing Ltd

IOP Conf. Series: Materials Science and Engineering 1159 (2021) 012052 doi:10.1088/1757-899X/1159/1/012052

change in the deformation and strength characteristics of plastic parts manufactured using additive technologies, depending on the connection method and operating conditions.

2. Objects and test procedure

The test specimens were printed on a Magnum Creative 2Pro 3D printer with print modes defined in earlier studies [8]. The material used was ABS-plastic made by REC, brand ABS M5. The samples are two butt-connected plates with a total overall dimension of 220 × 30 × 3 mm. The connection was made in three different ways: hot air welding, organic solvent bonding and 3D welding.

Welding with heated air was carried out at an air flow temperature of 300°C [9]. Air consumption during welding was 250 l/min.

The halves of the samples were glued on a flat surface. The ends of the plates were kept in an organic solvent acetone for 30 seconds, then half of the samples were placed on a glass surface, aligned relative to each other, and glued together with a counter axial force not exceeding 100 N [10].

To carry out the 3D welding process, half of the samples were placed on the printing table of the 3D printer and fixed on it coaxially with clamps. The extrusion temperature was 235°C, the speed of movement of the print head was 10 mm/min. The amount of filler plastic supplied was set by the program and was equal to 60 mm per side. Welding was carried out alternately on both sides. To minimize temperature deformations of the samples, they were heated to 105°C.

To compare the strength characteristics of the joints of plastic parts with solid parts, one-piece (reference) specimens with overall dimensions of 220 × 30 × 3 mm were made.

The appearance of the obtained samples is shown in Figure 1.

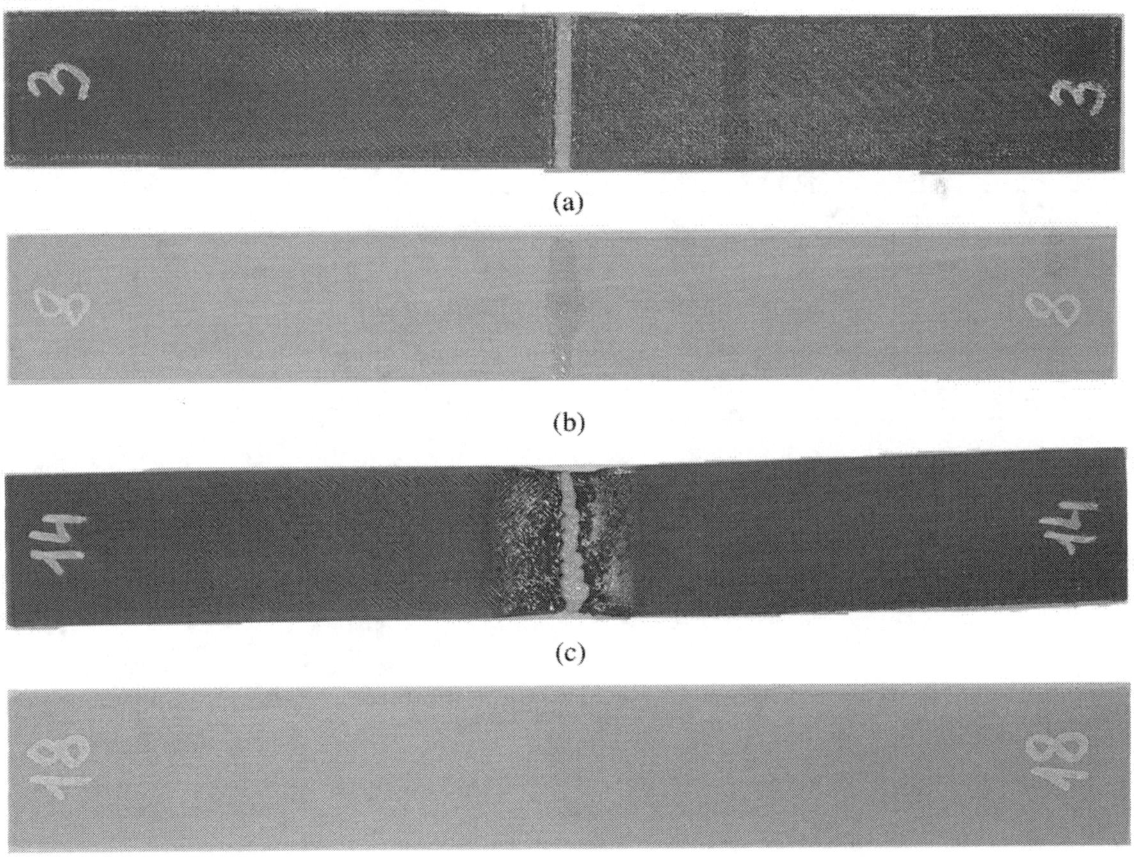

(a) (b) (c) (d)

Figure: 1. Samples connected by 3D welding (a), glued (b), welded with heated air (c), solid (d)

Since the operating conditions can differ significantly depending on the geographical location, the experiment was carried out at three different temperatures: + 20°C, -30°C, - 50°C [8]. To determine the effect of negative temperatures on the strength characteristics of the joints, the samples obtained were divided into three groups. Each of the groups of samples was kept for 30 days at the appropriate temperature. The first group is the control one. Holding temperature + 20°C (room temperature). The second group of samples was placed in a PozisFH-258-1 freezer with a temperature of -30°C (low temperature). The third batch of samples was loaded into a LOIPFT-311-80 cryostat with a temperature of -50°C (extremely low temperature).

After the end of exposure, the samples were tensile tested on a UTS-110M-50 testing machine using the temperature measurement system STI TS 3. The use of this temperature system allows you to provide the required temperature and maintain it during testing.

Tensile tests of plastic samples obtained by 3D printing were carried out in accordance with the requirements of GOST 11262-2017. The speed of movement of the grips of the testing machine during the tests was 2 mm / min.

3. Results and Discussion

Table 1 presents a summary of the test results for all three groups of samples.

Table 1. Results of tensile strength tests of specimens with different joining methods at + 20°C, -30°C and -50°C

Sample joining method	Strength characteristics of joints								
	Ultimate strength, MPa			Elastic modulus, MPa			Relative deformation at destruction,%		
	+20°C	-30°C	-50°C	+20°C	-30°C	-50°C	+20°C	-30°C	-50°C
3D welding	16.454	39.801	29.272	548.96	781.06	454.385	3.4	4.1	7.3
Hot air welding	1.794	13.65	17.521	622.38	752.69	473.925	0.3	1.6	3.45
Gluing	11.456	22.026	23.258	445.81	725.66	514.11	2.4	2.25	4.5
Whole (reference) samples	28.353	56.343	52.81	607.06	880.86	532.98	6.1	5.75	10.8

In figure 2 is a bar graph showing the change in tensile strength for each bonding method as a function of temperature.

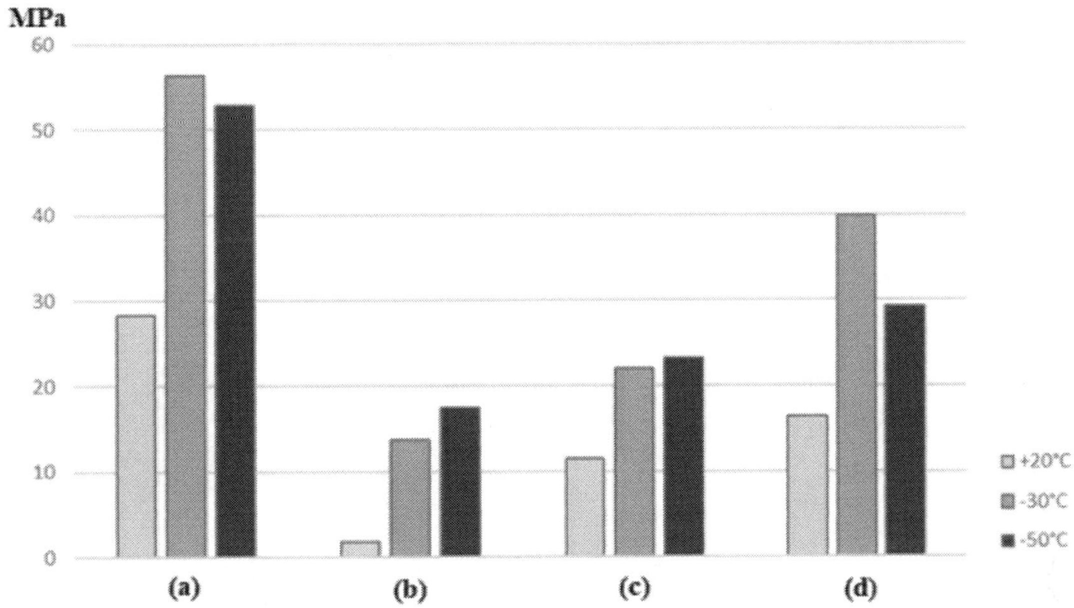

Figure 2. Change in tensile strength depending on the method of connection and ambient temperature: a – whole samples; b – hot air welding; c – gluing; d – 3D welding.

The histogram of the change in the elastic modulus at different temperatures, depending on the connection method, is shown in figure 3.

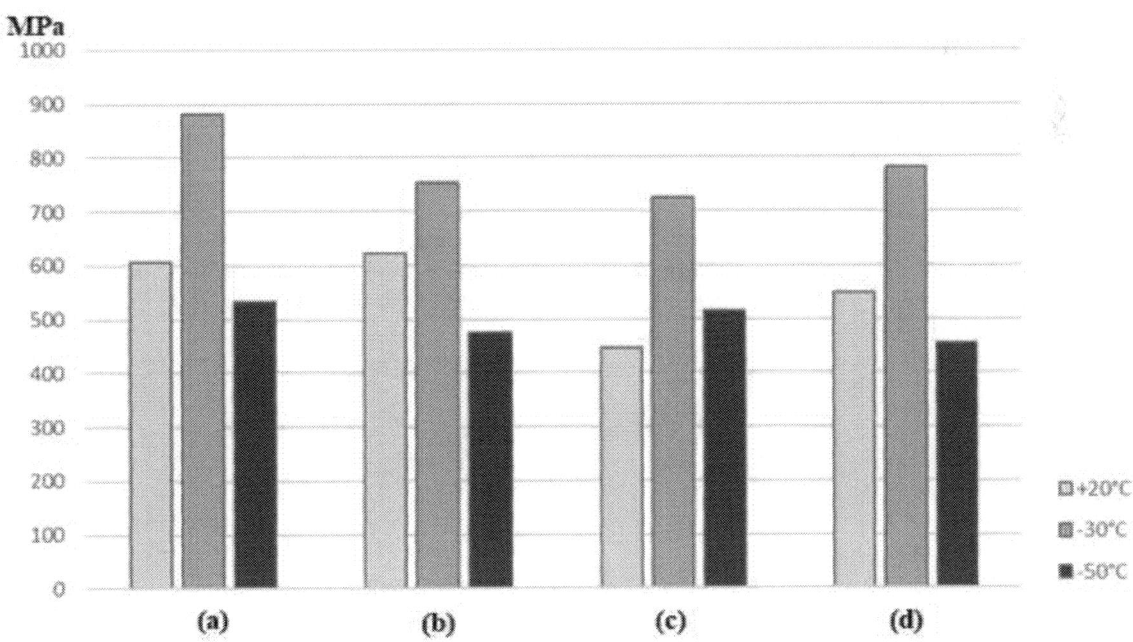

Figure 3. Change in modulus of elasticity in tension depending on the method of connection and ambient temperature: a – whole samples; b – hot air welding; c – gluing; d – 3D welding.

A comparative histogram of changes in the relative elongation of samples depending on the method of connection and temperature is shown in figure 4.

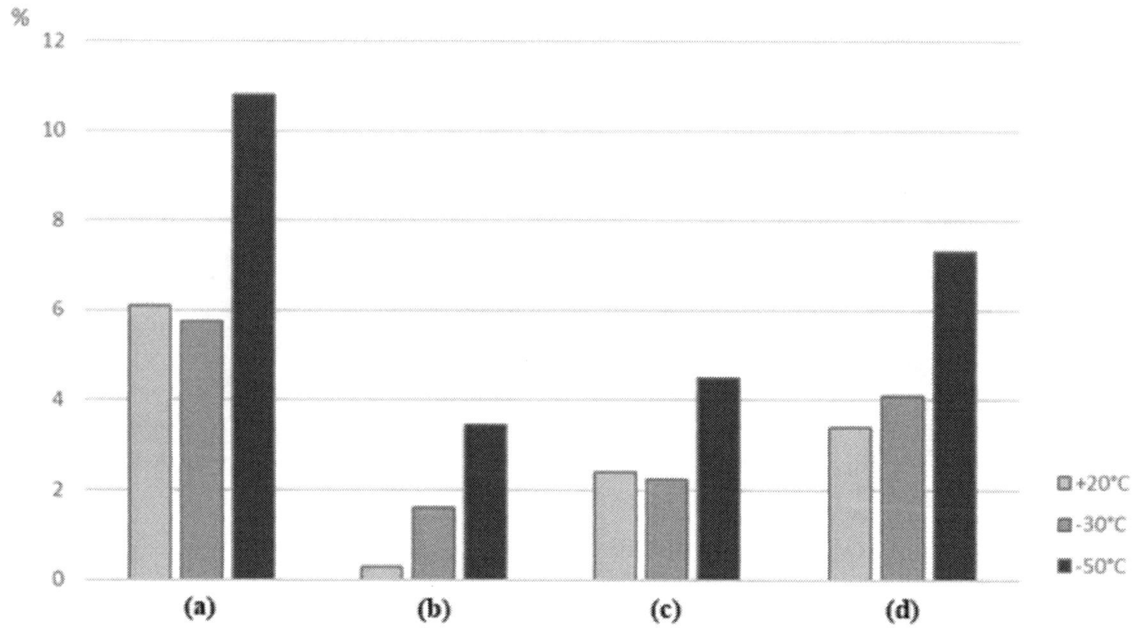

Figure 4. Change in the relative deformation of the samples under tension depending on the method of connection and the ambient temperature

As follows from the data obtained, negative temperatures generally have a strengthening effect on the joints of plastic parts. The method of joining proposed by the authors is characterized by the highest tensile strength in comparison with traditional joining methods.

4. Conclusion

The obtained test results make it possible to assess the prospects of the method proposed by the authors. An important role is played by the relatively low influence of the human factor on the quality of the resulting welded joint. It can be used in the conditions of repair enterprises. However, to obtain a more complete picture, it is necessary to carry out additional tests, including impact strength.

Acknowledgments

The material was prepared within the framework of scientific research under the project No. FSFM-2020-0011 (2019-1342), experimental studies were carried out using the equipment of the MADI collective use center.

References

[1] Valetov V A 2015 *Additive technologies (state and prospects)* (Saint Petersburg, Universitet ITMO)
[2] Zlenko M A 2013 *Additive technologies in mechanical engineering* (Saint Petersburg, SPbGU)
[3] Pavlov A P, Poluhin E V 2017 Methods for ensuring the efficiency of the operability of 3D printers in a production environment, *Repair, restoration, modernization* **3** 16–20
[4] Pavlov A P, Poluhin E V 2016 Application of 3D printers based on the basic principles of redundancy *Repair, restoration, modernization* **4** 27–31
[5] Zorin V A, Bocharov V S 2003 Machine reliability (Orel, OrelGTU)

[6] Baurova N I 2009 Deflected mode simulation of adhesive joints *Polymer Science - Series D* **2** 54–57
[7] Skryabin V A, Shirtladze A G 2017 Technological methods of using polymeric materials during repair work *Repair, restoration, modernization* **7** 14–19
[8] Nefelov I S, Baurova N I 2019 Evaluation of structural defects of articles obtained in various modes of 3D printing *Polymer Science - Series D* **12** 77–79
[9] Volkov S S 2001 *Welding and gluing polymeric materials* (Moscow, Chemistry)
[10] Korab G N, Shestopal A N *A reference dictionary for welding and bonding plastics* (Kiev, Science. dumka)

IOP Conf. Series: Materials Science and Engineering 1159 (2021) 012053 doi:10.1088/1757-899X/1159/1/012053

Investigation of the mechanical properties of polymer composite materials with various types of hybrid matrices in the extreme conditions of the Arctic

E A Kosenko[1], N I Baurova[1], V A Zorin[1]

[1] Department of Manufacture and repair of automobiles and road machines, Moscow Automobile and road construction state technical university, 64, Leningradskiy avenue, Moscow, 125319, Russia

E-mail: KosenkoKate@mail.ru

Abstract. The technology of manufacturing carbon plastics samples with various types of hybrid matrices is presented, one of the components of which is completely cured, and the second (technical wax, anaerobic and organosilicon polymer materials) retains its viscoelastic state after forming products, which ensures a change in the mechanical properties of structural materials. The possibility of using carbon plastics with various types of hybrid matrices for the manufacturing of mechanical engineering products that are used in extreme Arctic conditions is studied. A method for conducting mechanical tests to determine the tensile strength of carbon plastics samples with various types of hybrid matrices at temperatures t_1=20±2°C, t_2=-30°C, t_3=-50°C. The results of changes in the strength and deformation properties of carbon plastics with different types of hybrid matrices at negative and positive temperatures compared to the values of the same indicators of carbon plastics without viscoelastic components in the matrix are presented and analyzed.

1. Introduction

Currently, polymer composite materials (PCM) are one of the main structural materials used for the manufacture of large-sized and small body parts, but structurally complex, critical parts of machines and mechanisms [1]. The widespread use of PCM in various industries is due to the ability to create products from them with specified physicomechanical, thermophysical, dielectric, and other properties. High corrosion resistance, strength and low density of PCM make them the most preferred materials for the manufacture of cases, bodies, cabs, frames, fuel tanks, high-pressure cylinders (for cars running on gas fuel) and many other machine parts [1].

Increasing the pace of the development of the Arctic zone and achieving the goals of the state policy of the Russian Federation in the Arctic [2], among other tasks, require the attraction of significant production resources to improve the transport network, in particular, road construction equipment and technological equipment. Extreme natural and climatic conditions, including low air temperatures and strong winds, are features of the Arctic zone of the Russian Federation that influence the formation of state policy in the Arctic [3, 4].

Average sub-zero temperatures in the Arctic range from -4 to -25°C, and the minimum temperatures sometimes drop to -55... -60°C, which makes it necessary to test PCMs used in the production of machine parts and mechanisms at temperatures that allow working in these conditions.

A team of authors, for the production of machine parts operating under alternating loads, has

Content from this work may be used under the terms of the Creative Commons Attribution 3.0 licence. Any further distribution of this work must maintain attribution to the author(s) and the title of the work, journal citation and DOI.
Published under licence by IOP Publishing Ltd

proposed a technology for the production of PCMs with various types of hybrid matrices, one of the components of which is cured, and the second retains its viscoelastic state after the forming of products [5].

The viscoelastic components of the matrix, like the binder in materials of living nature (for example, in plants, bone, wood), can increase the deformation and strength properties of PCM [6]. The purpose of this study is to determine the dependence of the change in the strength properties of PCMs with different types of hybrid matrices, in particular, the tensile strength, on the ambient temperature: t_1=20±2°C, t_2=-30°C, t_3=-50°C.

2. Materials and manufacturing technology samples

Biaxial carbon fabric based on T300 carbon fiber was chosen as the reinforcing material for PCM samples with various types of hybrid matrices. Elan-Tech EC 57 epoxy resin was used as a binder (for creating products that require increased thermal stability) with a W61 hardener (in a ratio of 100:16 by weight), which ensures the viability of the resin + hardener compound of 70 ... 80 minutes. Based on the analysis of the curing kinetics, the following materials were selected as viscoelastic components of the hybrid matrices: technical wax, anaerobic (Loctite 638) and organosilicon (Unisil-9628) polymer materials.

The samples were made using prepreg technology by vacuum forming: 4 layers of carbon fabric were pre-impregnated with an epoxy binder and laid out on a prepared glass tooling; between 3 and 4 layers along the carbon fabric base thread, the material of a viscoelastic component (≈5 mm wide) was applied at a distance of 50 mm from each other; sacrificial and absorbent fabrics were laid on top of the laid layers, after which a vacuum bag was formed and a vacuum discharge was created. The samples were cured under the influence of infrared radiation using the Trommelberg IR3C Standard installation, which provides heating to a temperature of +80±2°C. The resulting carbon fiber sheets were cut into individual samples by mechanical means, followed by processing the edges with a binder.

Four series of flat rectangular carbon fiber samples were produced for mechanical testing:

Series 1: reference sample-carbon fiber without viscoelastic components in the matrix (sample size was 20x250x1.3);

Series 2: carbon fiber with a hybrid matrix, the viscoelastic component of which is an anaerobic polymer material (Loctite 638) (specimen size was 20x250x1,3);

Series 3: carbon fiber with a hybrid matrix, the viscoelastic component of which is a two-component organosilicon polymer material (Unisil-9628) (sample size was 20x250x1,3);

Series 4: carbon fiber with a hybrid matrix, the viscoelastic component of which is technical wax (sample size was 20x250x3,4).

Each series of samples was made in three sets:

- the 1st set of samples was not exposed to negative temperatures and was stored at a temperature of t_1=20±2°C;
- the 2nd set of samples was kept in a PozisFH-258-1 freezer at a temperature of t_2=-30°C for 60 days;
- the third set of samples was kept in a LOIP FT-311-80 liquid cryothermostat at t_3=-50°C for 30 days.

In the cryostat, medical ethanol was used as a working fluid (heat carrier), and air was used in the freezer.

The elastic, rheological and strength characteristics of PCM are determined not only by the mechanical properties and the type of components, but also by the conditions of interaction at the interface and the interaction between the elements [7]. The considered viscoelastic components in the composition of the hybrid matrix form an independent PCM phase, with the formation of boundary layers with a reinforcing material and a binder material.

3. Test procedure

The samples were tested in accordance with GOST RU 56785-2015 «Polymer composites. Tensile method for flat specimens» using the machine for testing structural materials «UTS 110M-50» and using the temperature testing system «STI TS 3» (for samples subjected to pre-exposure at negative temperatures). Before testing at temperature t_2=-30°C and t_3=-50°C, the samples were kept at the corresponding temperature for 60 minutes. The test speed was V=2 mm/min for all samples.

4. Discussion of test results

The tests performed allowed us to obtain the values of absolute breaking forces and tensile strength of the manufactured series of samples at temperatures: t_1=20±2°C, t_2=-30°C, t_3=-50°C The average values of these indicators are shown in table 1.

Table 1. Average values of mechanical properties of carbon plastics with different types of hybrid matrices in tensile testing

Holding and test temperature	Matrix type of carbon plastics samples			
	without viscoelastic components	with anaerobic polymer material	with organosilicon polymer material	with technical wax *
t_1=20±2°C	1803 / 46866	1907 / 49584	1288 / 33494	557 / 37900
t_2=-30°C	1356 / 35266	1371 / 35651	1159 / 30134	414 / 10783
t_3=-50°C	1269 / 32990	1274 / 33113	1226 / 31873	134 / 9125

Tensile strength, MPa / Absolute breaking strength, N

* In a comparative assessment of the mechanical properties of carbon plastics with various types of hybrid matrices, an adjustment should be made for the scale factor of the samples (samples of carbon fiber with technical wax in the matrix with the same width had a thickness of 2,6 times more than the other samples).

Carbon plastics samples with an anaerobic polymer material in the matrix have the highest average tensile strength at both positive and negative temperatures, which is explained by the nature of interfacial interactions at the «fiber-binder-anaerobic polymer material» interface. The somonomers included in the composition of anaerobic polymers, which are acrylic and methacrylic acids and amides of unsaturated acids, contribute to an increase in adhesion [8], which explains the higher average value of the ultimate strength for a given series of samples, as well as a small spread in the values of this indicator at a temperature t_1=20±2°C (table 2).

When the temperature decreases, the properties of the anaerobic polymer material change (the recommended storage temperature for Loctite 638 is in the range of 8 ... 21°C, and transportation is allowed at a temperature not lower than -16°C). Presumably, this factor negatively affects its adhesive properties. Reducing the adhesive strength in the «fiber-binder-anaerobic polymer material» system at temperatures t_2= -30°C, t_3=-50°C increases the value of the unbiased point estimate of the dispersion of the parameters monitored during tests by almost 2 times. However, the value of the ratio of dispersion to the mathematical expectation of ultimate strength remains within 5% (table 2). This is a good indicator and allows predicting the properties of PCMs with this type of hybrid matrix with a sufficient degree of reliability in the manufacture of products operating under negative temperatures.

An analysis of the data obtained during mechanical tests allows us to make an unambiguous conclusion that low temperatures lead to a decrease in the mechanical properties of PCMs both with various types of hybrid matrices and PCMs without viscoelastic components in the matrix. At the same time, it should be noted the intensity of the decrease in the studied indicators for various types of PCM. So, after holding in a freezer and testing at a temperature of t_2=-30°C for samples without viscoelastic components in the matrix, there is a decrease in the average value of the ultimate strength by 25% is observed. For samples with anaerobic polymer material and technical wax in the matrix, respectively, by 28 % and 26%. The smallest decrease in the average value of the ultimate strength in comparison with the samples not exposed to low temperatures is observed in PCM with an organosilicon polymer material in the matrix - by 10%.

After holding the samples in a cryothermostat and testing them at a temperature of t_3=-50°C, for carbon plastics without viscoelastic components and with an anaerobic polymer material in the matrix, a decrease in the average value of the ultimate strength by 30% and 33%, respectively, is observed in comparison with the samples of these series, which not exposed to low temperatures. The greatest decrease in the average value of the tensile strength compared to this indicator for samples that were not exposed to low temperatures is observed in carbon plastics with technical wax in the matrix – by 76%. Of particular interest are samples with an organosilicon polymer material in the matrix, since for these samples at a temperature of t_3=-50°C, there is a minimal decrease in the controlled indicator - by 5% compared to samples of this series that were not exposed to low temperatures. At the same time, in comparison with the samples of this series tested at a temperature of t_2=-30°C, at t_3=-50°C, an increase in the average value of the ultimate strength by 5% is recorded.

Various physical properties of polymers (mechanical, electrical, magnetic, optical, thermal) are determined by the processes of molecular mobility occurring in them. These processes depend on the structural features of their chains and structural organization [9].

Insignificant, in comparison with other series of samples, the change in the average value of the ultimate strength of carbon plastics samples with an organosilicon polymer material in the matrix with decreasing temperature can be explained from the standpoint of the structure of organosilicon polymers. In these materials, the main chain is made up of silicon and oxygen atoms, and the side groups contain atoms of carbon, hydrogen or other chemical elements, which makes organosilicon compounds have increased thermal and chemical resistance [9].

In the process of testing at a temperature of t_2=-30°C, samples of carbon plastics with an organosilicon polymer material in the matrix are destroyed in the most mechanically stressed places with lower breaking forces. However, the adhesive strength of organosilicon polymers increases during cooling [10]. This leads to an increase in the adhesion strength in the "fiber-binder-organosilicon polymer material" system and, as a consequence, to a slight increase in the average value of the ultimate strength of this type of samples at a temperature t_3=-50°C, as well as a decrease in the value of the unbiased point estimate of the variance of this indicator (table 2).

Table 2. Values of the unbiased point estimate of the variance of the tensile strength of carbon plastics with different types of hybrid matrices

Holding and test temperature	Matrix type of carbon plastics samples			
	without viscoelastic components	with anaerobic polymer material	with organosilicon polymer material	with technical wax *
t_1=20±2°C	217,01 / 12,04	35,95 / 1,89	158,77 / 12,25	109,18 / 19,59
t_2=-30°C	36,7 / 2,7	65,55 / 4,78	59,5 / 5,13	30,21 / 7,2
t_3=-50°C	85,54 / 6,74	61,13 / 4,8	53,26 / 4,35	22,98 / 17,13

Unbiased point variance estimate / Percentage of variance value to mathematical expectation, %

Technical wax with a carbon plastics impregnated with a binder forms a mechanical bond, which is carried out due to the mechanical engagement of the irregularities of the contacting surfaces and the pressure that is created by a vacuum bag when the formation of samples. At a temperature of t_2=-30°C, the technical wax in the composition of the carbon fiber matrix freezes, but the mechanical connection with the reinforcing material remains. The sharp decrease in the average value of the ultimate strength of specimens of this series at a temperature of t_3=-50°C (table 1) is explained by the complete loss of plasticity by the technical wax. After holding in a cryothermostat, the technical wax became brittle and crumbly, due to which the mechanical connection with the reinforcing material was completely lost. The absence of a plastic layer led to the loss of deformation properties and weakening of the material as a whole.

An important characteristic of the viscosity and deformation of a material during fracture is the elongation at fracture δ. The average values of the relative elongation in fracture of carbon plastics samples with different types of hybrid matrices and without viscoelastic elements in the matrix at temperatures t_1=20±2°C, t_2=-30°C, and t_3=-50°C are presented in table 3.

Table 3. Average values of elongation at fracture of carbon plastics samples with different types of hybrid matrices

Matrix type of carbon plastics samples	Elongation of specimens at fracture δ,%		
	Holding and test temperature t_1=20±2°C	Holding and test temperature t_2=-30°C	Holding and test temperature t_3=-50°C
without viscoelastic components	5,3	4,9	4,5
with anaerobic polymer material	5,7	5,6	4,6
with organosilicon polymer material	5,5	5	4,7
with technical wax *	7,1	4,4	4

The viscoelastic components in the PCM hybrid matrix form internal interfaces, the strength of which is less than 1/5 of the total adhesion strength of the material. This is the main condition for the implementation of the crack braking mechanism at the interface according to Cook-Gordon [11-13]. The presence of weak internal surfaces, contrary to expectations, does not weaken the material as a whole, but strengthens it, making it viscous like living wood or bone. This allows us to explain the higher values of the relative elongation of samples with different types of hybrid matrices at a temperature of t_1=20±2°C and decrease in the value of this indicator at temperatures t_2=-30°C, t_3=-50°C due to freezing of viscoelastic components and their destruction.

When assessing the mechanical properties of carbon plastics with different types of hybrid matrices and their changes under the influence of negative temperatures, not only strength and deformation characteristics, but also the character of the destruction of samples are important (table 4).

Table 4. Description of the nature of destruction of carbon plastics samples with different types of hybrid matrices

Matrix type of carbon plastics samples	Character of destructions		
	Holding and test temperature t_1=20±2°C	Holding and test temperature t_2=-30°C	Holding and test temperature t_3=-50°C
without viscoelastic components	Longitudinal splitting in the working area (found in the grippers), transverse fracture under grippers		
with anaerobic polymer material	Longitudinal splitting in the working area (found in the grippers), transverse fracture occurs under the grippers	Longitudinal splitting in the working area, lateral destruction under the jaws	Longitudinal splitting in the working area (found in the grippers), lateral destruction under the grippers
with organosilicon polymer material	Longitudinal splitting in the working area (found in the grippers)	Longitudinal splitting in the working area, lateral destruction under the grippers	Longitudinal splitting in the working area, lateral (oblique) fracture under grips
with technical wax *	Longitudinal splitting in the working area and in the grippers	Longitudinal splitting in the work area and in the grippers or slippage in the grippers	Longitudinal splitting in the working area and in the grippers

At low temperatures, PCMs are characterized by such fracture mechanisms as: cross-cracking of the matrix, disruption of the interlayer gap, peeling and destruction of fibers [14].

Figures 1-3 show photographs of carbon plastics samples tested at temperatures t_1=20±2°C, t_2=-30°C and t_3=-50°C. For samples (with the exception of carbon plastics with technical wax in the composition of a hybrid matrix), when tested for tensile tests at negative temperatures, severe cracking along the fibers of the base used in the manufacture of carbon plastics samples is characteristic.

a. b. c. d.

Figure 1. Carbon plastics samples after tensile testing at t_1=20±2°C: a – without viscoelastic components; b –with anaerobic polymer material; c – with organosilicon polymer material; d – with technical wax

a. b. c. d.

Figure 2. Carbon plastics samples after holding and tensile tests at t_2=-30°C: a – without viscoelastic components; b – with anaerobic polymer material; c – with organosilicon polymer material; d) with technical wax

Figure 3. Carbon plastics samples after holding and tensile tests at t_3=-50°C: a – without viscoelastic components; b – with anaerobic polymer material; c – with organosilicon polymer material; d) with technical wax

5. Conclusion
With a decreasing in temperature, the destruction of carbon plastics of all series occurs at lower stresses and is accompanied by intense cracking of the surface.

Carbon plastics with an anaerobic polymer material in the composition of the hybrid matrix, when passing to the region of negative temperatures, has the best tensile strength, stable and maximum (compared to the samples of the other series) elongation at fracture at a temperature of t_2=-30°C, and also comparatively high value of this indicator at t_3=-50°C (2% higher compared to carbon plastics without viscoelastic components in the matrix).

Tests of carbon plastics with an organosilicon polymer material as part of a hybrid matrix showed a slight decrease in the tensile strength (by 10% and 5% at temperatures t_2=-30°C and t_3=-50°C, respectively), while for this series of samples the maximum the value (in comparison with the samples of the other series) of the relative elongation at fracture at a temperature of t_3=-50°C (4% higher compared to carbon plastics without viscoelastic components in the matrix).

The samples of carbon plastics reinforced plastic with technical wax in the composition of the hybrid matrix at a positive temperature have a maximum relative elongation at fracture compared to the samples of the remaining series (20 ... 25% higher). However, in the region of negative temperatures, this type of specimen showed the highest rate of decline and the worst values of both deformation and strength properties.

Thus, the use of carbon plastics with anaerobic and organosilicon polymer materials in the composition of hybrid matrices in the design and production of mechanical engineering products operating in the Arctic, will provide higher strength and deformation properties of machine parts compared to carbon plastics without viscoelastic components in the matrix.

Acknowledgments
The material was prepared within the framework of scientific research under the project No. FSFM-2020-0011 (2019-1342), experimental studies were carried out using the equipment of the MADI collective use center.

References
[1] Timoshkov P N, Khrulkov A V and Yazvenko L N 2017 *Proceedings of VIAM* **6** 61–68
[2] Ob utverzhdenii gosudarstvennoj programmy Rossiiskoi Federacii «Social'no-ekonomicheskoe

razvitie Arkticheskoj zony Rossiiskoi Federacii. Postanovlenie Pravitel'stva Rossiiskoi Federacii ot 21.04.2014, № 366 (On approval of the state program of the Russian Federation" Social and economic development of the Arctic zone of the Russian Federation. Decree of the Government of the Russian Federation of 21.04.2014, № 366), Moscow, 2014.

[3] Osnovy gosudarstvennoj politiki Rossijskoj Federacii v Arktike na period do 2020 goda i dal'nejshuyu perspektivu. Utverzhden Prezidentom Rossiiskoi Federacii 18.09.2008,№ Pr – 1969 (Fundamentals of the state policy of the Russian Federation in the Arctic for the period up to 2020 and beyond. Approved by the President of the Russian Federation on 18.09.2008, № Pr-1969)

[4] Strategiya razvitiya Arkticheskoj zony RossiiskoiFederacii i obespecheniya nacional'noj bezopasnosti na period do 2020 goda. Utverzhdena Prezidentom Rossiiskoi Federacii (Strategy for the development of the Arctic zone of the Russian Federation and ensuring national security for the period up to 2020. Approved by the President of the Russian Federation)

[5] Kosenko E A , Baurova N I and Zorin V A 2020 *Polymer Science, Series D* **3** 341–344

[6] Kosenko E A, Baurova N I and Zorin V A 2020 *Tekhnologiya metallov* **2** 2–7

[7] Adamenko N A, Fetisov A V and Agafonov G V 2010 *Structural polymer composites* (Volgograd: VOLG) p 100

[8] Petrova A P and Malysheva G V 2017 *Adhesives, adhesive binders and adhesive prepregs* ed E N Kablov (Moscow: VIAM) p 472

[9] Bartenev G M and Zelenov Yu V 1983 *Physics and mechanics of polymers* (Moscow: Higher. School) p 391

[10] Ostec-materials, available at: https://ostec-materials.ru/tech_lib/publications_otm/proizvodstvo-mikroskhem-i-poluprovodnikovykh-priborov/opredelenie-nizhney-granitsy-rabochikh-temperatur-silikonovykh-elastomerov-dlya-proizvodstva-izdeliy1.php (accessed 3 august 2020)

[11] Gordon J E 2006 *Princeton University Press The New Science of Strong Materials: Or Why You Don't Fall Through the Floor* (Princeton: Princeton Science Library), p 287.

[12] Cook J and Gordon J E 1964 *Proceedings of the royal society A* **282(1391)** 508–520

[13] Nikolaeva E A 2010 *Fundamentals of fracture mechanics* (Perm: Perm State Technical University) p 103

[14] Egorov D V 2014 *Reshetnev readings* **1** 81–82.

Formation of minimum social standards in the field of public transport services for the provinces of Syria

D M Othman[1], A I Roshchin[1]

[1] Moscow Automobile and Road Construction State Technical University (MADI), Leningradsky Ave, 64, Moscow, 125319, Russia

E-mail: dareenothman1731991@gmail.com

Abstract. The problem of planning public transport in most Syrian cities is an essential problem since until now almost no one has addressed it. The reasons for this are the lack of adequate information about volumes of passenger traffic and the work of urban passenger transport, which affects on the quality of decisions made by the management of transport organizations, which, as a rule, choose either unsuitable systems for organizing urban passenger transport (UPT), or allow private operators to make their own decisions. Depending on the results of practical researches of passenger transportation in the province of Latakia, as well as after studying the social standards of different countries of the world, including the Russian Federation, authors have defined some parameters which that characterize the passenger experience before, during and after each trip. These materials have allowed to formulate recommendations in the field of public transport passenger service in the form of minimum social standards. These standards should become the basis for Improving public transport services for passengers not only in Latakia province, but also in Syria as a whole.

1. Introduction

1.1. Minimum social standards in the field of public transport services (MSSPTS)

The development of minimum social standards in the field of public transport has a very significant socio-economic importance because of the fact that the efficiency of social activities and various production sectors is greatly influenced by the level of transport services for the population, which in turn affects the economic situation of society [1].

The required transportation system should ensure, as far as possible, the mobility requirements of the population, and the budget allocated to activate this system should be directed towards improving the status of transportation system services in the community. According to this, the required transport system should:

- take into consideration all possible modes of transportation services (public, commercial, individual);
- apply the specified standards for the quality of the transportation service (safety, reliability, comfort, environmental protection and rational use of natural resources);
- be comfortable and accessible to all segments of the population.

The social norms taken in the field of transportation service for passengers are scientifically based on the quantitative and qualitative characteristics of the optimal state of the city environment, which depend on transportation and affect the working life of a person. [2].

Content from this work may be used under the terms of the Creative Commons Attribution 3.0 licence. Any further distribution of this work must maintain attribution to the author(s) and the title of the work, journal citation and DOI.
Published under licence by IOP Publishing Ltd

IOP Conf. Series: Materials Science and Engineering 1159 (2021) 012054 doi:10.1088/1757-899X/1159/1/012054

1.2. The objectives of the development and use of minimum social standards in the field of public transport services (MSSPTS) are:

- Long-term planning for the development of the place, taking into account the planning of buildings and the economic and social characteristics of this place.
- Creating a new mechanism to finance the development of the passenger transport system, with the aim of increasing the efficiency of the use of budget funds and the operation of the passenger transport system.

2. Examples of social standards in the field of public transport services in different countries and regions

Social standard of transport services for the population in the implementation of passenger and baggage transportation by road and urban land electric transport was approved by the order of the Ministry of Transport of the Russian Federation of January 31, 2017 No. NA-19-R [3]. An analysis of the use of MSSPTS on the territory of the Russian Federation showed that the system of minimum social standards for transport has been developed and operates in the following regions: the Republic of Bashkortostan, Kabardino-Balkaria, Karelia, Tatarstan, Sakha – Yakutia, Krasnodar territory, Arkhangelsk, Astrakhan, Kirov, Sverdlovsk regions and others. Each region in the Russian Federation has approved this standard on its own territory. Examples are given for the Krasnodar territory and the Moscow region in table 1 [4], [5].

Table 1. Minimum social standards in the sphere of public transport services in the Moscow region and Krasnodar territory.

Indicator	Moscow region	Krasnodar region
Reliability level of public transport modes, %	+	+
Comfort level of passenger transportation, people / m^2	+	-
Minimum number of trips public transport to sparsely populated towns and villages with the resident population, trip /day	+	+
Level of saturation of the region with bus fleet, units/1000 citizen	+	-
Level of saturation of the region with passenger seats, seats / 1000 citizen	+	-
Maximum allowable interval of urban transport, min	+	-
Distance of localities from public transport lines, km	+	+
Share of transport in environmental pollution, %	-	+
Specific lost free time fund (for 1 person per week), h	-	+
Road accident rate, units/100,000 trips	-	+
Annual mobility of the population with socio-cultural goals, %	-	+
Ratio costs between infrastructure and bus fleet, %	-	+
Share of public transport in passenger movements, %	-	+
Level of development of muscular modes of transport (such as bicycles) in urban and suburban communications, %	-	+
Effectiveness of financing of road economy	-	+

Note. "+" – the indicator exists in the district; "-" – the indicator does not exist in the district.

In foreign countries, much attention has also been paid to the current state of public transport in cities and to ensuring the convenience of a passenger travel route that suits all sectors of society - both the poor and the rich. In many of them, including the countries of the European Union, Latin America and others, studies were carried out on this problem, specific recommendations were proposed to improve passenger service and ways to implement some of them [6, 7, 8, 9, 10].

3. Evaluation of the significance of standards by experts in the field of passenger transport.
Having studied the experience of foreign countries in improving the quality of passenger transport on public transport and taking into account the peculiarities of the Syrian provinces and the current state of passenger traffic in them [11,12,13], the following indicators were proposed, which are important in terms of improving the quality of passenger service and improving the efficiency of the passenger transport system.

The researchers used a questionnaire method in the studied area, and during the survey, experts (12 experts from Syria working as transport specialists at Tishreen University and institutions affiliated with the Syrian Ministry of Transport) were interviewed to evaluate 16 indicators in the field of public transportation services (PTS). In the evaluation process, a scale consisting of 5 values was adopted. One value is given for each of the 16 indicators, depending on the importance of the indicator, as shown in the following lines:

- The indicator is not meaningful to provide the minimum acceptable level of PTS;
- The indicator is of little significance to ensure the minimum acceptable level of PTS;
- The indicator is unambiguously recognized as significant for ensuring the minimum acceptable level of PTS;
- The indicator is significant for ensuring the minimum acceptable level of PTS;
- The indicator is extremely significant for ensuring the minimum acceptable level of PTS.

Table 2 shows the results of experts' assessment of indicators established as minimum social standards (MSS) in the field of public transport services.

Table 2. Results of experts ' assessment of indicators in the field of public transport services

№	Indicator		Evaluation				
			1	2	3	4	5
1	Transport mobility of the population in accordance with socio-cultural activities	Frequency of assessments	1	2	2	**4**	3
2	Level of integrated transport availability	Frequency of assessments	0	0	4	2	**6**
3	Level of reliability of public transport	Frequency of assessments	0	0	0	4	**8**
4	The time spent for the movement of passengers	Frequency of assessments	0	0	1	3	**9**
5	Ratio of equipment costs and infrastructure, %	Frequency of assessments	1	**5**	**5**	1	0
6	Level of comfort of transportations of passengers, person/ m^2 (for city buses), the factor of filling microbus is no more 1. –Temperature of air in a vehicle; – A level height of a floor of buses, mm	Frequency of assessments	0	1	0	**9**	2
7	Efficiency of financing of a road facility	Frequency of assessments	0	2	**5**	4	1
8	Level of saturation of the region with passenger seats, seats/1000 citizen	Frequency of assessments	2	3	2	**4**	1
9	Level of saturation of the region with bus fleet, units/1000 inhabitants	Frequency of assessments	0	1	1	**6**	4
10	Minimum number of trips public transport to sparsely populated towns and villages with the resident population, trip /day	Frequency of assessments	0	2	3	**4**	3
11	Availability of schedules indicating working hours and frequency of public transport	Frequency of assessments	0	0	0	5	**7**
12	Availability of certain stops on the route	Frequency of assessments	0	1	1	**7**	3
13	Maximum allowable interval of movement:	Frequency of assessments	0	1	1	**5**	**5**
14	Percentage of buses operating in the Dispatcher control system	Frequency of assessments	0	2	2	**6**	2
15	Safety of movement: – speed of bus fleet km/h, – Reliability of functioning of vehicles.	Frequency of assessments	0	0	0	**6**	**6**
16	Number of transfers in movements:	Frequency of assessments	0	3	0	**7**	2

As a result of the questionnaire survey, it was found that 14 indicators were recommended as MSSPTS that received a good (4) and excellent (5) expert assessment (table 3)

Table 3. The list of indicators established as MSSPTS in Syria

Indicator name	Unit of measurement
Transport mobility of the population in accordance with socio-cultural activities	movement/ year
Level of integrated transport availability	travel time (hour)
Level of reliability of public transport	trip time (min)
Time spent for the movement of passengers	min
Level of comfort of transportations of passengers	person/m^2
Level of saturation of the region with passenger seats	seats/1000 citizen
Level of saturation of the region with bus fleet	units/1000 inhabitants
Minimum number of trips public transport to sparsely populated towns and villages with the resident population	trip /day
Availability of schedules indicating working hours and frequency of public transport	–
Availability of certain stops on the route	stop / 1km
Maximum allowable interval of movement	min
Percentage of buses operating in the Dispatcher control system	%
Safety of movement	number of road accidents / year
Number of transfers in movements	change / movement

4. Conclusion

The task of establishing minimum social standards for public passenger transport is a fundamental and priority task for improving public transport services, and many countries have paid great attention to it. Thus, our study is based on the establishment of minimum standards for passenger transport services in the Syrian Arab Republic. Further research will assess these indicators from the point of view of passengers traveling by public transport, then the adoption of standards in the field of passenger transport which are important to improve the process of mass passenger transport (based on the results of the matrix of indicators).

References

[1] Grishkevichene D 2008 Influence of the convenience of the transport system of passengers on the standard of living in the city of Vilnius *The 8th International Conference "Reliability and Statistics in Transportation and Communication* (Riga, Latvia) pp 73–80

[2] lezer V A, Semeryanova N A and Kopytova A V 2018 Public transport as a socio-legal phenomenon on the example of the city of Nizhnevartovsk (Khanty-Mansi Autonomous Okrug of Yugra)*MATECWeb of Conferences* **239** (TransSiberia) pp 201–210

[3] Social standard of transport services for the population in the implementation of transportation of passengers and luggage by road and urban land electric transport: order of the Ministry of Transport of Russia dated January 31, 2017 NA-19-r (as amended on April 13, 2018).

[4] Roshchin A I 2007 Development of a methodology for the formation of minimum social standards of the Moscow region in the field of transport services for the population: report on research MADI (Moscow)107 p

[5] Order of February 13, 2018 No. 42 "On approval of standards for the minimum provision of the

population with points for technical inspection of vehicles for the Krasnodar Territory and for its constituent municipalities" (as amended on April 27, 2018)
[6] *The role of the state in providing affordable public transport services for low-income residents* 26 p. 2011 Available at: www.internationaltransportforum.org
[7] Knupfer Sht and Pokotilo V 2018 Elements of success: urban transport systems of 24 cities of the world (mckinsey) 69 p
[8] Changhwan M O, Young-in K and Sangjun P 2014 *The current state of public transport in the megalopolises of Asia* (South Korea: The Korea Transport Institute)
[9] Yolandi R 2013 *Comparative Study of Public Transport Systems in Developing Countries (*South African: *U*niversity *of Cape* Town)
[10] Yves B *2010* Changing mobility in Asian cities. *South East Asia Geography Conference (SEAGA).* (Hanoi HAL Id: hal-00908986) p 21
[11] Roshchin A I and D M Usman 2019 Comprehensive survey of transport mobility of the population in Latakia province (Syria) *J.Vestnik RGUPS* **4** 149–154
[12] Roshchin A I and D M Usman 2019 Assessment of the level of development of the passenger transport system in Syria (on the example of the province of Latakia) *J. Universum* **5(62)** 28–30
[13] Aleksandrov S B and Alkhelu N 2019 Studying the problems of efficiency of public passenger transport between Gabla and Latakia and its development trends *Automobile. Road. Infrastructure* **4(22)** 14

IOP Conf. Series: Materials Science and Engineering 1159 (2021) 012055 doi:10.1088/1757-899X/1159/1/012055

Engine performance, emissions characteristics of the compression ignition engine using HHO gas (Brown's gas)

Nguyen Thin Quynh[1], Andrey Y. Dunin[2], Le Hoai Duc[1] and Mikhail G. Shatrov[2]

[1] University of transport and communications, 3 Cau Giay Strs., Ha Noi, 100000, Vietnam
[2] Moscow Automobile and road construction state technical university (MADI), 64, Leningradsky Prosp., Moscow, 125319, Russia

E-mail: thinquynh@utc.edu.vn

Abstract. HHO gas (Brown's gas) has recently been used as a new source of energy. The study presents the experimental results at 50% of the accelerator pedal when installed hydroxy (HHO) gas generator on 490 QZL diesel engines. HHO gas was provided into the engine manifold. Comparing the experimental results found that: the economic, technical and exhaust emissions of the diesel engine was improved. The brake power index increased by 2.8% in average compared to pure diesel and 4.2% was the percentage of the maximum engine power increasing at 2200 rpm. Especially, the amount of brake specific fuel consumption (*BSFC*) in pure diesel was always higher than one in diesel with HHO gas, which was 13.7% in average. At 1800 rpm, the difference of the minimum value of *BSFC* in two cases was 12.6%. The HC and CO emissions value reduced by 7.7% and 19.8% in average, respectively, while NO_x increased by 13.3% in average compared to pure diesel.

1. Introduction

The internal combustion engine (ICE) is widely used in the world because of their higher thermal and economic efficiencies than other types. However, it is one of the major sources of air pollution with toxic substances such as carbon monoxide (CO), sulfur oxides (SO_x), hydrocarbons (HC), especially particulate matter (PM) and nitrogen oxides (NO_x). In addition, fossil fuels (oil, coal, gas) are the primary fuel for ICE while their reserves in nature are being depleted. This problem becomes too exacerbate with the increasing of vehicle numbers. In recent years, many studies have been done to find solutions to the above problems. Including use of catalytic convert and particle filters; bio fuel; redesign and improvement the engine structure [1, 2, 3].

Regarding to diesel engines, the tightening environmental requirements and the need to improve their energy efficiency have led to the implement the following solutions:

- Increasing the injection pressure up to 3000 bar [4, 5];
- Ensuring the required shape of the tip edge of the injection characteristic [6, 7], taking into account the wave phenomena arising during fuel injection [8];
- Controlling of fuel distribution by the combustion chamber zones [9];
- Converting diesel into gas engine or gas-diesel [10, 11].

One of the most effective solutions is to use that using hydrogen as a fuel, which is characterized by high energy efficiency and without impurities.

There are different ways to use hydrogen. One of these is the production and storage of hydrogen

Content from this work may be used under the terms of the Creative Commons Attribution 3.0 licence. Any further distribution of this work must maintain attribution to the author(s) and the title of the work, journal citation and DOI.
Published under licence by IOP Publishing Ltd

IOP Conf. Series: Materials Science and Engineering 1159 (2021) 012055 doi:10.1088/1757-899X/1159/1/012055

[12, 13, 14, 15]. In this way, the hydrogen is compressed into the tanks and installed on the vehicle, therefore, vehicles need to redesign to have space for hydrogen tanks, and to ensure safety when operating. In addition, it is necessary to have a hydrogen distribution station system for vehicles [16]. As a result, it makes the use of hydrogen more complicated and increases costs. Another way is to produce HHO in a generator by the electrolysis of water, which is being studied and widely applied. The HHO gas can be injected with diesel fuel or supplied into the engine's manifold [16, 17, 18, 19]. The generator can be wet or dry type. This is the most basic method for making almost pure hydrogen [20]. The amount of hydrogen formed depends on the number of the electrode plates, the current supply as well as the electrolyte of solution [21, 22].

The resulting HHO gas is colourless and tasteless. The major components of HHO gas are H_2 and O_2, with H_2 accounting for 60.79%, O_2 accounting for 30.39% and a small amount of water vapor, O, OH ions and some other active elements [23]. Molar mass of HHO gas is 12.3 g/mol. At a pressure level of p = 0.1 MPa and at a temperature of T = 0°C, the density of oxygen gas is ρ_{O_2} = 1.43 kg/m^3, the density of hydrogen gas is ρ_{H_2}= 0.09 kg/m^3, and the density of the HHO gas is ρ_{HHO} = 0.54 kg/m^3 [26]. The energy in one-liter HHO is about 600J (± 34 J). The HHO gas is unstable and can live for about 11 minutes [24]. After the active elements of the HHO disappear, what it leaves is a mixture gas of common H_2 and O_2, with a volume ratio of 2:1. The lower heating value of hydrogen is almost three times that of diesel, but hydrogen has an auto-ignition temperature of 585°C, so hydrogen cannot be used directly in compression ignition engines without spark plugs or grow plugs [25].

In this study, the HHO gas generator will be manufactured and installed for the experiment. The HHO gas was fed into the engine's intake manifold. The values of engine power, brake specific fuel consumption and exhaust emissions were collected, analysed and compared in two cases that is pure diesel and diesel with HHO gas. The results received will be the basis for improvements of diesel engines.

2. Experimental setup

2.1. Principle of the HHO generation process

The HHO generator works on the principle of water electrolysis that generates hydrogen and oxygen. Diagram of the electrolysis principle is displayed in Figure 1. The electrolysis process uses direct current with inert electrodes. Reactions taking place at the electrodes are given below:

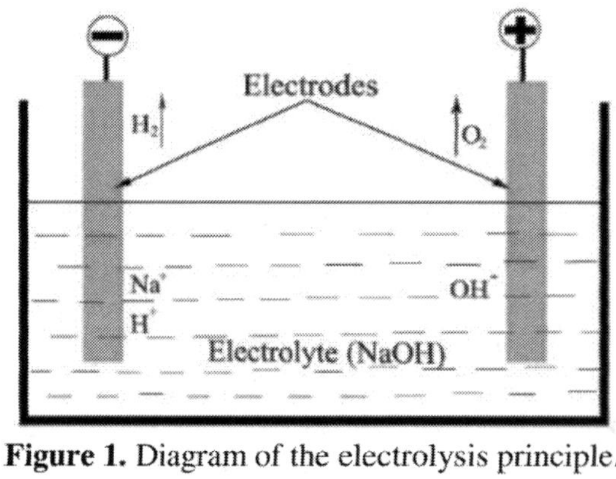

Figure 1. Diagram of the electrolysis principle.

At cathode: $2H_2O + 2e \rightarrow H_2 + 2OH^-$ (1)

At anode: $4OH^- \rightarrow O_2 + 2H_2O + 4e$ (2)

Overall reaction: $2H_2O \rightarrow 2H_2 + O_2$ (3)

The amount of gas from electrolysis is calculated following the second Faraday's law:

$$m = \frac{A \cdot I \cdot t}{z \cdot F}, \quad (4)$$

where:

m is the mass of the substance liberated at an electrode in grams
F = 96,485 C/mol is the Faraday constant
A is the molar mass of the substance in grams per mol
z is the valency number of ions of the substance (electrons transferred per ion).
t is the total time the constant current was applied.
I is electric current in Amperes
The volume of gas in ideal gas conditions is determined by the formula (5):

$$V = \frac{q \cdot R \cdot I \cdot T \cdot t}{p \cdot z \cdot F} \quad (5)$$

where:

V is volume of gas (l)
R is ideal gas constant (l.atm/(mol.K))
p is ambient pressure (atm)
T is temperature (K)
q is number of water compartments

2.2. *HHO generator*

The optimal surface of electrolyte needed to generate sufficient amount of HHO was 20 times the piston area and the volume of water needed in electrolysis cell was about 1.5 times that of engine capacity [27]. Moreover, the input voltage to the electrodes must be higher than the minimum dissociation voltage of ions in the water. For these reasons, the electrodes are made of stainless steel, which contains at least 7% nickel, 16% chromium and 0.08% (maximum concentration) of carbon.

In Figure 2 shows the HHO gas generator diagram. It is made with dry cells following the scheme – nnn + nnn–, where n is a plate of 304 stainless steels with dimensions 140 x 140 x 1 mm (width x length x thickness). In Figure 3 shows the experimental model of the gas generator.

In order to increase the efficiency of the electrolytic process, an electrolyte solution using concentration of NaOH is used.

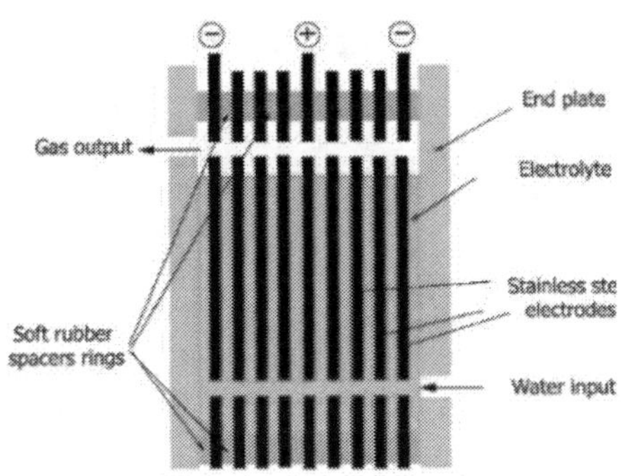

Figure 2. Diagram of a gas generator for the HHO gas production.

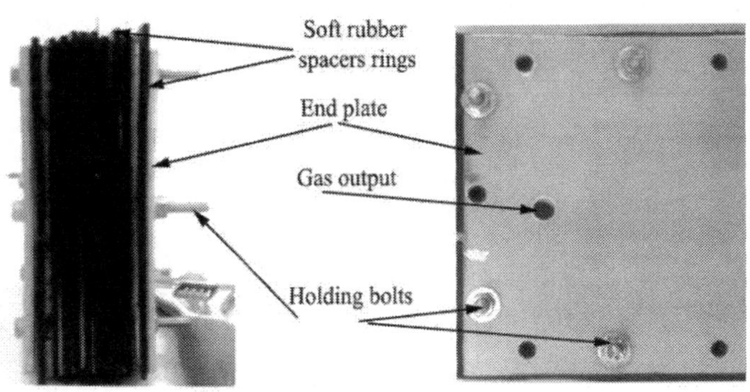

Figure 3. An experimental model of a gas generator for the HHO gas production.

The relationship between electrolytic efficiency for using different concentrations of NaOH at different engine speeds Figure 4, which can be seen that 4 g/L of NaOH concentration gave the highest efficiency of the electrolysis process compared to other NaOH concentration at different engine speeds [28].

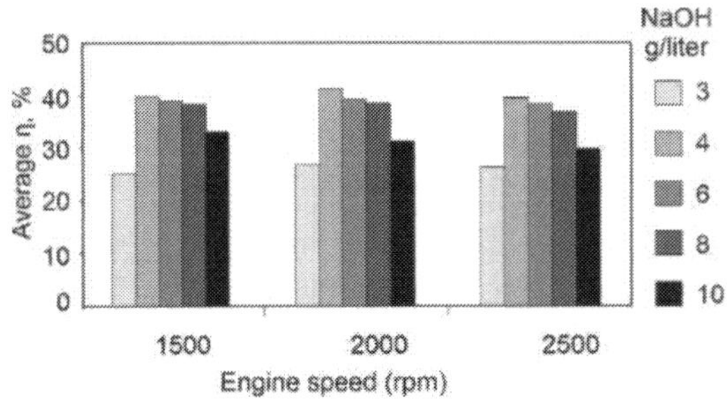

Figure 4. Average efficiencies for using different concentrations of NaOH at different engine speeds [28].

2.3. Experimental equipment

The experimental diagram is shown in Figure 5. Table 1 compares the properties of hydrogen and diesel [25, 29]. The detail of the test engine is listed in the Table 2.

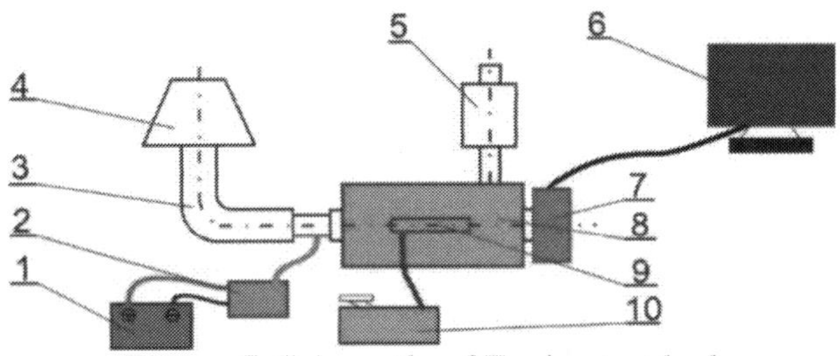

Figure 5. Schematic of Engine test bed
1 – Battery, 2 - HHO generator, 3 - Intake manifold, 4 - Air filter, 5 - Exhaust line, 6 - Display screen, 7 - Hydraulic brake, 8 - Engine, 9 – Fuel pump, 10 – Diesel fuel tank.

Table 1. Properties of diesel and hydrogen [25, 29].

Properties	Diesel	Hydrogen
Density (kg/m^3) at 16°C and 1.01 bar	840	0.082
Lower heating value (MJ/kg)	42.7	120.21
Higher heating value (MJ/kg)	45.58	142.18
Flame velocity (m/s)	0.3	2.70
Auto ignition temp. (°C)	254-285	585
Carbon residue (%)	0.1	0.0
Stoichiometric AFR on mass basic	14.7	34.3
Molar mass (kg/kmol)	200	2.02

Table 2. Experimental engine specification.

Stroke (mm)	100
Bore (mm)	90
Number of cylinders (-)	4
Connecting rod length (mm)	200
Maximum power at 3200 rpm (Hp)	80
Torque at 2200 rpm (N.m)	230
Diesel fuel injection	Inline Pump
Compression ratio (-)	17

To measure the parameters of the diesel engine during the experiment, Dyno max 2012 software is used. In addition, the 5-component exhaust analysis system - DYNOmite Exhaust Measuring System – is connected. Exhaust gas 5-components include HC, CO, NO_x, O_2, and CO_2.

3. Experimental results and evaluation

The economic, technical and exhaust emission parameters of the engine are measured when the engine runs at 50% of the accelerator pedal with different engine speeds and a jump of 200 rpm. The results are as follows:

3.1. Engine performance

3.1.1. Brake power

The variation of brake power when using diesel and diesel with HHO is demontrated in Figure 6.

Both of these cases, brake power value increases when engine speeds is from 1400 to 2200 rpm, while the engine speeds between 2200 and 2400 rpm, brake power value decreases. The engine performance is improved when HHO gas is added. The brake power index increases by 2.8% in average compared to pure diesel and 4.2% is the percentage of the maximum engine power increasing at 2200 rpm.

It can be explained because of the addition of oxygen concentration from the water electrolysis. At the same time, the additional gas component has OH^-, therefore, it increases the fire reaction area of fuel. In addition, due to the high burning rate and calorific value of hydrogen, the delayed combustion decreases, the burning time is shorter, and the heat loss is less. A mixture of diesel-hydrogen can be more ignited and quickly combusted than using the pure diesel. The combustion of the engine improves, the thermal efficiency increases, so the engine power increases with HHO gas.

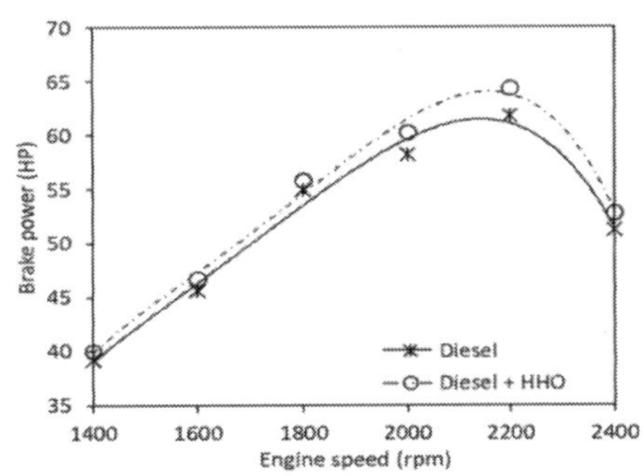

Figure 6. Variation of engine power with engine speed.

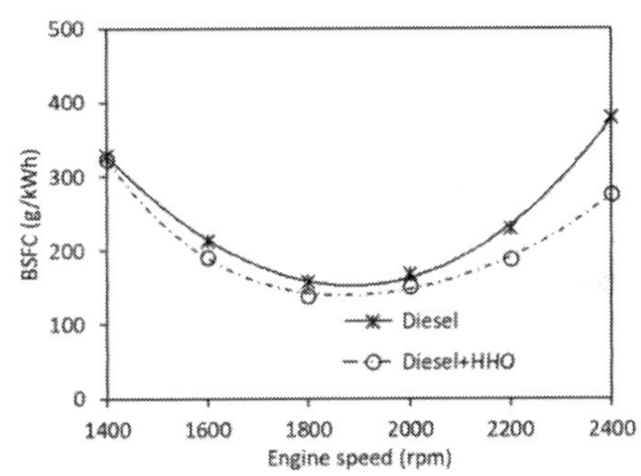

Figure 7. Variation of brake specific fuel consumption (BSFC) with engine speed.

3.1.2. Brake specific fuel consumption (BSFC).

Figure 7 shows the variation of *BSFC* with engine speed in both cases. The *BSFC* greatly reduces with speeds from 1400 to 1800 rpm. When speeds from 1800 rpm to 2000 rpm, *BSFC* fluctuates slightly and reaches the lowest point of value. *BSFC* increases rapidly as the speed between 2000 and 2400 rpm. The *BSFC* in pure diesel is always higher than one in diesel with HHO gas, which is 13.7% in average. At 1800 rpm, the difference of the minimum value of *BSFC* in two cases is 12.6%. *BSFC* decreased because the HHO is supplied on the intake manifold so times for diffusion, mixing HHO gas with fresh air increases. Gas mixture is better. At the same time, the oxygen value from the electrolysis process is also added to the mixture. OH ions have a large impact radius, which affects the C-H chains of the fuel, therefore, combustion reaction efficiency in combustion chamber is improved. In addition, with the high speed (□ 2000 rpm), the diesel-air mixture is difficult to completely burn at lean conditions because of the increase in the residual gas fraction and poor mixing. HHO has a wide ignition ability and high fire rate, which improves significantly the combustion at the high speed conditions.

3.2. Exhaust gas emissions

3.2.1. HC emissions

The relationship between HC and engine speed is illustrated in Figure 8. HC value is the unburnt fuel. When engine speed rises, which decreases the time of fuel combustion process, as the result the amount of unburnt fuel increases. However, when the HHO gas is added, the rate and burning area of the flame in the combustion chamber are improved, therefore, HC value goes down. HC decreases by 7.7% in average compared to pure diesel.

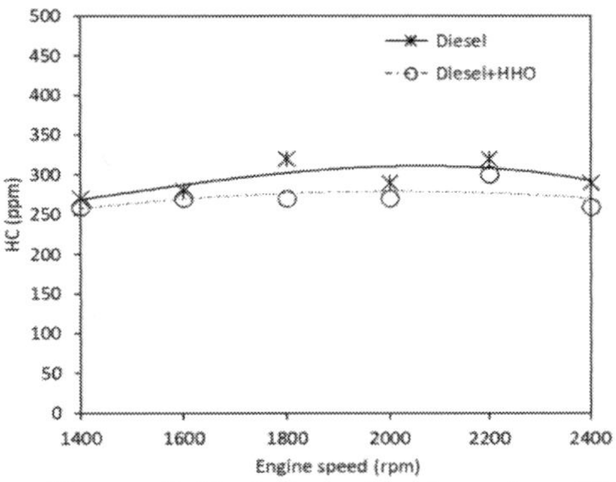

Figure 8. Variation of hydrocarbon (HC) emissions with engine speed.

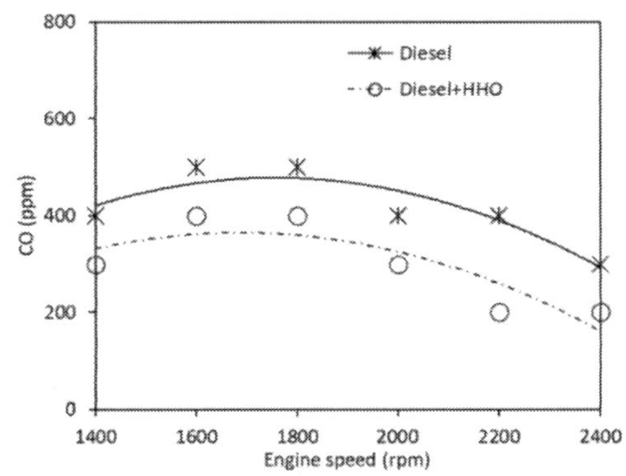

Figure 9. Variation of carbon monoxide (CO) emissions with engine speed.

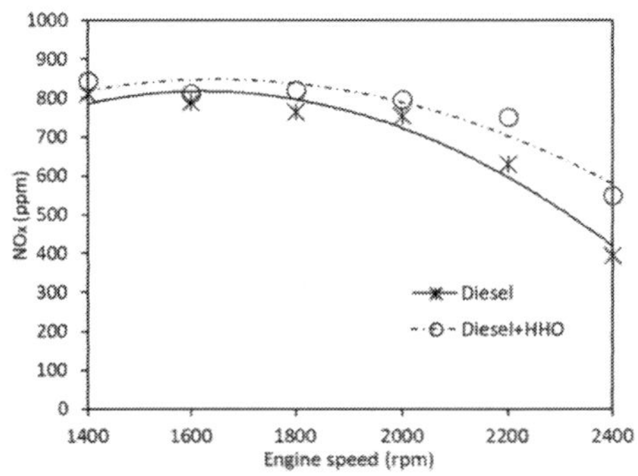

Figure 10. Variation of nitrogen oxides (NO_x) emissions with engine speed.

3.2.2. CO emissions

CO variation is similar to the change of HC. As the engine speed increases, the turbulence of the gas flow in the combustion chamber goes up, therefore, the amount of CO decreases in both cases. CO decreases with addition of HHO gas, which can be explained because HHO has a wide combustion range and a high burning rate, and the HHO component has oxygen and active OH ions. The OH ions have a powerful impact on H-C chain of diesel fuel. Moreover, the amount of oxygen is added to help the combustion reaction of the fuel completely. As a result, the CO value reduces by 19.8% in average. Experimental results are shown in Figure 9.

3.2.3. NO_x emissions

NO_x generation requires three simultaneous conditions, including high temperature, more oxygen concentration for reaction with nitrogen and large combustion area. When the engine speed increases, the burning time in each cycle decreases, therefore, the reaction time between oxygen and nitrogen reduces, which leads to the amount of NO_x declines. Figure 10 demonstrates the NO_x value when using pure diesel and diesel with HHO gas. Overall, there is an increase in the NO_x value when using HHO generator. It can be seen that the amount of NO_x increases by 13.3% in average, which is caused by adding the oxygen to the fuel combustion reaction. Besides, HHO gas has an impact on the burning area, and high temperature area in the combustion chamber. As a result, NO_x goes up at all measuring points.

4. Conclusions

In this study, HHO gas was provided into the engine manifold. The effects of HHO gas on the economic, technical and emission parameters of the diesel engine were presented. These values were compared to pure diesel with the corresponding values. Should be noted:

- The dry type of HHO generator is designed, manufactured and installed on engine test bed for experiment which have small diameter and easy to install on vehicles.
- The experimental investigation on diesel engine with the HHO gas generator at 50% of the accelerator pedal, the results showed that improvement of engine power, brake specific fuel consumption and exhaust emissions over the measuring speed range.
    - The brake power index increased by 2.8% in average compared to pure diesel and 4.2% was the percentage of the maximum engine power increasing at 2200 rpm. The *BSFC* in pure diesel was always higher than one in diesel with HHO gas, which was 13.7% in average. At 1800 rpm, the difference of the minimum value of *BSFC* in two cases was 12.6%.
    - The HC and CO emissions value reduced by 7.7% and 19.8% in average respectively, while NO_x increased by 13.3% in average compared to pure diesel.

References

[1] Nguyen Thin Quynh, Le Hoai Duc 2015 Experiments research effects of EGR to emissions of 490QZL diesel engine *Transport and Communications Science Journal Papers* 6

[2] Dunin A U, Nguyen Thin Quynh, Shatrov M G and Golubkov L N 2020 Analysis of the Nozzle Hole Diameter Effect to Common Rail Diesel Engine Characteristics using a Calculated Model of an Internal Combustion Engine *Int. J. of Emer. Trends in Engineer. Res.* **8(6)** 2301–08

[3] Shatrov M G, Dunin A U and Thin Quynh Nguyen 2020 Effect of pressure boost on the diesel performance and gaseous emission with fuel spray pressure up to 3000 bar *IOP Conf. Series: Materials Science and Engineering* **832** 012083

[4] Shatrov M G, Golubkov L N, Dunin A U, Yakovenko A L and Dushkin P V 2015 Influence of high injection pressure on fuel injection perfomances and diesel engine worcking process *Thermal Sci.* **19(6)** 2245–53

[5] Shatrov M G, Golubkov L N, Dunin A U, Yakovenko A L and Dushkin P V 2015 Research of the impact of injection pressure 2000 bar and more on diesel engine parameters *Int. J. of Applied Engineer. Res.* 10(20) 41098–102.

[6] Shatrov M G, Golubkov L N, Dunin A U, Dushkin P V and Yakovenko A L 2017 The new generation of common rail fuel injection system for Russian locomotive diesel engines Pollution Res. 36(3) 678–84

[7] Shatrov M G, Golubkov L N, Dunin A Yu, Dushkin P V and Yakovenko A L 2017 A method of control of injection rate shape by acting upon electromagnetic control valve of common rail injector *Int. J. of Mechan. Engineer. and Technol.* **8(11)** 676–90

[8] Shatrov M G, Golubkov L N, Dunin A U, Yakovenko A L and Dushkin P V 2016 Experimental research of hydrodynamic effects in common rail fuel system in case of multiple injection *Int. J. of Applied Engineer. Res.* **11(10)** 6949–53

[9] Shatrov M G, Malchuk V I, Dunin A U and Yakovenko A L 2016 The influence of location of input edges of injection holes on hydraulic characteristics of injector the diesel fuel system *Int. J. of Applied Engineer. Res.* **11(20)** 10267–73

[10] Shatrov M G, Sinyavski V V, Dunin A Y, Shishlov I G and Vakulenko A V 2017 Method of conversion of high- and middle-speed diesel engines into gas diesel engines *Facta Universitatis. Ser. Mechan. Engineer.* **15(3)** 383–95

[11] Sinyavski V V, Alekseev I V, Ivanov I Ye, Bogdanov S N, Trofimenko Yu V 2017 Physical Simulation of High- and Medium-Speed Engines Powered by Natural Gas *Pollution Research* **36(3)** 684–90

[12] Shirk M G, McGuire T P, Neal G L and Haworth D C 2008 Investigation of a hydrogen-assisted combustion system for a light-duty diesel vehicle *Int J Hydrogen Energy* 33(23) 7237–44.

[13] Verhelst S, Wallner T 2009 Hydrogen-fueled internal combustion engines, Prog. Energy Combust. Sci. **35(6)** 490–527

[14] Jorge Antunes, Rikard Mikalsen, Tony Roskilly 2009 An experimental study of a direct injection compression ignition hydrogen engine *Int J Hydrogen Energy* **34(15)** 6516–22

[15] Saravanan N, Nagarajan G 2010 Performance and emission studies on port injection of hydrogen with varied flow rates with diesel as an ignition source *Appl Energy* **87(7)** 2218–29

[16] Premkartikkumar SR, Annamalai K, Pradeepkumar AR 2014 Effectiveness of oxygen enriched hydrogen-HHO gas addition on direct injection diesel engine performance, emission and combustion characteristics *Therm. Sci.* **18(1)** 259–68

[17] Bari S, Mohammad EM 2010 Effect of H2/O2 addition in increasing the thermal efficiency of a diesel engine *Fuel* **89(2)** 378–83

[18] Samuel S, McCormick G 2010 Hydrogen enriched diesel combustion *SAE technical paper* 9

[19] Birtas Adrian, Voicu Iulian, Petcu Cristian, Chiriac Radu, Apostolescu Nicolae 2011 The effect of HRG gas addition on diesel engine combustion characteristics and exhaust emissions *Int J Hydrogen Energy* **36(18)** 12007–14

[20] Dincer I 2012 Green methods for hydrogen production *Int J Hydrogen Energy* **37(2)** 1954–71

[21] Kadir Aydin, Raif Kenanoglu 2018 Effects of hydrogenation of fossil fuels with hydrogen and hydroxy gas on performance and emissions of internal combustion engines *Int J Hydrogen Energy* **43(30)** 14047–58

[22] Ismail T M, Khaled Ramzy, Abelwhab M N, Elnaghi B E, El-Salam M A, Ismail M I 2018 Performance of hybrid compression ignition engine using hydroxy (HHO) from dry cell *Energy Conversion and Management* **155** 287–300

[23] Ruggero M S 2006 A new gaseous and combustible from of water *Int. J. Hydrogen Energy* **31(9)** 1113–28

[24] C. Eckman 2010 Plasma orbital expansion of the electrons in water *Natural Philosophy Alliance Proc.* **6(2)** 142–44

[25] Saravanan N, Nagarajan G, Dhanasekaran C, Kalaiselvan KM 2007 Experimental investigation of hydrogen port fuel injection in DI diesel engine *Int J Hydrogen Energy* **32(16)** 4071–80

[26] Alfredas Rimkus, Jonas Matijošius, Marijonas Bogdevičius, Ákos Bereczky, Ádám Török, An investigation of the efficiency of using O2 and H2 (hydrooxile gas -HHO) gas additives in a CI engine operating on diesel fuel and biodiesel *Energy* **152** 640–51

[27] Ammar A. Al-Rousan 2010 Reduction of fuel consumption in gasoline engines by introducing HHO gas into intake manifold Int. J. of Hydrogen energy **35(23)** 12930–35

[28] EL-Kassaby M M, Eldrainy Y A, Khidr M E, Khidr K I 2016 Effect of hydroxy (HHO) gas addition on gasoline engine performance and emissions *Alexandria Engineering Journal* **55(1)** 243–51

[29] Senthil Kumar M, Ramesh A, Nagalingam B 2003 Use of hydrogen to enhance the performance of a vegetable oil fuelled compression ignition engine *Int J Hydrogen Energy* **28(10)** 1143–54

Simulation of the force factors causing engine idling to vibrate

P V Safronov[1], I V Alekseev[1]

[1] Moscow Automobile and Road Construction State Technical University (MADI), 64 Leningradski prospekt, Moscow 124319, Russia

E-mail: pavel_safronov@mail.ru

Abstract. The article provides a diagram and a description of the physical reasons that cause the engine to vibrate on the suspension at idle. The substantiation of the assumption about the inertial nature of these reasons is given. The main dependences for modeling the force effects on the engine casing at idle modes are given. The above theoretical aspects of modeling oscillations at idle speed are confirmed by the results of experiments to determine the influence of the moment of inertia of rotating engine parts on the amplitude of its oscillations on the suspension.

1. Introduction

Internal combustion engine is active source of vibrations in wide frequency range due to cyclical working processes and kinematic features of crank mechanism. As you know, the main force factor causing vibrations of the engine on the suspension is the variable overturning moment, which is the result of the force interaction between the engine and the external load. Variable inertia forces are another significant force factor. These factors take place at high speed and load conditions [1]. In the modes of minimum idle speed, these factors are minimal or equal to zero [2]. However, as practice shows, the vibration amplitudes of the engine on the suspension at idle speed can reach significant values. The analysis and modeling of the effect of inertial forces and torque on the vibrations of the engine on the suspension is widely presented in the works of many authors. In this case, the influence of friction forces in the main bearings of the crankshaft and the moment of inertia of rotating parts on engine vibrations on the suspension is practically not considered. However, at idle and low loads, this force factor is the main one, disturbing the vibrations of the engine on the suspension. This article is devoted to the consideration of the factors causing vibrations of the engine on the suspension at idle modes.

2. Modeling of inertial action

For consideration and further mathematical modeling of oscillations of the engine on the suspension, consider the scheme of a single-cylinder engine and the forces and moments acting on its elements, shown in Fig. 1. Stages of creation of mass-inertial model of engine on suspension are considered in detail in operation [3], [4].

Content from this work may be used under the terms of the Creative Commons Attribution 3.0 licence. Any further distribution of this work must maintain attribution to the author(s) and the title of the work, journal citation and DOI.
Published under licence by IOP Publishing Ltd

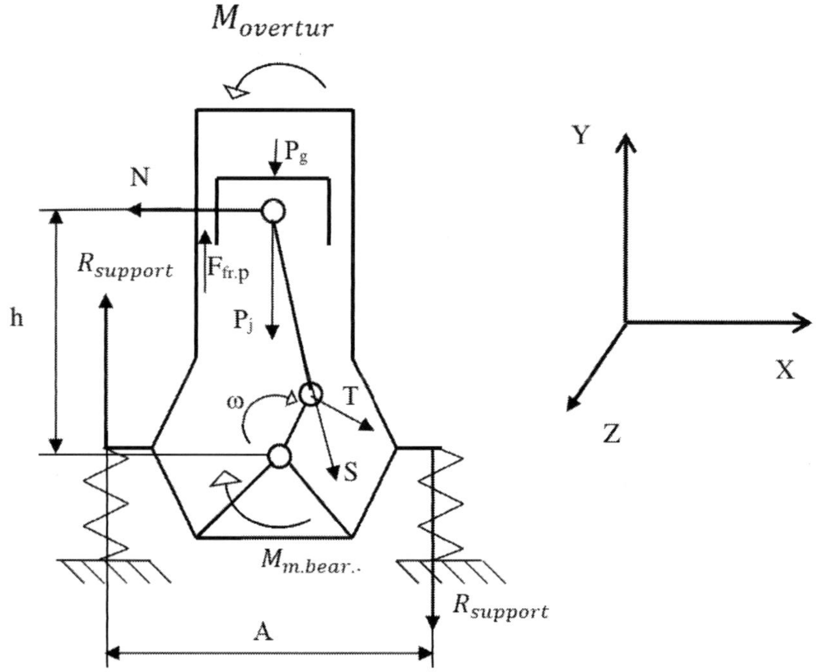

Figure 1. Forces and moments acting on the elements of the internal combustion engine.

The equation of angular motion of the engine body (Fig. 1) can be written in the following form :

$$M_R = M_{overturn} - M_{j\_body} \quad (1)$$

where $M_R = R_{support} \cdot A$ - moment from the elastic forces of the supports $R_{support}$, $M_{overturn}$ - overturning moment, $M_{I\_body} = I_{body} \cdot \frac{d\gamma}{dt^2}$ - instantaneous inertial moment of the case, I_{body}- reduced to the crankshaft axis moment of inertia of the housing, γ- angular movement of the body around the longitudinal axis Z.

The overturning moment acting on the body is the result of the action of two moments of forces:

$$\bar{M}_{overturn} = N \cdot h - M_{m.bear.} \quad (2)$$

where $M_{m.bear.}$ - moment from friction forces in the main bearings. Its value is approximately 1.5 ... 1.9 Nm for one main bearing at a speed of 1000 rpm at idle [5], N – lateral force, which is a projection onto an axis perpendicular to the cylinder axis, reactions from the connecting rod to forces acting on the piston $N = (P_g + P_j + P_{fr.}) \cdot tg(\beta)$, [6] h – arm of force N (distance from the axis of the crankshaft to the axis of the piston pin), β – is the angle of deflection of the connecting rod.

Substituting expression (2) into (1) and carrying out transformations, we obtain:

$$M_R = N \cdot h - M_{m.bear.} - I_{body} \cdot \frac{d\gamma}{dt^2} \quad (3),$$

Usually, when analyzing the dynamics of an internal combustion engine, friction forces are ignored. This approach is justified when considering the nominal regime. For this mode, the cycle average value $M_{m.bear.}$ is only a few percent of the average Nh value and practically does not affect the results of calculating the forces and moments acting on the motor housing. In this case, the instantaneous values of the loads on the engine mounts will depend on the value of the overturning moment Nh, as well as the moment of inertia of the body. As the engine load decreases, the average Nh value decreases and becomes equal to the average value $M_{m.bear.}$ at idle. However, the instantaneous value of the load on the engine mounts, in this mode, will depend on the ratio of Nh and $M_{m.bear.}$ at a particular moment of time, as well as the value of I_{body}. Despite the equality of the cycle mean values of Nh and $M_{m.bear.}$ at an arbitrary time instant Nh $\neq M_{m.bear.}$. This is due to the difference in the main factors influencing them. The magnitude of the frictional moment in the main bearings is mainly influenced by the angular velocity of the crankshaft ($M_{m.bear.}$ = f (ω)). The value of Nh is mainly influenced by the force of gas pressure Pg and the angle of rotation of the crank. Figure 2. presents the results of experimental determination of the indicated lateral force and angular speed of the crankshaft for a four-cylinder gasoline engine operating at idle speed. The lateral force was calculated from the results of measuring the pressure in each cylinder after one degree of rotation of the crankshaft and the calculated values of the inertia forces from the masses of parts moving back and forth. The measurement of the angular velocity and pressure in the cylinders was synchronized with each other and with the position of the pistons.

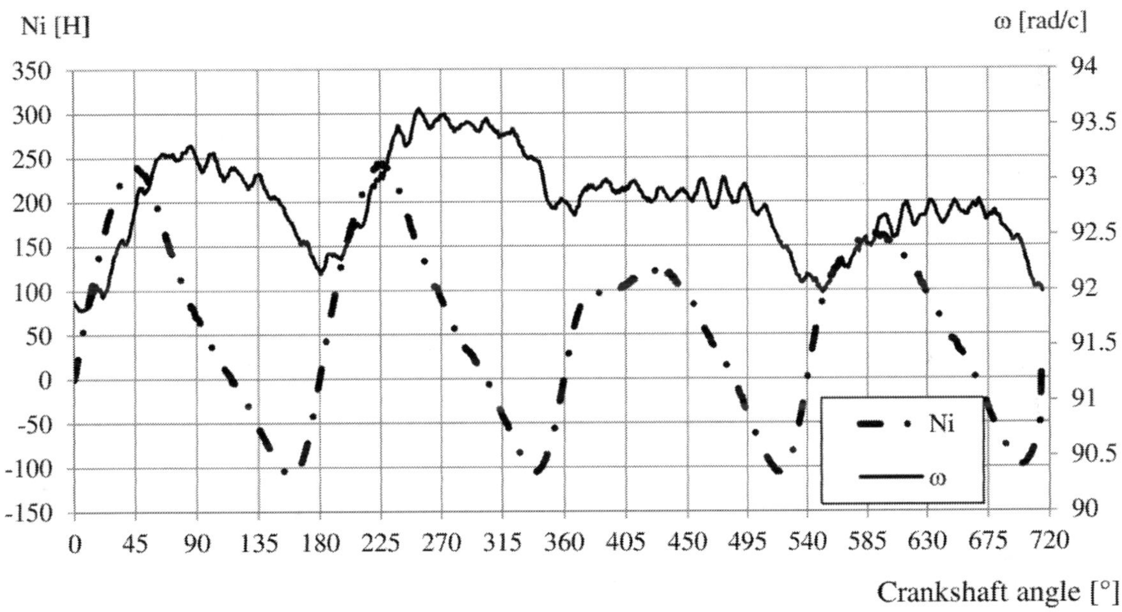

Figure 2. Results of simultaneous determination of the indicated lateral force Ni and the instantaneous angular velocity of the crankshaft ω.

Presented in Fig. 2. Ni force is called indicator force, since when determining it, the friction of the piston and rings on the cylinder wall was not taken into account.

The graph shows that the maximum lateral force is reached 40...50 degrees of crankshaft rotation earlier than the maximum angular velocity.

The movement of the crankshaft is generally described by the following equation:

$$M_{cr} = M_{rez} - I_{cr} \cdot \frac{d\omega}{dt} \quad (4),$$

where M_{cr} – moment created by force T (Fig.1) on the crankshaft, M_{rez} – total moment of resistance to crankshaft rotation, I_{cr} – the moment of inertia of the moving parts of the crank mechanism reduced to the axis of rotation of the crankshaft. Considering that the moment of inertia of rotating parts is usually 90 ... 95% [2] of the total reduced moment of inertia, I_{cr} = const. can be considered. I_{cr} = const.

$M_{cr} = T \cdot r$, where r – crank radius, $M_{rez} = M_{ext.rez.} + M_{m.bear.}$, where $M_{ext.rez.}$ – moment of external load.

Transforming expression (4), we obtain an expression for the moment of friction forces in the main bearings:

$$M_{m.bear.} = T \cdot r - M_{ext.rez.} - I_{cr} \cdot \frac{d\omega}{dt} \quad (5),$$

Substituting the resulting expression in (3) we get:

$$M_R = N \cdot h - T \cdot r + M_{ext.rez.} + I_{cr} \cdot \frac{d\omega}{dt} - I_{body} \cdot \frac{d\gamma}{dt^2} \quad (6).$$

From expression (6) it follows that the magnitude of variable loads on the engine supports, when operating under load, will be determined by the amplitude of the value Nh. When the engine is idling ($M_{ext.rez.}$= 0), the magnitude of variable loads on the engine mounts will be determined by the acceleration of the crankshaft ($\frac{d\omega}{dt}$), as well as the ratio of the moments of inertia of the engine body and the reduced moment of inertia of the moving parts I_{cr}.

3. Results of experimental determination of engine vibration parameters at different flywheel inertia moments

The influence of the flywheel moment of inertia on engine vibrations is confirmed by the experiments carried out to measure the vibration accelerations of the supports of a VAZ-2111 in-line four-cylinder engine installed on a test bench when it is idling with and without a flywheel. In this case, the moment of inertia of all moving parts, including the flywheel, reduced to the crankshaft axis, is I_{cr}=0,091 kg·m^2, and the moment of inertia of the flywheel itself is $I_{flywheel}$=0,075 kg·m^2. Moment of inertia of engine housing parts was I_{body}=0,128 kg·m^2. In order to separate the frequencies of the disturbing factors caused by the overturning moment and other force effects, the unevenness of the rotational speed was artificially increased by eliminating the ignition in one of the cylinders. At an average crankshaft rotation speed of 1000 rpm, the period of the disturbing factor caused by a misfire is two crankshaft revolutions and has a frequency of 8,4 Hz (Fig. 3 and Fig. 4). From the graphs in these figures it can be seen that the value of vibration accelerations j of the attachment point of the left engine support in the direction of the cylinder axis (Y-axis in Fig. 1), at the indicated frequency, in the absence of a flywheel, is almost 2 times less than in its presence. Thus, the moment of inertia of the flywheel creates an external inertial load on the engine. The variable nature of this load, which is a consequence of the unevenness of the rotational speed, causes the engine to vibrate on an elastic suspension. The amplitude of these vibrations is directly proportional to the moment of inertia of the moving parts, reduced to the axis of rotation of the crankshaft, and the value of the angular velocity irregularity.

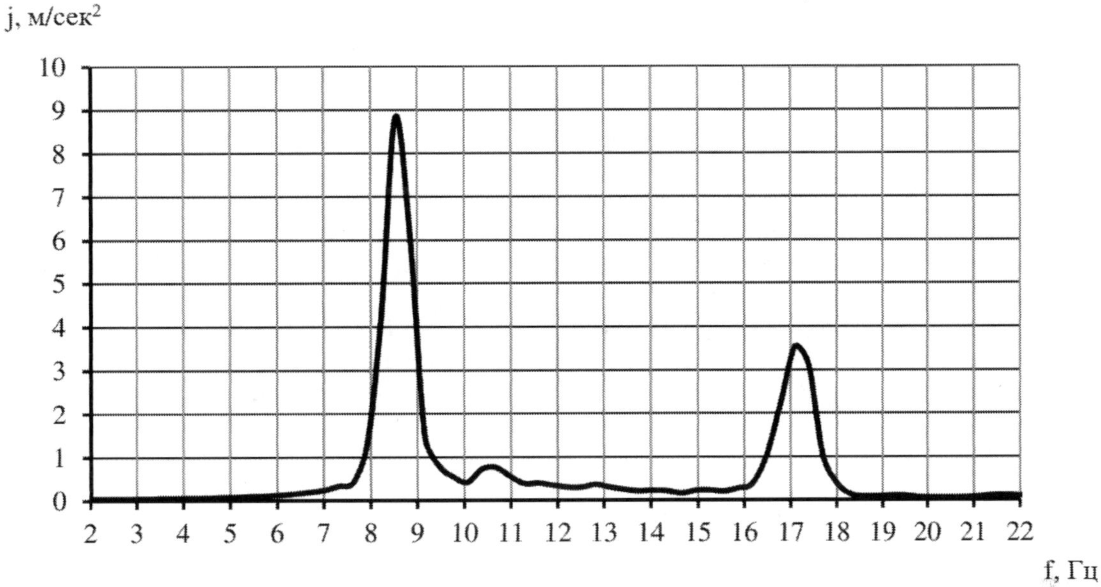

Figure 3. The spectrum of accelerations of the left engine support in the direction of the Y axis when the engine is running without ignition in one of the cylinders, complete with a flywheel (I_{cr} = 0.091 kg·m^2).

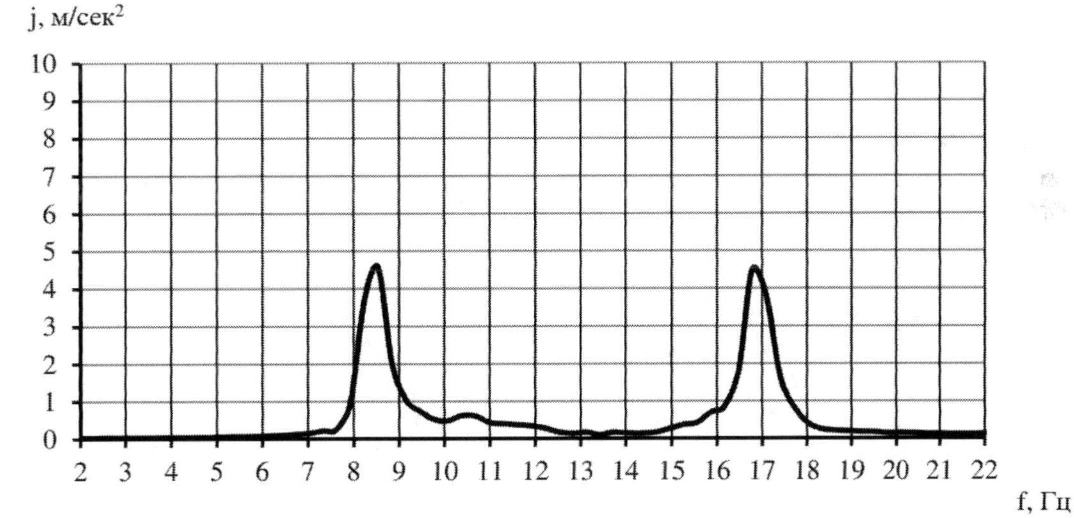

Figure 4. The spectrum of accelerations of the left engine support in the Y-axis direction when the engine is running without ignition in one of the cylinders in the configuration without a flywheel (I_{cr} = 0.016 kg·m^2).

4. Conclusion

1. The main reason for the occurrence of engine oscillations at idle is a mismatch in the time of change in the moment from the lateral force and friction coil in the main bearings.

2. The value of engine oscillations on the suspension in this case is directly proportional to the ratio of moments of inertia of the engine and its flywheel.

3. Reducing the reduced moment of inertia of the engine rotating masses from I_{cr}= 0.091 kg·m^2 to I_{cr}= 0.016 kg·m^2 resulted in a decrease in vibration acceleration of the engine support by almost two times (from 8,9 m/s^2 to 4,6 m/s^2).

4. When simulating engine oscillations on the suspension, it is necessary to take into account not only the frequency characteristics of power effects on the engine, but also their relative phase displacements.

References

[1] Tol'skij V E, Korchemnyj L V, Latyshev G V, Minkin L M 1976 *Fluctuations of the power unit of the car* (Moscow: Mashinostroenie)

[2] Lukanin V N, Shatrov M G, Alekseev I V, Morozov K A, Hachijan A S, Prishvin S A, Golubkov L N, Gorshkov Ju V, Chernjak B Ja, Sinjavskij V V, Trusov V I, Kamfer G M, Mahov V Z, Nazarov N I, Matjuhin L M 2010 *Internal combustion engines. Theory of working processes* (Moscow: Vysshaya shkola)

[3] Safronov P V 2018 Simulation of parametric disturbance of power unit oscillations on *Vestnik Moskovskogo avtomobil'no-dorozhnogo gosudarstvennogo tehnicheskogo universiteta* **2(53)** 38–45.

[4] Safronov P V, Karpov M A 2008 Procedure for calculation of dynamic behavior of four-cylinder engine on elastic supports during idling operation *Vestnik Moskovskogo avtomobil'no-dorozhnogo gosudarstvennogo tehnicheskogo universiteta* **3(14)** 26–32

[5] Sun L 2004 *PhD Thesis* 125

[6] Shatrov M G, Alekseev I V, Morozov K A, Hachijan A S, Prishvin S A, Golubkov L N, Gorshkov Ju V, Chernjak B Ja, Sinjavskij V V 2010 *Automobile engines* (Moscow: Akademija)

Dynamic tests in real vehicle conditions

Anton V Aleksandrov[1], Ivan A Dolgov[1], Timofey V Morozkin[1], Alan K Tsoriev[1] and Dmitriy S. Konyushkov[1]

[1] Moscow Automobile and Road State Technical University (MADI), 64, Leningradsky Prosp., Moscow, 125319, Russia

E-mail: madilab@mail.ru

Abstract. The report is devoted to the problem of building feedback with the car during its operation. It analyzes the problems that require feedback from the vehicle, clarifies the reasons why insufficient attention is paid to this issue, and discusses methods for obtaining the most informative data at the lowest cost. When working on the report, the authors relied on the results of their professional activities, which consists in the development and application in practice of diagnostic techniques for automobile internal combustion engines. The least costly and technically easy to implement is getting data from the engine management system controller. To increase the information content of such data, it is required to organize their correct processing and cataloging. You can get information about the progress of the working process in the engine cylinders by means of indication - direct or indirect. The implementation of direct indication is laborious and costly, therefore, with the help of direct indication, the authors of the report develop an indirect indication method based on the analysis of changes in the instantaneous crankshaft speed obtained as a result of processing the signal from the crankshaft position sensor. A promising complex of feedback with a car during operation should receive most of the data on engine operation from the ECM, but to calculate the generated torque, the complex must register a signal from the crankshaft position sensor using an ADC.

1. Introduction

When creating an engine and its control system, the conditions under which it will be operated are usually known [9, 10, 11]. Next, the prototype engine is installed on the engine stand and with the help of an engineering control unit, all its characteristics are checked for compliance with the specified ones. Then, as part of the car, the engine goes to road tests, during which, hung with a lot of additional sensors, it is checked in conditions as close as possible to real operational conditions. At the same time, based on the results of test tries, some changes are made to both the design and the management program. One of the tasks of such test trips is to develop a self – diagnosis system-how the control system should respond and how it reacts to the occurrence of malfunctions.

After that, the car moves to another stage of its life cycle. At this stage, most modern ECM as feedback, there is only a warning light when the malfunction occurs. Further tightening of environmental standards, together with the need to improve fuel efficiency [5] and the use of alternative fuels [6, 7, 8], creates the prerequisites for improving the methods of feedback with the car during its operation.

Content from this work may be used under the terms of the Creative Commons Attribution 3.0 licence. Any further distribution of this work must maintain attribution to the author(s) and the title of the work, journal citation and DOI.
Published under licence by IOP Publishing Ltd

2. Tasks of getting "feedback" from the car in the vehicle's operating conditions.

Is there any confidence that in a car, which, on the one hand, is a complex technological product, and on the other, is a mass-produced product, made, moreover, in a short time, the workflow is optimally managed? And if any malfunction occurs, then how adequately will the self-diagnosis system work in this case?

The self-diagnosis system of the car in general and the engine in particular is best suited to deal with obvious malfunctions - wire breaks or sensor values outside the set limits. In this case, the self-diagnostic messages are quite specific and on their basis it is easy to localize the malfunction that has occurred. In the case of more complex malfunctions - a violation of the mutual consistency of the sensor readings, that is, their signals go beyond the mathematical model - the developers of engine control systems offer various test blocks to help localize the malfunction. In the event of a malfunction, the accompanying information is also recorded - the driving mode, readings of some sensors and control parameters. But the greatest influence on the speed and correctness of diagnostic actions depends on the competence of the diagnostic engineer performing these actions.

To date, there is no comprehensive approach that allows you to get full feedback from the car during its operation. There are several reasons for this:

- The technical ability to receive data from the ECM controller, as well as the controller itself, as an integral part of control systems, appeared less than 30 years ago.
- "Development" and "operation" refer to different periods of the vehicle's life cycle, so it is not economically feasible to transfer the "feedback" methods used during development and Debugging to the period of operation.
- For widespread use of "feedback" techniques, there must be an appropriate level of technical personnel accompanying the vehicle during operation.

The automotive industry is developing very dynamically and what seemed to be a fantasy, such as self-driving cars, is now considered a not so distant prospect. In this regard, the methods of obtaining a stable "feedback" with the car during operation are becoming more and more relevant.

Let's look at what tasks this is relevant for:

- A set of tasks related to traffic management, traffic mode, and route planning. At the same time, the vehicle speed and dynamics of its changes (vehicle acceleration), engine operating mode (crankshaft speed and generated torque), as well as geolocation data are relevant for analysis.
- Clarification of vehicle and engine operating modes, analysis of the operation of the control system in these modes, as well as analysis of changes in the characteristics of the engine and control system components over time. This "feedback" is relevant for car manufacturers, as well as analytical centers of the "Car – Environment – Road" type.
- Monitoring the condition of cars of individual customers (subscribers) of service centers or the fleet of their own cars. This control allows you to detect and even predict the onset of failure of some elements of the engine and control system, as well as speed up the localization of the fault in the event of combined failures.
- Control of the vehicle operation in case of changes in the control program, as well as when the vehicle is operated in modes other than the calculated ones.

3. Analysis of the available opportunities for obtaining "feedback" with the vehicle in the vehicle operating conditions.

Information that can be obtained from the car as a "feedback" should be divided into two groups:

3.1. Information received from the ECM controller via the digital data exchange bus.

Today, the vast majority of cars produced have a standardized connector for connecting diagnostic equipment, through which you can also register-write to a file - some parameters of engine operation in real time. The trend in the global automotive industry is to expand the ability to access information received from the ECM - increasing the number of available parameters with the placement of these data on servers available to interested parties.

Receiving data from the ECM controller requires minimal technical costs, but a significant disadvantage of such data is its "subjectivity". In the case of various violations, the sensor signals are distorted, and the assessment of the operating mode and the control actions are determined incorrectly by the controller. This is especially important when determining the torque generated by the engine. The controller performs its calculation using a mathematical model based on sensor signals.

3.2. Information received via channels independent of the ECM controller. This information can be obtained from additional sensors placed on the engine, or as a result of independent signal processing of standard sensors.

Placing additional sensors on the car is very expensive and can only be justified if there are specific research tasks, or when conducting in-depth diagnostics in case of complex malfunctions.

Information about the flow of the working process in the engine cylinders is the most important information and diagnostic parameter, but directly obtaining it - indexing – is very expensive and time – consuming. Therefore, indication under operating conditions is rarely used. Various methods of indirect indexing are an alternative to direct indexing. The crankshaft senses the torques generated as a result of working processes in all cylinders of the engine, its angular velocity and acceleration depend on the nature of their flow. When developing methods for indirect indexing, it is necessary to take into account all the factors that affect the parameters of the crankshaft rotation and identify the influence of the working process against their background.

When conducting a number of research works and in-depth diagnostics, the noise and vibration of the power unit may also be of interest. Appropriate sensors are used to study them. If you process the signals of these sensors together with the signal of the crankshaft position sensor, then noise and vibration can be synchronized with the processes in the internal combustion engine.

4. Practical implementation of "feedback" with the car during operation.

More than 15 years ago, the development of a diagnostic complex (DC) for internal combustion engines began in the Problem Laboratory of Transport Engines of Moscow Automobile and Road Construction State Technical University. Interim results of work with practical examples of the use of diagnostic complex in vehicle diagnostics were reported at IASF 2007. The first version of the complex was based on a 16-channel ADC installed in the PCI slot of the PC, that is, it was stationary.

In the process of using the complex - diagnosing vehicles with complex malfunctions and carrying out research work - ideas for its further development appeared, and a list of tasks for the solution of which the use of the complex would be relevant.

It became clear that the complex must be mobile to solve many problems. Therefore, a mobile version of the diagnostic complex was created several years ago.

The number of sensors and actuators in a modern automobile engine has multiplied, therefore, in the mobile version of the complex, the number of simultaneously recorded data lines has increased.

To obtain information on the progress of the working process in our laboratory, a miniature pressure sensor was developed, for which a patent was obtained in 2018. Figure 1 shows: a - sensor design, b - sensor integrated into the spark plug.

IOP Conf. Series: Materials Science and Engineering 1159 (2021) 012057 doi:10.1088/1757-899X/1159/1/012057

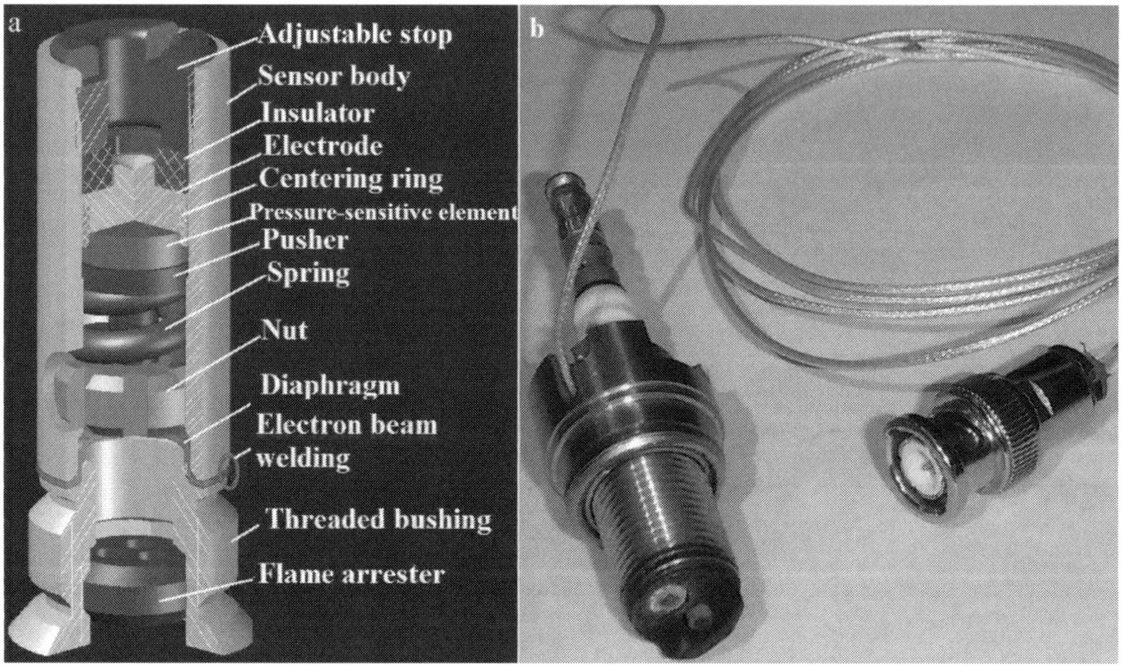

Figure 1.a - design of a miniature pressure sensor, b - pressure sensor combined with spark plug.

Now it is possible to carry out test trips on a car with a connected diagnostic complex, including indication of the internal combustion engine in road conditions. Now the ICE diagnostic complex is used when carrying out various studies of engine operation as part of a vehicle - Fig. 2.

Figure 2. A car with a connected diagnostic system.

A good correlation between the torque generated by the engine and the acceleration of the crankshaft has been known for a long time and is evident - Fig. 3, but to calculate the torque based on the change in the angular speed of the crankshaft, an indirect indication technique is required. Now we are working on such a technique using direct indexing data; more will be said about this later.

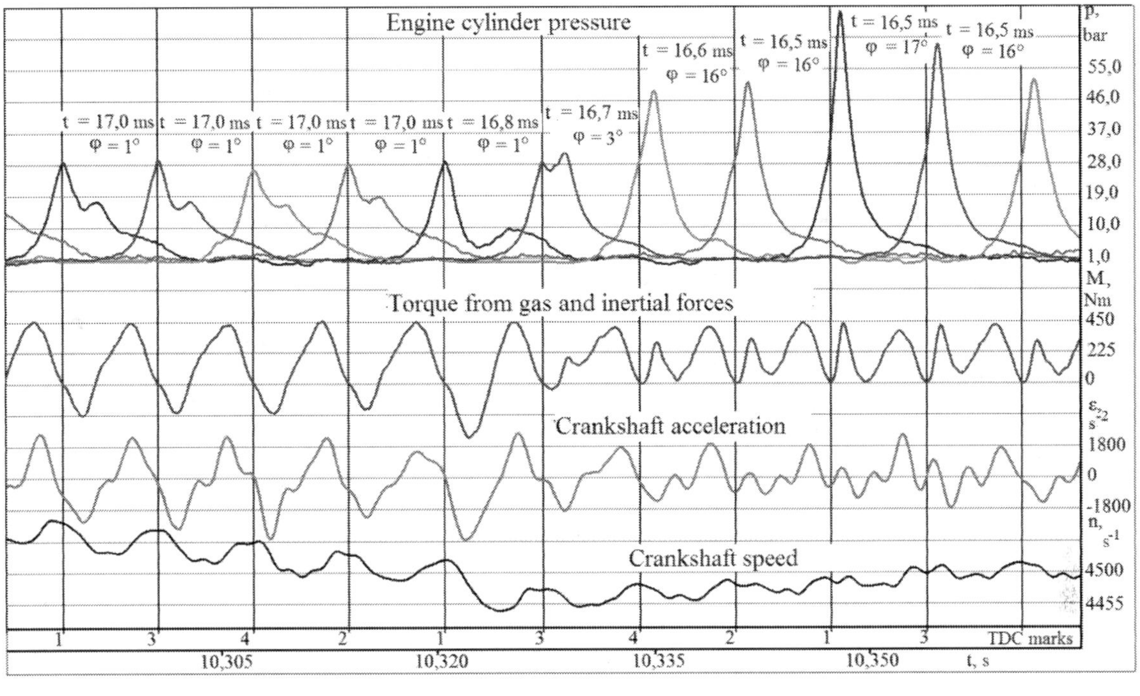

Figure 3. A fragment of the engine operation at the moment of a sharp change in the generated torque.

5. Processing data received from the ECM.

Receiving data from the ECM through the diagnostic connector is the simplest, from the point of view of technical implementation, way of obtaining "feedback" from the car. These data, even if they are analyzed without data of direct or indirect indexing, can be quite informative, while the way of their processing and systematization is of decisive importance. Depending on the research tasks, the data obtained from the ECM controller, it is advisable to analyze using various methods:

- As a function of time. For such an analysis, fragments of a relatively short duration - tens of seconds and a pronounced causal relationship - are suitable - when an event entails a change in the parameters of work - Fig. 4.

IOP Conf. Series: Materials Science and Engineering 1159 (2021) 012057 doi:10.1088/1757-899X/1159/1/012057

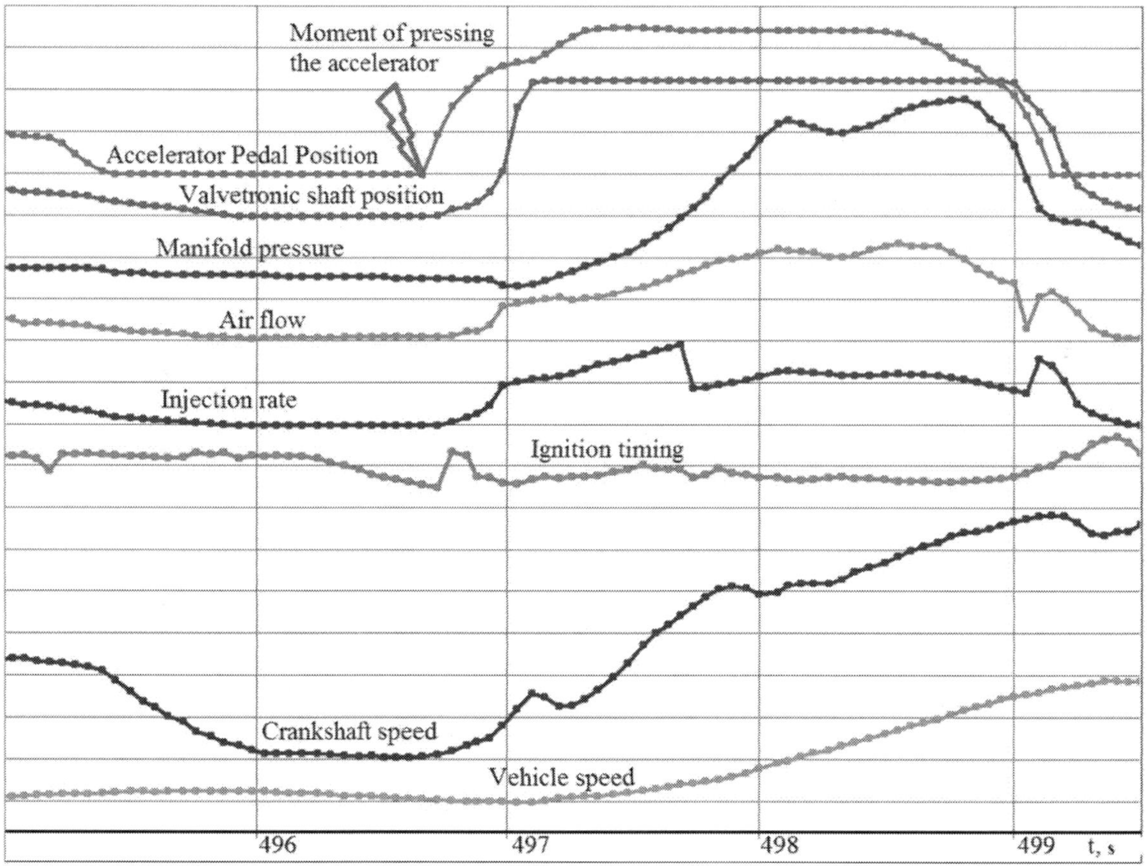

Figure 4. An example of a data fragment from the ECM as a function of time.

- In the function of the operating mode. For this analysis, the data recorded as a function of time must be converted to a function of the engine operating mode. To avoid "unnecessary" data in the mode field - formally, by mode parameters related to this mode, but by other parameters falling out of it and blurring the result, you need to use the filtering mechanism. To solve the problem of sorting large (several hours) time fragments of the engine operation in real operation conditions, as well as filtering them, the DataConvert program was written. This program allows you to assign any two of the registered or calculated on their basis parameters, as determining the mode, and the remaining parameters are considered in their function. The program also has several filters to exclude certain engine operating modes from consideration. A screen copy of this program is shown in Fig. 5. With this method of processing the obtained data, it becomes possible to compare the parameters of operation and control of the engine, obtained in different periods of time, as well as the engines of different cars of the same type.

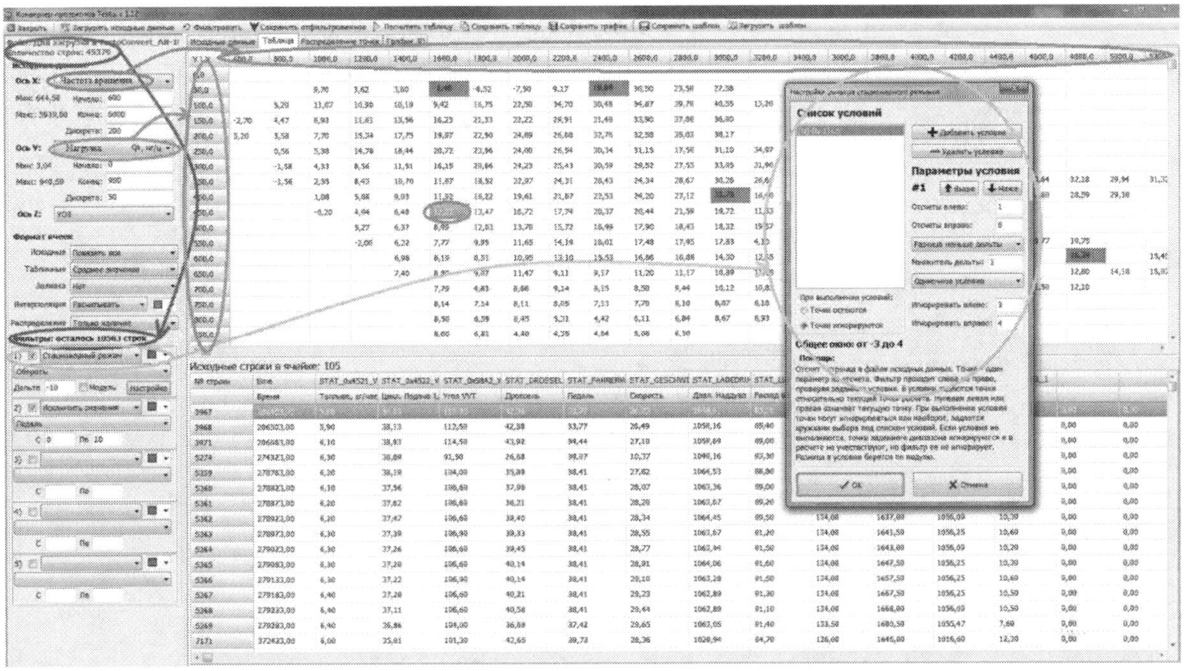

Figure 5. DataConvert program at the stage of setting the filter of the internal combustion engine operating modes.

Data from the ECM is an alternative to recording oscillograms of signals from sensors and actuators for an in-depth study of the operation of the internal combustion engine - direct or indirect indication.

6. Indirect indication technique based on crankshaft position signal processing and direct indication data.

When processing the signal from the crankshaft position sensor, you can get the angular velocity and its change - acceleration - of the crankshaft. But on the crankshaft, in addition to torques from external forces - gas forces, payload and resistance forces - internal - inertial ones act. They do not change the total kinetic energy of the moving parts of the engine, but change the angular velocity of the crankshaft. To exclude their influence from consideration, it is necessary to analyze not the angular velocity of the shaft, but the total kinetic energy of the moving parts of the internal combustion engine. When developing a method for indirect indication, it was decided to use two forms of connection between torque and shaft rotation - differential and integral. The differential form connects the instantaneous acceleration of the crankshaft and the total torque acting on it; integral - the change in the kinetic energy of the moving parts of the engine for a certain angular interval with the work of external forces at the same interval.

According to Newton's second law, the angular acceleration of the crankshaft must obey the relationship:

$$\varepsilon = \frac{\sum M}{J_k}, \qquad (1)$$

where:

$\sum M$ - total torque acting on the crankshaft;

$J_к$ - moment of inertia of rotating engine parts.

Let's list the components of the total torque:

$$\sum M = M_g + M_l + M_f + M_i, \tag{2}$$

where:

M_g - torque from gas forces, when indicated, it can be calculated with acceptable accuracy;

M_i is the moment of inertial forces from parts performing reciprocating and plane-parallel motion. With known geometric dimensions, the mass of the piston set and the connecting rod, this component of the total torque can also be calculated with sufficient accuracy;

M_f - the moment of internal losses;

M_l - the moment of the payload - in this work, the operation of the engine without external load is considered, therefore it is equal to zero.

Integral relationship between torque and crankshaft rotation:

$$A_e = \int_{\varphi_s}^{\varphi_e} \sum M_{out}, \tag{3}$$

where:

M_{out} - torque from external forces,

$$\sum M_{out} = M_g + M_l + M_f, \tag{4}$$

then:

$$\Delta E_{kin} = A_e, \tag{5}$$

where:

$$\Delta E_{kin} = \frac{J_k(\omega^2(\varphi_e) - \omega^2(\varphi_s))}{2} + \frac{m_p(v_p^2(\varphi_e) - v_p^2(\varphi_s))}{2}, \tag{6}$$

where:
ω is the angular speed of the crankshaft;
ω (φ_s) - angular velocity at the beginning of the considered section;
ω (φ_e) - angular velocity at the end of the considered section;
m_p is the mass of parts that reciprocate;
v_p - the speed of the parts making a reciprocating motion;
v_p (φ_s) - speed at the beginning of the considered section;
v_p (φ_e) - speed at the end of the considered section.

To calculate the kinetic energy of the moving parts of the internal combustion engine, it is required to know the kinematics of the crank gear the moment of inertia of the rotating parts, the mass of the piston, the mass of the connecting rod and its moment of inertia. Of all these parameters, the most difficult is the determination of the moment of inertia of rotating parts. This task - the experimental determination of the moment of inertia of rotating parts of an internal combustion engine based on the action of inertial forces from pistons and connecting rods - was solved in the process of working on a master's thesis. After determining all these parameters, the kinetic energy of the moving parts of the internal combustion engine is calculated as a function of the rotational speed and angular position of the crankshaft.

The next step in the developed indirect indexing technique is to determine the moment of drag forces in the internal combustion engine as a function of the speed, load, and angular position of the crankshaft.

Further, assuming that the moment of the payload in the angle of rotation changes insignificantly, it is necessary to obtain a relationship between the torque determined on the basis of direct indication and the change in the kinetic energy of the moving parts of the internal combustion engine determined based on the signal of the crankshaft position sensor. If, according to the results of the indication in road conditions, it is proved that the torque calculated on the basis of the signal from the crankshaft position sensor corresponds with the specified accuracy to the torque obtained during the indication, then the problem of indirect indication can be considered solved.

7. Prospects for creating a complex that allows you to receive feedback from the car during its operation.

The complex, which allows realizing "feedback" with the car during its operation, should combine simplicity, reliability, small dimensions and high information content. This can be obtained by combining data acquisition from the ECM and recording oscillograms of standard and additional sensors. A significant part of the engine operating parameters, and in a normalized form, can be obtained avoiding the laborious connection to data lines, directly from the ECM, to carry out their primary processing and remotely send them to the monitoring center, where they must form a database. The technical solution - an adapter installed in the diagnostic socket of a car, working wirelessly with a mobile device - has existed for a long time.

To fundamentally expand the capabilities of the complex, in addition to the digital communication line with the ECM, it requires several ADC channels to connect the signal from the crankshaft position sensor and additional external sensors - cylinder pressure, accelerometers. This is demonstrated in Figure 6.

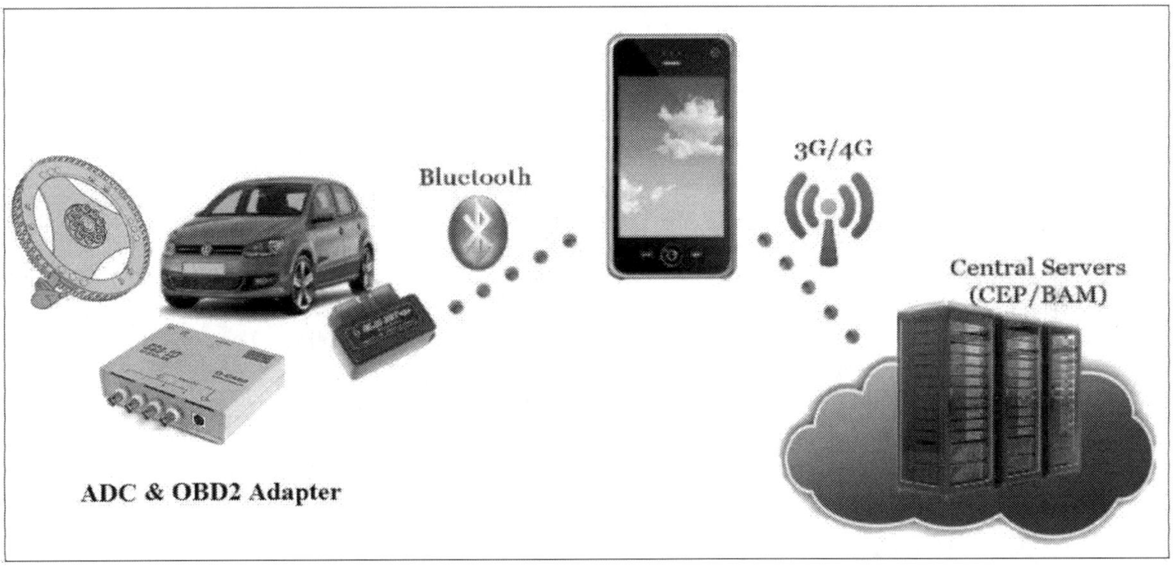

Figure 6. Diagram of a promising complex for obtaining feedback from a car.

Engine-specific calibrations are determined from a test drive with cylinder pressure sensors. Later, already without pressure sensors, using the method of indirect indication, it becomes possible to determine the instantaneous torque generated by the engine, that is, it becomes possible to monitor the progress of the working process in the engine cylinders.

At the moment, there is no such complex, but we are already working on its prototype.

The increasing complexity of engine management systems creates a demand for feedback mechanisms with it under operating conditions. The main task along this path is to create an optimal algorithm for collecting, processing and cataloging the data obtained.

References

[1] Alexandrov A V and Dolgov I A 2016 Relevance of internal combustion engine indication *Journal of Automotive Engineers: a periodic scientific journal* 5 (100) (Moscow: Publishing house AAE press) pp 19–23

[2] Morozkin T V 2020 *Investigation of the correlation of torque from gas forces and the dynamics of rotation of the crankshaft in various operating modes of the internal combustion engine* [Text]: dis. … Master: 13.04.03: defended 03.07.20. (Moscow: MADI)

[3] Alexandrov A V and Dolgov I A 2017 Mobile complex for registration and processing of the parameters of the car engine *Journal of Automotive Engineers: a periodic scientific journal* 2 (103) (Moscow: Publishing house AAE press) pp 22–28

[4] Alexandrov A V and Morozkin T V 2019 Indication of the engine in transient conditions *8th Lukanin Readings. Problems and prospects of development of the motor transport complex. Collection of proceedings of the International scientific and technical conference* 270–282

[5] Shatrov M G, Matyukhin L M, Sinyavski V V and Dunin A Y 2020 An alternative approach to the assessment of internal combustion engine filling and its technical and economic parameters *International Journal of Emerging Trends in Engineering Research* 8(6) 2805–2811

[6] Iakovenko A L, Dunin A Y, Dushkin P V, Savastenko E A and Shatrov M G 2020 Research of the influence of alternative fuels on diesel engine noise level *IOP Conference Series: Materials Science and Engineering* 832 012004 DOI:10.1088/1757-899X/832/1/012004

[7] Iakovenko A, Dunin A, Dushkin P, Savastenko E and Shatrov M 2019 The influence of mass composition of water-fuel emulsion on ecological characteristics of a diesel engine *Energies* 12(14) 2689 DOI:10.3390/EN12142689

[8] Sinyayski V V, Shatrov M G, Dunin A Y, Shishlov I G and Vakulenko A V 2019 Results of simulation and experimental research of automobile gas diesel engine *2019 systems of signals generating and processing in the field of on board communications. Proceedings Paper* 8706756 DOI: 10.1109/SOSG.2019.8706756

[9] Shatrov M G, Malchuk V I, Skorodelov S D, Dunin A Y, Sinyavski V V and Yakovenko A L 2019 Simulation of fuel injection through a nozzle having different position of the spray holes *Periodicals of Engineering and Natural Sciences* 7(1) 458–464

[10] Dunin A Y, Quynh N T, Shatrov M G and Golubkov L N 2020 Analysis of the nozzle hole diameter effect to common rail diesel engine characteristics using a calculated model of an internal combustion engine *International Journal of Emerging Trends in Engineering Research* 8(6) 2301–2308

[11] Shatrov M G, Malchuk V I and Dunin A Y 2020 A laboratory investigation into the fuel atomization process in a diesel engine for different configurations of the injector nozzles and flow conditions *Fluid Dynamics & Materials Processing* 16(4) 747–760 DOI:10.32604/fdmp.2020.08991

IOP Conf. Series: Materials Science and Engineering 1159 (2021) 012058 doi:10.1088/1757-899X/1159/1/012058

Factors affecting the formation of the established deceleration of two-wheeled vehicles

I S Brylev[1], S A Evtyukov[1], M M Blinder[1], T Y Cherepnina[2], V R Rogov[2]

[1] Saint Petersburg State University of Architecture and Civil Engineering, Saint Petersburg, 4, 2nd Krasnoarmeyskaya st., St. Petersburg, 190005, Russia
[2]Moscow Automobile and Highway State Technical University (MADI), 125319, 64, Leningradskiy prospect, Moscow, Russia

E-mail: ilya2104@mail.ru, maria.blinder@mail.ru

Abstract. The purpose of this study is to assess the factors that influence the formation of braking parameters for two-wheeled vehicles, specially bicycles, on dry and wet asphalt. Braking tests were carried out to determine deceleration using a decelerometer model "LWS-2MC" on dry and wet asphalt. Depending on the type of braking system, the method of braking, the mass of the two-wheeled vehicle and the cyclist, as well as the suspension design, graphs were built and regression equations were obtained. A quantitative assessment is given to the factors influencing the formation of the established deceleration and the deceleration rise time of a two-wheeled vehicle during emergency braking.

1. Introduction

In Two-wheeled vehicles (TWV), in particular bicycles, are active participants in road traffic, as well as in many road accidents. An important aspect of the reconstruction of road accidents with their participation is the assessment of the ability of the vehicle to brake.

Currently, there are very few published sources that include data on acceleration, braking and rolling resistance for bicycles. In the work of VA Puchkin [1], a table with the parameters of the steady-state deceleration of bicycles and motorbikes is presented, however, due to the active development of the design of the suspension, braking system and types of bicycles, updating and a more detailed consideration of the factors affecting the braking parameters are required, with the purpose of updating the databases and existing techniques for the reconstruction of road accidents. In the monograph Trofimenko Y.V., Shelmakova S.V., Zege S.O., Shashina E.V. [2] considered the advantages and disadvantages of cycling and the level of its development in different countries but did not provide data on road accidents involving cyclists.

2. Justification of the problem

Six cyclists took part in the tests on six bicycles with different frame materials, types of braking systems, suspensions, road vehicle geometry and its purpose. Also were collected the height and weight of cyclists, the weight of the bicycle and the total weight of the subjects (bicycle + cyclist) (Table 1). The road vehicle deceleration was measured using the LWS-2MC model decelerometer, which is designed for any vehicle category [3]. Braking tests were carried out on two sites: on smooth wet asphalt and on smooth dry asphal. The technique consists in the fact that the drivers of road

Content from this work may be used under the terms of the Creative Commons Attribution 3.0 licence. Any further distribution of this work must maintain attribution to the author(s) and the title of the work, journal citation and DOI.
Published under licence by IOP Publishing Ltd

IOP Conf. Series: Materials Science and Engineering 1159 (2021) 012058 doi:10.1088/1757-899X/1159/1/012058

vehicle applied emergency braking at a speed of 20-30 km / h, with the wheels locked, experiencing a certain type of braking: front, rear brakes and a combined method (front and rear brakes)[4].
According to the statistics of "New-retail" [5], at the moment the mountain bike is the most popular type of two-wheeled transport in Russia. Many road accidents involve mountain biking. According to the statistics of the official website of the State Traffic Inspectorate [6] for the period 2017 - 2020. the dynamics of road traffic accidents involving two-wheeled vehicles is traced (Figure 1).

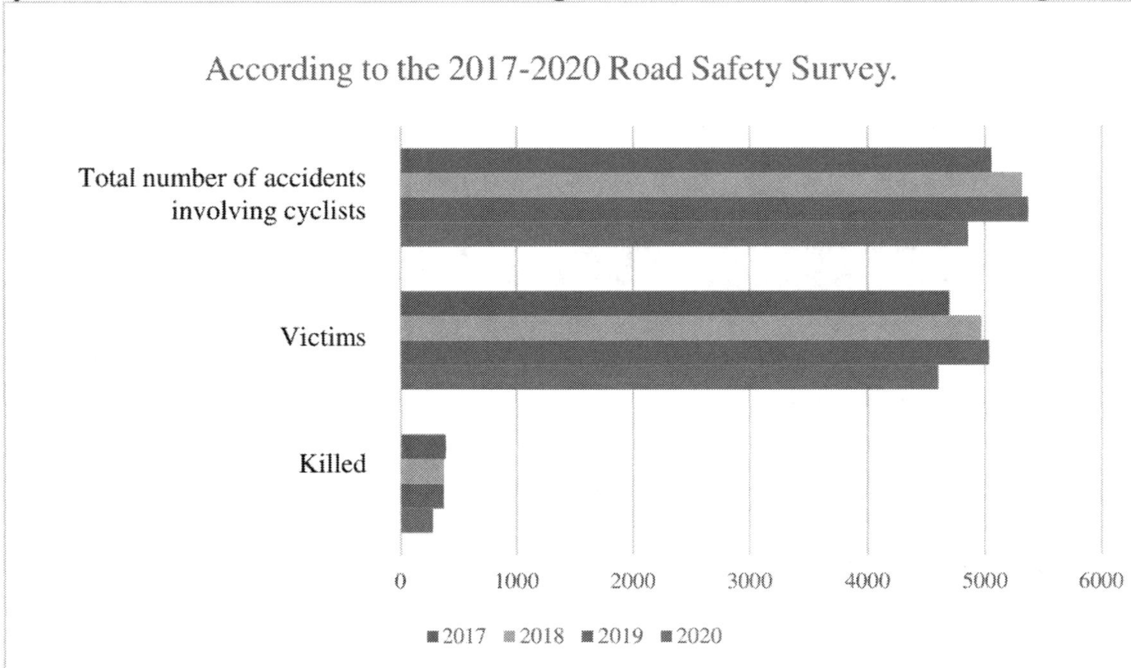

Figure 1. The dynamics of road traffic accidents with the participation of TWV with the dead and wounded in 2017-2020 in RF (people)

Establishing the true causes of traffic violations that led to the accident, and the circumstances contributing to them, is one of the important tasks of ensuring traffic safety and the operation of bicycles [4].

3. Method for assessing factors affecting the braking of two-wheeled vehicles

Reasoning An important aspect of road accident reconstruction is the ability of the vehicle to brake. There are 3 methods for braking TWV. The first method - front brake only - involved pulsating or intermittent locking of the front wheel during braking. The safer method is not to lock completely, but to quickly and repeatedly lock and unlock the front wheel, especially on uneven surfaces. By keeping the front wheel from locking, the driver can maintain control [7]. As in the case of cars not equipped with anti-lock brakes, after locking the front wheels, the driver loses significant control over the steering [8]. The second method involves locking the rear wheel and opposing the steering to maintain balance. The second method differs from the first by blocking the brake wheel. [9] The third method, in which both brakes are used (combined braking), is the most effective method of speed reduction, therefore, this article will consider the combined braking of the TWV [10,3].

IOP Conf. Series: Materials Science and Engineering 1159 (2021) 012058 doi:10.1088/1757-899X/1159/1/012058

Table 1. Characteristics of bicycles and cyclists

Bicycle	Bicycle weight (kg)	Brake-type	Purpose	Tire width front-rear (mm)	Tread depth front-rear (mm)	Driver' sweight (kg)	Gross weight (kg)	Height (sm)
Diamondback	13.6	Shimano 105	Mountain	52 52	2.5 1.7	90.1	103.7	189
Mongoose	12.1	Shimano BR-M400	Mountain	52 52	3.1 3.1	69.1	81.2	172
Cannondale	7.9	Shimano 105	Mountain	24 24	<0.9 <0.9	68.3	73.2	173
GT	12	SRAM Level TL	Mountain	55 49	2.4 2.4	75	87	171
Merida	11.4	TRP SpyreFlatmount	Mountain	29 29	0.9 <0.9	71.2	82.6	175
Specialized	9.4	Shimano BR-MT200	Mountain	47 52	3.2 4.8	96.6	106	191

Table 2. Braking tests carried out on smooth. wet asphalt. Testing brakes on a flat surface using combined braking

Bicycle	Steady-state deceleration (j), m/s^2	Rise time deceleration (t_3),s
Diamondback	4.0	0.15
Mongoose	4.2	0.2
Cannondale	4.2	0.16
GT	4.4	0.18
Merida	4.5	0.19
Specialized	4.4	0.17
Mean	4.3	0.18

Table 3. Braking tests carried out on smooth dry asphalt. Testing Asphalt Brakes Using Combined Braking

Bicycle	Steady-state deceleration (j), m/s^2	Rise time deceleration (t_3),s
Diamondback	4.9	0.1
Mongoose	5.1	0.13
Cannondale	5	0.11
GT	5.4	0.13
Merida	5.5	0.13
Specialized	5.6	0.12
Mean	5.25	0.12

Table 4. Average values of steady-state deceleration and deceleration rise time on smooth dry and wet asphalt

Act	Steady-state deceleration (j), m/s²	Rise time deceleration (t_3),s
Braking with isp. both brakes on wet asphalt	4.3	0.18
Braking with isp. both brakes on dry asphalt	5.25	0.12
Braking with isp. both brakes on dry asphalt	5.25	0.12

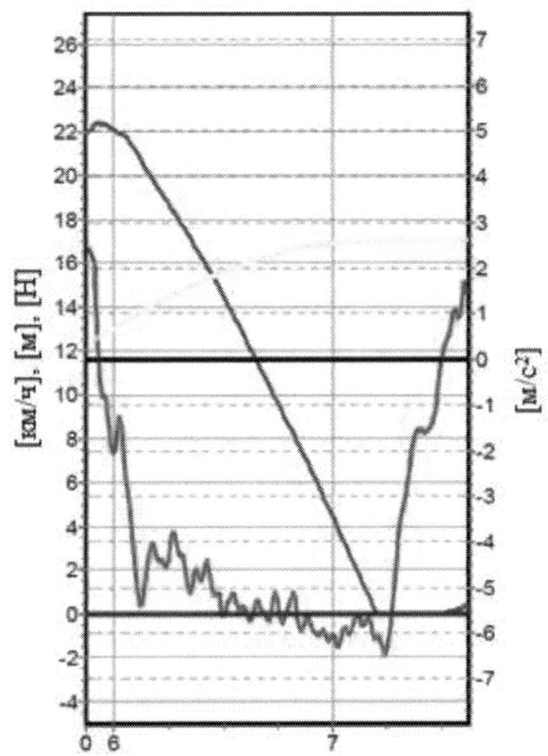

Figure2. Braking diagram in combined modebraking and coefficient of adhesionφ≈0.5

The nature of the change in the magnitude of deceleration and the time of its rise in the combined mode of braking, the coefficient of adhesion □≈0.5, in the loaded state, was obtained in the course of experiments and is presented in the form of a braking diagram (Figure 2).

- Maximum steady-state deceleration (j) = 6,47 m/s^2;
- Average steady-state deceleration (j) = 5,22 m/s^2;
- Maximumspeed = 22,38 km/h;
- Deceleration rise time (t3) = 0,15 s;

Based on the experimental data. for each investigatedtwo-wheeled vehicles. the nature of the change in the steady-state deceleration and deceleration rise time depending on the total mass was determined (Figure 3,4). The processing of the results of experimental studies made it possible to obtain polynomial regression equations.

IOP Conf. Series: Materials Science and Engineering 1159 (2021) 012058 doi:10.1088/1757-899X/1159/1/012058

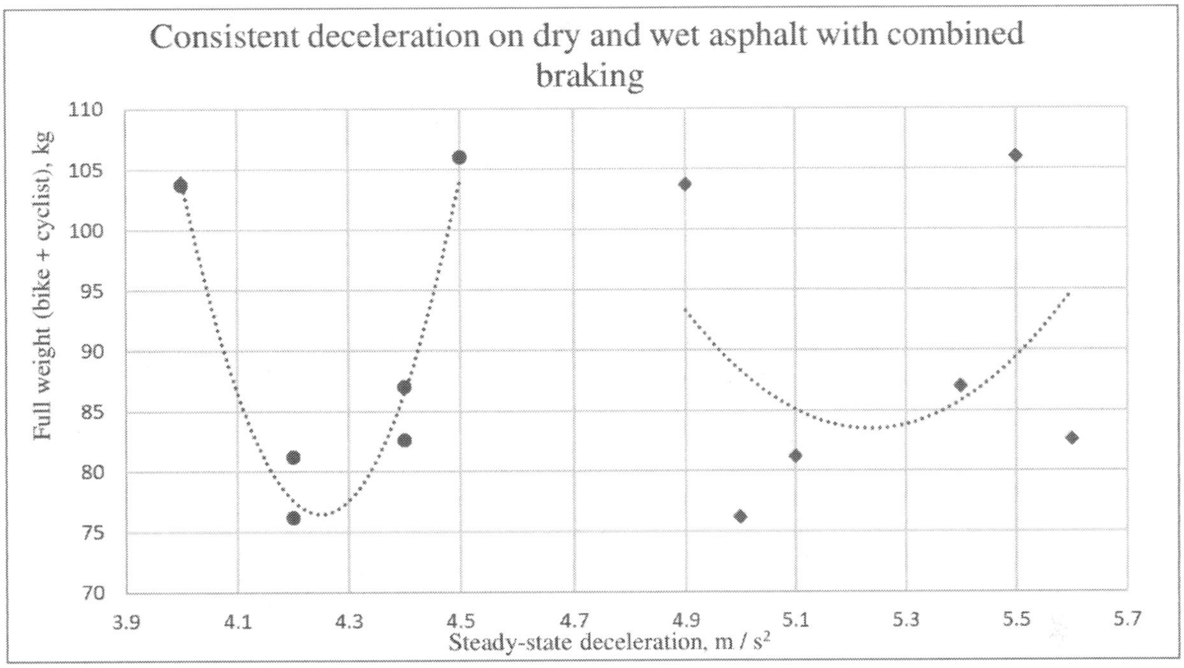

Figure 3. Plot of steady-state deceleration versus gross weight on dry and wet asphalt using combined braking

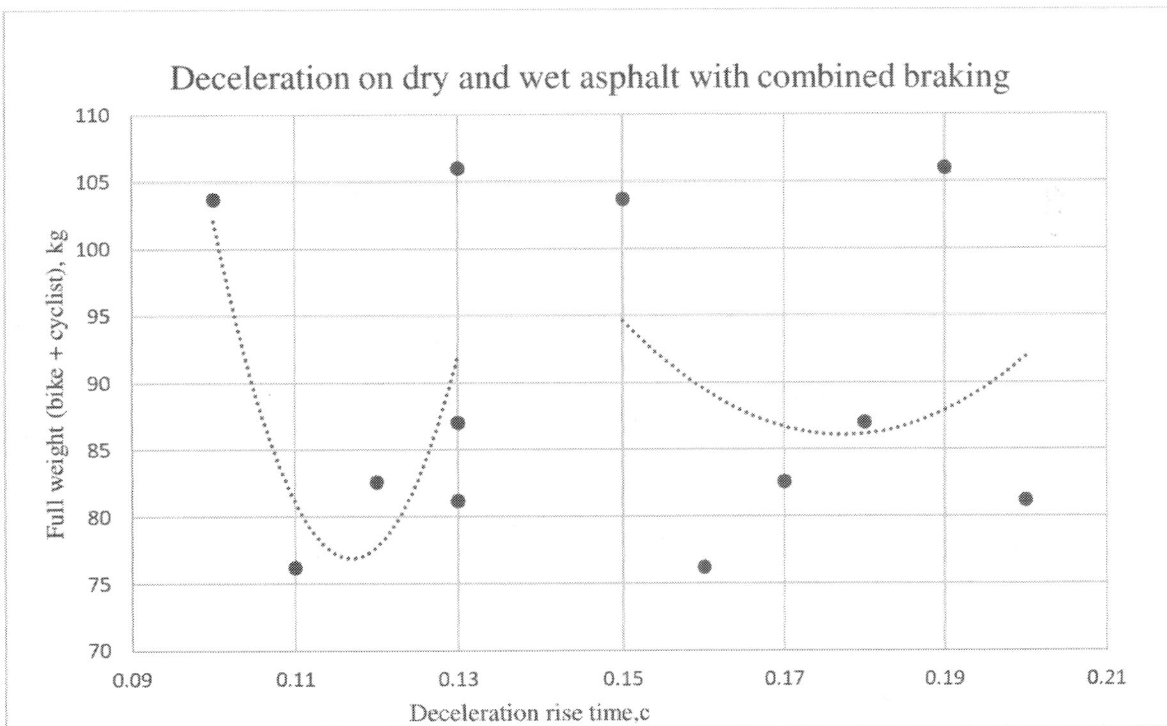

Figure 4. Rise time plot of deceleration versus gross weight on dry and wet asphalt using combined braking

The obtained regression equations describe polynomial curves (average values), with the help of which it is possible to obtain a unit value of the steady-state deceleration. for any TTP braking mode, depending on the state of the road surface, the weight of the bicycle and the cyclist (Table 5).

Table 5. Equations of the regressions of the steady-state deceleration values and the TTP deceleration rise time

№	Measurementparameters	Figure	Regression equations, y=f(x)
1	J steady-stateis on dry asphalt using combined braking	3	$y = 86.053x^2 - 901.3x + 2443.5$
2	J steady-stateison wet asphalt using combined braking	3	$y = 445.22x^2 – 3784.6x + 8119.4$
3	$t_3$is on dry asphalt using combined braking	4	$y = 41583x^2 – 9515.8x + 625.7$
4	$t_3$ison wet asphalt using combined braking	4	$y = 11411x^2 – 4047.2x + 444.93$

where x - is the total mass of the subject, kg, y - is the steady-state deceleration. m/s^2 (item 1.2), y - is the deceleration rise time, s (item 3,4)

4. Conclusion

The practical significance of the regression equations lies in the possibility of their use to determine the value of the steady-state deceleration and the deceleration rise time for various combined TTP braking modes. when research or conducting an investigative experiment to determine the steady-state deceleration value and the deceleration rise time is not possible due to the resulting accident damage to two-wheeled vehicles.

Thus, as a result of the research carried out, the following conclusions can be drawn:

1. The mass of the driver and TPA influences the formation of the steady-state deceleration parameter and the deceleration rise time on dry and wet asphalt;
2. The value of the steady-state deceleration during combined braking on dry asphalt pavement is 18% higher than the combined braking of TPA on wet asphalt pavement;
3. The value of the deceleration rise time during combined braking on dry asphalt pavement is 33% less than the combined braking of TPA on wet asphalt pavement;

Further research will be devoted to identifying the factors influencing the formation of the value of the steady-state deceleration and the time of its increase in other types of two-wheeled vehicles in other weather conditions with different design features, which will allow updating and expanding the database necessary to improve the methods of reconstruction of road accidents.

References

[1] Puchkin V A 2010 *Fundamentals of expert analysis of road traffic accidents: Databases. Expert technique. Solution methods* (Rostov-on-Don IPO SFU) 192-193

[2] Trofimenko Y V, Shelmakova S V, Zege S O, Shashina E V 2020 *Cycling transport in cities* (MADI) 13-35

[3] Evtyukov S A, Evtyukov S S 2018 Problems of calculating the speed of movement of two-wheeled motor vehicles during an accident *Transportation research procedure* **36** 84-89

[4] Brylev I S, Evtyukov S S Vorozheikin 2019 Estimation of the reaction time of the driver of a

two-wheeled mechanical vehicle when applying braking(Bulletin of civil engineers SPbSPBGASU 683) 277-283
[5] NEW RETAIL media portal. 2019 Available at: https://new-retail.ru/novosti/retail/velosipednyy_bum_v_rossii_za_god_spros_vyros_na_tret2897/
[6] Official site of the Ministry of Internal Affairs of the Russian Federation 2020the State Traffic Safety Inspectorate of Russia Available at: http://stat.gibdd.ru.
[7] Evtyukov S A, Evtyukov S S 2018 Problems of calculating the speed of movement of two-wheeled motor vehicles during an accident *Transportation research procedure* **36** 84-89
[8] Beck R F 2004 *Mountain Bicycle Acceleration and Braking Factors Casteel* (San Diego)
[9] Rutman J 2007 *How to do a Roll-down Test*
[10] Evtyukov S S, Ginzburg G 2016 Problem of accident reconstruction mechanism using parameters of braking process of two-wheeled motor vehicles *Architecture and Engineering* **31**

IOP Conf. Series: Materials Science and Engineering 1159 (2021) 012059 doi:10.1088/1757-899X/1159/1/012059

Computer materials science of electrical construction composite materials with specified properties

A V Ilyukhin[1], I I Zaripova[1] , M Y Abdulkhanova[1], E V Marsova[1]

[1] Moscow Automobile and Road Construction State Technical University (MADI), Moscow, 64, Leningradsky prospect, 125319, Russian Federation

E-mail: aviluhin@mail.ru, zaripovairinamadi@mail.ru

Abstract. This article is devoted to computer materials science of electrical construction composite materials for obtaining materials with specified properties. The idea, principles and distinctive features of computer materials science are given. An approach to solving a number of problems is described, which makes it possible to fully take into account the presence of random elements that are present in the process of forming composite materials. Implementation of algorithms is implemented, in which results obtained in previous stages affect progress of algorithms execution in subsequent stages. The radar absorbing material is considered in detail using the example of radar absorbing concrete. It has been proved that with multi-factor dependence, it is possible to optimize the composition only using an automated computer system that calculates the composition of the concrete mixture. One possible computer material science problem has shown that in order to increase absorption efficiency, it is necessary to substantially increase the effective absorption volume, which can be achieved by increasing the non-conductive layers in order to reduce the through conductivity of the radar absorbent concrete, i.e., as if to break the material into a large number of small electromagnetic screens. In this case, the volume of interaction of the material with the electromagnetic field extends deep into the material and is made commensurate with the size of the sample. In conclusion, a possible list of tasks and directions for the development of computer materials science is presented.

1. Introduction

Studies in the field of the use of computers in modeling the electro-physical properties of inhomogeneous dielectric materials in order to identify factors affecting the properties of such materials were begun in the mid-60s of the XX century. Unfortunately, due to the low performance of computers existing at that time, these works were not widely developed, although very interesting results were obtained.

Currently, when the performance of modern computers has reached the required level, it is possible to continue this work. The intensive development of computer technology and software has led to the fact that computers began to be used in areas where their use was not even intended before. The results of such work formed a new direction in materials science of electrical composite materials – «Computer materials science of construction composite materials». The following are some of the results of this work in MADI.

2. Problem setting

Since the widespread use of composite materials, many scientists have been trying to obtain analytical

Content from this work may be used under the terms of the Creative Commons Attribution 3.0 licence. Any further distribution of this work must maintain attribution to the author(s) and the title of the work, journal citation and DOI.
Published under licence by IOP Publishing Ltd

IOP Conf. Series: Materials Science and Engineering 1159 (2021) 012059 doi:10.1088/1757-899X/1159/1/012059

relationships linking the properties of the components and their concentration in the mixture with the properties of the finished composite. Such dependencies allow the calculation of the concentration-granulometric characteristics of mixtures for the manufacture of materials with predetermined properties. Still V.I. Odelevsky showed that many properties of the composites of the matrix structure (aggregate and binder matrix), depending on the concentration of the aggregate, are described in the same terms, which follows from the general theory of phase transitions. This work was the impetus for the appearance of a large number of formulas with the help of which they tried to describe the properties of composite materials. However, numerous experiments in physical models have shown that the fairness of such formulas extends to cases where there is relative proximity of the characteristics of the components and a small bulk concentration of aggregate. Such limitations are related to the difficulty of considering the mutual influence of individual closely spaced particles. Nevertheless, many authors continue to use such formulas, and to reduce the time for cumbersome calculations, they use computers. In this case, such calculations relate to computer materials science.

High hopes were placed by researchers on a relatively new theory of «percolation» - studying the properties of connected components of random graphs [1,2]. This theory, in relation to materials science, using the method of cluster structures, describes the behavior of heterogeneous materials depending on the concentration of components in them. By virtue of the universality and similarity hypotheses, the characteristic parameters of the «percolation» theory are considered independent of the particular nature of the phase transition, i.e., the various materials of the matrix and the aggregate. However, strict evidence in the framework of the theory of «percolation» is obtained for only a few statements. Most often, in such studies, the term computer materials science refers to the use of a computer to find numerical values in statistical processing of experimental data or Monte Carlo modeling of correct lattice problems.

Summing up the above-described tasks of the construction composite materials called by the authors, it can be stated that in all cases the computer acts as a means of computing, which, if especially necessary, can be performed manually. In our opinion, such calculations cannot be called computer materials science.

The computer, as a means of automating the process of searching for the optimal composition of composites, should be used to obtain the latter with predetermined properties. Programs should ensure that the final result is obtained using not algorithms that implement calculations according to predetermined analytical expressions, but only limiting criteria that should direct the calculations in the desired direction. This approach, as applied to construction materials, will be understood by the term computer materials science, which allows to fully realize the random elements necessarily present in the process of forming composite materials, since as the algorithms are executed, the results obtained in the previous stages affect the progress of the algorithm in subsequent ones. Therefore, the distinguishing feature of programs implementing such algorithms should be the absence of initial parameters and the presence of limiting criteria set by the researcher.

In our country, a new radio absorbing material has been developed, called «Radio Absorbing Concrete» (RAC). This material can be advantageously used both to prevent the effects of electromagnetic radiation on instruments, equipment and maintenance personnel, i.e. to ensure the radio tightness of buildings and structures, and to provide protection against electronic means of detecting and tracking objects of a terrestrial, long-term nature. Unlike traditional radio-absorbing and radio-shielding materials, RAC has a significant spread of efficiency absorption (EA) values from sample to sample, which is a consequence of the dependence of its electro-physical properties on a significant number of parameters [3].

3. Materials and methods.

As is known, the loss power of electromagnetic waves by nonlinear dielectrics that do not have magnetic properties (to which RAC belongs) is described by the Landau-Lifshitz formula

$$P = \frac{\omega}{2} V_{\text{кон}} \alpha''_{\text{E}} |E|^2 \qquad (1)$$

where: α''_{E} – imaginary component of electrical polarizability of phase interface of two-component material of matrix type;

$V_{кон}$ – volumetric concentration of conducting phase;

E – electric field of absorbed wave.

By analyzing the expression (1), it can be concluded that in order to increase the absorption efficiency of electromagnetic waves in the RAC, it is necessary to have the maximum content of the conductive phase in the material and only a small content of thin interlayer films, which ensure the bonding of the conductive phase particles into a single whole. However, such a conclusion is erroneous. The fact is that when the V_kon increases, the amount of dielectric gaps between the conductive particles decreases and, as a result, the value of potential barriers at the boundaries of these particles decreases. This leads to the fact that although cement layers are a dielectric with very low intrinsic conductivity, nevertheless, in the form of thin layers (on the order of several tens of angstroms) they can have great transparency for electrons due to the tunnel effect. This increases the homogeneity of the material, in which case the material must already be considered as an electromagnetic shield, and according to the theory of electromagnetic shielding, the volume in which the effective absorption of electromagnetic energy is observed is limited by the size of the skin-layer. As a result, the reflectance of the material is increased, which is unacceptable for RAC. To improve absorption efficiency, it is necessary to significantly increase the effective absorption volume, which can be achieved by increasing the non-conductive layers to reduce the through conductivity of the RAC, i.e., as if to break the material into a large number of small electromagnetic screens. In this case, the volume of interaction of the material with the electromagnetic field extends deep into the material and is made commensurate with the size of the sample.

According to the reasonings given above can seem that the internal contradiction consists in expression (1). Actually, there is no contradiction as with increase in V_kon layers between the carrying-out particles decrease that leads to reduction of interlaminar polarization, i.e., α_{E}'' and vice versa.

Thus, with the set properties it is necessary to make optimization for receiving RAC, both on conductivity, and on frequency characteristics of electromagnetic fields.

Besides, as RAC is composite material, its properties have to depend on its structure, concentration of the carrying-out phase and its particle size distribution.

Therefore, by drawing up concrete mix RAC it is necessary to consider a factor of conductivity of components, their particle size distribution, concentration in mixes and frequency of the electromagnetic field at which ready RAC has to work, and all these parameters are closely interconnected with each other, and change of one of them results in need to change also others under difficult laws [3].

At such multiple-factor dependence optimization of structure, perhaps to carry out, only using the automated computer system making calculation of composition of concrete mix RAC.

In [4-6] it is shown that electro-physical characteristics of conductive composites are well described by the theory of «percolation». However, application of this theory encounters the insuperable difficulties connected with impossibility of the calculations, due to the lack of ways of obtaining values of fundamental coefficients called by «critical indexes». The way out was for the first time offered in [7] where it is shown that there is no need to be engaged in search of values of «critical indexes» of the theory of «percolation», and it is possible to use some formulas of «the effective environment» within the theory of «the generalized conductivity», imposing on them restrictions of the theory of «percolation».

As part of the theory of «generalized conductivity» the properties of heterogeneous systems (composite materials) combined under the general name «generalized conductivity» (electrical conductivity, thermal conductivity, dielectric and magnetic permeability, etc.) are calculated. This combination is based on the known formal coincidence of differential equations of scalar and vector fields for stationary heat, electric current, electrical and magnetic induction flows. There are a large number of formulas describing the generalized properties of heterogeneous systems depending on the properties of the components and their concentrations (for example, Lawrence, Maxwell, Mosotti-Lorenz, Lichteneker, Odelevsky, Landauer-Bruggeman), all of them describe with approximately the same accuracy the concentration behavior of the properties of composites outside the critical region of aggregate concentrations for matrix systems (critical refers to the value of the concentration of electrically conductive aggregate at which a sharp, threshold change in the electrical conductivity of the entire composite occurs). None of these formulas includes absolute particle sizes, only volumetric phase concentrations appear in them. This circumstance is by no means accidental, and can be generalized to all cases in which surface, linear or point effects are absent or excluded from consideration. However, none of these formulas allows us to obtain values of the properties of composites in critical areas that mainly interest us, since it is here that the maximum dependence «composite property - the concentration of electrically conductive aggregate» comes.

Based on the methodology set forth in [3,4,7], it is possible to obtain a solution of the Landauer-Bruggeman formula in the framework of the theory of an «effective medium» when superimposing the conclusions of the «percolation» theory on it, given that, due to the theory of «generalized conductivity» the dielectric constant ε and the specific conductivity σ are described by the same expressions:

$$\begin{cases} \dfrac{\sigma_{ef}}{\sigma_y} \approx \dfrac{V_{кон}(1-V_{кон})}{V_{кон}^{к}(V_{кон}^{к}-V_{кон})} - \dfrac{k\cdot V_{кон}(1-V_{кон})}{(V_{кон}^{к}-V_{кон})^5\cdot 2(\omega\tau)^2} \\ \dfrac{\varepsilon_{ef}}{\varepsilon_c} \approx \dfrac{V_{кон}^{к}-V_{кон}}{V_{кон}^{к}} + \dfrac{V_{кон}(1-V_{кон})}{(V_{кон}^{к}-V_{кон})^3\cdot 2(\omega\tau)^2} \end{cases} \tag{2}$$

где: $k = 1/4 + V_{кон}(1-V_{кон})$

σ_{ef} – effective value of specific conductivity of composite (conductivity characteristic of the whole material);

ε_{ef} – effective value of dielectric constant;

σ_y – specific conductivity of carbon (aggregate);

ε_c – specific dielectric constant of cement stone (material of binding matrix);

$V_{кон}$ – current value of volume concentration of carbon in composite;

$V_{кон}^{к}$ – critical volume concentration value (percolation threshold);

ω – angular frequency of electromagnetic field acting on material;

τ – relaxation time of the double electric layer at the interface of carbon-cement substances $(\tau = \varepsilon_c/\sigma_y)$.

Using the proposed expression, it is possible to obtain «effective» values of conductivity and dielectric constant of the entire RAC by multiplying the right and left parts by σ_y and ε_c. Therefore, knowing the electro-physical characteristics of the carbon and cement used in the manufacture of the material, it is possible to select a value of the volume concentration of carbon that will provide a given loss (absorption) power of the electromagnetic field in the material.

On the basis of the given theoretical conclusions the «CritConc» program which allows to study influence of conductivity of carbon and dielectric permeability of a cement stone on extent of absorption, to vary particle size distribution and volume concentration of carbon, to define values «critical volume» concentration of filler of any composite material and also to accumulate data for the further statistical processing that allows to count optimum composition of material for specific objectives was developed.

Figure 1 shows the experimental and model average concentration relationships of the uptake efficiency of the RAC at the voltage of the electric component of the electromagnetic field E = 1 V/m and a frequency of 1 GHz. The gap in characteristics is due to the fact that in the zone of «critical» carbon concentrations during the experiment there is a rapid heating of the RAC, as a result of which there is a significant drift of readings and an increased probability of destruction of the material sample (carbon burnout). Experimental data were captured in an echo-free chamber. To obtain data, 150 RAC samples were produced with an increasing volume concentration. The value of EA for each volumetric concentration was taken as an average of 150 samples. Model dependence is obtained as the average for 35 implementations of the program.

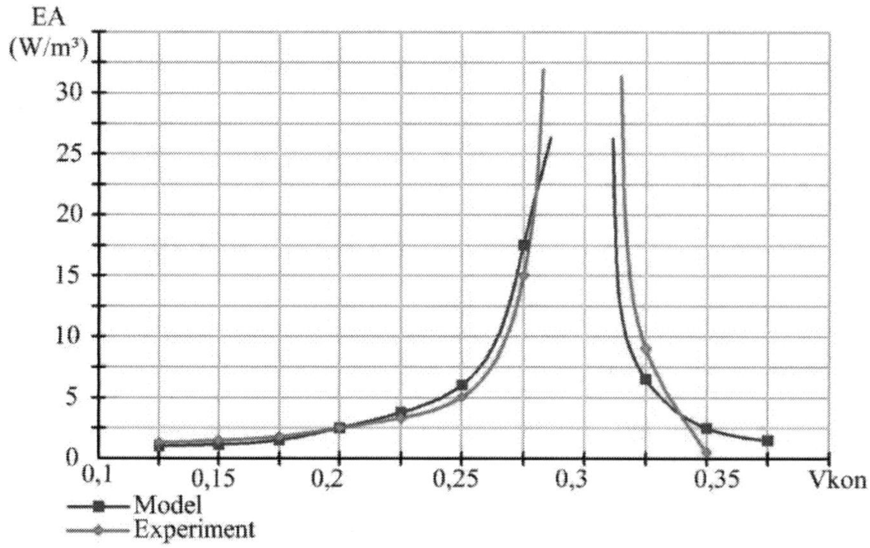

Figure 1. Experimental and model average concentration dependencies absorption efficiency of RAC

The sharp decrease in absorption efficiency in the «after critical» area is due to the increase in homogeneity of the material and its gradual transition from the radio absorbing material to the radio shielding one. In addition, the strength characteristics of the RAC in the «after critical» region are also sharply reduced, since the volume concentration of carbon increases so much that the cement matrix is not able to effectively bond its individual particles. For this reason, for real products from RAC, a volumetric concentration of carbon 0.25... 0.3 is used.

4. Results

Unfortunately, it is not possible to describe all aspects of the work carried out within the framework of the submitted materials. Note only that the values of the «critical» concentration themselves are determined using mathematical modeling using a bipolar (bispherical) coordinate system and are a function of variation in the particle size distribution of the aggregate (carbon).

Characteristics and dependencies similar to those shown in Figure 1 are given in [8-10] for aggregate compositions of nickel, chromium, niobium, molybdenum, etc. As can be seen from Figure 1, experimental and model dependencies have a good coincidence in the precritical region (divergence of not more than 15%) and a slightly worse coincidence in after critical, which can be explained by the absence of a mechanism in the model that allows modeling the reflection of an electromagnetic wave.

The results of the production tests showed that with the traditional (experimental) technology for selecting the composition of the RAC, only the 20% of products with a 10%-th variation in absorption efficiency from the expected value established by technical standards are absolutely suitable. When using the program «CritConc» the absorption efficiency of all products fits into the 10%-th range of the tolerance for scattering values, and the average maximum deviation from the expected values is 9.7... 9.8%.

5. Conclusion

Thus, it can be concluded that the use of the developed method for calculating the composition of the RAC allows us to provide the 100% yield of finished products suitable for use.

Further research in the field of computer materials science of electrical construction composite materials with given properties made it possible to determine the range of problems that can be solved by the methods described above:

• Modeling of composites structure,
• Modeling of composite structure with edge effects,

IOP Conf. Series: Materials Science and Engineering 1159 (2021) 012059 doi:10.1088/1757-899X/1159/1/012059

- Modeling of low-filled and highly filled composites,
- Study of clustering processes,
- Study of the mutual coordination and kinetics of the formation of the skeleton of composites,
- Modeling of strength characteristics of composites,
- Modeling of electro-physical properties of composites,
- Simulation of thermal conductivity of composites,
- Modeling of formation, development and quenching of cracks and pores in composites,
- Modeling of gas-liquid permeability of composites,
- Modeling of filtering properties of porous composites.

These are briefly the tasks of modeling and studying the properties of building composites in the framework of computer materials science.

Consider possible options for research and the use of a generalized problem of calculating and designing composites with optimization by their properties.

As already noted, the scope of these tasks is the development of a software product with a convenient interface, designed not for researchers, but for a practical user. Therefore, the purpose of such programs is to optimize the given properties of composites in order to obtain materials with specific parameters. The development of such applications can be based on two approaches. In the first approach, the initial data for the programs are only the specified properties of the composite and the data bank for existing source materials. In this case, the computer «informs» how much and what aggregate is to be taken, the range of its particle size distribution and the type of binder used. With the second approach, it is possible to realize the now popular trend in construction, aimed at using local raw materials or waste from other industries. In this case, a specific aggregate material and a specific binder type are specified, and the task of the program is to optimize only the concentration-particle size characteristics of the aggregate. This approach makes it possible to obtain a cheaper material, but at the same time the probability that it will not be possible to provide the given properties of the composite increases. Therefore, on a case-by-case basis, the user must decide whether or not to take a particular approach.

Based on all of the above, it is possible to formulate possible, at the moment, tasks of software development to optimize composites:

- Strength optimization,
- Electro-physical optimization,
- Thermal conductivity optimization,
- Minimization of crack formation,
- Porosity optimization,
- Optimization by moisture and gas saturations,
- Optimization by filtering properties.

References

[1] Grimmett G, Manolescu I 2013 Universality for bond percolation in two dimensions *The Annals of Probability, University of Cambridge. Institute of Mathematical Statistics* Vol 41 **5** 3261-3283

[2] Kesten H 2002 Some Highlights of Percolation *Probability Theory and Related Fields* Vol 122 **4** 345-362

[3] Ilyukhin A V 2004 *Automation of the technological process of preparing components of radio absorbing concrete with optimization by electrophysical characteristics of the conductive phase* (Moscow: dis.... Dr. Tech. sciences: 05.13.06)

[4] Bazhenov Yu M, Vorobyov V A, Ilyukhin A V at al. 2006 *Computer materials science of construction composite materials: a textbook* (Moscow: Publishing House of the Russian Academy of Engineering)

[5] Zaripova I I 2016 Influence of percolation threshold on properties of composite materials *All materials. Encyclopedic Handbook* **3** 13-18

[6] Zaripova I I 2014 Application of the theory of percolation for modeling of structure of composite material on the example of concrete *All materials. Encyclopedic reference book* **11**

25-30
[7] Vorobyov V A, Ilyukhin A V 1995 Strength of concrete and the theory of leaking *Izv. higher education institutions. Construction* **7** 60-63
[8] Kovneristy Yu K, Lazarev I Yu, Ravaev A A 1982 *Materials absorbing microwave radiation* (Moscow: Science)
[9] Ilyukhin A V, Chantieva M E, Gematudinov R A, Shukhin V V 2011 Cluster structures and percolation theory in computer materials science of building composites *Bulletin of Moscow Automobile and Road State Technical University (MADI)* **4** 97-101
[10] Vorobyov V A, Ilyukhin A V 1999 Mathematical modeling in computer materials science *Russian Academy of Architecture and Building Sciences. Bulletin of the Department of Construction Sciences* **2** 117-125

IOP Conf. Series: Materials Science and Engineering 1159 (2021) 012060 doi:10.1088/1757-899X/1159/1/012060

Assessment of the ecological and social potential of the organization of ecological zones to ensure sustainable transport development of territories

V V Donchenko[1], S V Shelmakov[2], M I Sharov[3], A V Lobikov[2] and V S Chizhova[1,2]

[1] «NIIAT», 24, st. Heroes of Panfilovtsev, Moscow, 125480, Russia
[2] MADI, 64, Leningradsky Prosp., Moscow, 125319, Russia
[3] Irkutsk national research technical University, 83, st. Lermontova, Irkutsk, 664074, Russia

E-mail: mavese@rambler.ru

Abstract. The article presents the results of an assessment of the impact on the environment and public health of measures to prohibit entry into a certain territory of the city of cars belonging to low environmental classes ("ecological zone" or "ecozone"). When justifying the choice of the city on the example of which the research was conducted (Novosibirsk), both environmental and socio-economic aspects were taken into account. The assessment of the environmental impact of the ecozone organization consisted in modeling traffic flows on the road network before and after the organization of the ecozone, estimating the emissions of pollutants from the totality of traffic flows, calculating the maximum and average annual surface concentrations of the considered pollutants on the territory of the ecozone. Based on the calculated average annual concentrations of pollutants, characteristics of the population density living in the ecozone, and scenarios of its interaction with pollutants, public health risk assessments were carried out before and after the organization of the ecozone. Studies have shown that the change in gross emissions of major pollutants during the organization of the ecozone in Novosibirsk will be 40...60%. Ground-level concentrations of these pollutants will decrease by about the same amount. Carcinogenic and non-carcinogenic risks to the population will be reduced by about half.

1. Introduction

One of the mechanisms aimed at reducing the impact of the city's transport system on the environment and public health is the introduction of restrictions on the movement of cars of low environmental classes in a certain territory. Such territories are called "ecological zones" (German: Umweltzone) or "low-emission zones" (English: Low Emission Zone).

"Ecozones" are traffic management schemes designed to restrict vehicle access to certain urban areas based on their environmental characteristics. They are mainly organized in cities with a population of more than 200 thousand people. Usually, restrictions are based on the compliance of vehicles with a certain certification level (environmental class). In many cases, vehicles of lower environmental classes are allowed to enter ecozones if they are upgraded (for example, if diesel vehicles are equipped with soot filters). Vehicles that do not meet the established emission requirements may be banned from entering the ecozone, or they may be allowed to enter the ecozone for a fee.

Now ecozones operate in more than four hundred cities around the world. In particular, in Europe, they are organized in Germany, Austria, France, Belgium, Denmark and Spain.

Content from this work may be used under the terms of the Creative Commons Attribution 3.0 licence. Any further distribution of this work must maintain attribution to the author(s) and the title of the work, journal citation and DOI.
Published under licence by IOP Publishing Ltd

Russian legislation provides for the powers of the constituent entities of the Russian Federation to impose restrictions on the movement of vehicles in localities, recreation and tourism areas, and specially protected areas in order to reduce the release of pollutants into the air. On the other hand, comprehensive action plans are currently being developed and implemented to reduce air pollution in major industrial centers, including Novosibirsk. Therefore, the question of the potential effectiveness of this measure in specific Russian conditions is very relevant.

2. Object of research

The following factors were taken into account when choosing a locality to evaluate the effectiveness of the ecozone organization. The organization of an ecozone makes sense in those localities where:

- there is a high degree of air pollution;
- there is a high contribution of motor transport to air pollution;
- a large number of people live;
- there is a high level of motorization;
- the urban planning structure and the structure of the road network allows you to select a zone with a high population density and the possibility of organizing a detour;
- public transport is well developed;
- it is possible to conveniently mark the boundaries of the ecozone for the administration of established restrictive measures;
- there is a favorable socio-psychological "background" that allows us to hope for the support of the organization of the ecozone from the majority of the population.

The following cities received the highest expert rating of desirability for ecozone organization: Moscow, Saint Petersburg, Novosibirsk, Samara, Omsk, Kazan, Yekaterinburg, Nizhny Novgorod, Voronezh, Ufa, Krasnodar, Krasnoyarsk, Vladivostok, Chelyabinsk, Perm.

Novosibirsk was chosen as an example from this list, since research for Moscow and St. Petersburg required significantly more labor. Therefore, any effective measures to improve the environment in the city center should be considered highly desirable.

The main source of emissions of polluting substances in atmospheric air in Novosibirsk and the region is road transport. However, its share in total emissions from all sources has tended to increase in recent years. In 2018, the share of emissions from road transport accounted for 68.5% [1].

The automobile fleet of Novosibirsk as of July 1, 2019 totaled 449.8 thousand vehicles, the population was 1618 thousand people, the level of motorization – 278 cars per thousand inhabitants, which is slightly lower than the national average (306 cars per thousand people) [2].

3. Description of the method

The assessment of the ecological and social effect of the ecozone organization in this work was of a predictive nature. This imposed specific conditions on the selection and formation of initial data, and, accordingly, on the methods used for evaluating intermediate and final results.

The environmental and social performance of the ecozone organization was evaluated in the following sequence:

1. Based on the analysis of urban planning information, the structure of the road network and transport characteristics, the territory that is potentially suitable for organizing an ecozone was selected.
2. The structure of the road network used for modeling traffic flows was formed for the selected city territory. In this work, the simulation was performed using the PT VISUM software.
3. Based on transport modeling, the maximum intensity of traffic flows was determined for all elements of the road network that is part of the ecozone.
4. The composition of traffic flows (i.e., the distribution of vehicles by type and environmental class) was determined based on the assumption that the composition of traffic flows is identical to the average composition of the car fleet in the region. The average composition of the fleet was determined based on the processing of traffic police data in the interpretation of the analytical Agency "AUTOSTAT".
5. The average coefficients of run-through emissions of pollutants for the traffic flow of a given composition were calculated. For this purpose, COPERT software was used, which

implements the standard pan-European method of inventory of vehicle emissions EMEP/EEA [3].

6. Based on the assumption that the composition of traffic flows on the entire road network in the ecozone is identical, the maximum single (g/s) and gross (kg/h) emissions of pollutants on each segment of the road network were calculated.
7. Gross emissions were added up for all sections of the road network to obtain total emissions for the entire ecozone.
8. Maximum one-time emissions were used as input data for modeling the dispersion of pollutants in the atmosphere to estimate the emerging surface maximum one-time concentrations throughout the ecozone. In this work, the unified program for calculating atmospheric pollution "ECO CENTER" was used for this purpose. The obtained surface maximum single concentrations were compared with the hygienic standards of air quality in populated areas in order to analyze their compliance on the territory of the ecozone.
9. The calculated maximum one-time concentrations (at several characteristic points located throughout the ecozone) were converted to average annual values based on the time heterogeneity of transport flows and climatic characteristics.
10. Average annual concentrations of pollutants were used as input data for assessing public health risk indicators.
11. Several possible scenarios for changing the initial data after the ecozone was established were developed, and one of them was selected as the most likely one to be used for further assessments.
12. For the selected scenario, paragraphs 3.10 were performed, taking into account changes in the initial data describing the intensity and composition of traffic flows associated with the organization of the ecozone.
13. A comparative analysis of the results obtained was carried out to determine the environmental performance indicators of the ecozone organization, and conclusions were formulated.

4. Features of estimation of average annual concentrations of pollutants

In this paper, the average annual concentrations were estimated as follows:

- several calculation points located near the largest residential areas were selected within the ecozone territory;
- at these points, the maximum single concentrations of pollutants that characterize the most unfavorable conditions were calculated both by the power of emissions (maximum traffic intensity on the entire road network within the ecozone) and by meteorological conditions (adverse weather conditions: calm + inversion);
- according to expert given variations in traffic density during the day, days and weeks per year was calculated the coefficient of reduction in average concentration caused by the change of intensity of a transport stream within a year;
- based on the analysis of publicly available weather parameters in the city under consideration (Novosibirsk) [4] were determined: a) the number of cases of adverse weather conditions in the year and b) the average duration of adverse weather conditions;
- based on these data, the coefficient of decrease in the average annual concentration was determined due to the probabilistic nature of the formation of adverse meteorological conditions;
- the total value of the average annual concentration was calculated as the product of the maximum single concentration and the two coefficients described above.

The first coefficient quite adequately corrects the change in concentration, since the relationship between traffic intensity, emissions, and concentration is linear in the first approximation. Although the second coefficient introduces an error (it seems to underestimate the results, since the binary model "there is an inversion – no inversion" is used), it allows us to take into account the very complex influence of weather parameters on the result.

5. Features of public health risk assessment

IOP Conf. Series: Materials Science and Engineering 1159 (2021) 012060 doi:10.1088/1757-899X/1159/1/012060

The public health risk assessment was carried out in accordance with the guidelines 2.1.10.1920–04 on public health risk assessment for exposure to chemicals that pollute the environment [5], which provides for the unification of methods and criteria for risk assessment.

To assess the exposure, the average characteristics of the population group exposed to traffic flow emissions in the ecozone under study were selected (table 1).

Table 1. Recommended standard values for exposure factors

Exposure factor	Value
Lifetime exposure (carcinogens)	70 years
Life expectancy	70 years
Exposure frequency, residential area scenario	350 days/year
The period of averaging of exposure, years	carcinogens: 70 years
The speed of the inhalation, adult, General characteristics	20 м³/ day
Body weight, adult, 18 years or more	70 kg

The amount of chemical intake into the human body was calculated using the formula:

$$LADD = \frac{C \cdot IR \cdot ED \cdot EF}{BW \cdot AT \cdot 365} \quad (1)$$

where LADD – average daily dose during life, mg/kg × day;
C – concentration of the substance in the air, mg /м³;
IR – average daily volume of inhaled air, m³/day;
ED – duration of exposure, years;
EF – frequency of exposure, days/year;
BW – human body weight, kg;
AT – the period of averaging of exposure, years;
365 – number of days per year.

The carcinogenic risk was calculated using the formula:

$$ICR = LADD \times SFi, \quad (2)$$

where LADD – average daily dose during life, mg/kg × day;
SFi – carcinogenic potential factor.

When exposed to multiple carcinogens, the total carcinogenic risk for this route of entry (inhalation) is calculated using the formula:

$$CR_T = \Sigma\, CR_j, \quad (3)$$

where CR_T – total carcinogenic risk for the route of entry of T;
CR_j – carcinogenic risk for the *j*-th carcinogenic substance.

Additionally, we calculated the:

Cri = LADD × SFi × 1 000 000 – carcinogenic annual population risk (per 1 million people).

Cri = LADD × SFi × 1 000 000 / 70 – carcinogenic annual population risk during life (per 1 million people).

Assessment of non-carcinogenic risk is carried out by calculating the hazard coefficient

$$HQ_i = \frac{C_i}{RfC} \quad (4)$$

where HQ_i – hazard ratio of substance i;
C_i – level of exposure to the substance i, mg/m³;

RfC – safe exposure level, mg/m^3.

If the calculated hazard coefficient (HQ) of a substance does not exceed one, then the probability of developing harmful effects in a person with daily intake of the substance during life is insignificant and such an impact is characterized as acceptable. If the hazard ratio exceeds one, the probability of harmful effects in humans increases in proportion to the increase in HQ, but it is impossible to specify the exact value of this probability.

In case of combined intake of several substances in any way, the hazard coefficients are added together and the total hazard index is calculated, which should also not exceed one.

6.Results and Discussion

Using the developed methodology, the ecozone configuration for the city of Novosibirsk was developed. The most heavily trafficked Central historical part of the city was taken as a basis, including all or part of the borders of Pisarev street, Hippodromskaya, Kamenskaya highway, Kirov street, Voskhod, Ob river embankment to the railway line.

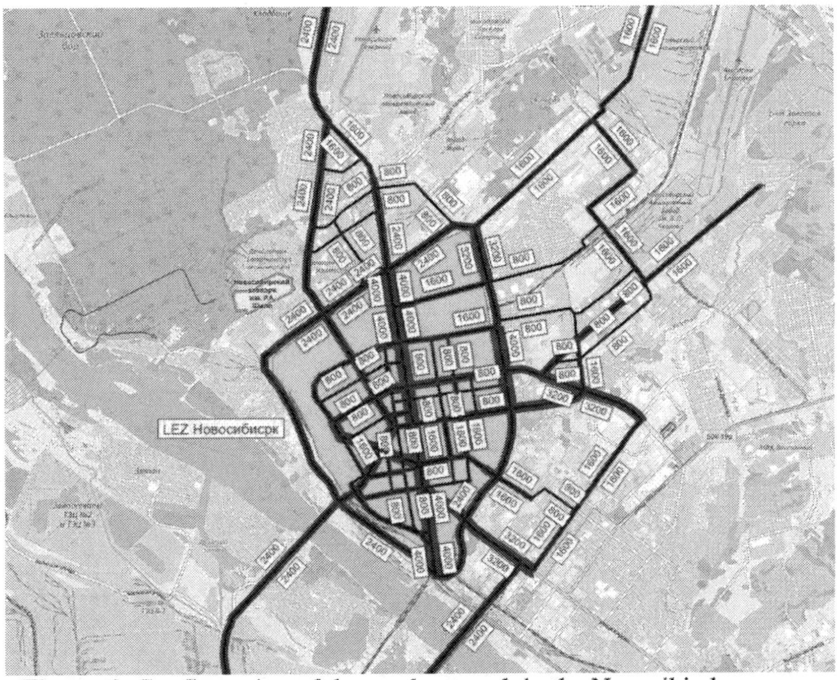

Figure 1. Configuration of the road network in the Novosibirsk ecozone

Description of the basic scenario: the structure of the vehicle fleet was adopted according to expert data processed by the traffic police for Novosibirsk. For vehicles that do not comply with Euro-3, it was 44.2%. The output coefficient for vehicles of environmental classes Euro-0 and Euro-1 is adopted in the range of 0.4...0.8, for vehicles of other environmental classes – the output coefficient is equal to 1. The coefficient of production of cars of ecological classes Euro-0 and Euro-1 was assigned by experts based on a visual analysis of the structure of real traffic flows on the street and road network of Novosibirsk. Traffic flow rates are assumed to correspond to the maximum capacity of network sections (saturation mode, typical for peak time intervals).

Description of the forecast scenario **«Prohibition of vehicles less than Euro-3»**: the share of buses in the traffic flow increased by 1%, which is enough to compensate for the ban on entry into the ecozone of passenger vehicles of environmental classes Euro-0...Euro-2; coef. output for vehicles of Euro-0 environmental classes... is reduced by 95% compared to the basic version, their share is reallocated to higher environmental classes in proportion to the existing structure of the vehicle fleet; traffic intensity on the road network has not changed.

Changes in gross emissions of major pollutants during the organization of an ecozone in Novosibirsk are shown in table 2.

IOP Conf. Series: Materials Science and Engineering 1159 (2021) 012060 doi:10.1088/1757-899X/1159/1/012060

Table 2. Changes in gross emissions of major pollutants during the organization of an ecozone in Novosibirsk

Pollutant	Emissions reduction, %
CO_2	2,3
CO	65,4
VOC	57,2
NMVOC	58,5
HC_4	34,5
NO_x	40,1
NO_2	20,6
NO	41,8
TSP	12,6
PM_{10}	17,0
$PM_{2,5}$	25,9
BC	50,3
OC	70,4
N_2O	37,5
NH_3	79,7
SO_2	1,4

Note: NMVOC – non-methane volatile organic carbons; TSP – total amount of suspended solids; BC – soot; OC – organic dispersed particles.

The results of calculating surface concentrations before and after the ecozone organization showed a decrease in CO by 65% (Figures 2 и 3), NO_2 – by 40%, polycyclic aromatic hydrocarbons (for benzo(a)pyrene) – on 7%, aldehydes (for formaldehyde) – on 31%, aromatic hydrocarbons (for benzene) – on 66%, persistent organic pollutants (by dioxins) – on 42%.

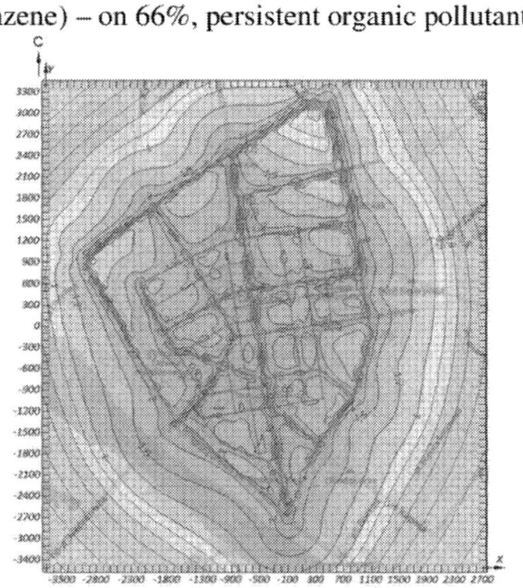

Figure 2. Estimated surface CO concentration (in fractions of the maximum permissible concentration) before the ecozone is established

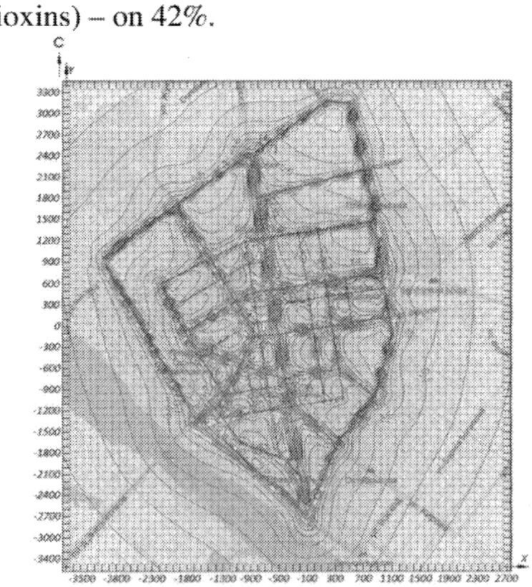

Figure 3. Estimated surface CO concentration (in fractions of the maximum permissible concentration) after the ecozone organization

A comparison of the results obtained for the baseline scenario and monitoring data [6] shows a satisfactory adjustment of the calculated models, with the exception of benzo(a)pyrene, for which monitoring shows significantly higher values. Apparently, this is due to the decisive contribution of stationary sources to atmospheric pollution with this substance, which were not taken into account in this work.

IOP Conf. Series: Materials Science and Engineering 1159 (2021) 012060 doi:10.1088/1757-899X/1159/1/012060

The organization of an ecozone is most effective in reducing emissions of those pollutants whose emissions are more determined by the availability and efficiency of exhaust gas treatment systems. These substances include CO, VOC, NO_x, $PM_{2.5}$, BC, OC, NO_2 и NH_3. Those substances whose emissions depend on both the efficiency of exhaust gas treatment systems and tire and brake wear (TSP, PM10) show a moderate decrease. Emissions of CO_2 and SO_2, which are largely determined by fuel consumption, are practically not reduced, since transport flows remain unchanged.

The obtained values of the maximum single concentrations of pollutants were recalculated to the average annual values using the approach described above. The resulting conversion coefficient (0.174) was slightly higher than the recommended value of 0.1 in [7] for the case of simplified estimation.

The calculation of the average carcinogenic risk for the Novosibirsk ecozone shows that this indicator will decrease from a value of $3.7 \cdot 10^{-5}$ (or 4 cases per 100 thousand population) to a value of $1.7 \cdot 10^{-5}$ (or 2 cases per 100 thousand population).

The results of calculating the total hazard index (non-carcinogenic risk) show a decrease in risk from the average current value (2.14) to a low value (1.08) after the organization of the ecozone.

7. Conclusion

Studies have shown that the change in gross emissions of major pollutants during the organization of the ecozone in Novosibirsk will be 40...60%. The ground-level concentration of these pollutants will decrease to about the same extent.

The results of the carcinogenic risk assessment show that the risk fluctuates close to the acceptable limits. The organization of the ecozone led to a noticeable decrease in the carcinogenic risk for all substances except benzo(a)pyrene, which is associated with a slight change in its concentrations.

The results of the assessment of the total hazard index (non-carcinogenic risk) show a decrease in risk from the average current value to a low value after the organization of the ecozone.

Thus, the study showed a fairly high ecological and social potential of the organization of ecological zones. Therefore, this measure of traffic management can be recommended for other large cities in Russia with a high share of motor transport in the structure of air pollution.

References

[1] On the state and environmental protection of the Novosibirsk region in 2018 Novosibirsk 2019 162 p

[2] Rating of Russian cities with millions of people in terms of car availability in 2019 available at: https://www.autostat.ru/press-releases/41923

[3] EMEP/EEA air pollutant emission inventory guidebook 2016

[4] Collecting weather data *RealMeteo Service* available at: https://realmeteo.ru/

[5] 2.1.10.1920–04 Guidelines for assessing public health risks from exposure to chemicals that pollute the environment Moscow Federal center of Gossanepidnadzor of the Ministry of health of Russia 2004 143 p

[6] Safonov V V 2012 Analysis of the state of atmospheric air in Novosibirsk for 1998-2011 *Interexpo Geo-Siberia* available at: https://cyberleninka.ru/article/n/analiz-sostoyaniya-atmosfernogo-vozduha-g-novosibirska-za-1998-2011-gg

[7] Methods for calculating the dispersion of emissions of harmful (polluting) substances in the atmospheric air. Approved by order No. 273 of the Ministry of natural resources of the Russian Federation dated 06.06.2017

[8] Mikhailov A Yu, Sharov M I, Levashev A G, Butuzova A B and Ovchinnikova N I 2019 Evaluating reliability of municipal public transport operation in the Russian Federation cities **632 (1)**

[9] Trofimenko Yu V and Chizhova V S 2019 Justification of measures to reduce the health risk from air pollution by suspended particles less than ten micrometers (PM10) on the urban road network *Ecology and Industry of Russia* **23 (7)** pp 48–51

[10] Pisoni E, Christidis P, Thunis P and Trombetti M 2019 Evaluating the impact of “Sustainable Urban Mobility Plans” on urban background air quality *Journal of Environmental Management* **231** pp 249–255

[11] Settey T, Gnap J and Beňová D 2019 Examining the impact of the deployment of low emission zones in Europe on the technological readiness of road freight transport *Transportation Research Procedia* **40** pp 481–488

[12] Hooftman N, Messagie M, Van Mierlo J and Coosemans T 2018 A review of the European passenger car regulations – Real driving emissions vs local air quality *Renewable and Sustainable Energy Reviews* **86** pp 1–21

[13] Bigazzi A Y and Rouleau M 2017 Can traffic management strategies improve urban air quality? A review of the evidence *Journal of Transport and Health* **7** pp 111–124

[14] Cyrys J, Peters A, Soentgen J and Wichmann H E 2014 Low emission zones reduce PM10 mass concentrations and diesel soot in German cities *Journal of the Air and Waste Management Association* **64 (4)** pp 481–487

[15] Dias D, Tchepel O and Antunes A P 2016 Integrated modelling approach for the evaluation of low emission zones *Journal of Environmental Management* **177** pp 253–263

[16] Ellison R B, Greaves S P and Hensher D A 2013 Five years of London's low emission zone: Effects on vehicle fleet composition and air quality *Transportation Research Part D: Transport and Environment* **23** pp 25–33

[17] Carslaw D C and Rhys-Tyler G 2013 New insights from comprehensive on-road measurements of NOx, NO_2 and NH_3 from vehicle emission remote sensing in London, UK *Atmospheric Environment* **81** pp 339–347

[18] Kurtenbach R, Kleffmann J, Niedojadlo A and Wiesen P 2012 Primary NO_2 emissions and their impact on air quality in traffic environments in Germany *Environmental Sciences Europe* **24** 21

Energy efficiency of autonomous vehicles on hilly roads

S Shadrin[1,2]

[1] BaseTracK Rus, 3-23, office 41, Kolodezniy Pereulok, Moscow, 107076, Russia
[2] Moscow Automobile and Road Construction State Technical University (MADI), 64, Leningradskiy Pr., Moscow, 125319, Russia

E-mail: dr.sergey.shadrin@gmail.com

Abstract. Energy efficiency is traditionally among the main development trends in the automobile industry. We suspected that intelligent speed control systems can provide economic and ecological benefits for implementing civil autonomous driving technologies. We introduced the approach for autonomous driving, which is based on the analysis of the precise Digital Terrain Model (DTM) and optimal vehicle control actions with respect to a requested transport task, which can be a task of fuel or energy economy in this particular case. We addressed the research question of fuel economy estimation with respect to the reference parameter that is applicable to compare different control algorithms. We demonstrated the use of roadway-related factors of DTM to affect fuel consumption. We contributed the algorithm for saving fuel while driving on a hilly road and examined the influence of vehicle load on possible economy values. We compared fuel and energy-saving values of gasoline and electric vehicles based on the same chassis with and without use of the proposed algorithm. Practical results of tests on public roads are presented.

1. Introduction

This study focuses on predicting the possible economic and ecological benefits of intelligent speed control systems of autonomous road vehicles. Thus, it is also concerned with the subject of autonomous driving system design and the subject of fuel (energy) economy.

The rest of the article is organized as described below.

In Section 2, we provide a description of state-of-the-art research in the field of autonomous driving design and in the field of fuel consumption analysis and optimisation.

In Section 3, we introduce an approach for fuel (energy) saving on hilly roads, describe the way in which we estimate possible economy values, and describe the process of mathematical model verification and validation.

Section 4 describes experimental research of real vehicles on public roads and represents results of fuel and energy-saving values for gasoline and electric vehicles based on the same chassis with and without use of the proposed algorithm of intelligent speed adjustment. An analysis of the vehicle's load influence on possible economy values is also presented.

The main results and novel contributions are highlighted in the conclusion.

2. State-of-the-art Research

Development of autonomous vehicle technologies has been ongoing in the air, marine [1, 2], ground, and space domains for a long time in different ways and with different results. We focused our attention

Content from this work may be used under the terms of the Creative Commons Attribution 3.0 licence. Any further distribution of this work must maintain attribution to the author(s) and the title of the work, journal citation and DOI.
Published under licence by IOP Publishing Ltd

on the ground domain only. Automation of road vehicles has been driven by approximately a century of research in different countries [3]. Several scientific studies [4-6] have addressed the issues of automatic vehicle control, such as the object component base, the mathematical apparatus and its implementation, and technical vision systems.

Developmental projects on autonomous vehicle technologies in the ground domain are currently being undertaken by all global motor manufacturers, especially those in the US, Germany, Japan, and China. Automobile groups such as Ford, Daimler, Volkswagen, Toyota, Honda, GM, Geely, Tata, and Tesla, as well as other major technological organisations such as Google, Continental, Delphi, Siemens, and Bosch are also involved in these projects. Other players in the field include defence departments and agencies of different countries, universities (e.g., Stanford University, Carnegie Mellon University, Technical University of Munich, University of Karlsruhe, Fraunhofer Institute, University of Minnesota, Universidad Politécnica de Madrid [7]), and many other institutions. Thus, autonomous driving is a global trend and no automakers, suppliers, or technological companies will be able to overlook it [8-11].

We suppose that one of the main reasons why autonomous driving technologies have not yet been widely deployed for everyday life use is that car manufacturers and technological companies have chosen the paradigm of variability of manoeuvres and driving modes at the tactical level of vehicle control. This paradigm makes autonomous vehicles essentially sensitive to the quality of the technical vision and decision-making systems. These systems are typically based on artificial intelligence solutions, and thus, achievement of 100% reliability seems ambiguous [12].

We suggest decreasing the number of degrees of freedom at the tactical level of vehicle control [13, 14] and excluding the variability of manoeuvres, while retaining the variability of driving modes. We offer a simplified approach for autonomous driving and a more reliable and less expensive solution. This approach can be compared with a virtual spatial railway for the vehicle, which acts like a train. Accordingly, in the simplest case, the vehicle should decrease its driving velocity and stop when it detects an obstacle or risk of any type.

Optimisation of a vehicle's fuel consumption is a complex subject, and numerous variables influence vehicle energy and emission rates. These variables can be classified into six broad categories [15]: travel-related, weather-related, vehicle-related, traffic-related, driver-related, and roadway-related factors.

Travel-related factors represent eco-routing systems, which are capable of providing the shortest-distance route and the shortest-duration route, and routes that minimise fuel consumption with respect to traffic conditions. The possible fuel economy improvements of eco-routing systems achieve values in the range from 8.73% to 42.15% [15].

Weather-related factors deal with parameters of ambient temperature, humidity, and wind effects. These factors affect fuel consumption through vehicle attachments, e.g., the air conditioner and the water pump. According to the U.S. Environmental Protection Agency, fuel economy improvements in this case can achieve values of up to 1%.

Vehicle-related factors belong to the core factors that influence fuel economy (e.g. engine size and type, its characteristics, loading, and aerodynamics). The fuel saving efficiency can reach any value, and it depends only on what type of vehicle and scenario will be used as a basis for comparison.

Traffic-related factors include traffic flow and traffic signalling. Several studies have indicated that the use of traffic signal information has significant potential to save fuel. Fuel consumption can be reduced by 22–50% depending on the application scenario considered [15].

Driver-related factors refer to driver behaviour and aggressiveness, which are typically identified by speed and acceleration profiles. Several studies have indicated that aggressive driving can cause 30–40% higher fuel consumption compared to calm driving [16].

Roadway-related factors refer to the physical characteristics of a road (roadway grade, roughness, curvature). In this case, an eco-driving system was based on model predictive control (MPC) to run a vehicle on roads with up and down slopes [17]; their simulation on a virtual road indicated fuel savings in the range of 5–7.04% when the eco-driving vehicle was compared with automatic speed control drive vehicle. This result was achieved by increasing the vehicle speed before entering the uphill slope and taking advantage of the down-hill slope. In another paper, the method of global and local optimisation was used [18], and a fuel reduction of 5.5% was obtained using simulation techniques. Researchers have also appraised the available efficiency of these systems as being in the range of 3–20% for fuel savings, according to [15], and up to 6.89–24.78%, according to [19]. The existing on-the-market Volvo "I-see" system is known, and it has reported fuel savings of about 5%, but it is not clear how this value can be confirmed.

We work with a precise Digital Terrain Model, so we will further discuss only roadway-related factors in this paper. The value of the fuel savings is critically affected by the value of fuel consumption that is used as a reference value for comparison. For example, a decrease in the average velocity on a route will lead to a reduction of air and rolling resistance factors and lead to fuel savings, but it will also increase the trip time. The next question is associated with the problem of conducting a road test on a long route because it is influenced by traffic and environmental conditions and it is difficult to repeat. Thus, we addressed the research question of estimating fuel economy with respect to the reference parameters that are required to compare different control algorithms and conduct a real road test on a long route.

3. Approach for Fuel (Energy) Saving on Hilly Roads

Fuel consumption of the road vehicle with a gasoline engine can be estimated using the following equation [20]:

$$B_{IC} = \frac{\int b_e \cdot \frac{1}{\eta}\left[\left(m \cdot f \cdot g \cdot \cos\alpha + \frac{\rho}{2} \cdot c_W \cdot A \cdot v^2\right) + m \cdot (a + g \cdot \sin\alpha) + B_r\right] \cdot v \cdot dt}{\int v \cdot dt}, \quad (1)$$

where B_{IC} is the fuel consumption per distance, b_e is the specific fuel consumption, η is the average transmission efficiency, m is the vehicle mass, f is the average rolling resistance, g is the acceleration of gravity, α is the longitudinal road inclination, ρ is the density of air, c_w is the coefficient of aerodynamic drag, A is the front vehicle area, v is the vehicle velocity, a is the vehicle longitudinal acceleration, B_r is the braking resistance, and t is time.

We consider comparing the fuel consumption of different driving modes with respect to the same driving time during all the tests. First, we set the nominal value for the driving velocity and lower and upper limits for speed regulation. Then, we analysed data from the Digital Terrain Model and estimated the function of longitudinal road inclination angles compared to the distance driven. We then estimated the optimal speed profile for the current route. We then estimated the fuel consumption and driving time with activated intelligent speed adjustment, determined the average velocity, and estimated the referenced fuel consumption with this constant speed. Thus, the reference parameter that is applicable to the fuel consumption comparison using different control algorithms is the driving time.

The energy consumption of electric vehicle is specified as an integration of the output power of the electric vehicle at the battery terminal. The total power flows at the battery terminal can be defined as [21, 22]:

$$P_{total} = P_{bat_{out}} - P_{bat_{in}} = P_{bat_{tractive}} + P_{bat_{aux}} - P_{bat_{in}}, \quad (2)$$

If we simplify the above equation and make assumptions about braking, regenerative braking, and auxiliary load absence, it will lead to the following equation:

$$B_{EV} = \frac{\int \frac{1}{(\eta_t \cdot \eta_m)} \cdot \left[\left(m \cdot f \cdot g \cdot \cos\alpha + \frac{\rho}{2} \cdot c_W \cdot A \cdot v^2\right) + m \cdot (a + g \cdot \sin\alpha) + B_r\right] \cdot v \cdot dt}{\int v \cdot dt}, \quad (3)$$

where B_{EV} is the energy consumption per distance, η_t is the average transmission efficiency, and η_m represents motor drive efficiency.

In the next stage, we need to find a way to verify the unknown parameters from equations (1) and (3). We suggest performing road tests on public roads using a predetermined DTM. This step will provide an advantage for the fuel consumption model validation in further simulations of fuel consumption on a long route with a known DTM.

4. Setup and Implementation

The fuel consumption experiment was performed on a highway between Moscow and Volokolamsk using a gasoline Chevrolet Cruze (experimental autonomous road vehicle). Figure 1 illustrates the altitudes of the highway compared to the travelled distance. The test road distance was 99.8 km, and the altitudes were obtained from a DTM with a height accuracy of 10 cm. The maximum difference in the altitude on the route was 163.3 m. To obtain a clear understanding of the fuel economy values, we estimated both driving directions: forward and then backward.

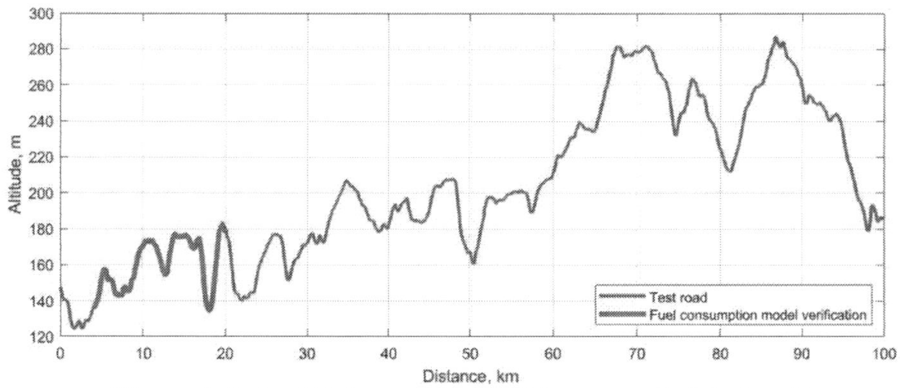

Figure 1. Federal highway altitudes (Moscow to Volokolamsk).

The unknown fuel consumption model parameters were determined by optimisation through comparison of the simulation results with those of road tests on the part of the test road that is identified with the red colour in Figure 1, for a total distance of 32 km while driving in both directions.

The specific fuel consumption b_e and the fuel consumption map of the experimental vehicle were obtained from the CAN bus (Controller Area Network [23,24]) by analysing OBD-II PIDs data [25] with the OBDLink MX device [26]; this is an imprecise method, but we consider it to be cost-effective and applicable for evaluation and rapid prototyping purposes. The adjusted parameters are as follows: b_e=7.9292e-05 g/(W·s); η=0.95; m=1500 kg; f=0.01; g=9.81 m/s^2; ρ=1.22625 kg/m^3; and c_w=0.3; A=2.3768 m^2. The model adequacy was examined by comparing the results of road tests with those of simulations, and further statistical analysis of the error functions. Figure 2 shows the comparison between experimental and simulation data. The relative error of the estimated fuel trip consumption was within 2.8%.

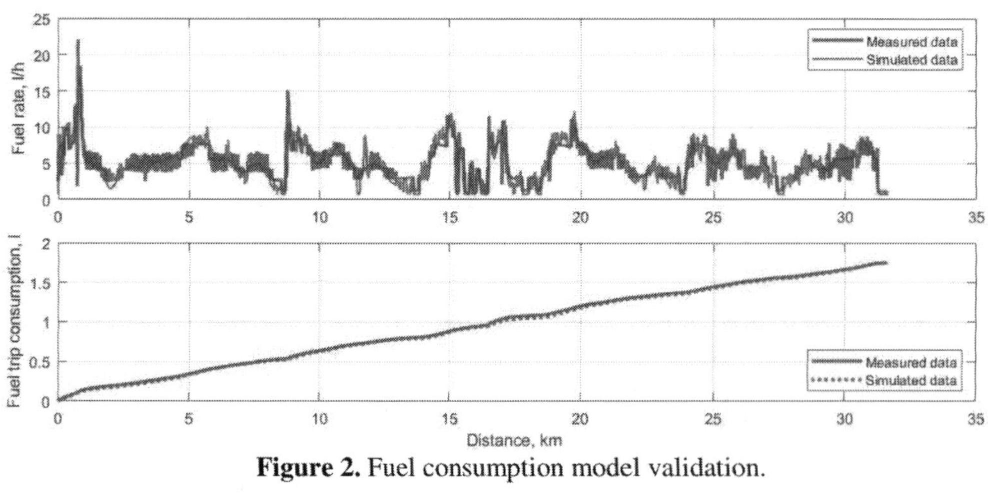

Figure 2. Fuel consumption model validation.

After validation of the fuel consumption model, we simulated driving on a whole route in both directions. The following assumptions were made: we simulated driving at the highest gear in the transmission (rates were detected by analysing the OBD-II PIDs data) and assumed that the gearbox torque converter is blocked. We did not use braking, there was no external wind load, and we neglected the manoeuvring effect. Furthermore, the road's and wheels' adhesion properties were unchanged, there were no other traffic participants, and weather conditions (temperature, humidity, pressure) were not taken into consideration. Such a scenario is not realistic, but we tried to show the possible values of fuel economy under ideal conditions, the proposed comparison method, and the practical use of precise geoinformational data.

Figure 3 illustrates the intelligent speed control regulation during the simulation process, according to the proposed algorithm.

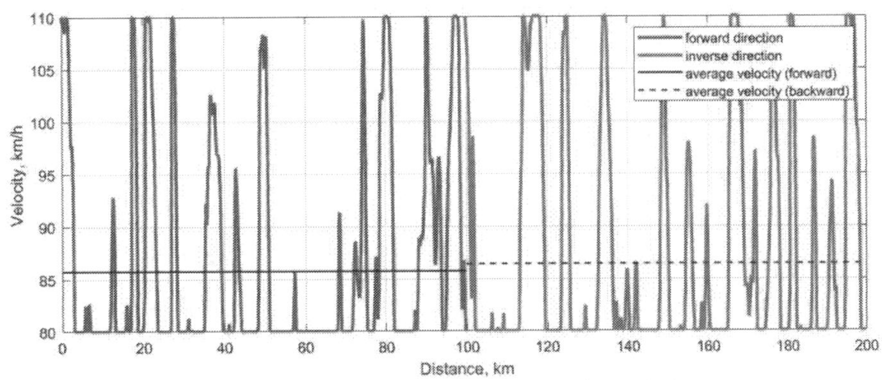

Figure 3. Intelligent speed control regulation.

The main results of the simulations are shown in Table 1. In both driving directions on the entire route, we compared two driving modes: 1) driving with intelligent speed control with a speed regulation algorithm at a nominal value of 95 ±15 km/h, depending on whether the route has an upward or downward slope; and 2) driving with a constant velocity as the average velocity in the first case (and therefore the same time of driving). Simulations of the fuel consumption were also performed for the case of a variable vehicle mass.

Table 1. Simulation results for gasoline vehicle fuel consumption while driving on a hilly road with and without intelligent speed control.

Veh. mass, kg	Forward direction (100 km)			Inverse direction (100 km)			Both ways (200 km)		
	Const.V km/h	80 to 110 km/h	Econ., %	Const.V km/h	80 to 110 km/h	Econ., %	Const.V km/h	80 to 110 km/h	Econ., %
	Fuel cons., l	Fuel cons., l		Fuel cons., l	Fuel cons., l		Fuel cons., l	Fuel cons., l	
1400	4.61	4.51	−2.30	4.41	4.32	−2.02	9.02	8.82	−2.16
1500	4.76	4.63	−2.76	4.54	4.42	−2.58	9.30	9.05	−2.67
1600	4.91	4.75	−3.21	4.67	4.53	−3.13	9.58	9.27	−3.17
1700	5.05	4.87	−3.65	4.81	4.63	−3.64	9.86	9.50	−3.64
1800	5.20	4.99	−4.08	4.94	4.74	−4.12	10.14	9.73	−4.10

An increase of the vehicle mass resulted in an increase of fuel economy for the intelligent speed control system. An increase in the driving velocity up to 150 km/h for the experimental car eliminates the efficiency of the proposed control algorithm. The optimisation of fuel consumption for autonomous driving mode on a hilly test road over a distance of 200 km resulted in up to 4.1% savings compared with constant-speed driving for the same time.

In the next stage, the experimental vehicle was replaced with an equivalent electric vehicle [27]. The motor drive efficiency was considered to be equal to 0.9 [28].

The main simulation results are shown in Table 2. The optimisation of energy consumption resulted in up to 2.4% savings compared with constant-speed driving with for same time using the full vehicle mass.

Table 2. Simulation results for electric vehicle energy consumption while driving on a hilly road with and without intelligent speed control.

Veh. mass, kg	Forward direction (100 km)			Inverse direction (100 km)			Both ways (200 km)		
	Const.V km/h Energy, kWH	80 to 110 km/h Energy, kWH	Econ., %	Const.V km/h Energy, kWH	80 to 110 km/h Energy, kWH	Econ., %	Const.V km/h Energy, kWH	80 to 110 km/h Energy, kWH	Econ., %
1400	13.28	13.13	−1.06	12.63	12.58	−0.37	25.91	25.72	−0.72
1500	13.68	13.48	−1.45	12.99	12.88	−0.84	26.68	26.37	−1.15
1600	14.09	13.84	−1.83	13.36	13.19	−1.30	27.45	27.02	−1.57
1700	14.51	14.19	−2.21	13.73	13.49	−1.77	28.24	27.68	−1.99
1800	14.93	14.54	−2.58	14.11	13.79	−2.23	29.03	28.33	−2.41

Figure 4 illustrates the possible economy values compared to the vehicle's load, which shows linear dependence.

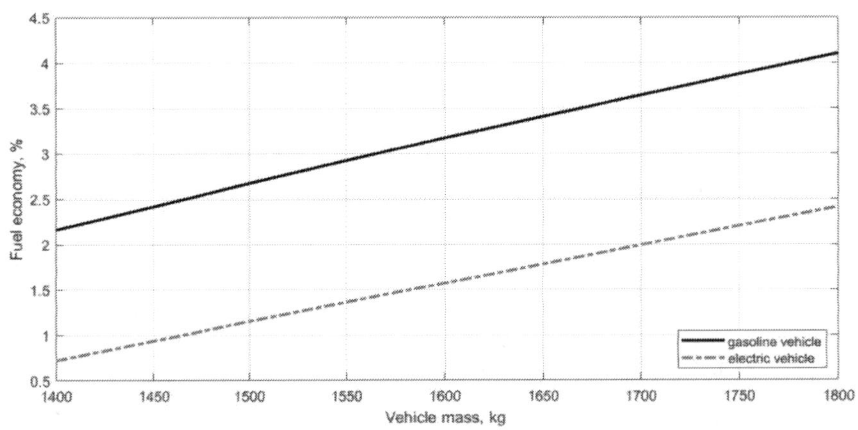

Figure 4. Vehicle's load influence on possible saving values using intelligent speed adjustment.

A road vehicle with an inner combustion engine demonstrated higher fuel saving values on a hilly road using intelligent speed adjustment compared with a similar electric vehicle.

5. Discussion and Conclusions

Estimating the fuel or energy consumption is not easy because there are many influencing parameters. Some papers on this subject showed values of savings for some routes without analysing the driving modes. We tried to compare fuel consumption of different driving modes and control algorithms with respect to the same driving time during all tests. We have demonstrated the practical use of the Digital Terrain Model in the task of autonomous driving and fuel consumption optimisation on hilly roads.

An approach for fuel (energy) saving on hilly roads, the process of mathematical model verification and validation, an experimental research method of testing the real vehicle on public road, and analysis of the vehicle's load influence on possible economy values have been introduced.

In addition, we have shown the possible values of fuel consumption (energy) savings for a gasoline and electric passenger car under ideal conditions. Fuel consumption optimisation for autonomous driving mode on a hilly test road over a distance of 200 km resulted in up to 4.1% savings for the experimental vehicle and up to 2.4% for an equivalent electric vehicle.

It is essential to solve the issues surrounding optimal vehicle control using the precise geoinformation environment.

References

[1] Singh Y et al. 2018 Optimal path planning of Unmanned Surface Vehicles *Indian J. Geo-Marine Sci.* **47(07)** 1325–34

[2] Zhang F 2016 Cyber-maritime cycle: Autonomy of marine robots for ocean sensing *Found. Trends Robot* **5(1)** 1–115

[3] Shadrin S, Varlamov O and Ivanov A 2017 Experimental autonomous road vehicle with logical artificial intelligence *J. Adv. Transp.* DOI:10.1155/2017/2492765

[4] Buehler M, Iagnemma K and Singh S 2009 *The DARPA urban challenge: Autonomous vehicles in city traffic* (Springer) p 651

[5] Özgüner Ü, Acarman T and Redmill K 2011 *Autonomous ground vehicles* (Artech House) p 289

[6] Nonami K et al. 2013 *Autonomous control systems and vehicles – Intelligent unmanned systems* (Springer Japan) p 306

[7] Naranjo J, Clavijo M, Jiménez F, Gómez O, Rivera J and Anguita M 2016 Autonomous vehicle for surveillance missions in off-road environment *IEEE Intelligent Vehicles Symposium (IV)* 98–103

[8] Ivanov A and Shadrin S 2018 Development of autonomous vehicles' testing system *2018 IOP Conf. Ser.: Mater. Sci. Eng.* **315** 012011

[9] Ruparel T, Yonten K and Eskandarian A 2014 Analyzing roadside safety implications of future vehicle designs *ASME Int. Mechanical Eng. Congress Expo., Proceedings (IMECE)*

[10] Eskandarian A 2012 *Handbook of Intelligent Vehicles* (Springer) p 1628

[11] Saykin A, Bakhmutov S, Terenchenko A, Endachev D, Karpukhin K and Zarubkin V 2014 Tendency of creation of "driverless" vehicles abroad *Biosci. Biotechnol. Res. Asia* **11** 241-246

[12] Saikin A, Buznikov S and Karpukhin K 2016 The analysis of technical vision problems typical for driverless vehicles *Res. J. Pharmaceut. Biol. Chem. Sci.* **7(4)** 2053–59

[13] Michon J 1985 A critical view of driver behavior models: What do we know, what should we do? *Human Behavior and Traffic Safety* 485–520

[14] SAE J3016: Taxonomy and Definitions for Terms Related to Driving Automation Systems for On-Road Motor Vehicles (U.S.A.) p 30

[15] Zhou M et al. 2016 A review of vehicle fuel consumption models to evaluate eco-driving and eco-routing *Transport. Res. Part D* **49** 203–218

[16] Sanchez M, Cano J and Kim D 2016 Predicting traffic lights to improve urban traffic fuel consumption *6th International Conference on ITS Telecommunications* Proceedings 331–336

[17] Kamal M, Mukai M, Murata J and Kawabe T 2011 Ecological vehicle control on roads with up-down slopes *IEEE Trans. Intell. Transp. Syst.* **12(3)** 783–794

[18] Wang J, Yu Q, Li S, Duan N and Li K 2014 Eco speed optimization based on real-time information of road gradient *J. Automot. Safe. Energy* **5(3)** 257–262

[19] Min Z, Jin H and Ding F 2017 Minimizing vehicle fuel consumption on hilly roads based on dynamic programming. *Adv. Mechan. Eng.* **9(5)** 1–8

[20] *Bosch Automotive Handbook 9th Ed.* p 1544

[21] Ye F, Wu G, Boriboonsomsin K and Barth M 2016 A hybrid approach to estimating electric vehicle energy consumption for ecodriving applications *2016 IEEE 19th International*

Conference on Intelligent Transportation Systems (ITSC) 719–724
[22] Qi X, Wu G, Boriboonsomsin K and Barth M 2018 Data-driven decomposition analysis and estimation of link-level electric vehicle energy consumption under real-world traffic conditions *Transport. Res. Part D: Transport Environ.* **64** 36–52
[23] ISO 15765-4:2016 Road vehicles -- Diagnostic communication over Controller Area Network (DoCAN) -- Part 4: Requirements for emissions-related systems
[24] ISO 11898-1:2015 Road vehicles -- Controller area network (CAN) -- Part 1: Data link layer and physical signaling
[25] ISO 15031-5:2015 Road vehicles -- Communication between vehicle and external equipment for emissions-related diagnostics -- Part 5: Emissions-related diagnostic services
[26] OBDLink MX Wi-Fi http://www.obdlink.com/mxwf/
[27] Saykin A, Buznikov S, Endachev D, Karpukhin K and Terenchenko A 2017 Development of Russian driverless electric vehicle *Int. J. Mech. Eng. Technol.* **8(12)** 955–965
[28] Fiori C, Ahn K and Rakha H 2016 Power-based electric vehicle energy consumption model: Model development and validation *Appl. Energy* **168** 257–268

Main trends in digital transformation of passenger automobile transport

N O Bludyan[1,2]

[1] Motor transportations, Moscow Automobile and Road Construction State Technical University (MADI), 64, Leningradsky prospect, Moscow, 125319, The Russian Federation
[2] Administration, The Association for the advancement of the transport industry "Transport Association of the Moscow Agglomeration" (The Association "TAMA"), 2, Denisa Davydova, Moscow, 121170, The Russian Federation

E-mail: np-tama@mail.ru

Abstract. Digitalization has entered many areas of the economy and life. The transport industry has not become an exception and is actively moving into the era of digitalization. However, the use of modern technologies requires regulation and definition of common standards. Unmanned vehicles must comply with safety standards and be capable of functioning. The regulation must be synchronized with the appearance of drones. Digitalization is a complex of devices and software products that ensure the functioning of such vehicles. Software products (applications) entered the sphere of transportation by passenger taxis and buses, providing passengers with new mobility. Despite all the positive aspects of digitalization, it also brought some difficulties. Therefore, mutual deliberate and timely regulation of digital technologies in transport is required.

1. Introduction

The number of urban population in the world is currently increasing by about 75 million people annually. According to UN forecasts this growth will continue in the coming decades, although at an increasingly slow pace. The world urban population from 4 billion in 2015 will reach 6.5 billion in 2050 [1]. The symbolic threshold of 50% of the world population living in cities was passed in 2007 [2]. In all regions of the world, urbanization is expected to grow in the coming decades.

Almost 90% of the urban population growth occurs in Africa and Asia. In Asia, this is mainly due to rapid urbanization, especially in East and Southeast Asia. Between 2015 and 2050, China, India and Nigeria together are expected to account for about 37% of the world's urban population growth [1]. In the same period, the urban population of India will increase by 390 million, China by 270 million and Nigeria by 210 million. [1]. Until recently, the largest urban areas were located in the more developed regions of Europe, Asia, and North America.

Today, the largest cities are concentrated in countries with transition and developing economies. Agglomerations with a population of 10 million are well known from various studies. The number is growing, today it is 28 against 10 in 1990. Megacities, however, account for only 12% of the world's urban population. About 50% of the world's urban population lives in small cities with population of less than 500,000 inhabitants. Although their proportion will gradually decrease, the role of these cities will remain significant. The fastest growing cities are those in Asia and Africa with population between 500,000 and 5 million people. Cities with less than 10 million inhabitants in countries with economies in transition or developing countries will account for almost 40% of the growth of the

Content from this work may be used under the terms of the Creative Commons Attribution 3.0 licence. Any further distribution of this work must maintain attribution to the author(s) and the title of the work, journal citation and DOI.
Published under licence by IOP Publishing Ltd

global economy by 2025. This is more than the entire developed world and the markets of megacities together

While urban sprawl has been observed in developed countries in the second half of the 20th century, there has been recent evidence that a reverse trend has begun (sometimes called «the great inversion»). For example, the density of the population in European megacities was increased by about 20% between 2001 and 2012.

In some cities, there is a concentration of population in the center. This phenomenon can be observed in Tokyo, where the population is growing faster within 10 km from the city center [3]. London has also seen steady growth in recent years. In North America, the real estate crisis has clearly accelerated the trend of moving back to city centers.

Sustainability and mobility limitations will be increasingly concentrated in cities, especially in lower-middle-income cities, where urbanization is the fastest. The International Transport Forum forecasts that from 2010 to 2050 the total world mileage of vehicles will increase from 117 to 233% [4].

Urban zones often extend beyond administrative and political boundaries, thus creating complex management challenges, especially in terms of mobility. In particular, the formation of polycentric configurations of megacities may lead to increased demand for space and equipment associated with the interaction of various sub-centers of agglomerations (Figure 1).

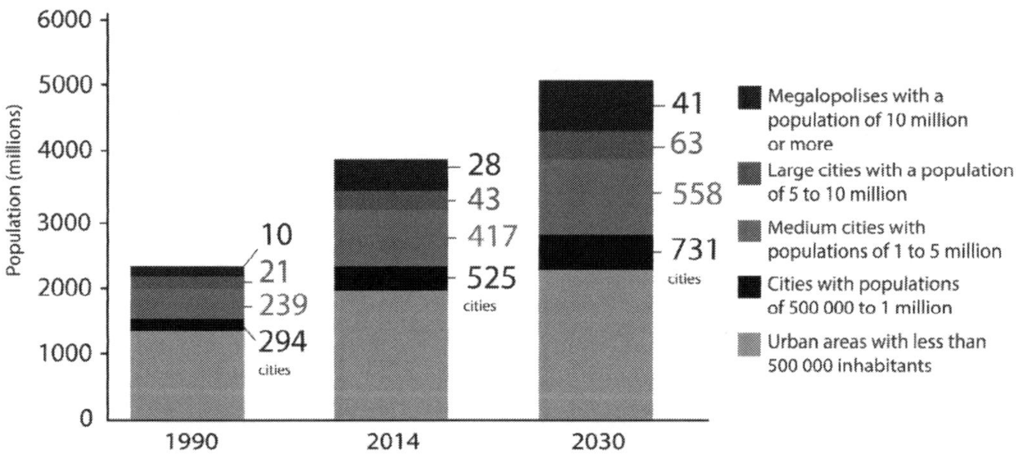

Figure 1. Global urban population growth is driven by the growth of cities of all sizes

The world economy has moved into the phase of the fourth industrial revolution, which is based on digital revolution. Digitization has enormous potential for a revolutionary leap in various industries and activities. At the same time, it can become a destructive force for certain industries, and even for states in general. In the universal understanding of the trends of the further development of civilization, the mutual agreement and common understanding of the objectives of power, business and society, among different states, is crucial. A fundamental trend of digital development in business, government, and society is the policy of synchronization, integration of "all with all". Within this trend, the main and most significant challenge is the ever-increasing number, quality and diversity of interrelationships between the state, business, consumers and socio-economic systems.

Cybersecurity is already at the forefront of the agenda of global and inter-state relations.

All of this is accompanied by a leapfrogging of data and transaction volumes.

Here are some predictions by well-known specialists of the patterns of global economic development as a result of its digitization [5]. By 2025, unmanned vehicles with a probability of 78.2% will account for 10% of the total number of US vehicles. With a probability of 67.2%, the number of commuting by vehicles will exceed the number of trips by private vehicles. With a probability of 63.7%, one can assume that the first city with a population of more than 50,000 without traffic lights will appear during this period.

IOP Conf. Series: Materials Science and Engineering 1159 (2021) 012062 doi:10.1088/1757-899X/1159/1/012062

As an example to justify explosive growth, table 1 lists the growth rates of companies before and during the digital revolution [5].

Table 1. Growth rates of companies

	Capitalization of the 3 largest companies	
1	Detroit, 1990	SiliconValley, 2014
2	36 billion $	1.09 trillion $
	Revenue (Gains)	
3	3.250 billion $	247 billion $
	Staff	
4	1.2 million people	137.000 people

The automobile is increasingly becoming a computer, with the cost of electronics now almost 40% of its total value. Apple and Google are gradually targeting the automotive industry. In the future, when the share of electronics is larger than the rest, it may be more strategically profitable not to produce cars, but to develop and sell a license for software for cars.

Not all sectors of the economy currently experience the same destructive effects of digitization. But they will all be drawn into the process of continuous, hard natural selection due to digitization. Everyone will have to operate in a mode of constant development.

Digital technologies have led to new ways of combining goods and services that have disrupted the order and erased traditional boundaries between industries. Amazon has evolved from a traditional bookstore into a retail conglomerate with an annual return of $100 billion. This is the result of understanding customer preferences and timely execution of customer orders.

2. Vehicles

Automated vehicles moving in unmanned mode have been in place for quite some time. Technologies, element base, operating principles were different. With the development of electronics and new technologies, the very technological phenomenon of unmanned transport has reached a new level. The International Automotive Engineers Union SAE has developed a classification according to which drones are divided into five levels of autonomy [6].

Level 1 implies a level of automobile autonomy, where the driver needs to keep his hands on the steering wheel, but control makes electronic auxiliary systems easy [6].

Level 2 means that the vehicle can be maneuvered off-road without the involvement of the driver when the hands can be removed from the wheel. Many cars now "park independently".

All the companies of the large Euroscope are working on the development of Level 3 "Platooning" technologies, and MAN has already transferred such vehicles into regular test operation of one of the transport companies in Bavaria. You cannot reconfigure from stripe to stripe in automatic mode. The management should take over the driver. That is, the main condition of level 3 is that the driver should still be in the cab.

Driving without a driver by remote control, that is Level 4. It's targeting a kind of plunging technology that's connected to a truck leader with Wi-Fi. The vehicle is able to disperse, follow a straight or long-range curve and slow down to 12 km/h [6]. This has already been implemented in trucks.

In July 2018, logistics services operator DB Schenker, car manufacturer MAN Truck&Bus and the Fresenius University of Applied Sciences launched a joint project to transport goods on unmanned trucks [7]. The convoy is a system of vehicles on the road, at least two of which move along the highway in a caravan. They are equipped with driver support systems and automatic controls. All the trucks in the convoy are electronically "connected" by means of which they are connected. The leading truck sets the speed and direction, and the others follow.

IOP Conf. Series: Materials Science and Engineering 1159 (2021) 012062 doi:10.1088/1757-899X/1159/1/012062

A convoy of highway trains departed from the DB SchenkerNeufarn near Munich for Nuremberg on the experimental section of the A9 motorway with a length of about 145 km. The first truck in this pair is driven by the driver, the second - by a computer. For safety in the cabin of the second truck there is still a person [7].

The use of self-propelled trucks in the transport of goods reduces fuel consumption and frees up space on motorways. The trucks are able to drive at a distance of 15 meters from each other instead of the 50 meters. This is the first case of the use of long-distance autotrains in a convoy in Germany, and it will set new standards in the logistics market [7].

There is no legal prohibition to use level 4 unmanned equipment in closed territories. In Norway, in the BronnoyKalk AS career, six Volvo FH freighters will transport the rock through a 5-kilometer route from the mine to the crushing complex [6].

The highest level of self-sufficiency Level 5 implies that the car has artificial intelligence. The unmanned truck should make its own decisions, analyze the situation. The first area of application of drones is military. It is the army's needs that have given impetus to the whole industry, the military sphere is in the vanguard of drone building. Volvo Construction has now launched its second generation of all-electric, unmanned Volvo HX1 dump trucks. But a fleet of a dozen of these trucks still needs to be monitored by a dispatcher. They are used in quarries with a high degree of gas and dust content. Another real application for unmanned trucks is warehousing and in-house logistics. Prototypes already exist, but in reality, it is not widely implemented [6].

Today's reality of unmanned trucks is a fourth level of autonomy and closed areas of industrial and mining industries, as well as the military sphere.

Currently, tractors under artificial intelligence control are being tested in California and Nevada (USA), and their serial production is planned for 2019 [8].

Mercedes-Benz company has developed a truck of the future Future Truck 2025. For unmanned control of trucks, Highway Pilot technology is used, which is responsible for automatic movement on the highway [8].

In 2016, Future Truck was tested: together with the European Caravan project, Mercedes delivered three unmanned trailer tractors to the road. The trucks driven by an autopilot drove more than 600 km from German Stuttgart to the Netherlands (Rotterdam) [8].

The column was 80 meters long. For comparison: in the case of conventional trucks with drivers, the length of the column is at least 150 meters. Fuel consumption during the experiment was reduced by 10% [8]. Trucks equipped with sensors and synchronized with each other via Wi-Fi, followed each other for a distance of 15-20 meters without any assistance of a driver.

Experts are confident that the technology of unmanned transport will be introduced more rapidly in the sphere of cargo transportation. Long and relatively straight routes are much easier in autonomous management, and light drones are much more difficult to operate in the urban cycle [8].

In February 2015, Cognitive Technologies and KAMAZ announced the start of joint work on the development of an unmanned truck. Cognitive Technologies started creating a brain for the car. In autumn 2015, the developers showed the first prototype of an unmanned truck [9]. In the Unmanned KAMAZ project, Cognitive Technologies used all the experience and achievements. The company's specialists taught the system of piloting to create a zone of interest - a narrow area of foveal vision that surrounds the observed object. In order to ensure safety from the point of view of reducing the number of accidents, the company involves external experts, creates a database of precedents, which allows learning from the mistakes of others. Many situations are processed by specialists at the virtual testing ground [10].

In PJSC KAMAZ in 2019, it is planned to launch an experimental route of intra-plant logistics transportation, in the framework of which the cabins from the press-frame to the automobile plant of the company will be transported on unmanned trucks [11].

This is just one of the unmanned projects that KAMAZ specialists are currently actively working on. It is assumed that serial production of unmanned KAMAZ trucks will begin by 2025 [12].

On May 21, 2019, the US Postal Service announced the use of unmanned trucks to deliver parcels and letters [13].

Volvo announced the commercial launch on 13 June 2019 of its firs telectric driverless trucks. The logistics giant DFDS started using them to transport goods between the distribution centre and the port in Sweden [14].

IOP Conf. Series: Materials Science and Engineering 1159 (2021) 012062 doi:10.1088/1757-899X/1159/1/012062

Electric truck from Einride has already started transportation on public roads. The T-Pod model can make short trips between the warehouse and the terminal in the industrial area in Jönköping, but at a speed of no more than 5 km/h (with a maximum of 85 km/h). The German Lidl supermarket chain, the Svenska Retursystem courier service and several Fortune500 retailers, in addition to the DBS Chenker logistics company, have already become interested in electric trucks. Even now, the t/Pod electric truck uses a platform from a third-party developer Nvidia to process information about the space surrounding the vehicle [15].

The research and development of unmanned vehicles continue to be actively worldwide. Large databases are being created that require government regulation. There is a need for unified hardware and software and for regulation to enable the movement of drones, including buses, on public roads.

A drone by itself will not create new mobility, but, together with technologies like "vehicle networks" and "smart road", this can be achieved. With smart networks, drones exchange different data among themselves, reporting and alerting each other about different road situations and fighting, allowing them to move at minimum intervals and at maximum speeds maintaining high levels of safety. Smart roads provide the vehicle's interaction with the road and the surrounding area. With this technology, it is possible to increase the throughput of roads by regrouping vehicles into a more efficient flow [16].

The transition to unmanned vehicles will lead to revolutionary changes and better transport management and environmental conservation. But there is a big transition problem. It is impossible to do so at the same time. New forms of mobility, such as drones, will overlap with traditional motorists on the roads. This could lead to controversial situations and the impossibility of applying all the technologies and possibilities built into the autonomous transport model. Resources and time will be required for the disposal refurbishment of the existing fleet of vehicles. It takes time for the production of unmanned vehicles to be mass-produced, and the government will need incentives from both automobile manufacturers and the public to switch to new modes of transport [17].

The substitution of the vehicle driver for automatic control systems (unmanned vehicles) radically-altering changes the transport economy toward important benefit for the service provider. Uber now estimates that between 60% and 80% of income covers driver's costs. That is why many transport service providers are seeking to move to unmanned technologies. However, with the arrival of drones, a review of emerging risk standards and security assessment criteria will be considered [18].

Frost and Sullivan (California, Moutain View) and the RBK forecast that by 2025, unmanned vehicles will account for 40% of the global car market. The State should amend the legal framework, reducing the deterrent to the use of autonomous systems on public roads. Business structures, particularly taxi services, will facilitate the distribution of unmanned transport. PwS predicts complete drones will be available in the world's largest cities only by 2040, while the first drones will be available in 2021 [19].

According to the research unit of Swiss Investment Bank UBS: UBS Evidence Lab, by 2030, the global market for unmanned taxis and related industries could reach $2 trillion. The assessment is based on the modeling of autonomous transport development in New York. The authors of the report analyzed the road situation, population projections, demand for taxi services, development of public transport, and also used computer simulations of the movement of test unmanned taxis on different days of the week, segments of the day and in different areas [20].

UBS estimates that by 2030 the penetration rate of such vehicles in the global auto market will be about 5% of total sales of new cars [20]. Researchers estimate that by this time the UAV will reach 11 million cars, or 1% of the world's total fleet.

Experts believe that the development of unmanned taxis will radically optimize traffic and passenger traffic in major cities. The authors believe that the introduction of unmanned taxis will reduce by two-thirds the number of taxis now available in New York and significantly reduce the burden on the roads. According to computer simulation data, in a city like New York City, replacing the taxis fleet with unmanned vehicles will cover up to 15% of the demand for all public transport services [20]. Travel will be 80% cheaper by reducing the costs of paying drivers and managers, as well as reducing travel time by choosing the best route and mode of movement. For example, for Manhattan, it is estimated that a route that currently takes 40 minutes will take up to 15 minutes. UBS experts believe that the development of unmanned taxis stimulates the growth of related markets — most of the software, technologies, and components. According to Global Market Insights, the LiDAR laser sensor market, which is actively used in autonomous transport, will grow to $10 billion by 2025 from its current volume of $1 billion.

The increase in the UAV fleet will lead to the development of the 3D-visualization market, microprocessors, new components from major automotive manufacturers, etc. The April Markets and Markets report notes that Europe's UAV market will be most advanced in Germany, France, Norway, and Holland. In these countries, the best way to develop infrastructure for drones is by now. In addition, they have great productive potential in the relevant industries. In total, there are about 180 automobile enterprises in the EU, with the EU automobile sector: One of the largest investors in research and development. Experts note that the development vector for unmanned taxis indicates that manufacturers will strive for more energy efficiency of cars [20]. By 2030, UBS estimates that unmanned taxis will account for up to 10% of global sales of electric vehicles.

"Markets and Markets" expect that by 2021, half of the robotics produced by Uber and Volvo will be electric. There is also a joint project between Daimler and Bosch to develop unmanned electric vehicles, which should reach public roads as early as the 2020s. A similar goal, but only in China, is a joint project between Volvo and Baidu [20].

In late April, at an event for investors, Ilon Mask reported that already in 2020, Tesla was launching a program to create unmanned electric taxis. Tesla's unmanned taxi service is expected to last for 11 years, with a total of 1.6 million kilometers [20].

The Zetta serial electric vehicle is prepared at the stage of the experimental-industrial production and certification procedure. The vehicle will be produced in several packages, totaling more than 10.000 units per year and distributed equally among the Russian and foreign markets [21].

In the Russian Federation, the first steps have already been taken to organize the work on the admission of passenger drones to the road network. According to the Russian Government Resolution, an experimental exploitation experiment is planned in Moscow and Tatarstan [22].

In late 2016, Yandex started work on unmanned vehicle management technology. The company tested the first prototypes of unmanned vehicles in spring 2017. There are already test zones in Skolkovo where you can take a taxi without a driver. In late 2018, Yandex received a license to test unmanned vehicles in Israel, and in January 2019 showed a drone at CES Nevada. In March, the company signed a Hyundai Mobis agreement to develop new prototypes of UAVs. Yandex plans to manufacture more than 100 drones in 2019 [23].

3. Taxomotor passenger transportation

American taxi aggregators Uber and Lyft have become leaders in the application of digital tools in passenger transport - they are vivid retailers.

Consumers needed a convenient, transparent, understandable and modern service that kept pace with the times. New Internet services responded and released the right product to customers. Are Uber and Lyft representatives of "new mobility"? Rather not, because their activities are very similar to those of classical taxis, but also to the use of digital technologies, both in interaction with the taxi application and in payment for it. This transparency with availability and has catalyzed the popularity of such applications [24].

People have been given an alternative to outdated, crowded, and late land transportation through new taxi services. However, this has resulted in an increase in the load on the already-filled road network and new traffic congestion [24].

It is clear from all this that online aggregators have created a dual product, on the one hand providing "new personal mobility", on the other hand having a serious impact on the road network and the city as a whole. All these aspects require attention from city authorities to ensure a compromise between convenience and consequences for cities [24].

There is another serious problem with the performance of aggregators: Often, Internet services violate drivers' rights by lowering their salaries. These circumstances lead to the creation of protective structures for drivers defending their rights (in particular the hourly rate). The city adjusts the rent for the car that a park can charge drivers. Still, the distribution of income between drivers and landlords has not been fully resolved. For these reasons, trade unions and organizations are being established to protect and defend the interests of drivers [25]. The drivers of Uber and Lyft went on strike even before the IPO of these firms. Drivers demanded higher tariffs and lower internet service fees from 25% to 15%, as well as better working conditions (weekend pay and job security) [26]. The results of such strikes and the work of the drivers' rights unions can already be seen, as on April 30, 2019, a New York Supreme Court judge, Andrea Masley, upheld the New York State Taxi Commission (TLC) rule on minimal payments to taxi drivers and rejected Lyft's protest against this rule [27]. Thanks to such events, drivers could count on a legal, decent hourly fee. In London, TFL in 2017 refused to renew Uber's license because it did not meet the security requirements required for private transport [28].

If operators are effectively and objectively controlled, a system can be established that: To solve transportation tasks in general and in certain areas, to plan transit between regions, to create zones required for the boarding and disembarkation of taxi passengers, to control compliance with the filling capacity of land transport [24].

4. Challenges and consequences of digitization of taxis

Providing regulatory and legal regulation of the new digital era of transport is a pressing issue at the current stage of industry development.

The problems of functioning in the modern taxi industry, based on the application of mobile applications, have become a laboratory for understanding what to expect in the future, when all passenger transport will affect digitization.

Should classic taxi rules be applied to "aggregators"? Should classic taxis and single-use vehicles be regulated to ensure fair conditions for all parties?

Unfortunately, up to now there is no clear understanding of the policy of regulating the activities of mobile application holders in transport.

Analysis of Bill No. 481004-7 of 28.03.2019 establishes the legal status of ordering services for light taxi aggregators. The task of these services is defined as "processing order information". The concept of "information processing", which does not contain anything concerning the formation of tariffs or the cost of carriage.

It is quite obvious that it is necessary to prohibit by the legislation of the Russian Federation "aggregators" of orders to make tariff regulations and to charge (offer) their tariffs on carriage of passengers. At present, taxi ordering aggregators are in fact implementing tariff regulation of transportation. The system of transport pricing shall be backwards: Taxpayers and individual entrepreneurs should send their tariffs on carriage to the aggregator(s) voluntarily. Aggregators are required to make available to potential charters the full "menu" on freight rates for all companies. The franchisor, i.e., the future passenger, when ordering the service for carriage, should have the right and possibility to choose from the general "menu" of the aggregator(s) an acceptable tariff. This will ensure real competition in the taxi industry, ultimately leading to improved transport safety and economic stability for freight operators.

The bill provides for notification procedures for the operation of the taxi service. In conditions where the "aggregator" is the main and the main "organizer" of transport. Obviously, it is necessary to introduce at the federal level the "Authorization" system of admission of services of order of passenger taxis (aggregators) to carrying out its activities. The authorities of the subjects of the Russian Federation should be empowered to establish requirements for taxi services and to organize their admission to carry out their activities.

New mobility systems should respond to the question of whether to encourage the carrier when the vehicle is full or not and when the limits are exceeded? The carpool system can help, but it requires incentives from regulators to allow passengers to travel on this system. Carpool transportation requires effective state control. To eliminate the operation of a dishonest carpool, you must introduce a limit on the cost of running. The driver shall not exceed the maximum permitted total cost of the journey, otherwise such carriage may include a taxi. There are questions concerning the safety and regulation of these transport operations, in the carpool system, as transportation is personal and is not currently regulated. Clearly, there is a need to create incentives and support for carshering and carpool development, such as parking lots, allocated lanes, and incentives for the purchase of electric vehicles.

The introduction of new systems and tools, common travel services, and unmanned vehicles require the cooperation of many structures. It is necessary to create smart streets, infrastructure suitable for the movement of autopilots.

The digitization of the passenger transport system in any sector should be based on the unconditional observance of the following principles:

1. Unconditional compliance with the existing regulations on passenger transport at the federal and regional levels;
2. Inadmissibility of the introduction of digital technologies and mobile applications that conflict with established regulations and regulations on passenger transport;
3. Avoidance of the creation of technologies for the digitization of parallel state regulation of passenger transport systems, including with regard to tariff and cost regulation.
4. Correction, as necessary, of the existing legal regulation of passenger transport by the legislative authorities in order to quickly introduce digital opportunities, improve the quality and safety of transport services for the population of the country.

References

[1] Organizatsiya ob"yedinennykh natsiy Perspektivy mirovoy urbanizatsii: Izmereniye ot 2014, Osnovnyye sobytiya, 2014
[2] Mezhdunarodnyy institut Makkinsi, Urbanisticheskiy mir: Ekonomicheskaya moshch' gorodov, 2011
[3] OECD, Novoye opredeleniye termina «urbanizatsiya»: novyy sposob opredeleniya razmerov megapolisov, OECD Publishing, 2012
[4] Proyekt Soveta korporativnogo partnerstva Mezhdunarodnogo transportnogo foruma «Gorodskaya mobil'nost': obnovleniye sistemy»
[5] Shvab K M 2018 The fourth industrial revolution Moscow Excmo p 288
[6] Gladkij D The unmanned reality of the truck world Available at: https://fleetguru.eu/ru/auto/03-12-2018_bespilotnaya-realnost-mira-gruzovikov
[7] Tadviser (State. Business. IT). Unmanned trucks
[8] BesPilot (Electric cars and drones). Unmanned trucks. Available at: https://bespilot.com/tip/bp-gruzoviki
[9] Agadzhanov, M. Cognitive Technologies together with KAMAZ will develop an unmanned truck Available at: https://habr.com/ru/post/366113/
[10] Tadviser (State. Business. IT). Unmanned KAMAZ is expected before the end of 2017
[11] KAMAZ vehicle. Kamaz launches unmanned transportation Available at: https://kamaz.ru/press/releases/kamaz_zapustit_bespilotnye_perevozki/.
[12] TASS news Agency. KamAZ will start serial production of unmanned trucks by 2025. Available at: https://tass.ru/ekonomika/6143950
[13] Tadviser (State. Business. IT). Amerikanskaya pochta nachala dostavlyat' posylki na bespilotnykh gruzovikakh
[14] Tadviser Tadviser (State. Business. IT). VolvoVera (unmanned truck)
[15] Bobrova, T. Unmanned electric truck Einride began daily transportation of goods in Sweden Available at: https://vc.ru/transport/67471-bespilotnyy-elektrogruzovik-einride-nachal-ezhednevnye-perevozki-gruzov-v-shvecii
[16] Pavlovskij V ETasks of driving an unmanned vehicle in the AutoNIVA project Ogol'cov V N, SpiridonovaI A 2015 The second All-Russian scientific-practical seminar Unmanned vehicles with elements of artificial intelligence pp 107–114
[17] Makov P V Transition to unmanned vehicles - Relevance and problems Научная библиотека 2016 pp 52–53
[18] Jonathan Matus. Understanding the Future of Mobility Available at: https://techcrunch.com/2015/08/08/understanding-the-future-of-mobility/
[19] Ol'hovikov D E Intelligent transport systems of Russia Information magazine № 7 2018 pp 19–20
[20] Kommersant Publishing House Unmanned taxi increased counter Available at: https://www.kommersant.ru/doc/3989470
[21] Ahireev V Serial production of electric vehicles starts in Russia Available at: https://rg.ru/2019/05/20/v-rossii-startuet-serijnyj-vypusk-elektromobilej.html
[22] Resolution of the government of the Russian Federation dated November 26, 2018 No. 1415 "on conducting an experiment on the operation of highly automated vehicles on roads". Available at: http://www.consultant.ru/document/cons_doc_LAW_311956/

[23] Network Edition RIA Novosti *Volozh drew Putin's attention to the importance of testing unmanned cars* Available at: https://ria.ru/20190530/1555112656.html
[24] Schaller B. The Uber and Lyft Revolution? Available at: https://gspp.berkeley.edu/news/news-center/the-uber-and-lyft-revolution
[25] Biju Mathew. Taxi: Cabs and Capitalism in New York City Available at: https://digitalcommons.ilr.cornell.edu/cgi/viewcontent.cgi?article=1047&context=books
[26] Rozhdestvenskaja Ja *Uber and Lyft are waiting for the strike* Available at: https://www.kommersant.ru/doc/3961777
[27] O'Neill P H *Lyft Fails to Overturn NYC Driver Minimum Wage Law* Available at: https://gizmodo.com/lyft-fails-to-overturn-nyc-driver-minimum-wage-law-1834453821.
[28] TASS. The court of London did not consider biased the decision to grant uber a temporary license Available at: https://tass.ru/ekonomika/6160380

Modern methods of the digital virtual analysis for design and tests of road vehicle restraint systems

I Demiyanushko[1]

[1] Department of Construction Mechanics. Automobile and Road State Technical University (MADI), 125319, Leningradsky pr., 64, Russia

E-Mail: demj-ir@mail.ru

Abstract. Widespread introduction of information technologies of virtual digital testing to reproduce vehicle impacts against vehicle restraint systems with simulation impact modeling road safety barriers, crash cushions, terminals, removable barrier sections, transitions and passive safety devices, is one of the main actions providing decline in mortality on roads as a result of the road accidents. On this basis the so-called "Virtual Digital Proving Ground" (Virtual Test House) for testing of road arrangement elements is developed. The researches, conducted for its development are based on use of modern world-class software products and numerical methods of mechanics. Digital modeling of crash-situations with arrival of vehicles on elements of road arrangement, allows to solve complex problems of the analysis of new designs now, to investigate behavior of vehicles at with arrival on the design. Implementation of intellectual systems of automated management of vehicles does not give the chance to hope for the exception of refusals, both the equipment, and control systems, that does the problem of development of the innovation passive safety systems extremely relevant. Research and development carried out MADI by the general research team together with LLC M&PQ allowed to take the major step in advance of works on increase in safety on roads, to create tools for design of new innovations, to develop reasonable standards in the field of the road systems of passive safety. In article prospects the development of these researches in scientific and practical directions and for the purpose of increase in competences of educational programs are considered.

1. Introduction

At the Moscow automobile and road state university (MADI), generally the staff of department construction mechanics and the small innovation enterprise - LLC M&PQ, conducts the scientific and practical developments on elements of road vehicle restraint systems and others road infrastructure elements (ERA-Elements of Road Arrangements - road safety barriers, crash cushions, terminals, removable barrier sections, transitions and passive safety devices, acoustic panels, etc.) directed to creation of the system of virtual digital tests, digital information modeling and the design analysis, and also development of the complex system of the expert assessment of safety of traffic, including development and improvement of the specifications and technical documentation in this area. Digitalization in the road-building industry – the trend now, and the conducted researches completely match the directions of development of the intellectual transport systems.

It is necessary to tell that similar systematic researches are conducted by research teams about the whole world, the closest approach is implemented in George Washington University, FHWA/NHTSA

Content from this work may be used under the terms of the Creative Commons Attribution 3.0 licence. Any further distribution of this work must maintain attribution to the author(s) and the title of the work, journal citation and DOI.
Published under licence by IOP Publishing Ltd

National Crash Analysis Center (NCAC), USA, close researches are conducted in Department of Mechanical Engineering, The University of Sheffield, Sheffield, UK, in University of Nebraska–Lincoln, Nebraska (USA) and in many other scientific centers.

Our works in this direction began much later – since 2010, when in MADI our collective began researches on application of the wire rope safety barriers. These researches, conducted in MADI together with LLC M&PQ, allowed to take the major step in advance of works on increase in safety on roads, to create for rather short time the tools, based on digital technologies for design of new innovation road restraint systems (frontal barriers, mobile frontal barriers, new parapet and rope systems), to develop reasonable standards in the field of road restraint systems, to provide significant reduction of costs of the enterprises of the industry for implementation of new samples of designs.

So far, these works created the basis for forming of "Virtual Digital Proving Ground" (VDPG) for testing of elements of road arrangement, including restraint systems, which reduce the impact severity of vehicles collisions with these systems and often shortly called "Virtual polygon", which can be used effectively, first of all, for the purpose of certified assessment of compliance of ERA, according to requirements of normative documents, in particular TR TS 014/2011 "Safety of Highways". These works within two years are successfully carried out with use of tools of the virtual testing within the " ERA Testing Laboratory ", accredited to carrying out virtual certification tests (Type Tests) as department of LLC M&PQ. VDPG is intended also for the numerous researches, connected with development of the innovation designs of ERA, expertise, which are carried out according to requests of the different organizations, and in some other cases. Below, some important aspects of the digital virtual analysis applications different ERA designs, examples of digital models, problems and perspectives of application for the virtual digital analysis of ERA in the road-building industry with implementations of new materials and intellectual technologies, will considered.

2. The models and methods used in the digital virtual analysis

Application of traditional engineering approaches for researches of mechanics of ERA's work does not give the necessary accuracy of the analysis now, that is connected with not accounting of elastoplastic deformation at impact, big deformations, contact processes. The solution of these tasks demands use of modern numerical methods of space nonlinear dynamic modeling. At impact process during collision of the car with ERA, parts of ERA and the vehicle, running with impact, are affected by dynamic loads of high level, because of what they begin to move and be deformed. ERA's geometry in the course of deformation at impact in each instant is unique and does not repeat in time. Such tasks belong to area of nonlinear dynamics of transition processes. Scientific and practical developments on creation of multidisciplinary 3D mathematical models of elements of road infrastructure - road arrangement, models of different types of the vehicles and also the processes of the movement and collisions, described by ratios of nonlinear dynamics with use numerical linearization methods, in particular, FEM, are aimed at providing high level of adequacy of digital doubles - Digital Twins.

Researches are conducted with use of modern program complexes of the high-speed processes dynamic analysis, based on the finite-element method (FEM), in the majority of models the licensed Ls-Dyna [1] and also license codes of MSC. Software used.

Examples of 3-dimensional final element (FE) models of ERA some types with arrival of the vehicles used within PDPG shown on figure 1 (a - g). Now we executed over one thousand virtual tests of specific objects of ERA of different application and designs, materials of researches are stated in numerous articles, in particular, [2-4].

When modeling work of ERA at impact process it is necessary to consider the number of problems which are connected with not triviality of the task: features of creation of FEM grids for different structural elements, the different material model descriptions, taking into account speeds of deformations, the description of nonlinear friction processes and losses in contacts. On the last costs will stop especially. The main result of calculation depends on the correct accounting of energy absorption in facing designs elements during impact. Accounting of friction losses in contacts: tires – the paving, taking into account climatic conditions and quality of the covering, in variable contacts of the impact and sliding interaction vehicle during collision with ERA, intrinsic losses at friction and deformation of structural elements, for example, in ropes, is defining for adequate creation of models, the correct accounting these factors is especially important when using nonmetallic materials in barriers – plastics and composites.

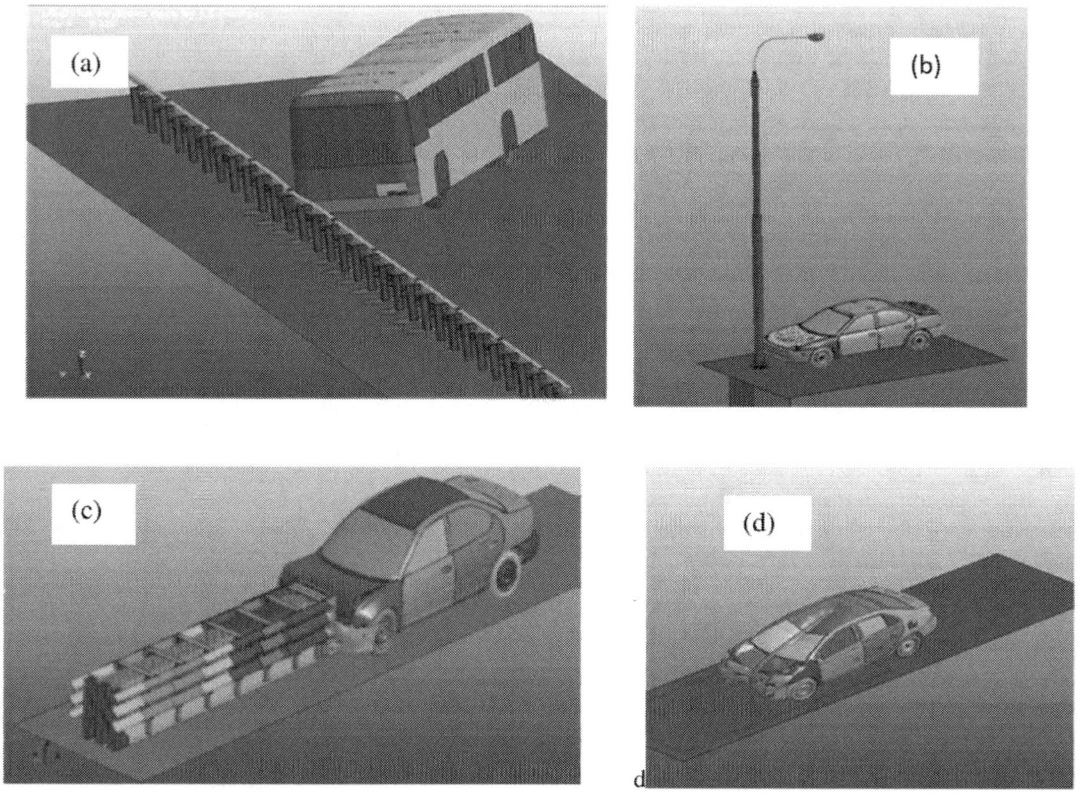

Figure 1. 3D digital FE models - doubles (Digital Twins) of digital virtual crash tests with the collision of various cars on the ERA (a - bus crash the safety road barrier, b – car crash the lighting support, c - car crash the front barrier, d - car crash the road sign.

Obligatory carrying out the validation analysis of models allows to consider the created models digital twins unlike the normal term "mathematical models". What there is "Digital Twin" and what it differs from mathematical model in - these disputes continue so far, see, for example, [5]. The term "Digital Twin" in our case is justified by need at each virtual testing to prove competency of estimates as they are used for certification of designs, and the received calculated parameters have to be compared to normative criteria, what safety of the designs established on roads depends on finally.

As well as in all cases of crash tests, accelerations in the fixed points of the vehicles, of arrival are subject to definition that allows to estimate Impact Severity and also the trajectory of vehicle, runout after impact. Now, because of the real test when determining severity assessment indices ASI of an errant vehicle value used [6], in the virtual analysis this index is also used. At the same time the virtual crash test allows to evaluate different indicators, in particular Theoretical Head Impact Velocity, index PHD (Post impact Head Deceleration), etc., and this direction needs even development.

Works on validation of the developed models and verification of decisions are conducted systematically. validation include comparative analysis of parameters and results of real tests and virtual tests of the same designs and also numerous bench and laboratory tests of critical structural elements and materials.

IOP Conf. Series: Materials Science and Engineering 1159 (2021) 012063 doi:10.1088/1757-899X/1159/1/012063

2.1 Validation of digital models by results of real tests

Comparison the results of full-scale real tests with virtual tests, on the one hand, is important as allows to estimate quality of full digital model of the crash test. On the other hand, in full-scale tests are defined very limited number of parameters - it is normal, these are the main standardized sizes, so-called, consumer characteristics. So, for road barriers [6,7], dynamic deflection (are often evaluated not authentically), lateral displacement, working width, containment level, the type of damages, behavior of the car, etc. are considered as consumer characteristics. These characteristics give in general the idea of the validity of the design, but at the same time, do not give the chance of studying of mechanics of her behavior in the impact course, and not fully can be used for optimization of designs, first of all, because of duration of full-scale tests and their high cost. Besides, in the analysis of test results there is the considerable subjective component.

At the same time validation of models demands the detailed research of all parameters that characterizes their main advantage – high informational content. Below in Figure 2 two normal episodes of comparison of road barrier behavior at bus impact at real tests (the data of the testing center of the Russian Federation SSUE NAMI used), which is carried out to identical instants are shown and allows visually work of model.

Figure 2. Fragments of the moments of the bus crash on the fence road barrier, on the left – real tests carried out of Testing Center RF SSUE NAMI, on the right - the FE model in Ls-Dyna, at the same moments of time – results of virtual testing

By the main consumer characteristics for road barriers on the basis of the numerous analysis practice determined the acceptable level of coincidence of parameters which are specified in standards, in particular state standard specifications on road safety barriers [6,7]. So, in the majority of parameters the digital difference of results should not exceed 10%. These requirements to results of virtual full-size tests are stated in the standardized technique [8] - LLC M&PQ.

IOP Conf. Series: Materials Science and Engineering 1159 (2021) 012063 doi:10.1088/1757-899X/1159/1/012063

2.2 Validation of digital models by results of bench and laboratory tests

At validation of models also the correct models of materials have to be applied. Usually in calculations use help characteristics of materials while in delivery the difference in characteristics can be considerable. The analysis shows that the difference in characteristics of the design at the same time can reach 10%, for example, on deflections of rope barriers experimental studies of materials of ERA and separate elements allow to choose more reasonably models of materials and the type of KE of models for structural elements. So, material tests from the ERA elements give the chance to consider real (true charts of deformation), but not engineering characteristics of steels, to consider the deformation speed in material model, that is important if to consider the shock nature of deformation at collision, both for materials of the ERA, and for materials of the contacting car elements.

FE models of separate critical elements allows correctly construct tests of posts, cables of barriers, elements of fastenings, that is important, as except requirements of ensuring accuracy, and because it is necessary to spend economically computing resources that is especially important, considering the iterative nature of decisions. In Figure 3 one types of the new validation dynamic tests of cables for wire road barriers developed – impact test, which have allowed to define the rational choice of FE model for cable at virtual tests, shown.

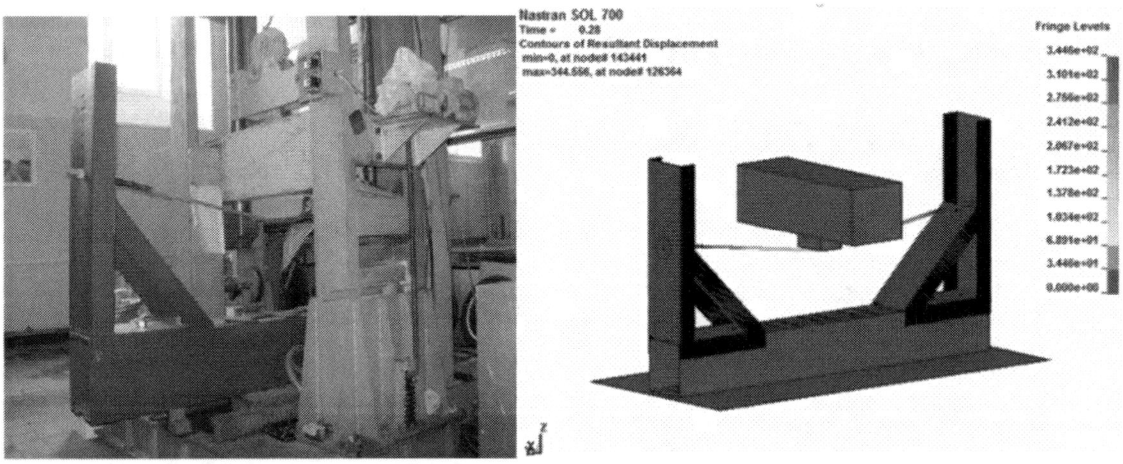

Figure 3. Bench impact tests of wire ropes and digital twin.

In Figure 4 the example of the scheme development of validation of barrier road protection element " post– anchor fastening" at bench virtual and real tests shown.

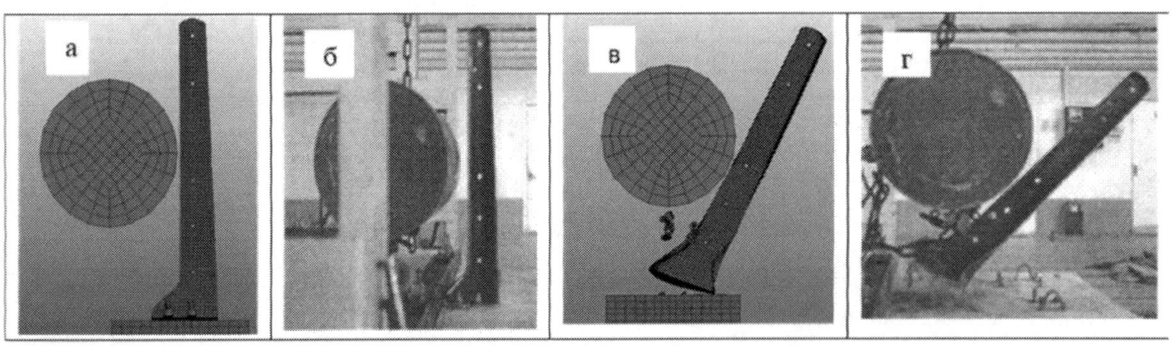

Figure 4. Example of Validation Analysis of Post-Anchor Model of Road Safety Barrier - Impact Test.

Here use of so-called "chemical anchors" for fastening of barriers posts, i.e., anchors with glue fastening were feature. Results of tests have shown compliance of behavior of models to natural samples.

One of the most important and indicative tasks is the problem of modeling of mobile frontal road restraint system (MFP - Mobile Frontal Protection) [9,10], which work is based on the predicted collapse, at the expense of which there is the main absorption of impact energy. Kinetic energy at impact to be spent for formation of free surfaces in the course of growth of crack, friction of special knife with tube surface, plastic deformation of sheets of tube and also plastic deformation of front frame and basic part of device (Figure 5 – type of design of the element of element destroyed directionally at impact).

Figure 5. The view of the structure of the Mobile Frontal Protection - a, the barrier beam with the knife device of the front absorber of the structure - b, the location of special concentrators – grooves, that determine the direction of crack propagation and make its opening predictable - c.

Validation was carried out by consideration of behavior of the cut-out element with bore at the movement in the course of impact (Figure 6).

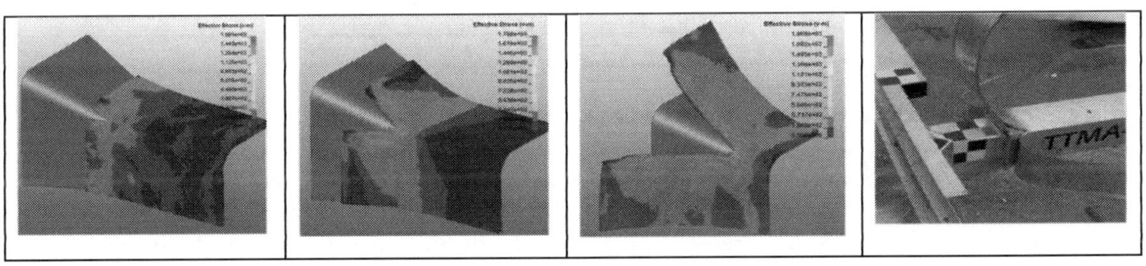

Figure 6. Simulation of directed destruction of energy absorbing element Mobile Frontal Protection.

The complex of works on creation and validation of the ERA models allows to consider that developed with use of the described technique, models can be fully considered as digital doubles of ERA.

3. Main advantages of digital twins and virtual analysis to ERA

The main advantage of the virtual digital analysis with use of 3-dimensional FE-models was already mentioned above - it is high informational content, that is important, first of all, at design of new ERA, especially innovative, on which there is still no sufficient experience of development and operation. So, advantages of the virtual analysis have been shown at creation of the first domestic samples of MFP, another name - "attenuator support vehicle". In detail problems which arose and have been solved during creation of models of crash tests of MFP described in [9]. Feature of models is the multibody of the general system here. All system of crash test model for the analysis of work of MFP consists of the maintenance car (usually, the truck), the hook-on device from MFP, and, at last, the running car. On

Figure 7 the developed virtual digital MFP model of all system of car's arrival on hook-on design is shown.

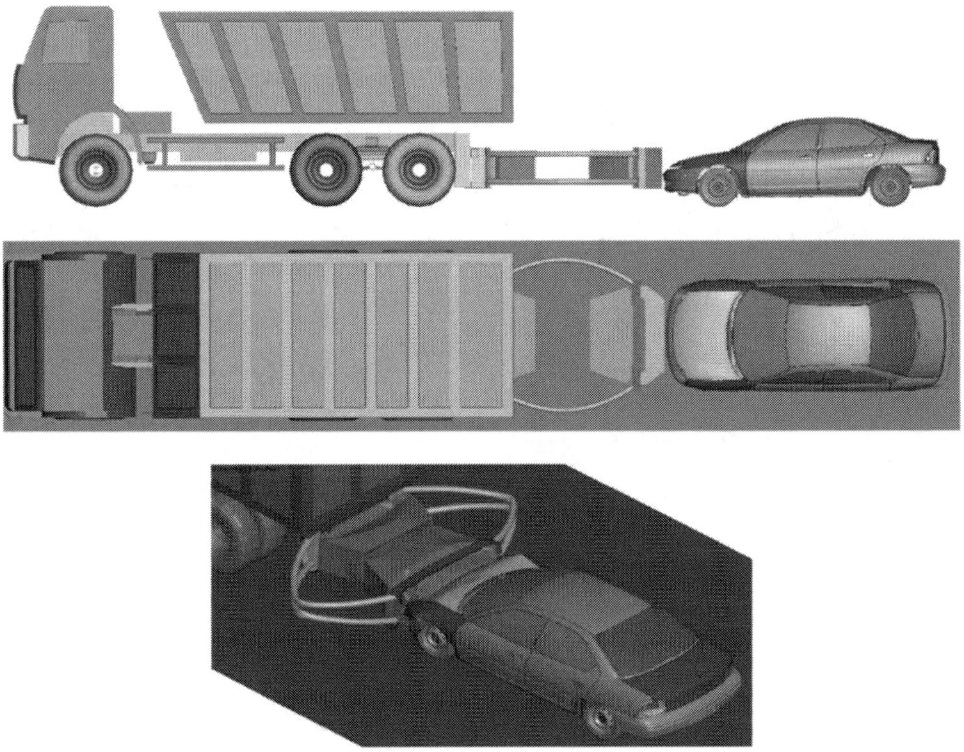

Figure 7. Virtual crash test of arrival of the car for hook-on Mobile Frontal Protection [6].

At virtual tests all characteristics of system behavior in time are defined in details, that opens considerable opportunities for the detailed analysis. So, in Figure 8 examples, received in calculations at virtual crashes various the vehicle with ERA of schedules of change of such characteristics of systems, important for the analysis, as:

a) acceleration in the vehicle center of gravity during the impact,

b) deflections in points of elements (here A- I is points of posts and its deflections near cables of wire barrier in the course of crash with the),

c) absorption of energy in system's elements of MFP and the car in the course of crash (different colors for different structural elements), and many others are given.

Especially these possibilities of the analysis of characteristics of change in time in the course of crash are important for structural design of ERA. So, the sequence change of absorption of energy elements of frontal and mobile frontal protections in the course of crash with the car gives the chance to operate of the absorbing structural elements characteristics. And change of sizes of acceleration in separate points of vehicle – to estimate time of approach of the most dangerous state at impact, etc.

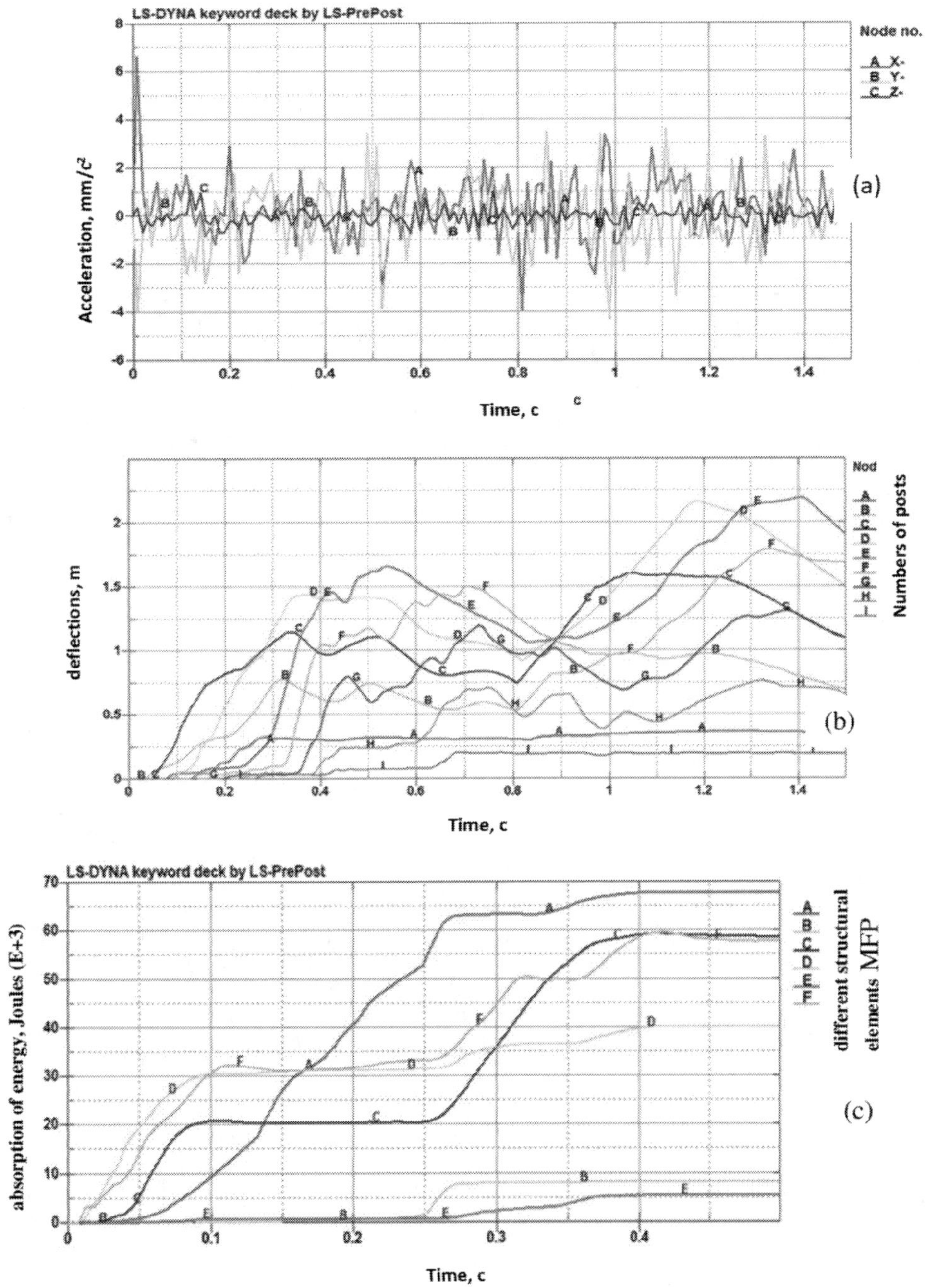

Figure 8. Example of change schedules of various characteristics important for analysis in time in case of virtual vehicle collisions with REA

Important advantage of the virtual analysis is also detailed visualization of impact processes and deformation of elements at crash, that equally belongs as to ERA, and the vehicle. It is obvious, that for the purpose of saving of time and computing resources it is necessary to sacrifice in some cases detail of models, so at probe of deformation of ERA in most cases in vehicle (car) model front part – bumper, front body details (side panels, cowl), front suspension, with wheels, windshield while, the rears of the car are modelled generally by solid bodies is in detail modelled. The vehicle shall be fitted with, as a minimum, one virtual accelerometer for measurement in three directions. That is, models are not in advance defined, and their design depends on analysis task. Below, in Figure 9 characteristic results of virtual test of road metal fence barrier are yielded by arrival of the car to separate timepoints (fragments of record of virtual tests), such simulation allows to look after all high-speed process of deformation and behavior of the facing objects, which usually at crash tests lasts 200 – 300ms.

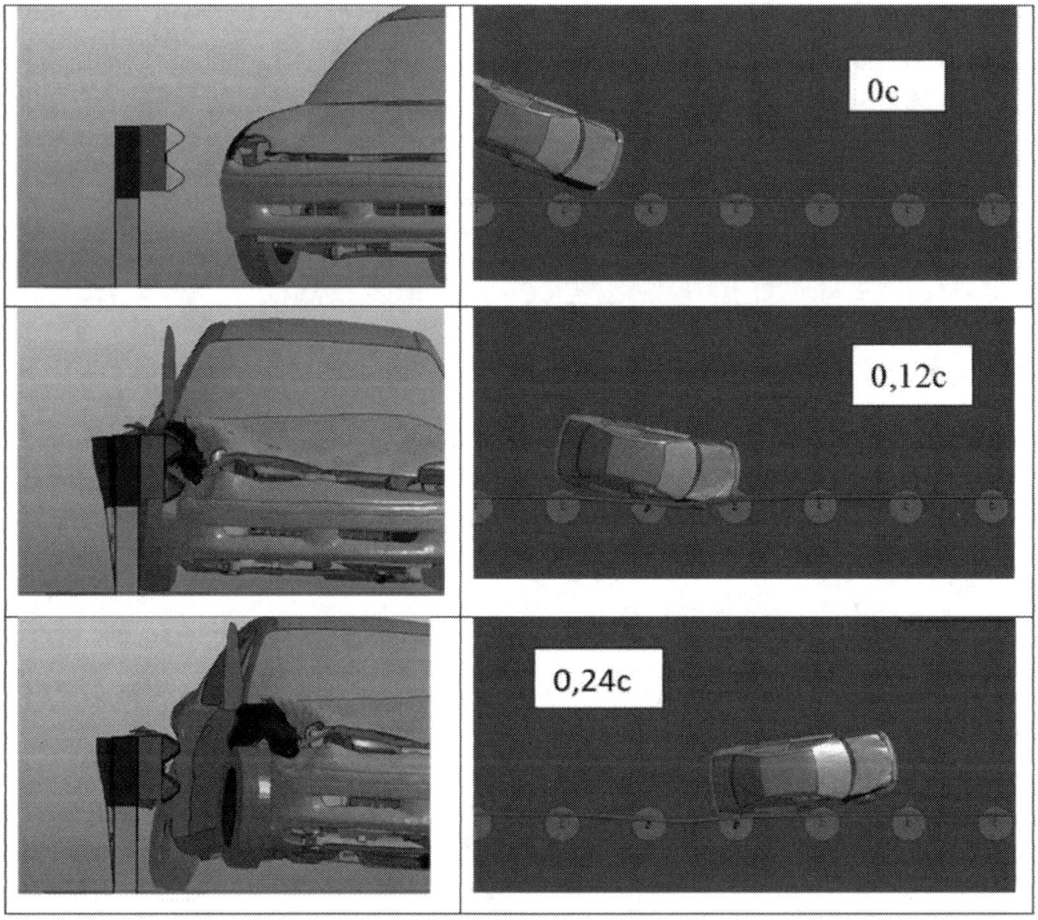

Figure 9. Fragments of a movie of a car crash a road barrier for virtual tests.

The same opportunity to estimate after-effect – residual deformation and collapse of designs is important. In Figure 10 the typical type of protection after arrival of the car is shown at natural and at virtual tests.

Essential advantage of the virtual analysis is economy of means when using these tools at design, at assessment of compliance to standards and the subsequent certification and also at examination and the analysis of the reasons of road accident.

Experience shows, that for one enterprise, issued the same road side barriers, the number of natural tests necessary for certification of these products decreases by 6-8 times, when using virtual certified tests, that leads to 25 ÷ 35-fold reduction of costs of certification.

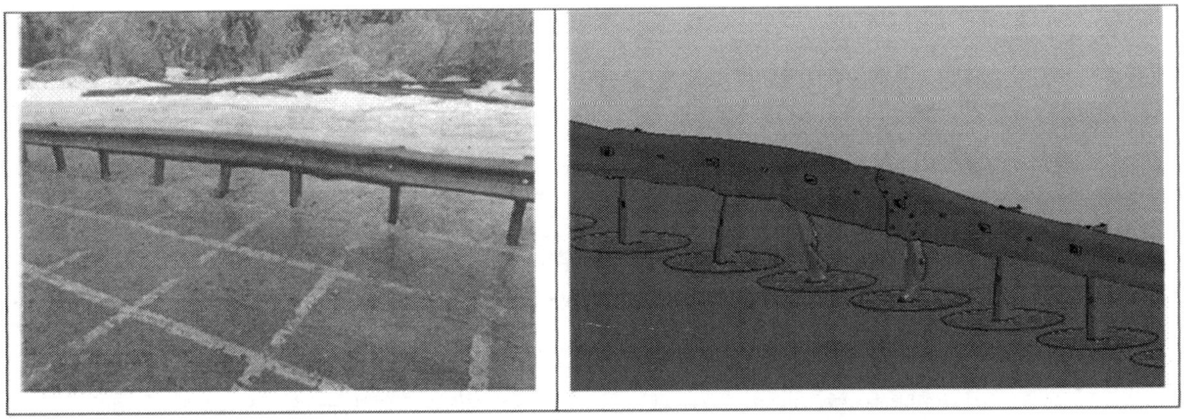

Figure 10. View of road barrier after full-scale and virtual tests.

4. Prospects of development of the virtual digital analysis of ERA

Of course, in article there is no opportunity to consider all aspects of the idea of creation of Virtual Digital Proving Ground (VDPG) for tests of ERA and further development of the virtual digital analysis of ERA. However, we will stop on the main, the directions which already to some extent are worked.

4.1. Use of the digital virtual analysis opportunities for reconstruction of the road accidents (RA)

Importance of this task is obvious to the road industry. All analysis of road accident, both applied (for examination), and scientific and practical, is connected generally with crashes the car, the behavior at the same time of the road ERA is practically not considered. Crash tests of cars (generally experimental) on crash with obstacles also pursue the purpose the analysis of deformability of the car, obstacle considered as rigid body. Thus, accounting of behavior of road designs at road accident in its analysis is practically not made. At obvious. importance of problem of reconstruction of road accident, the known probes carry, generally significantly the applied character, and their results are intended for the simplified examination, which methods rather in detail were already considered in a number of works [11,12]. The main program, applied even in profound examinations of cars crashes, use generally approaches of multibody dynamics, for example, of VCRASH4, when the facing bodies (generally cars) are considered in the movement as solid bodies. Few probes are devoted to assessment of road accident with penetration or colliding of the vehicle with elements of road arrangement, and they have concrete character, for example, crash with certain type of barriers [13,14], and probe comes to an end on determination of the car speed and trajectory after crash with barrier.

Digital modeling of possible situations of arrival of the car (initial speed, trajectory of the movement) on road protection and the alternative analysis allow to reconstruct rather precisely process of road accident and to define the reasons and, the main thing, share of road restraint systems (or any ERA), its elements and materials in absorption of impact energy, impact severity indexes for arrival cars and, that is important, for the cars which have got on the way of its movement on oncoming lane.

So, the digital twin of road accident with vehicle impact on road obstacle on the basis of full information on road accident is actually created. The example of such reconstruction with use of tools of the digital virtual analysis is given in a short form below in Figures 11 and12. Shooting from the video recorder (Figure 11), allowed to establish in initial position of the car running into wire road barrier, settlement scheme of simulation model, fragments of record of virtual arrival (Figure12) of the car on wire rope barrier, penetration on oncoming lane and arrival on the car, moving on oncoming lane, is shown. Final situational results well coincide with results of real road accident.

Figure 11. Filming from DVR and original impact diagram reproduced.

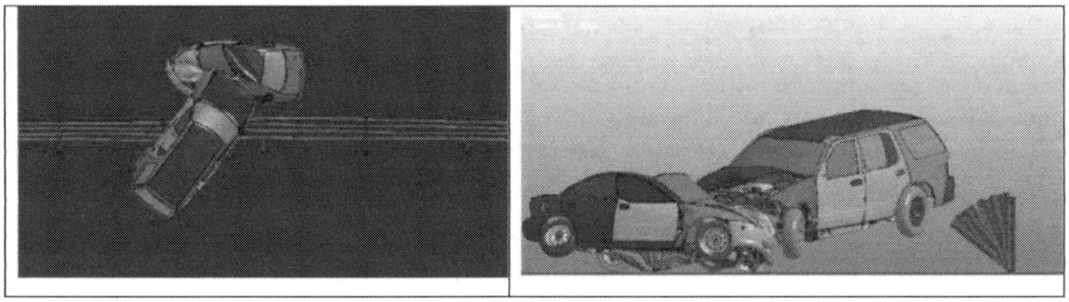

Figure 12. Fragment of a virtual collision of a car on a wire rope barrier and beyond, on a car after moving across wire rope barrier in the oncoming lane.

4.2. Use of the digital virtual analysis for new designs and development of regulatory requirements and data bases (DB)

Experience of application of the virtual analysis has turned out invaluable when updating state standard specifications on road protections and development of new standards, and, also, as it was already mentioned above, during creation of new innovative designs. Specific place among them is held by designs with use of elements from polymer and composite materials for road restraint systems. The detailed experiment allowing to create validating models of material and separate main structural elements first of all is necessary for these designs.

Experience of application of elements from modern high-strength polymer (Ultra High Molecular Weight Polyethylene - UHMWPE) for wire-rope barriers has shown, that without virtual modeling of tests it would practically impossible to find constructive solutions, particulars, for elements of fastening of cable – tape.

In figure 13 fragments of validation experiments with cable tape from Ultra High Molecular Weight Polyethylene (UHMWPE) shown: a) - on gap, b) – on impact, c) – FE model of attachment unit [15], used in the virtual analysis.

The considered analysis and especially, creation of Virtual Digital Proving Ground (VDPG), impossible and already becomes difficult in connection with variety of designs (both ERA, and vehicles), that obviously demands development of the databases (DB), which structure and the contents (maintenance) are based, first of all, on materials of VDPG.

In total, such bases can be united in uniform complex (BIG DATA), which can effectively be used, both in the mentioned applications, and at road accident reconstruction, and also to enter the general system of the databases organized when using BIM – technologies in the system of the digital road. The structure of DB, developed now, includes data on designs, 3D models, protocols of natural and virtual tests, reports on tests and a number of the specifications necessary for various users. It is natural that such bases are multilevel to meet requirements of various users.

Figure 13. Fragments of tests of innovative wire-rope barrier's element - tape from UHMWPE: at the left - tests of tape on gap, in the center-impact test, on the right - type of FE model of tape attachment unit in model of virtual tests.

5. Conclusions

The results of investigations, given in this work in the field of the digital technologies for the impact problems solution deformable complex systems on roads and called the road accidents are important, both from scientific, and from practical side. Complexity of researches for the solution of problems of impact of vehicles with obstacles, the accumulated considerable experience of simultaneously valid virtual analysis, define possibility of creation of the digital virtual proving ground (VDPG) for virtual tests of ERA, by analogy with known, including foreign, proving grounds for real tests. gives the chance of synthesis (creation DB) with the purpose of their analysis, not only for understanding of mechanics of difficult impact interaction of elements, but also for formulation of requirements to designs, design of new road safety structures.

As shows experience of probes, development of numerical methods of the nonlinear dynamic analysis of impact of deformable systems, must aimed, first of all, on the development of valid mathematical behavior models of structural materials at big deformations and collapse, at creation of the corresponding databases of materials and designs, and also their analysis. Development of methods of experimental studies for validation of the virtual digital models directed to analysis of damping in structural elements, creation of the specified mathematical models of materials and structural elements, is one of the major tasks for development of road restraint systems.

It is not necessary to think, that introduction of intellectual systems including the autonomous operating systems of cars, is panacea in the solution of question of ensuring safety – refusals of these systems are also probable as absolute reliability does not exist, both the virtual digital analysis and virtual testing are the necessary and effective tools in the system of safety on roads.

In conclusion it is also worth noting that the conducted researches are new step and in increase in competences of university's students of different level of training, including masters and graduate students of department are widely involved in these researches.

Acknowledgments
The author thanks all his students and colleagues, first of all, the staff of department of construction mechanics of MADI and LLC N&PQ (V.S. Nadezhdin, B.T. Tavshavadze, I.A. Karpov, P.S. Mikheyev, O.V. Titov and many others), whose results of researches and invaluable experience of discussions are used in this work. Work is promoted in by financing according to the Development — Digital Technologies Program with assistance of Russia "Fund of assistance to innovations" and also support of works on creation and processing of normative documents on road restraint systems from FDA RF (ROSAVTODOR).

References
[1] LS-DYNA® 2016 *Keyword User's Manual.* Livermore Software. Technology Corporation, Livermore, US
[2] Demyanushko I, Karpov I, Tavshavadze B 2017 Using a virtual experiment to determine the technical parameters of side retaining road barriers *J. Transport Construction* **6** 05–08
[3] Demyanushko I and Karpov I 2017 Virtual crash test *J. World of Roads* Nov. 49–50
[4] Demyanushko I 2017*Application of numerical methods of nonlinear dynamics to solving problems of collisions when vehicles hit road safety barriers* (IV School-NDM) M, IMASH RAS, p 482 - https ://elibrary.ru/item.asp id = 29130118
[5] Wright L, Davidson S, 2020 *How to tell the difference between a model and a digital twin* Adv. Model. and Simul. in Eng. Sci.) (7:13; https://doi.org/10.1186/s40323-020-00147-4
[6] GOST 33 128 - 2014. *Public roads. Road barriers. Technical requirements.* http://www.gostinfo.ru/catalog/gostlist
[7] GOST 33129 - 2014. *Public roads. Road barriers. Control methods.* http://www.gostinfo.ru/catalog/gostlist
[8] Standard paper 45029946-001-2018. *Procedure for virtual testing of side road retaining barriers.* LLC M&PC, RF, http//: www.niimech.ru
[9] Demiyanushko I, Tavshavdze B, Mikheev P, Karpov I Improving road safety during repair work *J, World of Roads* **131** 102–108
[10] TTMA-100 2010 *Innovative, Flexible and Affordable.* Gregory p 34.
[11] Niconov V 2018 *Reconstruction of accident, Introduction to modern methods of expert probes. Use crash-tests* Publishing decisions p 136 ISBN 978-5-4485-8175-5
[12] Fricke L1990 *Traffic accident reconstruction: Traffic accident investigation manual* (Vol. 2). Evanston, IL: Northwestern University Press.
[13] Coon, B and Reid J 2005 *Crash reconstruction technique for longitudinal barriers. J. of Transportation Engineering*, 131(1), 54–62. doi:10.1061/(ASCE)0733-947X (2005) 131:1(54).
[14] Asadollahipajouh M, Schmidt J, Lechtenberg K, Faller R 2017 Crash reconstruction technique for wire rope barrier systems *J. of Transportation Safety & Security*, DOI:10.1080/19439962.2017.1386251; https://doi.org/10.1080/19439962.2017.1386251
[15] Demyanushko I V, Ozerin A N, Golubev E K, Kurkin T.S., Putivsky I A, Romanov E V, Titov O V. 2019 *End capture for flat ropes and rope products from the high-strength films, threads from super molecular polyethylene, received by the solid-phase method* Patent for the invention of RU 2683045 C1, 26.03.2019

Navigation system with STV driver assistance function

V V Gaevskiy[1], S Ghith[1]

[1]Department of Automobiles, MADI, 64, Leningradskiy pr., Moscow, 125319, Russia

E-mail: vit-life@rambler.ru

Abstract. The article deals with the problems of application of modern driver assistance systems on single-track vehicles (STV). The design features of the STV that prevent the installation of such systems are presented. Variants of a motorcycle entering a turn are considered, with concerning the gyroscopic moment of the STV wheels and the causes of situations in the turn. A new navigation system with a driver assistance function with an analysis of its operation and advantages is proposed.

1. Introduction

The number of single-track vehicles (STV) is increasing every year. This trend is easy to explain. Overcrowded roads with cars in cities make people think about an alternative form of transport. The benefits of motorcycles (STV) to cars are obvious: mobility in traffic jams, ease of finding a parking space, economic efficiency, environmental friendliness.

However, the single-track vehicles have several disadvantages. The most significant of them is still its low safety. The challenge of enhancing safety in this type of means of transport will be an acute problem that will stand in the way of manufacturers for a long time.

Despite the general decrease in road accidents, the problem of road safety is urgent, since they remain the most dangerous means of transport in the world, STV drivers die 4-6 times more than car drivers if we compare the number of deaths per unit of vehicles. Low level of passive and active safety of STV, and the lack of widely used advanced driver assistance systems (ADAS), in particular those that are triggered when driving in a curve.

2. Main part

Active safety systems play an important role in the prevention of accidents and include: anti-lock braking system (ABS), vehicle stability control (ESC / ESP) and other driver assistance systems.

In recent years, the concept of driver assistance systems ADAS (Advanced Driver Assistance Systems) or in Russian Language – ISPV (intelligent driver assistance systems) has emerged and the number of electronic assistants has increased that support drivers in specific situations through information warning and /or intervention (if necessary) in driving control [1, 2, 3].

Initially, the vast majority of safety systems were designed for their implementation in road transport. For example, adaptive cruise control (ACC), autonomous emergency braking systems (AEBS) and others.

Due to the design features of the STV, practically none of the above systems can be applied on them without prior adaptation, and some systems are not used at all on the STV until now.

The main design features of the STV include:

Content from this work may be used under the terms of the Creative Commons Attribution 3.0 licence. Any further distribution of this work must maintain attribution to the author(s) and the title of the work, journal citation and DOI.
Published under licence by IOP Publishing Ltd

• the position of the motorcyclist (to control the motorcycle it is necessary to move the body, which affects the driver's fatigue);
• lack of a stable position when parking;
• due to the lack of a vehicle body in the motorcycle, the driver is exposed to negative external factors: wind, rain, dust, snow, foreign objects on the road, etc.
• mandatory requirement for the use of a protective helmet, as well as other elements of motorcycle equipment at the request of the driver;
• physical features of STV control when cornering;
• the specifics of the location of the controls.

Ultimately, the use of electronic systems is limited by the fact that the design and dimensions of the STV do not allow to accommodate all the necessary elements of these systems. In other cases, the use of systems is impossible due to the specifics features of the controls.

The list of systems that are still not commercially available on production of STV:
• "dead zones" warning system;
• parking assistance systems (PTS, AVM, TPMS);
• autonomous - robotic (partial or complete driver replacement) - adaptive cruise control (ACC), lane keeping system (LKSS), robot driver, etc.;
• assistance systems after an accident: eCall (emergency alert system launched), seatbelt (safety belt), active head restraints (AHR), airbags;
• active steering systems: direct adaptive steering (DAS), active front steering (AFS) [4];
• interior comfort systems (rain sensor, climate control, etc.) [5].

The analysis of the STV market shows that the following electronic safety systems, adapted for use on the STV, are now being installed on motor vehicles produced by popular and large motorcycle manufacturers (Yamaha, BMW, Kawasaki, etc.):
• anti-lock braking system (ABS);
• dynamic stability control systems (DSC);
• traction control system (TCS): Kawasaki traction control (KTRC);
• Hill-start Assist Control (HAC, HSA, HHC);
• front wheel detachment control system: Kawasaki launch control mode (KLCM);
• rear wheel slip level control system;
• cornering control system: cornering brake control (CBC);
• start control system.

It is important to mention that most of these systems are still found only on the most powerful and expensive modern motorcycles.

Thus, the applied list of electronic safety systems used in commercial large-scale production is small, and their field of application is limited by the powerful and overweight motorcycle. Most of the electronic systems are used in sports motorcycles, whose operating environment is a race track.

However, there are such marketing leaders as Bajaj, Racer, Motoland, which are highly popular due to their low cost, but they do not have any security systems. For example, the Bajaj brand has only one model with an ABS system - the Bajaj Dominar 400.

3. Causes of emergencies situations in turning a corner.

In view of the design of the STV and the peculiarities of its control, a section of the road with a curved trajectory is the most dangerous and requires special attention when driving along it.

Accident statistics show that the share of accidents involving associated with a motorcycle flying off the road in a turn without the participation of another vehicle is 41% of all fatalities involving motorcyclists. The percentage of accidents when turning a corner is twice the number of accidents for other reasons.

Driving at a speed above the critical speed or choosing the wrong trajectory leads to the fact that the driver tries to correct the situation directly during the cornering. In fear, an inexperienced driver may apply the brake or open the throttle too sharply. Such situations, when the STV moves with a certain roll angle, can negatively affect its stability, because at this moment, the mass of the motorcycle is redistributed and the area of the contact patch of the front or rear wheel with the road decreases. Loss of adhesion of one of the wheels to the surface leads either to the motorcycle skidding and falling in the turning direction, or the driver throws up and flies over the motorcycle. In this case, in addition to hitting the driver on the road surface, he also at risk of getting hit by the motorcycle itself which is moving behind him.

The reasons for this are:

1. Insufficient deceleration before entering a turn. High speed when entering a turn leads to the following consequences: driver fear and loss of control, attempts to reduce the speed by closing the gas and pressing the brake, an attempt to increase the roll angle by hanging the driver's body.

2. Errors in handling the motorcycle when exiting a corner. The transient process of movement, both entering and exiting a corner, is the most unstable and dangerous.

3. Incorrect reading of the traffic situation and the choice of the trajectory. A late reaction to a turn leads to the fact that the driver does not have time to select a safe driving mode, as well as to shift in the necessary direction within his lane, depending on the type of turn. Closed turns are especially dangerous and require an advance decision on safe traffic in this section (Fig. 1).

4. Changing the road surface. The presence of irregularities on the road, as well as oil, sand and other foreign objects leads to deterioration in handling, wheel traction and slipping.

Figure: 1. Dangerous trajectory in a closed corner

Cornering has a phase character: approach, entrance, movement in an arc, exit. In the approach phase, the driver reduces the speed by one of the braking techniques: engine, front brake, rear brake. In the best case, the driver tries to perform a preliminary maneuver - "approach", which allows him to build a trajectory of the maximum radius, which is terminologically referred to as "Trajectory Smoothing". Entering a turn is a transition from straight ahead to arc of a given radius. This process is the most dangerous, occurs because the center of mass displacement, as well as a shift in the contact spot patch of the tire with the road.

Movement along the arc presupposes ideally construction of trajectory smoothing, although it does not exclude additional maneuvers with forced change of trajectory during the competitive struggle. When moving along an arc, the methods of balancing techniques of maintaining stability and controllability are particularly important. The balance of stability is achieved by changing the tilting of body forward, backward, to the sides. Particular attention is paid to the throttling technique, since the complete cessation of traction provokes the motorcycle to fall after skidding, lateral side-slip and rotation. Depending on the configuration of the turn, the coefficient of adhesion, the individual characteristics of the rider, there are constant, variable, alternating throttling, as well as smooth, forced and sharp.

The movement along the arc of the smoothing trajectory involves a "cut", which is marked in the "apex" zone (the steepest part of the arc). When entering a turn, a rational gear is selected, which can change in the exit phase, or when stability is lost on the arc.

The phase of exit from the turn is associated with a decrease in the motorcycle roll, an increase in the load on the rear wheel, an increase in the radius of movement.

Consider the options for moving the STV in a turn, depending on the speed of movement. When cornering, it is important to take into consideration the gyroscopic moment of the STV wheels.

When driving at low speeds, for example, when maneuvering in a parking lot, the motorcycle turns in the usual way - turning the handlebar in the direction of movement (Fig. 2). The figure shows that the driver, moving around the site ground at low speed, turns the handlebar in the direction of the turn and goes in the direction in which the front wheel is moving.

Figure: 2. Maneuvering at low speed

Considering cornering at speeds where the gyroscopic moment has a large effect, the technique of entering the turn is fundamentally different. Since turning the handlebars at high speed towards the bend, contrary to the rider's expectations, his STV will tip over from the center of the turn, and not vice versa. Thus, in order to safely enter a corner at high speed, the driver must first shift the center of mass of the motorcycle by displacing the driver's body in order to tilt it, and only then turn the handlebars towards the turn (Fig. 3).

Figure: 3. Entry into a turn by shifting the driver's body towards the turn

There is a third special case of entering a turn when the driver turns the handlebar away from the turn. Before entering a turn, the driver deliberately turns the handlebar away from the center of the turn, knowing that the motorcycle will start to tip over to the same side which he turns, due to the gyroscopic moment on the wheels (Fig. 4).

Figure: 4. Entering a turn using a counter-steering technique

This way of cornering is extremely dangerous. As a rule, this technique of handling is used by motorcyclists on the racetrack. This method allows them to enter and move in a corner as quickly as possible and with the smallest radius. Also, motor sportsmen use the counter-steering technique at stunt riding competitions when performing drift on a motorcycle.

Thus, there are three main ways to entering the corner on a motorcycle:

1. at low speed: the absence of gyroscopic moment allows the driver to turn the handlebar in the direction of movement;
2. at high speed (safe): due to the influence of the gyroscopic moment, the driver is forced to first shift the center of mass of the motorcycle with his body, then smoothly turn the handlebar in the direction of the turn;
3. at high speed (unsafe): due to the influence of the gyroscopic moment, the driver intentionally turns the handlebar away from the center of Turning.

In everyday life, excluding extreme driving, the driver avoids critical driving modes in every possible way. But, moving along a curved trajectory, often the wheels of the motorcycle begin to slip. Of course, this can be caused by unforeseen circumstances, such as oil or sand on the road, but the most important reason for slipping the wheels in a corner is improper motorcycle control due to movement at speeds above critical. Every driver needs to clearly understand what speed is safe before an approaching turn.

As an example, let's calculate the critical slip speed for a specific turn, the radius of which is R = 48 m, for dry and wet asphalt. The coefficient of adhesion of the wheel to the road surface φ_y is 0.8 for dry asphalt and 0.3 for wet asphalt.

Let us calculate the critical slip speed in a turn using the following formula [6]:

$$Vcss = \sqrt{g\ \varphi_y R}$$

where g = 9.81 m / s^2 is the acceleration of gravity, is the coefficient of adhesion of the wheel to the road surface φ_y, R is the radius of curvature of the turn.

Vcss (dry asphalt) = 69.87 km/h

Vcss (wet asphalt) = 42.79 km/h

On the section for which the calculation was made, a speed limit sign of 50 km/h was installed. As it is observable, it is relevant for dry asphalt, because the design lateral slip speed exceeds the speed set by the SDA. However, if driving in the rain, following the 50 km/h limit may result in the rider falling. Therefore, speed limit signs can be objective for the safe movement of motorcyclists, moreover, one cannot be sure of the presence of such signs before each turn, since they may be insignificant for the motorist, while it will be useful for the motorcyclist to have information in advance that will allow him to enter safely into the turn.

IOP Conf. Series: Materials Science and Engineering 1159 (2021) 012064 doi:10.1088/1757-899X/1159/1/012064

4. Navigation system with STV driver assistance function

As previously mentioned, entering a corner incorrectly leads to a number of consequences that can be hardly dealt with either a very experienced driver or an intelligent driver assistance system (ISPV). However, these systems have a narrow application and are not yet available on most STV.

Intelligent driver assistance systems, which are implemented in motorcycle transport, affect the parameters of the motorcycle's movement when there is an immediate risk of falling. However, there are no such systems that could recognize turns in advance, read the traffic situation and give recommendations to the driver about the safest driving mode.

Therefore, one of the ways to deal with accidents in a corner is to prevent cases that the driver cannot cope with on his own, that is, the task is to inform the driver in advance about a possible change in the traffic situation. In order to minimize accidents and develop driving skills, we will consider a navigation system with an STV driver assistance function.

The navigation system in its usual sense is designed to determine the position of the vehicle, select and follow the route. In other words, the driver gets a general idea of how far to travel to the desired turn or final stop.

Navigation with a driver assistance function, theoretically considered in solving the problem, assumes that the driver will not only receive a signal about the presence of a turn, but also recommendations about its passage based on the following data:

- distance to the nearest turn;
- radius of curvature of the turn;
- speed limit on this section of the road;
- weather;
- technical data of the motorcycle.
- First, let's consider the navigation that is currently available using the example of the Yandex.Navigator application (Fig. 5a). When moving along the constructed route, the screen displays the following information:

- current speed of the vehicle;
- the speed set by the traffic rules for this section;
- distance to the nearest turn.

This navigation option does not take into consideration the peculiarities of traffic on the STV and the state of the road surface. But it is known that these factors are very important for safe movement and must be considered in navigation systems.

Also, the methods of entering the turn were considered depending on the gyroscopic moment on the wheels, which in turn depends on the speed of movement, mass and radius of the wheels. These factors should also be considered when adjusting of the navigation and offering recommended technique for the driver the to enter the corner.

Figures 5 b, c show examples of updated navigation. Its peculiarity is that in addition to the information that is available in the first version, the driver sees the following information on the screen:

- the recommended speed before the nearest turn, which will be relevant for entering the turn on the STV;
- recommendations on the technique of entering into the corner turn.

Thus, the driver will be able to reduce the speed in advance, select the desired gear, take the required position in the lane in which he is moving, and be ready to move in a dangerous area.

It is also assumed that in addition to visual information, prompts will be duplicated through a Bluetooth headset (Fig. 6) in order to distract the driver from the road as little as possible.

a) b) c)

Figure: 5. Variants of the navigation system: a - built route in Yandex.Navigator; b, c - proposed navigation options with the function of assistance to STV drivers

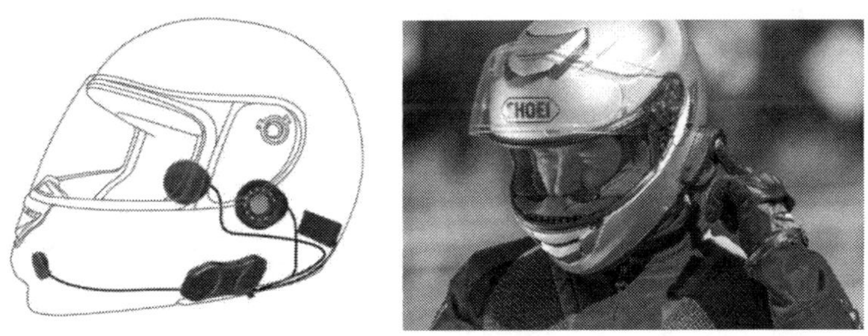

Figure: 6. Bluetooth headset

5. Algorithm of the modified navigation.

If the calculated critical speed is less than the one set by the traffic rules for this section of the road, then the program checks whether the driver exceeds the calculated critical speed. If the result is negative, then the driver receives only information about the cornering technique. If the result is positive, then in addition to the recommendation about the turning technique, the driver will be informed about the speed with which he needs to move before the next turn.

If the estimated critical speed is greater than or equal to that set by the traffic rules on this section of the road, then the program also checks whether the driver exceeds the traffic speed. Further, the information to the driver is received according to the principle of the first case: if the driver does not exceed the speed according to traffic rules, then information appears on the technique of cornering, if it does, then the motorcyclist is additionally informed that it is necessary to reduce the speed.

6. Conclusions

Features and benefits of improved navigation.

This function, of course, is not able to prevent an imminent fall, but it will help the driver to navigate and concentrate his attention, orientate himself more quickly in the road conditions and take the necessary actions for safe driving. The advantage of navigation with enhanced functionality is that it is accessible to any driver and will be especially useful for those drivers who are getting their first experience of driving a STV in an urban condition.

Moreover, standard navigation only informs the driver about those turns that have intersection of two or more roads. In other words, the turns formed by changing the direction of the road are not signaled in any way. This is of no importance to the motorist, but it is important for the STV driver to know in advance of any change in the direction of track motion.

References

[1] Shadrin S S, Ivanov A M 2016 Algorithm of Autonomous Vehicle Steering System Control Law Estimation while the Desired Trajectory Driving *[Text] ARPN Journal of Engineering and Applied Sciences* Vol. 11 **15** 9312–9316

[2] Shadrin S S, , Nevzorov D V 2015 Autonomous wheeled vehicle as part of intelligent transport systems *[Text] Natural and technical sciences* **6 (84)** 309–311

[3] Ivanov A M Shadrin S S 2013Development of a system of inter-object interaction of intelligent vehicles *Izvestiya VolgGTU. Series "Land Transport Systems". Issue 7: interuniversity. Sat. scientific. Art. / VolgSTU* (Volgograd) **21 (124)** 74–77

[4] Banks V A, Eriksson A, O'Donoghue J, Stanton N A 2018 Is partially automated driving a bad idea? Observations from an on-road study *Applied Ergonomics* Vol 68 138-145

[5] Gaevskiy V V and Ivanov A M 2018 Problems of the application of intelligent driver assistance systems on a single-track vehicles *IOP Conf. Ser .: Mater. Sci. Eng. 386 012021*

[6] Gaevskiy V V 2019 *Development of the theory of movement of single-track vehicles. Monograph* (Moscow Automobile and Road Construction State Technical University (MADI)) **64**

IOP Conf. Series: Materials Science and Engineering 1159 (2021) 012065 doi:10.1088/1757-899X/1159/1/012065

Modeling of diesel structure-borne noise in an unsteady operating mode

M G Shatrov[1], A L Yakovenko[1], I V Alekseev[1], S N Bogdanov[1], A O Glazkov[1]

[1] Moscow Automobile and Road Construction State Technical University (MADI), 64 Leningradski prospekt, Moscow 124319, Russia

E-mail: iakovenko_home@mail.ru

Abstract. The article considers some aspects of the modeling of the working process and structure-borne noise of a diesel engine in a transient mode. The main factors that cause differences in the diesel working process from similar stationary modes are described. A method for calculating the structure-borne noise from the diesel working process at the transient mode is presented. A study of the noise of a 6-cylinder (S=10.7 cm, D= 12,4 cm) diesel engine during acceleration was performed. The results of calculations showed that the noise of the 6-cylinder diesel engine is 2...3 dB higher in the entire range of engine speeds compared to similar speeds in steady-state for diesel full-load characteristics.

1. Introduction

In urban traffic conditions, most of the time the internal combustion engine operates in unsteady modes (USM). The experience of vehicles exploitation shows that these modes not only reduce efficiency, increase emissions of toxic components of exhaust gases, but also increase the engine noise level. When evaluating the external noise of the car according to GOST, noise measurements are made during acceleration of the car.

Therefore, development of methods and software tools for studies of the internal combustion engine noise at transient operation modes is an important and actual task.

2. The impact of the transient operation mode on the working process of a diesel engine

Among the sources of the engine structure-borne noise, the working process is crucial for diesels [1, 2] due to the high rate of pressure increase in the cylinder. Therefore, one of the tasks of the study was to determine the factors that cause changes in the working process of a diesel engine in acceleration mode compared to stationary modes.

It was found that the main influencing factors are: decline of the processes of mixing and combustion, an increase in the ignition delay period, the possibility of fuel sub-injection and other violations of the fuel supply process, thermal inertia of engine parts.

A sharp decrease of air-fuel ratio is most typical for supercharged engines. As the engine boost increases, this problem gets worse. This is explained by the fact that with a sharp increase in the load, the rate of pressure growth significantly lags behind the rate of increase in fuel supply due to the inertia of the supercharger. Also, increase of fuel in the working mixture may be affected by imperfections in the engine control system, its inability to respond to the acceleration mode.

Deterioration of the mixture formation at the USM leads to local re-enrichment of air-fuel mixture

Content from this work may be used under the terms of the Creative Commons Attribution 3.0 licence. Any further distribution of this work must maintain attribution to the author(s) and the title of the work, journal citation and DOI.
Published under licence by IOP Publishing Ltd

in the combustion chamber. This is caused by violation of the dynamics of gas exchange. The air velocity in the intake channels decreases, which affects the intensity of the vortex movement of the charge in the cylinder. This factor is especially important for diesels with volumetric mixing, in which it is extremely important to ensure the necessary charge turbulence.

The working process of a diesel engine is largely determined by its fuel system [3-5]. The deterioration of fuel evaporation due to low temperatures of the charge air and the surfaces forming the combustion chamber also negatively affects the mixture formation. Deterioration of the mixture formation leads to an increase in the ignition delay period. If the control system does not change the injection advance angle, this leads to a transfer of the main combustion phase to the expansion process and burnout at the outlet, which negatively affects the combustion efficiency.

A particular impact on the deterioration of diesel performance in USM is the possibility of a violation of the quality of fuel supply, for example, the possibility of fuel sub-injection in the absence of those in the steady state operation mode. One of the parameters of the fuel system operation is the residual pressure or residual volume in the high-pressure fuel line (HPL). The last parameter appears if there is a residual vacuum in the HPL. Residual pressures (or vacuum) determine the initial pressures or initial free volumes in the HPL before the next injection cycle. When the diesel engine is running, the residual vacuum of the previous cycle is equal to the initial pressure of the subsequent one. At USM, when there is a change in the speed or displacement of the fuel pump rail, there is a corresponding change in the residual and initial pressures. The initial pressure determines the speed of sound propagation and the compressibility of the fuel. Consequently, there is a difference in the initial pressure at SM and at USM and, accordingly, a difference in the working process of a diesel engine. This is due to changes in the fuel supply, changes in the start and end of the injection, and its duration. The appearance of fuel sub-injection in diesel USM cycles is associated with the appearance of the initial pressure on the USM significantly higher or lower than for SM, with the same speed and fuel pump rail position [6, 7].

One of the important factors in the difference between transient and stationary modes is the influence of thermal inertia of engine parts [8]. This is especially evident on diesel acceleration mode. Thermal inertia increases the ignition delay period and increases the harshness of operation of the diesel engine, which leads to an increase in structure-borne noise.

When using existing software tools for calculating the diesel working process (the Diesel-RK software, programs developed in MADI, the AVL FIRE software), it is necessary to supplement the initial data set for the stationary mode with amendments that take into account the transition mode. To solve this problem, special software tools can be used, for example, the MATLAB Simulink.

3. Method for calculating the diesel structure-borne noise in unsteady mode

After studying the influence of the acceleration mode on the working process of a diesel engine, a method for calculating its structure-borne noise from the working process was formed. It is based on the model of an engine as a structure-borne noise source developed at the Department of Heat Engineering and Automotive Engines of MADI [1] and is shown in Figure 1. Noise calculation is carried out in the software "Acoustics of internal combustion engines", developed in MADI.

At the first stage, a set of data on the engine and its three-dimensional model is formed. Next, the working process is calculated taking into account the unsteady operating mode, which results in an array of indicator diagrams for different engine speeds.

At the next stage, using the obtained indicator diagrams, the structure-borne noise from the working process is calculated and an array of sound power spectra is formed. Finally a curve for the dependence of the total sound power level on the crankshaft speed and time is obtained.

4. Results of calculating the structure-borne noise of a diesel engine in acceleration mode

Using the developed method, the structure-borne noise of the 6-cylinders diesel engine (D=10,7 cm, S=12,4 cm) in the acceleration mode was calculated.

The main parameters of the research object are shown in Table 1.

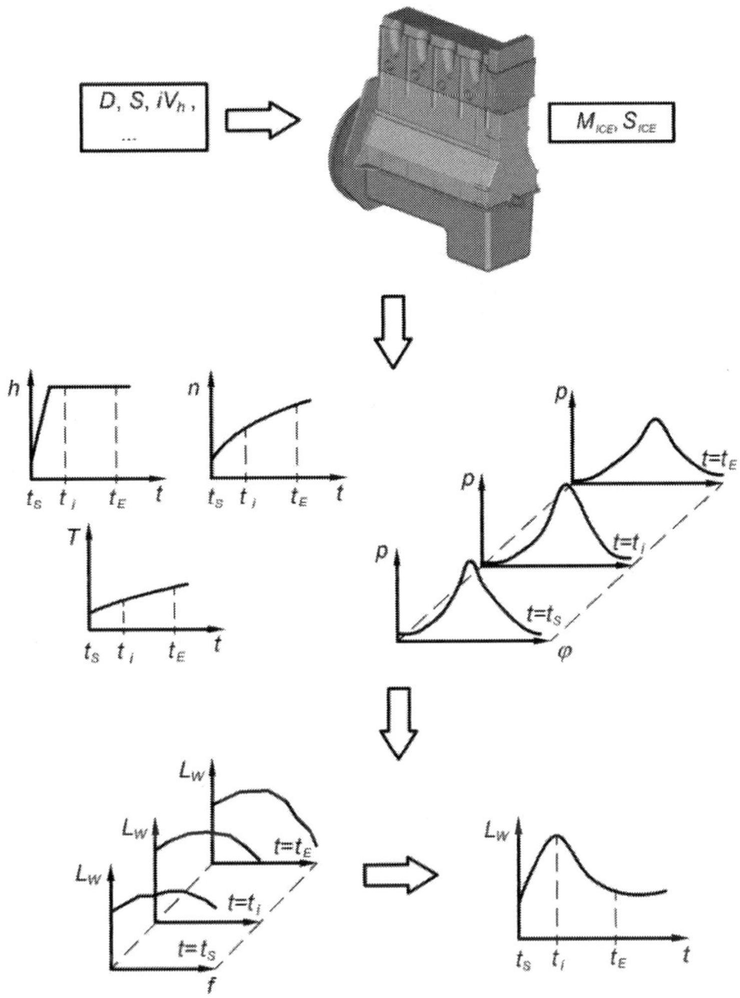

Figure 1. Method for calculating the structure-borne noise of a diesel engine in acceleration mode.

Table 1. Layout parameters of studied diesel

№	Name of parameter	Value
1	Engine type	diesel
2	Layout	in-line
3	Type of cooling system	liquid
4	Number of cylinders	6
5	Bore D, mm	107
6	Stroke S, mm	124
7	Compression ratio ε	17,3
8	Nominal speed n_{nom}, min^{-1}	2260

At the first stage, the working cycle was calculated for steady-state and non-steady-state operating modes. As a result, a set of indicator diagrams was obtained, a comparative analysis of which showed differences in the rate of pressure increase near the TDC (Figs. 2 и 3).

IOP Conf. Series: Materials Science and Engineering 1159 (2021) 012065 doi:10.1088/1757-899X/1159/1/012065

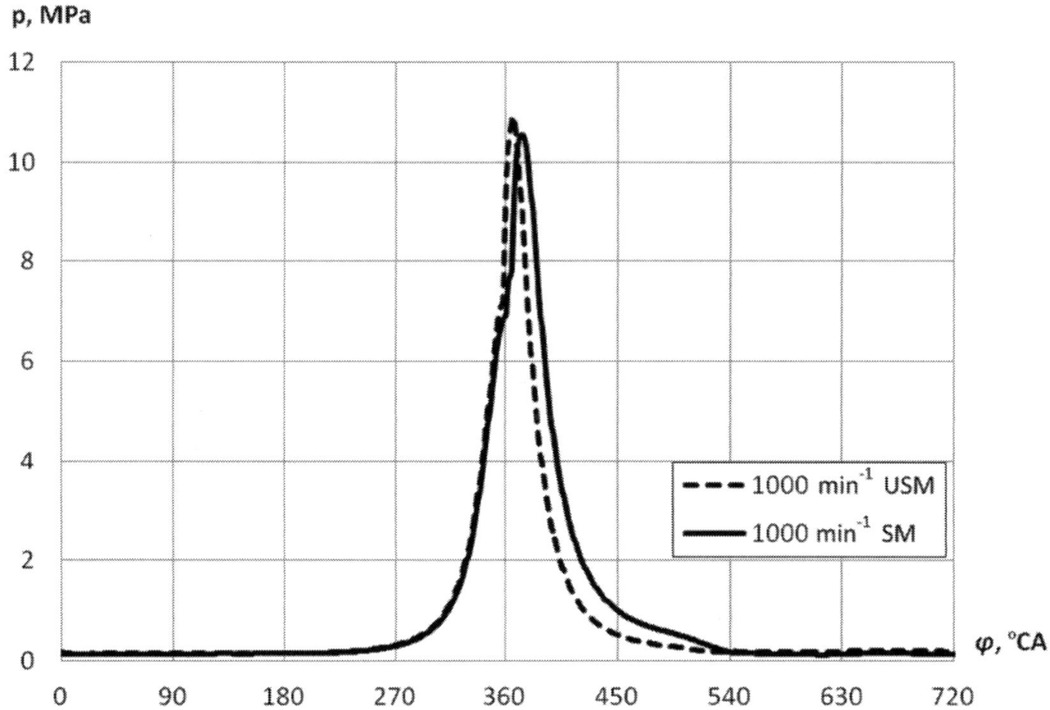

Figure 2. Comparison of indicator diagrams in steady-state and acceleration modes at n=1000 min^{-1}.

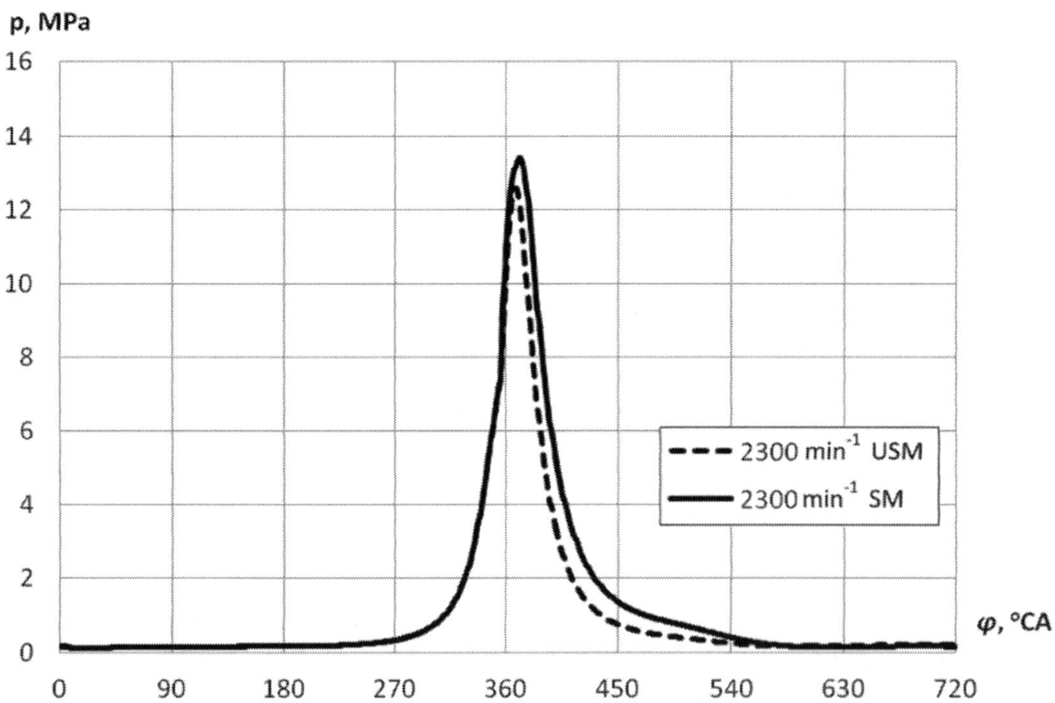

Figure 3. Comparison of indicator diagrams in steady-state and acceleration modes at n=2300 min^{-1}.

IOP Conf. Series: Materials Science and Engineering 1159 (2021) 012065 doi:10.1088/1757-899X/1159/1/012065

At the next stage, the spectra and total sound power levels of the diesel engine working process were calculated for all the studied operating modes.

To verify the calculations, an experimental determination of the noise level of diesel engine was performed. It showed good convergence of the results.

Some calculation results are shown in Figure 4 and Figure 5.

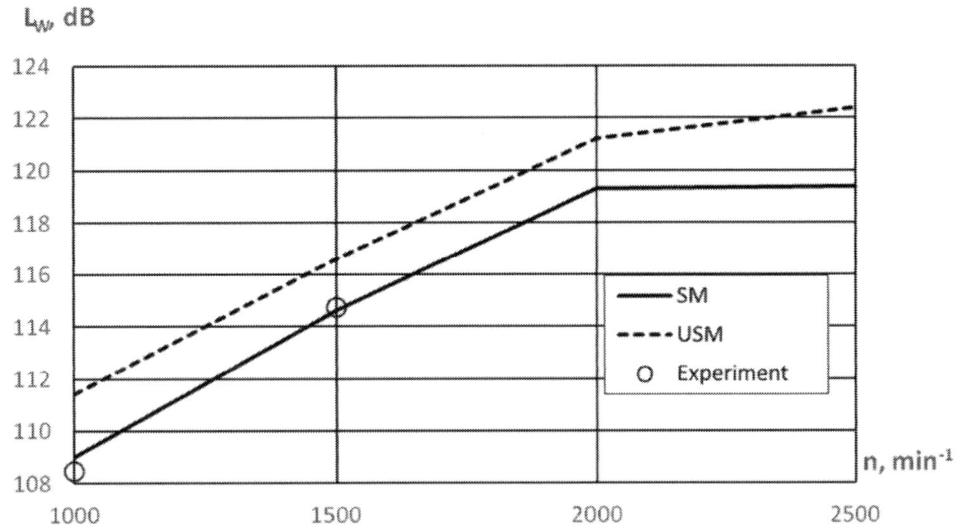

Figure 4. Changing the overall sound power level of diesel engine in different operating modes.

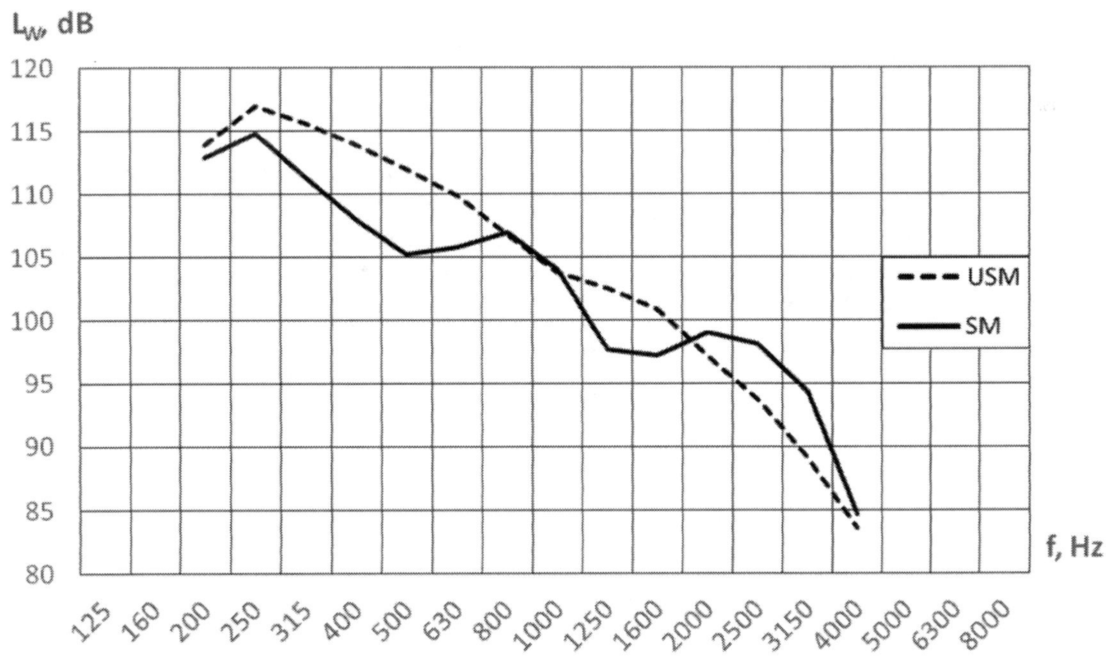

Figure 5. Sound power spectra of diesel from the working process at n=2300 min^{-1}.

Analysis of the sound power spectra showed that in the acceleration mode, the sound power level in the low frequency range of the spectrum is higher than in the stationary mode, over the entire engine speed range.

As a result, the overall sound power level of the diesel engine in the acceleration mode for all crankshaft speeds is 2...3 dB higher than in similar stationary modes.

5. Conclusion

1. The analysis of the features of modeling the working process and structure-borne noise in the unsteady mode for a diesel engine is made. A method for calculating the structure-borne noise of the engine in the acceleration mode is formed.

2. Simulation of the operating cycle of 6-cylinder diesel with D=10,7 cm, S=12,4 cm in the software "Diesel-RK" in steady-state mode according to the full-load characteristic is performed.

3. The structure-borne noise of the 6-cylinder diesel was calculated in the software "Acoustics of internal combustion engine" in steady-state mode based on the full-load characteristic. The difference in the obtained total sound power level for engine speed n=1000 min^{-1} and n=1500 min^{-1} compared to the experimental values was less than 1 dB.

4. The calculation of the working process and structural noise of the studied diesel engine in the acceleration mode from the minimum to the maximum speed is Performed.

5. The results of studying the structure-borne noise of diesel engine in different operating modes are compared. In the acceleration mode, the overall sound power levels of the studied diesel engine are 2...3 dB higher than in steady-state modes in terms of full-load characteristics over the entire range of diesel speeds. Thus, when developing measures to reduce the structure-borne noise of the designed diesel engine, it is necessary to take into account its operation in an unsteady mode.

References

[1] Yakovenko A L, Shatrov M G, Glazkov A O, Kuznetsov S M, Alekseev I V, Bogdanov S N 2020 *International Journal Of Online And Biomedical Engineering* **16 (7)** 105–113

[2] Chaynov N D, Markov V A, Savastenko A A 2018 *IOP Conference Series: Materials Science and Engineering* **327(2)** 022069

[3] Dunin A Ju 2018 *Avtomobil'naya promyshlennost'* 1 5–7

[4] Shatrov M G, Golubkov L N, Dunin A U, Yakovenko A L, Dushkin P V 2015 *International Journal of Applied Research* **10(20)** 41098–41102

[5] Gorbachevskij E V, Dunin A Ju 2018 *Vestnik Moskovskogo avtomobil'no-dorozhnogo instituta (Gosudarstvennogo tekhnicheskogo universiteta)* **4(55)** 38–46

[6] Solovyov D E, Patrahaltsev N N, Taipa Karlos Munares 2003 *Vestnik RUDN* **1** 19–22

[7] Shu Ge-qun, Wei Hai-qiao and Liang Xing-yu 2005 *SAE Paper №. 2005-01-2487*

[8] Gadir N 1991 *PhD Thesis* 133

Development of the integrated IKTS-MADI complex for design and research of common rail fuel systems of diesel engines with electronic control

A U Dunin[1], M G Shatrov[1], L N Golubkov[1], P V Dushkin[1] and A L Yakovenko[1]

[1] Department of Heat Engineering and Automotive Engines, Moscow Automobile and Road Construction State Technical University (MADI), Moscow, Russia

E-mail: a.u.dunin@yandex.ru

Abstract. Further tightening of environmental rules and standards regulating the content of toxic substances in diesel exhaust gases, accompanied by an increase in the requirements for their fuel efficiency, creates prerequisites for the creation of Russian common rail fuel systems (CR). The performance of this task is possible with the use of new engineering solutions in combination with computational and experimental research methods implemented on the basis of the integrated computational and experimental complex IKTS-MADI. The complex consists of computing and research components. The computing component of the IKTS-MADI allows calculating the working processes of the entire high-pressure line when the high-pressure fuel pump (HPFP), rail and electro-hydraulic common rail injectors (CRI) work together, taking into account the influence of wave phenomena on the distribution of injection parameters across diesel cylinders. Obtaining new data to expand the library of existing technical solutions and testing of fuel equipment samples designed using mathematical modeling are carried out on the research component of IKTS-MADI. The CR and its components are tested together with the control system (control unit, sensors and wiring harnesses). The injection characteristics required for calculating the diesel operating process are recorded.

1. Introduction

The further tightening of ecological standards specifying the content of toxic emissions in the exhaust gases of diesel engines [1], jointly with the need to improve fuel efficiency and use alternative fuels [2], creates prerequisites for perfection of the CR fuel systems of diesel engines. The fields of this perfection are:

- raising injection pressure up to 300 MPa [3, 4, 5];
- control of fuel distribution by the combustion chamber zones [6];
- ensuring the required shape of the front edge of the injection rate [7].

The desired fuel injection law at any operation mode of the engine is formed by variation of the control impulse duration and pressure in the common rail. It also depends on wave phenomenon originating in the high-pressure line and having a considerable impact on the fuel injection process in case of a multistage injection [8, 9].

These problems can be solved using the developed integrated computational and experimental complex IKTS-MADI.

Content from this work may be used under the terms of the Creative Commons Attribution 3.0 licence. Any further distribution of this work must maintain attribution to the author(s) and the title of the work, journal citation and DOI.
Published under licence by IOP Publishing Ltd

2. Description of the developed integrated computational and experimental complex IKTS-MADI

To develop the design of the CR and fine-tune its working process, an integrated calculation and experimental complex IKTS-MADI was created for cooperative research and adaptation of the working processes of diesels and their fuel systems. The complex consists of computing and research components.

The IKTS-MADI structure allows implementing the following strategies for research and adaptation of working processes of diesels and its fuel systems:

- exploratory research – after conducting initial experimental studies, additions are made to the mathematical models of the working processes of diesel and its fuel system. Then a search for a new technical solution is carried out, taking into account the specified restrictions;
- selection of a possible technical solution for the fuel system from those proposed by conducting a computational research, its experimental verification on a non-engine stand, and evaluation of the efficiency of it using to improve the performance of the diesel working process;
- parallel design of the fuel system with the diesel control system and debugging of their cooperative operation.

The computational component of IKTS-MADI consists of:

- non-engine stand B01.00.00.000 (figure 1) with tools for testing elements of direct-acting and common rail fuel systems elements;
- engine stand with single-cylinder engine M01.00.000 on the basis of a universal crankcase for studying the cooperative operation of diesel with experimental samples of fuel equipment with electronic control;
- non-engine stand PLTD. 387442.70.00 and measuring equipment (figure 2) for experimental research and testing of a full-size set of fuel systems with a diesel control system (figure 3).

The distinctive features of the research component: testing fuel system with pressure of 300 MPa; determine the power required to drive the fuel pump (up to 87 kW); a control system for the fuel system including the electronic unit with the necessary algorithms and sensors required to control the transport diesel.

To create a simulation of the operation of the electronic control system as part of the engine, disks (figure 3) of sensors position for the crankshaft 1 and camshafts 3 are installed on the test stand, similar to those installed on electronically controlled diesels.

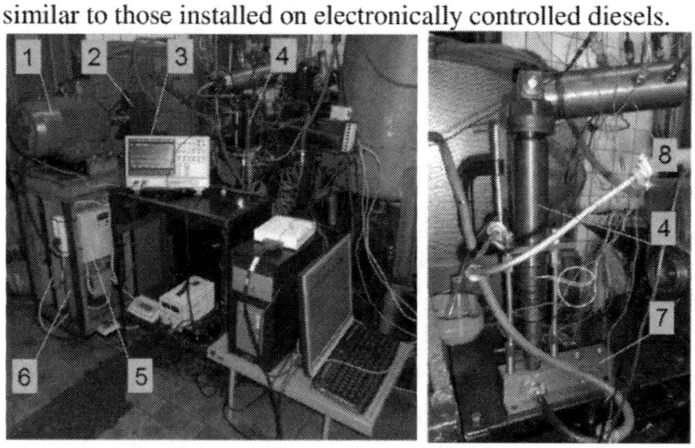

Figure 1. Non-engine stand B01.00.00.000 for testing elements of direct-acting and common rail fuel supply systems:
1 – electric motor; 2 – high-pressure fuel pump; 3 – oscilloscope; 4 – electrohydraulic injector; 5 – speed controller; 6 – bed; 7 – chamber for recording injection characteristics; 8 – rail.

When the test stend is running, the disks rotates, simulating the rotation of the engine crankshaft, and the control unit receives signals from the crankshaft position sensor located above the disk 4 having 60-2 teeth, which determines the engine speed, and from the camshaft position sensor located above the disk 2 having 6+1 tooth, which synchronizes the control pulses on the injectors relative to the conditional position of the top dead point of the first cylinder of the engine.

IOP Conf. Series: Materials Science and Engineering 1159 (2021) 012066 doi:10.1088/1757-899X/1159/1/012066

A mockup of the pressure control unit was created for conducting research. The development and production of the pressure control unit mockup was carried out in several stages.

- development of the PI-controller algorithm.
- development of the pressure control unit mockup.
- development of software and its transfer in the microcontroller of the control unit pressure.
- "on the table" debugging the developed system and determining of the pressure sensor characteristics.
- calibration of the integral and proportional parts coefficients of the controller on the test stend.

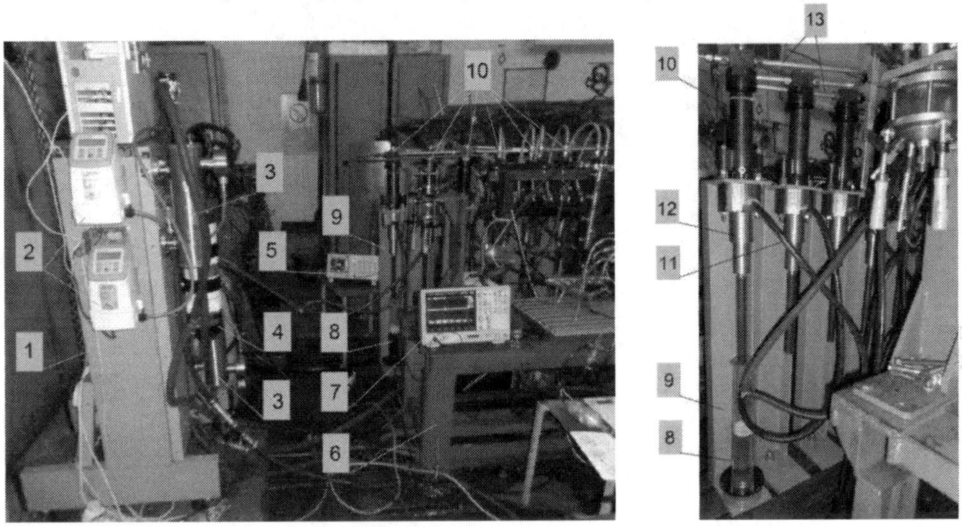

Figure 2. Non-engine stand PLTD.387442.70.00 and measuring equipment:
1 – measuring stand; 2 – registering parts of fuel flow meters; 3 – measuring parts of flow meters; 4 – fuel filter; 5 – oscilloscope for registering the control signal coming to the injector electromagnet; 6 – bed for mounting a high-pressure fuel pump; 7 – oscilloscope for registering changes in fuel pressure in the high-pressure line; 8 – measuring cup; 9 – stand for the injector mounting; 10 – electro–hydraulic injectors; 11 – hose for removing fuel spent on injector control; 12 – support for injectors mounting; 13 – fuel pipes.

Figure 3. Placement of diesel control system elements on the non-motor stand PLTD.387442.70.00:
1 – crankshaft position sensor; 2 – 6+1 disc; 3 – camshaft position sensor; 4 – 60-2 disc; 5 – control unit; 6 – electric brake machine.

Determination of the hydraulic characteristics of the CRI spray holes is carried out at a stationary fuel flow on the experimental stand shown in figure. 4. At the same time, the pressure at the inlet of the nozzle reaches 150 MPa, and in the working chamber – 16 MPa, which corresponds to the conditions of injection into the combustion chamber of a modern diesel engine.

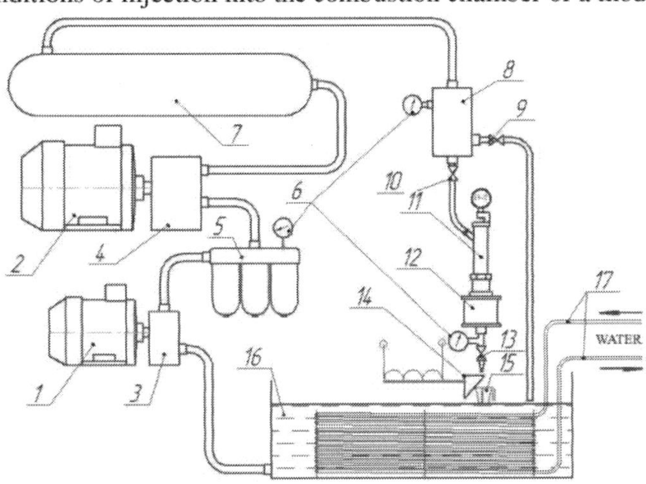

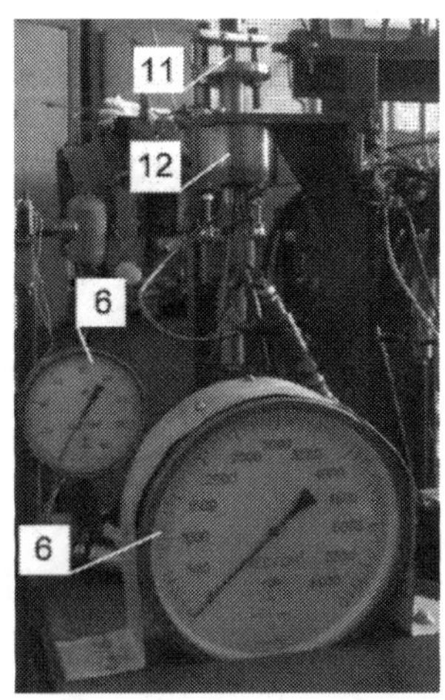

Figure 4. Structural diagram of an experimental stand for recording the hydraulic characteristics of nozzles:
1 – electric motor of the low-pressure fuel pump drive; 2 – electric motor of the high-pressure fuel pump drive; 3 – low-pressure fuel pump; 4 – high-pressure fuel pump; 5 – filter unit for fine fuel purification; 6 – arrow gauges; 7 – high–capacity rail; 8 – small rail; 9 – valve for adjusting the volume of fuel being passed; 10 – throttling valve for fine adjustment of the injection pressure; 11 – injector body with a nozzle and needle lift indicator; 12 – chamber; 13 – valve for adjusting the back pressure in spillage chamber; 14 – choke with electric actuator; 15 – measuring vessel; 16 – fuel tank; 17 – heat exchanger for fuel cooling.

Using complex, it is possible to carry out a full cycle of work related to testing and modification of fuel systems:

- checking the efficiency and effectiveness of the selected technical solutions of individual elements of the fuel system (element-by-element tests): high-pressure fuel pump, CRI, rail, high-pressure fuel lines, low-pressure lines;
- determining the characteristics of fuel system elements (hydraulic, speed, load and pressure characteristics);
- estimation of the power required to drive the fuel pump;
- full-size fuel system tests;
- debugging the cooperative operation of the control system and the test sample of the fuel system;
- study of the operation of the fuel system as part of a diesel engine with a primary assessment of its power, environmental and economic indicators;
- initial calibration of the diesel control system.

The calculation component of IKTS-MADI includes mathematical models and software packages based on them:

- hydrodynamic calculation of processes occurring in fuel systems;
- the diesel working process.

Working with the calculation complex provides a choice between modeling existing technical solutions and building a model focused on a new design, after making the necessary changes, obtained on the processing the results of experimental studies conducted using the research component (figures 1, 2 and 4).

The entire high-pressure line of the CR is represented by pipelines and cavities with control elements (valves, jets, etc.).

Figure 5 shows the main layout diagrams of the high-pressure line of the CR, which are presented in the IKTS-MADI complex as basic technical solutions: a fuel pump with a single CRI; a CRFS with a serial CRI connection; a CR with a parallel CRI connection.

The calculation of the fuel pump with one EHI is basic (figure 5, a). The design diagram of the high-pressure line (HPL) for this variant is shown in figure 6.

For figure 6: $F_1(t)$ and $W_1(t)$ – direct and reverse wave in the high pressure line in fuel pump rail; pакк – pressure in the main rail; ρ_0 is the density of fuel in the main rail; $F_1(t+\Delta t)$ and $W_1(t+\Delta t)$ – direct and reverse wave in the high pressure line in the CRI rail; $p_{acc\ CRI}$ – pressure in the CRI rail; $\rho_{acc\ CRI}$ – fuel density in CRI rail; $F_2(t)$ and $W_2(t)$ – direct and reverse wave in the channel from a CRI rail to the internal volume in the CRI body.

To calculate the effect of wave processes in a CR with a sequential connection of injectors (figure 5, b), the mathematical model of HPL with a single CRI was supplemented with the necessary algorithms that determine the nature of the connection of nodes and the formation of wave processes.

The fuel system was divided into six nodes. Each node consisted of an CRI and a fuel line connected to it (such a fuel line is the one whose opposite end is located closer to the main rail). Thus, three types of nodes were formed:

- the first type of node consists of an CRI, a connected fuel line that is joined to the main rail, and a fuel line that goes to the next CRI;
- the second type of node (4 pcs.) it consists of an CRI, a connected fuel line that is joined to the rail of the previous CRI, and a fuel line that goes to the next CRI;
- the third node consists of an CRI and a connected fuel line that is joined to the battery of the previous CRI.

The calculation scheme for the last (sixth) CRI has not changed, since it completely repeats the variant with a single CRI (figure 6).

IOP Conf. Series: Materials Science and Engineering 1159 (2021) 012066 doi:10.1088/1757-899X/1159/1/012066

The scheme for CRI from the first to the fifth takes into account the movement of forward and reverse waves in the high-pressure fuel line at the rail of the current injector moving to/from the next CRI.

In accordance with the scheme, the mass balance equation for the CRI rail, expressed in terms of density change, for a single CRI (1) is modified for CRI from 1 to 5 with two supply pipelines (2)

$$\rho^{i}_{acc\,CRI\,n} = \rho^{i-1}_{acc\,CRI\,n} + \frac{\Delta\tau}{6\cdot n_e \cdot V_{acc\,CRI}}\left(\left(\frac{f_{m1}\cdot\rho^{i-1}_{acc\,CRI\,n}}{a\cdot\rho_0}\right)\left(F_{1\,n}(m)\cdot e^{-\frac{k_1\cdot L_{1\,n}}{a}} - W_{1\,n}(m)\right) - \left(\frac{f_{m2}\cdot\rho^{i-1}_{acc\,CRI\,n}}{a\cdot\rho_0}\right)\left(F_{2\,n}(0) - W_{2\,n}(0)\cdot e^{-\frac{k_2\cdot L_2}{a}}\right)\right) \tag{1}$$

where: $\rho^{i}_{acc\,CRI\,n}$ – the current fuel density in the rail of the n-th CRI; $\rho^{i-1}_{acc\,CRI\,n}$ – fuel density in the rail of the n-th CRI at the previous integration step; $\Delta\tau$ – integration step; n_e – engine shaft speed; $V_{acc\,CRI}$ – volume of the CRI battery; f_{m1}, f_{m2} – cross-section area of the connected fuel line and channel to the volume in the CRI body; a – speed of sound in the fuel; ρ_0 – fuel density under initial conditions; $F_{1n}(0)$, $W_{1n}(0)$, $F_{1n}(m)$, $W_{1n}(m)$ – forward and reverse waves at the beginning and end of the pipeline connected to CRI; F2n(0), $F_{2n}(0)$, $W_{2n}(0)$, $F_{2n}(m)$, $W_{2n}(m)$ – forward and reverse wave at the beginning and end of the channel from the CRI rail to the volume in the internal CRI cavity.

$$\rho^{i}_{acc\,CRI\,n} = \rho^{i-1}_{acc\,CRI\,n} + \frac{\Delta\tau}{6\cdot n_e \cdot V_{acc\,CRI}}\left(\left(\frac{f_{m1}\cdot\rho^{i-1}_{acc\,CRI\,n}}{a\cdot\rho_0}\right)\left(F_{1\,n}(m)\cdot e^{-\frac{k_1\cdot L_{1\,n}}{a}} - W_{1\,n}(m)\right) - \left(\frac{f_{m2}\cdot\rho^{i-1}_{acc\,CRI\,n}}{a\cdot\rho_0}\right)\left(F_{2\,n}(0) - W_{2\,n}(0)\cdot e^{-\frac{k_2\cdot L_2}{a}}\right) - \left(\frac{f_{m1}\cdot\rho^{i-1}_{acc\,CRI\,n}}{a\cdot\rho_0}\right)\left(F_{1\,n+1}(0) - W_{1\,n+1}(0)\cdot e^{-\frac{k_1\cdot L_{1\,n+1}}{a}}\right)\right) \tag{2}$$

where $F_{1\,n+1}(0)$; $W_{1\,n+1}(0)$ – forward and reverse waves at the beginning of the pipeline going from CRI № n to № n+1.

When the CRI is connected in parallel (figure 5, b), two local rails are added to the HPL design scheme, to which three CRI with fuel lines are connected and two fuel lines. The first of it connect the main and first local rails, and the second connect the local rails to each other.

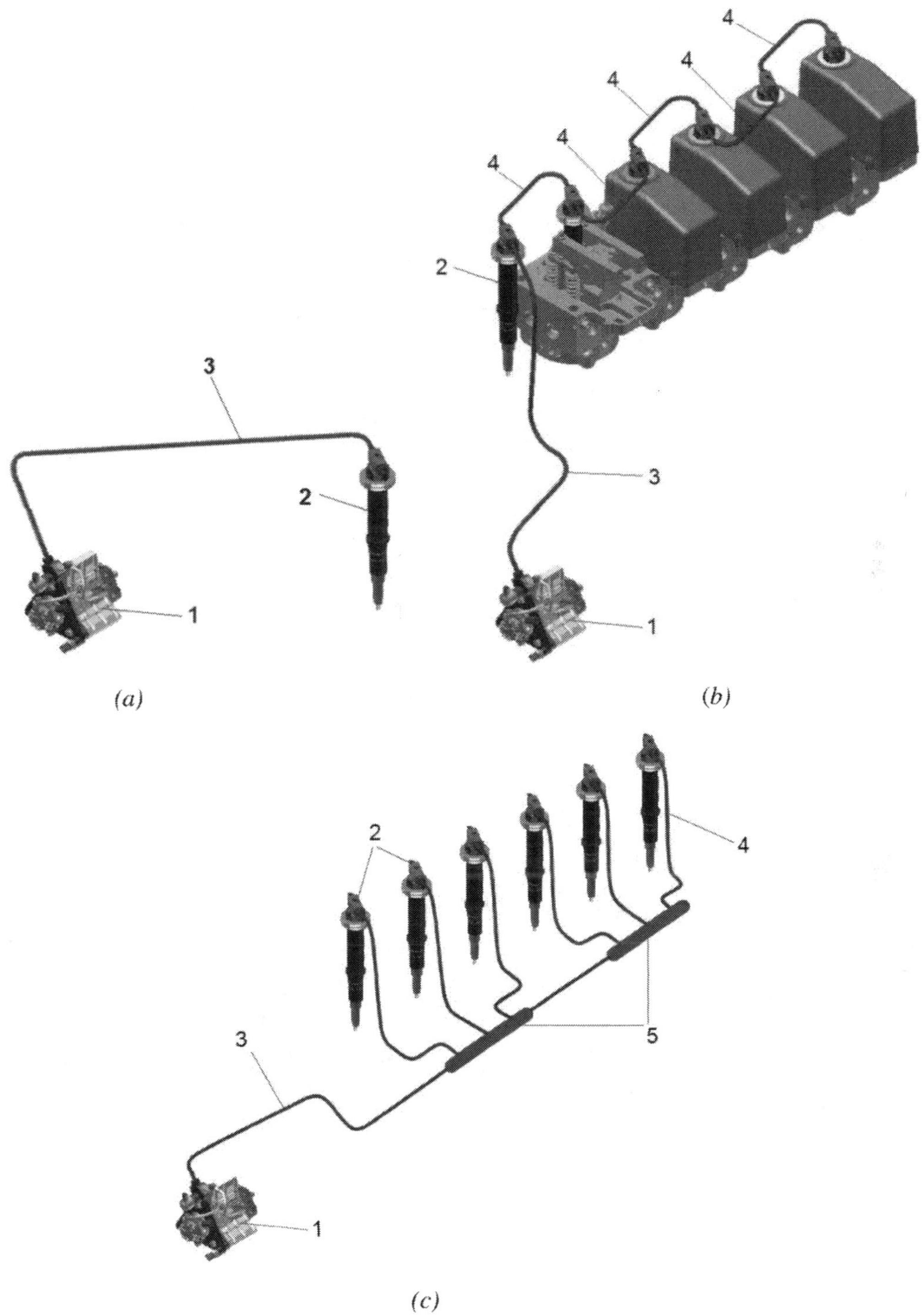

Figure 5. Layout diagrams of the CR high-pressure line used in the design component of the IKTS-MADI complex:
a – fuel pump with one CRI; b – serial connection of CRI; c – parallel connection of CRI;
1 – high – pressure fuel pump; 2 – electro-hydraulic injector; 3, 4-high-pressure fuel lines; 5 – rail.

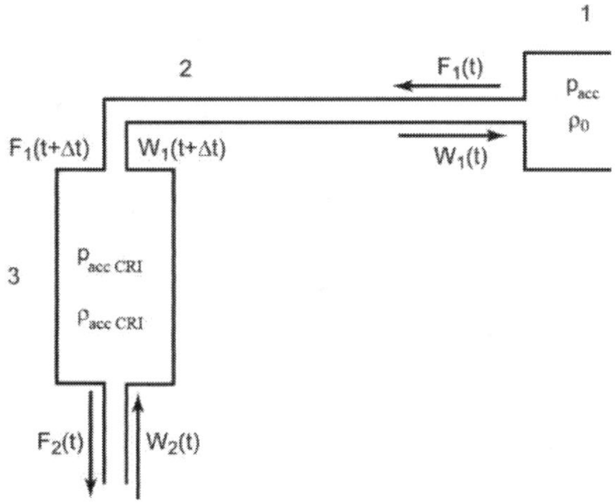

Figure 6. Calculation scheme of a high-pressure line with one CRI:
1 – rail; 2 – high-pressure fuel line; 3 – CRI rail.

The calculation scheme of the first rail is shown in figure 7.

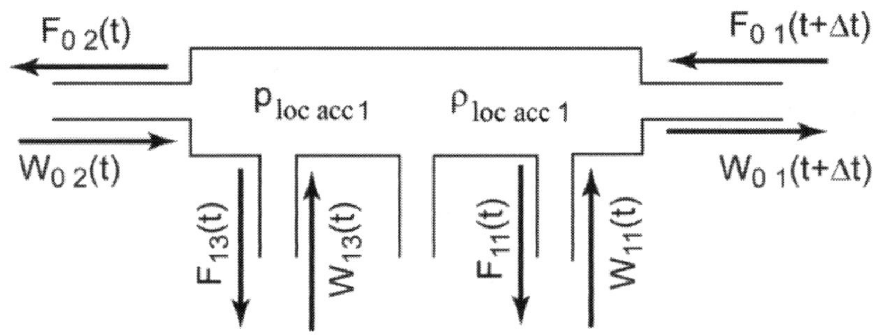

Figure 7. Calculation scheme of the first local rail with parallel CRI connection.

For figure 7: $F_{0\,1}(t+\Delta t)$ и $W_{0\,1}(t+\Delta t)$ – forward and reverse waves in the high - pressure fuel line (main rail– local rail) for the first local rail; $p_{loc\ acc\ 1}$– pressure in the first local rail; $F_{0\,2}(t)$ и $W_{0\,2}(t)$ – forward and reverse waves in the high-pressure fuel line for the CRI rail connecting the first and second local rails; $\rho_{loc\ acc\ 1}$ – fuel density in the first local rail; F1 1(t), $F_{1\,1}(t)$, $W_{1\,1}(t)$, $F_{1\,2}(t)$, $W_{1\,2}(t)$ (not shown), $F_{1\,3}(t)$, $W_{1\,3}(t)$ – forward and reverse waves in the fuel line connecting the first local rail and the CRI rail from Nos. 1 to 3.

Based on the presented scheme (figure 7), the equation for the fuel density changing in the rail has the form:

$$\rho^{i}_{loc\ acc\,1}=\rho^{i-1}_{loc\ acc\,1}+\frac{\Delta\tau}{6\cdot n_e\cdot V_{loc\ acc}}\left(\left(\frac{f_{m0}\cdot\rho^{i-1}_{loc\ acc\,1}}{a\cdot\rho_0}\right)\left(F_{01}(m)\cdot e^{-\frac{k_0\cdot L_{01}}{a}}-W_{01}(m)\right)-\right.$$
$$-\left(\frac{f_{m0}\cdot\rho^{i-1}_{loc\ acc\,1}}{a\cdot\rho_0}\right)\left(F_{02}(0)-W_{02}(0)\cdot e^{-\frac{k_0\cdot L_{02}}{a}}\right)-\left(\frac{f_{m1}\cdot\rho^{i-1}_{loc\ acc\,1}}{a\cdot\rho_0}\right)\left(F_{11}(0)-W_{11}(0)\cdot e^{-\frac{k_1\cdot L_{11}}{a}}\right)- \qquad (3)$$
$$\left.-\left(\frac{f_{m2}\cdot\rho^{i-1}_{loc\ acc\,1}}{a\cdot\rho_0}\right)\left(F_{12}(0)-W_{12}(0)\cdot e^{-\frac{k_1\cdot L_{12}}{a}}\right)-\left(\frac{f_{m3}\cdot\rho^{i-1}_{loc\ acc\,1}}{a\cdot\rho_0}\right)\left(F_{13}(0)-W_{13}(0)\cdot e^{-\frac{k_1\cdot L_{13}}{a}}\right)\right)$$

For the second local rail, the calculation scheme is shown in figure 8.

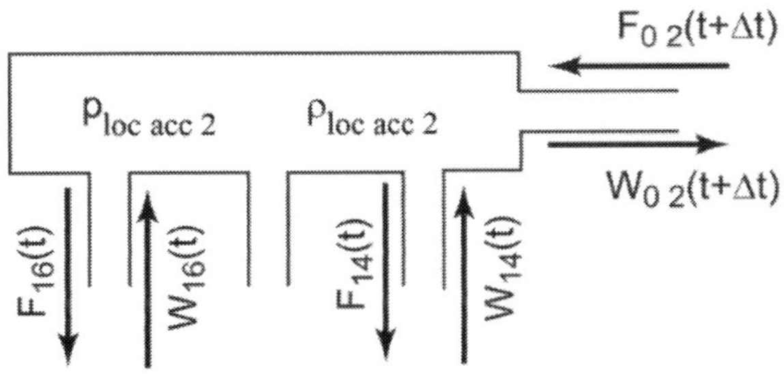

Figure 8. The Calculation scheme of a second local rail in parallel connection of CRI.

For figure 8: $F_{0\,2}(t+\Delta t)$ и $W_{0\,2}(t+\Delta t)$ – forward and reverse waves in the high-pressure fuel line (second local rail – first local rail); $p_{loc\ acc\ 2}$ – pressure in the second local rail; $\rho_{loc\ acc\ 2}$ – fuel density in the second local rail; $F_{1\,4}(t)$, $W_{1\,4}(t)$, $F_{1\,5}(t)$ (not shown), $W_{1\,5}(t)$ (not shown), $F_{1\,6}(t)$, $W_{1\,6}(t)$ – direct and reverse wave in the fuel line connecting the second local rail and the CRI rail from Nos 4 to 6.

The equation for the fuel density changing has the form:

$$\rho^{i}_{loc\,acc\,2} = \rho^{i-1}_{loc\,acc\,2} + \frac{\Delta\tau}{6\cdot n_e \cdot V_{loc\,acc}}\left(\left(\frac{f_{m1}\cdot\rho^{i-1}_{loc\,acc\,2}}{a\cdot\rho_0}\right)\left(F_{0\,2}(m)\cdot e^{-\frac{k_0\cdot L_{0\,2}}{a}} - W_{0\,2}(m)\right) - \right.$$
$$-\left(\frac{f_{m4}\cdot\rho^{i-1}_{loc\,acc\,2}}{a\cdot\rho_0}\right)\left(F_{1\,4}(0) - W_{1\,4}(0)\cdot e^{-\frac{k_1\cdot L_{1\,4}}{a}}\right) - \left(\frac{f_{m5}\cdot\rho^{i-1}_{loc\,acc\,2}}{a\cdot\rho_0}\right)\left(F_{1\,5}(0) - W_{1\,5}(0)\cdot e^{-\frac{k_1\cdot L_{1\,5}}{a}}\right) \quad (4)$$
$$\left. -\left(\frac{f_{m6}\cdot\rho^{i-1}_{loc\,acc\,2}}{a\cdot\rho_0}\right)\left(F_{1\,6}(0) - W_{1\,6}(0)\cdot e^{-\frac{k_1\cdot L_{1\,6}}{a}}\right)\right)$$

3. Conclusions

1. An integrated computational and experimental complex IKTS-MADI has been created. The structure of the complex allows to choose a possible technical solution for the fuel system, its experimental verification and evaluation of the it using efficiency to improve the performance of the diesel working process.

2. A method for calculating the high-pressure line of the common rail fuel system has been developed. The basic technical solutions in the IKTS-MADI complex are: fuel pump with one CRI; CR with a serial CRI connection; CR with a parallel CRI connection.

References

[1] Sinyavski V V, Alekseev I V, Ivanov I Ye, Bogdanov S N and Trofimenko Yu V 2017 Physical simulation of high- and medium-speed engines powered by natural gas *Pollution Research*. **36(3)** 684–690

[2] Markov V A, Kamaltdinov V G and Savastenko A A 2019 Use of rapeseed oil and water emulsion as diesel fuel *2019 International Science and Technology Conference "EastConf"*. 8725348

[3] Pflaum S, Wloka J and Wachtmeister G 2010 Emission reduction potential of 3000 bar common rail injection and development trends *CIMAC Congress*. № 195 1-12

[4] Wloka J, Potsch C and Wachtmeister G 2011 Injection spray visualization for 3000 bar diesel injection *24th Conference of the Institute for Liquid Atomization and Spray Systems*. 1–10

[5] Shatrov M G, Dunin A U and Nguyen T Q 2020 Effect of pressure boost on the diesel performance and gaseous emission with fuel spray pressure up to 3000 bar *IOP Conference*

Series: Materials Science and Engineering **832** 012083 DOI:10.1088/1757-899X/832/1/012083

[6] Shatrov M G, Malchuk V I and Dunin A Y 2020 A laboratory investigation into the fuel atomization process in a diesel engine for different configurations of the injector nozzles and flow conditions *Fluid Dynamics & Materials Processing.* **16(4)** 747–760; DOI:10.32604/fdmp.2020.08991

[7] Fink C, Drescher M, Rabe J R and Harndorf H 2013 Hydraulic measures to improve common-rail injection system performance - Impact of injection rate shaping on emissions of a medium speed diesel engine *CIMAC Congress 2013.* **248** 1–10

[8] Catania A E, Ferrari A, Manno M and Spessa E 2008 Experimental investigation of dynamics effects on multiple-injection Common Rail system performance *Journal of Engineering for Gas Turbines and Power, ASME.* **130** 032806

[9] Savastenko E A, Savastenko A A, Gorbachevskiy E V and Markov V A 2020 The use of reusable fuel injection with the aim of improving the economic and environmental performance of the diesel engine *IOP Conference Series: Materials Science and Engineering.* **832** 012083; DOI:10.1088/1757-899X/832/1/012083

IOP Conf. Series: Materials Science and Engineering 1159 (2021) 012067 doi:10.1088/1757-899X/1159/1/012067

Reduction and equalization of thermal loading of contact pairs of the friction unit

P A Polyakov[1], A E Litvinov[1], E A Polyakova[1], E S Fedotov[1], A A Golikov[1], A E Zadayanchuk[2]

[1] Kuban State Technological University, Krasnodar, 2a, Moskovskaya st., Krasnodar, 350020, Russia
[2] Kuban State University, Krasnodar, 149 Stavropolskaya, st., Krasnodar, 350040, Russia

E-mail: artstyleone@mail.ru

Abstract. Controlling the processes of interaction of the working pairs of the friction unit is possible when solving the problem of determining the parameters of influence. As the main criterion of the process, it is necessary to highlight the temperature on the surface of the friction unit. This paper presents a method for modeling the magnitude and distribution of temperature over the surface of a metal friction element depending on the nanogeometry of the surface profile of contact pairs of friction units. The ability to control the process of interaction of contacting pairs of the friction unit will further predict the life of the part and avoid such phenomena as burns and the formation of foci of microcracks.

1. Introduction

In the works of researchers [2,3,4], the effect of temperature on the tribocontact of various friction units is indicated. In the literary source [7], modified friction units are proposed based on new polymer materials used in braking devices of various types. The paper [6] proposes a method for controlling the energy-loading of the friction unit using the example of a drum-shoe type brake. The work [8] investigated the developed thermal model of the brake disc with the use of surface layers on the working and matt surfaces of the brake disc. The literary source [1] describes the interaction of the electric fields of the surfaces of the contact pads of the friction units. From the analysis of works [9, 10], devoted to the wear of the working surfaces of friction units, it can be seen that the specific contact pressures of the working surfaces of friction pairs are unevenly distributed over the area of the metal element. These studies allow us to say that the interaction of the working pairs of the friction unit is a controlled process that depends on many factors and can and should be influenced. In [5], the authors proposed the results of modeling the contact pressure on an elementary area of the friction unit depending on the nanogeometry of the surface. This paper provides further research on the effect of surface nanogeometry on other parameters such as the temperature of contacting pairs.

In this regard, to control the processes of friction and wear in the friction unit, it is necessary to study the effect of temperature and temperature gradient on the process of destruction of the material of the parts of the friction unit. The criterion for the thermal destruction of the material is the limiting value of the temperature T * for the materials of the friction lining and the metal element when cracking occurs. Consequently, the most important criterion is the maximum temperature reached in the material T *.

Content from this work may be used under the terms of the Creative Commons Attribution 3.0 licence. Any further distribution of this work must maintain attribution to the author(s) and the title of the work, journal citation and DOI.
Published under licence by IOP Publishing Ltd

IOP Conf. Series: Materials Science and Engineering 1159 (2021) 012067 doi:10.1088/1757-899X/1159/1/012067

Thermal destruction in the development of braking devices at the design stage can be controlled using the nanogeometry of the surface of the contact area of the rubbing elements of the friction unit. It is important to minimize the maximum temperature of the contact area of the friction unit during braking. In the process of braking, an external heat source arises on the contact area of the rubbing elements, caused by external friction of the non-metallic material, the working surface of the metal element. As a result of conductive heat exchange, the temperature of the non-metallic and metallic friction elements increases. The control variables are taken to be the parameters of the nano-geometry of the surface of the metal element. For a mathematical model to reduce the thermal state of the rubbing elements of the friction unit, using the example of a disc-shoe brake, we take the differential equation of thermal conductivity in a polar coordinate system:

$$\frac{\partial^2 T}{\partial r^2} + \frac{1}{r}\frac{\partial T}{\partial r} + \frac{1}{r^2}\frac{\partial^2 T}{\partial \theta^2} = 0. \qquad (1)$$

The metal friction element is assigned to the polar coordinate system r, θ, and the origin of coordinates is chosen at the center of concentric circles L0 and L with radii R0 and R, respectively. We simulate the brake disc as a homogeneous isotropic body.

The sought function H (θ), which describes the nanogeometry of the surface of an elementary area of the friction lining of the brake device, can be represented as a segment of the Fourier series:

$$H(\theta) = \sum_{k=0}^{\infty}\left(a_k^0 \cos k\theta + b_k^0 \sin k\theta\right); \qquad (2)$$

where a_k ^ 0, b_k ^ 0 are control parameters.

The function describing nano-geometry is assumed to be given. Consequently, the problem of optimizing the thermal state is reduced to determining the control coefficients of the Fourier series.

The first step in solving the optimization problem is the need to determine the thermal state of the elementary surface of the friction unit, and then to find the temperature on the contact surface.

t = T - T0 - is the temperature difference.

Based on this condition, we write down the heat conduction equation and its boundary conditions:

$$\Delta t = 0,\ A_{t1}\lambda\frac{\partial t}{\partial n} - A_{t2}\alpha_1 t = -Q_b(\theta),\ (r = \rho);$$
$$\lambda\frac{\partial t}{\partial r} - \alpha_1 t = 0,\ (r = R_0); \qquad (2)$$

where Δ is the Laplace operator;

At1 — heat absorption surface, m2;

At1 - cooling surface, m2;

Qb (θ) is the amount of heat supplied within the boundaries of the contact area, J;

λ — coefficient of thermal conductivity of the disc material, W / (m K);

α1 is the heat transfer coefficient between the outer surface of the brake disc and the environment (inside the perimeter of the contact pad), W / (m2K).

For the distribution of excess temperature on the surface of the brake disc, we use an expansion in a small parameter, in which we neglect the terms of ε to a power higher than the first:

$$t_{\text{пов}} = t_{|r=\rho} = t^{(0)}{}_{|r=R} + \varepsilon\left[\frac{\partial t^{(0)}}{\partial r}H(\theta) + t^{(1)}\right]_{|r=R}, \qquad (3)$$

where t (0), t (1) are the temperatures of the brake disc in zero and first approximations.

The temperature of the brake disc in zero and first approximations will be as follows:

$$t^{(0)}{}_{|r=R} = C_{10} + C_{20} lnR + \sum_{k=1}^{\infty}\left(C_{10}^{(k)}R^k + C_{20}^{(k)}R^{-k}\right)\cos k\theta$$
$$+ \sum_{k=1}^{\infty}\left(A_{10}^{(k)}R^k + A_{20}^{(k)}R^{-k}\right)\sin k\theta;$$

$$t^{(1)}|_{r=R} = C_{11} + C_{21}lnR + \sum_{k=1}^{\infty}\left(C_{11}^{(k)}R^k + C_{21}^{(k)}R^{-k}\right)\cos k\theta +$$
$$+\sum_{k=1}^{\infty}\left(A_{11}^{(k)}R^k + A_{21}^{(k)}R^{-k}\right)\sin k\theta ; \qquad (4)$$

where C_{10}, C_{20}, $C^{(k)}{}_{10}$, $C^{(k)}{}_{20}$, $A^{(k)}{}_{10}$, $A^{(k)}{}_{20}$, C_{11}, C_{21}, $C^{(k)}{}_{11}$, $C^{(k)}{}_{21}$, $A^{(k)}{}_{11}$, $A^{(k)}{}_{21}$- permanent;
(∂t ^ ((0))) / ∂r - temperature gradient along the radius of the brake disc, K / m.
The gradient can also be expanded in a Fourier series:

$$\frac{\partial t^{(0)}}{\partial r} = C_{20}\frac{1}{R} + \sum_{k=1}^{\infty} k\left(C_{10}^{(k)}R^{k-1} + C_{20}^{(k)}R^{-k-1}\right)\cos k\theta +$$
$$+\sum_{k=1}^{\infty} k\left(A_{10}^{(k)}R^{k-1} + A_{20}^{(k)}R^{-k-1}\right)\sin k\theta \qquad (5)$$

To determine the constants C10, C20, C (k) 10, C (k) 20, A (k) 10, A (k) 20, it is necessary to use the optimization boundary conditions in the zero approximation:

$$C_{10} = \frac{\Delta_1}{\Delta};\ C_{20} = \frac{\Delta_2}{\Delta};\ \Delta = -A_{t2}\alpha_1\left(\frac{\lambda}{R_0} + \alpha R_0\right) - \alpha_2\left(\frac{A_{t1}\lambda}{R} - A_{t2}\alpha_1 \ln R\right);$$
$$\Delta_1 = -\sigma f v \alpha_0^0\left(\frac{\lambda}{R_0} + \alpha_2 R_0\right);\ \Delta_2 = -\alpha_2\sigma f v \alpha_0^0;\ C_{10}^{(k)} = \frac{\Delta_{1k}}{\Delta_k};\ C_{20}^{(k)} = \frac{\Delta_{2k}}{\Delta_k};$$
$$\Delta_k = \left(A_{t1}\lambda k R^{k-1} - A_{t2}\alpha_1 R^k\right)\left(\alpha_2 R^{-k} - \lambda k {R_0}^{-k-1}\right)$$
$$+\left(\lambda k {R_0}^{k-1} + \alpha_2 {R_0}^k\right)\left(A_{t2}\alpha_1 R^{-k} + A_{t1}\lambda k R^{-k-1}\right);$$
$$\Delta_{1k} = -\sigma f v \alpha_k^0\left(\alpha_2 R_0^{-k} - \lambda k {R_0}^{-k-1}\right);\ \Delta_{2k} = \sigma f v \alpha_k^0\left(\lambda k {R_0}^{k-1} + \alpha_2 {R_0}^k\right);$$
$$A_{10}^{(k)} = \frac{\Delta_{1k}^0}{\Delta_k};\ A_{20}^{(k)} = \frac{\Delta_{2k}^0}{\Delta_k};\ \Delta_{2k}^0 = \sigma f v \alpha_k^0\left(\lambda k {R_0}^{k-1} + \alpha_2 R_0^k\right);$$
$$\Delta_{1k}^0 = -\sigma f v \beta_k^0\left(\alpha_2 R_0^{-k} - \lambda k {R_0}^{-k-1}\right);\quad p^{(0)}(\theta) = \alpha_0^0 + \sum_{k=1}^{\infty}\left(\alpha_k^0 \cos k\theta + \beta_k^0 \sin k\theta\right);$$
$$C_{11} = \frac{\Delta_1^1}{\Delta};\ C_{21} = \frac{\Delta_2^1}{\Delta};\ \Delta_1^1 = -\sigma f v A_0\left(\frac{\lambda}{R_0} + \alpha_2 R_0\right);\ \Delta_2^1 = -\alpha_2\sigma f v A_0;\ C_{11}^{(k)} = \frac{{\Delta^1}_{1k}}{\Delta_k};\ C_{21}^{(k)} = \frac{{\Delta^1}_{2k}}{\Delta_k};$$
$${\Delta^1}_{1k} = -\sigma f v A_k^0\left(\alpha_2 R_0^{-k} - \lambda k {R_0}^{-k-1}\right);\ {\Delta^1}_{2k} = \sigma f v A_k^0\left(\lambda k {R_0}^{k-1} + \alpha_2 {R_0}^k\right);$$
$$A_{11}^{(k)} = \frac{\Delta_{1k}^{01}}{\Delta_k};\ A_{20}^{(k)} = \frac{\Delta_{2k}^{01}}{\Delta_k};\ \Delta_{2k}^{01} = \sigma f v B_k^0\left(\lambda k {R_0}^{k-1} + \alpha_2 R_0^k\right);$$
$$\Delta_{1k}^{01} = -\sigma f v B_k^0\left(\alpha_2 R_0^{-k} - \lambda k {R_0}^{-k-1}\right); \qquad (6)$$

where α is the heat transfer coefficient between the outer surface of the friction pad and the environment, W / (m2K);
α2 is the heat transfer coefficient between the outer surface of the brake disc and the environment (outside the perimeter of the contact area), W / (m2K);
σ is the heat flow distribution coefficient;
f is the coefficient of friction;
v - average speed during the braking period, m / s;
p _ ^ ((0)) (θ) - specific pressure on the friction contact area of the friction unit, MPa;
α_0 ^ 0, α_k ^ 0 〖, β〗 _k ^ 0, A_0 ^ A_k ^ B_k ^ are the coefficients of the Fourier series.

Let us write down the heat conduction equation (2) and its boundary conditions in the form of a Fourier series:

$$\left[A_{t2}\alpha_1 \frac{\partial t^{(0)}}{\partial r} - \lambda A_{t1} \frac{\partial^2 t^{(0)}}{\partial r^2}\right] \mathrm{H}(\theta) - Q_b^{(1)}(\theta) = A_0 + \sum_{k=1}^{\infty}\left(A_k^0 \cos k\theta + B_k^0 \sin k\theta\right), \quad (r = R); \quad (7)$$

The amount of heat from the source is determined based on the dependence:

$$Q_b^{(1)}(\theta) = \sigma f v p^{(1)}(\theta). \quad (8)$$

For the function of the surface temperature of the brake disc, we find the maximum value:

$$t_{\text{пов max}} = t^{(0)}(\theta_1) + \varepsilon\left[\frac{\partial t^{(o)}(\theta_1)}{\partial r} \mathrm{H}(\theta_1) + t^{(1)}(\theta_1)\right]_{|r=R}. \quad (9)$$

The value of θ1 is determined from the following equation:

$$\frac{dt_{\text{пов}}}{d\theta} = 0. \quad (10)$$

The function H (θ) must be chosen so as to ensure the minimization of tпов max. Since the temperature on the surface tпов is an indicator of the quality of control, and also equation (9) linearly depends on the desired coefficients (2), the posed optimization problem is reduced to a linear programming problem. The main condition is the limitation:

$$t_{\text{пов max}} \leq [t]. \quad (11)$$

where [t] is the permissible excess temperature, which is a characteristic of the material, K.
In addition to constraint (11), the following condition must be satisfied:

$$\sum_{i=1}^{n} \mathrm{H}(\theta_i) = nR_a, \ a_0^0 \geq 0, a_k^0 \geq 0, b_k^0 \geq 0. \quad (12)$$

where n is the number of sites;
Ra is the arithmetic mean deviation of the nanogeometric profile of the contact area of the friction unit.

For solving the optimization problem, the simplex algorithm is the most effective method, because it leads to an optimal solution in a finite number of iterations.

In the expansion (2), the functions H (θ) were limited to k = 5 terms. The values of the control parameters (coefficients a_k ^ 0 and b_k ^ 0) are found depending on the thermophysical and geometric parameters of the contact area of the friction unit. The results of preliminary computer simulation are presented in tables 1 and 2.

Table 1. Calculation results (subject to v = 0.2 m / s)

a_0^0	a_1^0	a_2^0	a_3^0	a_4^0	a_5^0
0,642	0,598	0,522	0,355	0,248	0,218
b_1^0	b_2^0	b_3^0	b_4^0	b_5^0	
0,627	0,563	0,312	0,288	0,175	

Table 2. Calculation results (subject to v = 1.0 m / s)

a_0^0	a_1^0	a_2^0	a_3^0	a_4^0	a_5^0
0,788	0,734	0,622	0,545	0,423	0,312
b_1^0	b_2^0	b_3^0	b_4^0	b_5^0	
0,78	0,721	0,544	0,41	0,233	

In the optimization problem, the parameters of the brake disc diameter were determined. Modeling has shown that solving the optimization problem, i.e. the temperature on the surface of the elementary area of the contact surfaces of the friction unit depends on its nanogeometry of the surfaces of the metallic H (θ) and non-metallic H1 (θ) friction elements. The final stage after determining the nanogeometry of the surface of the rubbing elements of the friction unit, it is necessary to find the condition for the uniform distribution of the surface temperature. Thus, when choosing the nanogeometry of the friction surface, it is possible to reduce the temperature level. According to dependence (3), it can be concluded that the temperature on the surface linearly depends on the coefficients a_k ^ 0 and b_k ^ 0 of the Fourier series for the function H (θ) and on the coefficients a_k ^ 1 and b_k ^ 1 of the Fourier series for the function H1 (θ). The temperature function on the surface of the elementary contact pad will look like this:

$$t_{пов}(\theta, \tau) = F(\theta, \tau, a_0^0, a_k^0, b_k^0, a_0^1, a_k^1 b_k^1), (i = 1,2 \dots m). \quad (13)$$

where τ is the braking time, s.

The optimization task is to determine the values of unknown parameters that will provide the following condition:

$$U = \sum_{i=1}^{M} \left[F(\theta, \tau, a_0^0, a_k^0, b_k^0, a_0^1, a_k^1 b_k^1) - \bar{t}\right]^2 \to \min; \quad (14)$$

where t ̄is the optimal temperature value on the friction surface, K.

2. Materials and methods

The principle of least squares states that the most likely values of the parameters will be those at which the sum of the squares of the difference between the surface and optimal temperatures will be the smallest. The final solution to the optimization problem is presented in tables 3 and 4

Table 3. Calculation results (subject to v = 0.2 m / s)

a_0^0	a_1^0	a_2^0	a_3^0	a_4^0	a_5^0
0,685	0,601	0,584	0,331	0,206	0,185
b_1^0	b_2^0	b_3^0	b_4^0	b_5^0	
0,422	0,344	0,129	0,065	-0,173	

Table 4. Calculation results (subject to v = 1.0 m / s)

a_0^0	a_1^0	a_2^0	a_3^0	a_4^0	a_5^0
0,702	0,655	0,598	0,433	0,389	0,241
b_1^0	b_2^0	b_3^0	b_4^0	b_5^0	
0,549	0,533	0,322	0,211	0,095	

Figure 1 shows a graphical representation of the temperature distribution on the surface of an elementary contact pad of the friction unit at different sliding speeds. The solid line is shown for a smooth contour, and the dashed line is for the law of roughness distribution, described by the function H (θ). The modeling results clearly show that when choosing the nanogeometry of the surface of an elementary contact area, the surface temperature decreases and is distributed over the contact surface, in contrast to a smooth contour. The exception is the boundaries of the elementary area, where, as a result of the redistribution, the surface temperature of the elementary area increases in comparison with the smooth contour of the brake disc. The figure may indicate that the optimization problem on the example of an elementary area of the contacting surfaces of the friction unit of the brake disc performs its functions and minimizes and evens out the temperature over the surface.

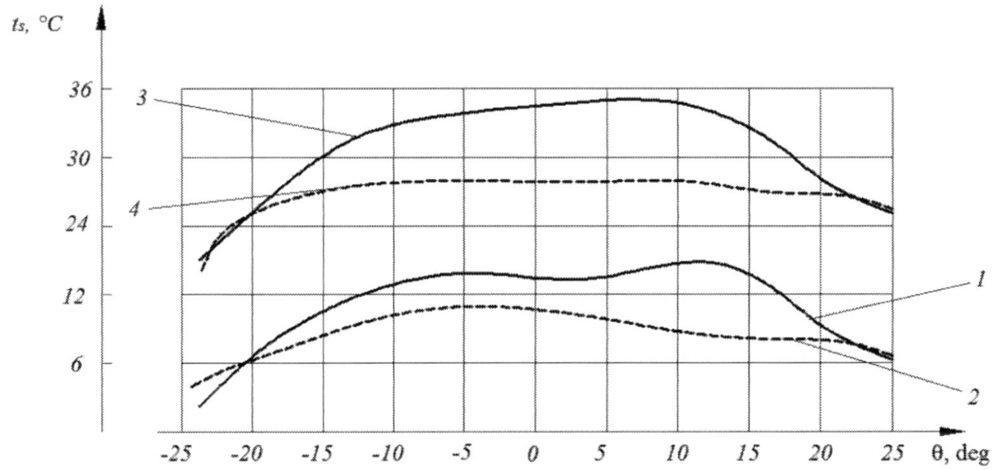

Figure 1. Dependences of the surface temperature of the contact pad of the friction unit on the value of the polar angle: curves 1 and 2 for a sliding speed v = 0.2 m / s, curves 3 and 4 for a sliding speed v = 1.0 m / s.

3. Conclusion

The optimization problem has shown that the surface temperature of an elementary area depends both on the nanogeometry of the working surfaces of the friction unit and on the contact pressure. The obtained control parameters make it possible to simulate the temperature distribution on the surfaces of the friction unit. Close to uniform temperature distribution on the surface allowed to reduce its value and distribute over the entire surface of the metal friction element. This fact is explained by the fact that with this choice of the nano-geometry of the profile, the entire surface of the contact area interacts and there are no foci of huge energy release on the surface.

The developed optimization problem makes it possible to design a friction unit with a predicted thermal load, which will further allow to control the process of minimization and distribution of the friction unit over the surface of the working platforms.

References

[1] Volchenko N A 2019 Pulse-contact frictional interaction of microprotrusions of friction pairs of brake devices *IOP Conf. Series: Materials Science and Engineering* 560 p

[2] Volchenko N A 2018 Electrochemistry with double electrical layers in frictional interaction metal-polymer tribolink *IOP Conference Series: Materials Science and Engineering* 327 p

[3] Kolesnikov I V 2014 Triboelectric phenomena at metal-polymer friction contact and their dependence on temperature *Engineering journal of Don* **4-1(31)** p 70

[4] Polyakov P A 2020 Design of surface profile of pairs of friction unit *IOP Conf. Series: Materials Science and Engineering* Vol 843

[5] Litvinov A E 2020 Drum brake energy control *Mechatronics, automation and robotics* **5** 74–78

[6] Kindrachuk M 2019 Polymeres with enhanced energy capacity modified by semiconductor materials *Functional Materials* **26-3** 629–634

[7] Polyakov P A 2012 Development of a thermal model of a brake disc frictional *unit Vestnik Irkutsk State Technical University* 24 **1(150)** 64–76

[8] Kolesnikov I V 2019 Analysis of the effect of the hardened surface layer properties of the tribocontact parts on their wear and stress-strain state Advanced Materials. *Springer Proceedings in Physics* Vol 224 pp 637–642

[9] Kolesnikov I V 2019 The Nonstationary Contact Problem of Thermoelasticity for a Ring in the Presence of a Reinforcing Layer on the Friction Surface *Journal of Friction and Wear* **4** 447–455

IOP Conf. Series: Materials Science and Engineering 1159 (2021) 012068 doi:10.1088/1757-899X/1159/1/012068

Evaluation of SVM modulation methods for indirect two-level three-phase frequency converter and direct frequency converter

Le Van Tung[1,2], Ngo Van An[3], Le Dinh Hieu[4,5], Pham Thanh Long[3]

[1] St. Petersburg Mining University, 2, 21 Line of Vasilyevsky Island, St. Petersburg, 199106, The Russian Federation
[2]Quang Ninh University of Industry, 18 street, Quang Ninh city, 208830, Viet Nam
[3]Thai Nguyen University of Technology, 666, 3/2 street, Tich Luong ward, Thai Nguyen city, 251750, Viet Nam
[4]National Research Technological University "MISiS", 4, Leninsky prospect, Moscow, 119049, Russian.
[5]Hue Instrial College (HueIC), 70 Nguyen Hue street, Vinh Ninh Ward, Hue City, Thua Thien Hue province, 49100, Vietnam

E-mail: vanankten@tnut.edu.vn

Abstract. In current practice, indirectly structured frequency converters are popular because of the simple switch and control circuit diagram, and pulse width modulation (PWM) can be used to control the semiconductor switches. The PWM method has the advantage of a simple control circuit but has large switching losses and generates a common-mode voltage that causes motor damage. The development of the space vector modulation (SVM) method takes advantage of the PWM method, reducing the switching loss of the switches and reducing the harmonic at the output of the inverter. With direct converters, there are many types depending on the circuit structure and semiconductor key structure, including matrix frequency converters which are widely studied. The classical modulation method according to the duty-cycle matrix approach for matrix converter has disadvantages such as complex computational algorithms, which require many trigonometric calculations. Based on the PWM and SVM modulation algorithms for indirect converters, a space vector modulation method for matrix converters will be developed. Compared with the indirect converter, the SVM modulation method for the matrix converter will be based on two separate modulation principles: control the input current vector and control the output voltage vector. The relationship between current and voltage at the converter input is closely related to the voltage and current at the output. The relationship between current and voltage at the converter input is closely related to the voltage and current at the output. Make sure at the input of the converter, the power factor cosφ = 1 and can be adjusted, it has the quality of sine input and outputs current and the exchange power in two directions.

1. Introduction

Technology to control frequency converters is growing, so the study and improving the structure of the converter, the control algorithms are important tasks in the manufacture of frequency converters. The study of modulation techniques is an important step to ensure the quality of input and output

Content from this work may be used under the terms of the Creative Commons Attribution 3.0 licence. Any further distribution of this work must maintain attribution to the author(s) and the title of the work, journal citation and DOI.
Published under licence by IOP Publishing Ltd

parameters and compliance with matching devices. Indirect frequency converter with a voltage source inverter (VSI), as shown in Figure 1, the rectifier circuits and inverter isolated transition sided circuit consisting of 2 elements L, C. The rectifier circuit uses unmanaged diodes; the inverter circuit uses managed keys IGBT modules, so management is easily done using the PWM and SVM methods [1]. The disadvantage of the indirect frequency converter is the presence of a diode s in the rectifier circuit. Because of, them harmonics appear in the network; it is not possible to regulate the input voltage of the inverter circuit; when working with a regenerative braking load, the energy will not return to the grid, but dissipated at the discharge resistor.

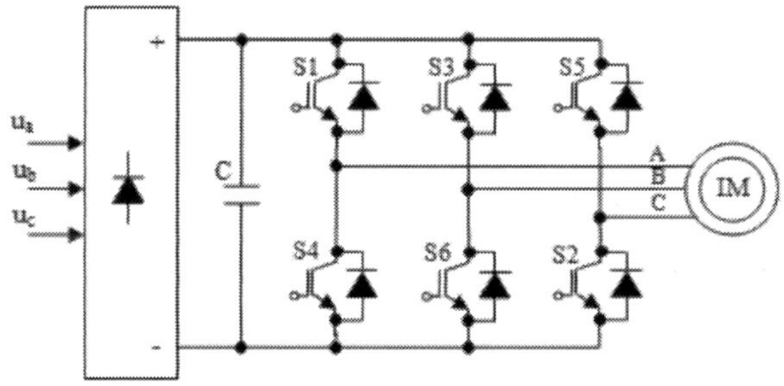

Figure.1. Diagram of a two-level converter

Direct frequency converter (AC-AC) is a AC direct converter in a variable frequency and variable voltage. Matrix converter (MC) is a direct converter based on semiconductor modules with with bidirectional switch [1, 2]. The block diagram of the electric drive with a matrix frequency converter is shown in fig. 2.

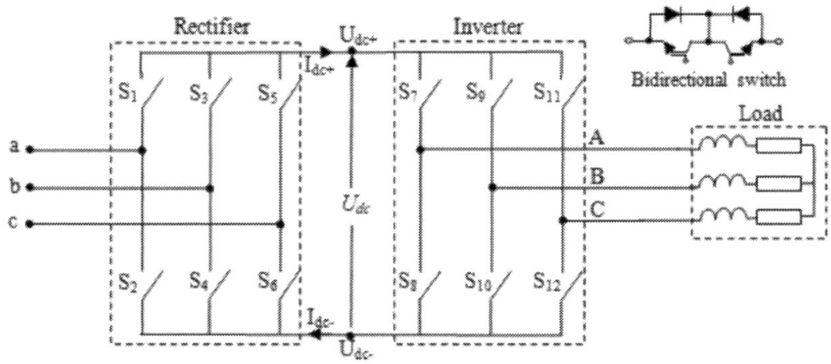

Figure.2. Diagram of a two-level converter

The main advantages of matrix converter are the possibility of bilateral exchange of energy, the regulation of the form of the input current, ensuring sinusoidal input current and voltage regulation, power factor cosφ=1 [3]. The main problem of the use of MC is more complex than the indirect converter structures, modulation control systems, as well as more complex methods of modification of the modulation system [4].Currently, there are several types of modulation control systems for matrix converters: Venturini - Alesina method; scalar modulation method; space vector modulation. However, the modulation system Venturini - Alesina has disadvantages such as the algorithm requires complex calculations; the algorithm requires several trigonometric transformations; the algorithm requires continuous measurement of the input voltage with high accuracy [5]. The Scalar modulation system is simpler, but it has the main drawback, which will consist in the high dependency of control quality on the accuracy of computational and measuring operations [6, 7]. In the article, the strategy of controlling matrix transformation via DC virtual link is given. This DC link is not physically present, but the switches are divided into the virtual rectifier and virtual inverter in figure 2. The control of these two converters circuits has provided a platform for analyzing and derive several extends PWM strategies. On the basis of the PWM and SVM methods for indirect converters, a space vector modulation method for matrix transformations will be developed by separate control for virtual rectifier circuits and virtual inverters [8, 9, 10].

2. Methods

2.1. SVM modulation principle in the indirect two-level three-phase frequency converter

Method SVM inherits the principle of the method the pulse-width modulation PWM and applies the theory of spatial vectors. SVM is a method of bundling closed states of semiconductor switches in which the value of the vector modulated output voltage close to the value of the vector of the desired output voltage [11, 12, 13].

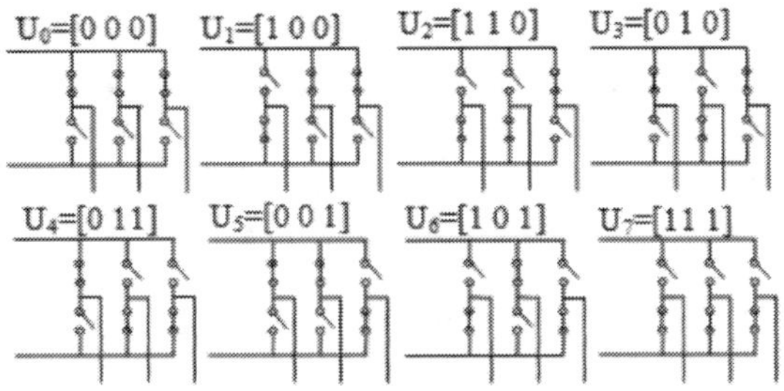

Figure.3. Status of keys in the inverter circuit

The inverter is considered as a single unit with 2^3=8 distinct closed States from 0 to 7 (Figure 3), according to each closed switch state will receive 1 unchanging voltage vector (both direction and magnitude) listed in table 1.

Table 1. Voltage values for different closed states of keys

Vector	Key			Phase voltage			Wire voltage		
	S_1	S_3	S_5	U_{An}	U_{Bn}	U_{Cn}	U_{AB}	U_{BC}	U_{CA}
U_0	0	0	0	0	0	0	0	0	0
U_1	1	0	0	2/3	-1/3	-1/3	1	1	1
U_2	1	1	0	1/3	1/3	-2/3	0	1	-1
U_3	0	1	0	-1/3	2/3	-1/3	-1	1	0
U_4	0	1	1	-2/3	1/3	1/3	-1	0	1
U_5	0	0	1	-1/3	-1/3	2/3	0	-1	1
U_6	1	0	1	1/3	-2/3	1/3	1	-1	0
U_7	1	1	1	0	0	0	0	0	0

We see that all of the vectors from the voltage U_1 to U_6 have the same value which is equal to $2/3U_{dc}$, phase angles deviate to $\pi/3$, other two vectors U_0, U_7 have zero value. Of these standard pairs of boundary vectors, vector space is divided into 6 uniform sectors with the angle between the vectors is $\pi/3$.

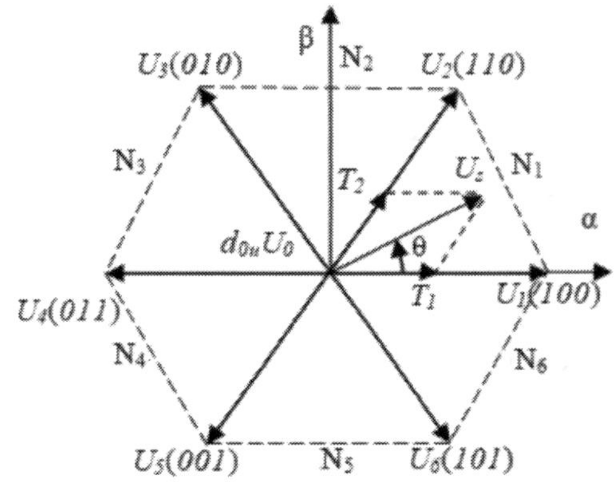

Figure.4. Output voltage vector diagram

Any voltage vector of a vector space should be in one of the six sectors, as shown in Figure 4, and will be represented by the sum of two component vectors or standard boundary vectors of this sector [14-16].If necessary to determine the output voltage when its vector is in sector 1, as Us (U_A , U_B , U_C) then by the formula to calculate the vector sum we have: $U_s = k.\ U_1 + l.U_2$. U_1 is formed by closing the keys S_6, S_1, S_2, and U_2 is formed by closing S_1, S_2, S_3. k , l is the closing coefficient of a cell in one pulse period. From here it is possible to develop a method of modulation of the vector U_s, by closing the key S_6, S_1 and S_2 (vector U_1) during time T_1(s), closing the valves S_1, S_2, S_3 (vectorU_2) within time T_2(s) period the hash pulses T_s(s) satisfy the condition: $k = T_2/T_s$, $l = T_1/T_s$. Since T_1, T_2 □ T_s, the values of k and l are in the range (0,1). The remaining time $T_0 = T_s$ - (T_1 T_2) is the time for vector modulation. Thus, the formula for determining output voltage (which should be achieved by modulation) U_s depending on the standard vector of boundary voltages and the closing time of the key:

$$U_s = \frac{T_1}{T_s}U_1 + \frac{T_2}{T_s}U_2 + \frac{T_{0/7}}{T_s}U_{0/7}$$

According to the geometrical calculation formula to calculate values for T_1, T_2:

$$T_1 = T_s.m.\frac{2}{\sqrt{3}}sin(\frac{\pi}{3} - \theta)$$

$$T_2 = T_s.m.\frac{2}{\sqrt{3}}.sin(\theta)$$

Where:$m = \frac{|U_s|}{U_{dc}}$ and $T_s = \frac{1}{f_s}$

After determining the times T_1, T_2, and T_0, the time of the closing all keys to the rack at work is determined. In each sector the switching time of each key is different and is described in [17-18].

2.2. Modulation for direct frequency converter

This section presents a space vector modulation method for the matrix converter, the structure of the power circuit described in figure 2. For the synthesis system modulation control based on the methods of space vector modulation, it is necessary to consider the matrix converter as a converter indirect with an intermediate DC link. The division of the MC to the rectifier and the inverter gives the possibility to perform synthesis of modulation control with the formation of the output voltage and controlling the input current.

2.2.1. Modulation of the spatial vector of the input current

The vector diagram of the MC input current is presented in Fig. 5. DC bus voltage and the input current can be represented as follows:

$$\begin{bmatrix} U_{dc+} \\ U_{dc-} \end{bmatrix} = \begin{bmatrix} S_1 & S_3 & S_5 \\ S_2 & S_4 & S_6 \end{bmatrix} \begin{bmatrix} U_a \\ U_b \\ U_c \end{bmatrix}$$

$$\begin{bmatrix} I_a \\ I_b \\ I_c \end{bmatrix} = \begin{bmatrix} S_1 & S_2 \\ S_3 & S_4 \\ S_5 & S_6 \end{bmatrix} \begin{bmatrix} I_{\text{dc+}} \\ I_{\text{dc-}} \end{bmatrix}$$

The given vector of the input current of the rectifier (grid current) can be represented by the combination of the base vectors of the two active vectors, between which there is a specified vector of the input current and the zero vector:

$$I_i = d_\gamma I_\gamma + d_\delta I_\delta + d_{0i} I_0$$

Where d_γ, d_δ, d_{0i} – interval modulation coefficients, which determine the duration of the base vectors of the current in the loop modulation.

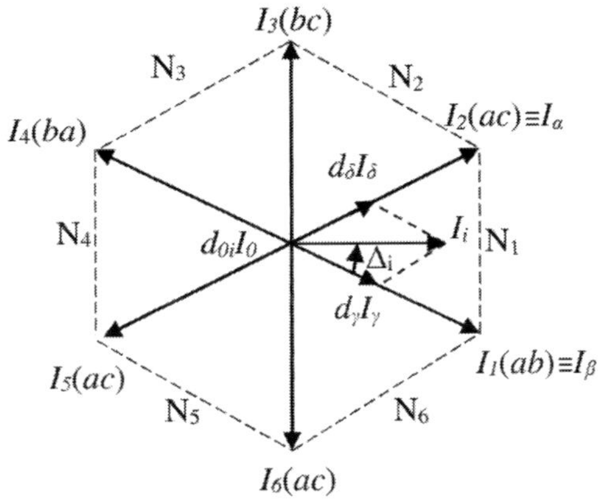

Figure 5. Vector input current diagram

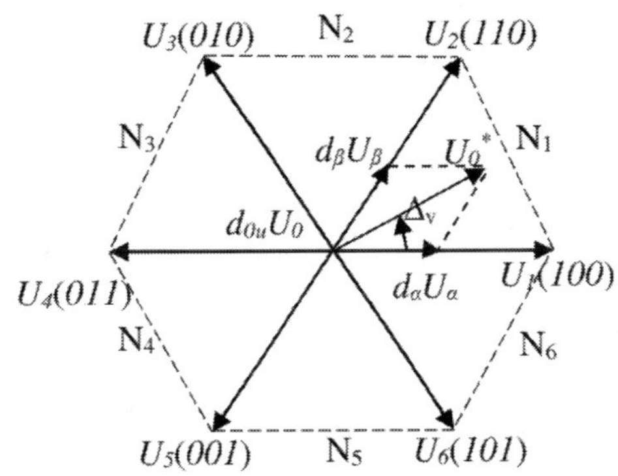

Figure 6. Output voltage vector diagram

Current modulation factor is calculated as follows:

$$m_i = \frac{|I_i|}{I_{dc}}$$

The interval modulation coefficients of the current are calculated as follows:

$$d_\gamma = m_i \sin\left(\frac{\pi}{3} - \Delta_i\right)$$
$$d_\delta = m_i \sin(\Delta_i)$$
$$d_{0i} = 1 - d_\gamma - d_\delta$$

2.2.2. *Modulation of the space vector of the output voltage*

The vector diagram of the MC output voltage is presented in Fig.7. Direct current and output voltage can be represented as follows:

$$\begin{bmatrix} U_A \\ U_B \\ U_C \end{bmatrix} = \begin{bmatrix} S_7 & S_8 \\ S_9 & S_{10} \\ S_{11} & S_{12} \end{bmatrix} \begin{bmatrix} U_{dc+} \\ U_{dc-} \end{bmatrix}$$

$$\begin{bmatrix} I_{dc+} \\ I_{dc-} \end{bmatrix} = \begin{bmatrix} S_7 & S_9 & S_{11} \\ S_8 & S_{10} & S_{12} \end{bmatrix} \begin{bmatrix} I_A \\ I_B \\ I_C \end{bmatrix}$$

The specified output voltage vector inverter a (load voltage) can be represented by a combination of base vectors as follows:

$$U_0^* = d_\alpha U_\alpha + d_\beta U_\beta + d_{0u} U_0$$

Where d_α , d_β , d_{0u} - interval modulation coefficients, which determine the duration of the base vector in the voltage in the modulation cycle.

The voltage modulation factor is calculated as follows:

IOP Conf. Series: Materials Science and Engineering 1159 (2021) 012068 doi:10.1088/1757-899X/1159/1/012068

$$m_u = \frac{|U_0|}{U_{dc}}$$

Interval coefficients of the modulation voltage are calculated as follows:

$$d_\alpha = m_u \sin\left(\frac{\pi}{3} - \Delta_v\right)$$
$$d_\beta = m_u \sin(\Delta_v)$$
$$d_{0u} = 1 - d_\alpha - d_\beta$$

2.2.3. Formation of switching functions using the indirect method

To ensure the symmetry of the input current and the output voltage in the modulation period, it is necessary to provide a certain switching sequence in the right mode (γ - δ -0) with the switching of the inverter (α - β -0) .Therefore, the modulation should provide the following switching sequence:

$$\alpha\gamma \rightarrow \beta\gamma \rightarrow \beta\delta \rightarrow \alpha\delta$$

Modulation factor for each stage:

$$d_{\alpha\gamma} = d_\alpha d_\gamma$$
$$d_{\alpha\delta} = d_\alpha d_\delta$$
$$d_{\beta\gamma} = d_\beta d_\gamma$$
$$d_{\beta\delta} = d_\beta d_\delta$$
$$d_0 = 1 - \left(d_{\alpha\gamma} + d_{\alpha\delta} + d_{\beta\gamma} + d_{\beta\delta}\right)$$

The coefficients of the modulation can be represented as follows:

$$d_\gamma d_\alpha = \frac{2|U_0|}{\sqrt{3}|U_i|} \sin\left(\frac{\pi}{3} - \Delta_v\right) \sin\left(\frac{\pi}{3} - \Delta_i\right)$$
$$d_\delta d_\alpha = \frac{2|U_0|}{\sqrt{3}|U_i|} \sin(\Delta_i) \sin\left(\frac{\pi}{3} - \Delta_i\right)$$
$$d_\gamma d_\beta = \frac{2|U_0|}{\sqrt{3}|U_i|} \sin\left(\frac{\pi}{3} - \Delta_i\right) \sin(\Delta_v)$$
$$d_\delta d_\beta = \frac{2|U_0|}{\sqrt{3}|U_i|} \sin(\Delta_v) \sin(\Delta_i)$$

These coefficients can be calculated for all phases of the MC. From the presented equations it can be concluded that the maximum voltage gear ratio for MC is equal to $\sqrt{3}/2 = 0{,}866$.

The sequence of formation of the base current vectors MFC should be constructed as follows:

- For rectifier:

$$I\gamma \rightarrow I_\delta \rightarrow I_0 \rightarrow I_\delta \rightarrow I\gamma$$

- For inverter:

$$U_\alpha \rightarrow U_\beta \rightarrow U_0 \rightarrow U_\beta \rightarrow U_\alpha$$

In MC output voltages should be formed in nine steps:

- If the sum of $N_V + N_i$ is odd value then:

αγ-βγ-βδ-αδ-0-αδ-βδ-βγ-αγ;

- If the sum of $N_V + N_i$ is even:

βγ-αγ-αδ-βδ-0-βδ-αδ-αγ-βγ.

The order of formation of the output voltage vectors for all sectors of the phase plane (Fig.5 and Fig.6) is presented in Table 2.

IOP Conf. Series: Materials Science and Engineering 1159 (2021) 012068 doi:10.1088/1757-899X/1159/1/012068

Table 2. The order of formation of the vector of the output voltage of the Matrix converter

Voltage and current sector N_v-N_i	Switch combination					Switching order
	p_1	p_2	p_3	p_4	p_0	
U1-I6	abb	aab	aac	acc	ccc	αγ-βγ-β -αδ - 0 - αδ-βδ-βγ-αγ
U2-I6	bab	aab	aac	cac	ccc	βγ-αγ-αδ-βδ- 0 -βδ-αδ-αγ-βγ
U3-I6	bab	baa	caa	cac	ccc	αγ-βγ-βδ-αδ- 0 - αδ-βδ-βγ-αγ
U4-I6	bba	baa	caa	cca	ccc	βγ-αγ-αδ-βδ- 0 -βδ-αδ-αγ-βγ
U5-I6	bba	aba	aca	cca	ccc	αγ-βγ-βδ-αδ- 0 - αδ-βδ-βγ-αγ
U6-I6	abb	aba	aca	acc	ccc	βγ-αγ-αδ-βδ- 0 -βδ-αδ-αγ-βγ
U1-I1	aac	acc	bcc	bbc	bbb	βγ-αγ-αδ-βδ- 0 -βδ-αδ-αγ-βγ
U2-I1	aac	cac	cbc	bbc	bbb	αγ-βγ-βδ-αδ- 0 - αδ-βδ-βγ-αγ
U3-I1	caa	cac	cbc	cbb	bbb	βγ-αγ-αδ-βδ- 0 -βδ-αδ-αγ-βγ
U4-I1	caa	cca	ccb	cbb	bbb	αγ-βγ-βδ-αδ- 0 - αδ-βδ-βγ-αγ
U5-I1	aca	cca	ccb	bcb	bbb	βγ-αγ-αδ-βδ- 0 -βδ-αδ-αγ-βγ
U6-I1	aca	acc	bcc	bcb	bbb	αγ-βγ-βδ-αδ- 0 - αδ-βδ-βγ-αγ
U1-I2	bcc	bbc	bba	baa	aaa	αγ-βγ-βδ-αδ- 0 - αδ-βδ-βγ-αγ
U2-I2	cbc	bbc	bba	aba	aaa	βγ-αγ-αδ-βδ- 0 -βδ-αδ-αγ-βγ
U3-I2	cbc	cbb	abb	aba	aaa	αγ-βγ-βδ-αδ- 0 - αδ-βδ-βγ-αγ
U4-I2	ccb	cbb	abb	aab	aaa	βγ-αγ-αδ-βδ- 0 -βδ-αδ-αγ-βγ
U5-I2	ccb	bcb	bab	aab	aaa	αγ-βγ-βδ-αδ- 0 - αδ-βδ-βγ-αγ
U6-I2	bcc	bcb	bab	baa	aaa	βγ-αγ-αδ-βδ- 0 -βδ-αδ-αγ-βγ
U1-I3	bba	baa	caa	cca	ccc	βγ-αγ-αδ-βδ- 0 -βδ-αδ-αγ-βγ
U2-I3	bba	aba	aca	cca	ccc	αγ-βγ-βδ-αδ- 0 - αδ-βδ-βγ-αγ
U3-I3	abb	aba	aca	acc	ccc	βγ-αγ-αδ-βδ- 0 -βδ-αδ-αγ-βγ
U4-I3	abb	aab	aac	acc	ccc	αγ-βγ-βδ-αδ- 0 - αδ-βδ-βγ-αγ
U5-I3	bab	aab	aac	cac	ccc	βγ-αγ-αδ-βδ- 0 -βδ-αδ-αγ-βγ
U6-I3	bab	baa	caa	cac	ccc	αγ-βγ-βδ-αδ- 0 - αδ-βδ-βγ-αγ
U1-I4	caa	cca	ccb	cbb	bbb	αγ-βγ-βδ-αδ- 0 - αδ-βδ-βγ-αγ
U2-I4	aca	cca	ccb	bcb	bbb	βγ-αγ-αδ-βδ- 0 -βδ-αδ-αγ-βγ
U3-I4	aca	acc	bcc	bcb	bbb	αγ-βγ-βδ-αδ- 0 - αδ-βδ-βγ-αγ
U4-I4	aac	acc	bcc	bbc	bbb	βγ-αγ-αδ-βδ- 0 -βδ-αδ-αγ-βγ
U5-I4	aac	cac	cbc	bbc	bbb	αγ-βγ-βδ-αδ- 0 - αδ-βδ-βγ-αγ
U6-I4	caa	cac	cbc	cbb	bbb	βγ-αγ-αδ-βδ- 0 -βδ-αδ-αγ-βγ
U1-I5	ccb	cbb	abb	aab	aaa	βγ-αγ-αδ-βδ- 0 -βδ-αδ-αγ-βγ
U2-I5	ccb	bcb	bab	aab	aaa	αγ-βγ-βδ-αδ- 0 - αδ-βδ-βγ-αγ
U3-I5	bcc	bcb	bab	baa	aaa	βγ-αγ-αδ-βδ- 0 -βδ-αδ-αγ-βγ
U4-I5	bcc	bbc	bba	baa	aaa	αγ-βγ-βδ-αδ- 0 - αδ-βδ-βγ-αγ
U5-I5	cbc	bbc	bba	aba	aaa	βγ-αγ-αδ-βδ- 0 -βδ-αδ-αγ-βγ
U6-I5	cbc	cbb	abb	aba	aaa	αγ-βγ-βδ-αδ- 0 - αδ-βδ-βγ-αγ

2.3. Results and analysis

To study the effectiveness of the use of the indirect method of modulation of matrix Converter, simulations were carried out in Matlab. The simulation model has the following parameters: input voltage 220(V), 50(Hz), load resistance R = 2(Ω), inductance L = 10(mH).

2.3. 1. *Two-level inverter VSI*Simulation

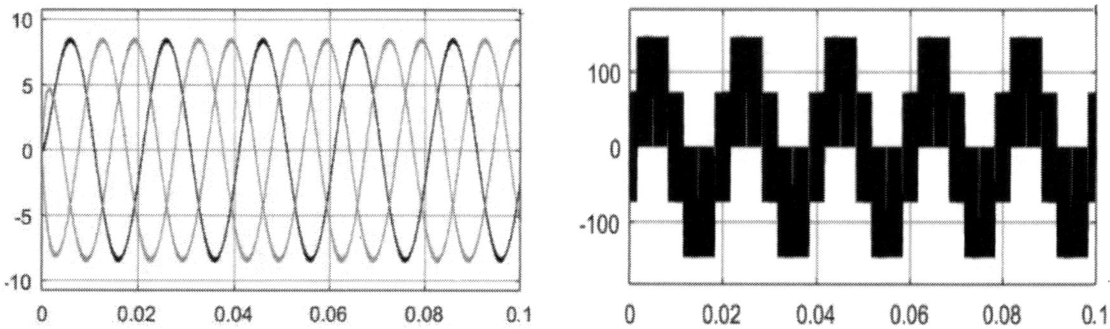

a) b)

Figure 7. SVM modulation results,(a) VSI load current, (b) the output voltage (f = 50 Hz)

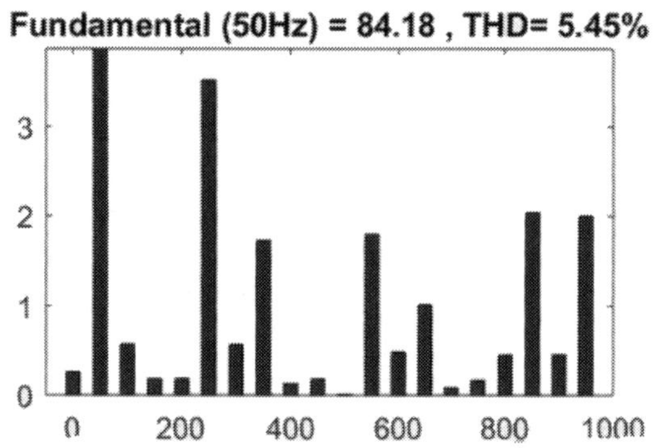

Figure 8. Output voltage total harmonic distortion *(THD=5,45%)* of the PWM method

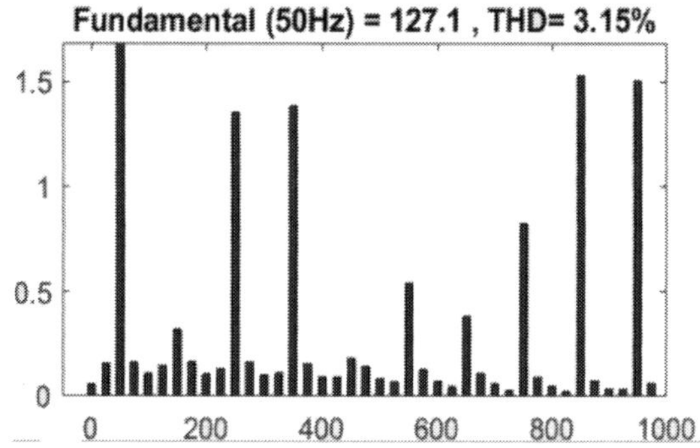

Figure 9. Output voltage total harmonic distortion *(THD=3,15%)* of the SVM method

2.3. 2. *Results when using a matrix converter*

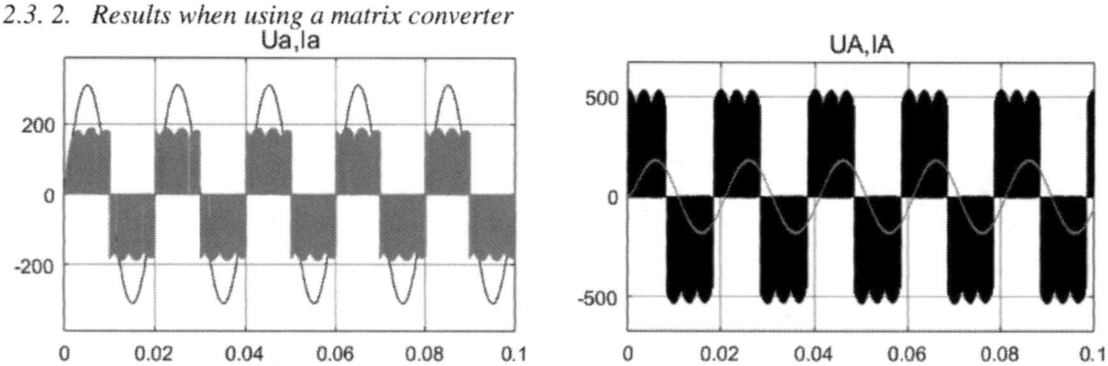

a) b)

Figure 10. The voltage and current at MC, (a) input MC, (b) output MC

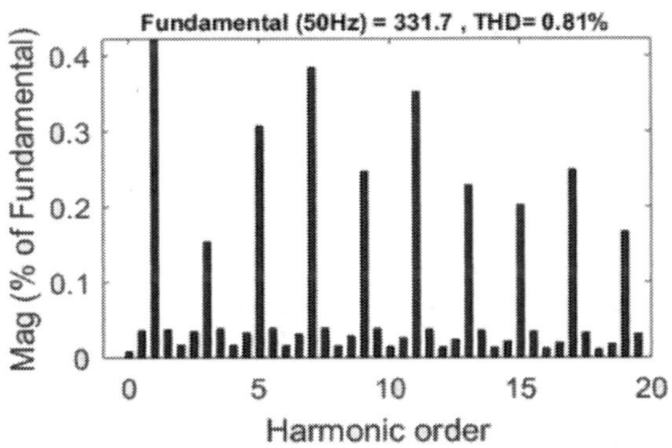

Figure 11. Output voltage total harmonic distortion *(THD=0,81%)*

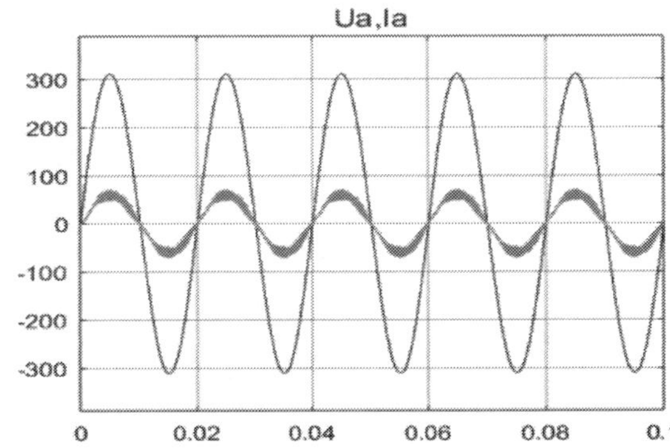

Figure 12. The voltage and current at the input when using filter circuit

The modulation method SVM for VSI shows the better output voltage and current than the PWM method (Fig.8 and Fig.9) because the SVM method creates a continuous shift of equivalent to a space vector voltage of the inverter on a circular path. With a uniform displacement of the spatial vector on a circular path, the higher harmonics are removed. There is a possibility to prepare SVM to optimize switching or optimization of harmonics depending on the choice of modulation standard boundary voltages. However, the quality of voltage and current still has the disadvantages of the VSI Converter. With modulation method SVM for matrix converter (Fig 10-12) the results show that the phase of the input current almost coincides with the phase of the input voltage, no voltage surge, the output voltage has a sinusoidal form. At the entrance there are no harmonics of voltage, indicating the correctness of the control logic switching.

3. Conclusion

The Based on the theoretical analysis and simulation of PWM and SVM modulation methods for indirect frequency converters, a space vector modulation method for 3x3 matrix converters has been developed. This modulation method has divided the bidirectional valves into two virtual rectifiers and virtual inverters to control the current vector and voltage vector separately. This ensures that in each cycle, it is only necessary to determine the phase angle of the voltage at the input with the time when the source voltage passes to the value of 0 without regard to the instantaneous value of the voltage. The output voltage is determined by the required frequency and the desired modulation factor. This will minimize the steps of calculating the modulation factor so the anti-interference is good, is a direct energy converter in two directions. The paper reviews the method of controlling semiconductor valves in indirect converters to develop methods of controlling two directions valves in matrix frequency converters, with the desire to develop the advantages of matrix converters in the electric drive control system.

Acknowledgments

The authors would like to express thanks to Thai Nguyen University of Technology for all supports and encouragements.

References

[1] Roshan Ghosh 2017 IEEE Trans.India 1281–1288
[2] Rmili L, Rahmani S, Vahedi H, Al-Haddad K 2015 *Journal of Renewable Energy and Sustainable Development (RESD)* 30–37
[3] Kokorin N V, Arakelyan A K 2010 *Journal Electricity* **1** 43–47
[4] Xiaohang Yu, Mengqi Wang 2016 *IEEE Trans.USA* 1120–127
[5] Satheesh, G, Chamdranth Naidu, Priyanka, P 2017 *Journal IJAICT* **3(11)** 1197–1205
[6] Mohamad Hosseini Abardeh, Reza Ghazi 2014 *Advances in Power Electronics* **14** 72–78

[7] Mohammad Ali 2017 *IEEE Transaction on power electronics* **32(6)** 4864–4877
[8] Cailhol Simon, Vidal Paul-Etienne 2018 *IEEE Transactions on Industry Applications* **54(5)** 4515–4522
[9] Yang Xia, Bin Gou, Yan Xu 2018*Protection and Control of Modern Power Systems* **33(3)** 2–9
[10] AbramovichB N 2018 *Zapiski Gornogo Instituta/Journal of Mining Institute* **229**31–40
[11] Zhukovskiy Y, Korolev N and Koteleva IN 2018 *IOP Conf. Series: Journal of Physics: Conf. Series* **1015(3)** 032158
[12] Tereshkin V M, Grishin D A, Makulov I A 2019 *Journal of Mining Institute* **240** 678–685
[13] Shashank Tiwari, Sanjaya Kumar 2017 *Research Journal of Engineering Sciences* **6(8)** 8–12
[14] Kozyaruk AE, Albert MK 2019 *Journal of Mining Institute* **39** 576–582
[15] Bogdan Y, Le Van Tung 2018 *IEEE Trans Rus* 80–87
[16] Zipan Nie, Matthias Preindl, Nigel Schofield 2016 *8th IET International Conference on Power Electronics, Machines and Drives* 2211–2217
[17] Omer Turksoy, Unal Yılmaz, Adnan Tan, Ahmet Teke 2017 *Natural and Engineering Sciences* **2(2)** 73–84
[18] Le Dinh Hieu and Temkin I O 2018 *IEEE Seventh International Conference on Communications and Electronics (ICCE)-Vietnam* 196-201

Kinematic and power characteristics of the elastic wheel while rolling with camber and convergence

T A Balabina[1], **M Yu Karelina**[1], **A N Mamaev**[2]

[1] Moscow Automobile and Road Construction State Technical University (MADI), 125319, Leningradskiy prospect, 64, Moscow, Russia
[2]Moscow Polytechnic University (Mospolitech), 107023, B. Semenovskaya street, 38, Moscow, Russia

E-mail: balabinata@gmail.com

Abstract. Vehicles steering wheels are installed in such a way that the plane of the wheel makes a certain angle with vertical plane (so-called camber angle) and at the same time is located at an angle to machine movement direction (toe-slip angle). These angles result in lateral force, increased contact slip and increased wear on the wheel tread. In the article, from phenomena consideration occurring in the contact of an elastic wheel with a rigid base, kinematics and rolling mechanics issues of elastic wheels installed with camber and convergence are considered; dependences are derived that allow determining the wheel power and kinematic parameters, toe and camber angles optimal ratio from the point of view of reducing the wheel running track rolling resistance and wear is determined.

1. Introduction

Vehicles steering wheels are installed in such a way that plane of wheel makes a certain angle with vertical plane (so-called camber angle) and at the same time is located at an angle to machine movement direction (toe-slip angle).

The presence of these angles leads to a side force, increased contact slippage, and increased wear on the wheel treadmill and rolling resistance. Various analytical [1-11] and experimental [12-17] methods were used to study these phenomena.

Despite the extensive research in this area, the problem remains unsolved about the optimal ratio of the angles of convergence and camber in terms of reducing the rolling resistance and wear of the treadmill.

To solve the problem, we will use motion reversal method, which is as follows. The system "wheel - support surface" is mentally imparted with an additional speed «-*V*», i.e. speed equal in magnitude and opposite in direction of wheel axis speed. As a result, the wheel stops (continuing to rotate at the same angular velocity) and supporting surface moves at a speed «-*V*». In this case, interaction mechanism of wheel with supporting surface does not change. A similar approach with interaction mechanics consideration of wheel with a rigid support surface was used, in particular, in [1-7].

Content from this work may be used under the terms of the Creative Commons Attribution 3.0 licence. Any further distribution of this work must maintain attribution to the author(s) and the title of the work, journal citation and DOI.
Published under licence by IOP Publishing Ltd

2. Materials and methods

Let us first consider separately the cases of wheel rolling with camber and with withdrawal in longitudinal tangential force absence in contact, and then a more general case when the wheel rolls with withdrawal and camber with simultaneous implementation of longitudinal tangential force.

When rolling with camber, in the contact between wheel and rigid base, treadmill lateral displacements occur due to wheel plane inclination.

Figure 1shows that magnitude of these lateral displacements can be found from geometric considerations. Consider a longitudinal section of a wheel 1-1. The point of the wheel surface located in this section, after coming into contact with supporting surface, moves (in reversed mechanism) along a straight line 1′-1′. At the same time, if there was no reference surface, the further trajectory of this point would represent a circle arc in plane 1-1. The projection of this trajectory onto the contact plane is described by equation:

$$W_y \sin\gamma = \frac{a^2 - x^2}{2r}\sin\gamma \approx \gamma\frac{a^2 - x^2}{2r},$$

where $W_y=(a^2-x^2)/2r$is wheel deflection in the section under consideration at a distance$x$from midpoint (in longitudinal direction), a s a contact area half-length.

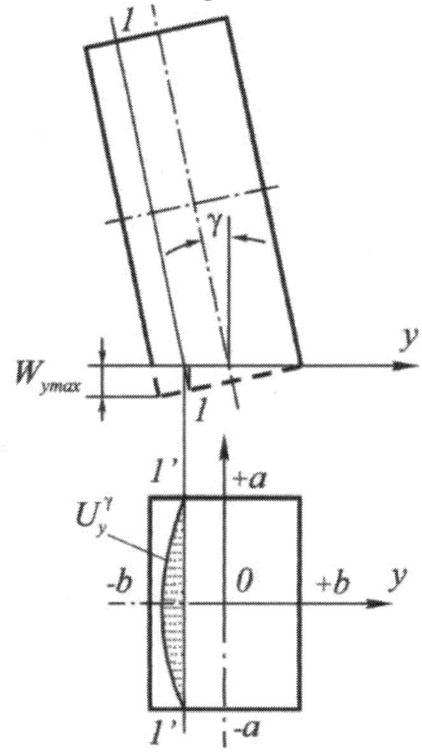

Figure 1. Wheel installed with a camber (γ-camber angle), and lateral tangential displacements diagram of wheel surface elements in contact zone arising in this case.

Thus, motion trajectory of point in a free state would have to be projected onto the reference surface at a distance $W_y sin\gamma$from straight line 1-1. However, as shown in the Figure 1, adhesion to supporting surface prevents this, resulting in point lateral displacement in opposite direction, i.e. lateral tangential displacement of wheel surface elements caused by camber in adhesion section changes according to the law:

$$U_y^\gamma = -\gamma(a^2 - x^2)/2r$$

Using this formula, we consider angle to be positive if distance between the planes of left and right steered wheels increases with distance from supporting surface.

Figure 2,a shows that in rectilinear rolling of an elastic wheel with withdrawal, the wheel plane is located at an angle δ to wheel axis speed vector. Due to this, wheel surface elements entering the contact zone in reverse motion with the wheel axle stopped, as a result of adhesion with the road, move in the contact zone along a straight line parallel to axis velocity vector in actual motion. In this case, as shown in the Figure 2, in adhesion area of wheel surface point contact zone, lateral tangential displacements are obtained, changing according to the equation:

$$U_y^\delta = (a-x)tg\,\delta \approx \delta(a-x).$$

When wheel rolls with withdrawal and camber, the total lateral displacement of a point on the wheel surface will be equal to the sum of lateral displacements caused by slip and camber angles, i.e.:

$$U_y = U_y^\gamma + U_y^\delta = (a-x)(\delta - \gamma\frac{a+x}{2r})$$

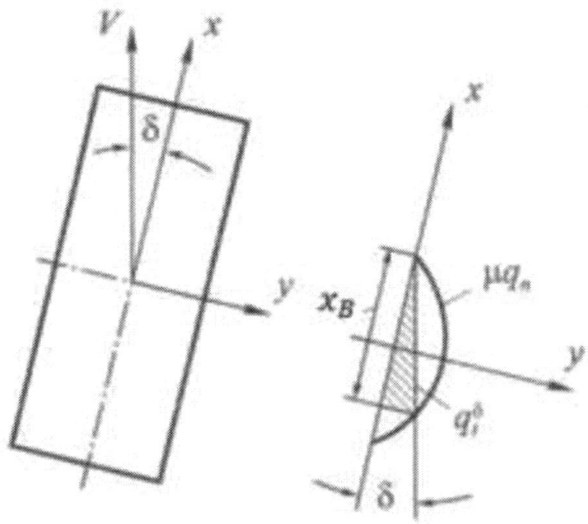

Figure 2. Wheel rolling with withdrawal (toe-in) and diagram of specific lateral tangential forces (tangential stresses) in contact caused by withdrawal (toe-in) of the wheel.

Taking [5,6] that specific tangential forces (tangential stresses) arising in adhesion site contact are proportional to tangential displacements of treadmill points $q_t = \lambda U$ (where λ is tangential stiffness coefficient), we obtain that specific lateral tangential forces value caused by camber and toe-in of wheel will be equal to

$$q_{t_y} = \lambda_y U_y = \lambda_y (a-x)(\delta - \gamma\frac{a+x}{2r}) = \lambda_y (a-x)\delta^*,$$

where
$$\delta^* = \delta - \gamma\frac{a+x}{2r},$$

λ_y- wheel tangential stiffness coefficient in the lateral direction.

It should be borne in mind that in the last expressions the angles δ and γ are substituted each with its own sign.

When longitudinal tangential force is realized in contact, longitudinal specific tangential forces appear, the values of which, as shown in [5, 6], are determined by the dependence

$$q_{tx} = \lambda_x U_x = \lambda_x \xi_x (a - x).$$

As the wheel surface point moves into the contact depth, specific lateral and longitudinal tangential forces increase until the resulting specific tangential force $q_t = \sqrt{q_{t_x}^2 + q_{t_y}^2}$ reaches the value μq_n, the limiting adhesion (here μ is sliding friction coefficient, q_n is normal contact pressure), after which wheel surface elements sliding in contact zone begins.

Figure 2,b shows that coordinate x_B of adhesion and sliding area boundary is determined from equality:

$$q_{t_{(x=x_B)}} = \mu q_{n_{(x=x_B)}}$$

Complexity of process describing under consideration when tangential stiffness coefficients of the wheel λ_x and λ_y in longitudinal and transverse directions are not the same, because in this case total tangents q_t in adhesion section do not coincide with relative velocity ΔV in contact, as shown in the Figure 3, and with shear stresses on slip section $q_t^{sl.} = \mu q_n$ The relative speed ΔV and total shear stresses in adhesion section are directed at angles θ and α to wheel speed. Wherein:

$$tg\theta = \frac{\xi_y}{\xi_x} \approx \frac{\delta^*}{\xi_x} \quad u \qquad tg\alpha = \frac{q_{t_y}}{q_{t_x}} = \frac{\lambda_y \xi_y}{\lambda_x \xi_x} \approx \frac{\lambda_y \delta^*}{\lambda_x \xi_x},$$

Where $\xi_y = \Delta V_y / V \approx \delta^*$ and $\xi_x = \Delta V_x / V$ - relative velocity difference components $\xi = \Delta V / V$i n the contact zone, respectively, in the lateral and longitudinal directions.

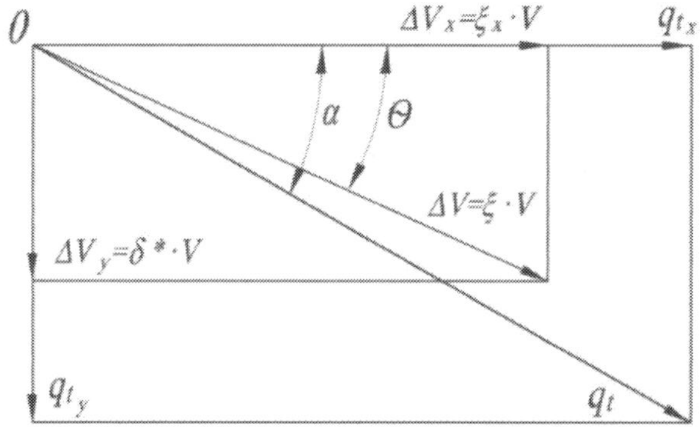

Figure 3. Relative velocities and tangential stresses in contact area adhesion section.

Depending on stiffness ratio, different ratios of θ and α are possible: for $\lambda_y > \lambda_x \alpha > \theta$; for $\lambda_y = \lambda_x \alpha = \theta$; for $\lambda_y < \lambda_x$ $\alpha < \theta$. On the slip section, the shear stress $q_t^{sl.} = \mu q_n$ and the relative velocity ΔV coincide in direction. Thus, at some point in time, there is a change in shear stress direction (the angle between shear stress direction and wheel axis speed changes from α to θ), and the beginning of this process begins after shear stress reaches maximum adhesion value, and a breakdown occurs. In view of the fact that duration of change in total shear stress direction after breakdown is unknown, we will proceed from assumption that simplifies problem of an instantaneous change in q_t t breakdown point - at adhesion and slip sections boundary.

As already noted, in sliding section, the shear stress $q_t^{sl.} = \mu q_n$ coincides in direction with sliding speed and its longitudinal and lateral components can be calculated using the following formulas:

$$q_{t_x}^{sl.} = \mu q_n \frac{V_x^{sl.}}{V^{sl.}} = \mu q_n \frac{\xi_x}{\xi} q_{t_y}^{sl.} = \mu q_n \frac{V_y^{sl.}}{V^{sl.}} = \mu q_n \frac{\delta^*}{\xi},$$

where

$$V^{sl.} = \xi V = V\sqrt{\xi_x^2 + \delta^2} V_x^{sl.} = \xi_x V V_y^{sl.} = \delta^* V$$

Longitudinal and lateral forces arising in the contact are defined as:

$$F_x = \int_{-b}^{+b} t_x dy = 2bt_x \qquad F_y = \int_{-b}^{+b} t_y dy = 2bt_y$$

At this time

$$\left.\begin{aligned} t_x &= \int_{-a}^{x_B} q_{t_x}^{sl.} dx + \int_{x_B}^{+a} q_{t_x} dx \\ t_y &= \int_{-a}^{x_B} q_{t_y}^{sl.} dx + \int_{x_B}^{+a} q_{t_y} dx \end{aligned}\right\}$$

- linear longitudinal and lateral forces, respectively.

Due to distribution asymmetry over the contact area of longitudinal and lateral tangential stresses in the contact plane, a moment arises:

$$M_z = M_{tx} + M_{ty},$$

where $M_{t_x} = \int_{-b}^{+b} t_x y dy$ - moment due to longitudinal stress distribution asymmetry; $M_{t_y} = \int_{-b}^{+b} m_y dy$ - moment due to lateral stresses asymmetry:

$$m_y = \int_{-a}^{x_B} q_{t_y}^{sl.} x dx + \int_{x_B}^{+a} q_{t_y} x dx$$

- elementary moment arising in a longitudinal section due to transverse stresses distribution asymmetry.

Power of losses due to sliding friction in contact will be equal to:

$$P_{fr.} = \int_{-b}^{+b} t\xi V dy = \xi V \int_{-b}^{+b} \sqrt{t_x^2 + t_y^2} dy,$$

where $t = \sqrt{t_x^2 + t_y^2}$ - linear force acting in wheel longitudinal section.

The choice of normal pressures distribution law q_n does not have a significant effect on calculated values of the force acting in contact, wheel speed relative loss, and friction power losses in the contact. In particular, with longitudinal tangential force $F_x < 0{,}5\mu F_z$ (F_z is vertical load on wheel), it is more convenient to use a simplified trapezoidal law (which, according to experiments, is closer to low-pressure tires). In this case (with ratio of upper and lower sides of trapezoid basec = 0,8) $x_B \approx -0{,}9a$, which greatly simplifies calculation formulas:

$$\left.\begin{aligned} t_x &= 1{,}8\lambda_x a^2 \xi_x t_y = 1{,}8\lambda_y a^2(\delta - \gamma a/3r) \\ F_x &= 3{,}6\lambda_x a^2 b\xi_x F_y = 3{,}6\lambda_y a^2 b(\delta - \gamma a/3r) \\ P_{fr.} &= 3{,}6a^2 bV\sqrt{\xi_x^2 + \delta^2}\sqrt{\lambda_x^2\xi_x^2 + \lambda_y^2(\delta - \gamma a/3r)^2} \end{aligned}\right\}$$

From the presented dependences it follows that even with equal values of camber angles γ and slip (toe) δ camber angle influence on wheel side track lateral force and wear rate is much less noticeable than influence of slip angle and ξ_x.

By choosing ratio of angles δ and γ it is possible to reduce lateral force values F_y and friction losses power P_{fr} in contact. In particular, at $\delta = \gamma a/3r$, lateral force becomes equal to zero, and friction losses power in contact is minimal and equal to $P_{fr.} = P_{fr_{min}.} = F_x V\sqrt{\xi_x^2 + \delta^2}$.

Derived dependencies use makes it possible to analyze influence of such factors as rolling mode and longitudinal and lateral stiffness ratio of wheel on resistance to lateral pull. In particular, this problem was solved in relation to low-pressure pneumatic tires.

It can be noted that lateral force increasing effect at δ=const with braking force application to wheel is more pronounced for tires with $\lambda_x/\lambda_y < 1$, i.e. with greater tangential elasticity in longitudinal direction, because in this case, longitudinal force of the same magnitude leads to a greater distortion of normal pressure diagram symmetry. This ratio is typical for radial tires.

3. Conclusion

1. Obtained dependences analysis shows that convergence (slip) angle influence on value of lateral force and treadmill wear intensity of wheel is much greater than camber angle influence.

2. By choosing slip and camber angles ratio, it is possible to reduce values of lateral force and friction losses in contact. In particular, at withdrawal angle $\delta = \gamma a/3r$, the lateral force becomes equal to zero and friction losses power in contact (and therefore wear) is minimal.

3. A simple dependence has been obtained that allows calculating lateral stiffness coefficient through lateral slip resistance coefficient found for the driven wheel, which significantly reduces experimental work amount to determine *F, M, P*.

4. The influence of rolling mode and longitudinal and lateral stiffness ratio of wheel on lateral withdrawal resistance is analyzed; lateral force increasing effect at a constant slip angle with braking torque application to wheel is more pronounced for tires with greater tangential elasticity in longitudinal direction, because in this case, longitudinal force of the same magnitude leads to a greater distortion of normal pressure diagram symmetry.

References

[1] Lorenz R 1928 *Schiene und Rad* VDJ-Z 2 6
[2] Freudenstein G 1961 Luftreifen bei Schrag und Kurvenlauf (Experimentale und theoretische Untersuchung an LKW-Reifen) *Deutsche Kraftfahrtforschung* H 152
[3] Fiala E 1954 Seitenkrafte am rollendenLuftreifen VDJ-Z Bd. 96, 29 973.79
[4] Clark S K 1983 *Tire rolling resistance Rubber Div ACS*
[5] Balabina T A, Brovkina Y I, Mamaev A N 2019 Elastic toroidal wheel rolling with side slip Lecture *Notes in Mechanical Engineering* (Springer, Cham) 2027-2035 DOI: 10.1007/978-3-319-95630-5_218.
[6] Balabina TA, Simonov D S, Rogov V R, Mamaev A N, Vorontsov A S 2020 Determination of Power and Kinematic Wheel Parameters when Rolling against a Drum *IOP Conference Series: Materials Science and Engineering* **832** 012077 doi:10.1088/1757-899X/832/1/012077
[7] Bakker E, Nyborg L, Pacejka H B 1980 Tyre modeling for use in vehicle dynamics studies *SAE Technical Paper* 870421
[8] Gim G 1990 An analytical model of pneumatic tyres for vehicle dynamic simulations, part I: pure slip *Int J Veh Des* 11:589-618
[9] Cho J, Lee S, Jeong HY 2015 Finite element analysis of a tire using an equivalent cord modal *Finite Elem Anal Des* 105:26-32
[10] Li J, Zhang Y, Yi J 2012 A hybrid physical-dynamic tire/road friction model *J Dyn Syst Meas Control Trans ASME* 135:359-370
[11] Ozaki S, Kondo W (2016) Finite element analysis of a tire traveling performance using anisotropic frictional model *J Terramech* 64:1-9
[12] Wang H, Al-Qadi I L, Stanciulescu I 2012 Simulation of tyre-pavement interaction for predicting contact stresses at static and various rolling conditions *Int J Pavement Eng* 13:310-321
[13] Kim S, Kondo K, Akasaka T 2000 Contact pressure distribution of radial tire in motion with camber angle *Tire Sci Technol* 28:2-32
[14] Guan YJ, Zhao GQ, Chenge G 2007 FEA and testing studies on static camber performance of the radial tire *J Reinf Plast Compos* 26:1921-1936
[15] Zhang Y, Yi J 2014 Static tire/road stick-slip interactions; analysis and experiments *IEEE ASME Trans Mechatron* 19:1940-1950
[16] Palanivelu S, Rao K V N, Ramarathnam K K 2015 Determination of rolling tyre modal parameters using finite element techniques and operational modal analysis *Mech Syst Sig Prosses* 64::85-402

IOP Conf. Series: Materials Science and Engineering 1159 (2021) 012069 doi:10.1088/1757-899X/1159/1/012069

[17] Zang LG, Zhao YQ, Li B, Wang J, Fu HX 2016 An experimental study on the ground contact characteristics of non-pneumatic mechanical elastic wheel *Automot Eng* 386350-355 (in Chinese)

IOP Conf. Series: Materials Science and Engineering 1159 (2021) 012070 doi:10.1088/1757-899X/1159/1/012070

Wheel rolling on deformable ground with slippage

T A Balabina[1], M Yu Karelina[1], A N Mamaev[2]

[1]Associate Professor, Moscow Automobile and Road State Technical University (MADI), 125319, Leningradskyave., 64, Moscow, Russia
[1]Professor, Moscow Automobile and Road State Technical University (MADI), 125319, Leningradskyave., 64, Moscow, Russia
[2]Professor, Moscow Polytechnic University (Moscow Polytech), 107023, B. Semenovskayastr., 38, Moscow, Russia

E-mail: balabinata@gmail.com

Abstract. A characteristic feature of wheeled earth-moving machines and tractors is their operation on deformable soils. This makes it necessary to consider the features of the interaction of elastic wheels with the ground, including such issues as the depth of the track and power losses for rutting, wheel movement along the laid track, rolling with pull, forces and moments acting on the wheel, the effect of wheel parameters and rolling mode etc. When a pneumatic wheel with a high internal air pressure rolls on a deformable soil, the nature of the interaction of the wheels with the soil is such that the wheel can be considered rigid, since its normal deformation is small. This greatly simplifies the calculations associated with the analysis of the operation of wheeled machines taking into account the transforming properties of the wheeled propulsion unit, in particular, the determination of the kinematic and power parameters of the wheel. In this paper, using the motion reversal method, the picture of physical phenomena in the contact of the wheel with the ground has been considered, which made it possible to obtain relatively simple expressions for calculating the circumferential thrust force and the coefficient of tangential elasticity of the pair "wheel - deformable soil".

1. Introduction

Wheeled tractors and wheeled earth-moving machines are widely used in road construction and municipal services. A large and varied group of wheeled vehicles in terms of purpose and design are general transport vehicles used for the transportation of various types of bulk cargo, and specialized vehicles that are designed to perform narrow target functions.

A characteristic feature of these types of wheeled vehicles is their ability to work not only on roads with asphalt-concrete pavement, but also on deformable soil surfaces.

When the wheels roll on the ground, the power consumption required to ensure the movement of the wheeled machine increases sharply, and questions arise related to the possibility of the machines, their controllability and loads in the transmission [1-9]. In this regard, it becomes necessary to consider various aspects of the movement of machines in these conditions, in particular, the determination of the forces realized in the contact of the wheels with the ground.

Content from this work may be used under the terms of the Creative Commons Attribution 3.0 licence. Any further distribution of this work must maintain attribution to the author(s) and the title of the work, journal citation and DOI.
Published under licence by IOP Publishing Ltd

IOP Conf. Series: Materials Science and Engineering 1159 (2021) 012070 doi:10.1088/1757-899X/1159/1/012070

In this work, to determine the circumferential force realized in contact, the authors used a proven methodology previously applied to the case of interaction of a wheel with a rigid support surface [10-12].

2. Materials and methods

Let us consider the rolling of a radially rigid wheel with circumferential compliance on deformable soil (which corresponds to the wheel model "elastic band on a rigid rim").

In this case, we will take the assumption about the elastic properties of the soil, i.e. we will assume that up to a certain limit zone there is a linear relationship between the tangential (with respect to the wheel surface) soil displacements and the limiting tangential forces in the contact.

For the convenience of solving the problem, we use an inverted mechanism (i.e., the wheel axis is stationary, and the base - the ground - moves at a speed -V in the direction opposite to the actual direction of the wheel movement, as shown in the Figure 1, similar to how it was used in the works [10-12].

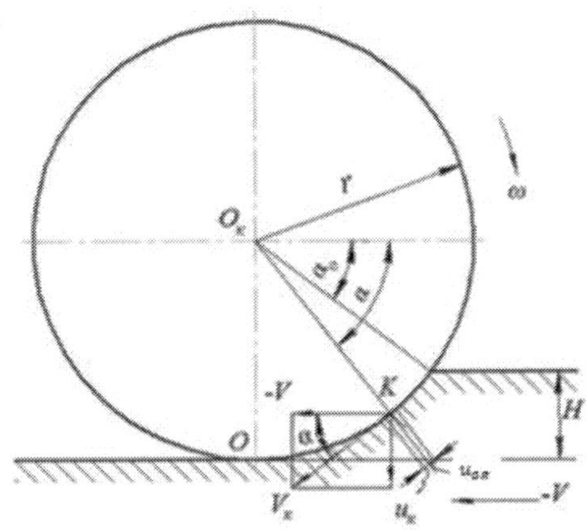

Figure 1. The rolling of a radially rigid wheel with circumferential compliance on deformable soil.

We will consider the movement of point *K*, which belongs to both the wheel and the ground in the contact zone at the adhesion section.

In the time corresponding to the rotation of the wheel from the moment the pointKenters contact with the ground to its position determined by the angle α, the tangential displacement of the ground point will be U_{GR}, and the tangential displacement of the wheel point (relative to the beam on which this point was located before entering the contact) - U_x.

The speed of the point *K* of the wheel surface in the reversed mechanism in the absence of tangential displacements is equal to $V_{K_{a.t.d.}} = -\omega r$.

The actual speed of the point under consideration in the presence of tangential displacement can be represented as follows:

$$V_K = V_{K_{a.t.d.}} + \frac{dU_K}{dt} \quad (1)$$

where dU_K - increment of the tangential displacement of the wheel point during the time $dt=dx/V$ (for the driving wheel it is negative, for the driven and brake ones - positive), dx - elementary movement (increment) of the base (ground) in the horizontal direction in the reversed mechanism.

Similarly, we can represent the same speed of the same point, referring it to the base:

$$V_{GR} = V_{GR_{a.t.d.}} + \frac{dU_{GR}}{dt} \quad (2)$$

IOP Conf. Series: Materials Science and Engineering 1159 (2021) 012070 doi:10.1088/1757-899X/1159/1/012070

Here dU_{GR} is the increment in the tangential displacement of a point on the ground surface over the same period of time;

$$V_{GR_{a.t.d.}} = -\frac{V}{\sin\alpha}$$

is the speed of a given base point in a reverse gear, i.e. in the absence of tangential displacements.

Since expressions (1) and (2) represent the same speed of point K, then, equating their right-hand sides, after transformations, we get:

$$d(U_{GR} - U_K) = \left(V_{K_{a.t.d.}} - V_{GR_{a.t.d.}}\right)dt = \left(\frac{V}{\sin\alpha} - \omega r\right)\frac{dx}{V} = \left(\frac{1}{\sin\alpha} - \frac{r}{r_k}\right)dx \qquad (3)$$

Considering that $\sin\alpha = \sqrt{r^2 - x^2}/r = \sqrt{1 - x^2/r^2}$, the expression 1/sinα, after expanding it in a power series and discarding quantities of the second and more orders of smallness, can be represented in the following form:

$$1/\sin\alpha \approx 1 + x^2/2r^2$$

Putting the last expression into (3), we obtain:

$$dU = d(U_{GR} - U_K) = \left(1 + \frac{x^2}{2r^2} - \frac{r}{r_k}\right)dx \qquad (4)$$

As a result, for the position of point K, determined by the x coordinate, in the adhesion area

$$U = \int_0^x dU = \left(1 - \frac{r}{r_k}\right)x + \frac{x^3}{6r^2} + c \qquad (5)$$

Taking into account the boundary conditions (at $x=a$ $U=0$), after finding the constant c, we will finally have:

$$U = \left(\frac{r}{r_k} - 1\right)(a - x) - \frac{a^3 - x^3}{6r^2} \qquad (6)$$

Specific tangential forces (tangential stresses) acting at a given point on the wheel and on the ground are equal in magnitude, but opposite in direction: $q_t = q_{t_k} = -q_{t_{GR}}$. Under the accepted assumption about the proportionality of the specific tangential forces to tangential displacements, the last equality can be represented as:

$$\lambda_k U_K = -\lambda_{GR} U_{GR}, \qquad (7)$$

where λ_k and λ_{GR} are coefficients of tangential stiffness of the wheel and soil.

Expressing the value U_{GR} from equality (7) and taking into account that $U = U_{GR} - U_W$, after transformations we obtain an expression for the tangential displacements of points on the wheel surface, and then the soil:

$$U_K = \frac{\lambda_K}{\lambda_K + \lambda_{GR}} U \qquad U_{GR} = \frac{-\lambda_K}{\lambda_K + \lambda_{GR}} U \qquad (8)$$

Thus, the specific tangential forces acting at the considered point of the adhesion section will be represented by the following dependencies:

$$q_{t_k} = \frac{\lambda_K \lambda_{GR}}{\lambda_K + \lambda_{GR}} U \qquad q_{t_{GR}} = \frac{-\lambda_K \lambda_{GR}}{\lambda_K + \lambda_{GR}} U$$

or

$$q_t = q_{t_k} = -q_{t_{GR}} = \lambda_{np} \qquad (9)$$

where

$$\lambda_{np} = \frac{\lambda_K \lambda_{GR}}{\lambda_K + \lambda_{GR}} \qquad (10)$$

is the reduced coefficient of tangential stiffness of the wheel-soil pair. Expressions (5) and (9) determine the tangential displacements and specific tangential forces, respectively, due to both the implementation of the circumferential force in the contact and the geometry of the contact zone.

With the adopted wheel rolling model, even with the implementation of significant forces close to the limiting adhesion forces, the coordinate of the boundary of the adhesion and sliding sections is $x_B=0$ (Figure 2 shows the stress distribution along the sweep of the contact arc), which greatly simplifies the calculations, in particular, the determination of tangential strength.

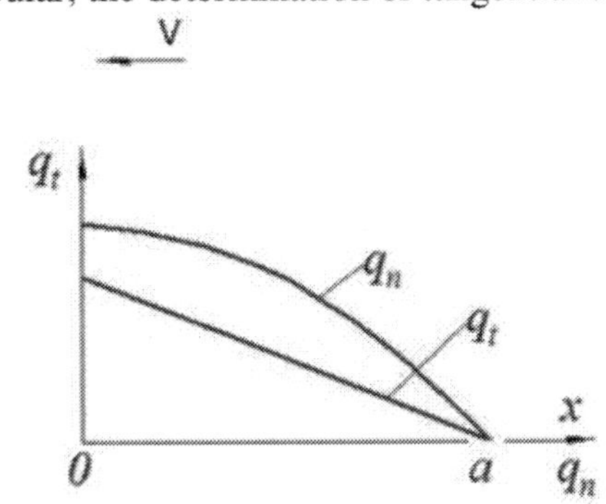

Figure 2. Distribution of stresses along the contact arc sweep.

Tangential stresses distributed along an arc of a circle in the contact zone create a moment relative to the wheel center:

$$M_t = \int_{\alpha_0}^{\frac{\pi}{2}} 2b\, q_t\, r r da = 2br^2\lambda_{GR} \int_{\alpha_0}^{\frac{\pi}{2}} U da \quad (11)$$

We will use this expression to find the rolling radius r_{k_0}, at which $M_t=0$. Substituting equation (5) into (11) and equating the expression obtained after transformations to zero, we will have:

$$r_{k_0} = r / (1 + a^2 / 4r^2) \approx r - a^2 / 4r = r - H / 2 \quad (12)$$

Taking this formula into account, from (5) we obtain a dependence that determines the law of change in tangential displacements in the contact zone, at which the torque on the wheel, due to the presence of tangential stresses, would be equal to zero:

$$U_0 = \frac{a^2}{4r^2}(a-x) - \frac{a^3 - x^3}{6r^2} = (\frac{r}{r_{k_0}} - 1)(a-x) - \frac{a^3 - x^3}{6r^2} \quad (13)$$

Subtracting the value U_0, from the total tangential displacement, we obtain a component due only to the action of the moment M_t:

$$U_t = U - U_0 = (\frac{r}{r_k} - \frac{r}{r_{k_0}})(a-x) = \frac{r}{r_{k_0}}(\frac{r_{k_0}}{r_k} - 1)(a-x) = \frac{r}{r_{k_0}}\xi(a-x), \quad (14)$$

where

$$\xi = \frac{r_{k_0}}{r_k} - 1 \quad (15)$$

As a result, the tangential stresses in the adhesion section, due to the implementation of the thrust force in the contact, will be represented by the dependencies:

$$q_{t_k} = \lambda_{np} U_t = \lambda_{np} \frac{r}{r_{k_0}} \xi(a-x) \quad (16)$$

$$q_{t_{IP}} = -\lambda_{np} U_t = -\lambda_{np} \frac{r}{r_{k_0}} \xi(a - x) \tag{16,a}$$

The corresponding torque on the wheel is found as:

$$M_t = \int_0^a 2bq_{t_k} r dx \tag{17}$$

After substitution of dependence (16) and subsequent integration, we obtain:

$$M_t = \lambda_{np} ba^2 r^2 \xi / r_{k_0} = F_t r, \tag{18}$$

where

$$F_t = \int_0^a 2bq_{t_k} dx = \lambda_{np} ba^2 r \xi / r_{k_0} \tag{19}$$

will be called the circumferential traction.

Elementary tangents $dF_\tau = q_t dx2b$, distributed along the arc of contact, have the resulting F_τ, located outside the arc of contact.

Let's find the magnitude and line of action of this force.

Figure 3 shows that each pair of elementary tangential forces «τ′» and «τ″», acting at points whose coordinates (abscissas) are symmetric about the midpoint with the coordinate

$$\alpha_{mid} = (\pi/2 + \alpha_0)/2$$

has its resulting:

$$\tau = \sqrt{(\tau')^2 + (\tau'')^2 + 2\tau'\tau'' \cos\varphi}, \tag{20}$$

passing through the intersection of the lines of action τ' and τ'', located at an angle φ relative to each other.

For points located on the edges, i.e. when $\alpha \to \alpha_0$ and $\alpha \to \pi/2$ the angle $\varphi \to \frac{\pi}{2} - \alpha$; for midpoints, i.e. when $\alpha \to \alpha_{mid}$, the angle $\varphi \to 0$.

We transform expression (20) by presenting it in the following form:

$$\tau = (\tau' + \tau'') \cdot \sqrt{1 + \frac{2\tau'\tau''}{(\tau' + \tau'')^2}(\cos\varphi - 1)} = (\tau' + \tau'') \cdot \sqrt{1 - \frac{\tau'}{\tau''} \frac{\varphi^2}{(1 + \frac{\tau'}{\tau''})^2}} \tag{21}$$

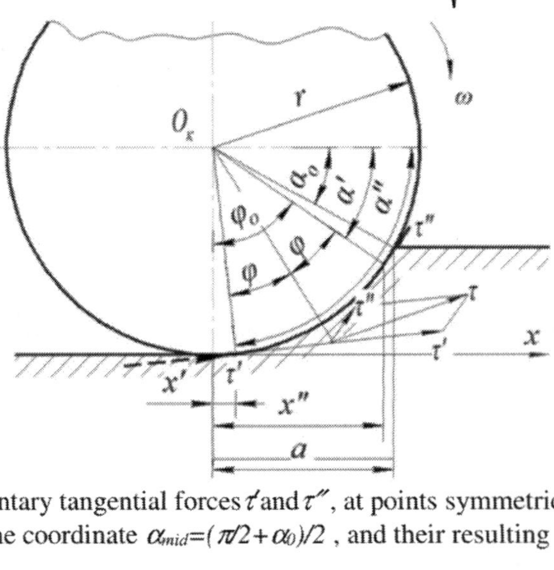

Figure 3. Elementary tangential forces τ' and τ'', at points symmetric about the midpoint with the coordinate $\alpha_{mid} = (\pi/2 + \alpha_0)/2$, and their resulting τ.

For the extreme and close to them points of the considered contact area $\tau'' >> \tau'$, therefore, even with a significant immersion depth of the wheel (for example, when $\varphi \approx 1$) the value of the square root in the last expression differs little from unity. For points located in the middle part of the contact, as before, $\tau'' > \tau'$ (except for the point with $\alpha = \alpha_{mid}$, for which $\tau'' = \tau'$ and, in addition (which is more important), $\varphi \to 0$ ($\varphi = 0$ at $\alpha = \alpha_{mid}$). Therefore, for the middle section of the contact, the value of the square root is close to unity. This allows us to take as a first approximation $\tau = \tau' + \tau''$.

Considering that $\alpha = \arccos\frac{x}{r} \approx \frac{\pi}{2} - \frac{x}{r}$, from which $x = (\frac{\pi}{2} - \alpha)r$ and $a = (\frac{\pi}{2} - \alpha_0)r$, dependence (18) can be represented as:

$$q_t = \lambda_{np}\xi(\alpha - \alpha_0)r^2 / r_{k_0} \tag{22}$$

Then the values of elementary tangential forces acting on symmetrically located elementary areas, as it seen at Figure 3, are found as:

$$\tau' = q_t' 2brd\alpha = \lambda_{np} 2br^3\xi(\alpha' - \alpha_0)d\alpha / r_{k_0} \tag{23,a}$$

$$\tau'' = q_t'' 2brd\alpha = \lambda_{np} 2br^3\xi(\alpha'' - \alpha_0)d\alpha / r_{k_0} \tag{23,b}$$

For the considered symmetrically located sites

$$\alpha' = \alpha \quad \alpha'' = \pi/2 - (\alpha - \alpha_0) = \pi/2 - \alpha + \alpha_0,$$

which gives us

$$\tau' = \lambda_{np} 2br^3\xi(\alpha - \alpha_0)d\alpha / r_{k_0} \tag{24}$$

$$\tau'' = \lambda_{np} 2br^3\xi(\frac{\pi}{2} - \alpha)d\alpha / r_{k_0} \tag{25}$$

As a result:

$$\tau = \tau' + \tau'' = \lambda_{np} 2br^3\xi(\frac{\pi}{2} - \alpha_0)d\alpha / r_{k_0} \tag{26}$$

The diagram of the resulting elementary tangential forces is shown in Figure 4.

Summing up these resulting forces, we can find the magnitude and direction of the total tangential force.

Since the diagram is a system of forces converging at point A,as shown in the Figure 4, then the total tangential force also passes through this point, being perpendicular to the line $O_к A$, i.e. at an angle ψ_t to the horizontal. From geometric considerations

$$\psi_t = \psi_t/2 = (\pi/2 - \alpha_0)/4 \tag{27}$$

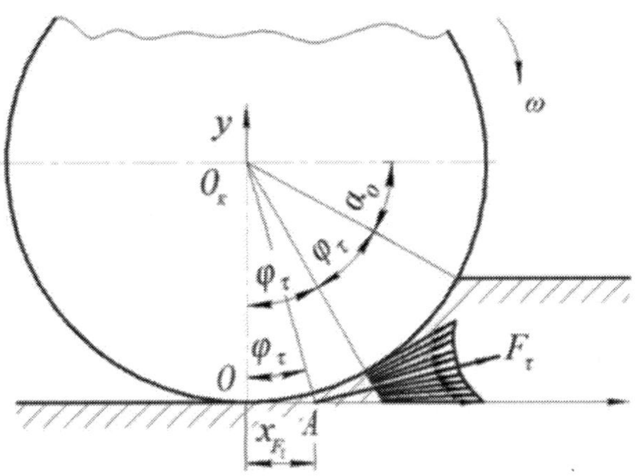

Figure 4. Diagram of the resulting elementary tangential forces.

Figure 4 shows that the moment arm $r_\tau = O_k A$ of force F_τ and the coordinate x_{F_t}, defining the point of application of this force will be found (), as:

$$r_\tau = O_k A = r / \cos\psi_t \approx r(1+\psi_t^2/2) \approx r \tag{28}$$

$$x_{F_t} = rtg\psi_t \approx r(\pi/2-\alpha_0)/4 \approx r\psi_t \tag{29}$$

When determining the magnitude of the total tangential force, summing up the resulting elementary tangential forces, one should take into account the angles between these resulting forces. However, due to the small value of these angles (which are in the range from 0^0 to $2\psi_t$) their cosines are close to unity (for example, if we take the average value of the indicated angle, equal to ψ_t, then at an angle $\alpha_0 = \pi/6$, corresponding to immersion of the wheel into the ground to a depth of 0.5r, the cosine of the angle ψ_t will be $\approx$0.97). This allows the shear force to be calculated using the following formula:

$$F_\tau = \int_{\alpha_0}^{(\pi/2+\alpha_0)/2} \tau = \int_{\alpha_0}^{(\pi/2+\alpha_0)/2} \lambda_{np} 2b\xi(\pi/2-\alpha_0) r^2 d\alpha / r_{k_0} = \lambda_{np}\xi b a^2 r / r_{k_0} \tag{30}$$

Neglecting the indicated angles gives a slightly overestimated value of the tangential force, calculated by the last formula, but this overestimation does not exceed several percent.

Thus, the actual value of the total tangential force F_τ is slightly less than the circumferential traction force F_t, calculated by formula (19), and its arm of force (see formula (30)) is slightly larger than the radius r. However, these differences are small, and in the future, in our calculations, we will assume $F_t = F_\tau$ and $r_\tau = r$.

Having a dependence for calculating the tangential force, one can obtain a dependence for finding the rolling radius.

It follows from the equilibrium condition of the driving wheel that

$$F_{t_x} = F_t \cos\psi_t = F_x + F_{GR} \tag{31}$$

Substituting formula (19) into this equality expressing the force F_t, and also taking into account that $\xi = (r_{k_0}/r_k - 1) r / r_{k_0}$, after the transformations, we will have:

$$r_k = r_{k_0} \frac{r_{k_0}^2}{\lambda_{np} b a^2 r} F_t = r_{k_0} - \frac{r_{k_0}^2}{\lambda_{np} b a^2 r \cos\psi_t}(F_{GR} + F_x) \tag{32}$$

or

$$r_k = r_k^c - \frac{r_{k_0}^2}{\lambda_{np} b a^2 r \cos\psi_t} F_x, \tag{33}$$

where

$$r_k^c = r_{k_0} - \frac{r_{k_0}^2}{\lambda_{np} b a^2 r \cos\psi_t} F_{GR} \tag{34}$$

- free rolling radius when there is no longitudinal force on the wheel axis.

The presented formulas have the same notation with the formula obtained by E.A. Chudakov [13]. In this regard, the expression

$$\frac{r_{k_0}^2}{\lambda_{np} b a^2 r \cos\psi_t} = \gamma_F \tag{35}$$

represents the coefficient of tangential elasticity of the pair "wheel - deformable soil".

Formulas (32) and (33) are also valid for the cases of wheel rolling in the braking or driven mode, if the forces F_x and F_t are substituted into the indicated formulas with a minus sign.

Realization of a tangential force in the contact leads to a change in the normal pressures in the contact by an amount $tq_t \cos\alpha = q_t x/r$. With a driving wheel, for which $\xi > 0$, the value of normal pressures increases, for a brake wheel - decreases (since $\xi < 0$):

$$q_n = C(H-h)^m + q_t x / r = C(H-h)^m + \lambda_{np}\xi(a-x)x / r$$

Here q_n is the normal contact pressure, determined by the Berstein-Letoshnev formula: $q_n = C(H - h)^m$, C and m are soil parameters determined experimentally ($C$is the deformation modulus, m is a parameter characterizing the law of change in soil resistance to indentation), H – track depth, as it sown in the Figure 1, h – depth of immersion of the wheel in the ground at a distance$x$from the vertical axis.

In the slip section, taking into account the above, the dependence for the distribution of normal pressures will have the following form:

$$q_n = C(H-h)^m \pm \mu C(H-h)^m x / r = C(H-h)^m (1 \pm \mu x / r)$$

The coordinate of the border of the adhesion and sliding sections can be found from the equality $q_t = \mu q_n / \sin\alpha$.

An increase in normal pressures in the contact of the driving wheel and a decrease in them in the contact of the brake wheel leads to the fact that at the same value of the relative speed loss at $x_0 > 0$ a greater tangential force is realized in the contact of the driving wheel than in the rolling of the brake wheel. Accordingly, there will be more friction losses in the contact of the drive wheel, and, therefore, larger power costs required for rolling the wheel on the ground. This is clearly confirmed by the dependence obtained experimentally by Yu.V. Pirkovsky [13], it is shown in Figure 5.

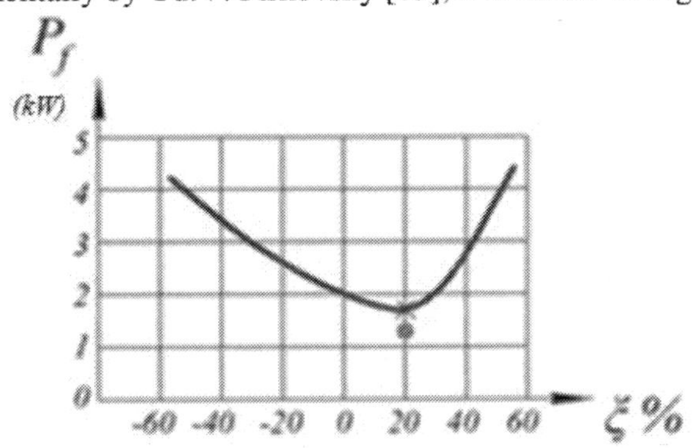

Figure 5. The graph of the change in the power of rutting during wheel rolling (r=0.5m, width B=0.3m) on loamy plowing (m=0.75, C=0.29).

The dependences obtained above are valid for the case when the shear stress in the soil is less than the shear stress value, at which the upper soil surfaces are sheared.

If the soil shear occurs earlier than the tangential stresses reach the limiting adhesion, then the picture of the phenomena occurring in contact and their mathematical description changes. When cutting the soil, it must move into the zone of increasing normal pressures. In this case, with an increase in normal pressures, the permissible stress increases. Due to the movement of the soil layers, there will be friction between them, while there will be no friction losses between the wheels and the soil. The value of the realized tangential force, with the same value of the relative speed loss, will be even less than in the previous case.

Denoting by $\tau_0 = C + q_n tg\rho$ (where ρ is the angle of friction between soil particles) the shear stress at which the shear of soil layers begins, we find the coordinate of the beginning of the shear from the equation:

$$q_t = \frac{\lambda_{np}\xi(a - x_0)r}{r_{k_0}} = \tau_0 \quad (36)$$

$$x_0 = a - \frac{\tau_0 r_{k_0}}{\lambda_{np}\xi r} \quad (37)$$

The realizable tangential force is found from the equation:

$$F_t = b[\tau_0(a - x_0)^2 + 2\int_0^{x_0} \tau \, dx] \quad (38)$$

At $q_i < \tau$ $x_0=0$ and

$$F_t = \lambda_{np} ba^2 \xi r / r_{k_0} \quad (39)$$

3. Conclusion

The studies of rolling on deformable ground of a rigid wheel in the radial direction performed in this work allow us to determine the circumferential traction force realized by the wheel using relatively simple expressions, taking into account the properties of the soil and wheel slippage.

References

[1] Becker M G 1973 Introduction to the theory of the "terrain-machine" system *Moscow Mechanical engineering* **520**
[2] Wong J 1982 Theory of Land Vehicles *Moscow Mechanical Engineering* **284**
[3] Clark S K 1983 Tire rolling resistance Ed. Rubber Div ACS
[4] Kosar F 1984 Wheeld armoured reconnaissance vehicles *Armada* **2** 36–56
[5] Apetrei Gh, Velicu D, Bobancu B 1978 Model petru analisa neconcordantei cinematice din sistemue de rulare al autotrenului cu semiremorca motoare la derlasarea in panta *Bul.Univ. Brascv* **20-A** 97–104
[6] Brylev I, Evtiukov S, Evtiukov S 2018 Problems of calculating the speed of two-wheeled motor vehicles in an accident *Transportation Research Procedia* **36** 84–89
[7] Evtiukov S, Golov E, Ginzburg G 2018 Finite element method for reconstruction of road traffic accidents *Transportation Research Procedia* **36** 157–165
[8] Repin S, Evtiukov S, Maksimov S 2018 A method for quantitative assessment of vehicle reliability impact on road safety *Transportation Research Procedia* **36** 661–668
[9] Marusin A, Ablyazov T 2019 Transport infrastructure safety improvement based on digital technology implementation *Atlantis Highlights in Computer Sciences 1 International Conference on Digital Transformation in Logistics and Infrastructure (ICDTLI 2019)* 353–357 DOI: 10.2991/icdtli-19.2019.61.
[10] Virabov R V 1967 Rolling an elastic wheel on a rigid base *Proceedings of universities Mechanical engineering* **4** 78–85
[11] Balabina T A, Brovkina Yu I, Mamaev A N 2019 Elastic toroidal wheel rolling with side slip *Lecture Notes in Mechanical Engineering* 9783319956299 2027–2035
[12] Mamaev A N, Balabina T A, Odinokova I V, Gaevskiy V V 2019 Interaction mechanics of the wheel with the drum *IOP Conference Series: Materials Science and Engineering* **632** 012083
[13] Pirkovsky Yu V 1974 *Rolling resistance of multi-drive vehicles and road trains on hard roads and deformable soil* Dissertation for the degree of Doctor of Technical Sciences Moscow MSTU named after N E Bauman **399**

IOP Conf. Series: Materials Science and Engineering 1159 (2021) 012071 doi:10.1088/1757-899X/1159/1/012071

The results of the analysis of the current state of the creation and operation of refuelling complexes of wheeled vehicles specially the big carrying capacity, taking into account their environmental, fire and explosion

A N Sova[1], A N Abramov[2], K R Korobeynikova[1], Y L Krasnobaev[1], Y V Trofimenko[2]

[1.] Department "Transportnyye ustanovki" Federal state budgetary educational institution of higher education " Moscow automobile and road state technical University (MADI), Moscow, 125319, Russia
[2.] Department "Tekhnosfernaya bezopasnost'" Federal state budgetary educational institution of higher education " Moscow automobile and road state technical University (MADI), Moscow, 125319, Russia

E-mail: korobeynikova909@mail.ru

Abstract. This article presents the results of the analysis of the creation and operation of multi-wheeled vehicle filling stations for the significance of creating a generalized methodological apparatus in order to ensure a sustainable reduction in the risk of accidents, explosions, man-made disasters and reduce the mortality of people at gas stations. More and more people are killed, injured and burned if fire safety measures are not followed when refueling large multi-axle wheeled vehicles. Accidents and explosions cause significant damage directly to the gas station itself, as well as to tankers carrying explosive fuel and substances. The substances used are mainly characterized by fire and explosion hazards. Therefore, in addition to pollution of environmental objects such as soil, air and water resources, and, as a result, changes in the ecosystem, further fire or explosion is possible. In this case, there is both an energy impact on neighboring objects, and their possible destruction. This article discusses the indicators of accidents in the event of accidents on vehicles, primarily on tankers, a preliminary assessment of the consequences of accidents and the possibility of taking measures to prevent the factors leading to them. The creation of a generalized methodological apparatus for assessing the consequences of accidents will make it possible to assess emerging emergencies without dividing them into classes of transported substances.

1. Introduction

To ensure comprehensive protection of facilities from emergencies, accidents and emergencies during the operation of refueling complexes for heavy-duty vehicles, it is of particular importance to develop a methodology and engineering measures aimed at evaluating and ensuring the comprehensive safety of the multi-axis wheeled vehicle refueling system.

Specific regulatory requirements for the prevention and prevention of emergencies are imposed on the territory of gas stations and dangerous areas, fuel filling columns, storage tanks for petroleum products, pipelines, and the site for draining petroleum products from a tanker. Gas stations (gas stations) and complexes (gas stations) are the most important part of the country's oil product supply

Content from this work may be used under the terms of the Creative Commons Attribution 3.0 licence. Any further distribution of this work must maintain attribution to the author(s) and the title of the work, journal citation and DOI.
Published under licence by IOP Publishing Ltd

system [6]. The object of research in this article is a complex of object gas stations of stationary and mobile type.

The life cycle of the research object includes the following stages:

1. The newly designed and built filling station was not in operation;
2. Newly built filling station, in operation for less than a year;
3. Filling station, in operation for more than a year, equipped in accordance with the project documentation;
4. Filling station, in operation for more than two years, having physically and functionally outdated operational characteristics of buildings and equipment, located, requiring investment to improve performance;
5. Filling station, not in operation as of the evaluation date (operation has been temporarily suspended or discontinued for any reason) [4].

Statistics show that today 43% of fires occur due to violations of operating rules and safety regulations, 22% due to malfunctions of electrical equipment and lighting devices, 13% due to fuel spills, and other causes are much less common [1].

Therefore, measures of environmental friendliness, fire and explosion safety at gas stations require special attention, both at the time of operation and at the time of construction.

Regular violation of these requirements and disregard for fire safety rules when refueling multi-axle wheeled vehicles by service personnel, violations during operation and maintenance of large-sized wheeled vehicles lead to irreversible consequences, such as the occurrence of emergency situations and, as a result, the death of people.

Unfortunately, the existing methods and measures to prevent emergencies are not effective enough, because accidents, explosions, fires, fuel spills are not a rare threat to human life and health at the present time.

Thus, the development of a new methodology and engineering measures aimed at assessing and ensuring the comprehensive safety of the multi-axle wheel refueling system is an urgent problem today.

One of the most common problems that arise when servicing and refueling vehicles with particularly heavy load capacity is the depressurization of the tank.

The substances used, which are transported on tankers, are mainly characterized by fire and explosion hazard. Explosions and fires can occur when the steam-air mixture ignites both inside the tank equipment and in an open area [3]. Therefore, in addition to pollution of environmental objects such as soil, air and water resources, and, as a result, changes in the ecosystem, further ignition or explosion of these substances is possible.

2. The study of the creation and operation of filling stations for wheeled vehicles of especially large carrying capacity, taking into account their environmental, fire and explosive properties

Oil spills may occur due to complete or partial destruction of storage tanks, pumping or pipeline equipment. Destruction is most often the result of corrosion processes, defects during installation work. According to statistics, the vast majority of gas stations in Russia do not meet the requirements of fire and environmental safety [5].

During the raids, massive serious violations were detected at gas stations in the Bryansk, Rostov regions, Krasnodar territory, as well as in other regions. Such statistics indicate that there are no working mechanisms in the industry to respond to violations.

Gas station owners in most cases downplay or completely ignore the facts of emergency situations in order to avoid liability, which significantly increases the fire hazard of stations. Analysis of statistics on accidents at gas stations in Russia and abroad, during which emergencies that could cause damage to the health of personnel and equipment of the facility were considered, showed that spilled oil products were most often ignited.

IOP Conf. Series: Materials Science and Engineering 1159 (2021) 012071 doi:10.1088/1757-899X/1159/1/012071

In about 18 percent of cases, there were explosions. Almost all serious emergencies at gas stations were accompanied by harm to the health of employees, as well as significant damage to property. In some cases, even complete destruction of objects was observed.

Such disappointing statistics indicate the urgent need to make decisions to eliminate existing violations in the field of fire safety at the design and operation stages of petrol stations.

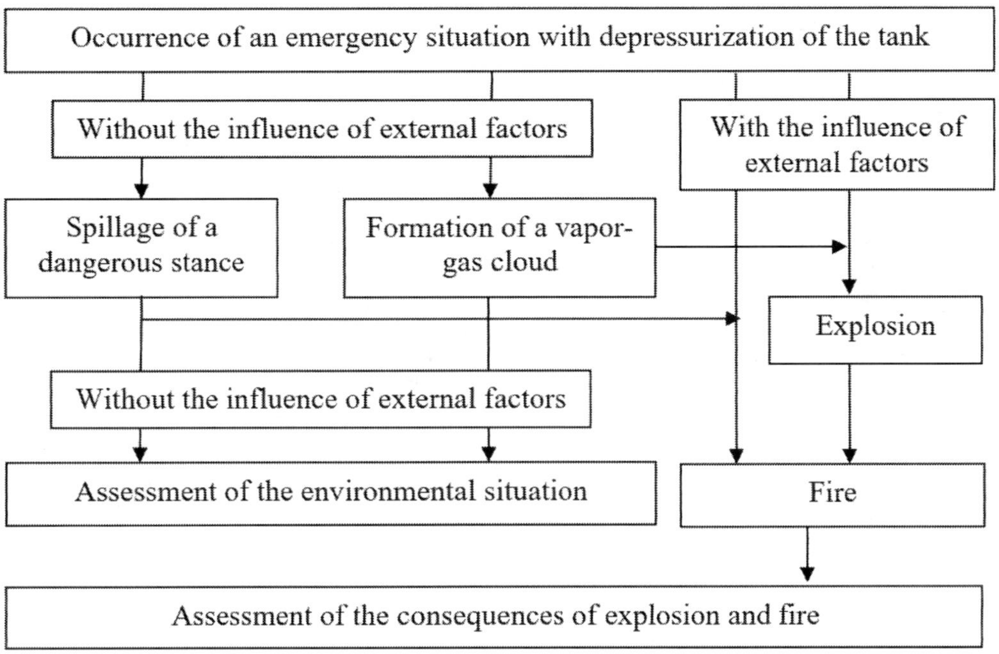

Figure 1. Diagram of the development of an emergency situation with depressurization of the tank

Based on this scheme, it is possible to create an algorithm for calculating the consequences of an emergency.

Input data required for assessing emergency situations on tankers:

a) characteristics of a chemically dangerous object (filled with a dangerous substance on a tank truck):

types of containers, amount of dangerous substance (S);

physico-chemical and Toxicological properties of OM or products of their destruction (taking into account the hazard class, mechanisms of Toxicological action of the substance, its speed of action, resistance, ways of impact on the environment);

finding environmental objects, including people, at the accident site;

presence of explosive and fire-hazardous substances at the accident site.

b) characteristics of a chemical accident:

place and time of the accident;

cause of the accident;

scale and nature of the accident;

the number of hazardous substances or products of their destruction, thrown out, spilled during the accident, their aggregate state.

Spillage of a dangerous substance is accompanied by contamination of a certain volume of soil. When dealing with the consequences, the issue of reclamation is costly because of the need to neutralize the entire volume of contaminated land [7]. The cost may increase significantly due to an increase in the hazard class of waste. To reduce environmental pollution, it is necessary to implement measures to reduce losses of petroleum products from evaporation, spills, reduce emissions of pollutants from wastewater and flue gases, and improve the quality of their treatment [2].

According to statistics from 2019, most often there were fires at gas stations, they accounted for 60% of the total number of accidents.

Depressing statistics show that the measures taken to date are not sufficiently developed and are not effective enough.

An external impact may cause a Strait fire and a fireball.

In emergency situations there is a simultaneous flow of high speed and low speed combustion processes. In the second case, a Strait fire or fireball may occur.

Assessment of the heat radiation intensity of the Strait fire

The intensity of thermal radiation q is determined by:

$$q = E \times F \times t,\ kWt/m^2 \quad (1)$$

where

E is the average surface heat radiation density of the flame, kW / m^2;

F - angular irradiance coefficient;

t is the atmospheric transmittance.

For real rocket fuel fireballs, the radius and time of heat exposure are determined by:

$$R = 17.6M^{0,32}, \quad (2)$$

$$R = 20.78\ m, \quad (3)$$

$$\tau = 2.71M^{0,32}, \quad (4)$$

$$q = E \times F \times t,\ kWt/m, \quad (5)$$

where

E is the average surface heat radiation density of the flame, kW / m^2; $E = 450\ kW / m^2$.

F - angular irradiance coefficient;

$0.012 = 100\ mt$ is the atmospheric transmittance.

We determine the intensity of thermal radiation at a distance of 100 m from the center of the fireball:

$$q = 76.5\ kWt/m^2 \quad (6)$$

The dose of heat radiation affecting people from the "fireball" is determined by:

$$Q = q \cdot \tau = J/m^2. \quad (7)$$

In order to monitor options for improving the design of gas stations and open-type structures located directly at the gas station, patent research was conducted.

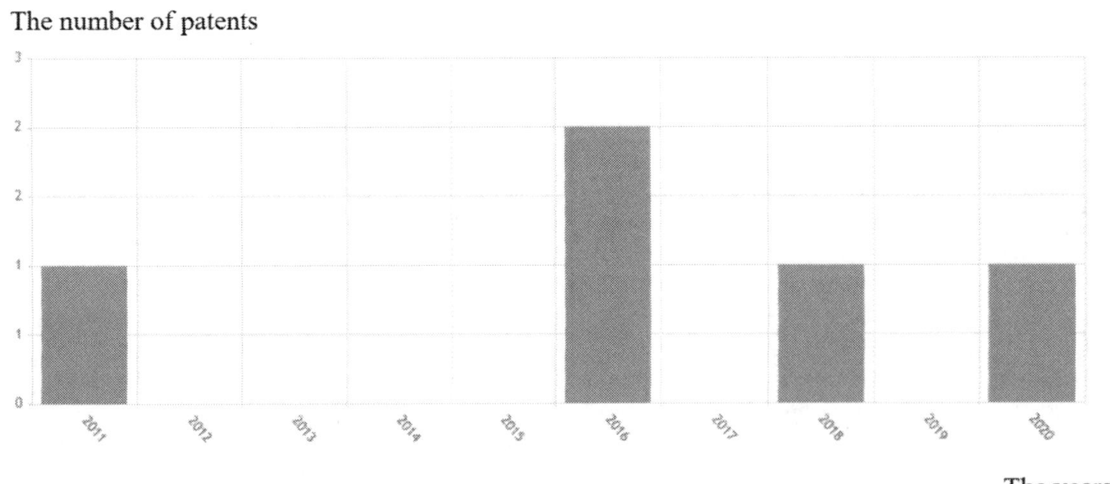

Figure 2. Diagram of patent applications filed for 2011-2020

Figure 3. Countries registering patents for inventions in this industry

In the course of patent research, it was revealed that there is a significant decrease in the volume of patents received for improving and analyzing designs and techniques in this area, which indicates a decline in the interest of specialists in this matter.

3. Conclusion

Thus, the use of a new generalized methodological tool will allow assessing the consequences of emergency situations on tankers, taking into account the environmental, energy and physical impacts on the surrounding objects, as well as help reduce the risks of emergencies, accidents and emergency situations and prevent the death of people.

References

[1] Efanov N V Expert, MAYUL LLC, Rostov, *Safety Requirements for the operation of lifting machines*

[2] Resolution of the Ministry of labor of the Russian Federation of May 6, 2002 *On approval of Intersectoral rules on labor protection in the operation of oil depots, fuel depots, stationary and mobile gas stations*

[3] Atapin E A, Shevtsov S A UDC 628.143 Ivanovo fire and rescue Academy of the Ministry of emergency situations of Russia. *Fire safety gas stations*

[4] GOST R 58404-2019 *Gas stations and complexes. Rules of technical operation*

[5] Safety and environmental friendliness of gas stations 01.06.2017 Available at: https://iz.ru/news/722108

[6] Kovalenko V G, Safonov A S, Ushakov A I, Shergalis V 2003 *Gas Stations: Equipment. Operation. Security* (SPb.: NPIKTS)

[7] Shevtsov S A, Kargashilov D V, Potekh S V, Bykov I A 2017 Evaluation of fire hazard of "big breaths" of ground reservoirs for oil products storage by numerical methods *Fire and explosion safety* **1** 43-51

A method for assessing the cross-country capability of a particularly heavy-duty vehicle with all-wheel control based on the compliance of the support surface during curved movement

G S Mazlumyan[1], R V Yushchuk[1], I O Sudarikov[1], T V Sevastyanova[1], A N Sova[1]

[1] Department "Transportnye ustanovki" Federal state budgetary educational institution of higher education " Moscow automobile and road state technical University (MADI), Moscow, 125319, Russia

E-mail: gmazlumyan@mail.ru

Abstract. The article presents a method for assessing the permissible speed of a particularly heavy-duty vehicle based on a road train using a combined hydraulic transformer in the tractor transmission, as one of the indicators of cross-country ability. In this paper, we consider one of the main ways to increase the cross-country ability of heavy-duty vehicles: the creation of heavy-duty vehicles based on a road train with an active trailer link. Increasing the passability of heavy-duty vehicles is primarily aimed at increasing the permissible speed of heavy-duty vehicles and the maximum angles of ascents to be overcome. When justifying the choice of an algorithm for controlling the voltage of a torque Converter inductor based on the speed of particularly heavy-duty vehicles, it is advisable to use the external characteristics of the torque Converter as evaluation indicators: the dependence of the efficiency coefficient, the dependence of the moment created by the pump, and the dependence of the transformation coefficient on the gear ratio of the torque Converter. As criteria for evaluating the feasibility of using a combined torque Converter for especially heavy-duty vehicles, it is advisable to use the maximum increase in efficiency, the moment created by the pump wheel, and the transformation coefficient of torque converters on magnetic fluid in relation to the efficiency, the moment created by the pump wheel, and the transformation coefficient of the torque Converter based on a conventional working fluid in the operational range of gear ratio changes. The external characteristic is calculated after calculating the blade system and selecting the design to determine whether the main parameters of the characteristic correspond to the parameters specified in the task for designing the torque Converter. Such calculations often have to be performed when fine-tuning torque converters with a known design, when the influence of geometric parameters of individual elements of the blade system (input and output angles, wheel radii, the number of blades in the impellers, etc.) on the transforming and loading properties of the torque Converter is investigated.

1. Introduction

The use of hydro-mechanical transmission has raised the transmission design of especially heavy-duty vehicles to a qualitatively new technical level in terms of reducing the load on the transmission and improving the driver's working conditions, but it has not solved the problem of individual power

Content from this work may be used under the terms of the Creative Commons Attribution 3.0 licence. Any further distribution of this work must maintain attribution to the author(s) and the title of the work, journal citation and DOI.
Published under licence by IOP Publishing Ltd

IOP Conf. Series: Materials Science and Engineering 1159 (2021) 012072 doi:10.1088/1757-899X/1159/1/012072

supply to the wheels depending on their rolling conditions. Various researchers have concluded that the achieved technical level of mechanical and hydro-mechanical transmission designs of vehicles with particularly high load capacity (VPHL) exhausted the possibility of further significant increase in the mobility of multi-axle chassis. Thus, when increasing the power of the power plant VPHL over 600 kW, it is difficult to create mechanical and hydro-mechanical transmissions with high technical and economic indicators. Therefore, with the growth of power, load capacity, and as a result of the number of axles, new technical solutions and qualitatively better methods of converting, distributing and transmitting energy in transmissions are needed VPHL. All this has led to the fact that at present the country and abroad are searching for new flexible types of transmissions for vehicles of particularly high load capacity for the automotive industry [1].

2. The study of a method for assessing the cross-country capability of a particularly heavy-duty vehicle

In this work, we consider one of the above main ways to increase the patency of a VPHL: creature of a VPHL based on a road train with an active trailer link. Increased cross-country ability of a VPHL it is aimed, first of all, at increasing the permissible speed of movement of a VPHL and the maximum angles of ascents to be overcome.

Significant increase in cross-country ability of a VPHL it is possible by improving the technical and operational characteristics of the hydro-mechanical transmission of the tractor transmission and the motor-wheels of the trailer device.

To improve the technical and operational characteristics of the hydro-mechanical transmission of a VPHL it is proposed to use in its composition a combined torque Converter controlled by an external rotating magnetic field, and as the drives of the wheels of the trailer link - valve-inductor drives [2].

Initial data for solving the problem of estimating the permissible speed of movement of a VPHL with a hydrodynamic transmission based on a combined torque Converter controlled by a rotating magnetic field and an active trailer link are divided into the following main groups:

when justifying the choice of an algorithm for controlling the torque Converter inductor strength by speed of a VPHL it is advisable as the evaluation indicators to use external characteristics of the Converter:

the dependence of the efficiency,

the torque generated by the pump,

the dependence of the transformation coefficient from the reduction ratio of the torque Converter [3,4].

As criteria for evaluating the feasibility of using a combined torque Converter for a VPHL, it is advisable to use the maximum increase in efficiency, the moment created by the pump wheel, and the transformation coefficient of torque converters on a magnetic fluid in relation to the efficiency, the moment created by the pump wheel, and the transformation coefficient of a torque Converter based on a conventional working fluid in the operational range of gear ratio changes [5].

Scheme of the algorithm for estimating the permissible speed of movement of a VPHL with a hydrodynamic transmission based on a rotating magnetic field controlled combined torque Converter includes the following main stages:

1) preparation and input of initial data;

2) modeling of system dynamics «internal combustion engines - hydrodynamic transmission based on a combined TC – WT VPHL» with speed control of a VPHL.

3) the feasibility of using a hydrodynamic transmission based on a combined torque Converter controlled by a rotating magnetic field is evaluated by comparing the combined transmission options with each other and with existing (known) transmissions of traditional design;

4) formation of requirements (General, special, special) for the developed structures by hydrodynamic transmission based on a combined torque Converter controlled by a rotating magnetic field of a VPHL.

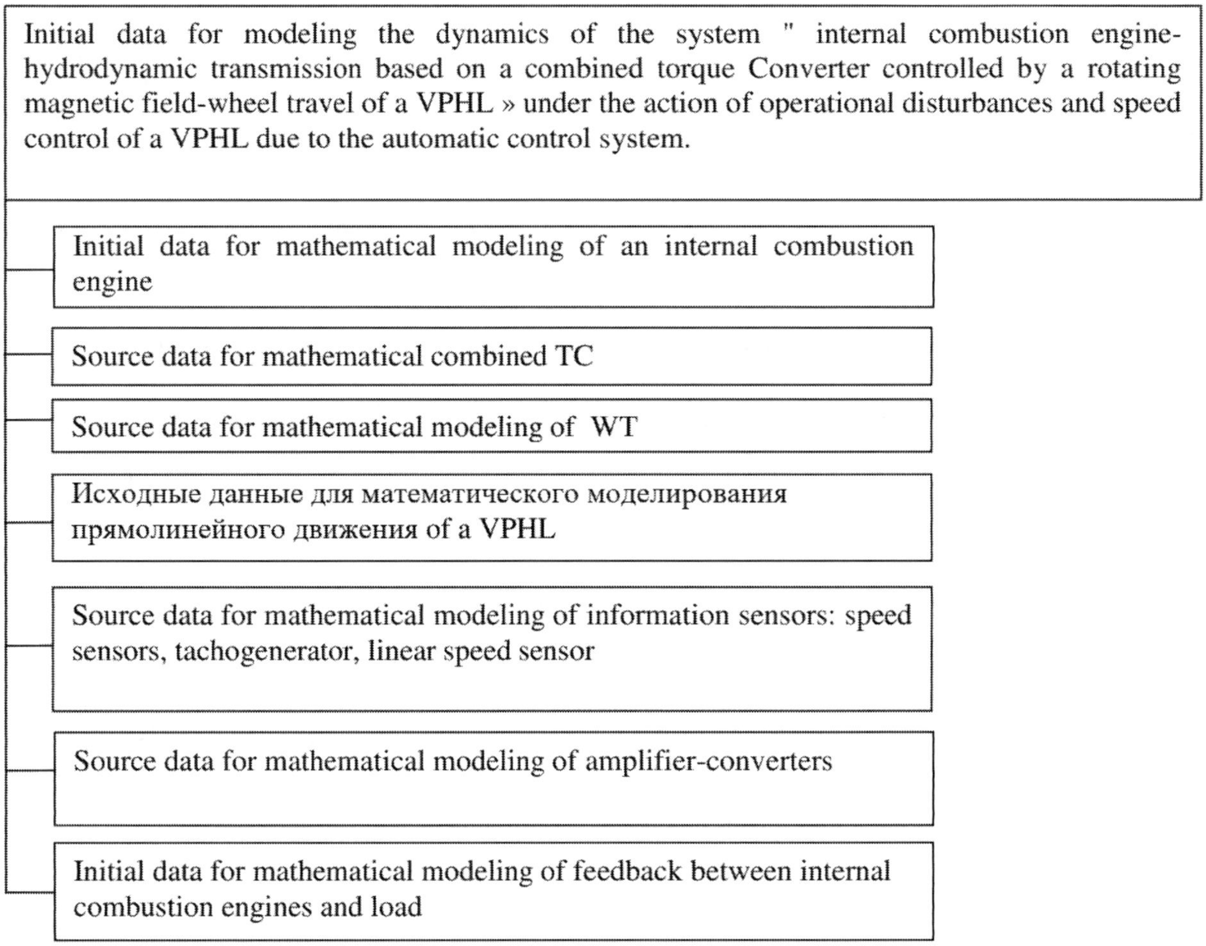

Figure 1. Structure of initial data for modeling the dynamics of the system "internal combustion engine-hydrodynamic transmission based on a combined torque Converter controlled by a rotating magnetic field-wheel travel of a VPHL" under the action of operational disturbances and speed control of a VPHL due to an automatic control system

The external characteristic is calculated after calculating the blade system and selecting the design to determine whether the main characteristic parameters (such as M, N, n, i, K0, P, d75, etc.) correspond to the parameters specified in the task for designing the torque Converter.

Such calculations often have to be performed when fine-tuning torque converters with a known design, when the influence of the geometric parameters of individual elements of the blade system (input and output angles, wheel radii, number of blades in the impellers, etc.) on the transformative-loading properties of the torque Converter is investigated.

Equations of moments. According to the theorem on the moment of the amount of movement, the second change in this moment is equal to the moment of external forces acting on this wheel. Second change in the moment of the amount of fluid movement in the impeller:

$$\frac{dL}{dt} = \rho Q(c_{u2}r_2 - c_{u1}r_1) \tag{1}$$

The moment of external forces on the pump MH is determined by the force action of the channel walls and blades driven by the motor on the fluid flow:

$$M_H = Q\rho(c_{uH2}r_{H2} - c_{uH1}r_{H1}) \tag{2}$$

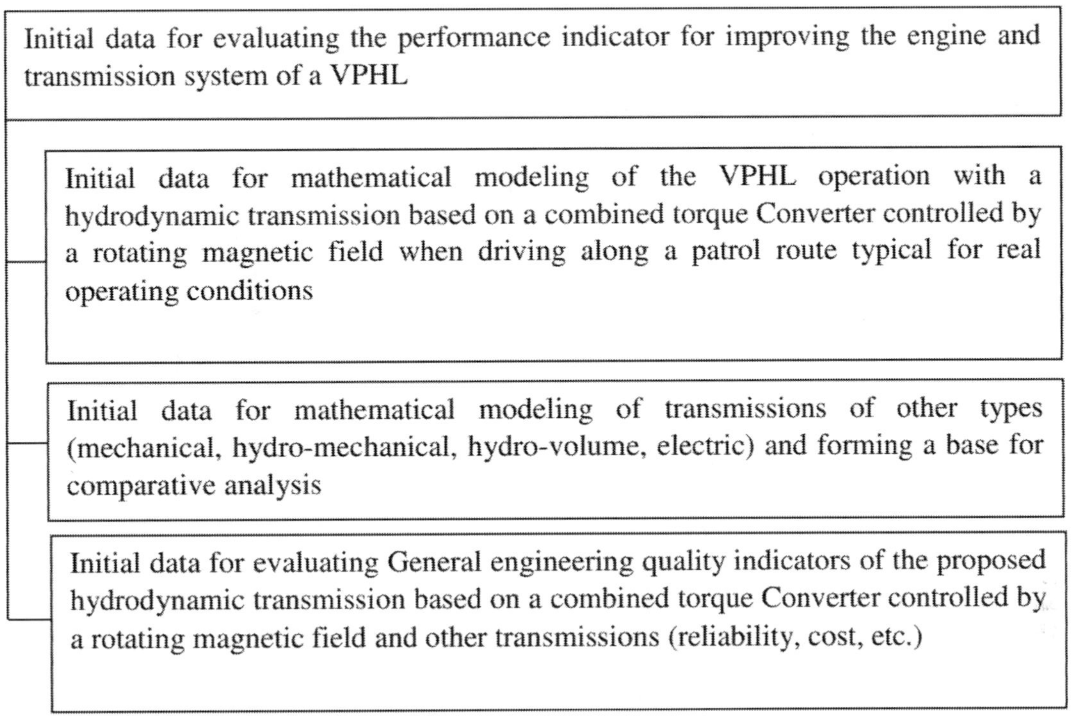

Figure 2. structure of the initial data for evaluating the efficiency indicator for improving the engine and transmission system of the VPHL

The moment of the amount of movement on the pump increases as the flow moves from the inlet to the outlet, and the torque is a positive value. The moment of the amount of fluid movement on the turbine is reduced, and the torque is a negative value:

$$M_T = Q\rho(c_{uT2}r_{T2} - c_{uT1}r_{T1}) \tag{3}$$

Torque generated at the reactor:

$$M_P = Q\rho(c_{uP2}r_{P2} - c_{uP1}r_{P1}) \tag{4}$$

This moment is transmitted to the body with which the reactor is rigidly connected, and can be positive or negative [6].

It follows from equations (3) and (4) that if the external moment is zero, then in the General case $r_2c_{u2} = r_1c_{u1} = rc_u = const$ that is, in the absence of energy transfer and in the absence of friction, the motion of the liquid is described by the law: $rc_u = const$ According to this law, the fluid moves in the inter-wheel gaps of torque converters and hydraulic couplings.

Determination of the flow rate Q in the working cavity of the torque Converter. The transfer of energy from the drive shaft to the slave in hydrodynamic transmission is accompanied by energy losses to overcome various resistances. Losses in hydrodynamic transmission can be divided into three types - hydraulic, mechanical and volumetric, which are taken into account by the corresponding efficiency: hydraulic PG; volumetric po and mechanical pmex.

Hydraulic losses are counted as specific energy or pressure losses. The viscosity of the working fluid causes a change in the H_{tH} and H_{tT} heads. So, pump head

$$H_H = H_{tH} - h_{\text{пот}.H} \tag{5}$$

where $h_{\text{пот}.H}$ - hydraulic losses in the pump.

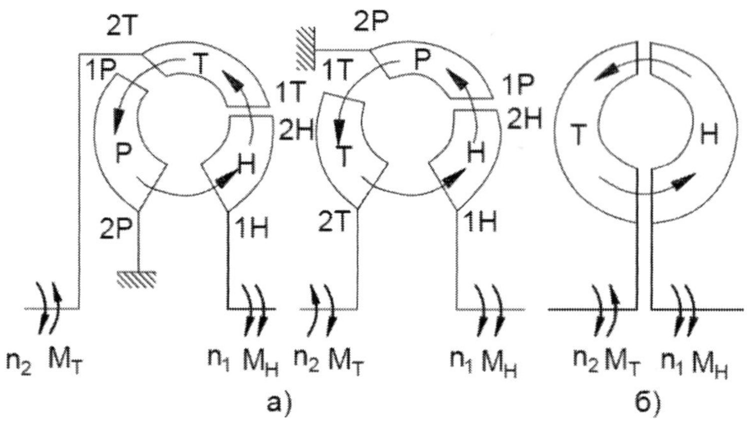

Figure 3. The scheme of movement of fluid in a hydrodynamic transmission:
a - in the torque Converter; b - in the fluid coupling

Turbine head

$$H_T = H_{tT} - h_{\text{пот}.T} \quad (6)$$

where is kpot. T - hydraulic losses in the turbine.

From the principle of operation of hydrodynamic gears, it follows that

$$H_H = H_T + h_{\text{пот}.P} + h'_{\text{пот.}} \quad (7)$$

Formula (7) expresses the balance of specific energy. From the energy balance equation in the torque Converter (7), the flow rate Q can be obtained for various operating modes. In hydraulic machines, the flow deviation is taken into account at the output of the impellers by the coefficient:

$$\mu = \frac{c_{u2}}{c_{u2\infty}} \quad (8)$$

where c_{u2}, $c_{u2\infty}$ - projections of the absolute velocity on the circumferential one, respectively, for a finite and infinite number of blades (see Fig. 4, a, b, v).

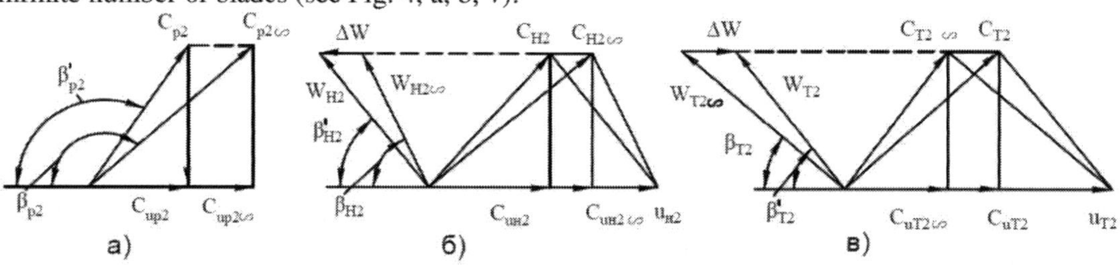

Figure 4. velocity Triangles taking into account the finite number of blades:
a-in the reactor; b - in the pump; v - in the turbine

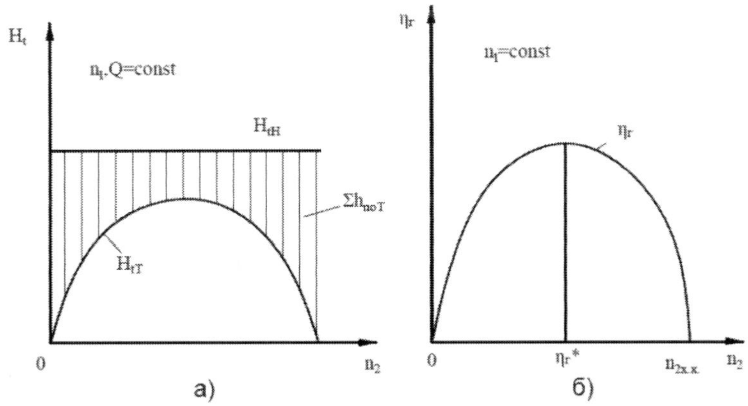

Figure 5. Pressure characteristic of the torque Converter: a-head balance, $H_{tH\infty}$ и $H_{tT\infty}$- theoretical pressure of the pump and turbine; shaded area-hydraulic losses in the torque Converter; b-hydraulic efficiency

where $c_{m2\ \text{пред}}$ и $c_{u2\ \text{пред}}$ the meridional and circumferential components of the speed at the exit of the previous wheel.

Flow rate value for different i:

$$\frac{Q}{\omega_H} = \frac{-2(b+ic) \pm \sqrt{4(b+ic)^2 - 4a(d+mi^2+2fi)}}{2a} \tag{13}$$

The calculation of the external characteristic is preceded by the calculation of the internal parameters of the torque Converter Q and H for various i. the flow Rate is determined by the equation. To determine the coefficients of the equation related to the geometric parameters of the blade system of the torque Converter, you must first determine the following calculated values: F_{H2}, F_{T2}, F_{P2}, - the cross-sectional area of the flow, normal to the direction of the speed cm, at the exit of the corresponding impellers.

The areas are determined taking into account the constraint of the flow by the blades according to the formula:

$$F = 2\pi r b \varkappa. \tag{14}$$

In addition, determine the coefficient of resistance of the wheel

$$k = \lambda \frac{l}{4R_{\text{г ср}}} \cdot \frac{1}{F^2_{\omega\ \text{ср}}}, \tag{15}$$

where λ - the experimental resistance coefficient is accepted within 0.06-0.08;

$R_{\text{г ср}}$ - the average hydraulic radius of the channel;

$F_{\omega\ \text{ср}}$ - average area of wheel channels, normal to the direction of relative flow rate.

The K coefficient for each wheel is calculated for several sections (see figure 6). figure 6 shows Δl_m - the distance along the midline in the Meridian section between the selected points; $\Delta lm = \frac{\Delta lm}{\sin\beta}$ - the actual distance along the average current line between the selected points; β - the current angle of inclination of the blade at the point (taken from the conformal diagram); $l = \Sigma\Delta l$ - length of the middle line of the blade profile; $t = \frac{2\pi r}{z}$ - current blade pitch at the given radius r; $a = t\sin\beta - \delta$ - the width of the channel between the blades; δ-the thickness of the blade at this point $f = ab$ - the area of the channel between the blades, normal to the direction of velocity ω (b is the width of the channel at a given radius r in the meridional plane); $f_{\omega\ \text{ср}} = \frac{\sqrt{n}}{\sqrt{1/f^2_{\omega 1} + 1/f^2_{\omega 2} + \cdots}}$ - average area of the channel (n is the number of selected calculation points); $F_{\omega\ \text{ср}} = f_{\omega\ \text{ср}} \cdot z$ - the average area of the wheel channels, normal to the direction of relative speed; $4R_{\text{г}} = \frac{2ab}{a+b}$ - the four-fold hydraulic radius of the channel at this point; $4R_{\text{г ср}} = \frac{4R_{\text{г}1} + \cdots + 4R_{\text{г}n}}{n}$ - the average value of the quadruple hydraulic radius.

It is advisable to calculate the value of K in tabular form. After calculating the geometric parameters, we determine the coefficients a, b, c, etc., and the flow rate Q according to the equation for different values of i. The known flow rates are used to determine the pressure values $H_{tH} = f(i)$ and $H_{tT} = f(i)$ and calculate the hydraulic moments on the impellers:

$$M_H = \frac{\rho g Q H_{tH}}{\omega_H} \tag{16}$$

$$M_T = \frac{\rho g Q H_{tT}}{\omega_H \cdot i} \tag{17}$$

Determine the hydraulic efficiency:

$$\eta_\text{г} = \frac{H_{tT}}{H_{tH}} = f(i) \tag{18}$$

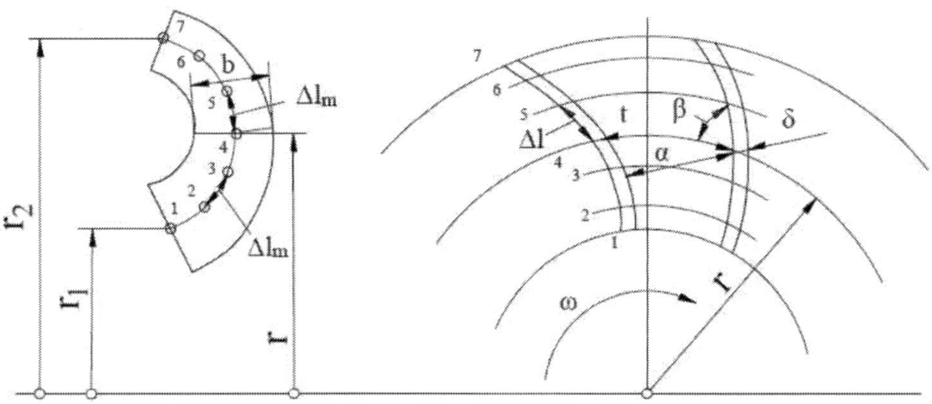

Figure 6. Calculation of the coefficient of resistance *k*
(points 1-7 correspond to the considered blade sections)

Torque converters that provide a constant mode of engine operation when the resistance to movement changes are usually called opaque. The load characteristic of an opaque torque Converter is represented by a single square parabola (Fig. 7, a).

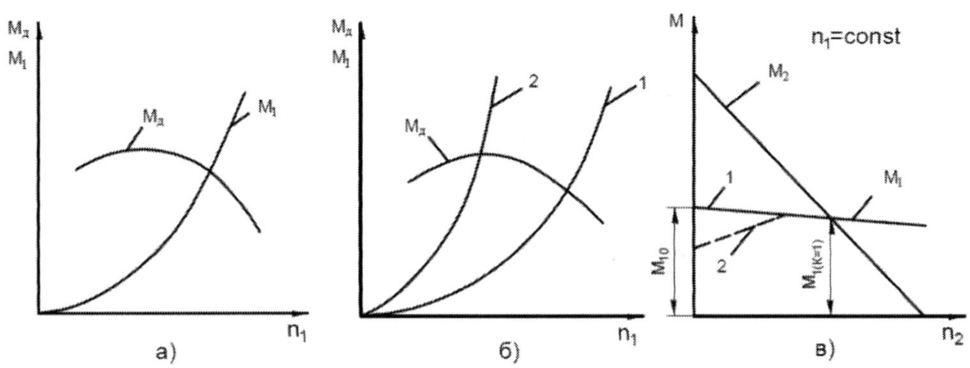

Figure 7. Transparency of torque converters:
a-load characteristic of an opaque torque Converter; b-load characteristic of a transparent torque Converter; C-external characteristic of a torque Converter: 1-direct transparency;
2 - reverse the transparency

In General, when the speed of the drive shaft and the torque on it change, the transparency coefficient is determined from the ratio:

$$\Pi = \frac{\lambda_{M1\,max}}{\lambda_{M1\,min}} \tag{19}$$

Where λ_{M1} max and λ_{M1} min are taken respectively M_{1max} and M_{1min}.

Determination of disk friction moments and overall efficiency.

The total efficiency of the torque Converter is calculated by taking n_0=0.95÷0.97 and determining the moments of disk friction and mechanical losses.

Moment of disk friction of adjacent surfaces $M_д = \frac{N_д}{\omega_д}$ We denote M_{1mex} and M_{2Mex} -respectively, the moments of mechanical losses in the seals and bearings of the leading and driven parts of the torque Converter.

For traction modes of operation of the torque Converter forward when i < 1.

$$\begin{cases} M_1 = M_Н + M_{д\,НТ} + M_{д\,НР} + M_{мех1}; \\ M_2 = M_Т + M_{д\,НТ} - M_{д\,ТР} - M_{мех2}. \end{cases} \quad (20)$$

If i > 1, the M_{dHT} is preceded by a minus sign.

For reversible operation modes:

$$\begin{cases} M_1 = M_Н + M_{д\,НТ} - M_{д\,НР} - M_{мех1}; \\ M_2 = M_Т + M_{д\,НТ} + M_{д\,ТР} + M_{мех2}. \end{cases} \quad (21)$$

If the supports of the torque converters are made in the form of rolling bearings, and the seals are labyrinth, then $M_{мех\,1} = M_{мех\,2} \cong 0$. If there are cast-iron sealing rings or cuffs on the shafts, the moments M_{mex1} and M_{mex2} should be determined experimentally for each type of torque Converter. With this in mind, we determine the overall efficiency and build the external characteristic of the torque Converter M_1, M_2, $\eta = f(i)$ for n_1 = const.

$$\eta = \frac{M_2}{M_1} \cdot \frac{n_2}{n_1} \quad (22)$$

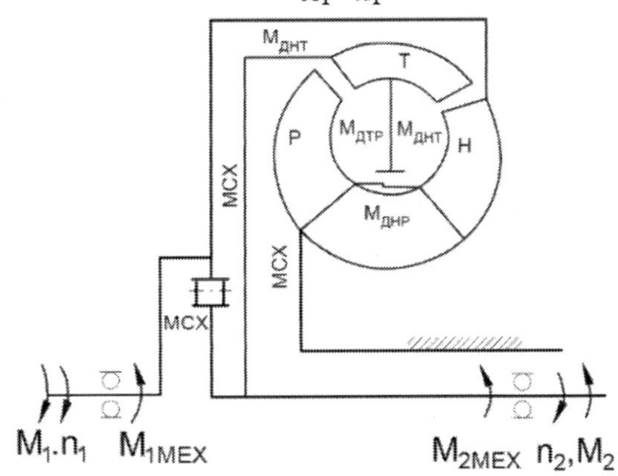

Figure 8. Diagram of the disk friction moments in the torque Converter

If the calculated parameters do not match the specified ones, we Refine the geometry of the blade systems theoretically or experimentally.

3. Conclusion

Thus, the proposed method makes it possible to assess the cross-country passability of wheeled vehicles of especially heavy carrying capacity with all-wheel control, taking into account curvilinear movement.

References

[1] Reimpell J and Sponagel P 1995 *Fahrwerktechnik: Reifen und Räder* (Vogel Fachbuch, Würzburg)

[2] Sova A N, Mazlumyan G S, Egorov O V, Eruslankin S A, Shadrin S S, Sova V A The results of modeling and evaluating the dynamics of a road train with an active trailer link based on valve-inductor electric machines *IOP Conference Series: Materials Science and Engineering*

[3] Sova A N, Mazlumyan G S, Egorov O V, Eruslankin S A, Shadrin S S, Sova V A The results of modeling and evaluating the relative slip coefficients during the interaction of the pneumatic tire with the supporting surface to determine the forces acting in the contact patch *IOP Conference Series: Materials Science and Engineering*

[4] Sova A N, Mazlumyan G S, Egorov O V, Eruslankin S A, Shadrin S S The results of modeling and evaluating the dynamics of the active semitrailer *IOP Conference Series: Materials Science and Engineering*

[5] *Dynamics of the "Road - tire - car - driver" system.* Under. ed. A A Khachaturova (Moscow: Mechanical Engineering, 1976)

[6] Mazlumyan G S, Trifonova G O, Trifonova O I, Presnyakov LA, Bulanov S V 2017 *Blade machines and hydrodynamic transmissions (textbook)* (Moscow: branch of FSUE "TSENKI" - KBTHM) ISBN 978-5-7962-0225-8

IOP Conf. Series: Materials Science and Engineering 1159 (2021) 012073 doi:10.1088/1757-899X/1159/1/012073

Modeling quality criteria for restoring parts of cars, vehicles and road-building machines

M Yu Karelina[1], A I Belyaev[2], A V Terentev[2], T Yu Cherepnina[1]

[1]Moscow Automobile and Road Construction State Technical University (MADI) Leningradsky Prospekt 64, Moscow, Russia
[2]Saint Petersburg State University of Architecture and Civil Engineering, 4 2nd Krasnoarmeyskaya str., Saint Petersburg, 190005, Russia

E-mail: karelinamu@mail.ru

Abstract. The article analyzes the types, methods and techniques of parts processing at restoration and quality criteria of parts of cars, other vehicles and road-building machines in terms of efficiency. Correlations between the quality criteria and the functionality of the part have been established, for which the requirements for machine safety and reliability after the restoration of individual parts have been taken. A mathematical model has been developed in the form of an algorithm for selecting the best variant of the restoration technology for a functionally important part, based on a typical restoration technological process, which consists in the development of a multi-criteria problem, the initial postulates of which are matrices: W "wear resistance → durability"; S "safety → reliability"; F "part fragility → machine safety". New parameters in the form of coefficients describing roughness, hardness and riveting of a part have been proposed. In particular, the authors define the coefficient of technological isotropy of roughness, which evaluates the roughness parameters in the process of technological changes of a part; the coefficient of isotropy of hardness, which compares the hardness parameters on the surface of the part and in the core of it; the coefficient of intensity of hardness, which is a complex parameter and in which there is a change of hardness by the depth of the part and the value of hardness depth (riveting). The purpose of introducing these coefficients is consistent with the model being developed to select the optimal technology for restoring parts of cars, other vehicles and road-building machines.

1. Introduction

The modern development of economy leads to an increase in construction volume in general and roads in particular. Road network development leads to intensification of their maintenance. This is where modern road construction technologies and use of modern materials intervene. All these factors together create conditions for modern cars development, vehicles and road construction machines, both quantitatively and qualitatively. To provide intensive construction technologies, the use wide range of purposes technological equipment is required.

The main features of operation of cars, vehicles and road-building machines are: high operation intensity; high price; isolation from service bases and complication of machines repair and restoration process in connection with this.

Parts reuse for cars, vehicles and road construction machines is, in fact, a reserve from the economic approaches point of view. An example is economic experience of all countries of the world, especially

Content from this work may be used under the terms of the Creative Commons Attribution 3.0 licence. Any further distribution of this work must maintain attribution to the author(s) and the title of the work, journal citation and DOI.
Published under licence by IOP Publishing Ltd

developed ones. As a specific example, we can name official data of the Tractor Parts Dealers Association (USA), on basis of which hundreds of enterprises carry out components and individual parts restoration that have gone out of service.

2. Research questions

Restoring parts technology typification for cars, vehicles and road-building machines. According to official statistics, 75% of the parts recognized as final marriage can in fact be restored [1]. Worn-out parts restoration, their double or even multiple use is economically feasible. Comparison of parts restoration with their new manufacture leads to the following conclusions: lower costs, lower labor intensity, high quality [2]. Rebuilding parts of cars, vehicles and road construction machinery is the best way to machines durability influence in general.

With the absolute correctness and relevance of these conclusions, it should be noted that issues related to restoring parts technology of cars, vehicles and road-building machines are not worked out. Each enterprise engaged in parts restoration solves technology issues independently, based on economic and material capabilities and considerations. At the same time, many questions in this area are well developed. According to the source, restoring parts technological process (TP) is a process that contains targeted actions to change a certain state of a part in order to restore its operational properties.But when organizing a production intended for restoring parts, it is recommended to use standard technological processes (STP). This makes it possible to reduce labor costs for technological process development, standardize parts nomenclature, and, as a result, make production more efficient from an economic point of view. It should be noted that the choice of the best route for specific part restoration of RBM depends on an understanding of various operations technological capabilities and their characteristic features.

Technological processes typification [3] raises technology level, reduces volume and shortens production preparation time.

Technological process schemes of restoring smooth shafts and axles are presented in the form of three most possible routes [4].

Technological processes typification implies the development of such technological process in which there would be a maximum number of operations varieties for a given type of parts.

A further algorithm for choosing the best technology option, based on the above STP, is to develop a multicriteria problem, the initial postulates of which are matrices: W "wear resistance → durability"; P "preservation → reliability"; F "part fragility → machine safety". A standard technological process is used to develop an optimal technological process with different objectives. Using multi-criteria dynamic programming and a typical technological process of part restoring, you can choose a sequence of operations that provide the best quality according to preferred criteria: either "durability" or "reliability" or "safety".

The authors [5] identified the main dependencies (three) for technological sequence choice for restoration of a car part or RBM part, where the difference was in different technological and, most importantly, economic indicators.

3. Methods and models

In most of the works devoted to this problem, criteria were used to select optimal technological process for cars parts restoring of vehicles or road construction or other machines, and mainly taking into account economic criteria.

To assess the choice of a method for part restoring, authors [6] proposed to select some evaluation criteria: shape, dimensions, coating thickness, surface hardness, part elements fatigue strength, and acting loads nature.

The author [6] proposes to solve the problem of choosing a method for part restoring that meets a complex technical and economic criterion, and technological process structure on the basis of a graph representation of possible options for combining technological operations that ensure objective function minimum value. Objective function value is put down at the top of graph.

The author [7] developed this system for choosing a method for recovering a detail by introducing graph arcs lengths values are given in their discontinuities.

Existing approaches to the choice of a restoring parts method for cars, vehicles and road-building machines inevitably face the need to solve problems in a multi-criteria formulation. In this case, as a rule, widespread methods of reducing multi-criteria problems to single-criteria ones (methods of using composite criteria) are used [8, 9].This approach has a significant drawback: when using composite criteria, the deficiency in one criterion is compensated for by another [10]. The most complete picture of possible solutions is provided by the method of effective plans set finding (Pareto set), since the use of this method allows you to compare all competitive options between each other [11]. The final choice of a solution from the set of Pareto sets can be carried out using the zoning method according to the principle of observing the hierarchical ratio of research environment possible states probabilities [12, 13].

When passing from a multicriteria problem to problems of probability theory, the probabilities of research environment states$p_j$are, in their meaning, adequate to criteria relative importance coefficientsc_j, i.e. $p_j \equiv c_j$.

Analytical solution is given in [12-11].

$$c_j = \begin{cases} \frac{1}{k}, if\ j = k \\ \frac{\lambda}{k}, \text{if}\ j < k,\ \text{where}\lambda = \frac{n-1}{n}, \\ \frac{1-\lambda}{n-k}, \text{if} j > k \end{cases}$$

where index$k$is determined from condition$a_{kj} = max_j\, a_{ij}$.

Fundamental difference between developed method is formalized relationship absence between relative importance coefficients according to individual criteria, which makes it possible to determine the values of solutions relative importance coefficients in multicriteria problems in accordance with the real conditions of system functioning under study.

4. Results and discussion

Consider machine durability in terms of the part properties. Durability is machine property to have the same performance for some time. Knowing the operating conditions, it is possible to determine the operability [13, 14]. Surface microrelief and correct lubricants are important [15]. According to other sources, the main reasons limiting reliability (and they are associated with increasing durability methods), in addition to surface wear, include the breakdown of the part itself, surface integrity violation due to contact stresses or work hardening, or corrosion [16].

Conclusion - machines durability is primarily manifested in the form of particular part function - wear. *Here the relationship "wear resistance → durability" is established.*

Reliability is determined by part property, while maintaining its characteristics, to perform specified functions. This occurs during the part running time period [17]. It can be concluded that relationship between durability and safety of the machine as a whole is obvious. *This is where the «persistence → reliability» relationship is established.*

Let us consider operation safety of cars, vehicles and road-building machines from the part properties point of view. It is possible to reduce cracking possibility by applying surface plastic deformation methods [18]. If safety indicators do not meet the requirements, then this can lead to catastrophic consequences in the field of life, human health, ecology and economic factors, and ultimately reduces machine efficiency to a minimum [19].*Here the link "part fragility → machine safety" is established.* A multicriteria dynamic model has been developed for choosing the best option for part restoring technological process, depending on priorities set in advance.

Surface layer of part has important characteristics, such as roughness, hardness (in the form of work hardening or heat treatment) and, as a result, residual stresses.

Roughness. Roughness is a very important and diverse quality criterion. This ambiguity, complexity and variety of relationships leads to the need to introduce a new parameter describing the roughness. It is proposed to introduce *roughness technological isotropy coefficient* for the partK_{RTI}.

IOP Conf. Series: Materials Science and Engineering 1159 (2021) 012073 doi:10.1088/1757-899X/1159/1/012073

Roughness parameters comparison in process of technological changes in part, that is, roughness comparison before and after operation, and will be desired value$K_{RTI} = \frac{RT_{before}}{RT_{after}}$,where:$RT_{before}$ – roughness parameter before any operation.

Hardness. It is proposed to introduce hardness isotropy coefficient for partK_{HIS}.Then hardness parameters comparison precisely on part surfaceH_{SUR}.and hardness of part coreH_{CORE}.and will be the required valueK_{HIS}:$K_{HIS} = \frac{H_{sur}}{RT_{core}}$.

Depth of hardness (work hardening). Surface hardness of part is changed not only by thermal or chemical-thermal treatment, but also by surface-plastic deformation methods. Enter *hardness intensity factor* for partK_{HI}. This is complex parameter comparison, in which there is a change in hardness along the depth of part and value of hardness depth (work hardening). For hardness, this isK_{HIS}= H_{SUR}/H_{CORE}.

5. Conclusion

After analyzing various indicators of hardness depth or work hardening, we came to conclusion that it is more preferable to evaluate hard layer depth during thermal or chemical-thermal treatment, during blade processing, or during SPD in absolute terms - in millimeters.

In general, proposed indicators: roughness technological isotropy coefficient, hardness isotropy coefficient and hardness intensity coefficient, can be summarized as integral quality indicators.

A mathematical model has been developed using dynamic programming method to obtain optimal trajectory of part movement from the initial state to the final state, based on the Bellman principle.

6. Recommendations

These integral quality indicators greatly simplify and facilitate the optimal selection process modeling of the best variant of technological process. Achieving maximum efficiency in technological process design of restoring parts of cars, vehicles and road-building machines according to the selected justified criteria (durability, reliability, safety) is possible by integrating zoning method using observing principle of probabilities ratio of research environment possible states into solving dynamic programming problem.

References

[1] Nurakova A S 2005 Economic efficiency analysis of technological options for worn parts restoration during machines repair *Bulletin of the Eurasian Humanitarian Institute* **3** 56–61

[2] Vinogradov V S 1999 *Equipment and technology of automatic and mechanized arc welding* (Moscow: Higher school)

[3] Popov V A, Shakhrai M L , Ivanov G A et al. 1952 Technology of heavy engineering [Text] (Moscow: Mashgiz)

[4] Molodyk N V , Zenkin A S 1989 *Restoring machine parts. Directory* (Moscow: Mashinostroenie)

[5] Filatov M I, Slavnenko V P 2011 Theoretical substantiation of restoring car parts method on the basis of information and energy assessment [Text] *Bulletin of OSU* **10 (129)** 151–155

[6] Panteleenko F I, Lyalyakin V P , Ivanov V P, Konstantinov V M 2003 *Restoration of machine parts: reference book* [Text] (M.: Mechanical Engineering)

[7] Bondareva G I 2010 System analysis of objects, functions and resources in the processes of restoring machine parts [Text] *Bulletin of the FSEI HPE MGAU* **2** 119–124

[8] Sosnin N A , Vichik B L 1992 *Plasma coatings* (Sankt-Petersburg)

[9] Taratuta N N , Sverchkov A A 1985 *Progressive methods of car repair* (Minsk: Urajay)

[10] Terentyev A V , Prudovskiy B D 2016 Decision-making methods under conditions of an uncertain state of the "external environment" / Transport planning and modeling: collection of articles. Materials of the International Scientific and Practical Conference (May 26-27, 2016) *SPbGASU* 145–149

[11] Terentev A V, Prudovsky B D 2015 Methods for determining the Pareto set in some linear programming problems [Text] *Notes of the Mining Institute.* (Sankt-Petersburg: National

Mineral Resources University "Mining") Vol 211 89–90

[12] Belyaev A I 2017 Optimization of road-building machines parts restoration using zoning method *Bulletin of civil engineers* **6**(65) 228–233

[13] Belyaev A I, Terentev A V, Pushkarev A E 2018 Assessment of methods for restoring parts of road-building machines in a multi-criteria setting *Bulletin of civil engineers* **1**(66) 123–127

[14] Feschenko V N 2016 *Constructor reference. Book 2. Design of machines and their parts: study guide. allowance* (Moscow: Infra-Engineering)

[15] Randolph M F 2003 Science and empiricism in pile foundation design. *Geotechnique* **53(10)** 847–875 doi:10.1680/geot.2003.53.10.847

[16] Zorin V A 1998 *Increasing the durability of road-building machines by improving the system of maintenance and repair: dis. Dr. Tech. sciences.* (Moscow)

[17] Karthigeyan S , Ramakrishna V V G. S. T., Rajagopal K 2007 Numerical investigation of the effect of vertical load on the lateral response of piles *Journal of Geotechnical and Geoenvironmental Engineering* **133(5)** 512–521

[18] Pulyaev S M, Stepanov M A, Kaitukov B A, Lukyanov N A et al. 2015 *Mechanical equipment and technological complexes* (Moscow: MGSU)

[19] Mandolini A, Viggiani C 1997 Settlement of piled foundations *Geotechnique* **47**(4) 791–816 doi:10.1680/geot.1997.47.4.791

[20] Rubailov A V, Kerimov F Yu, Dvorkovoy V Ya, Golovin S F et al 2007 *Operation of lifting and transport, construction and road machines* ed. E S Lokshin (Moscow: PublishingCenter "Academy")

IOP Conf. Series: Materials Science and Engineering 1159 (2021) 012074 doi:10.1088/1757-899X/1159/1/012074

Tensor analysis of uncertainty in freight transport ULS-systems

P F Yurchik [1], O I Maksimychev[1], V B Golubkova[1], A V Volosova[1]

[1] Automobile and Road Construction State Technical University, 64, Leningradsky prospect, Moscow, 125319, Russia

E-mail: upf.madi@mail.ru

Abstract. The article is devoted to the application of the tensor basis for the analysis of uncertainties in decision-making in freight transport ULS-systems. The use of tensors for constructing a mathematical model of the coordinator in the decision-making system is shown. Tensor equations make calculation possible the system parameters for any change in the structure of connections in the basic mathematical model. This provides additional advantages when making decisions in an uncertain environment. The authors also demonstrate the use of a tensor apparatus for constructing a mathematical model of deep uncertainty, which is impossible with traditional methods of analysis. The article shows that tensor analysis can be used to model decision-making under conditions of deep uncertainty in complex conditions, such as a catastrophes. Representing a fuzzy variable as a tensor makes it possible to explore additional properties of this variable. Tensors describing additional properties of the fuzzy variable (deviator, spherical tensor, symmetric tensor, and antisymmetric tensor) can be obtained from the fuzzy variable tensor. Applying special calculations to a fuzzy tensor variable makes it possible to obtain the membership function with high accuracy and find the defuzzified output value of the membership function. Tensor analysis significantly expands the scope of ULS-systems research.

1. Introduction

The tasks that arise when making decisions differ significantly by subject area, by the type of information received and presented, by the number and types of criteria, as well as by the problem situations that the decisions should resolve. The features of such tasks include a high degree of vagueness and uncertainty of information. Traditional methods of analysis are not sufficiently suitable for decision-making tasks in the implementation of transport dispatching, since in real conditions neither the initial data nor the ideas about the project and problem situations and the goal that was set are completely accurate and correspond to reality, which is compounded by the need to predict the development of the subject area parameters. So decision making in uncertainty is the main task in decision-making in freight transport ULS-systems [1]. Such systems should ensure decision-making in difficult conditions such as catastrophe.

The theory of catastrophes deals with processes in which a smooth change in the system parameters is interrupted by an abrupt change (catastrophes). As a result, the mode of existence of the system changes or it is destroyed. Thus, catastrophes can be called abrupt changes that occur in the form of a sudden response of the system to a smooth change in external conditions (see figure. 1).

Content from this work may be used under the terms of the Creative Commons Attribution 3.0 licence. Any further distribution of this work must maintain attribution to the author(s) and the title of the work, journal citation and DOI.
Published under licence by IOP Publishing Ltd

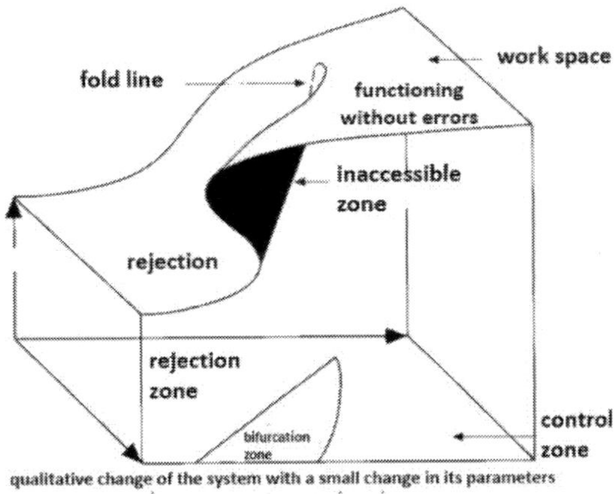

Figure 1. Surface of the catastrophe.

Let us consider the solutions $\Phi_1(t, x; c_a)$, $\Phi_2(t, x; c_a)$,... of a system of n equations defined in space R^N with coordinates $x=(x_1, x_2, ..., x_N)$,

$$F_i\left(F_i;\ c_a;\ t;\frac{dF_i}{dt};\frac{d^2F_i}{dt^2},\ \dots\dots\dots;\ x_l;\frac{dF_i}{dx_l},\frac{d^2F_i}{dx_l dx_m},\ \dots\dots\right)=0 \tag{1}$$

$1\leq i\leq n,\quad 1\leq l,\qquad m\leq N,\quad 1\leq a\leq k,$

The variables x_i and t can be considered spatial and temporal coordinates, respectively. Solutions of F_i describe the state of a certain system, so they are called variable States. The equations $F_i=0$ depend on k parameters of c_a, i.e. they can qualitatively affect the properties of solutions of F_i, so the parameters of c_a are the control parameters. Consequently, the theory of catastrophes considers the equilibrium state $F_i(c_a)$ of the potential function $U(F_i;\ c_a)$, which changes when the control parameters of the CA change. The state variables on which the function $U(F_i;\ c_a)$, depends are essentially generalized coordinates of the system under consideration. The main task of the dispatching service of a transport enterprise is to optimize traffic flows. The task of optimizing the decision-making process is a multi-criteria one.

$$Q(X)=\left(q_1(X),\dots,\ q_n(X)\right)\rightarrow max,\ X\in D, \tag{2}$$

D: $h_j(X)\geq 0$, $j=1..m$, X – desired solution,
$q_i(X)$ $(i=1..k)$ – solution quality function X,
$h_j(X)$ – restrictions that set the allowed range D of possible changes to the solution X.

The automated cargo transport control systemis proposed to be expressed as a two-level hierarchical system-coordinator. When the information support system is functioning, maximum coordination and coordination of its functioning processes should be achieved in order to achieve maximum efficiency of the system [2]. The coordination principle and the tasks of the control elements should be selected in such a way that the coordination principle is applicable and the transport information support system is coordinated based on the selected principle. The coordinator consists of the upper-level coordinating element C_0, which solves the problem of coordinating the entire system D_0, and the lower-level control elements $\{C^i,\ i\in M\}$, which solve the problems $\{D^i,\ i\in M\}$ of forming the optimal control signal for their functional subsystems (FS). The tasks $\{D^i,\ i\in M^{FS}\}$ of the control elements of the lower level of the coordinator are to select such a control signal $\tilde{m}^i\in M^i$ that ensures the maximum efficiency of each subsystem.

There are two types of conflicts that can occur in the management process:

– inter-level conflict, when achieving the maximum efficiency of one of the subsystems prevents achieving the maximum efficiency of the system as a whole;

– intra-level conflict, when an increase in the efficiency of one functional subsystem prevents the achievement of the maximum efficiency of another subsystem.

The coordinator consists of an top-level coordinating element C_o that solves the problem of coordinating the entire D_o system, and lower-level control elements $\{C^i,\ i \in M^{FS}\}$ that solve the tasks $\{D^i,\ i \in M^{FS}\}$ of generating the optimal control signal for their functional subsystems. The condition for the end of the iterative process in the coordinator is the choice of the optimal coordinating signal, i.e.:

$$A_k^i : \xi_g^i \rightarrow \xi_{g+1}^i,\ g = 1, 2, 3, \dots, g_z \tag{3}$$

The condition for the end of the iterative coordination process when optimizing parameters is the choice of the optimal control signal:

$$A_0^i : m_l^i \rightarrow m_{l+1}^i,\ l = 1, 2, 3, \dots, l_z \tag{4}$$

Tasks of the higher-level (coordinating) element and lower-level elements of the SIPD are formalized as tasks of predicting the state and maximizing the efficiency *(E)* of functional subsystems and dispatching objects. The main idea of the proposed approach is to introduce two levels of description of dispatching control systems. The macro description level of the system corresponds to the description of the implemented distribution:

$$Q = \{q(x_P, y_E),\ x_P \in X_P,\ y_E \in Y_E\}, \tag{5}$$

that is the probability *(P)* with which the required transport flow x_P will perform the y_E task. Thus, many of the microstates leading to the same distribution, or macrostate. The most likely distribution Q corresponds to a situation where there is the greatest uncertainty about the microstates of the system, since in this case there is the largest possible number of microstates leading to this distribution, assuming an equal probability of microstates.

2. Tensor basis for the system modeling of deep uncertainty

As noted above, the coordination process is complicated by the presence of various kinds of uncertainty. The computational complexity of the problem of processing fuzzy knowledge can be reduced through the use of tensor methodology [3]. The control process within the coordinator may be considered as a projection of the abstract control process in the space-structure of control flows. The dimension of space corresponds to the number of local control systems. The connection of local subsystems defines the coordinate system. The figure 2 shows a two-level management process, which is basic for building similar multi-level processes. The following designations are used in figure 2 below:

Z	– set of information signals from the managed process
Ξ	– set of signals from coordinator
M	– set of control actions on the process
ϑ	– set of control flows,
$coal_{\alpha\beta}$	– coalition interaction between control flow α and control flow β,
z_α	– signal of the controlled process to the control subsystem α,
ξ_α	– coordinating effect of the top-level coordinator on the control subsystem α,
m_α	– control influence of the subsystem α on the process

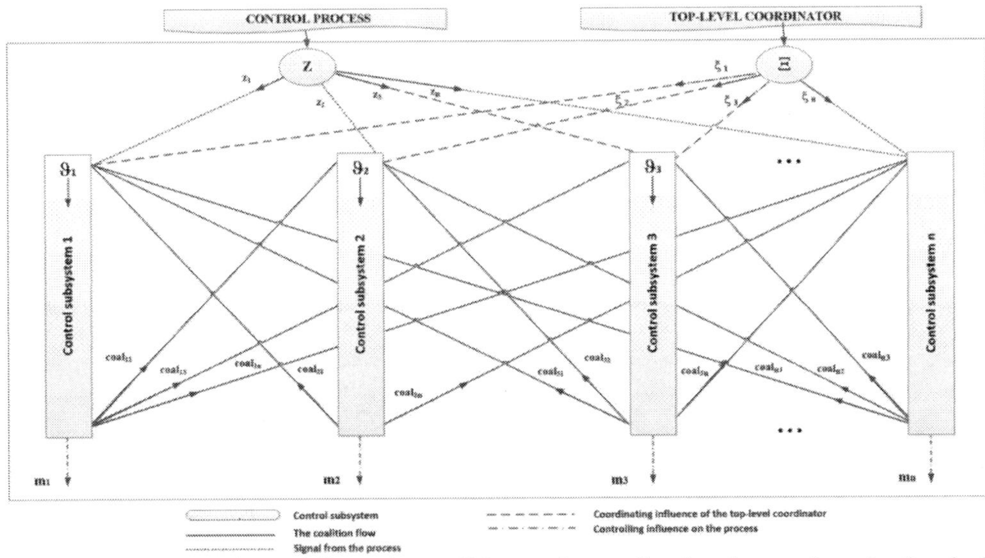

Figure 2. The distribution of flows in coalition and coordinating interactions in the decision-making process

The equations of the system can be written in tensor form:

$$\vartheta_\alpha = m_\alpha + \sum_\beta coal_{\alpha\beta}, \tag{6}$$

ϑ_α – control flow

$$coal_{\alpha\beta} = \chi_{\alpha\beta} * \vartheta_\beta, \tag{7}$$

$\chi_{\alpha\beta}$ – the value of the coalition flow of the control subsystem α, required to form a unit ϑ_β

$$z_\alpha = \mu_\alpha * \vartheta_\alpha, \tag{8}$$

μ_α- coefficient that characterizes the value of the signal of the controlled process to the control subsystem α necessary for forming the unit ϑ_α

$$\xi_\alpha = \eta_{\kappa\alpha} * \vartheta_\alpha,$$

(9)

$\eta_{\kappa\alpha}$- coefficient that characterizes the amount of coordinating influence on the subsystem α, necessary for the formation of a unit ϑ_α

$$m_\alpha = (\delta_{\alpha\beta} - a_{\alpha\beta}) * \vartheta_\beta, \left(\sigma_{\alpha\beta} - \chi_{\alpha\beta}\right) = (I - A), A - matrix\ of\ \chi_{\alpha\beta} \tag{10}$$

$$\vartheta_\alpha = \sum_\beta coal_{\beta\alpha} + \sum_\alpha (z_\alpha + \xi_\alpha) \tag{11}$$

Let's change the connection of local subsystems in any way (see figure 3). Each new connection is a new coordinate system. The invariant of the equation transformation is the value of the control action. Since there is a linear relationship between $coal_{\alpha\beta}$, z_α, ξ_α and ϑ_α we can find the optimal value of the control action for various connections of control subsystems.

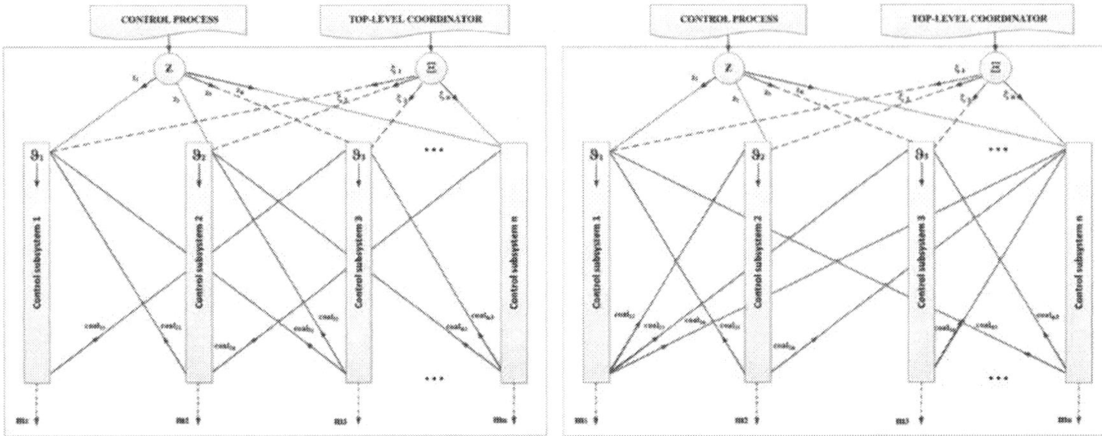

a) b)

Figure 3. The examples of changing the link structure

So we got the equations for the new connection of control subsystems:

$$\vartheta_{\beta} = G_{\beta}^{\beta'} * \vartheta_{\beta'} \tag{12}$$

$$G_{\beta}^{\beta'} - \left(\delta^{\alpha\beta} - \chi_{\alpha\beta}\right)^{-1} \cdot \left(\delta^{\alpha'\beta} - \chi^{\alpha\beta'}\right) \tag{13}$$

$$\vartheta_{\alpha'} = G_{\alpha'}^{\alpha} \cdot \vartheta_{\alpha} \tag{14}$$

$$\vartheta_{\alpha'} = G_{\alpha'}^{\alpha}\left(\sum_{\beta} coal^{\beta\alpha} + \sum_{\alpha}(z^{\alpha} + \xi^{\alpha})\right) \tag{15}$$

These equations describe transformations for tensor equations (6), (7), (8), (9), (11). These equations can be used to calculate parameters for any option of connecting subsystems.

We use a tensor basis [4] to determine the fuzzy variables in the presented system. The set of values of a fuzzy variable m on a certain universe U can be represented as a matrix: $m = \{m_i\}, i \in N, m_i \in U$, $\mu_F(m)$ is a function of the membership of elements U to the fuzzy set F.

$$F = \{< m, \mu_F(m) >\},\ m \in U \tag{16}$$

We define the matrix (17) and the tensor product T (18) (19) for the matrices m and $\mu_F(m)$:

$$\mu_F(m) = \{\mu_F(m_i)\}, n \in N,\ i = 1, n;\ \mu_F(m_i) \in F \tag{17}$$

$$T = m \otimes \mu_F(m) \tag{18}$$

$$T = \{m_1, m_2 ... m_n\} \otimes \begin{Bmatrix} \mu_F(m_1) \\ \mu_F(m_2) \\ \dots \\ \mu_F(m_n) \end{Bmatrix} = \begin{Bmatrix} m_1 * \mu_F(m_1) & m_2 * \mu_F(m_1) & \dots & m_n * \mu_F(m_1) \\ m_1 * \mu_F(m_2) & m_2 * \mu_F(m_2) & \dots & m_n * \mu_F(m_2) \\ \dots & \dots & \dots & \dots \\ m_1 * \mu_F(m_n) & m_2 * \mu_F(m_n) & \dots & m_n * \mu_F(m_n) \end{Bmatrix} \tag{19}$$

The dyad T defines the fuzzy variable m. By converting from a tensor variable, we can get attached tensors, the analysis of which allows us to determine additional properties of the fuzzy variable.

3. Results of applying the tensor basis for the system modeling of deep uncertainty

Let's assume that for some system the following fuzzy values (m_α, α={1,2,3,4}) of control actions of the subsystem α on the process are known: m_1 = *fuzzy 15* = {12, 15, 18}, m_2 = *fuzzy 26* = {23, 26, 29}, m_3 = *fuzzy 34* = {31, 34, 37}, m_4 = *fuzzy 7* = {4, 7, 10}. We heuristically define the membership functions μ_{F_α} for m_α and get a graphical representation of fuzzy sets (see figure 4).

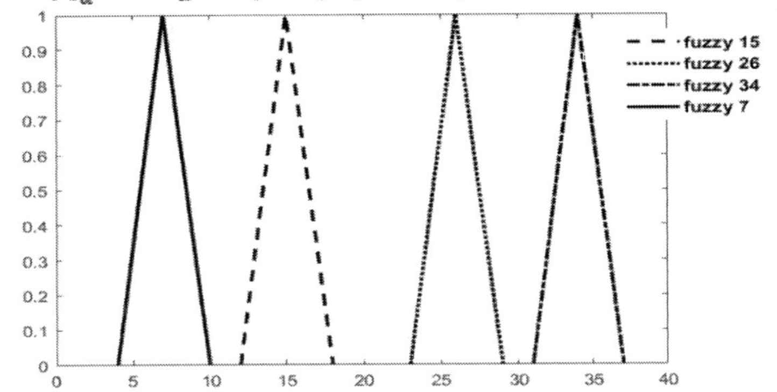

Figure 4. The fuzzy variables m_α

Using equation (18) we obtain a tensor representation of fuzzy variables (see figure 5).

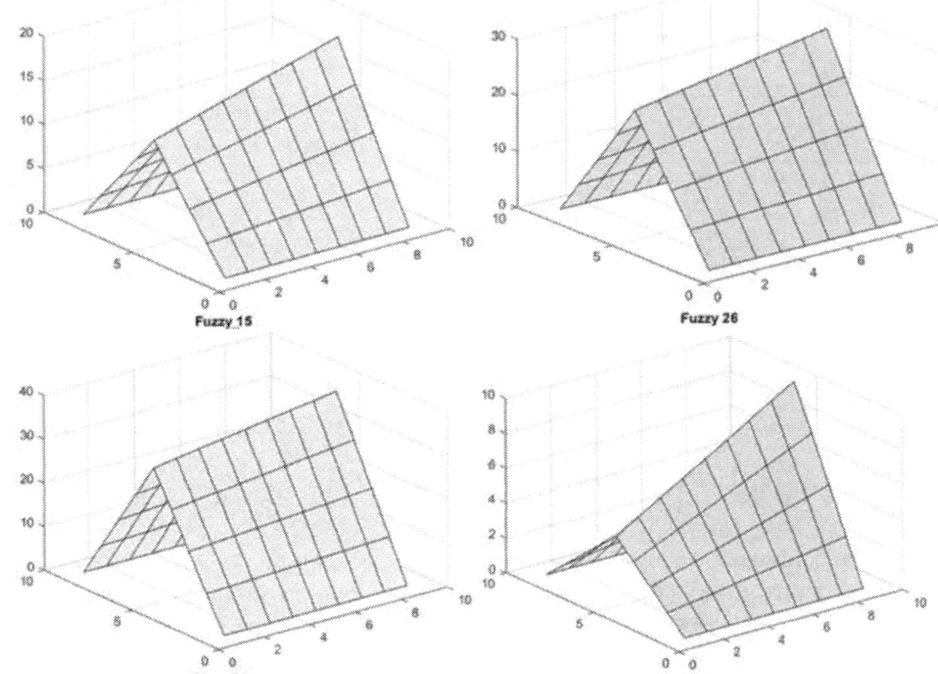

Figure 5. The tensors of fuzzy variables m_α

Finally, for each tensor we obtain by transformations the set of ordered pairs (SOP_{F_α}). We calculate the Frobenius norms for SOP_{F_α} and defuzzified values for SOP_{F_α}.We perform a singular decomposition of the T_α. Based on the results of the singular decomposition we define the SOP_{T_α}. We calculate the Frobenius norms for SOP_{T_α} and defuzzified values for SOP_{T_α}. The results of calculations are shown in table 1. Functions *trimf, kron, transpose* in table 1 are MatLab's functions.

Table 1. Combined table of calculation results

Fuzzy 15 = {12, 15, 18}	$F_1=\{m_1, \mu_{F_1}\}$, $m_1\in$ [12, 18], $\mu_{F_1}=$ trimf([12:0.75:18],[12 15 18]) SOP_{F_1}=[transpose([12:0.75:18]), transpose(μ_{F_1})] $T_1=m_1\otimes\mu_{F_1}$=kron([12:0.75:18], transpose(μ_{F_1}))	**Fuzzy set F_1**	**Tensor T_1**
	Frobenius norm	45.4037	45.4037
	Defuzzified value	15	15
Fuzzy 26 = {23, 26, 29}	$F_2=\{m_2, \mu_{F_2}\}$, $m_2\in$ [23, 29], $\mu_{F_2}=$ trimf([23:0.75:29], [23 26 29]) SOP_{F_2}=[transpose([23:0.75:29]), transpose(μ_{F_2})] $T_2=m_2\otimes\mu_{F_2}$=kron([23:0.75:29]), transpose(μ_{F_2})]	**Fuzzy set F_2**	**Tensor T_2**
	Frobenius norm	78.2336	78.2336
	Defuzzified value	26	26
Fuzzy 34 = {31, 34, 37}	$F_3=\{m_3, \mu_{F_3}\}$, $m_3\in$ [31, 37], $\mu_{F_3}=$ trimf([23:0.75:29],[31 34 37]) SOP_{F_3}=[transpose([31:0.75:37]), transpose(μ_{F_3})] $T_3=m_3\otimes\mu_{F_3}$=kron([31:0.75:37], transpose(μ_{F_3})]	**Fuzzy set F_3**	**Tensor T_3**
	Frobenius norm	102.1788	102.178 8
	Defuzzified value	34	34
Fuzzy 7 = {4, 7, 10}	$F_4=\{m_4, \mu_{F_4}= \}$, $m_4\in$ [4, 10], $\mu_{F_4}=$ trimf([4:0.75:10], [4 7 10]) SOP_{F_4}=[transpose([4:0.75:10]), transpose(μ_{F_4})] $T_4=m_4\otimes\mu_{F_4}$=kron([4:0.75:10], transpose(μ_{F_4})]	**Fuzzy set F_4**	**Tensor T_4**
	Frobenius norm	21.8518	21.8518
	Defuzzified value	7	7

From table 1, we see that the Frobenius norm of the F_α and Frobenius norm of the T_α completely coincide. Defuzzified value of the F_α and defuzzified value of the T_α also match. So the tensor fully reflects all the properties of the fuzzy variable. The representation of a fuzzy variable as a tensor allows us to investigate additional properties of uncertainty [5]. Since deviator, spherical tensor, symmetric tensor and antisymmetric tensor can be obtained from this tensor. Thus the representation of a fuzzy variable as a tensor is a representation of deep uncertainty [6] and opens up more possibilities for research.

4. Conclusion

The results obtained by the authors in the course of the study allow us to draw the following conclusions:

1. The mathematical model of the coordinator of the decision-making process under deep uncertainty can be implemented on the basis of a tensor basis. Tensor equations can be used to calculate system parameters for any change in the structure of connections in the basic mathematical model.

2. Representing a fuzzy variable as a tensor makes it possible to explore additional properties of this variable. Tensors describing additional properties of the fuzzy variable (deviator, spherical tensor, symmetric tensor, and antisymmetric tensor) can be obtained from the fuzzy variable tensor. Applying special calculations to a fuzzy tensor variable allows us to obtain the membership function with high accuracy and find the defuzzified output value for the membership function.

IOP Conf. Series: Materials Science and Engineering 1159 (2021) 012074 doi:10.1088/1757-899X/1159/1/012074

3. Tensor analysis can be used to model decision-making under conditions of deep uncertainty in complex conditions, such as catastrophes.

4. Tensor analysis significantly expands the field of research of ULS-systems.

References

[1] Subbotin B S, Ahmetzhanova E U 2019 Research on the integration of intelligent transport systems *IOP Conference Series: Materials Science and Engineering, International Conference on Digital Solutions for Automotive Industry, Roadway Maintenance and Traffic Control (DS ART 2019)* 1 November 2019 (Cholpon-Ata, Kyrgyzstan) **832**

[2] Yurchik P F, Maksimychev O I, Golubkova V B, Vinogradov V A, Zav Zav Aung 2018 The Integration of a Set of Functions for the Product Life Cycle of a Virtual Enterprise *Advances in Business and Finance. 2018 3rd BEM International Conference on Education, Sociology and Humanities (BEM-ESH 2018)* 14-16

[3] Volosova A, Matiukhina E, Asimov D2020 The use tensor method of dual networks for analysis of the transport and tourist components of sustainable development of territories First Conference on Sustainable Development: Industrial Future of Territories (IFT 2020) *E3S Web Conf* **208**

[4] Volosova A V 2019 Representation of fuzzy knowledge by means of the tensor algebra apparatus *Computational nanotechnology* **1**

[5] Minaev Y N, Guzii M, Filimonova O, Minaeva J I 2019 Alternative Methods of Analyses and decision-making in uncertainty conditions based of tensor decompositions *International Journal, Information Technologies & Knowledge* **13** (1)

[6] Kochenderfer M J 1980 *Decision making under urcentainty: theory and application*

IOP Conf. Series: Materials Science and Engineering 1159 (2021) 012075 doi:10.1088/1757-899X/1159/1/012075

Study of changes in the side clearance between vehicles and cable barriers installed along the highway centre

A V Kostsov, N A Lushnikov, A A Zuikov

Moscow Automobile and Road Construction State Technical University, Leningrad Avenue, 64

E-mail: kostsov_msfs@bk.ru , lab10@mail.ru, art-zuikov@yandex.ru.

Abstract. In the years ahead, one of the most important tasks of the government of the Russian Federation is to bring the condition of highways of various categories to standard parameters and ensure their safety. This is reflected in the decrees of the President of the Russian Federation and in national projects, for example, in the national project on "Safe and High-Quality Roads". In order to ensure safety and reduce damage from road traffic accidents on highways, various measures are used, including the installation of technical means of road traffic organization. These measures include cable barriers, which also have become popular recently. According to numerous studies, this kind of barrier has a positive effect on the passive safety of highways and it reduces the number of accidents with serious consequences that occur as a result of vehicles driving in the oncoming lane. However, the incorrect installation of these barriers (non-compliance with the minimum safety lanes at barriers, improper installation of barriers when the number of traffic lanes changes, etc.) and imperfection of the modern regulatory framework can negatively affect both a driver and the traffic flow in general, leading to the increase in the number of road traffic transport accidents. This is explained by the fact that a driver of a car, being afraid of close movement along the cable barriers at high speed, tries to stay away from the barrier as far as possible, increasing the side clearance between the vehicle and the barrier. However, at the same time, the side clearance between vehicles passing in the right lane (where a driver has limited visibility) or any obstacle on the right (for example, cars standing on the side of the road) sharply decreases. The article considers the results of the study of changes in the side clearance between the centre line of a road in the presence of a cable barrier and in its absence. The measurement results were compared. According to the results of the measurements, the authors gave the recommendations to improve road conditions, safety and the recommendations for further study of this issue.

1. Introduction

The improvement of the quality and safety of highways is one of the priority directions for the development of the road transport industry, which is reflected in the Decree of the President of the Russian Federation No. 204 of May 7, 2018. As a part of this Decree, the national project on "Safe and High-Quality Roads" (NP SHQR) [1] is currently being implemented, according to which it is necessary to reduce the number of places where road traffic accidents (RTA) are concentrated by 85% by 2025 [2].

In order to solve this problem, comprehensive measures are widely implemented such as bringing the network of federal and regional roads to the standard indicators as well as their installation.

Content from this work may be used under the terms of the Creative Commons Attribution 3.0 licence. Any further distribution of this work must maintain attribution to the author(s) and the title of the work, journal citation and DOI.
Published under licence by IOP Publishing Ltd

According to the legal document [3], one of the ways to reduce the accident rate in the places where road accidents are concentrated is the installation of technical means of traffic management (TMTM). These include the installation of road signs, marking, installation of barriers, etc. The installation of barriers is one of the most effective passive safety measures. Along the highway centre concrete and cable barriers are installed [4]. Barriers and concrete barriers are installed on roads with a minimum width of a central median strip of 5 meters. Cable barrier can be installed with a much narrower safety strip.

Cable barriers have several advantages over other barriers. These barriers need less space, and the cost of repairing them in case of damage is lower. They can be installed both on the highway centre and on road shoulders. In the first case, they prevent vehicles from entering the oncoming lane and prevent serious consequences of an accident associated with an oncoming collision of vehicles. In the case of installing barriers on the road shoulders, they prevent vehicles from overturning when they leave the roadbed. However, due to the sufficient minimum width of the unreinforced part of the shoulder, barrier fences are preferred.

The additional advantage of the installation of such a fence along the centre is convenient operation. In the case of a road traffic accident, the elements can be quickly changed as soon as possible as well as only the damaged segment. During winter operation there are no snow drifts of fences, even with regular heavy snowstorms. However, at the same time, it is difficult to clean the roadway surface with operational vehicles, but this problem arises for any type of fence installed along the centre of highway.

The disadvantages of cable barriers include a lower impact capacity. However, due to the impossible alternative caused by the limited conditions, this disadvantage is compensated by advantages.

On the road shoulders, the installation of cable barriers is carried out in the same way as the installation of barrier fences, taking into account the requirements [5] [6] [7]. However, the rules for the installation of barriers along the centre of the highway are insufficiently studied and are considered only from the point of view of possible deformations (dynamic deflection) after an accident. In this case, the safety strip along the barrier fence should be at least 0.5 meters on each side of the fence (Figure 1) [8].

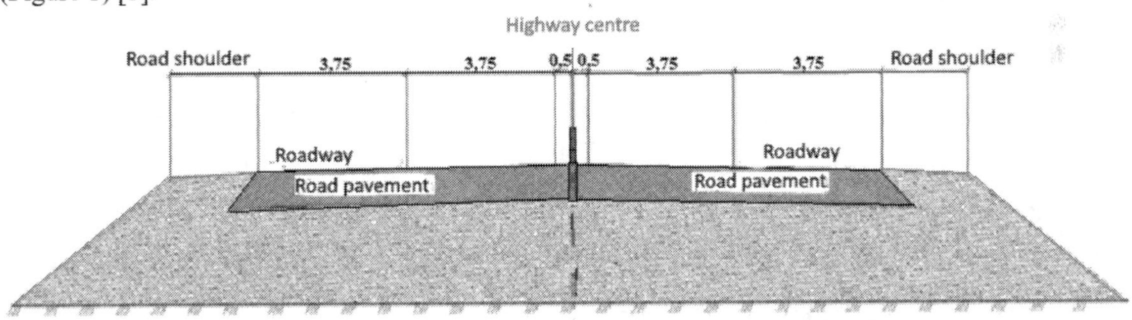

Figure 1. Installation diagram of cable barriers on the highway centre

As the analysis of the literature shows, this value is not justified from the point of view of the psychological perception of this fence by drivers. If the safety lane is insufficient, emergency situations may arise, leading to the increase in the number of road accidents. In addition, on highways, the technology of both posts setting (often the distances between the posts and the degree of tension of the cables are not observed) and the technology of the application of markings along the highway centre at the fence are not generally observed. In addition, there are problems of the installation of a cable barrier on the highway centre when the number of traffic lanes changes (an additional lane for turning left, advance lane, etc.). As a rule, on existing roads of the 1st and 2nd categories, due to the limited width of the roadway, the width of the roadway is not increased during the installation of cable barriers. It is necessary to take into account that it is recommended to install road signs "obstruction" and "detour right" in front of the barrier fence according to the recommendations of legal documents [3, 8]. All this affects the side clearance while the vehicle is moving.

2. Research

Side clearance is the required distance between a vehicle and an obstruction (or other vehicle) which allows free maneuvering on the road. The non-compliance results in the increase in the number of road accidents and reduces safety on the road in general.

In this paper the authors try to find out how does the installation of the barrier fence affect the side clearance between the vehicle and the cable barrier on two sections of the four-lane road: without a barrier fence and with installed cable barrier. The observation of the movement of cars was carried out using a drone quadcopter. In order to collect data from a transport facility, a section of a road with a minimum load level and the same permitted speed was studied. The work was carried out by an unmanned aerial vehicle PHANTOM-4. The control of this drone was carried out in manual mode. Figure 2 shows fragments of shooting road sections.

Figure 2. Measurement diagram of side clearance in the area with the installed cable barrier (left) and with a double solid line of marking (right)

3. Research results

Based on the research results, cumulative curves were created for the accumulated frequencies (proportional frequencies). As a result of statistical processing of the data, the distribution curves were obtained (Figure 3) from which it follows that:

• with cable barriers, the side clearance was 1.53 meters with 85% coverage, 1.20 meters with 50% coverage, 0.89 meters with 15% coverage;

• in areas with a double solid line, the side clearance was 1.27 meters with 85% coverage, 0.98 meters with 50% coverage, 0.72 meters with 15% coverage.

IOP Conf. Series: Materials Science and Engineering 1159 (2021) 012075 doi:10.1088/1757-899X/1159/1/012075

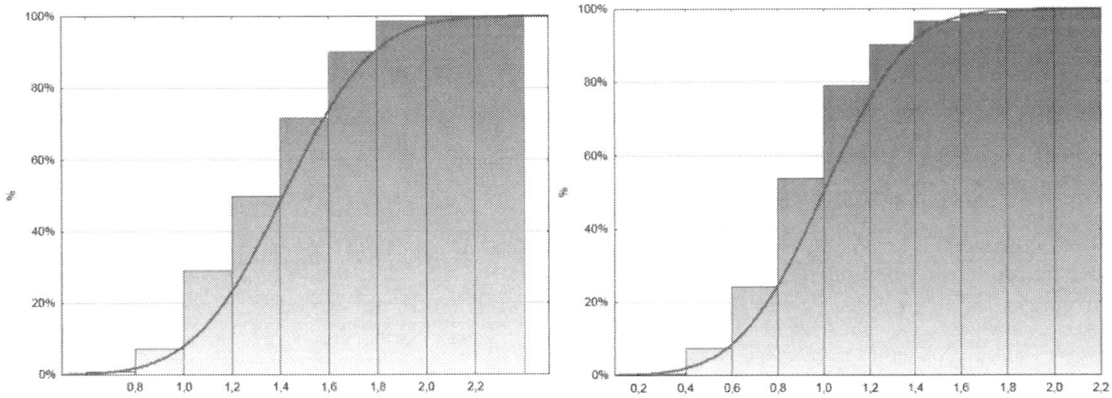

Figure 3. Processed measurement results with installed cable barrier (left) and with double solid line (right)

The increase in the side clearance at different was:

- 0.26 meters with 85% availability;
- 0.22 meters at 50% safety;
- 0.17 meters at 15% safety;

4. Conclusion

Thus, according to the results, the following conclusions can be made:

- With the installed cable barriers, the side clearance between the vehicle and the cable barrier increases by 15 - 30 cm. This indicates a different perception of the road conditions by a driver. It is necessary to note that with the increase in the left side clearance, the right one decreases by the same value, which can cause a possible accident, since vehicles moving on the right fall into a zone of less visibility. In addition, the possibility of collisions of vehicles with obstructions on the right, standing vehicles, etc. increases.
- In the presence of axial barriers, it is necessary to increase the width of the roadway with a central dividing strip or reduce the permissible speed. It is necessary to take into account that widening of the central dividing strip should be provided only by the increase of the width of the roadbed. It is unacceptable to consider the decrease in the width of the hard shoulder and other parameters of the transverse profile to the smaller dimensions provided in this legal document [7]. In the case of the reduction of the width of the shoulder, road accidents are possible which are associated with emergency stop of vehicles. In the case of the decrease in traffic lanes, it is necessary to reduce the permitted speed of vehicles and limit the advance of cars by trucks.
- Further studies of the traffic flow with the installed cable barriers are required. It is also necessary to find out how does the trajectory of vehicles change when advancing and does the number of advancing vehicles increase or decrease.
- It is necessary to consider additional measures related to road safety with the installed barriers along the highway centre, for example, such as "noise" markings, reflectors built into the markings, etc. Additional technical means of traffic management will help drivers to feel more confident when driving along cable barriers at high speeds and performing various maneuvers.
- It is necessary to analyze the accident rate on the road sections with installed cable barriers, to find out the nature of the accident, the damage and the severity of injuries in order to take them into account during the determination of safety measures on the road.

References

[1] Record of September 21, 2016. No. 2, PR-1919 of October 6, 2016, order of the Government of the Russian Federation of October 15, 2016 №DM-P6-6192) regarding the implementation of the priority project "Safe and high-quality roads"

[2] Passport of the national project "Safe and high-quality roads"
[3] GOST R 52289-2019 Technical means of traffic management. Rules for the use of road signs, markings, traffic lights, road barriers and guiding devices
[4] Road facilities - one of the priority areas of economic development, *Roads Journal* **8** [1065] 2020
[5] Road industrial methodical document 218.*6.004-2011 Guidelines for the construction of cable road barriers to ensure safety on highways*
[6] GOST R 52398-2005 Classification of highways. Basic parameters and requirements
[7] GOST R 52399-2005 Geometric elements of highways.
[8] Road industrial methodical document 218.6.018-2016 Recommendations on the rules of use, construction and operation of cable and combined road barriers on public roads

IOP Conf. Series: Materials Science and Engineering 1159 (2021) 012076 doi:10.1088/1757-899X/1159/1/012076

Electrolytic alloying of iron-chromium during deposition of coatings from a sulfate-chloride electrolyte

A F Sinelnikov[1], E V Bomeshko[1], N I Korneychuk[1], A S Ianuta[1]

[1] Moscow Automobile and Road Construction State Technical University (MADI), 64, Leningradsky Prospekt, Moscow, 125319, Russia

E-mail: ianyta_anton@mail.ru

Abstract. The paper presents the results of studies of the process of electroreduction of alloy iron-chromium from a sulfate-chloride electrolyte, influence of electrolyte composition and electrolysis (acidity, electrolyte temperature and cathode current density) on the process kinetics, the output of the alloy current efficiency and physico-mechanical properties of the coatings (microhardness, morphology). There were determined such parameters as the optimal ratio of concentrations of electrolyte components – 150 g/l of iron chloride (2) and 50 g/l of chromium sulfate (3) and modes of electrical recovery of metals – current density 40 A/dm^2; the temperature of the electrolyte is 50°C, the acidity of the solution is 0.5 pH units, which contribute to obtaining a high-quality coating with high tribological characteristics. Assumptions are made about the mechanism of the alloy formation by electrolysis from a mixed electrolyte.

1. Introduction

Currently, when repairing cars, the cost of spare parts is more than 50% of all costs. Restoration of worn parts can significantly reduce these costs, reduce the cost of repair and maintenance of machines, free up significant production capacity of the industry that manufactures spare parts and save metal used for these purposes.

As evidenced by the practice of renovation of machinery and equipment, the most cost-effective way to restore worn-out machine parts is galvanic metal build-up and, first of all, iron plating. Electrolytic iron, due to the high productivity of the process and the possibility of obtaining a coating with pre-defined physical and mechanical properties, is becoming increasingly common in the repair industry. However, the coating of electrolytic iron is inferior to chrome due to its insufficient microhardness and wear resistance. At the same time, iron plating allows the restoration of parts from environmentally friendly electrolytes in comparison with the use of toxic electrolytes based on hexavalent chromium compounds for chrome plating.

Iron-based alloys are a promising compromise between high deposition rate of electrolytic iron and wear-resistant chromium coating. Moreover, electrolytic alloys, in contrast to alloys obtained by metallurgical methods, have fine-grained properties and phase stability [1]. However, when selecting components for the deposition of a future alloy, it is necessary to take into account that the joint deposition and crystallization of two or more metals with the formation of alloys is possible in cases where their equilibrium potentials differ slightly from each other [2, 3].

The most widespread in the practice of repair production are ferric chloride electrolytes, from which coatings with high physical and mechanical properties and current output are deposited. At the same

Content from this work may be used under the terms of the Creative Commons Attribution 3.0 licence. Any further distribution of this work must maintain attribution to the author(s) and the title of the work, journal citation and DOI.
Published under licence by IOP Publishing Ltd

IOP Conf. Series: Materials Science and Engineering 1159 (2021) 012076 doi:10.1088/1757-899X/1159/1/012076

time, solutions of iron (2), chromium (3) and other metal chlorides have the property of unstable composition due to rapid oxidizability, aggravated by the use of high temperature during electrolysis. Sulfate electrolytes are largely devoid of these disadvantages, but their widespread industrial use is hindered by the relatively low productivity of the deposition process.

In our opinion, the use of mixed sulfate-chloride electrolytes is a compromise solution to the problem of applying a high-quality coating that has reliable operational properties and high process productivity. The analysis of the literature data on the compositions of mixed electrolytes used to produce the iron-chromium alloy [4] shows that the kinetics of the Fe-Cr alloy deposition process, the effect of various additives and modes of co-deposition of metals in the alloy are insufficiently studied.

The purpose of our research was to search for and optimize the composition of mixed sulfate-chloride electrolyte, electrolysis modes, and identify the mechanism of deposition of high-quality iron-chromium alloy coatings that meet the operational requirements of the surface to be restored.

2. Materials and methods

To study the deposition process of the iron-chromium coating, a sulfate-chloride electrolyte was prepared by mixing solutions of chromium sulfate and iron chloride in various proportions. The ferric chloride solution prepared by known methods [4], a solution of chromium sulfate – by dissolving a solid of hydrated chromium sulfate (3) in acidified water followed by bringing the acidity of the electrolyte to a predetermined pH by using 1 n HCl solution. The content of Fe^{2+}, Fe^{3+}, and Cr^{3+} cations in the electrolyte was monitored by potentiometric titration according to the approved method [5]. For the preparation of electrolytes, reagents of the "H" and "HCH" qualification were used.

Coatings were applied at direct current to cylindrical samples made of steel 20 with a surface area of 0.001 m^2. Before electrolysis, the samples were degreased with Viennese lime, and then anodically etched in a 30% aqueous solution of sulfuric acid, followed by washing in running water for 20 seconds.

To power the electrolyzer, a DC source was used, a VU-42/70A rectifier with a two-half-period current rectification circuit. in the studies, the current density was changed using the RNT voltage regulator. The current in the circuit was measured by an ammeter M–104 CL. 0.5. the temperature of the electrolyte was changed from 30 to 50°C, and the acidity – from 0.2 to 1.2 pH units.

Heating and temperature control of the electrolyte was carried out in a thermostat with an accuracy of 0.1°C using a thermorelet. The working volume in the cell was 2 liters. The acidity (pH) of the electrolyte was controlled using a pH meter Smart Sensor model AS218 with a measurement accuracy of 0.01 units. calibrated with standard buffer solutions.

To ensure a strong adhesion of the galvanic Deposit to the sample surface, the cathodic current density at the beginning of the electrodeposition process was gradually increased from 2 A/dm^2 to the set value for 10 minutes. The deposition time of the coating was 1 hour, taking into account the "acceleration".

The current yield of the metal was determined by gravimetric method. The value of the electrochemical equivalent of deposition of the iron-chromium alloy (at a chromium concentration in the coating of 18 %) is 0.971 g/A·h [6]. Samples were weighed before and after coating on a VLR–200 scale with an accuracy of 0.01 mg. The deposition rate of the iron-chromium alloy was determined by measuring the coating thickness with a micrometer MK 0-25 mm (GOST 6507-60) by the diameter of the sample in its middle part before and after deposition. When calculating the speed of the process from the deposition time, taking into account the loss of sample thickness during etching, the time to enter the operating mode was not taken into account.

The morphology of the coating was studied using a metallographic microscope (MMU–3) with magnification multiplicity of 250, 620, 990, with information output to a personal computer and image processing by S-EYE software version 1.6.0.11. the microhardness of the coating was measured on a PMT-3 microhardness meter with an indenter load of 100 g (1H), followed by measuring the size of the print diagonal and calculating the microhardness using a well-known method [5].

Polarization studies were performed in galvanostatic mode on a P5848 potentiostat in an electrochemical cell YASE–2 with platinum indicator and auxiliary electrodes and a silver-chloride reference electrode. The current sweep time was 750 s, the potential of the working electrode was measured with an electronic multivoltmeter with an accuracy of 0.01 mV and a fixation frequency of 1 Hz, with constant transfer of information to a personal computer. Registration of multi-voltmeter readings was performed by the software "UT61C_D" V3.03 .

3. The study of the structure of the modified lead-tin-base bronze

At the first stage of research, it was necessary to select the composition and concentration of the base electrolyte. According to the analysis of the literature data, all previously studied electrolytes used ferrous iron chloride or sulfate and trivalent chromium chloride or sulfate to produce the Fe-Cr alloy. Our research on the stability of electrolytes of different compositions to oxidation and their electrical conductivity allowed us to conclude that a mixed sulfate (chromium) – chloride (iron) electrolyte is the most effective and stable in quality composition. To determine the quantitative ratio of the components, their concentrations were varied widely: $FeCl_2$ – 50÷200 kg/m^3, $Cr_2(SO_4)_3$ – 10÷100 kg/m^3. Moreover, in all electrolytes, one of the components was always present in the predominant amount: it could be either a sulfate compound or a chloride one. Based on our research (resistance to oxidation, to the formation of colloidal precipitation, electrical conductivity of solutions), we adopted a mixed sulfate-chloride electrolyte with a predominant concentration of ferrous iron ($FeCl_2$) as the basis. To determine the optimal quantitative ratio of Fe(2):Cr(3) components in a mixed electrolyte, polarization studies were conducted to assess the effect of the component concentration, acid balance, and electrolyte temperature on the reduction potentials of chromium (3) and iron (2) ions in pure and mixed electrolytes (figures 1, 2, and 3).

Figure 1 shows the dependences of the electrode potential on the current density obtained in chromium sulfate electrolytes with a chromium sulfate concentration of 25-100 g/l and an interval between the studied concentrations of 25 g/l (curves 1-4, respectively); in iron chloride electrolytes with iron chloride concentrations of 100-200 g/l and an interval of 50 g/l (curves 5-7, respectively), as well as in a mixed electrolyte of the composition: 150 g/l $FeCl_2$ and 50 g/l $Cr_2(SO_4)_3$ (curve 8). Studies were performed at a temperature of 40°C and pH=0.6.

Analysis of figure 1 shows that the potential of the chromium electrode (curves 1-4) has shifted to the negative region relative to its equilibrium value (φ (for Cr^{3+}) = – (0.35÷0.45) mV, while in a highly acidic environment the equilibrium potential of the platinum electrode for chromium $\Delta\phi_{equil}$(Cr3+/Cr) = 50 mV) and practically does not depend on either the concentration of the metal cation or the current density. A simple calculation shows that the electrode was polarized by the value $\Delta\phi_{pol}$(for Cr^{3+}) = 85÷95 mV.

The potential of the electrode for iron (curves 5-7) turned out to be more positive in comparison with its equilibrium value (φ(for Fe^{2+}) = –(220÷260) mV, and in a strongly acidic medium $\Delta\phi_{equil}$(for Fe^{2+}/Fe) = -320 mV), and there are noticeable potential fluctuations both depending on the concentration of the Fe^{2+} ion and when the current density changes. The electrode polarization for iron $\Delta\phi_{pol}$(for Fe^{2+}) was – (60÷100) mV, bringing it closer in value to the opposite sign of the polarization for chromium, which indicates the presence of a common factor of displacement of metal potentials from their equilibrium values. With the combined presence of iron and chromium ions in a mixed electrolyte (curve 8), the potential of the electrode is further shifted to the positive range, indicating their mutual influence and participation in the electrolysis process, and at the same time practically does not react to changes in the current density.

From our point of view, the common factor of electrode polarization in strongly acidic electrolytes of chromium and iron plating is the discharge of hydrogen ions (H3O+) combined with metal cations. We believe that the behavior of the electrode in the chromium electrolyte is due not only to the step discharge of the chromium cation Cr^{3+} and its intermediate hydrate (due to hydrolysis) compounds together with the hydrated hydrogen ion H_3O^+, but also to the preferences for the H3O+ cation under these conditions (due to the low hydrogen overvoltage on platinum). The following process scheme is possible:

$$Cr^{3+} + H_3O^+ + \bar{e} = Cr(OH)^{2+} + H_2;$$
$$Cr(OH)^{2+} + H_3O^+ + \bar{e} = Cr(OH)_2^+ + H_2;\ Cr(OH)_2^+ + H_3O^+ + \bar{e} = Cr + H_2 + 3OH^-. \quad (1)$$

As can be seen (curves 1-4), at low current densities (2-30 A/dm^2), the release of hydrogen at the cathode from the chromium electrolyte prevails with a slight fluctuation in the concentration of Cr^{3+}, and subsequently, the chromium surface is also deposited on the chromium surface for the same reason of low hydrogen overvoltage on chromium.

At the same time, during electrolysis from a strongly acidic iron electrolyte (curves 5-7), even with a step discharge of Fe^{2+} and its hydroxo compounds together with H_3O^+, metal wins in their competition (due to the high overvoltage of hydrogen on the deposited iron after a current density of 10 A/dm^2). The scheme of the process is already different: $Fe^{2+} + H_3O^+ + \bar{e} = Fe(OH)^+ + H_2$;

$$2Fe(OH)^+ + \bar{e} = Fe \cdot Fe(OH)^+;\ Fe \cdot Fe(OH)^+ + \bar{e} = 2Fe + OH^- . \quad (2)$$

A shift to the region of more positive values of the electrode potential in a mixed electrolyte (figure 1, curve 8) indicates the formation of associative hydrolysis complexes of iron and chromium with a solvent in their joint presence according to the scheme:

$$Fe^{2+} + Cr^{3+} + 8H_2O <=> Fe[Cr(OH)_4]^+ + 4H_3O^+. \quad (3)$$

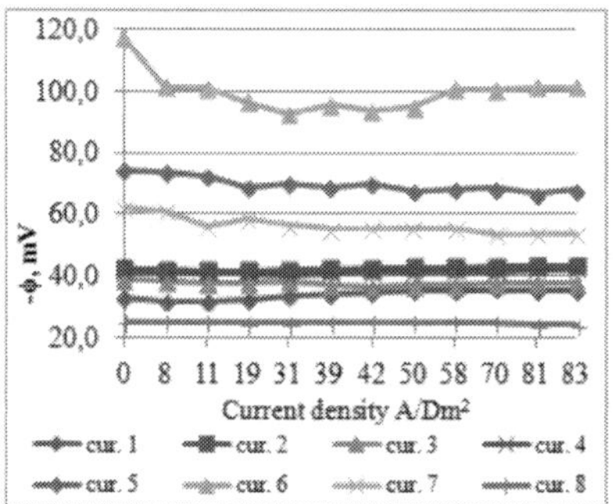

Figure 1. Effect of component concentration at tel=40°C and pH=0.6 on the value of the potential of the platinum electrode depending on the cathode current density in the chromium sulfate electrolyte (curves 1-4); in the iron chloride electrolyte (curves 5-7) and in the mixed sulfate–chloride electrolyte (curve 8) at concentrations (g/l): 1) $Cr_2(SO_4)_3$–100; 2) $Cr_2(SO_4)_3$–75; 3) $Cr_2(SO_4)_3$–50; 4) $Cr_2(SO_4)_3$–25; 5) $FeCl_2$–150; 6) $FeCl_2$–100; 7) $FeCl_2$–200; 8) FeCl2–150, $Cr_2(SO_4)_3$–50.

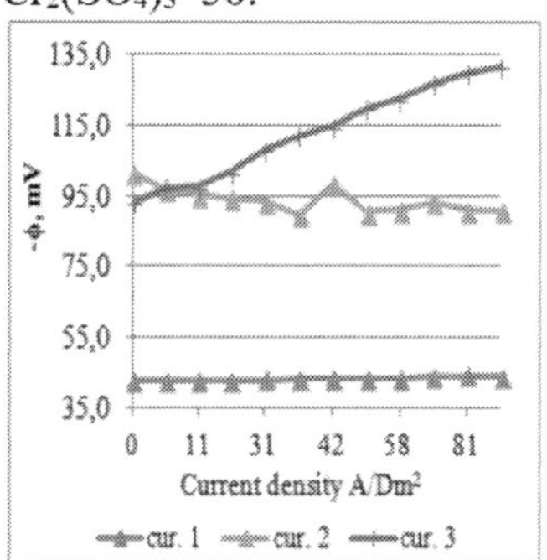

Figure 2. Changes in the potential of the platinum cathode in the ferric chloride electrolyte (curve 1), the chromating sulfate electrolyte (curve 2), and the mixed chloride-sulfate electrolyte (curve 3) at pH=0.8, temperature 40°C, and optimal component concentrations (g/l): 1) $FeCl_2$–150;
2) $Cr_2(SO_4)_3$ – 50;
3) $FeCl_2$–150, $Cr_2(SO_4)_3$–50

Figure 3. Influence of the cathode current density and temperature of the sulfate-chloride electrolyte (composition, g/l: 150 $FeCl_2$ and 50 $Cr_2(SO_4)_3$) at pH=0.4 on the polarization of the platinum cathode: curve 1 – t=30°C, curve 2 – t=50°C.

It is likely that the activation energy of the reduction of metal ions from hydrate complexes decreases, which facilitates the process of their joint discharge. However, testing the stability of the mixed electrolyte over time showed that the concentration of iron salt in it should not be increased to 200 g/l due to an increase in the viscosity of the highly concentrated solution and the formation of an even denser cathode film on the electrode surface, which slows down the ion discharge process [6].

To prove our opinion about the mechanism of mutual influence of iron, chromium and hydrogen ions on the electrolysis process, figure 2 shows the dependence of the cathode potential of the platinum electrode on the current density at the electrolyte temperature of 40°C and the acidity of the solution of 0.8 units. pH in pure electrolytes with concentrations (g/l): $FeCl_2$ – 150 (curve 1), $Cr_2(SO_4)_3$ – 50 (curve 2) and in mixed electrolyte 150 $FeCl_2$ + 50 $Cr_2(SO_4)_3$ (curve 3).

IOP Conf. Series: Materials Science and Engineering 1159 (2021) 012076 doi:10.1088/1757-899X/1159/1/012076

As follows from figure 2, the increase in the polarization of the electrode in the iron electrolyte with an increase in pH to 0.8 units. (curve 1) leads to a significant dependence of the discharge potential on the current density. The reason lies in the increased hydrolysis of iron cation in a less acidic environment and an increase in the concentration of iron hydroxide ions according to the Le Chatelier principle in the reaction: $Fe^{2+} + 2H_2O <=> Fe(OH)^{+} + H_3O^{+}$, (4)
this requires more energy to restore the metal from the oxygen-containing ion, which means that the dependence of the potential on the current density is increased.

The offset potential of the polarized cathode in a mixed electrolyte with increasing pH in the zone of more negative values compared to strongly acidic electrolyte (curve 3 in figure 2 compared to curve 8 in figure 1) brings it closer to the equilibrium value of the potential discharge of pure chromium, which facilitates a joint recovery of Cr and Fe, and the independence of polarization on current density suggests the idea of forming a more stable associative complexes of iron and chromium with the participation of not only solvent, but also anions, particularly chlorine anion Cl^{-}, this compacts the adsorption layer on the surface of the electrode and increases its concentration polarization. In this case, the joint discharge of iron (2) and chromium (3) cations is facilitated, in our opinion, at higher operating current densities (50-80 A/dm^2). The following process scheme is possible:

$$Fe[Cr(OH)_4]^{+} + \bar{e} = Fe[Cr(OH)_4];\ Fe[Cr(OH)_4] + H_3O^{+} + \bar{e} = Fe \cdot Cr(OH)_3 + 2H_2O;$$
$$Fe \cdot Cr(OH)_3 + H_3O^{+} + \bar{e} = Fe \cdot Cr(OH)_2 + 2H_2O; Fe \cdot Cr(OH)_2 + 2H_3O^{+} + 2\bar{e} = Fe \cdot Cr + 4H_2O. \quad (5)$$

This step-by-step mechanism for the formation of the cathode film and the discharge of Fe and Cr cations from it increases the speed of the process, increases the quality of higher-thickness wear-resistant coatings that provide high tribological reliability of the restored parts and meet production requirements for wear surfaces.

In General, the analysis of the research results presented in figures 1 and 2 allows us to conclude that it is advisable to mix iron and chrome electrolytes in a ratio of 3:1 or 150 g/l $FeCl_2$: 50 g/l $Cr_2(SO_4)_3$.

It was also of interest to find out the effect of the electrolyte temperature on the mechanism of the electrolysis process at the optimal ratio of component concentrations in it.

As a result of studies of the effect of the cathode current density and temperature of the sulfate-chloride electrolyte on the polarization of the electrode when coating with Fe-Cr alloy, it was found (figure 3) that with increasing temperature, the potential of the electrode shifts to more negative values, increasing its concentration polarization. At the same time, an increase in the current density over 20 A/dm^2 with an increase in temperature to 50°C stabilizes the potential of the electrode, indicating that the maximum discharge current has been reached. From our point of view, the observed effect is that the temperature rise, on the one hand, strengthen the hydrolysis of metal cations according to reactions described by equations (3) and (5), and on the other accelerates the diffusion of cathode film, its rebuilding as a result of partial loss of adsorbed ions and solvent molecules, and hence, the relative loosening of the film, thereby producing a positive effect on the process of extraction of metals iron and chromium in the alloy.

Note that the increase in temperature to a lesser extent than the acidity of the electrolyte affects the polarization of the electrode (for comparison, curve 8 in figure 1 at pH=0.6, curve 3 in figure 2 at pH=0.8, and curves 1-2 in figure 3 at pH=0.4).

In support of the above, the dependences of the current yield of the alloy (figure 4) and the deposition rate (figure 5) on the acidity and temperature of the mixed electrolyte at a constant current density were studied. The alloy was deposited from a mixed electrolyte with an optimal concentration of components ($FeCl_2$:$Cr_2(SO_4)_3$=150:50) at a cathode current density of 40 A/dm^2 and electrolyte temperatures of 30, 40 and 50°C.

Figure 4 shows that, regardless of temperature, the current output of metals increases with decreasing acidity of the solution (curves 1, 2, 3). The maximum current output effect (78 %) for a given current density is achieved in the deposition range pH=0.9–1.04 and t= 40°C (curve 2). It also corresponds to a high deposition rate of 0.25 mm/h (figure 5 curve 2). However, the appearance and quality of the coating at the initial acidity of the solution of 0.9 units with a subsequent increase in pH during electrolysis do not meet the specified properties. The explanation for this lies in the features of the mechanism of co-reduction of Fe^{2+}, Cr^{3+}, and H_3O^+ cations depending on the acid-base balance in the electrolyte.

In our opinion, when the concentration of hydrogen cations is insufficient (pH≥0.8÷0.9), the near-cathode layer is alkalinized by the reactions described by equations (1) and (2), colloidal iron and chromium hydroxides are formed with an intermediate degree of oxidation and their incomplete reduction from the film by hydrated electrons. Ingress of oxygen-containing components into the coating worsens the structure and properties of the sediment. When the temperature drops to 30°C (figure 4, curve 1), the current output becomes unstable even at lower pH values (0.4 – 0.9 units), varying in the range of 18-45 %. The same situation is observed with the deposition rate (figure 5, curve 1), which varies in the range of 0.12 – 0.24 mm/h, which confirms the assumption of incomplete reduction of iron and chromium cations in the entire pH range above 0.4 units. from cold electrolyte.

When the temperature rises to 50°C, the situation improves (figure 4, curve 3) both in terms of current output and sediment quality in the pH=0.4÷0.8 range . Under these conditions, the deposition rate of metals gradually increases from 0.15 to 0.28 mm/h (figure 5, curve 3). With a further decrease in acidity, the process becomes unstable again. The most stable in terms of current efficiency and deposition rate is the electrolysis process at temperatures of 40 and 50°C in the pH range of 0.6-0.8.

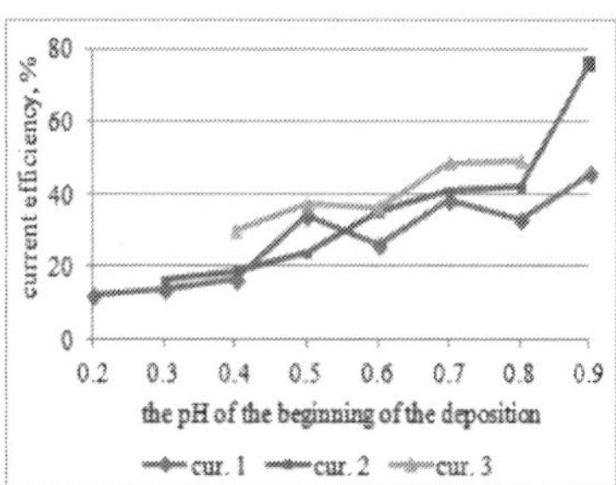

Figure 4. Dependence of the current output of the Fe-Cr alloy from a sulfate-chloride electrolyte with component concentrations (g/l) $FeCl_2$ – 150; $Cr_2(SO_4)_3$ – 50 and current density 40 A/dm^2 on the pH of the beginning of deposition and the temperature of the electrolyte: 1 – t=30°C; 2 – t=40°C; 3 – t=50°C.

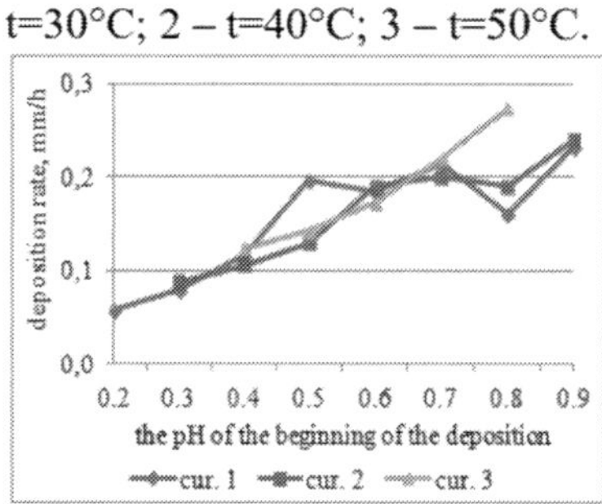

Figure 5. Dependence of the deposition rate of Fe-Cr alloy from a mixed electrolyte (150:50) at a current density of 40 A/dm^2 on the pH of the deposition start and temperature: 1 – t=30°C; 2 – t=40°C; 3 – t=50°C.

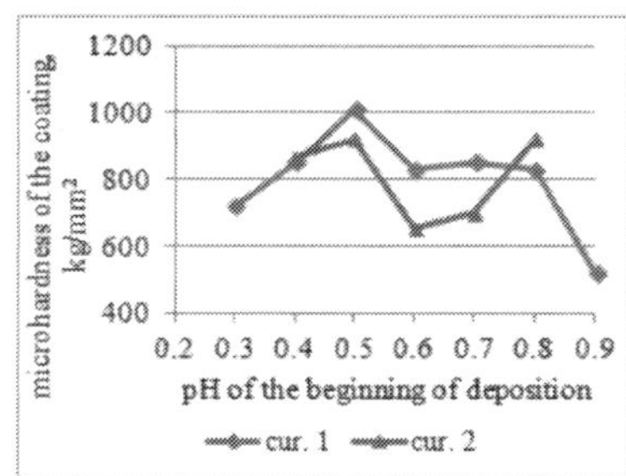

Figure 6. Dependence of the microhardness of the coating on the pH of the beginning of deposition and the temperature of the electrolyte: curve 1 – t=40°C; curve 2 – t=50°C.

The dependences of the alloy deposition rate and current output on the pH and temperature of the electrolyte described in figures 4 and 5 during electrodeposition on steel (ST20) electrodes indicate the extreme complexity of the mutual influence of ions on the kinetics of their recovery and correlation with polarization studies (figures 1-3). The discharge potentials of iron and chromium cations converge under certain conditions, providing their mutual support during recovery (see equations 4). At the same time, intermediate iron complexes formed under mild conditions accelerate the reduction of chromium (3) and its crystallization in a structure favorable to iron atoms, ensuring the formation of the alloy as an intermetallic compound or solid solution, which prevents the appearance of defects in the crystal lattice.

Physical studies of the microhardness and surface morphology of deposited metals from the $FeCl_2$–$Cr_2(SO_4)_3$ mixed electrolyte were performed)3 with an optimal concentration ratio (g/l) of 150-50 at a current density of 40 A/dm^2, depending on the acidity and temperature of the electrolyte. The research results are shown in figures 6 and 7.

Studies of the microhardness of coatings depending on the modes of electrolysis indicate that the surface formation is inhomogeneous when the acidity of the medium changes and when the electrolyte is heated (figure 6, curves 1 and 2). The most solid precipitation (900-1010 kg/mm^2) was obtained in the lower pH range (0.4–0.5) at Dk=40 A/dm^2 and te=40-43°C (curve 1) and at te=50-52°C (880-920 kg/mm^2), curve 2. Increase the pH of the electrolyte to 0.6 units. dramatically reduces the hardness of the coating at both temperatures, remaining stable at 40°C up to pH=0.8 (820 kg/mm^2) and rising in the same pH range to the initial value of 900 kg/mm^2 at 50°C. Further reduction of acidity to pH=0.9 and higher reduces the microhardness of the precipitate in an electrolyte with a temperature of 40°C to 500 kg/mm^2 and actually destroys the coating at a temperature of 50°C. Such changes in the microhardness of coatings indicate that the acidity of the electrolyte significantly affects the structure of the coating, shifting the equilibrium of the metal reduction process in the direction of iron, then in the direction of chromium. The choice of the electrolysis mode in favor of the hot electrolyte (t=50°) becomes quite obvious with an initial precipitation acidity of 0.8 units. pH and warm electrolyte (t=40°C) with an initial precipitation acidity of 0.5 pH units.

Studies of the morphology of coatings obtained from a mixed sulfate-chloride electrolyte at a current density of 40 A/dm^2, depending on pH and temperature, allowed us to see both cracked and uncracked coatings with fewer or more globules on the surface to be coated (figure 7, a, b, c, d).

The results obtained are in good agreement and are a real confirmation of our earlier assumptions about the mechanism of joint electrical recovery of iron and chromium metals into an alloy. For example, figure 7a provides an understanding of the high microhardness of the surface in a cold electrolyte at low current densities due to the formation of fine-crystalline Fe-Cr cellular complexes in the form of solid solutions with a predominance of iron or chromium in them, and therefore the presence of interfaces between these complexes.

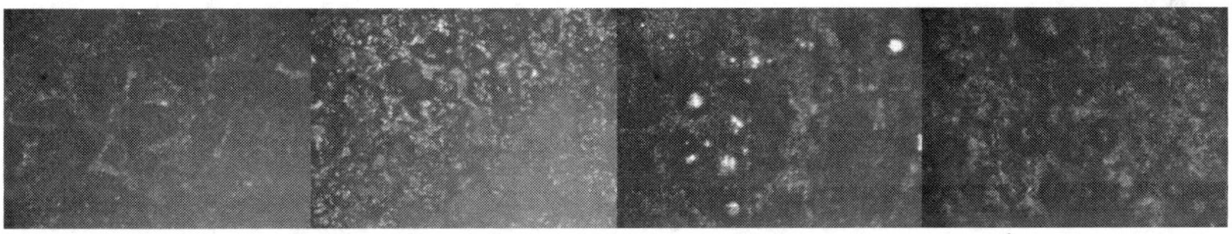

a b c d

Figure 7. Morphology of the iron-chromium alloy coating from a mixed sulfate–chloride electrolyte depending on the acidity and temperature deposited at a cathodic current density of 40 A/dm^2 Fe-Cr (x250): a – pH=0.5 $t_{el.}$=30°C; B – pH=0.5 $t_{el.}$=40°C; b – PH=0.5 $t_{el.}$=50°C; g – pH=0.8 $t_{el.}$=50°C.

The same cellular globular morphology is observed in a strongly acidic electrolyte at temperatures of 40 and 50°C (figure 7 b, c). moreover, during deposition from a hot electrolyte, the formation of larger globules on the surface and the growth of crystals of a solid solution of chromium in iron is observed. It is only when the pH is increased to 0.8 units that the morphology becomes more uniform (figure 7 d), an intermetallic compound of iron and chromium is formed, and the qualitative and quantitative characteristics of the coating are improved.

4. Conclusion

The article proposes a mechanism for the joint reduction of iron and chromium cations to an alloy by an electrolytic method from a mixed sulfate-chloride electrolyte. It defines optimal ratios of component concentrations (150 g/l iron chloride $FeCl_2$ and 50 g/l chromium sulfate $Cr_2(SO_4)_3$) and modes of electrical recovery of metals (current density 40 A/dm$_2$; pH of the electrolyte at the beginning of deposition 0.8 and 0.5; temperature 40±2°C, which contribute to obtaining a high-quality coating with high current output (49% and 42%, respectively), deposition rate (0.18 and 0.12 mm/h, respectively), microhardness of the sediment (820 and 1010 kg/mm$_2$, respectively).

References

[1] Petrov Yu N, Guryanov G V, Bobanova Zh I, Sidelnikova S P, Andreeva L N 1990 *Electrolytic deposition of iron* (Chisinau: Shtiintsa) 196 p

[2] Povetkin V V, Kovenskiy I M 1983 Structure of electrodeposited alloys of iron subgroup metals,

Izv. USSR ACADEMY OF SCIENCES. Metals **3** 108–111

[3] Serebrovsky V V, Serebrovsky V I, Pavlova E V, Bogomolov S A 2012 Increasing the strength of electrolytic iron with alloying additives of molybdenum *Izvestiya Yugo-zapadnogo gosudarstvennogo universiteta* **2-2** 143–145

[4] Blinkov B S, Serebrovsky V V, Kalutsky E S 2016 Electrodeposition of iron-based alloys *Bulletin of the Kursk state agricultural Academy* **2** 67–71

[5] Fadeeva V I, Shekhovtsova T N, Ivanov V M et al 2001 *Fundamentals of analytical chemistry. Practical guide: Textbook for higher education institutions*, ed. by Yu A Zolotov (Moscow: Higher school) pp 312–314

[6] Bomeshko E V, Korneychuk N I 2019 Electrodeposition of double and triple alloys based on iron and chromium: theoretical concepts and practical recommendations *Tiraspol; Bulletin of the Pridnestrovian University. Ser.: Physical,mathematical and technical Sciences* **3 (63)** 153–165

IOP Conf. Series: Materials Science and Engineering 1159 (2021) 012077 doi:10.1088/1757-899X/1159/1/012077

Increasing the level of automation of dispatch control when regulating at the end stops of public transport routes

P O Voronov[1], **A A Kudryavtsev**[1], **N A Filippova**[1]

[1]Chair of transport telematics, Moscow automobile and road construction state university (MADI), 64, Leningradsky prospect, Moscow, 125319, Russian Federation

E-mail: v.p.o.1998@rambler.ru, kudryavtsevaa@transnavi.ru, umen@bk.ru

Abstract. The subject of research is automated dispatch control of public transport. Smart systems application on transport started in the second half of the 20th century. Application of such systems aims to improve every aspect of public transportation. However, despite constant development of telematics systems in public transport, this has not been achieved yet. The article discusses the use of telematics in public transport and the existing and actually functioning systems: ASK, NimBus and Navitrans. Authors suggest algorithm for public transport control that should increase the level of automation of dispatch control. The result of such approach is the improved regularity of public transport which is one of the key points to public transport efficiency and comfort to passengers.

1. Introduction

Public transport is an indefeasible part of life and modern society, providing urban mobility to the population. At the same time, it is extremely important to ensure the timely solution of management problems, the availability and quality of transport services, as well as to meet the growing requirements for the operation of urban public transport. All these things are available due to the introduction of telematic systems for public transport.

In recent years, automated dispatch control systems based on satellite navigation have been created in cities across the country, which allow solving the problems of automated control of urban public transport, scheduling and routing of public traffic. The operation of these systems is based on a discrete analysis of the movement of public transport by checkpoints. The effective use of satellite navigation in organizing the work of public transport results in the creation and application of scientific, methodological and technological approaches to improve the quality of information support of the telematic systems.

Enhancing the efficiency of the dispatcher is one of the most important tasks in improving the dispatch management of public transportation. As a result of the efficient operation of dispatching systems, the quality, reliability, availability and comfort of public transport are increased. That is why it is so important to improve the formation of approaches to further automation of dispatch control of public transport, as this is what determines the relevance of the research topic.

Content from this work may be used under the terms of the Creative Commons Attribution 3.0 licence. Any further distribution of this work must maintain attribution to the author(s) and the title of the work, journal citation and DOI.
Published under licence by IOP Publishing Ltd

IOP Conf. Series: Materials Science and Engineering 1159 (2021) 012077 doi:10.1088/1757-899X/1159/1/012077

2. Review of domestic and foreign experience in the development and use of transport and telematic systems in public transport.

Currently, the concept of transport telematics is constantly expanding [1,2,3]. This is due to scientific and technological progress, which let us solve new practical problems. It is obvious that modern information systems are replacing traditional technologies. Telematics on automobile transport is aimed at developing and improving the transport infrastructure, as well as at ensuring the productive work of participants of the road transport process. The concept of telematics includes the application of information technologies based on the use of stationary and mobile systems, as well as on processing, accumulation and use of information in real time [4].

In the 60s of the last century, Europe, the USA and Japan introduced much more advanced transport systems than the previous traffic management. In the 60s and 70s were formed the basic principles of their functioning. And then in the 80s, the rapid development of electronics began, which made possible to introduce the first telematic systems DRIVE, ROMANSE, PROMETEUS in Europe, UTMS, ASV, ARTS in Japan, MOBILITY 2000 and IVHS systems in the USA [5].

In most of the systems we talk about, the vehicle position on the Earth surface determinating with the global navigation systems GPS(Global Positioning System, USA) and GLONASS (Global Navigation Satellite System, Russia) [6,7,8]. The result of the European Union and The European Space Agency - the Galileo system is also functioning now. The launch of the 5G-based Beidou system made it possible to start development of new innovative transport and infrastructure programs in China in 2020. The Development of the autonomous NavIC system (the system was renamed to IRNSS with the launch) was completed in India. The system is similar to the GLONASS and GPS systems in many ways, but IRNSS is regional. It covers the territory of India and an area up to 1500 kilometers to the Indian border [10].

Dispatch systems with global navigation systems based on collection and processing of operational information about the positioning of vehicles. This information makes it possible to control the transportation process and implementation of transport work, and to manage the transportation process.

Navigation systems consist of several main parts: planning of transportation processes, monitoring the operation of tracking objects, dispatch control, and the formation of various reporting forms.

Automated dispatch system (ADS) based on satellite navigation systems implements the following technological components: obtaining navigation marks, storing them onboard, and transmitting primary information to dispatching points. It also includes software and hardware for processing primary information. The scheme of work of the automated dispatching systems based on satellite navigation is shown on Fig.1.

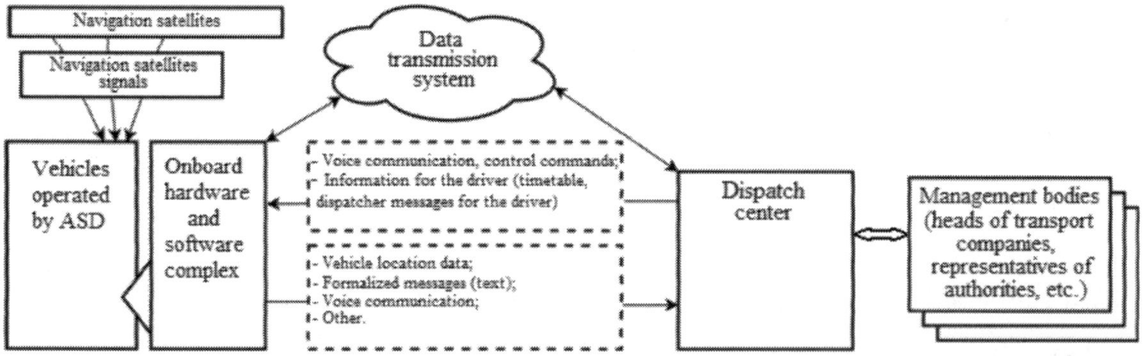

Figure 1. Schematic diagram of the operation of an automated control system based on satellite navigation

Below are some automated dispatch systems that applied at public transport earlier:

1) System «ASDU-A». In the late 70s of the last century, SPKB "Promavtomatica" in the city of Omsk developed the first automated dispatch buses control system for large cities. Control point

devices (CPD) are installed on the route network. And these points received marks from vehicles as soon as they entered the checkpoint area using an electromagneticloop, which was built under the road surface near the stopping point. Each route had to have at least two UCP. For each vehicle of the system, a mobile unit device (MUD) was installed. When the vehicle passed the checkpoint, signal that contained the vehicle number and the timestamp was generated. After that, this signal was transmitted to the dispatch center using a dedicated telephone line.

2) System «NEZHAN». The principle of the system, developed in the city of Nalchick in the 80s of the 20thcentury, was to install radio beacons and radio repeaters at the control points of the GPT routes. At the same time, a mobile unit device was installed on each vehicle. Mobile unit, being within a radius of several tens of meters, receives signals from radio beacons, which they send to the radio broadcast at regular intervals.Received signals are stored in the internal memory,they contain the beacon number and the time when the mark was received. The repeaters, in turn, receive the accumulated signals from mobile unit device and, using city telephone communications transmit them to the dispatch center. At the same time, the system made possible the voice communication between the dispatcher and the driver, but only in the zone of the repeater.

3) System «ASU-Reis». Also in the 80s of the last century, the State Unitary Enterprise "MosgortransNIIProekt" was developed a dispatching control system for buses. The main principle of the system was the installation of radio beacons at the control points of the public transport routes, which send information about their own identification number on the air. A receiver that receives these signals, a VHF-radio station and radio modem was installed on the vehicle. Together with the identification number, the current time value is also transmitted to the dispatch center by radio channel. Then, based on the received data, the dispatch center calculates the deviation from the schedule and sends this information back to the vehicle. This system, due to the VHF radio station, made it possible to organize dispatcher-driver radio communication along the entire route of the vehicle.

The main disadvantages of the listed systems, obviously, are the lack of a continuous monitoring mode, as well as the features associated with the need to install and operate the equipment of control points.

3. Review and analysis of existing technologies of dispatch control

1) ASK system – Public transport is a software package for tracking the movement of an object along a route, passing control points by this object in a certain or arbitrary sequence according to a schedule or without it.

Application of this system implements the following features: formation and editing of routes, control over the implementation of specified traffic routes and compliance with the schedule by the driver, control of the driver's working day schedule.

System consists of the following functional modules: module "Technologist", module "Dispatcher", module "Reports", and module "References".

A distinctive feature of this system is the service of stop terminals for informing the public. Specifically: informing riders about the arrival of all types of transport, providing background information (weather, news, traffic jams), alerting citizens about emergencies, if they arise, and advertising information.

2) NimBus system is a system for public transport control. The system is the most modern solution in terms of design among those considered in this work. Nimbus app helps public transport dispatchers plan and monitor bus transit. It can control urban, suburban and intercity public traffic [12].

The service combines the main functionality of satellite monitoring systems, as well as special tools for controlling route vehicles: distribution objects (vehicles) by routes, real-time movement tracking, control over adherence to schedules, tracking the location of objects on routes, creating a notification about events on routes, formation of reports on trip performance.

The following modules are declared in the system : stations (creating or import stations for each type of transport, with the ability to search for a specific one, as well as the ability to add and edit a schedule), routes (formation a route of movement, drawing up a timetable for each route, as well as creating schemes of action), trips (assignment of transport objects to routes with grouping by route and object), tracking (visual interface for controlling schedule execution, with the ability to track deviations from the route, as well as replace vehicles on the route).

3) *"Navitrans" system* is a modular system for dispatching control and monitoring of the public's flow. It has been introduced and operated in the enterprises of the State Unitary Enterprise "Mosgortrans" in Moscow, and in more than 80 cities located in 27 regions of Russia. The system allows to organize the management of public transport, control and accounting for implementation of transport work, promptly determine the locations of road accidents and emergencies, increase the speed of response to provide medical and evacuate victims, carry out measures in the line of the Ministry of Emergencies and mobilization [9,13,14,15,16]. A distinctive feature of the system is that the composition of the vehicle's on-board devices can be different and depends on the specific tasks assigned to the system.

The system also makes it possible to generate various reports, both on the work of the dispatch center, and on the results of the transport work of the vehicle. For instance: reporting data on the drivers, vehicles (schedule execution, total mileage, linear mileage, operating time on the line, downtime), reporting data on the work of system dispatchers(recordings of voice communication between the dispatcher and the driver, recording of control actions).

The sub-system ASM-PP is designed to optimize public traffic for a number of parameters in medium and long-term planning, based on the technology of automatic collection and analysis of information about public flows. The system also obtains characteristics of public traffic in the format of tabular survey data [14].

Basic functionality of the system: operational control of the number of transported persons to set daily plans for collecting revenue, optimization of the route network based on the analysis public flows on stations on each route.

Additional functionality of the system: calculation of profitability for each surveyed route, taking into account zoning of fare payment. Application of geoinformatics tools allows you to create and edit passports of routes, as well as adapt them to specific tasks of public transport, drawing up a timetable taking into account the calculated norms for mileage on any section of the route with a division for periods of the day.

The system uses contactless public count systems, with both domestic and Imported equipment, which is installed on buses, streetcars.

In general, we can say that the "Navitrans" system has the greatest functionality of all systems considered. It is obvious that the basic functions of all three systems are similar;however the "Navitrans" system, as a result of collecting and analyzing data of public flows, provides a much wider functionality.

4. The main problems of providing the quality of service of service on public transport

It is important to understand that the assessment of the quality of service is carried out according to an integrated indicator – the value of the quality factor. It contains estimates for individual particular indicators, for instance: filling of the bus, the time spent on the trip, the regularity of the public transport movement on the routes, the safety of thepublic during the ride.

Obviously, it is impossible to accurately assess the current state of transportation, or determine the population's need for this type of service without providing the management of public transport up-to-date analytical information of the actual capacity and dynamics of public flows on urban routes.

Fig. 2 shows an example of the possible types of supply and demand curves for transport services. It clearly shows that without an assessment of public flows, the curves do not correspond to each other.That means that riders either wait a long time for their transport, or the public transport is almost empty. The red line in the figure indicates the mode of operation that should be achieved at organizing traffic to provide a balanced filling of the cabin, while taking into account the carrying capacity of the route, efficient use of rolling stock, free entry and exit of people, waiting time at a stopping point, as well as the fastest transportation the public.

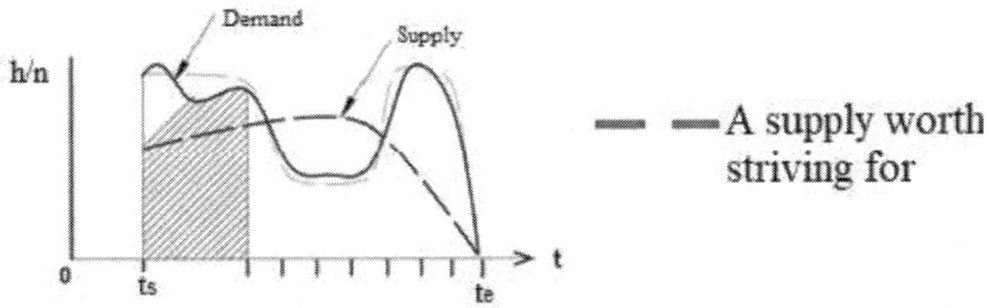

Figure 2. Supply and demand curves of transport services on the public transport routes

A similar situation occurs when the dispatcher system does not work correctly, or when an emergency situation occurs and there is no prompt action from the dispatcher. The regularity of the movement of rolling stock is an indicator of the quality of transport work,and it is the activity of dispatch systems that directly affects the regularity. Consequently, enhancing the automation of dispatching systems, and hence their efficiency, also increases the quality of services provided.

It is important to note that automated dispatch systems can improve another aspect of the quality of services provided, namely the safety of publics during transportation. Indeed, in case of an emergency situation, an accident, etc., when using automated dispatching systems, it is possible to take measures both in automated mode, e.g. calling emergency services, and using the driver-dispatcher voice communication. Hence automated systems can reduce the response time to emerging problems in the field of transport security.

One of the most effective ways to reduce traffic congestion in cities is to increase the proportion of citizens using public transport services and reduce the proportion of drivers and publics in private cars. It is necessary to increase the attractiveness of public transport by improving the quality of public flows to do this.

Despite the fact that the quality of public traffic is assessed in a comprehensive manner according to several parameters, one of the most obvious ways to improve the comfort of public traffic is to organize regular traffic of the public transport. This will solve the main problem faced by the public.Namely, the problem of long waiting times for the vehicle, which may be caused by the non-optimal spread of the rolling stock along the routes or non-compliance with the planned intervals of the vehicle. As a result, there is a problem of low ride comfort associated with overcrowding and increased waiting times for vehicles at stopping points.

Unfortunately, feedback from the public is not enough for competent adjustment of schedules and distribution of rolling stock on routes,which is why it is so important to work with operational data from the vehicle. The processing of the initial data on the passing time of the route stops will help to not only to determine the most problematic routes, but also the period of the day where the summary schedule on routes is not executed. The analysis of processed navigation data is a tool for improving the organization of transportation on this route by revising the existing timetables. Moreover, the results of the calculation in real time can be the basis for operational dispatching actions. For instance, dispatchers can put additional vehicles on the route to increase the carrying capacity riders at peak times. In turn helping improve the comfort of travel for passengers.

To obtain operational information about the quality of the transportation process on the route, it is necessary to be able to process the primary data coming from the rolling stock in real time.

IOP Conf. Series: Materials Science and Engineering 1159 (2021) 012077 doi:10.1088/1757-899X/1159/1/012077

5. The control algorithm of public transport

This calculation uses a recursive algorithm. When calculating the departure time of the current vehicle, the interval of the vehicle in front is not taken into account, since it is taken into account when calculating the departure time of the vehicle in front, when it was current.

5.1. Real-time simulation, end of the trip event processing.

1) Get the data from the retrospective data array with two requests for each direction.
2) Saving the data in the corresponding entities containing the following variables: stop time, trip number, stop number, vehicle number. The data is saved into an array, which stores the numbers of adjacent stops in the order of their passage. The associative array is used where the "key" is the time of the stop arrival, and the "value" is the duration interval between adjacent stops. At this moment the program can determine the time of the end of the trip.

5.2. Predicted arrival time calculation for the vehicle behind

This calculation is necessary in order to subsequently determine the value of the time interval for which the program needs to "hold" the current vehicle at the end stop to maintain a uniform interval, if the vehicle behind is late. This situation is shown in Fig.3. The program already stores the time of passage of the current vehicle in RAM. The program also receives in real-time simulation the navigation marks from the vehicle behind. So, when the current vehicle arrives at the end stop, the algorithm determines the last stop passed by the vehicle behind. We assume that the vehicle behind has the same time left to go until the end of the trip. That means that the interval between vehicles will remain unchanged. Further, the algorithm compares, according to formula 1, the interval, between the current vehicle and the vehicle behind,at the stop, the navigation mark about passage of which by the vehicle behind was received last, and the time of the specified interval for the vehicle's release on the trip.

$$\Delta t = (t_c - t_b) - t_i \quad (1)$$

where Δt – the time difference between the interval, between the current vehicle and the vehicle behind, and the interval for the vehicle's release on the trip;

t_c – the time the current vehicle passed the stop, the navigation mark, about passage of which by the vehicle behind was received last;

t_c – the time the vehicle behind passed the stop, the navigation mark about passage of which was received last;

t_i – the interval for the vehicle's release on the trip.

So, if $\Delta t > 0$, the vehicle going behind it is late.

5.3. Departure time calculation of a vehicle on a trip.

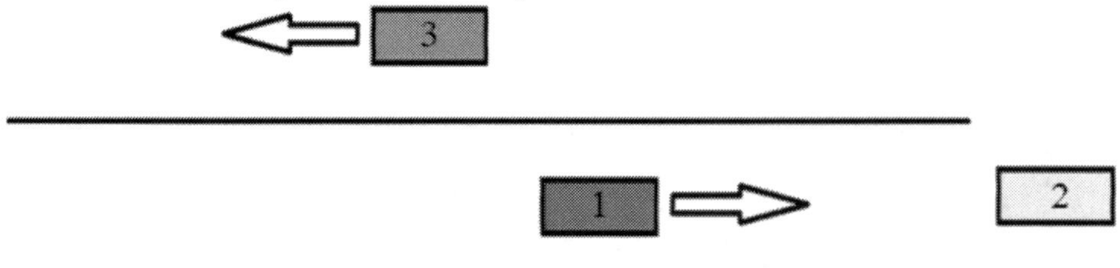

Figure 3. The situation when the vehicle at the end stop must be "held"

If $\Delta t < 0$, in further calculations we take $\Delta t = 0$. If $\Delta t > 0$, the left time of the trip is longer than the specified interval for releasing the vehicle on the trip. That means that the interval of release of the current vehicle on the trip must be increased by half of Δt. In this case, when calculating the adjusted interval, the value of half of Δt must be saved in the program's RAM in order to add it when calculating the adjusted interval for the vehicle behind when it becomes current. The adjusted interval is determined by formula 2:

$$t_{ad} = t_i + \frac{\Delta t}{2} + t_h \quad (2)$$

where t_{ad} – adjusted release interval of the current vehicle on the trip;
t_h – the half value of half of Δt saved in the program memory when calculating the adjusted interval for the previous vehicle.

6. Conclusions

The calculated departure time can also be used to inform thepublic. Moreover, the methods of informing can be completely different.

First of all, information can be displayed on the stop information display.Adjusted intervals can be displayed on stop information display, in a mobile application or web interface. If the installation of stop information display is impossible, a QR code can be installed on the stop. When this QR-code is scanned, the currently calculated interval will be displayed, or in case of improving the public information system, on the basis of the developed system, it is possible to display the time of arrival of the vehicle at a particular stop.

When considering the issue of informing passengers about the time of arrival of the vehicle at the stop terminals, it is important to understand that there are people who do not have access to Internet resources or the mobile platform. Therefore, to organize informing these citizens, a free telephone line can be launched. Operators of this line will provide the information of the movement and waiting time of public transport at the indicated passenger stop.

References

[1] Vlasov V M, Efimenko D B 2007 Coordinate-time and navigation support (KVNO) as a unified information basis for automation of basic technologies in transport (theses) *On Sat. reports "Fundamental and applied coordinate-time and navigation support". 2nd All-Russian Conference* (SPb .: IPA RAN.)

[2] Vlasov V M, Zhankaziev S V, Nikolaev A B, Prikhodko V M 2003 *Telematics in road transport* (Moscow: MADI)

[3] Gerami V D 1994 *Organization and management of urban passenger road transport* (Moscow: MADI)

[4] Federal Target Program "Global Navigation System", approved by the Government of the Russian Federation of 20.08.2001, №. 587.

[5] Decree of the President of the Russian Federation of 17.05.07 No. 638 "On the use of the GLONASS global navigation system in the interests of the social and economic development of the Russian Federation"

[6] Przhibyl P, Svitek M 2003 *Telematics in transport* ed. Prof. V.V. Silianova (Moscow: MADI)

[7] Vlasov V M, Bogumil V N 2000 All-seeing eye of the "GLONASS" system *Autocarrier* **3** pp 68-69

[8] Decree of the President of the Russian Federation of 17.05.07 No. 638 "On the use of the GLONASS global navigation system in the interests of the social and economic development of the Russian Federation"

[9] *Official website of the newspaper "People's Daily"* Available at: http://russian.people.com.cn/n3/2020/1023/c31517-9772507.html (accessed: 20.10.2020)

[10] *Official site of the interdisciplinary journal of navigation technologies "Bulletin GLONASS"* Available at: http://vestnik-glonass.ru/stati/obzor-indiyskoy-regionalnoy-navigatsionnoy-sputnikovoy-sistemy-irnss/ (accessed: 1.11.2020)

[11] *The official website of the "ASK" system*Available at: https://ask-glonass.ru/solutions-passager (accessed: 20.10.2020)

[12] *Official site of the "NimBus" system* Available at: https://gurtam.com/ru/nimbus (accessed: 20.10.2020)

[13] Vlasov V M, Ruzhilo A A 2002 *The effectiveness of the dispatcher of the automated radio navigation control system in critical situations on urban and suburban transport in*

Novokuznetsk (article) / Collection of scientific papers of MADI. "Actual problems of organization and management of road transport" (Moscow: MADI)

[14] *Official site of JSC "NPP Transnaviatsiya"* Available at: http://www.transnavi.ru (accessed: 10.25.2020) - Title. from the screen

[15] *The official website of the State Unitary Enterprise "Mosgortrans"* Available at: http://mosgortrans.ru (accessed: 01.11.2020)

[16] *The official website of the Mostransavto State Unitary Enterprise* Available at: http://www.mostransavto.ru/ (accessed: 10.27.2020)

[17] *Transport site of the city of Kazan* Available at: http://www.kazantransport.ru (accessed: 03.11.2020)

Synthesis of oil-soluble corrosion inhibitors from raw materials of plant origin

S M Gaidar[1], M Yu Karelina[2], Hoang Duc Quang[3] and A A Akulov[4]

[1] Russian State Agrarian University - Moscow Timiryazev Agricultural Academy, Russia, Moscow, Timiryazevskaya str., 49, 8 (916) 789-25-76
[2] Federal State Budgetary Educational Institution of Higher Education «Moscow Automobile and Road State Technical University (MADI), Russia, Moscow, Leningradskyave. 64, 8 (499) 155-03-74
[3]Southern Branch of the Russian-Vietnamese Tropical Research Technology Center, Vietnam, Ho Chi Minh City, Street 3/2, 3
[4]Federal State Budgetary Educational Institution of Higher Education «Moscow Automobile and Road State Technical University (MADI), Russia, Moscow, Leningradskyave., 64

E-mail: ter34t@mail.ru, quanghoang1510@gmail.com

Abstract. The use of low-soluble corrosion inhibitors based on raw materials of plant origin is one of the most promising methods of metalwork processing to prevent corrosion. Carbon acids with carbon radicals are used for the synthesis of such inhibitors, which participate in the formation of hydrophobic part of the molecule.It is known that the hydrophilic part of the molecule causes solubility in water, and the hydrophobic part - in hydrocarbons. The study found that by changing the structure of the hydrophilic and hydrophobic parts it is possible to obtain a substance that will dissolve in water as well as in hydrocarbons. The obtained organic compounds can be used as corrosion inhibitors, antiwear and antiadhesive additives. Therefore, by synthesizing carboxylic acids with various radicals and using them for treating metal surfaces, we can obtain not only anti-corrosion properties of the metal, but also anti-wear and anti-adhesion properties, which makes it possible to use inhibitors not only for protection against corrosion, but also for extending the service life of loaded or rubbing metal surfaces.

1. Introduction

Currently existing corrosion inhibitors are based on synthetic raw materials. Obtaining corrosion inhibitors synthesized on the basis of raw materials of plant origin for the purpose of processing metal structures by them is the most promising research in this field of science. Obtaining low-soluble properties of plant-derived inhibitors is the main objective of this research study.

Research into the synthesis of low-soluble corrosion inhibitors based on plant-based raw materials in science is one of the least studied ways to create surfactants to protect metal structures from environmental influences and further oxidation of the metal surface.

2. Materials and methods

To obtain oil-soluble organic compounds, molecules of which have chemisorption capacity, carboxylic acids are used as an active ingredient. In the synthesis of corrosion inhibitors, carboxylic acids with hydrocarbon radicals are used; in the synthesis of antiwear additives, acids with

Content from this work may be used under the terms of the Creative Commons Attribution 3.0 licence. Any further distribution of this work must maintain attribution to the author(s) and the title of the work, journal citation and DOI.
Published under licence by IOP Publishing Ltd

hydrocarbon or perfluorinated radicals are used. Radicals are involved in the formation of the hydrophobic part of the molecule.

Figure 1 shows the structural formula of fatty acid amide (FAA) obtained as a result of the polycondensation reaction of carboxylic acids with ethanolamine borate.

In the synthesis of FAA were used:

- carboxylic acids$CH_3(CH_2)_nCOOH$; C $F_3(CF_2)_nCOOH$

- diethanolamine

$$HN\begin{cases} CH_2-CH_2-OH \\ CH_2-CH_2-OH \end{cases}$$

- boric acid H_3BO_3.

As a result of the synthesis, organic compounds (Figure 1) with an asymmetric molecular structure were obtained, containing a hydrophobic radical (for compounds with a perfluorinated radical, it is lyophobic) and a hydrophilic part.

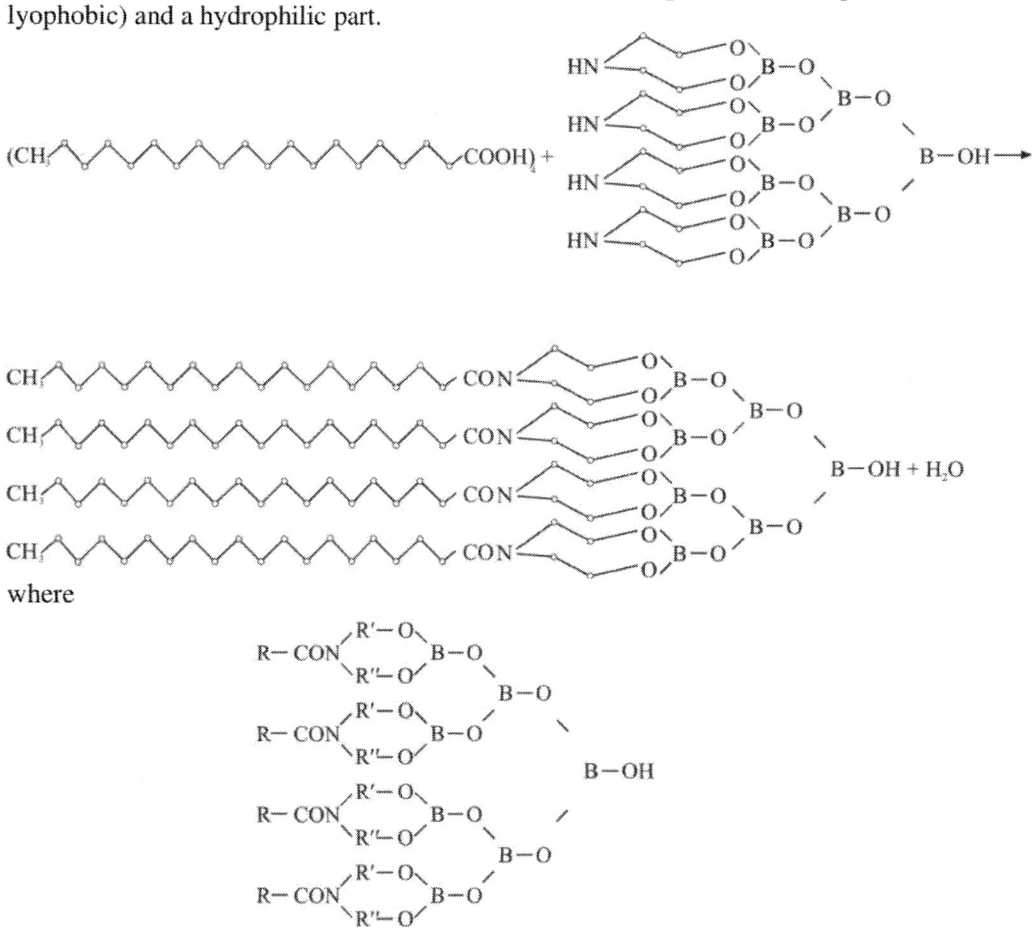

where $R - CH_3-CH_2 \cdots - CH_2-$
or
$CF_3-CF_2 \cdots -CF_2-$

R′, R″ $-CH_2-CH_2-$

Figure 1. Structural formula of fatty acid amide

The hydrophilic part of the molecule contains two functional groups:

amideand –OHhydroxyl, which cause chemisorption activity. Such molecules belong to non-ionic surfactants that do not dissociate in aqueous solution with formation of ions.

It is known that the hydrophilic part of the molecule causes solubility in water, while the hydrophobic part causes solubility in hydrocarbons. By changing the structure of the hydrophilic and hydrophobic parts, it is possible to obtain a substance that will dissolve in water as well as in hydrocarbons. Molecule structure depends on the ratio of ingredients in the synthesis, in this case there are three: carboxylic acid, diethanolamine and boric acid.

The molecules of the resulting compound are adsorbed on the metal surface due to the hydrogen bond, which is provided by hydrogen of the hydroxyl group.The hydrogen bond (7...15 kcal/mol) is energetically more than 10 times higher than Van der Waals forces. It easily displaces adsorbed water from the metal surface, which is very important in the preservation of wet surfaces. Undivided electrons of the nitrogen atom in the amide form complex compounds with metal cations.

In the course of a field study of all compounds it was found that the organic compounds obtained could be used as corrosion inhibitors, antiwear and anti-adhesion additives.

To obtain polyfunctional properties, benzotriazole is included in the fatty acid amide composition.

3. Results and discussion

As a result of the research it was proved possible to synthesize insoluble corrosion inhibitors based on raw materials of plant origin, which is a new method of metal processing to prevent oxidation and further rusting.

Also, by means of a field study, the anti-wear and anti-adhesion properties of organic compounds were proved, which allows their use to extend the life of loaded and rubbing surfaces of metals.

4. Conclusions

Researches on the synthesis of low-soluble corrosion inhibitors based on raw materials of plant origin have shown the possibility of giving anticorrosive properties to metal constructions. Other properties of these surfactants have also been identified, which allows their scope of application to be expanded.

Metal inhibitors based on synthetic raw materials do not have such a wide area of application, which makes them less in demand, as compared to low- soluble inhibitors based on plant-based raw materials.

As a result of the research, the possibility of synthesizing oil-soluble corrosion inhibitors from vegetable raw materials has been shown. This will protect the metal surface from corrosion by displacing water molecules from it. This technology will protect metal structures in regions with high humidity, which will increase their service life. It also means a complete opportunity to abandon the processing of metals using petroleum products.

Acknowledgments

The article was prepared as part of the research work on the topicT-1.16 «Development of technologies for obtaining highly effective protective materials against atmospheric corrosion using raw materials of plant origin and fluorine-containing surfactants» of the Program of scientific research and technological works of the Joint Russian-Vietnamese Tropical Research and Technology Center for 2020-2024.

IOP Conf. Series: Materials Science and Engineering 1159 (2021) 012078 doi:10.1088/1757-899X/1159/1/012078

References

1. Gaidar S M 2015 *Planning and analysis of the experiment: a textbook* (Moscow: Publishing house of thefederal state scientific institution Rosinformagrotech)
2. Karelina M Yu, Petrovskaya E A, Pydrin A V 2015 Optimization of inhibited composition to ensure the preservation of agricultural machinery *Proceedings of GOSNITI* Vol 121 pp 89–93
3. Karelina M Yu, Pydrin A V 2015 Optimal inhibited composition for anticorrosive treatment of agricultural machinery and equipment *Innovative directions of development of technologies and technical means of agricultural mechanization. Proceedings of the International Scientific and Practical Conference devoted to the 100th anniversary of the Department of Agricultural Machinery of the Agroengineering Faculty of the Voronezh State Agrarian University named after Emperor Peter I. The Russian Federation Ministry of Agriculture; Voronezh State Agrarian University named after Emperor Peter I.* 110-115
4. Kuznetsov Yu I 2002 Current state of the theory of metal corrosion inhibition *Protection of metals* Vol 38 **2** 122–131
5. Semenova I V, Florianovich G M, Khoroshilov A V 2002 *Corrosion and corrosion protection* (Moscow: FIZMATLIT)
6. Gaidar S M 2017 *Technology of agricultural machinery storage: training manual* (Moscow: publishing house of the federal state scientific institution Rosinformagrotech)
7. Gaidar S M 2011 *Theory and practice of creating corrosion inhibitors for conservation of agricultural machinery. Monograph* (Moscow: publishing house of the federal state scientific institution Rosinformagrotech)

IOP Conf. Series: Materials Science and Engineering 1159 (2021) 012079 doi:10.1088/1757-899X/1159/1/012079

Environmental safety diagnosis and maintenance specific of heavy vehicles and buses engines functioning with compressed natural gas

A P Boldin[1], Yu V Panov[1], A A Solntsev[1], P V Aksenov[2]

[1] Moscow Automobile and Road Construction State Technical University (MADI), 64, Leningradsky prosp, Moscow, 125319, Russia
[2]State Unitary Enterprise MOSAVTOSANTRANS, 2, Bol'shaya Filyovskaya, Moscow, 121087, Russia

E-mail: boldin1940@yandex.ru, panovyur@mail.ru, solntsev@madi.ru, pva@suntrans.ru

Abstract. The results of a statistical survey of the content of toxic components CO, CH and NOX in the exhaust gases of 38 "LiAZ" buses (4th and 5th ecological classes with foreign-made engines) at the nominal and increased engine crankshaft rotation speeds in idle running at compressed (compressed) natural gas in Moscow and adjacent areas of the region. It has been established that, as a first approximation, these modes provide information on the deterioration of environmental safety, primarily in relation to nitrogen oxide (NOx) emissions, the control of which in ordinary operation by the existing technical regulations for all types of vehicles, including heavy-duty gas vehicles are not provided. Statistical processing, with an assessment of the maximum permissible value of this diagnostic parameter in two idle modes (no more than 50 M-1 for the best attainable states of buses of the 5th ecological class), showed that the absence of such a simple control leads to approximately 75% of faulty states. These indicators can be significantly improved with the introduction of monitoring of operational monitoring of NOx and the development of more complex maintenance technologies. Herewith there is a significant environmental safety improvement of gas motor buses in comparison with diesel versions (approx. 6-8 times) and the technological and economic feasibility of their use for many regions of the country. Preliminary studies of the structure of such a technology have been carried out on a "KAMAZ" gas vehicle, which has an engine control different from "LiAZ" buses. At the same time, the higher NOX emissions level could be considered as allowable for intercity bus transportation and for activities outside the city, however, they can be reduced by about 3...5 times due to the use of the latest versions of pre-chamber devices "Ball Lightning".

1. Introduction

The Order of the Government of the Russian Federation No. 767-r [1] dated May 13, 2013 gave a new impetus to public transport gasification. According to this Order up to 50 % of buses and municipal heavy vehicles in million cities should be converted to the operation on natural gas (methane) as a motor fuel within the period of 7 years. It was planned that for settlements over 300 thousand inhabitants the share of buses and municipal heavy vehicles running on natural gas would be 30%, and for settlements over 100 thousand inhabitants – 10%. The reason of the issuance of the Order was not

Content from this work may be used under the terms of the Creative Commons Attribution 3.0 licence. Any further distribution of this work must maintain attribution to the author(s) and the title of the work, journal citation and DOI.
Published under licence by IOP Publishing Ltd

only economic criteria (the cost of methane for various regions of Russia is 2,5 - 3 times lower compared to the cost of gasoline and diesel fuel), but also a considerable contribution to the solution of the problem concerning to the protection of environment from the harmful effect of automotive transport (i.e., elimination of exhaust smoking, this phenomenon is unavoidable for diesel engines running under load. In recent years, there has been a trend to switch light motor vehicles to natural gas. First of all, it relies to taxi, but in southern regions of the country many private cars' owners are already using natural gas as a motor fuel. Conversion of vehicles to natural gas using gas-diesel process did not give a high replacement rate in urban traffic conditions. By the time the Order of the Government of the Russian Federation [1] was issued, this problem had been begun to solve successfully by factory production of specialized mono-fuel gas engines for buses LiAZ, NefAZ, Volgabus, PAZ, KavZ and trucks KamAZ and Ural. Other disadvantages that hinder the widespread use of methane, experts attribute the insufficient number of gas refuelling stations. According to the latest data, for more than 200 thousand motor vehicles running on compressed natural gas (CNG) in Russia there are slightly more than 500 CNG stations (automobile gas-filling compressor stations) and refueling sites for cars equipped with mobile auto-gas stations (PAGZ). At the same time for cars and light commercial vehicles using liquefied petroleum gas (propane-butane) the number of gas refuelling stations is 3500 units. In addition, maintenance and repair processes become more complicated due to the need to maintain gas equipment. Technogenic safety performance also deteriorates due to the need to store the filled gas under high pressure (20 MPa) [3]. However, the intensity of the public transport gasification processes, in spite of the noted problems, has recently increased significantly, primarily for Moscow, which is the largest city in Europe and has a circular layout that is not the most convenient for natural ventilation. State Unitary Enterprise "Mosgortrans" increases purchases of new LiAZ gas buses with foreign engines with higher environmental performance [7]. Simultaneously there is the growth of the fleet of more expensive electric buses for operation in the central regions of Moscow. In accordance with the Resolution of the Moscow Government dated April 22, 2019 No. 408-PP "On Approval of the General Gas Supply Scheme for the City of Moscow for the Period up to 2030 and the Scheme for External Gas Supply for the City of Moscow for the Period Until 2030" [2] in a short time (until 2020 year) 11 new CNG refuelling stations were built, and 9 more CNG stations are operating in the Moscow region. At present, there is a situation with an insignificant work load of Moscow CNG refuelling stations. All this contributes to the further growth of the methane car fleet in the near future. This poses an additional task of developing processes for organizing diagnostics and maintenance of heavy trucks and buses gas engines in order to maintain operation environmental safety at the certification requirements level set by the manufacturer, or even higher, in accordance with the state-of-the-art achievements of scientific and technical progress (STP).

2. Statement of the problem and analysis of possible methods of its solution

In connection with the latter, it becomes obvious to revise the previously established ideas about the environmental safety of heavy vehicles and buses, and technical regulations in force in the field of technical operation to ensure it. In particular, one of the authors of this publication [11] showed the feasibility, in addition to smoke, of additional control for city buses of the content of nitrogen oxides NO_X in the exhaust gases, which carry independent information about the ecological state of the rolling stock and specify the already known list of technological operations to improve it ... Somewhat earlier [10] it was shown the feasibility of an additional operational check of diesel passenger cars and light commercial vehicles (LDDVs) for their compliance with the EURO environmental classes using a simplified method of simulating driving cycles on unloaded running drums, which was also based on NO_X measurement. To do this, for a start, it is necessary to provide in the technical regulations for equipping motor transport enterprises (MTE) and service stations with five-channel gas analyzers, which have long been produced in Russia by commercial organizations at lower (compared to Western) prices and with better adaptation to our operating conditions. Direct measurements focus on certain harm to public health and the environment (*eutrophication*), which can increase when the normative values for NO_X emissions are exceeded, established by the standards of different countries (for the EU, the EURO standards). The relevance of emission control is confirmed by the so-called "Dieselgate" in the United States, when Volkswagen AG underestimated the actual values of exhaust gases toxic parameters for cars during certification. In the West, various types of studies are constantly being carried out related to the assessment of the harmful affect from excess NOx emissions from LDDVs, which give a significant scatter of values when using different approaches (for example, 900-500-300 [10] when assessing the total harm).

However, for the three noted approaches, there is a practical coincidence of the estimates of the comparative ranking according to the main indicators of pollution: for example, for Germany, the annual emission of particulate matter and soot was last estimated at 2800 conventional values, ozone (causing *smog*) at 1550, and NO_X at 930 [12], i.e., the latter accounted for more than 20% of the total pollution damage for diesel passenger cars, which is unacceptable.

There is no doubt that for heavy-duty vehicles and buses with diesel engines, damage assessments may have also significant indicators (according to conventional estimates "10" for Germany is about 30% less compared to LDDVs), therefore, all measures of the Moscow Government on the intensive transfer of city buses to the use of methane are justified. However, for these machines, the significance of NO_X emissions increases significantly due to the increase in compression ratio compared to gasoline engines, and the worse operating conditions for classic three-way catalytic converters due to excessive depletion of the combustion mixture. Therefore, along with the recommendations for diesel buses, it is necessary to monitor NOx emissions level in exhaust gases when using methane as motor fuel, which is not provided for by the current technical regulations [6]. It is also advisable to supplement the complex of operational diagnostics by checking the operation of spark plugs, the role of which may turn out to be even more significant in comparison with gasoline cars (due to an increase in breakdown voltages), and fuel consumption, at least at idle. For these purposes, it is possible to use both recently appeared and practically well-mastered, but little-known diagnostic equipment: the latter should include the *USB Autoscope* (models *II-III-IV*), which can be used to check the operation of the spark plugs on the oscillograms of the secondary voltage. Convenient for measuring fuel consumption during stationary diagnostics are *ELM 327* adapters, which are connected to the *universal diagnostic* connectors *EOBD* and *OBDII* of the car and display the fuel consumption (l / h) and ignition timing on the mobile phone, and the intensity of braking and acceleration of the car when driving; in the long term (after equipping the car with appropriate sensors), the content of toxic NO_X and greenhouse CO_2 components in exhaust gases, and other indicators can be determined. All of the above will make it possible to move to a higher level of maintenance of the environmental safety of gas buses (and heavy vehicles), and, according to the authors, in the social and possibly economic direction, will justify the complication and rise in the cost of maintenance technology. Optimization of these processes will become possible after a complex of laboratory and industrial research, the initial stage of which is devoted to this work. The accumulated experience shows that the necessary stage of such research in the field of technical operation of vehicles (TOV), providing an important source of information, are statistical surveys of the condition of vehicles in the MTE [4, 8, 9].

3. Experimental research.

The collection of statistical information was carried out at the Levoberezhnaya operational site of the Zelenogradskiy branch of the Mosgortrans State Unitary Enterprise (the former 11th Bus Depot in Moscow), which operates LiAZ gas buses equipped with various engines. To measure the composition of exhaust gases at idle speed at a warm engine at the diagnostic site, a five-channel gas analyzer of the 2nd accuracy class of Russian Federation production was used, which was provided by the Department of Operation and servicing of motor vehicles «MADI». Measurements were an integral part of the educational process of preparing the attestation work of the master's student of the department, which provided the necessary level of reliability in the further processing of the results obtained. The indicators of the content of CO, CO_2, CH and NO_X obtained for 38 buses at two idle modes are given in Table 1 and can be used with repeated independent processing by interested specialists.

Table 1. Environmental indicators of "LiAZ" gas buses when working on idling modes at n_1 = 700 rpm (1) and n_2 = 1600 rpm (2)

CO, % (1)	CO_2, % (1)	CH, M^{-1} (1)	NOx, M^{-1} (1)	CO, % (2)	CO_2, % (2)	CH, M^{-1} (2)	NOx, M^{-1} (2)	Env. class
0.318	10.64	56	1630	1,271	10.28	64	1052	5
0.32	10.82	72	1140	0.265	11.35	60	870	5
0.005	10.41	82	117	0.017	11.29	53	258	5
0	10.76	145	340	0	11.56	31	919	4
0	11.12	126	463	0	11.48	21	373	5
0	10.91	21	297	0.004	11.28	20	569	5
0.002	10.97	29	352	0	10.93	20	625	5
0.004	11.08	10	594	0.002	11.2	7	287	4
0	10.82	71	429	0.007	11.33	11	37	5
0	11.56	0	411	0	11.44	0	6	5
0	11.47	11	548	0	11.55	9	28	5
0.001	10.21	45	1445	0.001	10.81	18	1918	4
0.009	11.48	42	476	0.014	11.52	14	129	5
0.001	11.02	80	134	0.01	11.64	35	375	4
0	10.95	41	353	0	11.24	27	2053	4
0	9.255	0	336	0	9.075	1	469	5
0.011	8.318	9	26	0.006	7.934	5	13	5
0	7.589	5	211	0	6.539	5	49	5
0.002	7.22	0	8	0.003	7.529	4	51	5
0.002	7.24	25	262	0.001	7.005	11	1136	4
0,001	6,704	4	12	0,001	6,257	6	10	5
0,001	7,064	0	7	0,001	6,794	2	101	5
0,008	8,801	10	333	0,007	8,686	12	627	4
0	7,25	1	4	0,008	6,961	2	11	5
0,001	7,133	0	13	0,001	6,757	1	9	5
0	6,111	14	256	0	6,916	20	348	4
0	6,93	21	478	0,001	6,476	13	1171	4
0	11,43	1	754	0	11,53	1	26	5
0,059	11,33	3	226	0,025	11,56	3	211	5
0,013	11,5	0	420	0,013	11,61	0	8	5
0	11,27	0	510	0	11,45	3	450	5
0,001	11,44	0	428	0	11,57	2	587	5
0	11,43	13	1402	0	11,45	8	2124	4
0	10,08	138	459	0	10,35	38	3074	4
0	11,64	2	1202	0	11,63	2	0	5
0,002	11,41	1	50	0,002	2	11,47	6	5
0,005	11,2	7	409	0,005	11,35	10	424	5
0	11,46	1	3	0	11,51	1	9	5

The analysis was carried out using the Russian and the English versions of program STATISTICA-10-12,5. All indicators were transferred to a specialized spreadsheet (see below). The figures below show the results of processing the data from Table 1 obtained using these programs.

At the same time, it is convenient to set the automatic execution of calculations by comparing two independent samples based on t-tests (Fig. 1), from which one can clearly see the significant environmental advantage of buses of the 5th ecological class (in relation to the 4th) in terms of NO_X emissions, especially on increased frequencies at which load modes are in operation. (From Table 1 it follows that *CO* and *CH* emissions for gas buses are practically insignificant for environmental safety).

Another important conclusion that can be made after the *regression* analysis (according to the spreadsheet in Fig. 2) is the identification of a fairly close probabilistic (*stochastic*) effect of NO_X emissions at nominal idle modes on emissions at increased frequencies. (According to the numerical estimates on the information panel in Fig. 2, the correlation coefficient exceeded the critical value R> 0.5 and the determination R2> 0.25; Fisher's criterion, reflecting the *informativeness* of the regression dependence, reached a significant value F> 12 at a high level of significance, reflecting the probability obtaining an erroneous conclusion p<0.00148, therefore the line for the influence coefficient B1 by the STATISTICA program is highlighted in red). Hence, an important conclusion for practice follows that it is possible (and advisable) to use both the first and second indicators of NO_X emissions at idle modes as independent diagnostic parameters, and when they are in the range of maximum permissible values, consider a gas vehicle in an ecologically sound state. However, it must be remembered that this conclusion does not correspond to the specifics of checking gasoline cars, for foreign models of which, at the XXth century, *zeroing* of NO_X values for environmental classes EURO-3 and higher is already observed (sensitivity of an inexpensive gas analyzer of the 2nd accuracy class is not sufficient); here it is necessary to carry out speed tests on unloaded running drums, which already show NO_X [5] emissions.

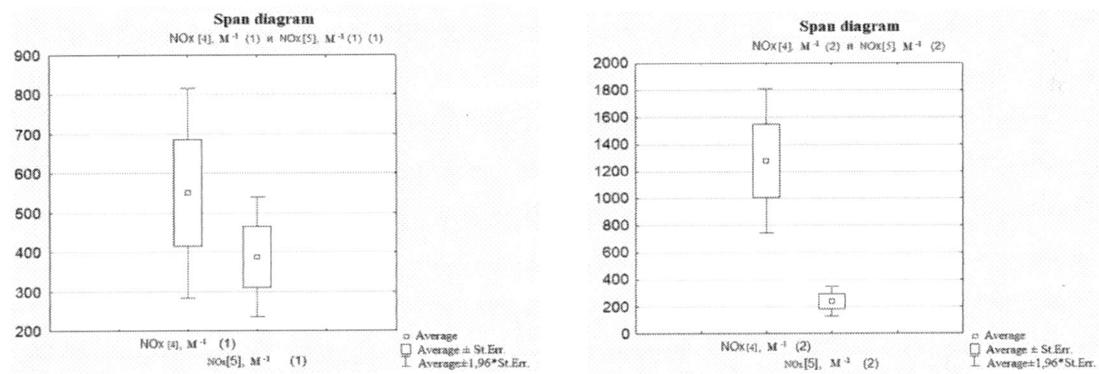

Figure 1. Span diagrams for assessing the average NO_X emissions of «LiAZ» gas buses of EURO-4 and EURO-5 ecological classes at idle mode at nominal (left) and increased frequencies (right) of engine crankshaft rotation

In a mass inspection of the state of vehicles, it is advisable to determine the maximum permissible values of diagnostic parameters on the basis of a statistical approach, gradually excluding the worst values from the initial array, until the remainder of the *best* and already achieved in practice indicators are obtained, the dispersion of which can be approximated with the greatest probability by unimodal theoretical laws. Such laws can be normal, exponential, gamma distribution with positive and negative skewness, and sometimes Weibull, although the latter was developed for assessing reliability indicators. The validity of this approach can be seen from the histograms in Fig. 3, built in accordance with the statistics requirements (Sturges criterion), the left of which shows the entire range of values of NO_X emissions at an increased crankshaft speed for "LiAZ" buses of the 5th ecological class, where the first column clearly reflects the presence of a zone of good health, and the remaining five with practically the same probability density - faulty. Subsequent analysis of the data for the 1st column in the range up to 65 M^{-1} (set by a special command) allowed the STATISTICA program to establish the only possible approximation of the distribution by the Weibull law (middle part of Fig. 3), the limitation of which with a probability of 0.95 in the first approximation allows set the maximum allowable value of the diagnostic parameter (no more than 50 M-1 according to the calculations on the probabilistic calculator). The same value is given by calculations for the nominal idle mode. From data analysis of table 1 it can be seen that even for buses of the 5th ecological class, about 25% of cars are formally serviceable in two parameters (informally, in terms of the main parameter for increased frequency - about 45%), and performing routine maintenance operations on them would lead to a decrease in emissions. Although these standards turned out to be impracticable for 4 class of buses, such statistical gradations will also help to reduce environmental pollution. In general, in terms of NOX emissions at nominal idle modes, there is a noticeable advantage of "LiAZ" 5th ecological class gas buses compared to diesel buses (Fig. 4): here, for the first 15% of cars, emissions above 800 M-1 could be easily reduced to average values of 400 M-1.

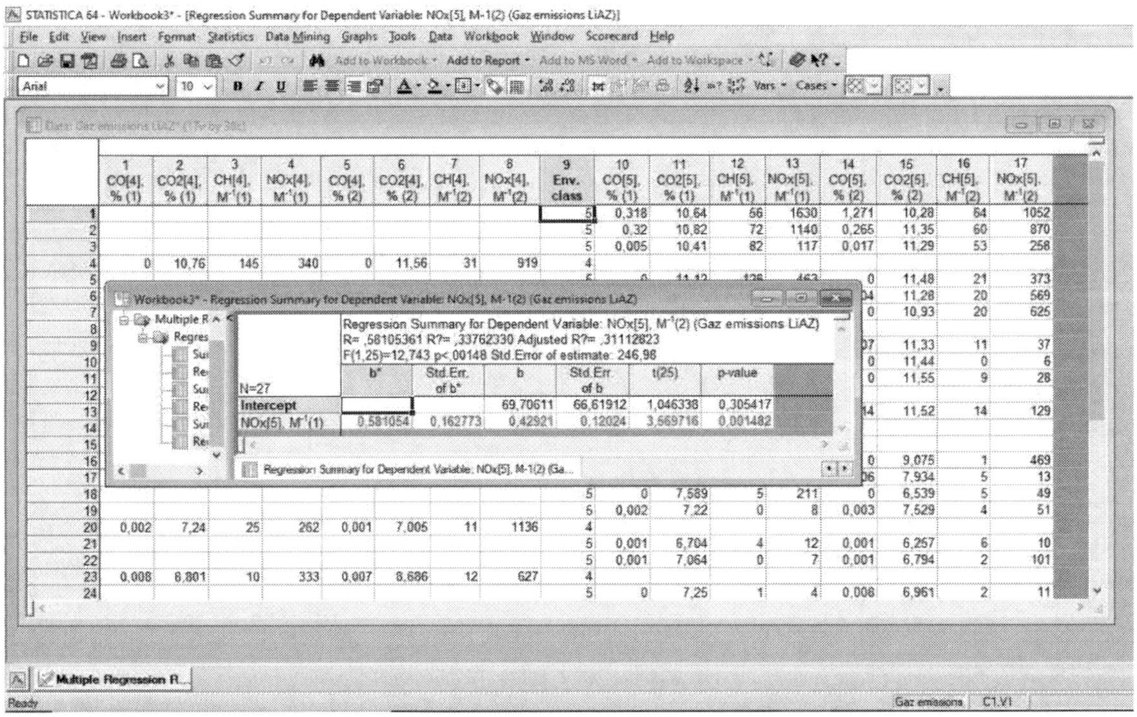

Figure 2. STATISTICA-12,5 program window with the estimation of the parameters of the linear model of the regression dependence of NO_X nitrogen oxide emissions at increased crankshaft speeds on emissions at nominal frequencies for "LiAZ" gas buses of the EURO-5 ecological class

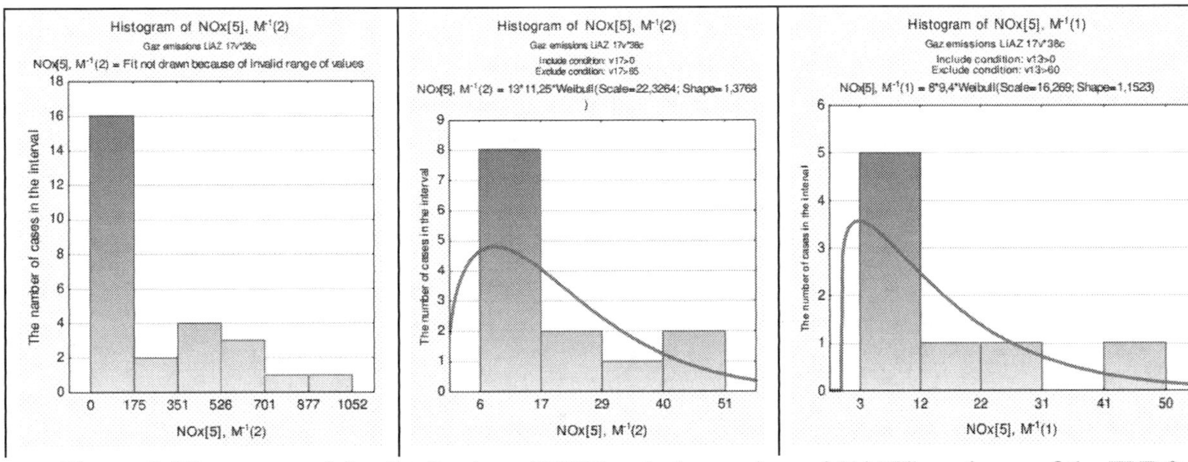

Figure 3. Histograms of the distribution of NOX emissions values of "LiAZ" gas buses of the EURO-5 ecological class at idle modes at increased engine crankshaft speeds for a mixed sample (serviceable and faulty condition) (on the left, fitting the law is impossible); after excluding the faulty state data based on a statistical approach (middle); at nominal frequencies after excluding faulty state data (right)

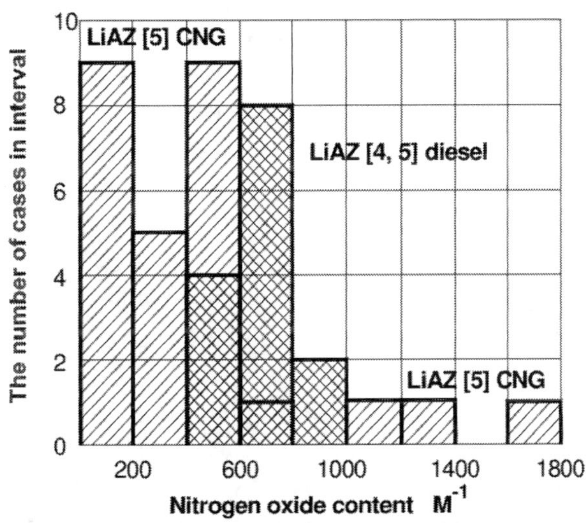

Figure 4. Histograms of the distribution of the content of nitrogen oxides NOX in the exhaust gases of "LiAZ" city buses running on compressed natural gas (CNG) of the declared EURO-5 ecological classes, and "LiAZ" city buses with diesel engines of the EURO-4-5 [11] declared ecological classes, on nominal speed of the crankshaft at

The need (and expediency) of using a more complex technology for servicing systems to ensure the operability of high-power gas engines (in this case, ignition systems, methane supply and exhaust gas neutralization) was confirmed by the example of diagnosing a "KAMAZ" vehicle equipped with EControls engine control system. First of all, it was carried out to assess the feasibility of using prechamber devices "Ball Lightning" (Fig. 5, left [8]) on such engines, the use of which on gasoline cars led to a decrease in NO_X emissions by 1.5 times (when imitating an urban driving cycle (ECE test) on the running drums without load [5]). In addition, the modernized pre-chamber device made it possible to use ordinary *"Zhiguli" spark plugs* with increased gaps, since the breakdown voltage was reduced by about 30% (Fig. 5, right), which also turned out to be relevant for methane engines with an

IOP Conf. Series: Materials Science and Engineering 1159 (2021) 012079 doi:10.1088/1757-899X/1159/1/012079

increased compression ratio using foreign plug models. Unfortunately, the *USB Autoscope* device, which allows observing the secondary voltage oscillograms through a laptop, has the ability to sequentially check the operation of the spark plug for only one cylinder (by reinstalling the capacitive sensor on the high-voltage wires), however, with two operators at the diagnostic site, this does not cause complications.

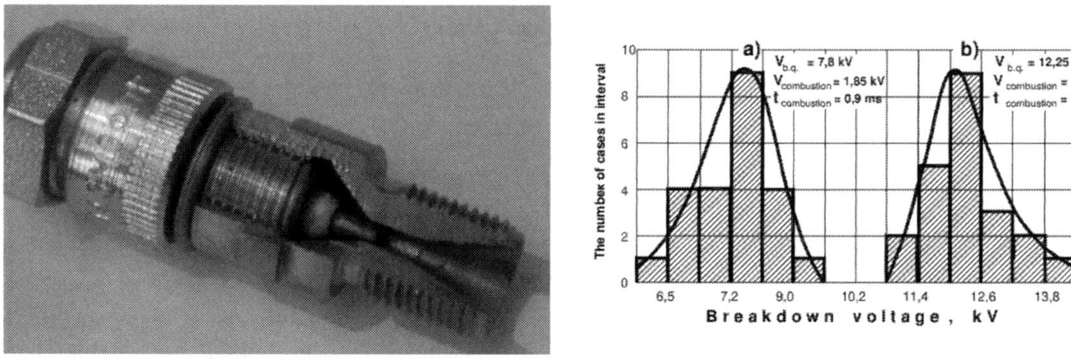

Figure 5. Section of the prechamber device "Ball Lightning" (for flare ignition of a combustible mixture in the cylinders of gasoline engines) with an experimental spark plug with an extended central electrode (left) [8, 9], and the distribution of breakdown voltage indices of gapless spark plugs BRISK PREMIUM LR15ZC when using a prechamber device a) and when working without it (b) on a VAZ-2107 car engine (right)

The measurements of the diagnostic indicators of the "KAMAZ" vehicle were carried out when it was parked in an open area at a temperature of 17^0 Celsius, which ensures the operability of the devices. At the same time, it turned out to be possible to record the relative changes in fuel consumption (without preliminary calibration of the measuring system) when the *ELM 327* adapter (for cars) is connected to the universal *OBDII* diagnostic connector of the "KAMAZ" car and displays the instantaneous fuel consumption (l / h) and the ignition timing on the mobile phone (and a large set of other indicators in the process of movement). The results of three consecutive stages of measurements (initial state, using the pre-chamber device "Ball Lightning" with gapless BRISK candles, repetition of the initial state) are given in tab. 2. The excess air coefficient α for an excessively lean mixture was calculated from CO, CO_2 and CH using the Porsche formula [5], since it went beyond the gas analyzer settings. From the operational analysis table 2, it can be seen that the environmental performance of the "KAMAZ" vehicle as a whole is somewhat worse compared to "LiAZ" buses of the 4th environmental class, due to a sharp, more than 100 times, increase in NO_X emissions at an increased rotational speed of the KV, but they can be reduced by 3 … 5 times when using pre-chamber devices (the latter can be assumed for "LiAZ" buses). In addition, during servicing, you can additionally correct the breakdown voltage of the spark plugs by individually adjusting the gap (to increase the spark power), and the ignition timing, since it was observed to change by (-5) degrees when using pre-chamber devices. According to the data obtained, an increase in fuel consumption for prechamber devices was not revealed, although (as it was mentioned above) when a passenger car was running on liquefied natural gas (propane-butane), efficiency deteriorated by 5%, and on gasoline - up to 10%. Thus, the tested prechamber devices are an ideal means of significantly increasing the environmental safety of gas engines of public transport vehicles, and with an increase in the efficiency of the NO_X neutralization systems, they can provide an increase in their service life. A higher level of estimation of NO_X emissions can be obtained with the additional use of a stationary stand with unloaded treadmills, on which a *classic imitation* of an urban driving cycle (ECE test) for passenger cars with high-speed sites of 15, 30 and 50 km / h, typical for urban traffic, is possible. The development of a model for recalculating the results obtained in indicators of specific emissions by mass of matter in *g / km* of run per 1 *liter* of the working volume of the engine, as was already justified for diesel buses "LiAZ" [9].

Table 2. Changes in the environmental and technical performance of the gas engine of a "KAMAZ" vehicle at nominal and increased idle speed when using gapless plugs BRISK PREMIUM LR15ZC in combination with prechamber device "Ball lightning" [8, 9]

Device, operating mode: *n, rpm*	Emissions *CO*, %	Emissions CO_2, %	Emissions *CH*, M^{-1}	Emissions NO_X, M^{-1}	Fuel consumption G_T, l/h	Excess air ratio [5] *α*, rel.units
Without device, 700	0,044	8,45	144	28	13,0	1,52
Without device, 1600	0,124	9,00	240	2850	25,0	1,41
Prechamber, 700	0,032	8,54	136	45	12,4	1,51
Prechamber, 1600	0,124	9,70	131	410	22,2	1,35
Without device, 700	0,034	8,55	140	12	13,1	1,51
Without device, 1600	0,124	9,25	274	1614	22,8	1,37

4. Conclusion

The absence in operation of checking the composition of exhaust gases from heavy-duty gas vehicles and buses with five-channel gas analyzers (for *CO*, CO_2, *CH*, NO_X and O_2) does not allow timely detection of the deterioration of their environmental safety and high-quality maintenance (MO) in order to reduce emissions into the environment the main polluting component - nitrogen oxides NO_X. At the same time, the probability of a malfunctioning state, according to preliminary estimates obtained by the authors, can reach values of the order of 55 ... 75%, which reduces the effectiveness of the current program of the Moscow Government to transfer 50% of public transport to methane, with already developed infrastructure for the supply of this type of fuel. Achievements of scientific and technological progress of the last 2-5 years makes it advisable and allows the use of more complex maintenance technologies based on complex diagnostics of ignition systems, fuel supply, neutralization devices and other necessary elements in the chains of the technological process, ensuring the maintenance of environmental safety of buses when working on the line on the level of significantly more expensive electric buses. The complication and rise in the cost of service, according to the authors, in the future will be multiplied by an increase in social and economic efficiency. The latter is also facilitated by the specific use of prechamber devices "Ball Lightning" for high-power gas engines, which reduce NO_X emissions by 3 ... 5 times (due to changes in the combustion processes of the fuel mixture) and increase the service life of the neutralizers in operation.

References

[1] Decree of the Government of the Russian Federation of May 13, 2013 No. 767-r *"On the regulation of relations in the use of gas motor fuel, including natural gas as a motor fuel."* Available at: https://auto-ally.ru/pravo/52569/index.html (date of access: 02.12.2020).

[2] Resolution of the Moscow Government dated April 22, 2019 No. 408-PP dated April 22, 2019 *"On approval of the General Scheme for Gas Supply to the City of Moscow for the Period until 2030 and the Scheme of External Gas Supply for the City of Moscow for the Period until 2030* Available at: https://www.mos.ru/upload/documents/docs/408-PP.pdf (date of access: 02.12.2020).

[3] *Gas-powered vehicles: environmentally friendly, economical, safe. Why is there so little of it in Russia?* AUTOSTAT, analytical agency, 2020. Available at: https://tass.ru/opinions/6974728 (date of access: 02.12.2020).

[4] Boldin A P and Bohumil V N 2016 *Use of STATISTICA program modules and computational models for solving scientific problems in the field of technical operation: textbook. manual*

for students, undergraduates and graduate students of institutions of higher education (Moscow: Tekhpoligrafcenter) 154 p

[5] Boldin A P and Aksenov P V 2017 *Features of the operational methodology for diagnosing passenger cars for compliance with EURO ecological classes. (Bulletin of MADI)* **1 (48)** 27-34

[6] GOST R 54942-2012. *Gas vehicles with spark engines. Vibrations of harmful (polluting) substances with exhaust gases. Norms and methods of control when assessing technical condition*, 2013

[7] *State Unitary Enterprise "Mosgortrans" transfers city buses to the gas modification [Electronic resource]. State Unitary Enterprise "Mosgortrans"* Official website, 2019 Available at: https://mosgortrans.ru (date of access: 07.10.2020).

[8] Kotov N M Boldin A. P Aksenov P V and Rizhov M B 2012 Environmental capabilities of the pre-chamber ignition system "Ball lightning" *Engine magazine* **4**

[9] Kotov N M and Aksenov P V 2010 *Patent for utility model 97869, IPC H01T 13/46. Prechamber device "Ball lightning" for an internal combustion engine* utility priority dated May 12

[10] Boldin A P Sarbaev V I and Aksenov P V 2017 *Diagnostics of passenger cars and minibuses with diesel engines for compliance with Euro emissions standards. International Jornal of Mechanical Engineering and Technology* Vol 8 **12** 933-943

[11] Boldin A P 2020 Analysis of some indicators of environmental safety of buses with diesel engines (on the example of Moscow) *International Conference on Digital Solutions for Automotive Industry, Roadway Maintenance and Traffic Control (DS ART 2019) 1 November 2019, Cholpon-Ata, Kyrgyzstan / DS ART 2019 IOP Conf. Series: Materials Science and Engineering* **832** 012001 doi:10.1088/1757-899X/832/1/012001

[12] Jonson J E, Borken-Kleefeld J, Simpson D A, Nyıri A, Posch M and Heyes C 2017 *Impact of excess NOxemissions from diesel cars on air quality, public health and eutrophication in Europe* 3-11

IOP Conf. Series: Materials Science and Engineering 1159 (2021) 012080 doi:10.1088/1757-899X/1159/1/012080

The concept of structural steels surface hardening for tribo-joint elements

O V Chudina and Yu M Luzhnov

Moscow Automobile and Highway State Technical University (MADI), 64, Leningradskiy prospect, Moscow, 125319, Russia

E-mail: Chudina_madi@mail.ru

Abstract. The concept of surface hardening of structural steels is proposed on the basis of the structural theory of strength and the Charpy principle at the macro- and microgeometric levels for the purposeful formation of a structure that can effectively resist wear and fatigue. An intellectual approach to the rational choice of technological combinations capable of maximally realizing the dislocation, substructural, grain-boundary, solid solution and dispersion mechanisms of steel hardening, which makes it possible to create optimal structures on the surface of tribo-interface elements, is described. As an example of such an approach, the surface laser alloying of structural steels with nitride-forming elements with subsequent nitriding was investigated; an increase in surface hardness, strength, wear resistance by 15 times and fatigue strength by 1.5 times compared with the initial parameters of steels was shown.

1. Introduction

The service life of the friction unit is determined by the optimal design solution, accurate calculation of real loads, as well as the selected material and the method of hardening of the tribo mating parts operating in specific operating conditions. Depending on the operating conditions, complex and interdependent multifactorial phenomena occur on the surface of the tribo-interface elements.

In the zone of frictional contact, under the influence of loads, a friction force arises, and, due to the rupture of metal bonds in places with the highest stress concentration, free energy is released. Moreover, with an increase in the friction force, the activation energy increases and it is spent primarily on the release of heat and change in the shape of the body. As the temperature rises, diffusion phenomena are also accelerated, which leads to significant changes in the structure of surface layers, up to phase transformations. At the first stage of the friction process, an increase in the dislocation density leads to some hardening of the metal surface. However, with the further development of plastic deformation, critical accumulations of dislocations are observed in some areas of the surface, which are the cause of the appearance of surface and subsurface submicrocracks, which, while developing, block the channels of dislocation advance and form a branched system of cracks, which ultimately leads to fatigue failure.

Since in the process of friction these surfaces interact with each other directly or through a lubricating medium, the energy released in the frictional contact zone also activates physicochemical processes, such as oxidation and reduction reactions, adsorption of elements from the environment, electrification and others related to them. phenomena. In real frictional contact, as a rule, all these processes proceed simultaneously [1].

Content from this work may be used under the terms of the Creative Commons Attribution 3.0 licence. Any further distribution of this work must maintain attribution to the author(s) and the title of the work, journal citation and DOI.
Published under licence by IOP Publishing Ltd

IOP Conf. Series: Materials Science and Engineering 1159 (2021) 012080 doi:10.1088/1757-899X/1159/1/012080

A decrease in the adverse effect of these factors is associated with the creation of dissipative compatible structures capable of adaptability, self-organization and efficient energy dissipation [2, 3]. The available information about such structures is rather empirical. In particular, research was carried out on some pairs of friction, which, with a certain type of lubricant in the process of mutual movement, could adapt to each other, providing a given durability without damaging the friction surface.

Analysis of literary sources indicates the absence of a systematic approach to the purposeful creation of materials for tribosystems.

An important role in reducing the effect of friction processes on the reliability and durability of machine parts is assigned to designers who perform calculations to quantify the critical stresses arising during operation. However, they cannot consider the entire spectrum of dislocation processes and related problems of the structural strength of materials.

Therefore, the solution of such a complex problem is possible on the basis of the joint use of the achievements of metal physics and fracture mechanics through scientific design and purposeful formation of dissipative structures capable of efficiently dissipating the energy generated in the friction contact zone and resisting its destructive effect on the surface of parts.

In this regard, an urgent problem is the development of the concept of surface hardening of structural steels for tribosteel elements based on the structural theory of strength by purposeful application of known hardening technologies or their combinations in order to increase the structural strength of materials operating under wear conditions.

Structural strength is expressed by indicators of mechanical properties, which are conventionally combined into two groups: a reliability group and a durability group.

The criteria for the reliability of materials are impact strength, fracture toughness (survivability, critical crack opening, etc.), and the criteria for durability are wear resistance, fatigue strength, contact endurance, corrosion resistance, etc.

Many machine parts operate under conditions of wear and alternating loads, at which maximum stresses arise in the surface layers, where stress concentrators are concentrated [4]. The sensitivity to them is sharply reduced when residual compressive stresses are created on the surface by hardening by induction, laser, electron-beam heating or chemical-thermal treatment, vibration rolling, including ultrasonic or in some other way.

All characteristics of structural strength are structurally sensitive, and therefore the way to increase the operational properties is associated with the development of such strengthening technologies that would lead to the creation of heat-resistant structures with high hardness and strength, providing the design level of reliability and durability of structures. The formation of the optimal structural state of metallic materials, which provides a high level of reliability and durability of products, is most fully realized on the basis of the application of the principles of dislocation hardening theory [5].

According to the structural theory of strength and fracture mechanics, strength is understood as the ability of a material to resist the movement of dislocations. This means that in order to increase the strength of metallic materials, it is necessary to slow down the movement of dislocations, that is, to create barriers in the path of their movement. Other dislocations, microstresses arising in solid solutions in the presence of alloying elements, grain boundaries, dispersed particles, coherent or incoherent with the matrix lattice, can be such barriers in the structure.

An analysis of the existing dislocation theories shows that the main hardening mechanisms that provide an increase in the plastic flow stress are hardening by dissolved interstitial or substitutional atoms, dislocations, grain and subgrain boundaries, and dispersed particles. Then the strengthening of real metals is determined by the combined action of the listed mechanisms:

$$\sigma_T = \Sigma\sigma_i = \sigma_0 + \Delta\sigma_{s.h.} + \Delta\sigma_d + \Delta\sigma_g + \Delta\sigma_s + \Delta\sigma_{p.h.}.$$

where σ_0 is the frictional stress of the crystal lattice (Peierls-Nabarro force); $\Delta\sigma_{s.h.}$ - an increase in the yield strength due to solid solution hardening; $\Delta\sigma_\partial$ - increase in yield strength due to dislocation (strain) hardening; $\Delta\sigma_g$ - increase in the yield point due to grain-boundary hardening; $\Delta\sigma_s$ - increase in the yield stress due to substructural hardening; $\Delta\sigma_{p.h.}$- an increase in the yield strength due to precipitation hardening.

The most favorable mechanisms that provide a combination of high strength with a sufficient margin of plasticity are grain boundary hardening σ_g, solid solution hardening $\sigma_{s.h.}$ (if alloying elements refine the grain) and substructural hardening σ_s. An increase in the density of unorganized dislocations (σ_d), increasing strength, to the greatest extent reduces the fracture toughness. Dispersion hardening ($\sigma_{p.h..}$) effectively increases the strength characteristics, while the negative effect of particles on the fracture toughness characteristics can be minimized by adjusting the structure parameters by technological methods.

On the basis of the implementation of certain strengthening mechanisms by purposeful formation of the structure by technological methods, it is possible to obtain the required characteristics of structural strength for parts operating under specific operating conditions. Using various technological combinations, you can use the maximum number of strengthening mechanisms and form a structure with high and ultra-high physical and mechanical properties [6, 7]. The main technological methods for the implementation of the most important hardening mechanisms are plastic deformation, especially with the use of ultrasound, quenching for martensite, laser processing, mainly with fusion and addition of carbide and nitride-forming alloying elements, chemical-thermal treatment, especially nitriding, etc. with surface fusion, dislocation, solid solution, grain boundary and substructural mechanisms are realized. However, calculations show that the greatest contribution to the overall level of hardening is made by the dispersion mechanism, especially by particles of nitrides of alloying elements coherent with the matrix. Such a mechanism can be realized by nitriding if laser alloying with nitride-forming elements is carried out beforehand [7]. At the same time, heating during nitriding to a temperature of 540 ° C relieves unfavorable tensile stresses arising at the border with the main one after laser exposure, leads to polygonization and the formation of a cellular substructure and, unlike other types of chemical-thermal treatment, does not lead to deformation of the product, while maintaining core structure unchanged.

An important advantage of a nitride-hardened surface is its high heat resistance (up to 600 ° C), which is important for products operating at elevated temperatures or under conditions of wear at high speeds, while martensitic structures obtained by traditional heat treatment (quenching and tempering), disintegrate with a loss of strength already at 250 ° C.

Thus, the combination of two mutually complementary technologies that increase the efficiency of each other, consisting in laser alloying (LA) of carbon steels with nitride-forming elements followed by nitriding, will make it possible to use the maximum possible number of strengthening mechanisms and purposefully create a structure that can withstand wear and fatigue.

The aim of this work is to implement the proposed concept of surface hardening of structural steels for tribo-joint elements using the example of a combined technology, which consists in laser alloying of the surface with nitride-forming elements and subsequent nitriding, and also to study its effect on the main parameters of structural strength characterizing the reliability and durability of parts, such as surface hardness, wear resistance and crack resistance.

2. Materials and methods

Carbon steels were treated with elements V, Cr, Mo, Al, etc. using laser radiation both in a pulsed mode with a radiation power of 24 J and in a continuous mode with a power of 1 kW at a beam travel speed of 2 to 30 mm / s. Nitriding was carried out in an ammonia atmosphere at a temperature of 540 ... 570 ° C for 3 ... 6 hours. Tests for wear resistance were carried out on an installation for studying tribological properties according to the "roller-block" scheme under dry friction conditions with the determination of the stabilized coefficient of friction. Crack resistance tests under low-cycle loading were carried out on a ZD-10 installation, and under high-cycle loading - on a URS-20/30000 installation. Fractographic studies were carried out using a Jeol-U3 electron microscope.

3. Results and discussion

In order to implement the maximum possible number of strengthening mechanisms, when choosing a matrix material, the type of alloying elements and technological parameters of laser alloying and subsequent nitriding, we were guided by the fact that the alloying elements had a high affinity for nitrogen, and also increased the solubility of nitrogen in ferrite. With these requirements in mind, V, Cr, Mo and Al were selected. Their optimal concentration in the alloying zone was determined depending on the ability to uniformly distribute in the iron matrix during laser alloying (Fig. 1).

The thickness of the alloyed zone is determined by the maximum possible thickness of the nitrided layer, which in most cases for alloyed steels does not exceed 600 μm (Fig. 2).

Steels with a carbon content of 0.2 ... 0.3% C are preferred as the matrix material. With a higher carbon content, the efficiency of the combined technology is somewhat reduced, since the alloying elements are partially bound into carbides and do not participate in the formation of nitrides.

The conducted studies of the influence of laser alloying parameters made it possible to establish the optimal technological modes of both pulsed and continuous radiation, at which the above requirements are satisfied.

Metallographic, X-ray diffraction and micro-X-ray spectral analysis showed that under optimal technological conditions in the zone of laser alloying with vanadium, chromium and aluminum, a structure of doped ferrite with an extremely fine grain of 2 ... 5 μm is formed (Fig. 3), which leads to an increase in microhardness up to about 6000 MPa. Doping with molybdenum in the structure may precipitate intermetallic phases, which increase the microhardness of the hardened layer to 11000 MPa. After nitriding at T = 540 ° C for 3 hours of the laser-doped surface of V, Cr and Mo, in the hardened layer, as a rule, a single-phase structure of a α -solid solution doped with nitrogen is formed without inclusions of nitride particles.

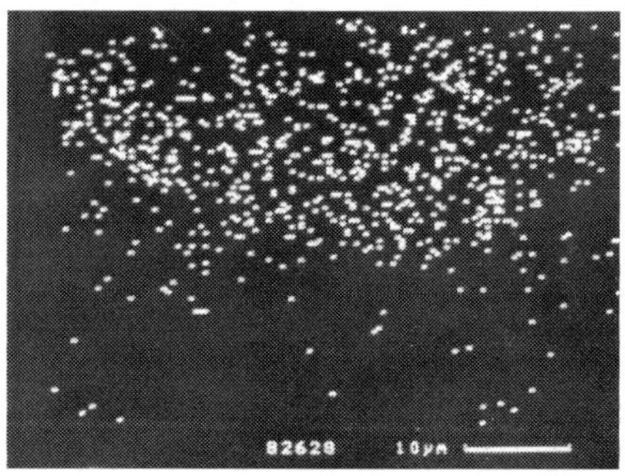

Figure 1. Distribution of vanadium in the laser alloying zone

Figure 2. Microstructure of steel 20 after laser alloying with chromium, x100

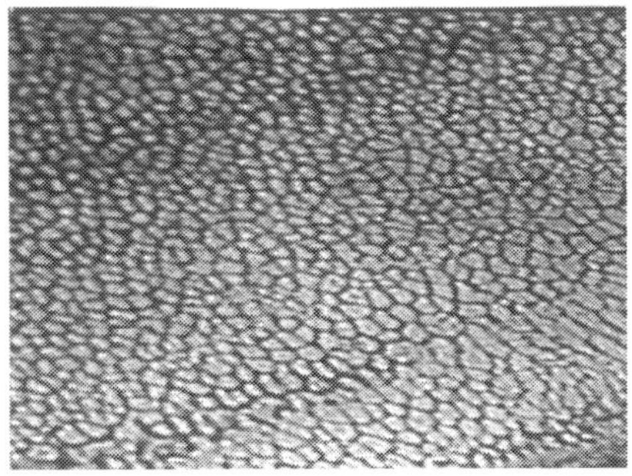

Figure 3. Microstructure of steel 20 after laser alloying with chromium, x1000

In zones doped with Al, a two-phase structure $\alpha + \gamma'$ - phase $(Fe, Al)_4N$ is formed. Such processing provides a significant increase in microhardness: LA (Al) + N up to 21000 MPa, LA (V) + N up to 18500 MPa, LA (Cr) + N up to 18000 MPa; LA (Mo) + N up to 12000 MPa. The increase in hardness is explained by the fact that after nitriding, the mechanism of solid solution hardening with nitrogen is additionally activated. When nitriding steels laser-alloyed with aluminum, the microhardness is maximum, since in this case the mechanism for increasing the strength combines the components of both solid solution and precipitation hardening with the γ' -phase (Fe, Al) 4N.

An increase in the duration of the nitriding process to 6 hours or aging leads to the release of nitrides of alloying elements fully or partially coherent with the matrix and, as a consequence, to an additional increase in hardness. It has been experimentally established that the maximum level of hardening is achieved after aging at a temperature of 250 ° C for 0.5 ... 1 h due to the release of dispersed particles of the optimal degree of coherence.

Calculations show that the most significant contribution is made by two mechanisms: solid solution - by nitrogen and dispersion - by particles of nitrides of alloying elements. The contribution of solid solution hardening with nitrogen is especially high in Fe-Cr alloys, which is explained by the strong influence of chromium on increasing the solubility of nitrogen in ferrite.

The proportion of solid solution hardening with an alloying element is very small: no more than 5%. Grain-boundary hardening is more significant in Fe-Mo alloys (7 ... 10%) in comparison with other alloys, which is explained by a finer-grained structure. The contribution of dislocation hardening, as well as grain-boundary hardening, significantly decreases after nitriding to 1 ... 6%.

The role of precipitation hardening in combined processing is maximal. It is the greater, the higher the thermodynamic stability of the nitride (in the series $Mo_2N \rightarrow Cr_2N \rightarrow VN$); upon strengthening with vanadium nitrides, its contribution reaches 83%. Moreover, the level of hardening by coherent particles is significantly higher than by incoherent ones.

Therefore, to achieve the maximum level of hardening, it is necessary to ensure the formation of a structure with dispersed particles of nitrides of alloying elements, which are in the optimal degree of coherence with the matrix, which is controlled by optimizing the technological parameters of the combined processing at each stage. Combined processing allows you to increase a whole range of structural strength indicators.

Wear tests have shown that the wear resistance of steel 20 after combined treatment is 15 times higher than that of normalized steel 20, and 1.5 ... 3 times higher than the wear resistance of nitrided nitralloy 38Kh2MYuA. High indicators of wear resistance are explained not only by the high hardness of the surface layer, but also by the formation of a Charpy-type relief on the surface, which significantly reduces the coefficient of friction. Moreover, the Charpy principle is realized both at the macrolevel with 50% filling of the surface with hardened zones, and at the micro- and nanolevel due to the creation of a structure inside the crystal, consisting of a soft and plastic α -solid solution and solid fine particles of nitrides of alloying elements. Well-known low friction antifriction materials such as babbits and cast bronzes have a similar structure. The best results are achieved by alloying with aluminum, vanadium and chromium, followed by nitriding (Fig. 4).

Cyclic tests under high-cycle loading with a frequency of 200 Hz for samples subjected to laser treatment showed a decrease in the growth rate of a fatigue crack and an increase in the threshold value of the stress intensity factor K_{1c}, below which the crack does not develop. Subsequent nitriding shifts the kinetic diagrams of fatigue fracture even more to the right.

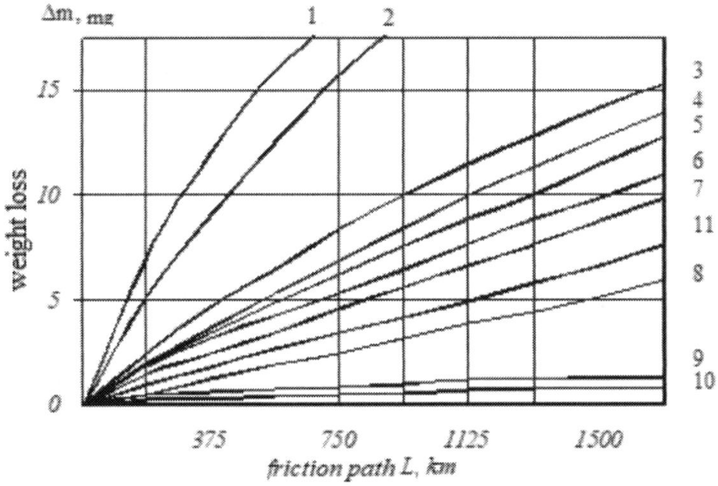

Figure 4. Wear of steel 20 after: 1 - normalization, 2 - nitriding, 3 - LA (Al), 4 - LA (Cr), 5 - LA (V), 6 - LA (Mo), 7 - LA (Mo) + N, 8 - LA (Cr) + N, 9 - LA (V) + N, 10 - LA (Al) + N, 11 - steel 38X2MЮA after nitriding

Tests under low-cycle loading with a frequency of 0.1 Hz when laser "tracks" are applied to a steel surface with overlapping, the rate of growth of a fatigue crack increases, and subsequent nitriding improves this characteristic to the level of the original sample. When laser "tracks" are applied without overlapping, followed by nitriding, the crack resistance of low-carbon steels increases by 1.5 times. The kinetic diagrams in this case have an unusual form (Fig. 5), which, however, fully corresponds to the internal mechanisms of fatigue failure of such heterophase samples. An unstable crack growth is observed, the accelerations of which alternate with periodic deceleration. The best results under high-cycle loading are achieved by alloying with molybdenum, and under low-cycle loading - with vanadium and chromium.

Fractographic studies have established that in the fracture of steels hardened by the combined LL + N technology, viscous components are observed (Fig. 6, b), while after laser alloying, the fractures of steel 20 have a pronounced brittle character (Fig. 6, a), which due to the significant contribution of the dislocation mechanism to the overall level of hardening.

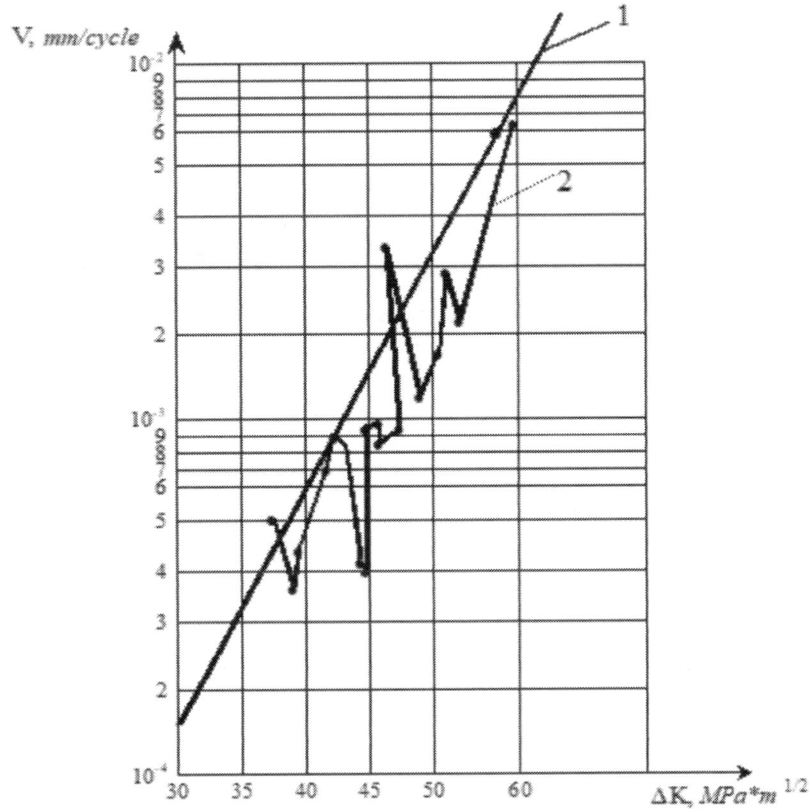

Figure 5. Kinetic diagrams of fatigue failure under cyclic loading with a frequency of 0.1 Hz for steel 20 after: 1 - normalization; 2 - LA (V) + N, laser lines applied without overlapping

During laser alloying and subsequent nitriding, a structure is formed on the steel surface, strengthened by dispersed particles of nitrides of alloying elements, which complicates crack initiation and promotes its effective deceleration, especially at the early stage of growth, and the contribution of the dislocation hardening mechanism decreases significantly.

Thus, on the basis of experimental studies of the influence of the modes of each stage of the combined technological process and the tests of the samples on wear resistance, crack resistance, technological modes have been determined that make it possible to increase the service life of products by 1.5 ... 1.8 times operating under conditions of intense wear, dynamic and cyclic loads at elevated temperatures (shafts, gears, tools, dies, punches for cold and hot deformation, spindles of rolling bearings, lead screws, sleeves, parts of the caliper group, etc.).

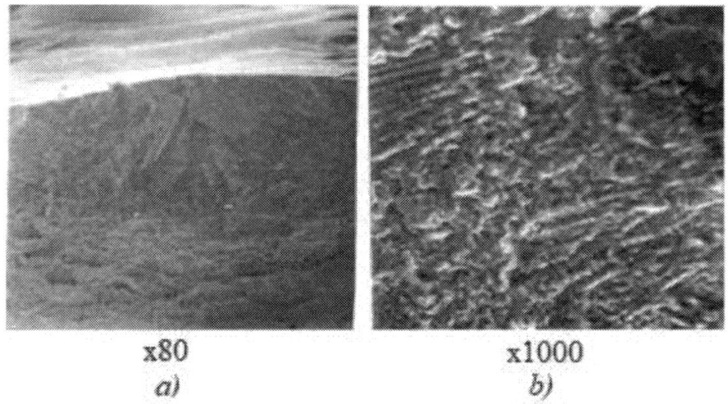

Figure 6. Fractograms of fatigue failure of steel 20 after: a) LA (V); b) LA (V) + N

4. Conclusion

1. The concept of surface hardening of structural steels is proposed, which consists in applying the principles of strength theory for the purposeful formation of structures capable of effectively resisting complex multifactorial phenomena occurring in the frictional contact zone, increasing wear resistance and crack resistance.
2. An intellectual approach to the rational choice of technological combinations capable of maximally realizing dislocation, substructural, grain-boundary, solid solution and dispersion mechanisms of steel hardening, allowing to create optimal structures on the surface of tribo-interface elements, is described.
3. Using the example of the technology of surface laser alloying of structural steels with nitride-forming elements followed by nitriding, an increase in surface hardness, strength, wear resistance by 15 times and fatigue strength by 1.5 times compared with the initial parameters of steels is shown.

5. Acknowledgments

The authors express their gratitude to the laboratory of mechanical properties of structural materials of the Institute of Metallurgy and Materials Science named after V. A.A. Baykov of the Russian Academy of Sciences, its head professor V.F. Terentyev and A.G. Bochvar for help in carrying out fatigue tests, fractographic studies and interpretation of the results.

References

[1] Kragelsky I V, Dobychin M N, Kombalov V S 1977 Fundamentals of Friction and Wear Analysis *Moscow: Mechanical Engineering* 526 p
[2] Boucher N A, Kopytko V V 1981 Compatibility of rubbing surfaces *Moscow Nauka* 126 p
[3] Gershman I S, Bushe N A 1995 Realization of dissipative self-organization of friction surfaces in tribosystems *Friction and wear* **16** **1** 61–70
[4] Lakhtin Yu M, Arzamasov B N 1985 Chemical heat treatment of metals *Metallurgy* 256
[5] Tushinsky L I 1990 Theory and technology of strengthening metal alloys *Novosibirsk Science Sib department* 306
[6] Petrova L G, Chudina O V 2010 Application of the methodology of structure formation management for the development of strengthening technologies *Metallurgy and heat treatment of metals* **5** 31–41
[7] Chudina O V, Aleksandrov V A, Brezhnev A A 2010 Combined surface hardening of carbon steels by laser alloying with subsequent nitriding. *Strengthening technologies and coatings* **4** 33-36

IOP Conf. Series: Materials Science and Engineering 1159 (2021) 012081 doi:10.1088/1757-899X/1159/1/012081

Software package for evaluation of efficiency of motor vehicle fleet operation

P I Smirnov[1], **B S Subbotin**[2], **D A Ptitsyn** [2], **A O Volkov** [2]

[1] Vologda State University,15, Lenina st., Vologda, 16029, Russia
[2] Moscow Automobile and Road Construction State Technical University (MADI), 64, Leningradsky Prosp., Moscow, 125319, Russia

E-mail: petrsm@bk.ru

Abstract. The article describes an improved approach to assessing the efficiency of using the road transport of enterprises based on the assessment of the technical readiness factor of the fleet using the average daily mileage and specific idle time in maintenance and operation. The software implementation of this method in MS Excel is shown, the results of application and the results obtained are analyzed approbation at the enterprise of the metallurgical industry.

1. Introduction

The problem of increasing the efficiency of motor vehicles operation in modern economic realities is one of the most frequently encountered tasks for motor transport enterprises of all forms of ownership [1-3].

At the same time, the key task is to fulfill the transport task on transportation.

The transportation task, with other equal factors, is ensured by:

1. Increase of the productivity of motor vehicles.
2. Increase of time of the vehicle being in serviceable condition.

At the majority of enterprises the considerable potential of increase of efficiency of work of motor vehicles consists in an increase of level of working capacity of cars (improvement of technical operation) which is scaled up by an indicator "coefficient of technical readiness" (CTR). There are several ways to determine a CTR [5] and, using one or the other methods, it is possible, deliberately or accidentally, to obtain significantly different CTR values.

For effective management of the motor transport fleet it is necessary to have objective information about its current condition. With proper application and consideration of all factors and conditions of motor vehicles, calculation by any method should give equally correct result. Unfortunately, this does not always happen in practice.

2. Materials and methods

The use of the standard formula for calculating CTR as the ratio of the number of serviceable cars (car-days) to the list - often leads to a significant error in the calculations. To eliminate it, first of all, it is necessary to take into account the idle time in the defective state taking into account possible presence of a vehicle in the correct or defective condition during one shift (i.e. in hours or fractions of

Content from this work may be used under the terms of the Creative Commons Attribution 3.0 licence. Any further distribution of this work must maintain attribution to the author(s) and the title of the work, journal citation and DOI.
Published under licence by IOP Publishing Ltd

IOP Conf. Series: Materials Science and Engineering 1159 (2021) 012081 doi:10.1088/1757-899X/1159/1/012081

shift). Secondly, in order to correctly calculate the CTR, it is necessary to take into account all the conditions when the vehicle is inoperable: delivery from the line, waiting for spare parts, waiting for posting, time of maintenance and repair works, etc. Taking into account only the labor intensity of maintenance and repair in the calculation unreasonably overstates the CTR values and is unacceptable. In addition, using the CTR value alone to assess the performance of the technical service is not enough, one should use the value of specific idle time in maintenance and repair (man-hour/1000 km mileage) [5]. Meeting these requirements allows one to receive reliable objective values for the analysis of efficiency of motor vehicle fleet operation and work of technical service of an enterprise.

Thus, the methodology for calculating the CTR taking into account these comments is as follows. To calculate the CTR value it is necessary to differentiate two possible vehicle conditions:

a) in good working order - it is in service and may be allowed for use;

b) faulty (including routine maintenance) - it is in the technical service and cannot perform transportation.

The moment of transition between states should be accurately recorded by a responsible person with documentary evidence. The sum of time when the vehicle is in all possible (two) states is equal to 100%, and this is used to control the correctness of accounting. Formula for calculating CTR:

$$\alpha_T = \frac{1}{1 + B_\Sigma \times L_{CC}} \qquad (1)$$

where B_Σ is the total specific idle time of the analyzed vehicles in maintenance and technical repair (being in faulty condition), shift/1000 km for the analyzed period of time;

L_{CC}– average daily mileage of the analyzed vehicles for the analyzed period of time, thousand km.

The total specific idle time of a vehicle in maintenance and technical repair B_Σ is defined as the sum of the i-th specific idle time for certain types of technical actions:

$$B_i = \frac{t_i}{X_i}, \text{ shift/1000 km} \quad (2)$$

where t_i — average vehicle idle time during the i-th technical effect, shifts,

X_i —mean time to failure, the i-th technical action, thousand km.

The mean time to failure X_i is determined by the following formula:

$$X_i = \frac{L_\Sigma}{n_i}, \quad (3)$$

where L_Σ is the total mileage of all analyzed motor vehicles for the calculation period, thousand km;

n_i is the number of failures for each calculated group of the i-th technical actions, units.

As the duration of the shift is taken the time of the vehicle in hours per day, i.e. for vehicles working 8 hours a day, 1 shift is equal to 8 hours, for vehicles working around the clock, 1 shift is equal to 24 hours.

It is recommended to calculate the total specific idle time in a faulty state by the following types of technical actions/conditions:

a) actual time of the vehicle staying in the maintenance and repair area;
b) waiting for evacuation;
c) evacuation to repair site;

d) diagnostics;
e) paperwork;
f) waiting for spare parts;
g) waiting for a post;
h) other unaccounted vehicle time.

To implement the abovementioned method of calculating CTR values for the motor transport fleet, a software macro has been developed in MS Excel environment with the possibility of synchronization and automatic data download from the systems of telematic control of equipment, as well as the working environment of enterprise fleet management (for example, 1C software). The appearance of the working variant of the CTR calculation macro is shown in Figure 1. An example of CTR calculation together with the data analysis block for CTR calculated values and specific idle time in maintenance and repair is shown in Figure 2.

Car group mileage 08.02.20 41 771 km

Average daily mileage of 1 car 163,168 km 0,163168 thousand km

	Number of orders	Total loss of time (hours)	Average downtime (hours)	Mileage to failure, km	Average idle time, shifts	Specific downtime in maintenance and repair, shifts /
1	2	3	4	5	6	7
1 service	4	19	4,75	10442,8	0,198	0,0190
2 service	3	72	24	13923,7	1,000	0,0718
Waiting for repair	6	247	41,166667	6961,8	1,715	0,2464
Emergency repair	1	12	12	41771,0	0,500	0,0120
Request repair	12	56	4,6666667	3480,9	0,194	0,0559
Repairs	31	376	12,13	1347,5	0,505	0,3751
					Total	**0,7800**

CTR **0,8871**

Fillable cells

Figure 1. Example of the working screen of the CTR calculation macro

	A	B	C	D	E	F	G	H	I
2	Reference and variable data				Period	01.04.2019	30.04.2019		
3	Group	67564W (VOLVO)	column	2					
4	Number of cars in the group	Environment category	Average age of cars, years	Average mileage since the beginning of operation, km					
5	6	-	-	-					
6		Average daily mileage in a group, thousand km	0,078						
7		Duration of shift, hours	12						
8		Car group mileage	14447						
9	Estimated data								
10				Targets					
11								Specific blank idle	
12	№	Zone / site / service	Number of refusals / arrivals	Mean time between failures, thousand km	Actual idle, hour	Average idle actual, hour	Average downtime, shifts	shift / 1000 km	%
13	1	1 service	0	0,00	0	0,0	0,00	0,00	0%
14	2	2 service	0	0,00	0	0,0	0,00	0,00	0%
15	3	2 service small	0	0,00	0	0,0	0,00	0,00	0%
16	4	Emergency repair	13	1,11	46	3,5	0,29	0,27	10%
17	5	Request repair	31	0,47	117	3,8	0,31	0,67	26%
18	6	Service maintenance	0	0,00	0	0,0	0,00	0,00	0%
19	7	Urgent repair	0	0,00	0	0,0	0,00	0,00	0%
20	8	Repairs	81	0,18	291	3,6	0,30	1,68	64%
21	Total		125		454,0			2,62	100%
22	ysis of the CTR ator	Actual value of CTR	0,830		lock of proposals				
23		Standard value of CTR	0,94						
24		Difference by CTR	-0,110						
25		Difference in % target	-11,66						
		Standard value of specific downtime in maintenance	0,82						

Figure 2. Example of operation of the analysis block of calculation data

The obtained values are automatically transferred to the summary table by divisions, models, subordinates etc. Subsequently, the calculated values are analyzed and compared in a month, quarter, year, and also compared with previous periods. An example of a summary analytical table is shown in Figure 3 (data of vehicles are hidden for reasons of data non-disclosure).

2	Work in February 2020																				
3																					
4	Gar. No.	tate numbe	Vehicle	Work Schedule No. / Scheduled Work Hours	Operating time on p / l, hours	1 service		2 service		Request repair		Urgent repair		Request repair		Repairs		Without driver			
5						number	time	number	time	number	time	number	time	number	time	number	time	number	time		
6	xxxxx	xxxxx	HYUNDAI HL760-7A	35	75,50											1	3				
7	xxxxx	xxxxx	HYUNDAI HL 7607 A	35	34,50			1	38												
8	xxxxx	xxxxx	HYUNDAI HL730-7A	35	23,00													3	36		
9	xxxxx	xxxxx	KOMATSU WA 470-3A	35	36,00					1	12	1	4								
10	xxxxx	xxxxx	KOMATSU WA 470-3	35	57,50													2	24		
11	xxxxx	xxxxx	KOMATSU-WA-420-3	35	47,00													2	24		
12	xxxxx	xxxxx	KOMATSU WA600-6	79	84,00																
13	xxxxx	xxxxx	KOMATSU WA600-3A	79	156,00	1	4			1	43			2	11						
14	xxxxx	xxxxx	KOMATSU WA600-3A	79	168,00			1	6												

Figure 3. Example of a summary table for analysis

The developed software tool for collection, calculation and analysis of information allows (under the condition of integration into the existing electronic document flow environment) to calculate CTG values and specific idle time in an automated mode in maintenance and repair, to carry out the primary analysis of reasons of fulfillment/non-fulfillment of planned indicators, to monitor the achievement of key performance indicators and thresholds in an automated mode, as well as to minimize the subjectivism of evaluation and human factor in the calculation and analysis of indicators.

The application of the developed software tools within the framework of activities of one particular enterprise of the metallurgical industry in 2019-2020 allowed to reveal a number of work features:

1. The calculation method is sensitive to the accuracy of accounting for changes in technical conditions, it requires prompt fixation and timely data transfer to the central server.
2. Operation of road construction and utility vehicles with high idle time while waiting for work results in extremely low CTR values with the overall good condition of the machines.
3. The analysis of efficiency indicators of motor transport fleet is only possible at a complex estimation of CTR values, specific idle time in maintenance and repair, and also average speed of movement that demands participation of an expert.
4. It is clearly shown that the discrepancy between the working hours of a vehicle and the work of the maintenance and repair service leads to an increase in specific idle time and a decrease in CTR.
5. The proper objectivity and reliability of data can be achieved only with automated and centralized data collection from motor vehicles using remote telematic transport control systems.

3. Conclusion

The given methodology sets the goal - to assess the efficiency of the technical service.
Having received the time that the car is in a faulty condition, we get the time that the car is in good working order. Comparison of this time with the operating time of cars on the line gives an indicator of the efficiency of the operation (commercial operation) service: how effectively the time spent by cars in good condition is used. This indicator can be specified in hours, shifts or percentages.

Thus, we have 2 indicators of the technical service performance and 3 indicators of the operation service.

Technical service (technical operation of vehicles):
1. Specific downtime in TO and TR, shifts / 1000 km.
2. CTR.

Maintenance service (commercial vehicle operation):
1. Clock in good condition / clock on line.
2. Productivity, tons (tkm).
3. Cost of transportation, USD / km (USD / tkm).

The widespread introduction of these performance indicators by production will help identify weaknesses in the technical and commercial operation of vehicles and improve key performance indicators.

References

[1] Khayyam H, Bab-Hadiashar A 2014 *Adaptive intelligent energy management system of plug-in hybrid electric vehicle* (Energy)
[2] Muratori M, Moran MJ, Serra E, Rizzoni G 2013 *Highly-resolved modeling of personal transportation energy consumption in the United States* (Energy)
[3] Whyte K, Daly H E, O Gallach oir B P 2013 *Modelling HGV freight transport energy demand in Ireland and the impacts of the property construction bubble* (Energy)
[4] Zhao H, Burke A, Miller M 2013 Analysis of Class 8 truck technologies for their fuel savings and economics *Transp Res Part D Transp Environ*
[5] Kuznetsov E S 2001 *Management of technical systems: a training manual for the specialty 150200 - Cars and automotive industry* (Moscow: MADI)
[6] Nepomnyaschiy E.G 2003 *Investment design: a training manual.* (Taganrog: Publishing house of TRTU)

[7] Ulyanov V N , Markosyan L A 2015 *Prospects of lean production development at Russian industrial enterprises* // Management systems **4 (29)**
[8] Smirnov P I 2017 *Telematic data flow analysis from trucks* International scientific and practical conference (Orenburg)
[9] Smirnov P I 2017 *Use of data obtained through satellite monitoring channels to adjust the frequency of car maintenance* (Vladimir)

IOP Conf. Series: Materials Science and Engineering 1159 (2021) 012082 doi:10.1088/1757-899X/1159/1/012082

Establishing the requirements for the minimum values of stopping sight distance on the road surface from the condition of vehicle braking safely

Yu V Kuznetsov[1], Ola Ismail Ghassan[1]

[1] Moscow Automobile and Road Construction State Technical University (MADI), 64, Leningradskiy pr., Moscow, 125319, Russia.

E-mail: dil_madi@mail.com, albylasan1988@gmail.com

Abstract. The article is devoted to determine the stopping sight distance from the condition of the possibility stopping a car that moving at design speed. Determination is based on experimental studies of the dependences the coefficient of friction on the vehicle's speed, when braking happens on the road surfaces that have the minimum permissible coefficient of friction value, established by the current regulatory literature. Braking of vehicles was studied on two surfaces with a minimum value of the coefficient of friction, but they were differing in surface roughness. Based on theoretically calculated braking distances, requirements for stopping sight distance with different design speeds have been developed. The obtained values of stopping sight distances are compared with the values contained in the normative documents of Russia and foreign countries.

1. Introduction

When calculating the required distance of visibility of the road surface, providing the possibility of stopping, two terms are usually considered - the distance traveled by the car in the time required for the driver to perceive an unexpected obstacle and the start of braking (the estimated driver's reaction time) and the braking distance - the distance traveled by the car from the start of braking until it stops completely [1,2,3,4]. The calculated visibility distance S is calculated by the formula:

$$S= ((t_r*V)/3,6) + ((K*V^2)/(2g*(\Phi\pm i)) \quad (1)$$

where V is the speed before braking,

t_r is driver reaction time,

K is the coefficient of the operational state of the brakes,

Φ is coefficient of friction,

i is longitudinal slope

The analysis of formula (1) shows that the coefficient of friction of the wheel to the road surface and the estimated speed of movement have the greatest influence on the visibility distance from the stopping condition.

2. Background

Since the braking distance of a vehicle largely determines the magnitude of the visibility distance, the justification of this length for various values of the design speeds should be given primary attention. The main parameter that determines the length of the braking distance, all other things being equal, is

Content from this work may be used under the terms of the Creative Commons Attribution 3.0 licence. Any further distribution of this work must maintain attribution to the author(s) and the title of the work, journal citation and DOI.
Published under licence by IOP Publishing Ltd

the coefficient of friction, which largely depends on the condition of the road surface. The road surface can be in various states - dry, wet, dirty, snowy, icy. Obviously, in snowy, icy and dirty conditions, the vehicle cannot move at the design speeds due to a decrease in the friction coefficient. Moreover, the driver has the ability to visually determine these conditions and to reduce the speed in a timely manner. Therefore, such pavement conditions should not be considered when justifying the visibility of the road surface. With a dry and clean surface, the grip qualities of the road surface enable the driver to drive at speeds permitted by the traffic regulations. As studies show [5,6,7,8], the coefficient of friction of dry surfaces to a small extent depends on the speed of movement, its values are very high on both rough and smooth surfaces, which allows, when justifying a safe distance, the visibility of the road surface do not consider driving the vehicle on a dry surface at the design speed.

Traffic rules do not limit driving speeds when the roadway is wet. And in the rain, the driver has the right to move along the road at the same speeds as in the dry state of the roadway. At the same time, the adhesion qualities of the road surface during wetting can sharply decrease to an unacceptable level for safe driving. Studies show that wet adhesion coefficients are largely determined by the macro- and micro-roughness of the road surface. The driver in the rain under a layer of water does not have the technical ability to assess the state of roughness, and the traffic rules do not require this from him. Therefore, in the rain, the driver is able to drive at the same speeds as in dry conditions. To get an answer to the question - does the wetting of the pavement affect the speed of movement, we analyzed the data of the speed of movement recorded by video surveillance cameras on the toll road "Northern Bypass of Odintsovo". In the summer of 2018, two days were selected, one of which was clear, sunny, and the other rainy, and the rain was characterized by a low intensity. The results showed that neither the average speeds, nor the 95 percent rainfall rates significantly changed. Similar results were obtained by Ph.D. N.A. Lushnikov [9] in the course of observing the speed of movement on wet slippery and non-slip sections of the road surface. Observations show that the speed of movement does not depend on the state of the roughness of the road surface, including when it is wet.

The braking distance of a car is largely determined by the grip of the road surface. Today in Russian Federation the value 0.3 (GOST 50597 p. 3.1.4.) is taken as the minimum admissible coefficient of friction in the wet state of the pavement [10]. In this case, the coefficient of friction should be measured at a speed of 60 km/h using a full-size automobile tire with a smooth tread with a landing size of 13 inches on a dynamometer trailer. As the results of studies carried out by MADI (STU) for many decades show, road surfaces characterized by values of the coefficient of friction is close to the minimum permissible can differ significantly in surface roughness. Studies show that macro-roughness - the roughness created by particles protruding on the surface of the pavement, determines the ability of the pavement to quickly drain water from the contact zone of a tire sliding on a wet road. Micro-roughness - the intrinsic roughness of protruding particles - contributes to the rapid destruction of liquid films present on the protrusion after removing its layer. The road surface can be characterized by a coefficient of 0.3 with a satisfactory macro-roughness, but minimal micro-roughness. The same coefficient can be characterized by a coating with a minimum macro-roughness in the presence of a satisfactory micro-roughness. Studies show that the graphs of the dependence of the coefficient of friction on the speed of movement on these surfaces differ from each other. On a macro-rough surface, at speeds up to 60 km / h, the curve is lower than a similar curve obtained on a surface with less macro-roughness, but at higher speeds the curves change places. This is due to the fact that at low speeds in contact between the tire and the road, the dry interaction zone has a large area of direct contact. At the same time, the frictional qualities created by the microroughness play a decisive role. At higher speeds with insufficient microroughness, the area of wet interaction in the contact of the tire decreases due to a decrease in drainage capacity, which leads to a more rapid decrease in the coefficient of friction on such surfaces.

3. Experiments and results

Studies carried out at MADI (STU) have established that macrorough coatings characterized by coefficient values close to the minimum allowable are, for example, coatings made of rubbled-mastic asphalt concrete (RAC) with an excess of binder immediately after their construction. As the binder wears out during intensive traffic in the first months of operation, the adhesion qualities of such coatings rapidly increase to values of 0.55 - 0.6. Rough coatings with the minimum allowable coefficient values were recorded by the MADI laboratories in the 70s - 80s on the Rostov-Baku road, where surface treatments were performed using river pebbles, which did not have good microroughness. Macro-rough surfaces with a long service life on roads with heavy traffic also have coefficients close to the minimum permissible value.

Coatings with low macro-roughness, which have the minimum admissible value of the friction coefficient, are much more common than macro-rough ones. As a rule, these are freshly laid asphalt concrete pavements made of mixture B, but sometimes also A, as well as pavements with a significant service life, made of asphalt concrete on crushed stone of low wear resistance.

Since pavements characterized by the minimum admissible values of the coefficient of friction have a right to exist, it is obvious that the calculated visibility distances should be determined considering the stopping distance on these pavements. In order to determine the length of the braking distance on such surfaces, experimental curves of the dependence of the coefficient of friction on the speed of movement were obtained (Fig. 1), one of which was recorded on the lower layer of the M-11 road during its construction, and the other - on rubbled asphalt concrete, made with excess bitumen on the surface.

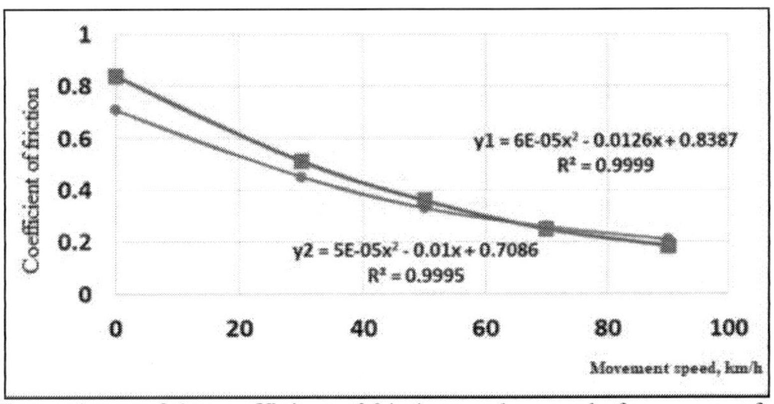

Figure 1. Dependence of the coefficient of friction on the speed of movement for a tire with a smooth tread (row 1 - micro-rough, row 2 - macro-rough).

The dependencies (Fig. 1) were obtained using a modified PKRS-2 device, the irrigation system of which was made with water supply under the action of excess air pressure in the water tanks, which made it possible to provide the flow rate required to create a water film of 1 mm at a speed of up to 90 km/h. Dependences similar in configuration were once obtained during mass tests of tires carried out by MADI (STU) in the 70s - 90s using the MADI-8 dynamometric laboratory. The watering system of the laboratory made it possible to carry out measurements at speeds up to 120 km/h.

Due to the fact that the Road Traffic Regulations allow the operation of tires with a residual tread depth of up to 1.6 mm, it becomes necessary to recalculate the dependences shown in Fig. (1) for the case of using tires with the maximum permissible tread wear on a car. Obviously, due to the presence of a pattern, the graphs of the ratio of the coefficient on the speed of movement for tires with a pattern will be located above the given ones. However, as shown by the studies carried out during tire tests, carried out in the 70s - 90s by order of the Research Institute of the Tire Industry, the ratio of the coefficients for tires with permissible wear and smooth depends on both the surface roughness and the speed of movement. So, on fine-rough surfaces at low speed, the ratio of the coefficients can be taken equal to 1.07, while at a speed of 90 km/h this ratio increases to 1.24. The ratios of the coefficients show that on fine-rough surfaces, the role of the tread pattern with an increase in the speed of movement increases significantly, since on such surfaces the drainage capacity of the coating is low and the pattern begins to play a significant role in reducing the area of wet interaction in the contact of the tire. On a macro-rough surface, the drainage of the contact zone is mainly carried out due to roughness, while the role of the pattern in water drainage is significantly reduced. Experiments show that on such surfaces, for all speeds, the same coefficient of 1.07 can be applied to recalculate the coefficients.

Considering the indicated coefficients, we calculate the dependence of the friction coefficients on the speed of movement and extrapolate them to a speed of 150 km/h (Fig. 2)

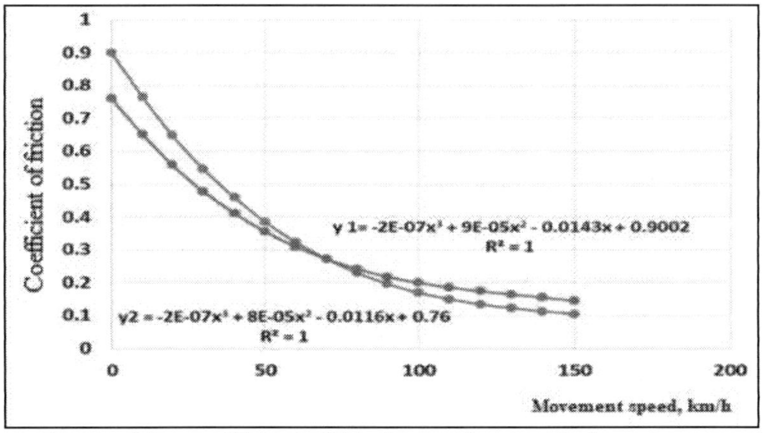

Figure 2. Dependence of the coefficient of friction on the speed of movement, calculated for tires with a tread pattern of 1.6 mm (row 1 - micro-rough surface, row 2 - macro-rough surface).

The obtained dependences clearly show that on roads with design speeds below 80 km / h, it makes no sense to arrange macrorough surfaces. On surfaces with less macro-roughness on such roads, the braking distance will be much shorter. Experiments on measuring braking distances during tire tests in braking mode, carried out in the past years, showed that on rough surface treatments at a braking speed of 60 km/h, there was always more braking distance recorded on rough surface treatments.

When comparing the obtained graphs, it should be borne in mind that the braking distance when the speed decreases, for example, by 10 km/h, is proportional to the square of the speed differences. Therefore, to reduce the speed of movement from 110 km/h to 100km/h, it is necessary to spend 8 times more work than when reducing the speed from 20 km/h to 10 km/h. Therefore, a visual examination of the curves creates an erroneous impression of their identity. Calculations based on the dependences obtained, given in (Table 1.) and shown in (Fig. 3), indicate that due to the seemingly insignificant difference in friction coefficients at speeds of more than 80 km/h at high speeds, a large difference in braking distances is obtained.

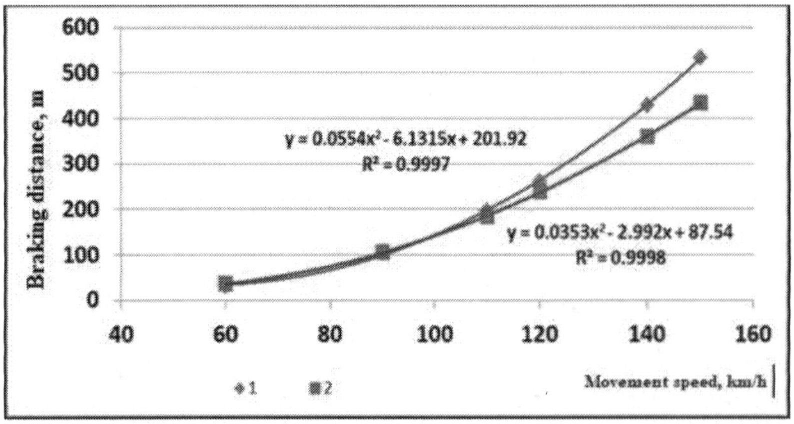

Figure 3. Dependence of the stopping distance on the speed of movement (row 1 - micro-rough surface, row 2 - macro-rough surface).

Table 1. Sliding condition of the locked wheels of cars

Braking start speed, km/h	Braking distance, m	
	micro-rough coating	macrorough coating
60	31.8	35
90	104.4	105
110	197.2	183.8
120	261.1	239.6
140	427.8	358.6
150	532.3	434.6

Table (1) is calculated based on the sliding condition of the locked wheels of cars. However, today almost all cars, including light trucks, are equipped with anti-lock braking systems that allow emergency braking without locking the wheels, with their partial slipping. As shown by studies carried out in MADI on passenger cars (Nissan, Peugeot) and light trucks (Volkswagen Crafter, Mercedes Sprinter), on snowy, icy and wet surfaces, braking distances during braking with full blocking and using anti-lock braking systems are characterized by similar values. However, braking with full blocking at speeds above 60 km/h always ends in a skid, while when the systems are on, emergency braking is possible at significantly higher speeds.

4. Conclusion

The study showed that even without considering the distance traveled by the car during the driver's reaction time, the values of the visibility distances adopted today at the minimum admissible value of the coefficient of friction cannot guarantee driving safety at speeds above 110 km/h. At a speed of 150 km/h, the braking distance on surfaces with the minimum admissible coefficient of friction becomes commensurate with the braking distance of a steam locomotive! To solve the problem, it seems advisable to increase the minimum admissible value of the friction coefficient for roads with an estimated speed of 120 km/h to 0.35, and for roads with an estimated speed of 150 km/h to 0.40.

The given values of the coefficients are technically achievable today without the transition to a ubiquitous surface treatment device. The experience of their widespread use in the 70s and 80s showed that due to the low quality of their device in those years, the coverage was slippery for 30% of the network length - it was characterized by friction coefficients below 0.3. The main defect of such coatings is bitumen on the rolling strips. Today, surfaces with adhesion values below 0.3 are rare. Another measure aimed at reducing the braking distance may be to increase the minimum permissible tread depth of the tires. However, special research is required to justify this value.

References

[1] AASHTO 2018, *American Association of State Highway and Transportation Officials* (Washington:DC 20001, GDHS-7,ISBN: 978-1-56051-676-7)
[2] Pooja Lonare, Prof. Apurwa Kahalkar, Prof. Lemonisha Uikey. 2017 *IJSTE - International Journal of Science Technology & Engineering*. **3** 25–28
[3] Ali Abdi Kordani, Omid Rahmani, Amir Saman Abdollahzadeh Nasiri, Sid Mohammad Boroomandrad 2018 *Civil Engineering Journal* **4** 46–57
[4] Burmistrova O N, Plastinina E V, Timohova O M 2015 *Technical science* **2** 2074–2078
[5] Kuznetsov Yu V 1978 *study of the main factors that affecting on the skid resistance between car tires and road surface* (Moscow: MADI)
[6] Nemchinov M V 2004 *Nauka i tekhnika v dorozhnoj otrasli* **4** 12–15

[7] Shutov A I, Kotuhov A N, Zagorodnij N A 2006 *New materials and technologies in mechanical engineering* **6** 119–122
[8] Nikiforovich A K, Ponurovskij V A 2015 *Actual directions of scientific research in the XXI century: theory and practice* **8-2(19-2)** 321–324
[9] Lushnikov N A 1989 *Development of requirements for adhesion qualities of road surfaces.* (Moscow:MADI)
[10] GOST R 50597-93 1994 *Automobile Roads and Streets. The Requirements to the Level of Maintenance. Satisfied the Traffic Safety* (Moscow:state standard of the russian federation)

Problems of introducing capacitive fuel level sensors on city buses

V A Maksimov[1,2], N V Pozhivilov[1,2], A E Andrianov[2*]

[1] Department of Operation of Motor Transport and Car Service, MADI University, Leningradsky prospect 64, 125319, Moscow, Russia
[2] Research Centre, Mosgortrans, Raushskaya naberezhnaya, 22/21, building 1, 115035, Moscow, Russia

E-mail: andrianov_ae.yz@mosgortrans.ru

Abstract. This paper substantiates the need for a procedure for warming up buses in motor transport enterprises engaged in passenger transportation, and the procedure for its implementation. The necessity of regulation of fuel consumption for heating is determined. A comparative analysis of existing methods and means of measuring fuel consumption for use in the regulation of fuel consumption for heating in motor transport enterprises engaged in passenger transportation is carried out. A schematic diagram of a device of its own design based on a fuel level sensor for measuring fuel consumption for heating is proposed. Methodical recommendations on the regulation of fuel consumption for heating with the help of the device are given. The procedure for processing statistical data is described in detail. The process of standardizing fuel consumption for warming up the LiAZ-429260 bus at the State Unitary Enterprise "Mosgortrans" is described. The results of standardization of fuel consumption for heating are shown.

1. Introduction

Warming up the vehicle engine consists in starting the internal combustion engine and then waiting for it to warm up to operating temperature. When carrying out transport work immediately after starting the engine, the wear of some of its parts can significantly increase. The use of heating allows to ensure the gentlest transition from the moment of starting the engine to the beginning of the transport work [1,7].

Despite the fact that the manufacturers of most modern vehicles have ceased to recommend mandatory warm-up of the internal combustion engine before driving for a moderate climatic region of operation, this procedure is still relevant in many cases. In particular, this applies to road transport enterprises in the territory of the Russian Federation engaged in urban passenger transportation.

The main factors that make it necessary to warm up city buses in MTE are:

1. Natural and climatic conditions. In the middle zone of the Russian Federation, the average monthly temperature in winter can drop to -10 ° C, which is much lower than in Germany (0 ° C), France (+5 ° C), Great Britain (+4 ° C) and other European countries ... At negative temperatures, operating fluids are not immediately able to reveal their properties [3, 8], working gaps between structural elements are violated, their relative position, shape, etc. change.

2. While the engine is warming up, the interior of the bus is also warmed up and its individual systems (air suspension, pneumatic brake drive, etc.) are brought up to operating condition, which ensures environmental and transport safety and passenger comfort when working on the line [2].

Content from this work may be used under the terms of the Creative Commons Attribution 3.0 licence. Any further distribution of this work must maintain attribution to the author(s) and the title of the work, journal citation and DOI.
Published under licence by IOP Publishing Ltd

IOP Conf. Series: Materials Science and Engineering 1159 (2021) 012083 doi:10.1088/1757-899X/1159/1/012083

3. Additional control of the technical condition of the bus by the driver and mechanic due to the detection of extraneous noise, leaks, leakage of the pneumatic systems of the bus, etc. during engine warm-up.

4. Injection of working pressure in the pneumatic systems of the bus for their reliable and safe functioning.

The start of the bus engine is usually carried out after pre-heating with a pre-heater to the minimum temperature for accepting the load (+ 40 ° C). Temperature control is carried out from the driver's workplace using a standard temperature sensor [4,5].

After the temperature arrow of the internal combustion engine cooling system deviates from the lower mark according to the standard sensor on the instrument scale corresponding to a temperature of + 40 ° C, the bus engine is started. In the process of joint operation of the preheater and the bus engine, the bus interior is heated, as well as the working pressure in the bus pneumatic systems is pressurized. When the arrow of the standard temperature sensor reaches 80 ° C, the heating stops and the bus moves to the line [2].

At the same time, in the conditions of motor transport enterprises, the issues of fuel economy occupy a leading position, therefore, it is necessary to pay special attention to instruments and methods for measuring fuel consumption for heating the rolling stock at low temperatures for the subsequent development of fuel consumption rates for heating at the enterprise.

2. Review of existing approaches to measuring fuel consumption by car

To select a device and / or method that is optimal for use in measuring fuel consumption for heating, a comparative analysis of existing solutions in this area was carried out [6].

2.1. Measurement by injector pulses

On vehicles equipped with engines with electronically controlled injectors in the fuel injection system, fuel consumption can be measured from injector pulses using a proximity reader. This method involves the removal of information from the internal measuring instruments of the standard information and diagnostic system of the vehicle about the operation of the fuel equipment. Signals are read without electrical contact with the injector wires and without interfering with the vehicle's electronic systems.

The advantages of the method:

- high accuracy;
- the ability to timely identify malfunctions of the vehicle fuel system;
- no need to use additional sensors.

Disadvantages of the method:

- the need for complex and time-consuming calibration of fuel equipment for each vehicle and determination of the dependence of the unit volume of injected fuel by the injector on the engine operating mode [2];
- high requirements for manufacturability and design perfection of standard fuel equipment and means of diagnosing and taking indicators of its work.

This method of measuring fuel consumption is the cheapest from the point of view of the required equipment, but extremely laborious, especially in the conditions of large motor transport enterprises with a diverse fleet of vehicles. The cost of using this method also needs to add fleet downtime during the equipment calibration process.

This method is not suitable for measuring fuel consumption for heating for large motor transport enterprises.

2.2. Measurement according to the data of the vehicle's on-board information buses

There are many sophisticated electronic devices aboard a modern car. For data exchange, they are united into a single network through onboard information buses - CAN (SAE J1939 standard) or J1708 (SAE J1587 standard). On-board data buses transmit information about the parameters of the transport, including data on fuel consumption.

The data that can be obtained from on-board information systems contains information about the level and amount of fuel consumed, as well as hourly and track fuel consumption.

The advantages of the method:

- no need to use additional sensors.

Disadvantages of the method:

- the accuracy of data on fuel consumption is lower than that of other measurement methods and differs depending on the standard sensors for measuring the fuel level in the fuel tank of the bus.

The measurement accuracy of this method differs depending on the measured vehicle. Increasing the measurement accuracy by calibrating standard sensors is impossible.

This method is also not suitable for rationing fuel consumption for heating due to the low accuracy of the results obtained.

2.3. Measurement of the fuel level in the fuel tank using fuel level sensors.

Electronic fuel level sensors (FLS) can be used as a tool for measuring the fuel level in the tanks of any vehicles.

The use of such devices allows:

- receive reliable information about the current volume of fuel in the car's tank;
- to determine the volume of refueled fuel;
- detection and prevention of fuel theft;
- control fuel consumption.

The advantages of the method:

- low cost;
- no interference with the fuel system of the car, installation on warranty vehicles is possible;
- measuring the level of various types of liquid fuel;

Disadvantages of the method:

- the measurement accuracy depends on the operating conditions of the vehicle, the shape and volume of the fuel tank;
- when installing the fuel level sensor, it is necessary to perform the tank calibration operation [2].

The use of such a device for measuring fuel consumption for heating when installed directly into the vehicle tank does not allow obtaining accurate data due to the large total volume of the tank. The use of the device with direct installation into the vehicle tank does not allow obtaining sufficiently accurate data on fuel consumption in the warm-up mode.

2.4. Measurement of fuel consumption in the fuel line of the engine

This measurement method measures the actual fuel consumption of the vehicle engine. Instantaneous fuel consumption can be accurately determined by direct measurement using a flow meter installed in the engine fuel line.

The use of flow meters allows the fleet owner to provide:

- accounting of actual fuel consumption;
- accounting of the time of equipment operation;
- testing of engines in terms of fuel consumption.

The advantages of the method:

- high measurement accuracy;
- fuel consumption is correctly determined in any engine operating modes, including short runs;
- it is possible to identify engine malfunctions, leading to increased fuel consumption, and violation by the driver of the optimal operating modes of the machine;

Disadvantages of the method:

- installation of a fuel flow meter is a complex procedure that requires appropriate qualifications and intervention in the design of the vehicle fuel system;
- on some machines the installation of a flow meter may change the operating conditions of the engine;

IOP Conf. Series: Materials Science and Engineering 1159 (2021) 012083 doi:10.1088/1757-899X/1159/1/012083

- volumes and time of refueling (draining) of fuel are not controlled [2].
- high cost of the device;

A more accurate determination of the fuel consumed in the transport mode of ATS operation in comparison with other devices and methods, however, the flow meter is not able to determine the fuel consumption for warming up small buses, due to the fact that such a device has a limitation in the form of a minimum volume of liquid passing through it (minimum spillage).

3. Description of the created installation

Due to the fact that none of the existing devices can fully meet all the accuracy requirements for measuring fuel consumption for heating city buses at the State Unitary Enterprise "Mosgortrans", the Research Centre has created its own solution based on the fuel level sensor R- 7 for accurate automated measurement of fuel consumption for heating the bus.

The device consists of two high cylindrical interconnected containers of different volumes.

The small tank contains the R-7 fuel level sensor, the large tank has an external marked glass tube, two holes at the bottom (for draining the fuel) and at the top (for connecting the fuel return line) (Fig. 1).

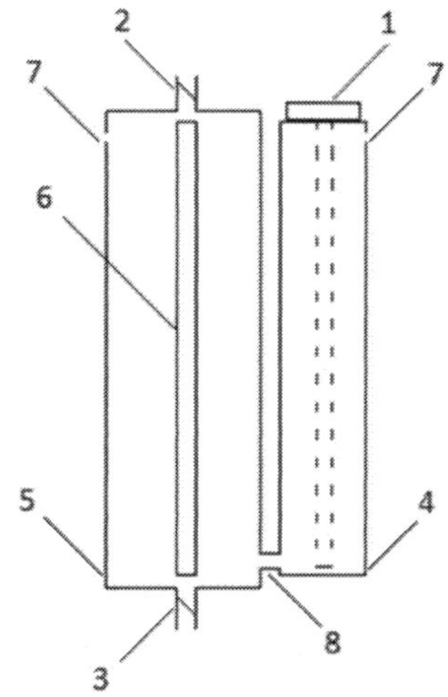

Figure 1. Schematic diagram of the installation:
1 - fuel level sensor; 2 - filler neck with a connector for connecting the return fuel line of the vehicle; 3 - drain neck with a connector for connecting the vehicle fuel supply; 4 - container for installing the fuel level sensor; 5 - main fuel tank; 6 - dimensional scale; 7 - openings for communication of containers with the external environment; 8 - the place of communication between the containers.

A device created according to this principle has the following advantages:
- low cost;
- measuring the level of various types of liquid fuel;
- high measurement accuracy due to the shape of the containers;
- low labor intensity when preparing the device for operation, due to the absence of the need for calibration before each measurement;
- automation of the measurement process.

Disadvantages of the method:
- the need to intervene in the vehicle's fuel system to connect the device.

The device, made according to the proposed schematic diagram, combines the advantages of using a fuel level sensor, as well as eliminating its main disadvantages - low accuracy and the need for individual calibration for use in each vehicle.

It should be noted that the degree of accuracy depends on the quality of the calibration procedure of the installation after assembly.

IOP Conf. Series: Materials Science and Engineering 1159 (2021) 012083 doi:10.1088/1757-899X/1159/1/012083

4. Guidelines for the regulation of fuel consumption for heating using the installation

Fuel consumption for warming up is the amount of fuel consumed from the moment the bus engine starts until it reaches the operating temperature.

To measure the consumed volume of fuel for heating using the installation, it is necessary to fill it with fuel through the filler neck (2) in Figure 1, connect it to the fuel system of the bus instead of the fuel tank (the fuel supply line of the bus is connected to the drain neck (3), and the fuel return line to the filler neck (2) It is important to install the device strictly vertically to exclude the influence of inclination on the measurement results.

After filling and connecting the unit, the heating procedure is carried out directly, consisting of:

1. Starting the engine preheater and its operation until the temperature reaches 40 ° C according to the standard temperature sensor;

2. Starting the bus engine and running it together with the heater for 10 minutes or more until the operating temperature is reached.

The volume of consumed fuel is measured in automatic mode by the model R-7 fuel level sensor and recorded in its memory. During the measurement process, it is necessary to keep track of the time to determine both the total warm-up time and its components.

After the end of the warm-up procedure, the bus is shut down, the fuel is drained from the installation and dismantled from the vehicle.

The next step in determining the norms is the processing of the results obtained. For this, data is extracted from the installation using specialized software and presented in a tabular form for further processing.

Depending on climatic conditions, it is possible to adjust the fuel consumption rates for warming up over the temperature range to increase reliability. Calculation of norms should be made separately for each temperature range.

The most important factor affecting the reliability of the measurement data is the sample size (number of measurements required). Fuel consumption for heating obeys the law of normal distribution, so the following formula is used to determine the sample size:

$$N \geq \frac{t^2 V^2}{\varepsilon_0^2} \quad (1)$$

where V is the coefficient of variation;

t is the value of the significance level, which at the five-one percent significance level is 1.96 at q = 0.05 and 2.58 at q = 0.01

q is the level of significance;

ε_0 is relative accuracy.

Depending on the required relative accuracy ε_0 and the level of significance q, as well as the coefficient of variation V, calculated on the basis of the results of the measurements, it is possible to flexibly adjust the required number of them, which is an important economic factor for MTE.

Table 1. Sample size for normal distribution

Significance level q	Coefficient of variation V	Relative accuracy ε_0			
		0.05	0.10	0.15	0.20
0.05	0.1	16	4	2	1
	0.2	62	16	7	4
	0.3	138	36	16	9
	0.4	245	62	28	16
0.01	0.1	27	7	3	2
	0.2	107	27	12	7
	0.3	240	60	27	15
	0.4	697	174	78	27

The heating time of individual rolling stock models may differ, therefore, the most convenient value for accounting for fuel consumption within the MTE is the hourly fuel consumption for heating (l/h).

The hourly fuel consumption of the preheater is calculated using the following formula:

$$Q_{ph} = Q_{p10} \times 6 \tag{2}$$

where Q_{ph} is the hourly fuel consumption of the pre-heater, Q_{p10} is the fuel consumption of the pre-heater for 10 minutes of operation.

The hourly fuel consumption of the engine in both modes is calculated using the following formula:

$$Q_{eh} = (Q_{eh10} + Q_{p10}) \times 6 \tag{3}$$

where Q_{eh} is the hourly fuel consumption of the engine, Q_{eh10} is the fuel consumption of the engine and the pre-heater during 10 minutes of joint work, Q_{p10} is the fuel consumption of the pre-heater for 10 minutes of operation.

Further, on the basis of the hourly rates of fuel consumption by the engine and pre-heater, as well as the time of their operation during engine warm-up, the calculation of the fuel consumption rate for warm-up is made.

5. Implementation of methodological recommendations and main results of fuel consumption data processing

In the autumn-winter period of 2019/2020, at the Research Centre of the State Unitary Enterprise "Mosgortrans", fuel consumption standards for heating buses were developed and, in particular, for the modification of LiAZ-429260 in the temperature range from -10 to 5 ° C according to the described method with using the described setting. Data from each measurement of fuel consumption for heating using the installation were recorded in Excel tables (electronic protocols). For each protocol, the hourly fuel consumption by the pre-heater and the engine was calculated according to formulas 2 and 3. In total, 11 measurements of the fuel consumption for warming up were made for this bus model. Table 2 presents the data collected.

Further processing of the data was carried out in the Statistica 10 program. When processing the data on the hourly fuel consumption, the calculations were performed on the upper limit of the mathematical expectation with a confidence level of 95%. Table 3 summarizes the post-processing data.

The hourly fuel consumption of the preheater is 1.7 l/h. The time of its operation when the engine warms up to the starting temperature is 20 minutes, and the total time of its operation during the warm-up until the engine reaches the operating temperature is 30 minutes.

Table 2. The results of measurements of fuel consumption on the LiAZ-429260 bus

№	Bus modification	Garage №	Fuel consumption by the engine, l/h	Fuel consumption by pre-heater, l/h	Warm-up time before starting the engine, min.	Ambient temperature °C
1	LiAZ-429260	11181	4.06	1.54	17	0
2	LiAZ-429260	11275	3.7	1.4	13	-1
3	LiAZ-429260	11285	3.16	1.32	19	-9
4	LiAZ-429260	11275	4.6	1.6	19	-7
5	LiAZ-429260	10981	3.8	1.1	17	5
6	LiAZ-429260	11091	2.7	*	*	5
7	LiAZ-429260	11175	2.8	1.9	10	4
8	LiAZ-429260	11281	3.5	1.02	17	4
9	LiAZ-429260	11284	3.6	1.6	19	3.1
10	LiAZ-429260	11276	3.5	1.6	12	3.1
11	LiAZ-429260	11093	2.6	*	*	5

* fuel consumption and heater warm-up time were not measured.

IOP Conf. Series: Materials Science and Engineering 1159 (2021) 012083 doi:10.1088/1757-899X/1159/1/012083

Table 3. The results of the post-processing data of hourly fuel consumption on the LiAZ-429260 bus

Bus modification	Preheater operation time, min	Fuel consumption by pre-heater, l/h	Engine running time, min	Engine fuel consumption, l/h	Total warm-up time, min
LiAZ-429260	30	1.7	10	3.9	30

Respectively:

$$(1.7 \div 60) \times 30 = 0.85 \text{ l}$$

The hourly fuel consumption of the engine for warming up is 3.9 liters, and the total operating time until the operating temperature is reached is 10 minutes, respectively

$$(3.9 \div 60) \times 10 = 0.65 \text{ l}$$

The total consumption of diesel fuel for heating is:

$$0.85 + 0.65 = 1.5 \text{ l}$$

Table 4 shows the obtained fuel consumption rates for heating the LiAZ-429260 bus in the temperature range from -10 to 5 ° C.

Table 4. Rates of fuel consumption for warming up the engine of the LiAZ-429260 bus

Bus modification	Diesel fuel consumption value for heating, l.		
	Preheater	Engine	Total
LiAZ-429260	0.85	0.65	1.5

6. Conclusion

The research shows the necessity of carrying out the procedure for warming up the buses engine was substantiated in connection with the necessity to warm up the interior of buses to ensure the comfort of passengers, to build up working pressure in the pneumatic systems of the bus for their reliable and safe operation and to additionally monitor the technical condition of buses in motor transport enterprises engaged in passenger transportation. The necessity of regulation of fuel consumption for heating has been substantiated.

In the course of a comparative analysis of existing methods and means of measuring fuel consumption for use in standardizing fuel consumption for heating, it was revealed that there are no ready-made solutions for use at MTE during the operation of modern low-floor buses performing passenger transportation.

A schematic diagram and design of a home-made installation based on a fuel level sensor for measuring fuel consumption for heating, suitable for use in standardizing fuel consumption, was proposed. The key advantages of this device are the ability to measure fuel consumption for warming up without additional setup and calibration on any vehicle, automation of measurement, low labor intensity and high accuracy of results. The disadvantage of the device is the need to change the fuel system of the vehicle by connecting the device instead of the fuel tank.

Methodological guidelines have been developed for standardizing fuel consumption for heating city buses using a specially designed installation, which consists in measuring fuel consumption for heating based on a fuel level sensor, the procedure for obtaining data, calculating the required number of measurements, statistical data processing and calculating the final fuel consumption rate for warming up for the vehicle.

The practical result of the research carried out at the Research Centre of the State Unitary Enterprise "Mosgortrans" is to determine the rate of fuel consumption for warming up the LiAZ-429260 bus in the temperature range from -10 to 5 ° C, equal to 1.5 liters.

A further area of research is the automation of the process of transferring, storing and statistical processing of data on fuel consumption for heating from the device in order to reduce labor costs and measurement time and improved usability.

References

[1] Liu Z, Han S, Yin X, Huang R, Sun Z, Yu X 2012 An experimental study on the measures for improving cold start/warm-up process of engine *Qiche gongcheng* **Vol 7** 596-600
[2] Chalgren R 2004 Thermal Comfort and Engine Warm-Up Optimization of a Low-Flow Advanced Thermal Management System *SAE Technical Paper* **2004-01-0047**
[3] Torregrosa A J, Broatch A, Olmeda P et al. Assessment of the influence of different cooling system configurations on engine warm-up, emissions and fuel consumption *Int.J Automot. Technol* **9** 447–458 2008
[4] Samhaber C, Wimmer A and Loibner E 2001 Modeling of Engine Warm-Up with Integration of Vehicle and Engine Cycle Simulation *SAE Technical Paper* **2001-01-1697**
[5] Shayler P, Christian S and Ma T A 2020 Model for the Investigation of Temperature, Heat Flow and Friction Characteristics During Engine Warm-Up *SAE Transactions* **102** 1588-1597
[6] Axel M, Krahl J and Speckmann H 2002 A fuel sensor for biodiesel, fossil diesel fuel, and their blends *ASAE Annual Meeting. American Society of Agricultural and Biological Engineers*
[7] Jarrier L, Champoussin J, Yu R and Gentile D 2000 Warm-Up of a D.I. Diesel Engine: Experiment and Modeling *SAE Technical Paper* **2000-01-0299**
[8] Andrews G, Harris J, and Ounzain A 1989 SI Engine Warm-Up: Water and Lubricating Oil Temperature Influences *SAE Technical Paper* **892103**

Methodology of assessing the regulated crossing throughput with a dedicated lane for ground public transport based on a probabilistic model

P I Pospelov[1], Le Duc Long [1], A G Tatashev [1,2], M V Yashina [1,2]

[1] Moscow Automobile and Highway State Technical University (MADI), 64, Leningradskiy prospect, Moscow, 125319, Russia
[2] Moscow Technical University of Communications and Informatics (MTUCI), 8A, Aviamotornaya St., Moscow, 125319, Russia

Email: ppospdom@gmail.com; lelongbg@gmail.com

Abstract. One of the problems of traffic management in the allocation of lanes for ground public transport (GPT) is the organization of the right turn at regulated and unregulated intersections. High traffic intensity at a regulated intersection on the main carriageway and a significant proportion of right-turn traffic leads to numerous violations of the rules of passage when making a right turn. A method for assessing the throughput of a controlled intersection with a dedicated lane before joining is presented on the basis of a deterministic stochastic model.

1. Introduction
The scheme of intersection planning and traffic management of a transport intersection provides, first of all, to ensure minimization of the number of road accidents and temporary delays of vehicles [2]. A decrease in the speed of traffic at regulated intersections with a high flow density occurs due to the prolonged phase of the prohibitory signal of the traffic light and the organization of right-turn traffic. At regulated intersections, the intensity of traffic of cars turning to the right makes it difficult for the NOT vehicles to move along the designated lane and causes a queue to form in the second lane. Cars turning from the second lane create a queue, which leads to the impossibility of performing a maneuver for several cycles and the formation of a congestion at the intersection. Vehicle infiltration into the designated lane can also be difficult due to the high traffic volume of vehicles in the designated lane. The queue of GPT funds on the designated lane also increases the time delays for public transport vehicles passing through the intersection [1].

2. Description of traffic at the intersection
To assess the characteristics of the traffic flow at a regulated intersection with a dedicated lane for GPT, field observations were carried out on the road network of Moscow. Figure 1 shows the intersection - the object of study, located at the address: Leningradskiy Prospect, 40.

Content from this work may be used under the terms of the Creative Commons Attribution 3.0 licence. Any further distribution of this work must maintain attribution to the author(s) and the title of the work, journal citation and DOI.
Published under licence by IOP Publishing Ltd

(a) (b)

Figure 1. The intersection of the study: (a) - the diagram of the intersection; (b) - type of vehicle

The main carriageway of the intersection (Leningradskiy Prospect) has four lanes, including:

- the first (rightmost lane) lane is a dedicated lane for the movement of GPT funds. The lane is designed for vehicles to move only in the forward direction and make a right turn.
- there is a forward movement in the second lane, but a small part of the traffic flow slows down the forward direction while waiting for the possibility of changing lanes to the right to perform a right turn.
- on the third lane there is movement only in the forward direction.
- from the fourth lane, a left-turn manoeuvre is being carried out, and only a small part of the traffic flow is moving in the forward direction.

The intensity of traffic in the fourth lane does not have any effect on the formation of the features of traffic in the dedicated GPT lane. This simple consideration leads to the conclusion that the main component of the traffic flow is formed on the 1st, 2nd and 3rd lanes.

Traffic intensity studies were carried out on weekdays at different times of the day during the busiest days and hours to obtain the characteristics of the traffic flow at different traffic intensities, with a dry surface. The results of measuring the traffic intensity on the lanes that affect the movement of the GTP on the designated lane are presented in Table 1.

Table 1. Traffic intensity on crossing lanes

Lane parts	Vehicle type	Traffic intensity (vehicle / hour) by hours of the day			
		05.00-07.00	07.00-09.00	14.00-16.00	19.00-21.00
1	Bus	11	29	29	36
	Truck	02	02	12	08
	Car	37	120	178	597
2	Bus	06	17	17	13
	Truck	11	21	26	29
	Car	78	245	388	617
3	Bus	0	0	0	0
	Truck	08	12	19	27
	Car	135	413	627	961

3. Analysis of the traffic intensity data obtained as a result of measurements

Each vehicle passing through the intersection was assigned to one of three types: buses (type 1), trucks (type 2) and cars (type 3). Here are the measurement results for the following four-time intervals with a duration of 2 hours: from 5 a.m. to 7 (interval 1), from 7 a.m. to 9 (interval 2), from 2 p.m. to 16 (interval 3), from 19 hours to 21 (interval 4). Let us denote by q_{ijk} - the intensity of movement of means of the i-th type on the j-th lane for the k-th mecasurement time interval.

IOP Conf. Series: Materials Science and Engineering 1159 (2021) 012084 doi:10.1088/1757-899X/1159/1/012084

As a result of measurements, the following values were obtained:

$$q_{111} = 11;\ q_{112} = 29;\ q_{113} = 29;\ q_{114} = 36;\ q_{211} = 2;\ q_{212} = 2;\ q_{213} = 12;\ q_{214} = 8;$$

$$q_{311} = 37;\ q_{312} = 120;\ q_{313} = 178;\ q_{314} = 597;\ q_{121} = 6;\ q_{122} = 17;\ q_{123} = 17;\ q_{124} = 13;$$

$$q_{221} = 11;\ q_{222} = 21;\ q_{223} = 26;\ q_{224} = 29;\ q_{321} = 78;\ q_{322} = 245;\ q_{323} = 388;\ q_{324} = 617;$$

$$q_{131} = q_{132} = q_{133} = q_{134} = 0;\ q_{231} = 8;\ q_{232} = 12;\ q_{233} = 19;\ q_{234} = 27;$$

$$q_{331} = 135;\ q_{332} = 413;\ q_{333} = 627;\ q_{334} = 961;$$

The resulting measurements can be used as input data and settings for the parameters of the mathematical model of movement on a multi-lane carriageway of various types of vehicles [3].

Then

$$q_{ij} = \frac{1}{2}\sum_{k=1}^{4} q_{ijk} \tag{3.1}$$

q_{ij} is an average over measurement intervals intensity of vehicles of the i -th type on the j -th lane, $i;\ j = 1;\ 2;\ 3;$

$$q_{i.} = \sum_{j=1}^{3} q_{ij}\ ;\ q_{.j} = \sum_{i=1}^{3} q_{ij} \tag{3.2}$$

$q_{i.}$ - the average over the intervals of measurements, the total intensity of vehicles of the i-th type on all lanes, $i = 1;\ 2;\ 3;$

$q_{\cdot j}$ is the total intensity of all types of vehicles on lane j averaged over the measurement intervals; $j = 1;\ 2;\ 3;$

$$q = \sum_{i=1}^{3} q_{i.} = \sum_{j=1}^{3} q_{.j} \tag{3.3}$$

q is an average over measurement intervals intensity of vehicles of all types on the lanes;

$$\alpha(j) = \frac{q_{.j}}{q};\ \beta(i) = \frac{q_{i.}}{q} \tag{3.4}$$

$\alpha(j)$ is the ratio of the traffic intensity of all types of vehicles on lane j averaged over the measurement intervals to the traffic intensity averaged over the measurement intervals of the vehicles of types on all lanes, $j = 1;\ 2;\ 3;$

$\beta(i)$ is the ratio of the flow rate of vehicles of the i -th type averaged over the measurement intervals in all lanes to the rate averaged over the measurement intervals of the flow rate of vehicles of all types in all lanes $j = 1;\ 2;\ 3;$

$$\alpha_{ik}(j) = \frac{q_{ijk}}{\sum_{s=1}^{3} q_{isk}};\ \beta_{jk}(i) = \frac{q_{ijk}}{\sum_{s=1}^{3} q_{sjk}} \tag{3.5}$$

$\alpha_{ik}(j)$ is the ratio of the traffic intensity of vehicles of the i -th type on the j-th lane for the k-th time interval to the intensity of the flow of vehicles of the i -type in all lanes for the k -th time interval, $i;\ j = 1;\ 2;\ 3;\ k = 1;\ 2;\ 3;\ 4;$

$\beta jk\ (i)$ is the ratio of the flow rate of vehicles of type i on lane j to the flow rate of all types of vehicles on lane j for interval $k;\ i;\ j = 1;\ 2;\ 3;\ k = 1;\ 2;\ 3;\ 4;$

$$\alpha_i(j)=\frac{\sum_{k=1}^{4} q_{ijk}}{\sum_{s=1}^{3}\sum_{k=1}^{4} q_{isk}}\;;\;\beta_i(j)=\frac{\sum_{k=1}^{4} q_{ijk}}{\sum_{s=1}^{3}\sum_{k=1}^{4} q_{sjk}} \tag{3.6}$$

αi (j) is the ratio of the intensity of the flow of vehicles of type *i* averaged over measurement intervals in lane *j* to the intensity of the total flow of vehicles of type *i* averaged over measurement intervals in all lanes; *s*; *i*; *j* = 1; 2; 3; *k* = 1; 2; 3; 4;

βj (i) is the ratio of the intensity averaged over measurement intervals of the total flow of vehicles of type *i* in lane *j* to the intensity averaged over measurement intervals of the total flow of vehicles of all types in lane *j*; *s*; *i*; *j* = 1; 2; 3; *k* = 1; 2; 3; 4.

There are some calculated values:

$\alpha_1(1) = 0.665$, $\alpha_{11}(1) = 0.647$; $\alpha_{12}(1) = 0.630$; $\alpha_{12}(1) = 0.630$; $\alpha_{14}(1) = 0.735$;

$\alpha_2(1) = 0.135$; $\alpha_2(2) = 0.492$; $\alpha_2(3) = 0.373$;

$\alpha(1) = 0.22$; $\alpha(2) = 0.31$; $\alpha(3) = 0.47$; $\beta(1) = 0.03$; $\beta(2) = 0.04$; $\beta(1) = 0.93$;

Based on the results of measurements and calculations performed, the following conclusions can be drawn:

- GPT funds move mainly along lane 1, $\alpha_1(1) = 0.665$; 66.5% of the total traffic intensity of GPT vehicles corresponds to traffic on lane 1, and the value of this share at two-hour intervals varies from 63.0% in the time intervals of 7.00 - 9.00 hours and 14.00 - 16.00 hours to 73.5% in the time interval of 19.00 - 21.00. Of the measurement time intervals, the interval 19.00 - 21.00 corresponds to the highest traffic intensity. With such an intensity, GPT vehicles are less likely to enter lane 2 and this can explain the fact that in this time interval the share of buses on lane 1 is maximum;
- the intensity of the flow of trucks averaged over the measurement intervals is: 13.5% in lane 1, 49.2% in lane 2, 37.3% in lane 3;
- Distribution of vehicles by lanes: the number of moving vehicles in lane 3 amounted to almost half (47%) of the total traffic intensity. The corresponding figures for the second and first bands are 31% and 22%, respectively. This can be explained by the fact that most cars will continue to move in the forward direction when passing the intersection, while part of the traffic intensity in lane 1 and lane 2 will turn to the right. Therefore, in order to minimize the influence of cars turning to the right, drivers prefer to change lanes in advance in the third lane of the carriageway. The intensity of traffic on the first lane outside rush hour is relatively low (from 50 to 250 vehicles / h), since this lane is intended for the movement of GPT vehicles. Its value increases only during peak hours, when the traffic intensity of the remaining lanes is close to the capacity.
- The results of the analysis also show that in the composition of the traffic flow, cars make up an absolute majority - 93%, and trucks and GPT vehicles are quite a small share (4% and 3%). those. In the course of the analysis, we can assume that in the composition of the traffic flow during modelling, it is possible to assert with a high degree of reliability only cars and GPT vehicles within the range of permissible error.

4. Model of movement on two lanes of the carriageway of two types of vehicles, differing in speed and size

Suppose there are two types of particles moving along a two-lane carriageway. Particle types differ in the direction of movement at the end of the section under consideration. Particles of the 1st type move along one strip and do not tend to move to another strip. Each type 2 particle tends to move to the far-right lane. The length of the section and the intensity of traffic flows on the lanes at the beginning of the section are given. (Fig. 2).

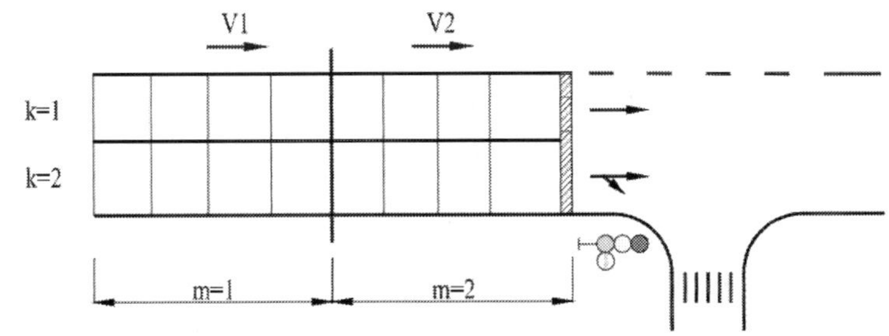

Figure 2. Cellular two-zone segregation field

The position of the particle is characterized by the point (k, y): where k is the traffic lane index, k = 1,2. The points with the values of the abscissa $k = 1$ will be called the points of the left (first) strip, and the points with the values of the abscissa $k = 2$ - the points of the right (second) strip; y - zone index; y = 1,2.

The site is divided into M segments. Each segment contains "l" cells. The size (length) of each cell located on the "m", segment is dm, d_m, $m=1,2...M$. The d_m value corresponds to the dynamic dimensions of the vehicle [3,5,6]:

$$d_m = 5.7 + 0.5v_m + 0.03v_m^2 \tag{4.1}$$

The rules for individual movement of a particle in the m-th zone, m = 1, 2, are as follows [4]:

- If a particle of the 1st type is on the strip k, then it will continue to move along the section on this strip. In one quantum of time, a particle of the 1-st type located on the k-th strip and the m-th section, with the probability $\lambda_{1,m}$□t, tends to move one cell in the direction of motion.
- If a particle of the 2-nd type is on the cell $(1, j)$, and the neighboring cell $(2, j)$ is free, then with probability $\lambda_{2,m}$□t the particle goes to the cell $(2, j)$.
- If a particle of the second type is on the 1-st lane, and the cell located in front of this particle is free, and the neighbouring cell on the 2-nd lane is occupied, then the particle does not move.

Let us make the following assumptions: in each cell located in the m-th zone on the k-th strip, with the probability $r_i(k, m)$ there is a particle of the i-th type, i, k = 1, 2; m = 1, 2. With probability $1 - r(k, m)$, this cell is free, where $r(k, m) = r_1(k, m) + r_2(k, m)$. The probabilities of the states of points do not depend on each other.

Let a particle of the i-th type be in the m-th zone on the 1-st lane in the cell $(1, ld_m)$ at time s□t. Then we denote by $p(1, m)$ the probability that the particle at the time instant $(s+1)$□t will be in the cell $(2, ld)$, m = 1,2:

$$p(1,m) = \left[1 - r(2,m)\right]\lambda_{2,m}\Delta t \tag{4.2}$$

Let $r_i(k,m)$ be the flux densities of a particle of the i-th type on the k-th strip of the m-th zone:

$$r_i(k,m) = d_m.\rho_i(k,m) \tag{4.3}$$

where $r_i(k,1)$ are flux densities of a particle of the i-th type on the k-th strip of the 1-st zone, are given.

The values $q_i(k,1)$ of the flow rates of particles of the i-th type on the k-th strip at the beginning of the 1-st zone are calculated by the formulas:

$$q_i(i,1) = r_1(i,1)\left[v_1 + \left\{1 - r(i,1)\right\}\lambda_{1,1}\Delta t\right] \tag{4.4}$$

$$q_i(k,1) = r_i(k,1).v_1, \quad \text{i} \neq \text{k} \tag{4.5}$$

$$q_1(k,2) = q_1(k,1) \tag{4.6}$$

Let $\beta(1,m)$ be the probability that a particle of the 2-nd type, which began to move in the m-th zone on the 1st lane, will have time to move to the 2-nd lane during its movement in this zone, $m = 1, 2$. Then the intensities $q_2(k, 2)$ the flux of particles of the 2-nd type on the k-th band in the 2-nd zone are calculated by the formulas:

$$q_2(1,2) = q_2(1,1)\left[1-\beta(1,1)\right] \quad (4.7)$$

$$q_2(2,2) = q_2(1,1)\beta(1,1) + q_2(2,1) \quad (4.8)$$

The probabilities that the points of the 2-nd zone are occupied by particles of the corresponding types at fixed $k = 1, 2$ are found from the relations:

$$q_i(i,2) = r_i(i,2)\left[v_2 + \{1-r(i,2)\}\lambda_{1,2}\Delta t\right] \quad (4.9)$$

$$q_i(k,2) = r_i(k,2).v_2, \ i \neq k \quad (4.10)$$

$$r(k,m) = r_1(k,m) + r_2(k,m) \leq 1; \ r_1(k,m) > 0, \ r_2(k,m) > 0 \quad (4.11)$$

This system of equations for each value $k = 1, 2$ is reduced to a quadratic equation. If the system has more than one solution, then we take the smallest values $r_1(k, 2)$ and $r_2(k, 2)$.

Let the random v199ariable $\eta(1,m)$ be the increment in the stationary system of the ordinate of a particle of the 2-nd type, which began to move along the m-th zone in the 1st lane, until the moment the particle transitions to the 2nd lane under the assumption that $L_m = \infty$, $m = 1, 2$.

We estimate the required probability $\beta(1,m)$, assuming that the random variable $\eta(1,m)$ can be considered distributed according to the normal law with the corresponding parameters, $m = 1.2$:

$$\beta(1,m) = P\left[\eta(1,m) < L_m\right] = \frac{1}{2} + \Phi\left[\frac{L_m - a(1,m)}{\sigma(1,m)}\right] \quad (4.12)$$

Where $\Phi(x)$ is the Laplace function: $\Phi(x) = \frac{1}{\sqrt{2\pi}}\int_0^x e^{-\frac{z^2}{2}}dz$

$$a(1,m) = \frac{v_m \Delta t}{p(1,m)}, \ \sigma^2(1,m) = \frac{\left[1-p(1,m)\right](v_m\Delta t)^2}{\left[p(1,m)\right]^2} \quad (4.13)$$

Suppose that the α probability that a particle of the 2-nd type passes into the right lane at the end of the section under consideration is:

$$\alpha = \frac{q_2(2,2)}{\sum_{k=1}^{2} q_2(k,1)} \quad (4.14)$$

5. Numerical examples

We consider the movement of cars with three different values of the parameters in the far zone and with the same values of the other parameters of the model. The values of the input parameters are as follows:

Δt =0.8 (s), $\lambda_{1,1} = \lambda_{1,2} = 0.5$ (1/s), $\lambda_{2,1} = \lambda_{2,2} = 1.0$ (1/s),

$\rho_1(1,1) = \rho_1(2,1) = \rho_2(1,1) = \rho_2(2,1) = 0.01$ (1/m), $L_1 = L_2 = 30$ (m),

Let's carry out calculations for three values v_m: v_m = 8 (m/s), v_m = 10 (m/s), v_m = 12 (m/s). The calculation results are given in Table 2:

Table 2. α vs v_m dependencies

v_m (m/s)	α
8	0.999
10	0.986
12	0.939

Thus, in the considered speed range, with a decrease in the deterministic component of the flow rate, the probability of a timely transition of the car to the right lane increases.

6. Conclusion

The presented methodology for assessing the throughput of a regulated intersection section on an element of a road network in the presence of a dedicated lane for GPT on the roadway on the basis of a stochastic segregation model and the results of field studies can be used to determine the throughput of traffic segregation sections, which will increase the validity of design decisions in the design regulated intersections and will help improve the quality of transport services.

References

[1] Babkov V F 1967 *Road conditions and modes of movement of vehicles* (Moscow: Transport)
[2] Lobanov E M 1990 *Transportation planning of cities* (Moscow: Transport)
[3] Inosse H, Tanaka T 1983 *The road traffic control* (Moscow: Transport)
[4] GOST 33100-2014 *Automobile roads of general use. Rules of road projecting* (Moscow: Informavtodor [Federal Agency for Technical Regulation and Metrology]), 2014
[5] Barnes H A 1955 *Operation Research* **3(4)** 536-544
[6] Nagel K, Wolf D E, Wagner P, Simon P M 1998 *Physical Review E.* **58(2)** 1425-1437

IOP Conf. Series: Materials Science and Engineering 1159 (2021) 012085 doi:10.1088/1757-899X/1159/1/012085

Theoretical aspects of assessing the residual engine oil resource

A V Laushkin[1], A A Khaziev[1]

[1] Department of Operation and servicing of motor vehicles, Moscow Automobile and Highway State Technical University (MADI), Leningradskiy prospect, 64, Moscow, 125319, Russian Federation

E-mail: lav82@mail.ru, madi-chim@mail.ru

Abstract. The article reflects issues related to changes in the properties of engine oil during operation. Proposals are considered for creating a methodology for assessing the residual resource of engine oil by the example of changing the kinematic viscosity.
The operating conditions of cars vary greatly, which affects the change in the engine oil resource. Automotive manufacturers set the engine oil change interval in kilometers. The actual resource of the engine oil may differ significantly from that recommended by the manufacturer of the equipment and declared by the manufacturer of the oil. The article presents the factors that determine the resource of engine oil and theoretical aspects of assessing the residual resource of engine oil. To assess the residual life of the engine oil, it is proposed to select a diagnostic indicator that comprehensively evaluates the state of the operating oil, assign limit values for the diagnostic indicator, fix the values of the diagnostic indicator during daily maintenance of the car, replace the dependence by which the diagnostic indicator changes with an approximating straight line, determine the coefficients for the approximating direct based on the last and previous values of the diagnostic indicator and operating time, calculate the engine oil operating time to the limit state according to a linear relationship The results of testing and forecasting the residual resource of engine oil for a motor transport enterprise are presented. Calculations made it possible to establish that the frequency of engine oil change can be increased relative to that adopted at a motor transport company by 28.3%.

1. Introduction

An important reserve for saving material resources during the technical operation of vehicles is the most complete use of the resource of engine oils. The reliability and service life of a car engine directly depends on the quality and condition of engine oil. Car manufacturers set engine oil change intervals in kilometers (miles). The scientific community and specialists of car service enterprises are finding alternative approaches to the appointment of change intervals.

2. Engine oil life and its determining factors

There are several factors that determine the life of an engine oil. A modern car engine operates under extreme conditions that change the properties of the engine oil. Given this, the manufacturer of the equipment recommends regular replacement after a certain mileage with some reservations to consider difficult operating conditions.

Content from this work may be used under the terms of the Creative Commons Attribution 3.0 licence. Any further distribution of this work must maintain attribution to the author(s) and the title of the work, journal citation and DOI.
Published under licence by IOP Publishing Ltd

IOP Conf. Series: Materials Science and Engineering 1159 (2021) 012085 doi:10.1088/1757-899X/1159/1/012085

In many ways, the resource of engine oil depends on the operating conditions of the car, among which the following should be noted:

- Driving the vehicle over short distances when the engine (and engine oil) does not reach optimum operating temperature and the power unit is running on a rich fuel mixture.
- long-term operation of the engine in conditions of traffic jams at idle;
- «sporty» driving style, aggressive driving style;
- prolonged idle time of the car without movement, when the owner rarely uses the vehicle;
- use of low-quality consumables (filter elements of the lubrication system, air purification, etc.);
- low quality of motor fuel;
- overheating of the engine and engine oil due to excessive loads, etc.

Along with this, the technical condition of the automobile engine and its systems affects the resource of the engine oil: the power supply system, the ignition system (for gasoline engines), the crankcase ventilation, the engine control unit, etc.

Thus, it can be stated that the resource of engine oil depends on a large number of factors that are difficult to consider. Under these conditions, as a general recommendation, you can either change the oil according to its actual state, or reduce the recommended oil change period by one and a half to two times.

3. Theoretical aspects of assessing the residual resource of engine oil

Residual resource of engine oil is the operating time of the lubricant from the current to the limit state.

To assess the residual resource of engine oil, you must:

- choose a diagnostic indicator that comprehensively evaluates the condition of the operating oil;
- assign limit values for a diagnostic indicator;
- measure the values of the diagnostic indicator several times with different operating time;
- determine the mathematical relationship by which the diagnostic indicator changes;
- determine the coefficients for the selected mathematical relationship;
- calculate the operating time of the engine oil to the limit state.

The actual resource of the engine oil may differ significantly from that recommended by the manufacturer of the equipment and declared by the manufacturer of the oil.

Scientists and manufacturers of automotive and road construction equipment recommend evaluating the condition of engine oil by the following indicators: kinematic viscosity, flash point in an open crucible, alkaline and acid numbers, water content, content of mechanical impurities, incl. soot, silicon, iron, copper, etc [1, 2, 5, 6]. The dependences of changes in the quality indicators of engine oil are described in sufficient detail in the literature [1, 3, 4, 7, 8].

Under stable operating conditions of the car, it is enough to measure from 2 to 5 values of the diagnostic indicator of engine oil and build a linear or non-linear regression model describing the change in the diagnostic indicator. Instead of a mathematical model of the influence of operational parameters, you can use empirical dependencies described by linear, polynomial, exponential and other dependencies. A large number of diagnostic points allows you to more accurately build the regression equation and predict the residual resource. If the operating conditions of the car change: there is an instability of the quality of the car fuel, several drivers are assigned to the car, the car is operated on different routes and with different loads, then the coefficients of the regression model describing the state of the engine oil must be adjusted.

By increasing the frequency of measurements of the diagnostic indicator, it is possible to more accurately record changes in the properties of engine oil in real time. With an infinite number of discrete measurements, the function of changing the diagnostic parameter of the engine oil will turn into a continuous one. At each point of a continuous function, the trend of its change can be described by a tangent. On a small segment of the engine oil operating time, the tangent function approaches the function itself. Thus, with sufficiently frequent discrete measurements, the function of changing the

IOP Conf. Series: Materials Science and Engineering 1159 (2021) 012085 doi:10.1088/1757-899X/1159/1/012085

diagnostic indicator of engine oil can be replaced with tangents. When the operating conditions of the car change, the intensity of the change in the diagnostic indicator of the engine oil, and, accordingly, the angle of inclination of the tangent will change.

By extending the tangent to the last point of the engine oil diagnostic indicator graph, you can determine the point of its intersection with the boundary line. Replacing the function of changing the diagnostic indicator of its tangent allows you to go to a linear relationship. Evaluation of the linear dependence of the change in the resource of the engine oil seems to be a simpler task compared to the recalculation of the coefficients of the regression model. In this case, the assumption is introduced that the change in the diagnostic indicator will be linear until the limit state of the oil is reached. With a large residual engine oil resource, this assumption will reduce the prediction accuracy. As the properties of the oil approach the limiting state, the prediction accuracy will be increased.

With a low labor intensity of oil analysis (less than 5 man-min.), It is possible to recommend measuring the diagnostic indicator of engine oil during daily car maintenance. With an average daily mileage of 200 km and an engine oil change interval of 10 - 20 thousand km, 50 - 100 measurements will be carried out. During a work shift, the operating conditions of the car can be considered stable. Thus, sudden changes in the properties of the engine oil will be recorded at the end or before the start of the work shift. Information about a change in the properties of engine oil may indicate a violation of operating conditions and correct its residual resource.

The residual resource of the engine oil will be defined as the operating time before the intersection of the straight line approximating the function of changing the diagnostic indicator with its boundary value. The formula for the approximating straight line is:

$$\alpha(l) = k \cdot l + b, \tag{1}$$

where $\alpha(l)$ is the calculated diagnostic indicator of the engine oil;
l is engine oil operating time, km;
k and b are the coefficients of the linear equation.

The slope k is equivalent to the rate of change in oil properties and is calculated as the ratio of the change in the value of the diagnostic indicator to the change in operating time:

$$k = \frac{\alpha_l - \alpha_{l-1}}{l_l - l_{l-1}}, \tag{2}$$

where α_l is the value of the diagnostic indicator at the current time;
l_l is operating time at the current time
α_{l-1} *is* value of the diagnostic indicator at the previous measurement;
l_{l-1} is operating time during the previous measurement

The free term is calculated by substituting the current values of the diagnostic indicator and operating time into the equation of the straight line:

$$b = \alpha_l - k \cdot l_l, \tag{3}$$

To determine the residual life of the engine oil lo, the operating time must be expressed from formula (1). The current values of the coefficients k from the formula (2) and b from the formula (3) are substituted into the formula (1). The boundary value αmin, is used as the value of the oil diagnostic indicator if the diagnostic indicator decreases (k <0) and αmax if the diagnostic indicator increases (k> 0) during maintenance:

$$l_0 = \begin{cases} a_{\min} - b_l / k_l & when \quad k < 0 \\ a_{\max} - b_l / k_l & when \quad k > 0 \end{cases}. \tag{4}$$

If it is known that the change in the diagnostic oil quality indicator is monotonous, then one equation and one boundary value of the diagnostic oil quality indicator can be limited.

4. Example of predicting the residual life of engine oil

Let us consider the application of linear extrapolation by the example of changing the kinematic viscosity of engine oil. The kinematic viscosity of an engine oil varies in terms of operating time under the influence of various factors. The nature of the change in viscosity is random. To predict the residual life, the values of the kinematic viscosity by operating time were recorded on a Scania P114GA4x2NA340 vehicle. During the test period, 78 measurements of the kinematic viscosity of Mobil Delvac MX Extra 10W-40 engine oil were processed (Fig. 1).

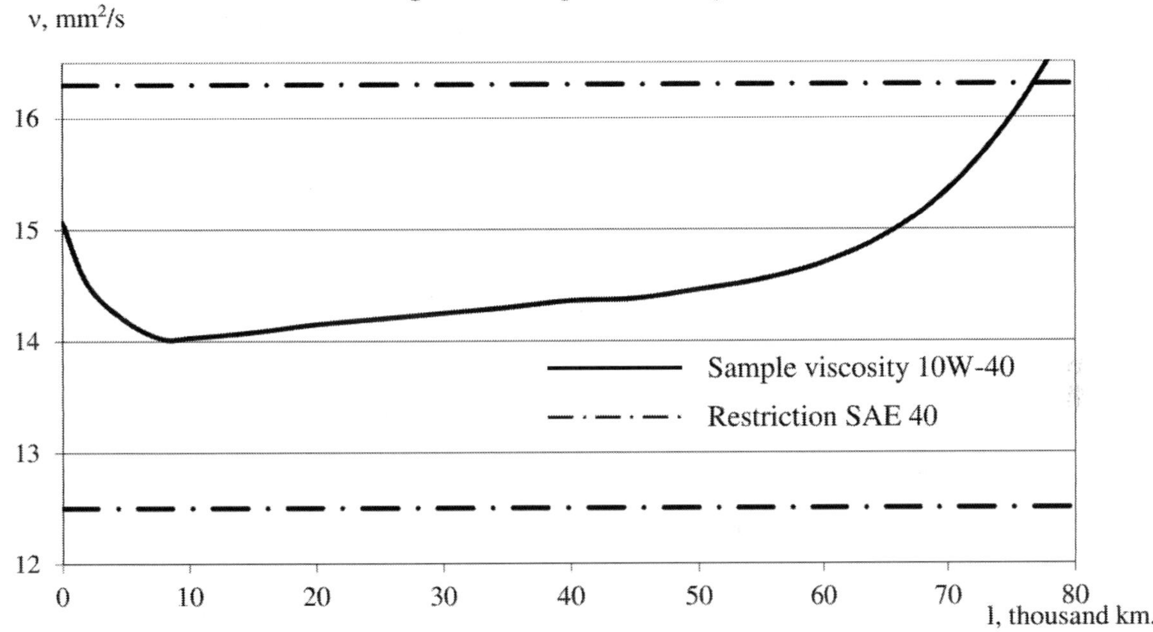

Figure 1. Change in engine oil viscosity by operating time

After maintenance for 8 thousand km, the kinematic viscosity of the engine oil decreased from 15.1 mm^2/s to 14.0 mm^2/s. Then the change in viscosity stopped, and after 10 thousand km of oil production, a slight thickening of the oil began to occur. With the service interval of 60 thousand km adopted at the enterprise, it was proposed to increase the operating period until the operating time, at which the kinematic viscosity of the oil at 100°C would go beyond the SAE 40 viscosity class. At the same time, in accordance with the proposed theoretical aspects, the residual resource of the engine oil was predicted (Table 1). The results of tests and predictions of the residual resource of the engine oil showed the need to replace it with operating time of 77 thousand km.

Table 1. An example of predicting the residual resource of engine oil

Oil production, thousand km	0	2	5	8	10	15	20	30	40	50	60	70	73	76	77
Residual resource, thousand km	10	14	23	72	70	65	60	50	40	30	20	10	5	1	0
Kinematic viscosity, mm^2/s	15.1	14.5	14.2	14.0	14.0	14.1	14.15	14.3	14.4	14.5	14.7	15.4	15.5	16.1	16.3

When predicting the resource of engine oil by kinematic viscosity, it was possible to observe erroneous values of the forecast up to operating time of 8 thousand km. If the tendency of viscosity changes at the beginning of the experiment persisted, the prediction would be correct. However, the nature of the change in kinematic viscosity has changed. According to the calculation according to the proposed formula, the residual resource of the engine oil with operating time of 8 ... 10 thousand km tends to infinity. After operating 50 ... 55 thousand km, the rate of change in viscosity began to grow. After operating 70 thousand km, the predicted residual resource began to decrease. The deviation of the real resource of the engine oil from the predicted one was 3.75%. At the same time, the increase in the resource of the engine oil in comparison with that adopted at the motor transport enterprise was 28.3%.

5. Conclusion

Based on the results of the study, assessing the residual resource of engine oil, it was proposed:

- use the viscosity of engine oil as a diagnostic indicator that comprehensively evaluates the condition of the oil;
- assign viscosity limit values in accordance with the SAE J300 viscosity grade limit values;
- fix the values of the diagnostic indicator during daily maintenance of the vehicle;
- replace the dependence according to which the diagnostic indicator changes with an approximating straight line;
- determine the coefficients for the approximating straight line based on the last and previous values of the diagnostic indicator and operating time;
- calculate the operating time of engine oil to the limit state according to a linear relationship.

The advantages of this method for assessing the residual resource of engine oil are:

- no need to build complex mathematical models of changes in the diagnostic indicator;
- the possibility of using the characteristics of engine oil as diagnostic indicators, the change in which by operating time has been little studied;
- quick correction of the values of the residual resource of the engine oil when operating conditions change.

References

[1] Vasilyeva L S 2014 *Operational materials for the rolling stock of motor transport* (Moscow: Nauka)
[2] Grigoriev M A 1981 *The quality of engine oil and reliability of engines* (Moscow: Publishing house of standards)
[3] Zarochintsev Yu M 1992 *Motor oils. Causes of motor oil pollution* (Chelyabinsk: ChSTU)
[4] Kovalenko V P 1978 *Pollution and cleaning of oil oils* (Moscow: Chemistry)
[5] Futsita H 1984 *Ishimaru M. Junkatsu* **29(12)** 889-892
[6] Khaziev A A 2014 *Bulletin of MADI* **4(39)** 11-17
[7] Khaziev A and Laushkin A 2020 *IOP Conference Series: Materials Science and Engineering* **832(1)** 012080

IOP Conf. Series: Materials Science and Engineering 1159 (2021) 012086 doi:10.1088/1757-899X/1159/1/012086

Methods of electrochemical research of corrosion resistance of metal treated with corrosion inhibitors, created on the basis of plant-based raw materials

M Y Karelina[1], O N Didmanidze[2]

[1]Federal State Budgetary Educational Institution of Higher Education «Moscow Automobile and Road State Technical University (MADI), 64 Leningradsky ave, Moscow, 125319 Russia
[2]Corresponding Member of the Russian Academy of Sciences, Professor,
Russian State Agrarian University - Moscow Agricultural Academy named after K.A. Timiryazev, 49, Timiryazevskaya str., 49, Moscow, 127550, Russia

E-mail: Karelinamu@mail.ru, Didmanidze@rgau-msha.ru

Abstract. The use of corrosion inhibitors based on raw materials of plant origin has been a poorly studied method of metal processing so far. To show the degree of corrosion protection, it is necessary to examine it. It is possible to do this by means of electrochemical research, the method of which consists of several methods: the extrapolation method; the polarization resistance method; the related methods. As a result of this method, a significant advantage of samples treated with corrosion inhibitors based on raw materials of plant origin was revealed.

1. Introduction

There are many ways to determine the corrosion resistance of metal, used in different conditions for different types of treatment. One of the most promising treatment methods is metal coating with corrosion inhibitors based on vegetable raw materials [1]. The method of investigation of corrosion resistance of metal samples processed with these materials requires a detailed study.It is necessary to achieve the most accurate results in determining corrosion resistance, as all research in this area of science will depend on it.

2. Materials and methods

The electrochemical methods for determining corrosion rate are based on the results of polarization measurements, most often obtained with potentiostats or various types of corrosion meters. The advantage of these methods is manifested in the ability to evaluate the mechanism of processes [2]. Their disadvantage is the incomplete adequacy of the results obtained and the theoretical concepts used in their interpretation, which may not always take into account the complexity of phenomena occurring on the surface.

To date, most methods for evaluating corrosion process parameters have been based on the analysis of the kinetic equation:

Content from this work may be used under the terms of the Creative Commons Attribution 3.0 licence. Any further distribution of this work must maintain attribution to the author(s) and the title of the work, journal citation and DOI.
Published under licence by IOP Publishing Ltd

$$I = I_{cor}\left\{\exp\left(\frac{2,3\Delta E}{b_a}\right) - \exp\left(\frac{-2,3\Delta E}{b_k}\right)\right\} \tag{1}$$

where$b_a$and$b_к$are respectively the anodic and cathodic Tafel constants;
$\Delta E = E - E_{cor}$ –electrode polarization, measured from corrosion potential (E_{cor}).

3. The extrapolation method

The classical method of polarization curve analysis is the extrapolation method [3]. Figure 1 schematically shows anode and cathode polarization curves in a medium with and without corrosion inhibitor. In semi-logarithmic coordinates, these curves turn into straight lines at a sufficient distance from the corrosion potential. Extrapolation of straight sections of anode or cathode polarization curve to the stationary potential allows determining the corrosion current density (i_{cor}, A/m^2) and calculating its speed (K, g/m^{2}h):

$$K = \frac{i_{cor} * A}{n * F}, \tag{2}$$

where A is theatomic weight of metal (55,85 foriron);
n – metal ion valence (2 for iron);
F– Faradayconstant, A·h/g-equivalent.

Comparison of corrosion current values in the presence of the inhibitor $\left(i^i_{cor}\right)$ and without it $\left(i_{cor}\right)$ allows to calculate the protective effect (Z, %) and the protection factor (γ):

$$Z = 100\frac{i_{cor} - i^i_{cor}}{i_{cor}} \tag{3}$$

$$\gamma = 100\frac{i_{cor}}{i^i_{cor}} \tag{4}$$

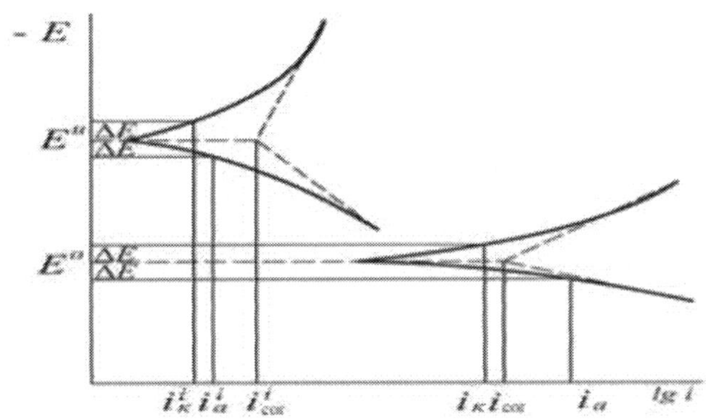

Figure 1. Anodic and cathodic polarization curves with and without corrosion inhibitor

The slope of rectilinear sections of polarization curves is related to the degree of polarizability of anode or cathode components of the corrosion process, respectively. It allows to calculate Tafel constants and to find out the nature of inhibitor action by type of control (cathode, anode, mixed) and to determine its degree.

The tests related to the polarization curves are carried out according to the instructions for the potentiostat used. Electrolyte (e.g. formation water or 3% solution of NaCl + 250 mg/lCH_3COOH) with aggressive gases dissolved in them or mixtures of electrolyte - liquid hydrocarbon (hydrocarbon condensate, kerosene) are used as the test medium.

Consistent testing of a sample in a solution with and without an inhibitor (after exposure in an inhibited solution) allows us to evaluate the effect of the inhibitor.

The disadvantage of the extrapolation method is the need for significant polarization of the electrode (~ ±0,5…1,5 V) so that the Tafel sections on the curves can be clearly detected. This also determines the limitations of the method, which are associated with possible irreversible changes in the electrode system, resulting in irreversibility and hysteresis of potential-current curves.

To evaluate the effect of the inhibitor separately on the cathode or anode process, you can use the polarization of the electrode in the appropriate direction by some constant value, for example, in the range of 5…30 mV. At the selected potential shift from the stationary value, the current value is determined. Such determinations are performed for pure and inhibited solutions either from both pairs of polarization curves, or (preferably due to less electrode exposure) directly by measuring the current value at a corresponding potential displacement.

As a result of such definitions, two pairs of current values will be found (see Figure 6.2): i_k, i_k^{u} and i_a, i_a^{u} – mainly related to the rate of cathode and anode processes in the initial and inhibited environment. Thecomparisonofvalues i_a and i_a^{u} characterizes the inhibitor effect on the anodic component of the process, while i_k and i_k^{u} refer to the cathodic one.

4. The polarization resistance method

This method is often associated with the names of Stern and Giri, who contributed most to its development. It is based on the assumption that the relationship between potential and superimposed current at a potential close to corrosion potential does not correspond to equation (6.5), but is linear. This assumption leads to a simple ratio between the corrosion current and polarization resistance (R_p):

$$I_{cor} = \frac{b_a \cdot b_k}{2,3(b_a + b_k)} \cdot \frac{1}{R_p} = \frac{B}{R_p}, \tag{5}$$

where - $R_p = \frac{dE}{dI}$

$$B = \frac{b_a b_k}{2,3(b_a + b_k)} \tag{6}$$

- constant depending on the Tafel constants

The polarization resistance method eliminates most of the problems associated with intense electrode polarization. In practical measurements, the polarization of the electrode rarely exceeds ±20 mV and is most often limited to ±10 mV. The method involves the use of conventional laboratory equipment for polarization measurements. In addition, the simplicity of the method stimulated the development of corrosion meters based on polarization resistance measurements by many companies [4]. When working with them, one should follow the instructions attached to them, taking into account the disadvantages of the method.

One of the disadvantages of the method is the assumption of the linearity of the basic kinetic equation leading to systematic errors [5]. The greater the polarization value, the more these errors become, they can reach several tens of percent. Another disadvantage of the method is related to the uncertainty of the B constant, the value of which cannot be obtained directly by measuring R_p and is set a priori or determined in a separate experiment, the use of one or another of its quantitative value must be reasoned in each particular case. Theoretically, the B value can vary from 6.5 to 51.11. Additional difficulties in the interpretation of the results obtained occur in low conductive media, as well as in the deposition of corrosion products on corrosive surfaces, such as hydrogen sulfide corrosion.

IOP Conf. Series: Materials Science and Engineering 1159 (2021) 012086 doi:10.1088/1757-899X/1159/1/012086

The use of corrosion meters based on polarization resistance measurements, taking into account the reservations made, can be recommended when it is necessary to obtain a large number of measurements in a short period of time, for example, when selecting an inhibitor from a large series of samples to obtain the dependence of the change in corrosion rate over time, to control the effectiveness of the inhibition with a constant nature of the corrosion process, etc.

5. Related methods

The related methods of polarization curve analysis largely combine the advantages of both previous methods, with almost no disadvantages [6]. Like the polarization resistance method, they provide for relatively small polarization, but, in contrast to this method, being based on the nonlinearity of equation (6.5), they allow all three parameters of the corrosion process to be determined in one experiment (b_a, b_k, I_{cor}). This only requires the determination of several current values with negligible polarization of the electrode near the corrosion potential. The distinctive feature of these methods, developed by various authors, is the nature of the necessary polarization and its direction. It can be defined as cathodic, anodic, or cathodic-anodic. The most convenient are the adjacent Beleevsky-Kudelin methods, which allow, depending on the choice of the experimenter, to choose either cathode or anode measurements, or to use the polarization of both signs, depending on the expedient need [7]. In particular, cathode measurements are preferred for systems where anode currents cause surface etching. Anodic polarization has an advantage when cathode shift of potential leads to deposition of corrosion products on the electrode or to restoration of the oxide formed during corrosion, as well as when another electrode process is applied (cathode emission of hydrogen in the study of corrosion with oxygen depolarization).

For the analysis of experimental data corresponding to the ratio (6.5), two variants can be used, based on the precise analytical solution of the system of initial equations at a certain choice of variables I and ΔE.

In particular, for points of any branch of the polarization curve selected with equal pitch (ΔE, $2\Delta E$, $3\Delta E$) or ($-\Delta E$, $-2\Delta E$, $-3\Delta E$) the required values are determined by the following formulas:

$$I_{cor} = \frac{|I_1|}{\sqrt{\frac{4I_3}{I_1} - 3\left(\frac{I_2}{I_1}\right)^2}}, \quad (7)$$

$$b_a = \pm\Delta E / \lg 0,5\left[I_2/I_1 \pm \sqrt{4I_3/I_1 - 3(I_2/I_1)^2} \right] \quad (8)$$

$$b_k = \pm\Delta E / \lg 0,5\left[I_2/I_1 \pm \sqrt{4I_3/I_1 - 3(I_2/I_1)^2} \right] \quad (9)$$

At other choice of variables, namely at ΔE, divisible by two: (ΔE, $2\Delta E$, $4\Delta E$) and ($-\Delta E$, $-2\Delta E$, $-4\Delta E$) corrosion current and Tafel constants are found from:

$$I_{cor} = \frac{|I_1|}{\sqrt{\frac{2I_4}{I_2} - \left(\frac{I_2}{I_1}\right)^2}}, \quad (10)$$

$$b_a = \pm\Delta E / \lg 0,5\left[I_2/I_1 \pm \sqrt{2I_4/I_2 - (I_2/I_1)^2} \right] \quad (11)$$

$$b_k = \pm\Delta E / \lg 0,5\left[I_2/I_1 \pm \sqrt{2I_4/I_2 - (I_2/I_1)^2} \right] \quad (12)$$

where the signs «+» and «–» in (3.11), (3.12), (3.14), (3.15) refer to anodic currents, while «–» and «+» – to the cathodic ones;

I_1, I_2, I_3, I_4 – currents respectively at polarization ΔE, $2\Delta E$, $3\Delta E$, $4\Delta E$.

Under the conditions of the applicability of equation (6.16), which is a special case of general equation (6.5), provided that $b_k \to \infty$ (diffusion control of cathode reaction in case of corrosion with oxygen depolarization) or$b_a \to \infty$ (metal passivity):

$$I = I_{cor}\left\{\exp\left(\frac{2,3\Delta E}{b_a}\right) - 1\right\} \quad (13)$$

or

$$I = I_{cor}\left\{1 - \exp\left(\frac{-2,3\Delta E}{b_k}\right)\right\} \quad (14)$$

The calculation of b_a, b_k, $I_{cor}$is possible by (6.10)…(6.15). It is reasonable to determine the parameters of such electrode systems by two values of currents I_1 and I_2, corresponding to anode (cathode) polarization of the electrode ΔE and $2\Delta E$ (i.e. polarization of one sign). In this case:

$$I_{cor} = \frac{|I_1|}{\frac{I_2}{I_1} - 2}, \quad (15)$$

$$b = \Delta E / \lg(I_2 / I_1 - 1) \quad (16)$$

The research was conducted in a three-electrode electrochemical cell made of Pirex glass with anode and cathode spaces separated by grinding. The electrochemical measuring complex by Solartron (UK) was used, consisting of an SI 1255 impedance analyzer and a potentiostat SI 1287 (Figure2).

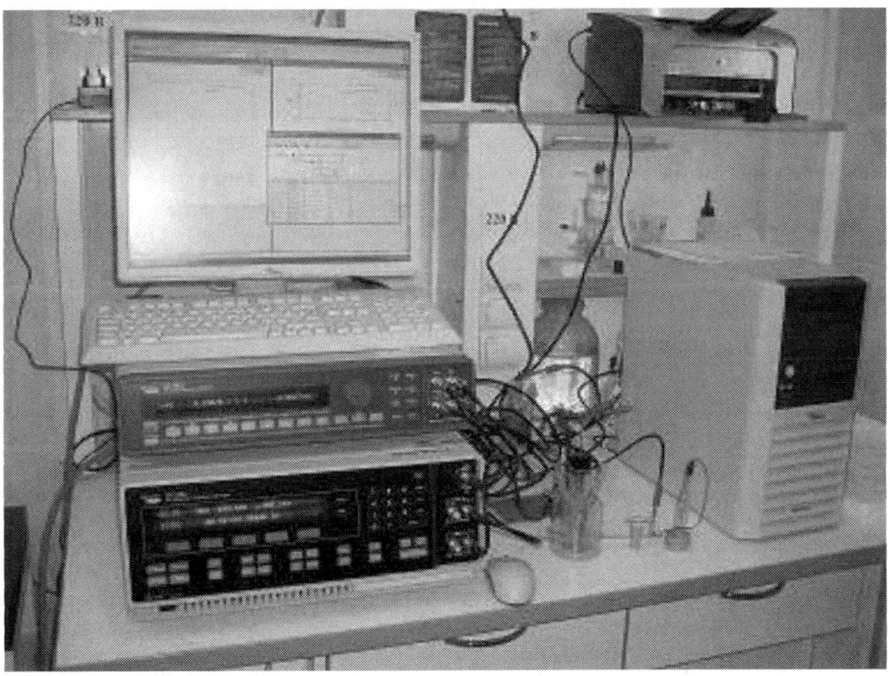

Figure2. Electrochemical measuring complex by Solartron (UK)

Working electrode - carbon steelSt.3 (composition, %: Fe – 98,36; C – 0,2; Mn – 0,5; Si – 0,15; P – 0,04: S – 0,05; Cr – 0,3; Ni – 0,2; Cu – 0,2) ($S_{el.}$ = 0,14 cm^2), reinforced into a mandrel made of epoxy resin ED-5 with polyethylene polyamine hardener. Auxiliary electrode - smooth platinum. Silver chloride electrode was used as a comparison electrode. After immersion the electrodes were kept in the solution to establish quasi-stationary potential (10...15 min). The samples were polished up to purity class 6 and degreased with acetone and alcohol in series.

IOP Conf. Series: Materials Science and Engineering 1159 (2021) 012086 doi:10.1088/1757-899X/1159/1/012086

6. Results and discussion

In the course of the research, three methods were analyzed to study the corrosion resistance of metal treated with corrosion inhibitors based on plant origin raw materials. As a result of the research carried out with these methods, a significant advantage of metals subject to processing has been revealed.

7. Conclusion

Anticorrosive properties of metals are the main assessment indicator in studies of corrosion inhibitors based on raw materials of plant origin. The received method of electrochemical research of anticorrosive firmness of the metal processed by the corrosion inhibitors created on the basis of raw materials of a plant origin, allows to define an indicator of anticorrosive firmness with high accuracy that makes it advantageous in researches of this direction.

Acknowledgements

The article was prepared as part of the research work on the topic T-1.16 «Development of technologies for obtaining highly effective protective materials against atmospheric corrosion using raw materials of plant origin and fluorine-containing surfactants» of the Program of scientific research and technological works of the Joint Russian-Vietnamese Tropical Research and Technology Center for 2020-2024.

References

[1] Gaidar S M 2015 *Planning and analysis of the experiment* (Moscow: publishing house of the federal state scientific institution Rosinformagrotech) 548 p
[2] Tsygankova L E, Vygdorovich V I, Pozdnyakov A P 2001 *Metal corrosion inhibitors* (Tambov: publishing house of the TSU) 190 p
[3] Kuznetsov Yu I 2002 Currentstateofthetheoryofmetalcorrosioninhibition *Protectionofmetals* Vol 38 **2** 122-131
[4] Gaidar S M 2017 *Technology of agricultural machinery storage: training manual* (Moscow: publishing house of the federal state scientific institution Rosinformagrotech) 216 p
[5] Gaidar S M 2011 *Theory and practice of creating corrosion inhibitors for conservation of agricultural machinery. Monograph* (Moscow: publishing house of the federal state scientific institution Rosinformagrotech) 304 p
[6] Gaidar S M, Pydrin A V, Karelina M Yu 2015 Technology of preservation of autotractor diesel engines with working-conservation composition *Rural machinery and equipment* **12** 38-42
[7] Gaidar S M, Karelina M Yu, Pydrin A V et al. 2016 *Metal corrosion inhibitor.* Patent for invention. № 2015113929/02 from 15.04.2015

IOP Conf. Series: Materials Science and Engineering 1159 (2021) 012087 doi:10.1088/1757-899X/1159/1/012087

Analysis of hydrogen use as an energy carrier in transport

A V Keller[1], K E Karpukhin[1,2], A F Kolbasov[2], V N Kozlov[2]

[1] MADI, 64, Leningradsky ave, Moscow, 125319, Russia
[2] Central Scientific Research Automobile and Automotive Engines Institute «NAMI», 2, Avtomotornaya str., Moscow, 125438, Russia

E-mail: K.Karpukhin@nami.ru

Abstract. The subject of the analytical review is the power units of electric vehicles using hydrogen fuel cells. A prerequisite of the research is the ever-increasing urgency of developing vehicles with zero emission of carbon dioxide and other harmful substances. Materials are taken both from open sources of information (Internet) and from scientifically reliable sources of information (peer-reviewed publications). The scientific methods of performing this analysis include an informative review of sources. Originality lies in the choice of the topic and the findings from the analysis. As a result of the analytical review the tendencies of the automobile market development have been revealed; the technical advantages of hydrogen fuel cells as compared to batteries have been given; the conclusions have been drawn about the availability of the programs of carbon dioxide emission reduction in the European Union, the USA, Japan, China, Korea as well as about the readiness of the technology of the fuel cells use in the traction electric drive for the introduction, including the patenting of the technologies by the leading automobile manufacturers; the most promising segments of transport for this purpose have been described. The results of the analysis are applicable to the design of electric vehicles and their power plants by vehicle designers and manufacturers.

1. Introduction

Currently, one of the main global trends in the development of the automotive industry is the development of environmentally friendly vehicles, since emissions from vehicles are a serious environmental problem, especially in megacities. The content of CO_2 has now reached 0.042% in the vicinity of St. Petersburg and 0.083% in Moscow and continues to increase. In Moscow, cars are the main source of carbon dioxide emissions (85% of all emissions from CO_2 in 1998 and about 92% in 1999). The Intergovernmental Panel on Climate Change (IPCC) has repeatedly reported on the link between global warming and increasing concentrations CO_2 (1990, 1995, 2001, 2007, 2014). In 1992, the Kyoto Protocol was adopted - an international agreement, an additional document to the UN Framework Convention on Climate Change, adopted in Kyoto (Japan). It commits developed countries and countries with economies in transition to reduce or stabilize greenhouse gas emissions.

In this situation, one of the measures to partially reduce the content of CO_2 and other harmful substances, including carcinogenic ones, is the transition of motor vehicles from internal combustion engines to electric drives or combined power plants (CPP) in large cities. However, while electric cars have really zero emissions, cars with a CPP have only partially reduced emissions. In addition, safety standards for cars with internal combustion engines are constantly revised towards stricter standards for harmful emissions in exhaust gases. For example, a number of environmental standards Euro 0 → Euro

Content from this work may be used under the terms of the Creative Commons Attribution 3.0 licence. Any further distribution of this work must maintain attribution to the author(s) and the title of the work, journal citation and DOI.
Published under licence by IOP Publishing Ltd

1 → Euro 2 → Euro 3 → Euro 4 → Euro 5 → Euro 6. Our country has adopted the Euro 5 standard and is mastering Euro 6 fuel production technologies.

One of the technical solutions with zero emission of CO_2 and other harmful substances is the use of hydrogen fuel cells (HFCs) on electric vehicles.

2. Cars on HFCs

Hydrogen fuel cells as an electric power source for motor transport are one of the main directions of energy efficient motor transport development in the world [1]. Modern and future-oriented intelligent transport is increasingly in need of power through the introduction of additional systems such as ADAS (Advanced Driver Assistance Systems) or multimedia [2,3]. In the USSR and Russia the development of the following prototype vehicles on hydrogen fuel cells was carried out:

- hydrogen minibus KVANT-RAF developed by "Kvant" Research and Production Association and "RAF" plant, with an alkaline fuel cell battery with capacity of 2 kW and nickel-zinc battery with a capacity of 5 kWh. (1982),
- models of passenger cars "AvtoVAZ" (based on VAZ-2131 "Niva" and VAZ-2111), "Electric Transport Technologies" LLC with the participation of specialists from the Center of Competences of National Technical Initiative Institute of Chemical Physics of the Russian Academy of Sciences and "InEnergy" LLC,
- commercial truck "NAMI" with a 25 kW fuel cell battery, developed in St. Petersburg (Central Research Institute of Marine Electrical Engineering and Technology).

Then the demonstration of passenger cars on hydrogen fuel cells became a frequent phenomenon. The total number of passenger car models on hydrogen fuel cells in the period 1990-2017 is 141 pcs. (Figure 1). Detailed information on the models can be found on the site at the link [4].

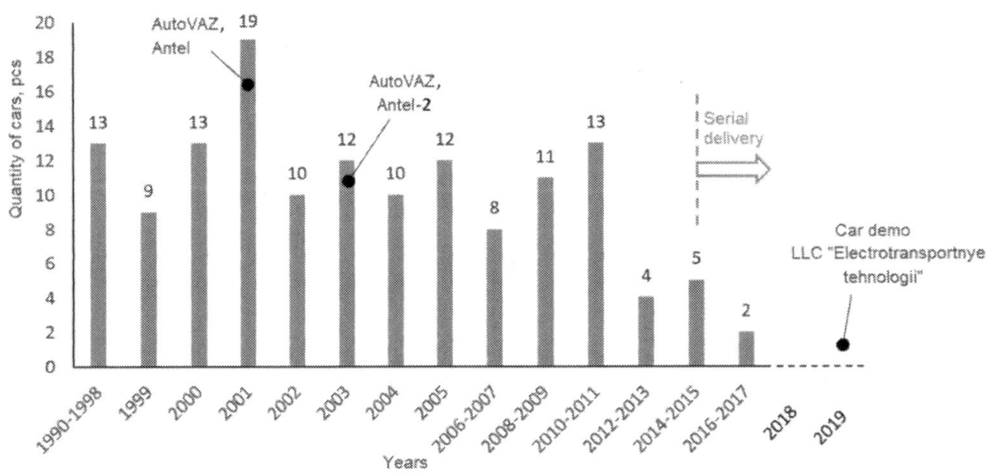

Figure 1. Number of hydrogen fuel cell passenger car models by years. Almost all models are demo models. Exceptions are the models presented in Table 1, which are also taken into account in the figure.

Almost all leading car manufacturers presented samples of passenger cars on fuel cells at different times. 2014-2015 can be considered as the beginning of commercialization (beginning of serial shipments of Mirai in Japan). The companies that have transitioned to serial production are presented in Table 1.

IOP Conf. Series: Materials Science and Engineering 1159 (2021) 012087 doi:10.1088/1757-899X/1159/1/012087

Table 1. Mass production of passenger cars on fuel cells

Manufacturer, model	Status
Honda FCX Clarity	Manufacture in Japan, 2016 – Series shipments to USA (California)
Toyota. Mirai FCHV-adv (2008)	2010 – Beginning of a three-year demonstration trial of one hundred specimens on US roads. 2014 – Mirai serial supply in Japan. 2017 – Mirai serial supplies to USA. 2017 – global sales amounted to 5300 Mirai (the USA – 2900 units, Japan – 2100 and Europe – 200)
Mercedes-Benz GLC F-CELL	2010 – limited edition production (500 copies) of Mercedes-Benz A-class F-Cell for European consumers. 2019, spring – market launch.
Hyundai ix35 FCEV	2013 – production start

Other leading automobile companies have developed fuel cell car concepts. Among them are: Alfa Romeo, Audi, BMW, Chang'an Automobile (Group) Co., Ltd., FAW Group, Chrysler, Daimler, Fiat, Ford, General Motors, Kia, Mazda, Mitsubishi, Peugeot, Renault, Subaru, Suzuki, Volkswagen.

In Russia, fuel cell cars are represented by demonstration models developed by JSC "AVTOVAZ": Antel based on VAZ-2131 "Niva" (2001) and Antel-2 based on VAZ-2111 (2003), in which alkaline fuel cells were used. One can also add a demonstration of the car in October 2019 at the forum "Open Innovations" (Figure 2).

(a) (b)

Figure 2. Fuel cell car (a), developed by " Electric Vehicle Technologies" LLC, 2019. Label on Car: «Electric Car on Hydrogen». The fuel cell system and hydrogen cylinders are located in the trunk of the car (b).

The car was developed by "Electric Vehicle Technologies" LLC with the participation of specialists from the Center of Competences of National Technical Initiative Institute of Chemical Physics of the Russian Academy of Sciences and "InEnergy" LLC [5]. The basis for this is the electric car Lada Ellada, produced in 2012 by AvtoVAZ in the amount of 100 pieces. The car was re-equipped with a hybrid power unit including a 24 kWh battery and a fuel cell battery. The fuel cell system with hydrogen cylinders occupies the entire trunk, which will not be approved by the consumer. This solution is unlikely to have a future. The reconfiguration will mean the development of a new car.

One of the promising segments for commercialization of hydrogen passenger cars is the taxi sector. For example, the European division of Toyota supplied 35 Mirai electric cars for the passenger transport company Noot Personenvervoer. Thus, in the Netherlands (The Hague) the first hydrogen taximotor park appeared [6]. Eco Taxi Deutschland GmbH in Wiesbaden has added two Hyundai ix35 FCEV on fuel cells to its park. Commercialization through the taxi segment is justified, as in the case of one owner it is easier to organize the operation of a new type of car, especially if the owner has the competence to service the car.

3. Trucks

Some truck manufacturers also have concepts working on HFCs (Hyundai, Mercedes-Benz, Ford, Skoda, Tata, Toyota). At this stage, manufacturers have not yet entered the stage of serial production and are conducting trial operation of the vehicles under development. The number of samples to be tested is calculated in tens in each case. At the same time, there is information about immediate plans for their wider deployment. For example, Norway aims to deploy 1,000 hydrogen fuel cell trucks by 2023.

The French project ROAD is testing a refrigerator on fuel cells. The project presented the world's first hydraulically driven refrigerated semi-trailer (coupled truck), which entered French roads for the first time on 4 July 2019 for testing [7].

Hydrospider intends to produce "green" hydrogen in Hessen, using hydroelectric power to supply the first 50 fuel-cell trucks for use on Swiss roads by 2025. The total planned number of trucks is 1600. Trucks will be produced by Hyundai Hydrogen Mobility, a joint venture between South Korean car manufacturer Hyundai Motor Company and H2 Energy [8].

The Singapore-based company Horizon Fuel Cell Technologies has signed a Memorandum of Understanding for the supply of 1,000 fuel cell batteries with a capacity of 100 kW or more (each battery) for heavy-duty trucks over the next three years, beginning in 2019. This is one of the largest deployments of heavy fuel cell vehicles in the world [9].

4. Forklifts and load carts

The most dynamic direction of fuel cell commercialization is loading and unloading equipment and machinery for moving cargoes in large logistics centers and warehouses. Such equipment includes 3- or 4-wheel forklifts or small load carts for transporting goods only horizontally. The power range of power units for this equipment is relatively small and amounts to 5-30 kW.

The first prototype of a forklift truck on a hydrogen fuel cell was created in 2000. In 2013 more than 4000 fuel cell forklift trucks were already used in the USA. In 2019, 23,000 forklifts were used. This segment is also developing in Europe and Japan.

(a)

(b)

Figure 3. Photo of forklift (a) and traction controlled cart (b).

IOP Conf. Series: Materials Science and Engineering 1159 (2021) 012087 doi:10.1088/1757-899X/1159/1/012087

The size of the fuel cell battery is almost identical to the size of the storage batteries. That's why HFCs can easily fit in the loader design.

The cost of a separate fuel cell forklift is higher than that of a battery-operated forklift. Economic efficiency is manifested when comparing the cost of maintaining the fleet of a certain number of units. An evaluation by the U.S. Department of Energy's National Renewable Energy Laboratory (note: evaluation was conducted in 2013) shows that with intensive operation (2-3 shifts per day, 6-7 days per week), the cost of maintaining 60 fuel cell loaders is lower than maintaining the same number of battery-powered loaders with comparable characteristics [10].

To fill the batteries of fuel cells with hydrogen, a hydrogen filling station shall be installed at the place of their use.

Advantages of fuel cell technology (as compared to batteries):

- filling time is 3-5 minutes;
- one refill provides a full shift (8-10 hours);
- longer service life than batteries (approximately twice as long);
- filling space is 75% smaller;
- 15% increase in productivity;
- power saving up to full fuel consumption;
- maintaining power at low temperatures (declared to -30 °C), which is important when working in open space (at low temperatures, battery capacity drops by 50% or more);
- no risk of acid spill.

The use of forklifts is appropriate in large logistics centers, where the fleet of forklifts is relatively large.

In July, the famous American giant Walmart acquired 55 million shares of one of the pioneers of hydrogen energy - the company Plug Power, announcing plans to equip its 30 centers with hydrogen fuelling stations for loaders. Now 22 American Walmart centers are equipped with such filling stations. In April this year, Amazon.com also bought more than 50 million shares of Plug Power and began to equip its warehouses with hydrogen fuelling [11].

5. Buses

In recent years, the world has been actively developing prototypes and small-scale models of buses that use the energy unit based on fuel cells as the main source of energy for driving the engine and bus auxiliary systems.

The hydrogen bus H2 City Gold (Figure 4) was developed by two companies. Toyota, actively developing hydrogen engines, was responsible for the power plant, and the rest was done by the Portuguese company CaetanoBus, engaged in the design and manufacture of such equipment.

Geely brand subsidiary, New Energy Commercial Vehicle Group (GCV), Yuan Cheng, introduced its F12 hydrogen fuel cell city bus and the C11 rechargeable public transit bus. Both are zero emission vehicles, with the hydrogen bus being the first for Geely.

As for the bus Toyota SORA Fuel Cell Bus 4K 2 HD, the acronym SORA symbolizes the water cycle in nature: Sky, Ocean, River and Air. Compressed hydrogen is in 10 cylinders with a total capacity of 600 liters under pressure of 70 MPa in the front of the roof. Electricity is stored in nickel-metal hydride batteries, which are located on the roof, just behind the cylinders.

Xcelsior CHARGE H2. It goes up to 300 miles after refueling without recharging the onboard electricity. Fast filling time varies from 6-20 minutes (depending on the bus model and operating conditions). It is powered by Ballard fuel cells. Regenerative deceleration, which recharges the batteries and reduces power consumption, as well as extends the range of action.

Wrightbus Street deck FCEV. The bus works on fuel cell Ballard FCveloCity, Siemens transmission and traction battery with a capacity of 48 kW. The fuel cell system on Streetdeck FCEV is manufactured by Ballard Power Systems. Operating range is 200 miles and the extended storage option extends the range up to 265 miles. Filling the bus with hydrogen takes about 7 minutes.

Figure 4. Hydrogen bus H2. City Gold

6. Other examples

Abundance of scooters in the Asian region is a source of environmental pollution. In Taiwan, the company "APFCT" has successfully implemented a program of fuel cell scooters production. The Taiwanese Ministry of Economic Affairs has issued reliability and safety standards for hydrogen fuel cell motorcycles to regulate mass production of such vehicles. The park of 80 scooters was tested in Pindong County. Twenty scooters were meant for the local government, the remaining 60 were distributed among 17 hotels in Kenting National Park to provide visitors with rental. Each site is equipped with a point of replacement of hydrogen tanks. Company "APFCT" is working on a scooter project in Hawaii. Hawaii is 90% dependent on imported fuel, and vehicles using fuel derived from solar electrification are an attractive option [12].

7. Conclusion

The development of hydrogen transport in Europe is carried out within the framework of the European Hydrogen Road Map [13], which considers the structural transformation of the energy sector as a whole. According to the road map, transport makes a major contribution to climate change by emitting 32% of emissions CO_2 in the European Union. According to the scenario, the region should reduce CO_2 emissions from transport by 72% by 2050, which is approximately 82 Mt.. This requires a transition from hydrocarbon fuel to renewable energy sources, which not only requires the introduction of new energy units in vehicles, but also a fundamental change in the entire energy supply chain.

There are similar programs in the United States, Japan, China and the Republic of Korea.

Deployment of fuel-cell vehicles in the near future is the most attractive for commercial taxi- and truck fleets used within municipalities, where their advantages are most relevant and the obstacles to infrastructure are lower than in other segments. Commercialization of forklifts is already underway.

Intensive and continuous investment in R&D has made hydrogen transport technologies ready for implementation. Currently there are five models of fuel cell cars on the market, and another 25 models will be announced within the next five years. Some manufacturers announced mass production of trucks with fuel cells. Two of them announced their entry into the European market.

Fuel cell power plants appear on the railway, river and sea transport.

Leading companies have the necessary package of patented technologies both in the field of vehicles and filling stations. The leaders are Japan, USA, China, Europe and Republic of Korea.

References

[1] Tuan N Kh, Karpukhin K E, Terenchenko A S, Kolbasov A F 2018 World trends in the development of vehicles with alternative energy sources *ARPN Journal of Engineering and Applied Sciences* **13 #7** 2535-2542

[2] Karpukhin K, Terenchenko A, Kolbasov A 2018 The creation of modern electric vehicles with additional source of energy *Earth and Environmental Science* 012043

[3] Poddubnaya A A, Keller A V 2019 "Forecast" - adaptive forecast diagnostic intelligent system for vehicles *Materials Science and Engineering. International Automobile Scientific Forum, IASF 2019 "Technologies and Components of Land Intelligent Transport Systems"* (Moscow: FSUE NAMI) 012008

[4] Hydrogen Cars Now on http://www.hydrogencarsnow.com/

[5] Lada Kalina rides on hydrogen on https://rg.ru/2019/10/29/lada-kalina-poehala-na-vodorode.html

[6] Ballard buys BC's first fleet of FCEVs with Toyota Mirai purchase 2019 (Singapore: Fuel Cells Bulletin) **Vol.2019** Iss.8 p 2 DOI: 10.1016/S1464-2859(19)30311-6

[7] French ROAD project tests fuel cell powered refrigerated semi-trailer 2019 (Singapore: Fuel Cells Bulletin) **Vol.2019** Iss.8 p 3-4 DOI: 10.1016/S1464-2859(19)30315-3

[8] Hydrospider plans first Swiss commercial hydrogen production 2019 (Singapore: Fuel Cells Bulletin) **Vol.2019** Iss.8 p 9 DOI: 10.1016/S1464-2859(19)30333-5

[9] Horizon to power 1000 fuel cell heavy vehicles in port operations 2019 (Singapore: Fuel Cells Bulletin) **Vol.2019** Iss.8 p 10 DOI: 10.1016/S1464-2859(19)30336-0

[10] Ramsden T An Evaluation of the Total Cost of Ownership of Fuel Cell-Powered Material Handling Equipment 2013 (Denver: National Renewable Energy Laboratory)

[11] Nosyrev I Hydrogen deficiency: what prevents the promotion of cars on light gas on https://www.rbc.ru/own_business/02/08/2017/5979d1ac9a7947802604085f

[12] Berzigiyarov P K, Tarasov B P etc. Development and creation of hydrogen system of reserve power supply and energy storage 2014 (Chernogolovka: Institute of Problems of Chemical Physics of Russian Academy of Sciences)

[13] Hydrogen roadmap Europe. A sustainable pathway for the European energy transition 2019 *Fuel Cells and Hydrogen 2 Joint Undertaking ed.1* (Luxembourg: Publication Office of the European Union) DOI: 10.2843/341510

Russian national non-commercial vehicle safety rating system RuNCAP

A M Ivanov[1], S R Kristalniy[1], N V Popov[1]

[1] Moscow Automobile and Road Construction State Technical University (MADI), Moscow

E-mail: ivanov-am@madi.ru, sportauto@rambler.ru, niko-popov@yandex.ru

Abstract. The article is devoted to the need to introduce an independent rating system for assessing the safety of cars RuNCAP in the Russian Federation. The introduction of this system will make it possible to more fully take into account the Russian operating conditions when assigning ratings to passive and active safety. RuNCAP will contribute to the wider adoption of intelligent driver assistance systems and their efficiency. The development of RuNCAP is proposed to be carried out taking into account the analysis of the features of the accident statistics for the Russian Federation. The article also presents the development strategy of RuNCAP for the coming years.

1. Introduction

The car is a source of increased danger. Therefore, it is subject to mandatory certification. In different countries, certification is carried out in different ways, but the basic principle remains the same: certification sets the **minimum** requirements for the safety of the structure and includes checks of cars for compliance with international and national regulatory documents.

In addition to mandatory certification, since 1997, independent rating systems for assessing vehicle safety have appeared. First of all, they began to check and evaluate the achieved level of passive safety. With the advent of electronic systems for optimizing tire adhesion, such as anti-lock braking system (ABS), electronic stability control (ESC) or traction control system (TCS) etc., they began to take into account the active safety aspects of vehicles in their overall rating.

The most famous safety rating systems are EuroNCAP (The European New Car Assessment Programme [1]), ANCAP (assessment program for new vehicles sold in Australia and New Zealand), LatinNCAP (assessment program for new cars sold in Latin America and the Caribbean), ASEANNCAP (new car assessment program for Southeast Asia), JNCAP (Japanese new car assessment program), GlobalNCAP (international platform for collaboration between new vehicle assessment programs). It is obvious that there is no single rating system in the world. National and regional ratings have some similarities, but, at the same time, they are purely individual, suitable for the respective market.

In addition to safety assessment systems, there are also complex systems for assessing the consumer properties of cars, used, in particular, by journalists of popular automotive publications. However, the results obtained significantly depend on the conditions of the tests, methods, techniques and applied comparison criteria.

For different countries, these conditions, methods and criteria may differ significantly. In particular, the test conditions for India and Russia clearly cannot be the same. The most popular

Content from this work may be used under the terms of the Creative Commons Attribution 3.0 licence. Any further distribution of this work must maintain attribution to the author(s) and the title of the work, journal citation and DOI.
Published under licence by IOP Publishing Ltd

European rating system for assessing the safety of new cars EuroNCAP [1] also cannot be applied in Russia or can be applied with a large number of limitations.

2. Organization of RuNCAP

Attempts to create Russian methods and test methods have been made for a long time. In particular, attempts are known to create techniques for the publication of "Autoreview" to determine the level of passive safety of cars (ARCAP) [2], resource, fuel consumption and other parameters.

In November 2018, an agreement was signed between the Central Scientific Research Automobile and Automotive Engines Institute NAMI, the Federal Agency for Technical Regulation and Metrology (Rosstandart), "Autoreview" LLC and the Federal State Budgetary Educational Institution of Higher Education "MADI" on the creation of a consortium to form a working group to develop a Russian rating system for assessing vehicle safety. The new structure, called RuNCAP, will have to eliminate the shortcomings inherent in the existing assessment systems, as well as take into account the Russian characteristics of operation and factors of interest to the Russian consumer.

At the same time, the European and American experience of such tests will in no way be denied, but will be used in full, taking into account the adjustment of the conditions and the elimination of the identified shortcomings [9, 10].

The development of RuNCAP is supposed to start with systems that characterize the safety of a car, without pretending at the first stage to assess its consumer qualities.

Work will be carried out on a new alignment of emphasis and taking into account the weight of the elements that affect the overall security picture, which are specific to Russia. In particular, special attention will be paid to the functioning of certain safety elements in the dark, at negative air temperatures, on unlit roads, in mud, rain, snowfall, fog etc.

At the same time, it is unacceptable to sacrifice the objectivity and reliability of these assessments. And this is the main difficulty in creating test procedures. Striving for absolute objectivity leads to costs for the development of the testing and metrological base. The development of RuNCAP will require a revision and significant expansion of the testing base of the NAMI test center.

3. Analysis of accident statistics in Russia

Russia has completely unique statistics of road accidents, not at all the same as in Europe or the USA. Based on the statistics of road accidents, which began to be published on the website www.stat.gibdd.ru [3], methods and test conditions will be developed that are specific to Russia, and also, possibly, to its different regions separately. An example of the analysis of such statistics, carried out using a specially developed technique on a representative sample, is shown in Fig. 1. It can be noted that a relatively large number of accidents with injured and dead occurs in severe road and climatic conditions.

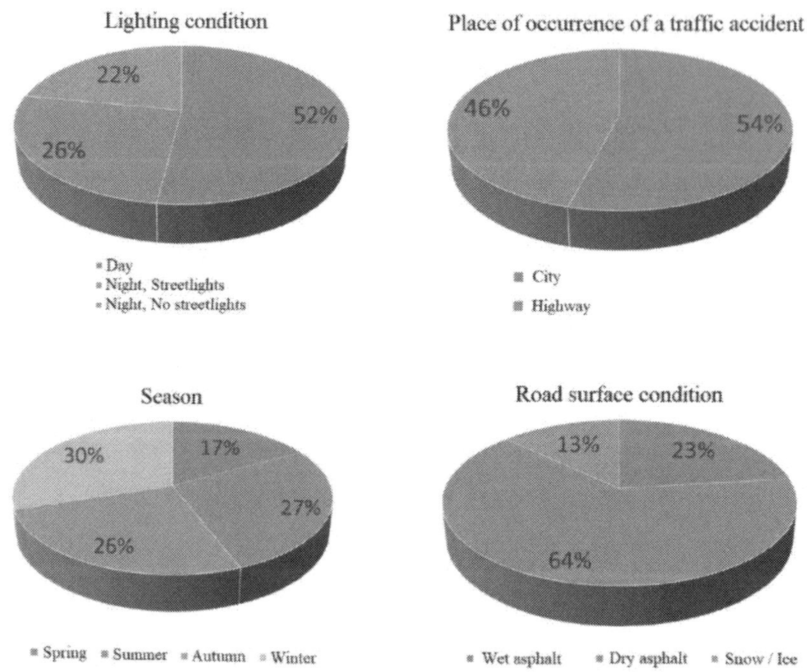

Figure 1. Analysis of traffic accident statistics on the territory of the Russian Federation in 2018 (type of accident - hitting a stationary obstacle)

4. Test procedure of AEBS by RuNCAP

The first fruit of the activities of the RuNCAP working group was the emergence of a method for determining the effectiveness of the automatic emergency braking systems (AEBS). Research tests were carried out to refine the technique. [4, 5, 6, 7]. The effectiveness of the AEBS is checked in the daytime and at night using a special target that simulates the rear of the car (Fig. 2). The test result is evaluated in credit points. During the test drive, the vehicle's movement parameters are recorded, video recording of collision warning signals and video recording of braking are also carried out (Fig. 3). This technique can be found in detail on the website of the Association of Automotive Engineers (AAE) [8].

Figure 2. Test maneuver according to RuNCAP method (day lighting conditions)

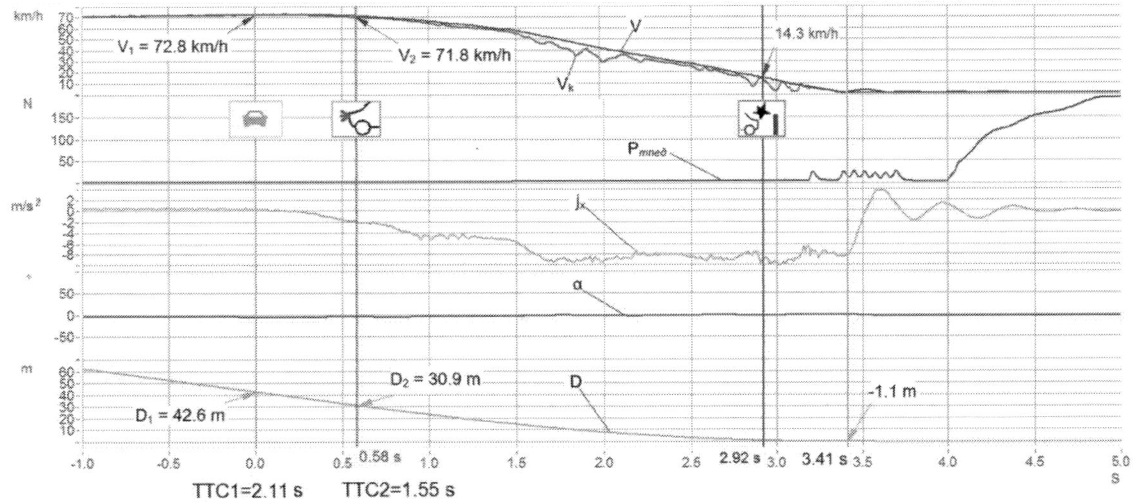

Figure 3. An example of parametric record of test run (the condition of area surface- is dry asphalt, the speed of approach is 70 km / h, light conditions are "Day" type):

V_1 - is longitudinal velocity of an object at time of possible collision warning moment;
V_2 - is an object speed at time of collision with the "target";
D_1 - the distance between the object and the "target" at warning time of a possible collision;
D_2 - the distance between the object and the "target" at the time of automatic braking start;
TTC1 - time before collision at possible collision warning moment;
TTC2 - time before collision at the moment of automatic braking beginning;
 - the moment of visual signal switch on during the detection of an obstacle (orange bar);
 - the moment of switching on brake signals on test object (red vertical line);
 - the moment of collision of an object with a "target" (blue vertical bar).

5. Development strategy of RuNCAP

In the future, the RuNCAP rating system will be supplemented by taking into account the effectiveness of the AEBS on other types of targets (pedestrians, cyclists), as well as, in the longer term, in difficult road conditions (surfaces with a low adhesion coefficient, rain, snow, mud, fog, glare headlights, etc.). In the future, it is planned to evaluate the effectiveness of parking assistance systems, adaptive cruise control systems etc.

Also, a crash test (a series of crash tests) will be introduced into the RuNCAP structure, the parameters of which are currently being discussed and agreed by the interested parties. The conditions for conducting crash tests are selected taking into account the statistical data published by the traffic police in open sources [3]. Also, additional, specially collected statistical data are involved in the work to form and substantiate the parameters of the passive safety tests.

In contrast to the certification system, the RuNCAP rating system shows not only that the vehicle has passed the minimum safety threshold of the structure. It will allow one to compare the actual achieved level of safety of different vehicles. In the future, using RuNCAP, it will also be possible to assess the consumer properties of vehicles.

With the development of vehicle design, RuNCAP will improve in harmony with the entire automotive industry and science. With the advent of new driver assistance systems, as well as unmanned technologies [6], the independent rating system will also be improved. Also, as the road situation changes, the emphasis in safety assessment will shift, the weight of certain elements will change. RuNCAP will become a "living" system that will constantly be changing.

Significant investments are required to launch this system. The most likely source of these investments may be the state, since the widespread use of RuNCAP meets the goals and objectives of the state. However, one cannot but note the contradiction, which consists in the lack of independence of the assessment due to the participation of the state in the management of RuNCAP. Here the role of professional associations comes to the fore, and, in particular, the Association of Automotive Engineers (AAE). At first, RuNCAP will exist as a committee at the AAE.

6. Discussion

The first RuNCAP test programs are published on the AAE website [8] (including in English), and everyone is invited to participate in their discussion and change.

The first test results of the AEBS with the assignment of a rating score according to RuNCAP were published in the popular Russian magazine "Autoreview" [11].

Manufacturers considering RuNCAP test conditions will naturally have some advantage.

Automakers and importers may oppose RuNCAP because it will force them to carry out additional tests and, accordingly, incur additional financial burden. These costs, in turn, will inevitably be transformed into an increase in the price (prime cost) of the car.

However, we encourage carmakers to familiarize themselves with the programs and methodologies and engage in constructive discussions. A broad constructive discussion of RuNCAP will enable its developers to take into account the opinion of manufacturers, as well as to promptly eliminate possible errors and inaccuracies.

Manufacturers could also share their experience in conducting their assessments of the safety and consumer properties of cars, for example, in the framework of AAE conferences.

The development of RuNCAP will require an expansion of the test base, including for the evaluation of unmanned vehicles, ADAS systems and new passive safety systems. And the main technical service for this will be the NAMI test center.

7. Conclusion

The creation of RuNCAP is already a fait accompli, and the only question remains: in what form will its development take place. RuNCAP will allow one to get objective ratings of car safety and create conditions for reaching a new level of marketing development in the automotive industry. Thus, RuNCAP will claim its role in the distribution and self-regulation of the non-commercial vehicle market. RuNCAP development is definitely worth the planned costs.

References

[1] *About Euro NCAP* Available at: https://www.euroncap.com/en/about-euro-ncap/
[2] *ARCAP, Autoreview Car Assessment Program* Available at: https://autoreview.ru/arcap/about
[3] *Road safety indicators* – Available at: http://stat.gibdd.ru
[4] Ivanov A M, Shadrin S S, Kristalniy S R, Popov N V 2019 *IOP Conference Series: Materials Science and Engineering* 4–12
[5] Malinovsky M P 2017 *NAMI Works* **1 (268)** 53–59
[6] Ivanov A M., Kristalny S R , Popov N V, Toporkov M A, Isakova M I 2018 *Proceedings of the NSTU. R.E. Alekseeva* **2(121)** 146–155
[7] Ivanov A M, Kristalniy S R, Popov N V, Toporkov M A, Isakova M I 2018 *Their application // IOP Conference Series: Materials Science and Engineering* Vol 386 **1** 012019
[8] *RuNCAP Method* – Available at: http://www.autoengineer.org/runcap1/metodika.html
[9] OCAS-DRI-CIB-16-08 *New car assessment program. Crash imminent braking system confirmation test. 2016 Nissan Rogue. Final Report* U. S. Department of transportation 108
[10] CAS-DRI-CIB-17-06 *New car assessment program. Crash imminent braking system confirmation test. 2017 Mitsubishi Outlander. Final Report.* 2017 U. S. Department of transportation 120
[11] *RuNCAP: reloading our rating ARCAP* Available at: https://autoreview.ru/articles/supertest/runcap

Modeling of systems for restoration of parts of cars, other vehicles and road construction machines

M Yu Karelina[1], A I Belyaev[2], A V Terentev[2], V R Rogov[1]

[1]Moscow Automobile and Road State Technical University (MADI), Leningradsky Prospekt 64, Moscow, Russia
[2]Saint Petersburg State University of Architecture and Civil Engineering, 4 2nd Krasnoarmeyskaya str., Saint Petersburg, 190005, Russia

E-mail: karelinamu@mail.ru

Abstract. The article deals with the application of process typification in the restoration of the most worn-out parts of cars, other vehicles and road construction machines from the point of view of application of optimization of the best quality parameters selection. A typical technological process of part recovery using the solution of a multi-criteria dynamic problem by the zoning method has been proposed with the introduction of the best variant of the matrix technology into the algorithm of selection: W "wear resistance → durability"; S "safety → reliability"; F "part fragility → machine safety". The goal is to achieve maximum efficiency in structuring the technological process of restoration of parts of cars, other vehicles and road construction machines according to a number of reasonable criteria. While solving the task, the analysis of the existing methods of modeling for solving such tasks has been performed. The authors proposed their own solution in the form of an algorithm of the zoning method based on the principle of compliance with the hierarchical correlation of probabilities of possible states of the external environment in order to select an optimal variant of the required solution. As a result, it has been found that obtaining the maximum efficiency of the technological process of shaft-type parts restoration in cars, other vehicles and road construction machines can be solved by the task of dynamic programming as determining the optimal trajectory of the part transitions during the restoration process.

1. Introduction

In the analysis of operating and maintenance conditions of automobiles, other vehicles and road construction machines has identified a number of reasons why the relevance of the development of metrology for the formation of a system of technology selection that increases the efficiency of restoration of road and construction machines is justified. Restoration and repair facilities are focused on poor quality of restoration, as they are limited by the typical set of equipment for restoration or lack of equipment. When restoring parts, the choice of technology is based on the availability of equipment and technology. Faulty parts are very expensive and sophisticated.

Since cars, vehicles and road construction machines are operated under special conditions, the parts of these machines also experience special conditions. The nature of failure of these parts has a specific appearance, and methods of their restoration should also be specific.

The development of methodology for the formation of the technological process of restoration of automobiles, other vehicles and road construction machinery, considering the dominant group of

Content from this work may be used under the terms of the Creative Commons Attribution 3.0 licence. Any further distribution of this work must maintain attribution to the author(s) and the title of the work, journal citation and DOI.
Published under licence by IOP Publishing Ltd

IOP Conf. Series: Materials Science and Engineering 1159 (2021) 012089 doi:10.1088/1757-899X/1159/1/012089

factors determined by their nature, becomes relevant. In turn, the nature of the factors is determined by the area of their influence on quality indicators (reliability, safety of operation). At present, the importance of determining the sustainable characteristics of these areas is being actualized by the specifics of the cars, vehicles and road construction machinery themselves, the area of their use and infrastructure, and the production and technical base of services.

2. Analytical model of optimization of the route selection of the technological process of restoration of functionally important parts of cars, vehicles and road construction machines

Let us present the methodology developed on the basis of the decisions made [1-10]. This method has been tested in the form of three routes with different content of operations. This study examines a technological process that, according to the model, has the following hierarchy:
In block 1 the formation of source data, the variants of routes for restoring a specific functionally important part - the shaft - are compiled.

An important role in the technological process is played by the detection and evaluation of wear and tear – fault detection of parts. As a result of fault detection, some parts are recognized as fit and sent for assembly, and some are to be repaired or replaced.

These variants are chosen on the basis of the rules of construction of technological processes of restoring machine parts. Usually the restoration technology is developed only on the basis of available equipment and technologies of the given enterprise or plant. It is generally accepted that restoration of details at the enterprises is carried out according to the developed technological process which should provide high quality of carried out works at the least expenses. The proposed developed model allows designing the restoration technology very flexibly.

The mathematical model makes it possible to consider the following when developing the options: workload or inaction of the equipment and technologies of a particular restoration facility; possibility to attract and use the equipment and technologies of the enterprises, which can be used by cooperation. Therefore, hereinafter we present variant routes based on the preconditions of various possibilities. The assumptions in route construction boil down to the fact that the number of operations selected for the example is quite diverse, and the number of operations selected is equal for ease of understanding the example. However, in real production the number of operations and the number of routes can be any, depending on the capabilities of enterprises.

In the abovementioned control system, when choosing the optimal solution of the technological process, the tasks are presented as a multilevel hierarchical system. At each level, different tasks are solved according to their criteria. The information generated at the *k*-th level, goes to the (*k*+1)-th level, where it is used to solve other tasks by other criteria etc. As a rule, when moving from level to level, the number of tasks to be solved decreases, but their importance and complexity considerably increases.

The criteria of the tasks of the *k*-th level must be consistent with the interests of the (*k*+1)-th management level. This fact often helps in formulating recommendations in a multi-criteria environment. Transition to a higher level of management allows to solve one more problem - correct evaluation of coefficients of relative importance in composite criteria. Let us present the system as a whole

$$K = \sum_{j=1}^{m} c_{m_k n_k}^{k_p} k_{m_k n_k}^{k_p} \rightarrow opt \qquad (1)$$

where $c_{m_k n_k}^{k_p}$ is the "weight", or the importance of some performance indicator;

$k_{m_k n_k}^{k_p}$ – performance indicator value.

The lowest level usually contains the simplest operational control tasks (the requirement to obtain the necessary properties - roughness, hardness, etc.), reflected by the index:

$j = (1,\ldots,m)$ for each individual route;

Block 2 contains the digitization of the source data, here occurs the transition from the content part of the technology to the values in quantitative estimates, subsequently determining the desired properties of the part.

We will apply expression (1) to solve a specific problem by the method that confirms the chosen solution

$$\sum_{j=1}^{m} c_{m_k n_k}^{k_p} k_{m_k n_k}^{k_p} \rightarrow \max, \tag{2}$$

where in the criterium $k_{m_k n_k}^{k_p}$ the coefficients of the level index have the following correspondences.:

in the coefficient k_p the value k is the route number of the restoration technology; the index p of the coefficient k indicates the property of the part.

The essence of the further solution is described below. We know the solution from the source [11-21]. Let the matrix of efficiency have the following form:

$$\left\| a_{ij} \right\| = \begin{pmatrix} a_{11} & a_{12} & \ldots & a_{1n} \\ a_{21} & a_{22} & \ldots & a_{2n} \\ \ldots & \ldots & \ldots & \ldots \\ a_{m1} & a_{m2} & \ldots & a_{mn} \end{pmatrix} \tag{3}$$

If it is necessary to present everything in relative units and at the same time to reflect the fact that part of the indicators strive for the maximum and part - for the minimum, a new matrix B (b_{ij}) is formed from the numerical values calculated by the following formulas:

$$b_{ij} = \begin{cases} \dfrac{a_{ij}}{\max_{1\langle\langle i \langle\langle m} a_{ij}}, if \text{ the } j-th \text{ indicator is maximised} \\ \dfrac{\max_{1\langle\langle i \langle\langle m} a_{ij}}{a_{ij}}, if \text{ the } j-th \text{ indicator is minimised.} \end{cases} \tag{4}$$

Block 3 is the data aggregation block.

The criterion $k_{m_k n_k}^{k_p}$ of the coefficient of level index from the above expression (2) has the following correspondences in this calculation: in the coefficient m_k the quality criteria are numbered with the value m and cover their integral value (1-4); in the coefficient n_k the value n determines the number of the operation in the sequence, in the given route.

The ratio of mathematical expectation to the criterion $k_{m_k n_k}^{k_p}$ introduces a new criterion $K_{m_k n_k}^{k_p}$ necessary for transition to the II level of the hierarchy.

Thus, we will calculate the average of a random X value, which we will define as $M[X]$:

$$M[X] = \frac{x_1 p_1 + x_2 p_2 + \ldots + x_n p_n}{p_1 + p_2 + \ldots + p_n} = \frac{\sum_{i=1}^{n} x_i p_i}{\sum_{i=1}^{n} p_i} \tag{5}$$

or considering that $\sum_{i=1}^{n} p_1 = 1$

$$M[X] = \sum_{i=1}^{n} x_i p_i . \tag{6}$$

We calculate by formula (5), substituting our values.

$$M[\mathrm{I}_1] = \frac{\sum_{i=1}^{n} P_{Ii} p_{Ii}}{\sum_{i=1}^{n} p_i} = \frac{P_{I1} p_{I1} + P_{I2} p_{I2} + P_{I3} p_{I3} + P_{I4} p_{I4} + \ldots + P_{In} p_{In}}{\sum_{i=1}^{n} p_i}$$

$$M[\mathrm{I}_1] = \frac{\sum_{i=1}^{n} P_{Ii} p_{Ii}}{\sum_{i=1}^{n} p_i} = \frac{0,51125 \cdot 0,98 + 1,35125 \cdot 0,89 + 0,642 \cdot 0,91 + 0,8675 \cdot 0,88}{0,98 + 0,89 + 0,01 + 0,88} =$$

$$= \frac{3,05}{3,66} = 0,83(3)$$

$$M[C_1] = 1,83$$

$$M[\mathrm{L}_1] = 1,12$$

Thus, we formalized or obtained (prepared) the initial analytical basis for the transition to the second level (II level) of the hierarchical system for solving the multi-criteria problem:

$$K = \sum_{j=1}^{m} c_{m_k n_k}^{k_p} (k \rightarrow K)_{m_k n_k}^{k_p} \rightarrow opt, \tag{7}$$

3. Results and discussion

Technological process of parts, units and assembly unit's restoration is a complex of recovery methods and means with a regulated sequence of technological operations. Technological processes of restoration should be designed on a scientific basis considering changes in production factors, renewal of the range of materials, development of new methods of restoration, processing and peening. It gives the chance to receive variant decisions with the subsequent choice of a rational one.

The evaluation of the result was carried out by the method of ranking and the method of expert evaluations in combination with standard instrumental control by quality indicators. Positive results were obtained. The content of operations in the restoration technology is reflected in the subsequent operating properties of the part and the road construction machine as a whole. This type of processing allowed to improve quality performance of parts after restoration. We have noted the increase of indicators related to the properties of reliability, safety and durability of the machine with various advantages by 10%.

4. Recommendations

The analysis of methods for solving multi-criteria problems by bringing them to single-criteria models typical for restoring functionally important parts of road construction machines was carried out. It was found that most of these methods had many objective flaws. In such tasks, the solution is performed by simplifying the source data and depends on the choice of applying a number of methods or mathematical models to solve multi-criteria problems, but may lead to wrong decisions.

A general methodology has been developed to study the process of restoring functional-important parts of cars, vehicles and road-building machines with a number of sequential operations in the form of a multi-level hierarchical system, for which the conclusion is important, that if we move to a higher level, it becomes clear which criteria should be preferred and the coefficients of relative importance in the composite criteria should be correctly evaluated. It is proposed to solve the problem by setting the coefficient of relative importance in accordance with each variant of the solution, which comes down to determining the numerical values of these coefficients.

References

[1] Bellman R 1960 *Dynamic Programming* (Moscow: Inostrannaya literatura)
[2] Bellman R, Dreyfus S 1965 *Applied Dynamic Programming* (Moscow: Nauka)
[3] Belyaev A I 2018 *Methodology of forming the optimal structure of road and construction machinery renovation technology: monograph* (SPSUACE. St.Petersburg)
[4] Belyaev A.I. S.A. Yevtyukov, A. V. Terentyev, A.E. Pushkarev, I.V. Arifullin 2019 *Maintaining the performance of cars. Methods of calculation of the production program for current maintenance and current repair of cars: a training manual* (SPSUACE. St.Petersburg: LLC Publishing house "Petropolis")
[5] Bondareva G I 2010 System analysis of objects, functions and resources in machine parts restoration processes *Bulletin of the Moscow State Agricultural Engineering University named after V.P. Goryachkin* **2(41)** 119–124
[6] Vavilov A V, Kotlobay A A, Kotlobay A Ya 2013 *Design of construction and road machines: a training manual* (Minsk: BNTU)
[7] Vasilyev K A, Nikolaev A K 2003 *Transport vehicles: training manual* (St.Petersburg: Saint Petersburg Mining University (technical university))
[8] Vasilyev V I 2013 *Engineering technology, restoration and repair of machine parts [Text]: UMO textbook* (Tomsk: publishing house of the Tomsk State University of Architecture and Construction)
[9] Kapustin A A 2015 Automobilization and car *Energy and Industry of Russia: journal* **(05)265** 38–39
[10] Kravchenko I N, Gataullin R M 2005 Basics of system approach to optimizing machine parts recovery techniques *Repair. Restoration. Upgrade* **4** 34–37
[11] Potapov Yu S, Korostil A P 1979 *Repair of car parts and agricultural machinery* (Chişinău: KartiaMoldavenyakse)
[12] Samarskiy A A 1979 Mathematical modeling and computational experiment *Bulletin of the Academy of Sciences of the USSR* **5** 38–49
[13] Terentyev A V, Yefimenko D B, Karelina M Yu 2017 Zoning methods as methods of motor transport process optimization *Bulletin of Civil Engineers* **6 (65)** 291–294
[14] Terentyev A V, Prudovsky B D 2015 Methods of Pareto set determination in some linear programming tasks *Notes of the Mining Institute* (St.Petersburg: Saint Petersburg Mining University) Vol 211 89–90
[15] Terentyev A V, Prudovsky B D 2014 Investigation methods for «current repairs labour–intensiveness» factor for a vehicle *Life Science Journal* **11(10s)** 307–310
[16] Chelpan L K, Vasilyev B S, Du A, Savin S V 2005 *Technology for repairing cars, road and construction machinery: lecture notes* (Moscow: LLC "Tekhpoligraphcentr")
[17] Shadrichev V A 1976 *The basics of automotive technology and car repair* (Leningrad: Mashinostroenie, Leningrad department)
[18] Arabani M, Haghi A K, Amani B 2009 Making a decision between the rehabilitation and reconstruction of asphalt pavements using the rough-set theory *ScientiaIranica* **16(2A)** 116–125
[19] Belyaev A I, Pushkarev A, Kuzmichev V 2018 Vehicle reliability and safety through simulation of the sequence of restoration of motor vehicle components *Transportation Research Procedia* **36** pp 50–55
[20] Biruk S, Jaśkowski P, Czarnigowska A 2017 Updating linear schedules with lowest cost: A linear programming model *Paper presented at the IOP Conference Series: Materials Science and Engineering* **245(7)** doi:10.1088/1757-899X/245/7/072011
[21] Bo L, Sminchisescu C, Kanaujia A, Metaxas D 2008 Fast algorithms for large scale conditional

3D prediction *Paper presented at the 26th IEEE Conference on Computer Vision and Pattern Recognition, CVPR* doi:10.1109/CVPR.2008.4587578

[22] Bošnjak S, Petković Z, Zrnić N, Simić G, Simonović A 2009 Cracks, repair and reconstruction of bucket wheel excavator slewing platform *EngineeringFailureAnalysis* **16(5)** 1631–1642 doi:10.1016/j.engfailanal. 2008.11.009

Crushing and screening digital system

A V Ostroukh[1], N G Kuftinova[1], C B Pronin[1], Y E Vasiliev[1], K.A. Barinov[1]

[1] Moscow Automobile and Road State Technical University (MADI), 64, Leningradsky prospect, Moscow, 125319, Russia

E-mail: ostroukh@mail.ru, caesarpr12@gmail.com

Abstract. The article proposes the structure of the Crushing and Screening Digital System (CSDS). A software implementation of a digital system for a cone crusher has been completed, which allows integrating a set of interfaced technological units in the control of one hardware equipment as part of the controller and installed at the operator station of the SCADA system. The most important functions of the CSDS are the continuous monitoring of the equipment condition of the complex of crusher mechanisms and aggregates in the process of work and the formation of a signal to regulate the productivity of the crusher. CSDS allows you to integrate in the control of one hardware as part of the selected controller and the SCADA-system server crusher installed at the operator station, a set of interfaced technological units, the efficiency of joint work of which is largely determined by the degree of coordination and the speed of two-way information exchange. In the event of malfunctions or abnormal situations, it will notify the operator. All accidents and crusher operation parameters are recorded in an electronic archive. The developed system provided a significant increase in the operator's comfort, providing him with a single user interface for controlling the cone crusher from one operator station and an additional amount of information and services.

1. Introduction

Modern crushing and screening production of stone materials for constructions products and structures is a complex with a large set of technological operations [1, 2]. Technological operations include the delivery of the source material, its purification and preliminary sorting, crushing in several stages, with the sorting of the required commodity fractions by crushing stages, storage and transportation to the final consumer. Technological processes of crushing and sorting are among the most critical processes in the technology of obtaining building materials, but they are not highly efficient enough.

To increase the efficiency of technological processes for crushing and sorting stone materials, it is required to develop new methods and scientific approaches to the creation of a Crushing and Screening Digital System (CSDS), which is a complex multi-level system.

2. Materials and Methods

A set of CSDS technological equipment (Fig. 1) is a branched flow-transport system, in its simplest version, consisting, for example, of a storage hopper, a screen, a crushing plant and several belt conveyors, and in a more complex version, it has several dozen units of equipment linked into a single technological process and operating in automatic mode [3, 4].

Content from this work may be used under the terms of the Creative Commons Attribution 3.0 licence. Any further distribution of this work must maintain attribution to the author(s) and the title of the work, journal citation and DOI.
Published under licence by IOP Publishing Ltd

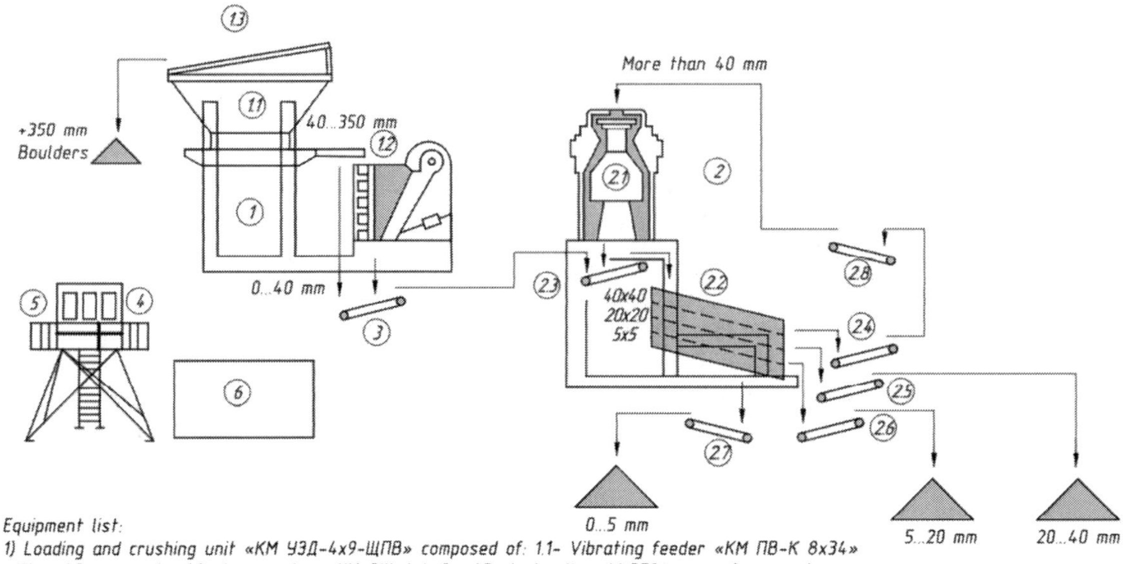

Equipment list:
1) Loading and crushing unit «KM УЗД-4x9-ЩПВ» composed of: 1.1- Vibrating feeder «KM ПВ-К 8x34» with a 40 mm. grate; 1.2- jaw crusher «KM ДЩ-1-4x9»; 1.3- hydraulic grid 350 mm. on frame - 1 pcs.;
2) Sorting loading and crushing unit «KM УДС-1200» composed of: 2.1 - cone crusher «KMД-1200Гр»;
2.2 - Sieving machine «ГИС-53» with meshes in 40, 20 and 5 mm; 2.3 - loading conveyor 1000 mm; 2.4 - 2.6 dispensing conveyors 650 mm; 2.7 - conveyor 1000 mm; 2.8 - conveyor 650 mm. on frame - 1 pcs.;
3) Conveyor «KM КЛО,65/15» on supports with a left-sided service platform, with a flexible coupling drive - 1 pcs.
4) Control cabin - 1 pcs.
5) Electrical equipment set (switchboard, control panel, cables) - 1 set;
6) Diesel generator «AKSA AD-330» with cover for «AKSA APD330», with «ЗИП» kit - 1 pcs.

Figure 1. Technological scheme of CSDS

The choice of the CSDS technological scheme depends on the type of processed raw materials, its physical characteristics, quality requirements and the purpose of the finished product. In addition, the technological scheme of the crushing and sorting plant also depends on the required ratio of cuboid grain, the planned capacity, as well as the readiness of the owner of the DSK to bear a certain share of the costs of restoring the operability of its working bodies.

Aggregates with jaw crushers (Fig. 2) are used for coarse and medium crushing of hard and abrasive rocks (granite, diabase, sandstone). Crushing is carried out by means of two "cheeks" - plates, one of which makes oscillatory movements.

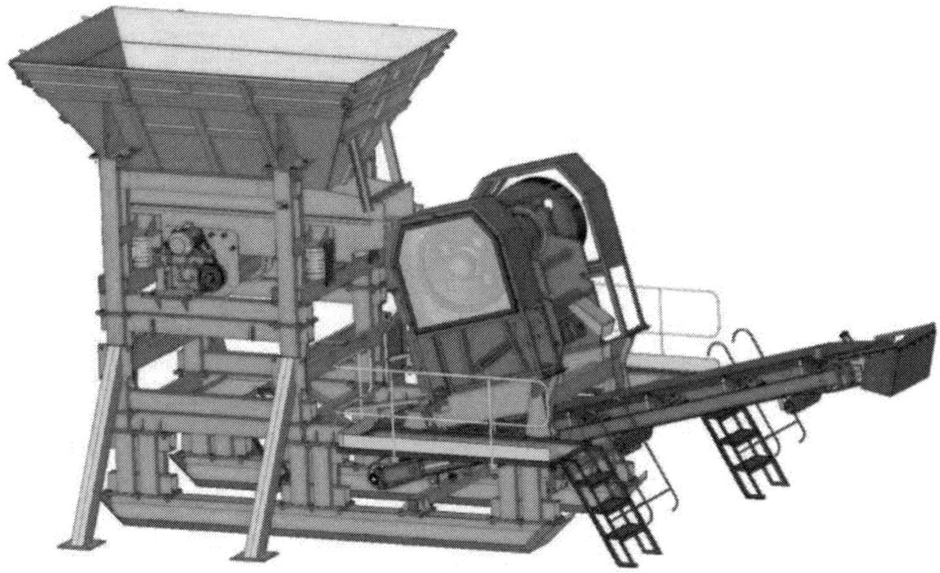

Figure 2. Unit for loading sorting and crushing KM UZD-4x-ShchPV

Aggregates with cone crushers (Fig. 3) are most effective for medium and fine crushing of strong and highly durable rocks. In them, crushing is carried out continuously due to the rotation of an eccentric conical rotor - a "crushing cone" inside an external stationary cone.

Figure 3. Unit for loading sorting and crushing KM UDS-1200

In cone crushers, instead of vibrations, continuous rotation is used, and the gap is changed, in contrast to jaw crushers, not simultaneously over the entire width of the gap, but by its alternating change along its length. The working gap in a cone crusher does not change simultaneously over the entire width, as in a jaw crusher, but continuously around a circle, contributing to an improvement in crushing quality. The inlet and outlet slots in the cone crusher are concentric rings. The maximum and minimum size of the exit slot is set by the adjusting device.

The size of the cone crusher product depends mainly on the size of the exit slot and the strength of the rock.

3. Results and Discussions

The power supply and automation system is built on a modular basis, which makes it possible to reconstruct or modernize production. Remote and local modes of operation are debug and repair modes. Visualization of the technological process is provided in various ways - by installing a control panel with a mnemonic diagram of the technological process, an operator's touch panel or a computer as part of an automated workstation (AWP) of a dispatcher [5, 6].

At the operator station of the SCADA-system [7, 8], a visualization project was implemented with reference to the main video frame for monitoring the state of the crusher equipment. In the right part of the screen there is a tool-information panel (Fig. 4)

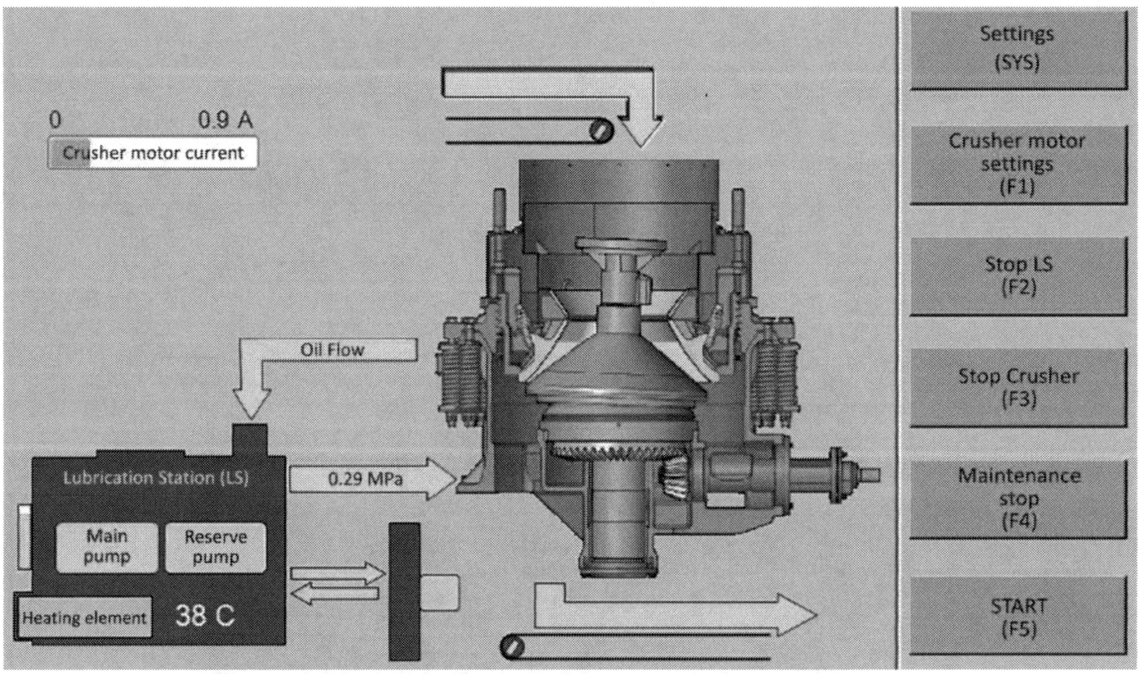

Figure 4. Mnemonic diagram of the cone crusher control system

In the process of adjusting the system software, it was revealed that it was impossible to remotely control the hydraulic unit to rebuild the gap in automatic mode due to failures in the adjustment of the sensors of the hydraulic unit pressure switch.

When rebuilding the unloading slot in the local operating mode of the hydraulic unit, the maintenance personnel must proactively enter information about the size of the unloading slot from the operator station. The system monitors the fact of operation of the hydraulic unit and informs about the need to enter information on the actual size of the gap in a special information window.

The function of "optimal control of loading ore into the crusher" is realized only from a programmable logic controller (PLC) of Russian production OWEN PLC 150 (Fig. 5) and provides control of the drives of the feeding conveyor depending on the calculated value of the indicator "current value of the unloading gap", automatically generating a task for the amount of stone material fed into the crusher (Q, t / h).

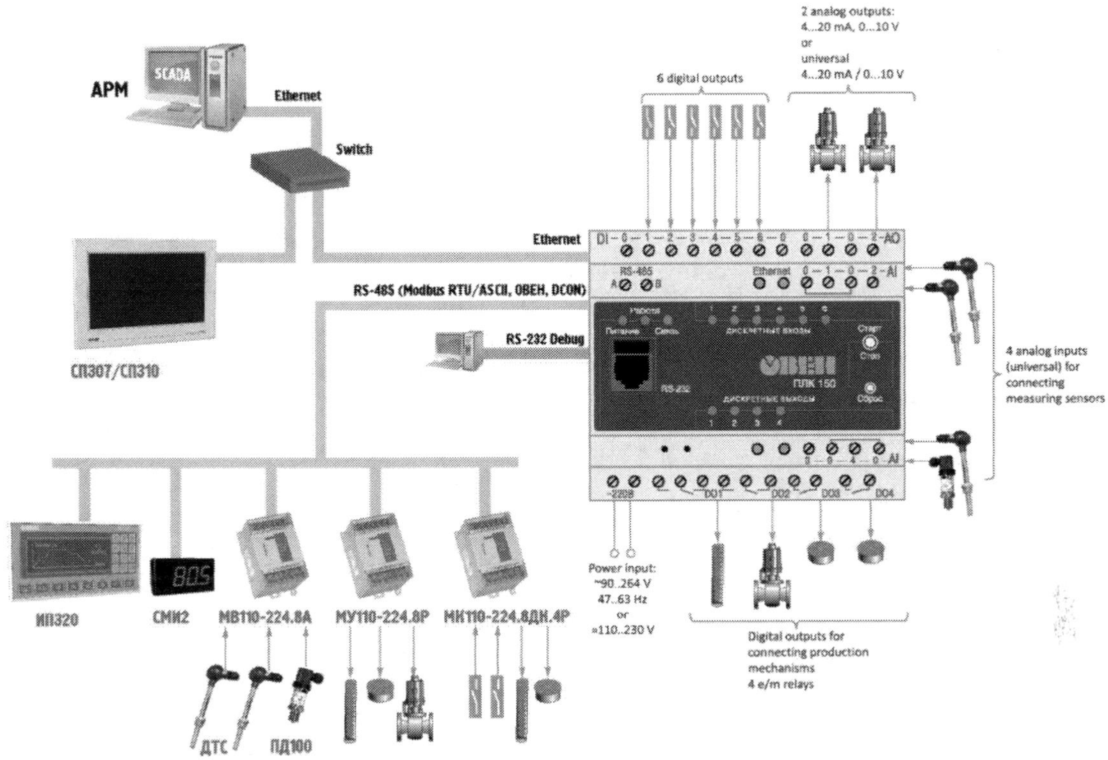

Figure 5. Programmable logic controller OWEN PLC 150

Optimization of material supply as a function of the size of the discharge gap is performed only in the recommended technological range of values of this gap (8.5 - 10.5 mm). When the crusher is operating with a slot outside this range, the regulator processes the algorithm setting in a gentle mode. At the crusher operator's station, information messages are generated about the need to tighten the gap.

The regulator implements control of the change in the supply of stone material to prevent the crusher blockage in the event of interruptions in the supply of material on the conveyor according to the following algorithm: in the case of a sharp decrease in the supply of ore (more than 70% of the steady-state value for 3 seconds), not due to the formation any control action from the control system, the setpoint of the material feed task is reduced.

4. Conclusion

A structural diagram of a CSDS has been developed, consisting of a set of elements, taking into account possible relationships; the features of the functioning of the elements of the scheme are determined, due to the need to automate the crushing and screening production.

The software implementation of the automated control system for the cone crusher in the SCADA system has been completed. This made it possible to integrate in the control of one hardware and technical complex as part of the selected controller and the SCADA-system server crusher installed at the operator station, a set of interfacing technological units, the efficiency of joint work of which is largely determined by the degree of coordination and the speed of two-way information exchange.

The developed system provided a significant increase in the operator's comfort, providing him with a single user interface for controlling various technological units from one operator station and an additional amount of information and service services, and also confirmed its economic efficiency.

Acknowledgments
The research was funded by a grant from the Russian Science Foundation for the event "Research by world-class scientific laboratories in the framework of the implementation of the priorities of scientific and technological development of the Russian Federation" of the President's program of research projects implemented by leading scientists, including young scientists.

References

[1] Ostroukh A V, Moroz D G, Vorobiev A I, Gavriluk M V, Akhterov A V 2017 Mobile Crushing and Screening Machine Control System Journal of Engineering and Applied Sciences 12 (23) 7165-7169 DOI: 10.3923/jeasci.2017.7165.7169.

[2] Ostroukh A, Surkova N, Varlamov O, Chernenky V, Baldin A 2018 Automated process control system of mobile crushing and screening plant Journal of Applied Engineering Science 16 (3) 343-348 DOI: 10.5937/jaes16-15586.

[3] Tikhonov A and Velichkin V 2017 Automation for Crushing and Screening Equipment to Produce Graded Paving Crushed Stone In IOP Conference Series: Earth and Environmental Science Vol 90 1 p 012194

[4] Mattisson A and Alsmarker D 2018 Forming an Inventory Control Policy in a Stochastic Environment: A Case Study at Sandvik Crushing and Screening

[5] Baurova N I and Konoplin A Y 2020 Visualization of the Dynamics of Measuring Processes of the Quality Indicators of Engineering Products during Production, Repair, and Operation Polymer Science, Series D 13 193-196

[6] Konoplin A, Baurova N and Abrakov S 2020 Visualization models of technological heredity in mechanical engineering In IOP Conference Series: Materials Science and Engineering Vol 971 2 p 022017

[7] Teixeira M A, Salman T, Zolanvari M, Jain R, Meskin N, Samaka M 2018 SCADA System Testbed for Cybersecurity Research Using Machine Learning Approach Future Internet 10(8) 76 https://doi.org/10.3390/fi10080076

[8] Aghenta L O and Iqbal M T 2019 Development of an IoT Based Open Source SCADA System for PV System Monitoring IEEE Canadian Conference of Electrical and Computer Engineering (CCECE) (Edmonton, AB, Canada, 2019) pp 1-4 doi: 10.1109/CCECE.2019.8861827

Protection of steel constructions against wear and corrosion in difficult climatic conditions by new techniques of zinc metallization

L G Petrova, P E Demin, A S Sergeeva and G Y Timofeeva

Moscow Automobile and Road Construction State Technical University (MADI), Russia, 64, Leningradsky prospect, Moscow, 125319, Russia

E-mail: Petrova_madi@mail.ru

Abstract. In the paper modernized techniques of zinc metallization of steel in combination with nitriding are considered. Two methods of combined surface modification of low-carbon steel are studied: application of zinc coatings by painting technique and subsequent gas nitriding; simultaneous thermo-diffusion saturation by zinc and nitrogen. Microstructure and microhardness of steel samples were examined; forming of a transition zone was determined under visually observed surface layer/coating with increased concentrations of Zn and N. Combined saturation by zinc and nitrogen forms diffusion layers with increased wear resistance and reduced friction coefficients together with improved corrosion resistance in salt fog. Nitriding of Zn coatings protects the steel both in salt and acid mediums.

1. Introduction

Corrosion and wear of steel components of bridge constructions are the main damages limiting their lifetime in difficult climatic conditions. Both phenomena are the consequences of metal interaction with external factors; their appearances depend on composition and structure of the metal's surface. Techniques of coatings application and surface modification are widely used for protection against wear and against corrosion. The problem is that, as a rule, the structure of the surface layer increasing wear resistance simultaneously decreases corrosion resistance and vice versa. Can we receive coatings or layers which influence positively both wear and corrosion resistance? Some previous results suggest promising solutions on forming of multi-component layers in steels by combined techniques of metallization and nitriding [1, 2].

It is known that zinc-based coatings application provides effective corrosion protection of steel parts and constructions working in a corrosive environment of atmosphere, precipitations, humidity, condensed moisture, etc. There are several methods of Zn-based coatings and layers application which can be categorized into two main groups according to the factors of existence or absence of diffusion processes.

The first group includes processes of thermodiffusion metallization by zinc with forming of diffusion layers of gradual structure [3, 4]. Corrosion resistance of such layers is the consequence of the protective properties of different Fe-Zn phases. We have supposed that the combination of diffusion saturation by zinc and nitrogen may influence the structure and phase composition of the steel surface layer and may amend wear-corrosion relation.

Processes of application of so-called zinc-rich coatings belong to the group of "diffusion-free"

Content from this work may be used under the terms of the Creative Commons Attribution 3.0 licence. Any further distribution of this work must maintain attribution to the author(s) and the title of the work, journal citation and DOI.
Published under licence by IOP Publishing Ltd

processes [5, 6]. Such coatings are applied by painting techniques, for example, by cold zinc-plating, and provide two possible mechanisms of corrosion protection: electrochemical (cathode) mechanism and barrier mechanism of painted films. Protective Zn coatings have a distinct boundary with a steel substrate without enough adhesion. The origin of low adhesion lays in the sharp change of residual stresses at the interface of a coating with a metal substrate. To a certain extent, the residual stress profile corresponds to the microhardness gradient along the transition zone. Therefore, the control of microhardness gradient across the surface layer may be a tool for the prevention of coatings exfoliation. Gradual microhardness profiles are attributes of diffusion layers receiving by thermochemical treatment (TCT). Thus, nitriding is one of the basic TCT processes for wear resistance increase for different steel components and tools. The nitriding process is based on the principles of diffusion saturation; and it forms diffusion layers with gradual structural changes [7, 8].

The purpose of the study consists in the investigation of the nitriding effect on corrosion and wear resistance of Zn coatings and in the examination of properties of modified layers received by combined processes of zinc diffusion metallization with nitriding.

2. Methods of Study

Low-carbon steel 09Mn2Si was used for experimental study (Table 1).

Table 1. Chemical composition of 09Mn2Si steel according to the GOST RF 19281-89

Steel grade	ASTM analog	Elements concentration, weight %						
		C	Mn	Si	Cr	Ni	Cu	Fe
09Mn2Si	*A515-55*	<0.12	1.30-1.70	0.50-0.80	<0.30	<0.30	<0.30	Bal.

Two techniques were applied for surface modification of steels:

A. Diffusion metallization by zinc and nitriding. Slip suspension is the source of diffusion saturation of steel by zinc. It contains zinc powder (25%), quartz sand (70%) as a filler, ammonium chloride NH_4Cl (5%) as a transportation agent, and PVA-glue as a binder. Diffusion saturation of covered specimens occurred in the ammonia atmosphere at 600^0C for 4 hours.

B. Zinc coatings application and nitriding. We used a painting technique for the zinc-coatings application. Paint mixtures contained fine-dispersed Zn powder and film-forming substances: organic binder (modified polyurethane) or silicate binder (sodium liquid glass with a density 1.2 g/sm^3). Zinc concentration was up to 85 weight % in organic-based mixtures and up to 75 weight % in silicate-based mixtures. Nitriding of painted specimens occurred in the ammonia atmosphere at 540^0C for 24 hours.

We used the following methods and tests: metallographic examinations and Auger-spectroscopy for the study of the structure of coatings and diffusion layers, microhardness tests, two methods of corrosion tests: standard salt fog tests in 5%-solution of sodium chloride (NaCl with pH=6.5) and electrochemical corrosion tests in 0.5M solution of sulfuric acid H_2SO_4, and standard wear tests with friction coefficients measurement and detection of volume wear parameter.

3. Results and Discussion

3.1. Microstructure of diffusion layers and nitrided zinc coatings

The combined process of diffusion saturation of steel by zinc and nitrogen forms modified layers with the thickness up to 100 microns. The diffusion layer contains an external surface case enriched by zinc (up to 80–92%Zn directly at the surface) and a transition zone (Fig. 1). The thickness of the surface zone is about 30 microns, and it consists of intermetallic compounds δ–$FeZn_7$ and Γ–Fe_3Zn_{10} in Fe_α(Zn, N). Specific of the combined metallization-nitriding technique consists in the consumption of previously applied suspension during the saturation process that leads to the diffusion of zinc atoms (from suspension) into the steel substrate, their solution in ferrite, and forming of Fe-Zn intermetallic compounds. Parallel diffusion of nitrogen from the gas atmosphere results in forming of a nitrogen-enriched solid solution. Since the diffusivity of nitrogen (interstitial element) in metals is significantly higher than the diffusion ability of zinc (substitution element) then nitrogen deeper penetration to the steel substrate is evident. The data in the Table 2 confirms the decreasing concentrations of zinc and nitrogen in different areas of the layer. Thus, we suggest considering the nitrogen-enriched $Fe_{\alpha}(N)$ sub-layer as the transition zone; the thickness of this zone is 30…40 microns.

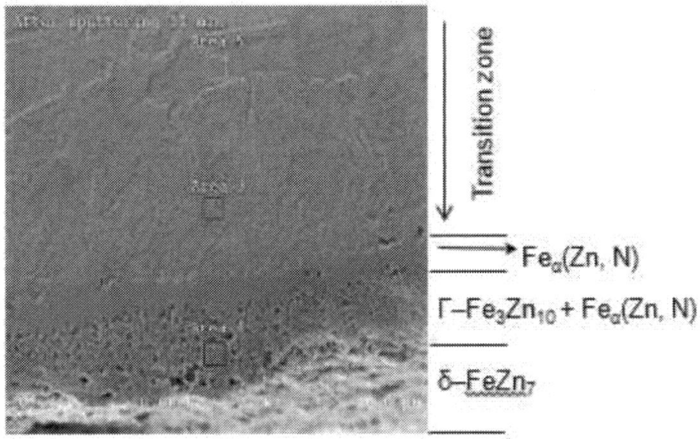

Figure 1. SEM-image of diffusion layer after combined diffusion saturation by zinc and nitrogen.

Table 2. Concentrations of elements in areas of Zn-N diffusion layer (according to the Auger-spectroscopy shooting points at the Fig. 1)

Shooting point	Concentration of elements, [at.%]	
	N	Zn
Area 4	10.8	4.5
Area 3	7.1	0
Area 5	3.3	0

The diffusion layer has increased microhardness compared to the steel core; this is the result of the formation of intermetallic compounds in the surface zone and of solid solution strengthening in the transition zone. Due to the gradual changes of structure and saturating elements' concentration in the diffusion layer the microhardness depth profile is quite smooth (Fig. 2).

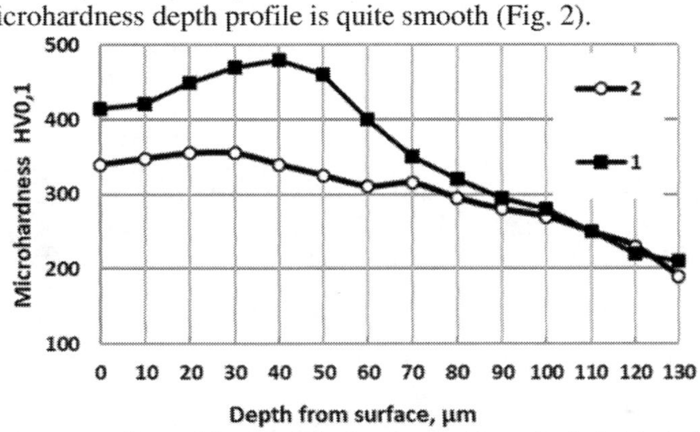

Figure 2. Microhardness profiles of Zn-N diffusion layer (1) and of nitrided zinc coating (2).

We considered the technique of Zn-coated steel nitriding as an alternation to the combined diffusion metallization-nitriding. The expectation was to receive a similar transition zone under zinc coating due to the diffusion processes activation. The possibility of nitrogen diffusion through Zn coating was never studied before; moreover, zinc was used as a protective covering against nitriding.

After gas nitriding of previously applied zinc coating, the film of about 50 microns thickness is formed at the surface having a visible boundary with steel substrate. Auger-spectroscopy detects both zinc and nitrogen under Zn coating after nitriding (Fig. 3). This confirms that nitriding activates zinc diffusion from the coating and nitrogen diffusion through the coating to the steel substrate.

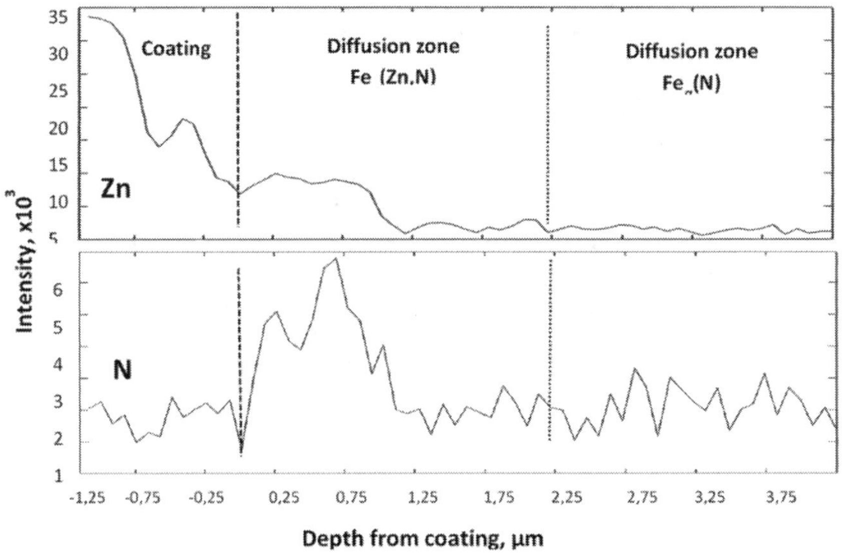

Figure 3. Auger-profiles of zinc and nitrogen concentrations in nitrided zinc coating and in diffusion zone.

Examination by microspectral X-ray analyses proves that during heating in ammonia Zn penetrates into the steel substrate under the coating (Fig. 4a); besides the analysis detects iron directly in Zn-coating (Fig. 4b) that is the consequence of Fe atoms counter diffusion from the substrate.

Phase composition in areas of the coating and of the diffusion layer was estimated on the basis of detected elements' concentrations (Table 3). Near the surface the composition of the coating is quite close to ZnO oxide (Spectrum points 1 and 2). The diffusion layer contains Zn in Fe_{α}-solid solution (Spectrum points 4 and 5). The diffusion transition layer consists of the "Zn+N solid solution zone" Fe_{α}(Zn,N) and of epy "nitrogen solid solution zone" Fe_{α}(N). This diffusion layer softens the microhardness profile (see curve 2 in Fig. 2) that promotes better adhesion of the coating. Increased microhardness of the internal part of the layer is the consequence of the solid solution hardening by nitrogen.

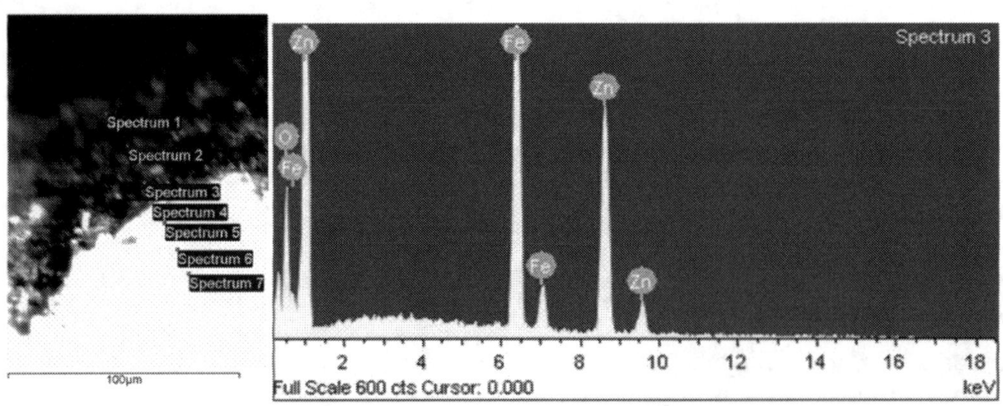

a b

Figure 4. SEM-image of nitrided Zn coating (spectrum points 1-3) and the diffusion layer below (spectrum points 4-7) (a); spectrum example at "Spectrum 3" point (b).

Table 3. Microspectral X-ray analysis in spectrum points according to the Fig. 4

Spectrum point	Localization	Elements concentration, weight %					Estimated phases
		Fe	Zn	Si	Mn	O	
1	Coating	-	80.15	-	-	19.85	ZnO
2	Coating	0.89	82.31	-	-	16.80	ZnO
3	Coating	33.76	54.57	-	-	11.67	ZnO+Fe_{α}(Zn)
4	Diffusion layer	93.82	3.75	1.56	0.87	-	Fe_{α}(Zn)
5	Diffusion layer	95.67	1.85	1.31	1.16	-	Fe_{α}(Zn)
6	Steel substrate	97.88	-	0.99	1.14	-	Fe_{α}
7	Steel substrate	97.98	-	0.90	1.12	-	Fe_{α}

3.2. Corrosion tests results

Electrochemical corrosion tests by measuring of potentiodynamic curves (PD) and passivation curves detect parameters of local corrosion: stationary corrosion potential E_c (the potential of anodic dissolution), and anodic current density of corrosion i_{MAX}. The lower is corrosion current the better is the protection. Both methods of surface treatment of Zn metallization and nitriding increase the corrosion resistance of steel in a diluted saltwater solution: the decrease of i_{MAX} value indicates this (Table 4).

Table 4. Electrochemical corrosion parameters of 09Mn2Si steel in 16.5 g/l NaCl water solution

Treatment	E_c, mV	i_{max}, μA/sm^2
As-recieved	-412	12.37
Nitrided Zn coating	-460	0.58
Diffusion saturation Zn+N	-403	3.7

Salt fog chamber tests have shown that combined diffusion saturation of the steel by zinc and nitrogen increase its resistance against corrosion in salty atmospheres: the protection time parameter (τ) is higher, and the damage area parameter (S) is less compared to the parameters of untreated samples (Fig. 5).

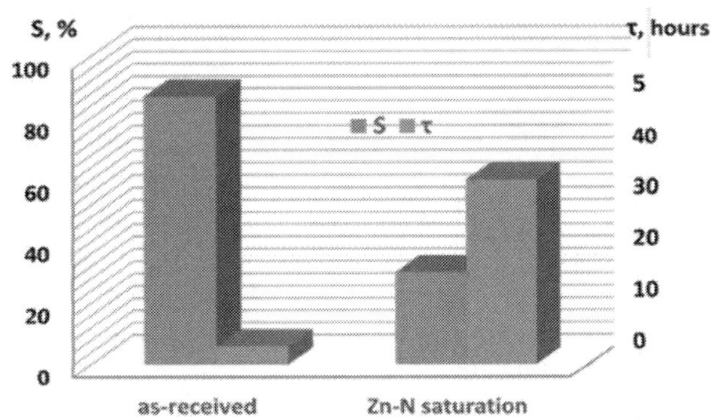

Figure 5. Results of salt fog chamber 36-hour tests for as-received steel specimens and for speciments after Zn+N diffusion saturation.

The corrosion resistance of steel in acid solution after diffusion saturation by zinc and nitrogen is less than the resistance of as-received Zn coating: current density i_{MAX} value is about 3 times higher

(Table 5). At the same time, the passivation interval E_p^c- E_p^s remains about the same as for the steel with Zn coating (Fig. 6). Passivation parameters of Zn+N saturated steel are much better comparing to the parameters of the steel nitrided in the same process mode (Table 5).

Table 5. Electrochemical corrosion parameters of 09Mn2Si steel in 0.5M solution of sulfuric acid H_2SO_4

Treatment	E_p^c- E_p^s, mV	i_{max}, μA/sm^2
Zn coating	1106	55
Diffusion saturation Zn+N at 600^{0}C/NH_3, 4 h	1089	157
Nitriding at 600^{0}C/NH_3, 4 h	54	26

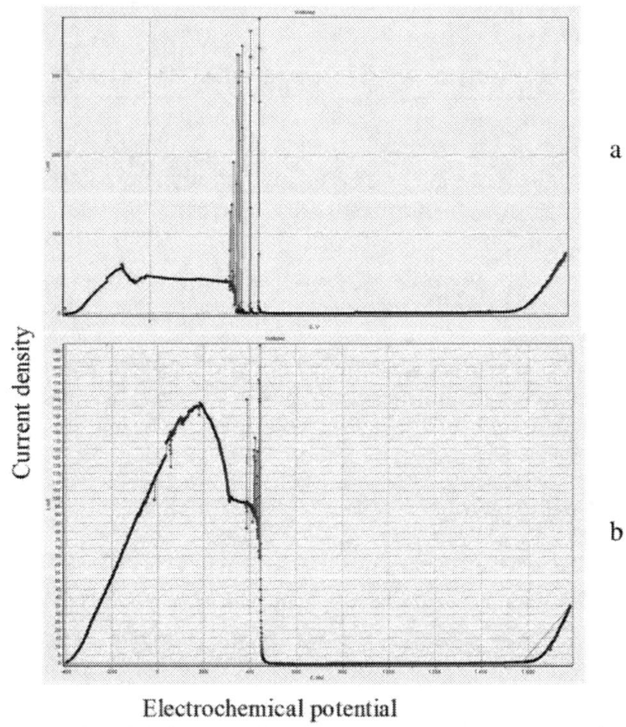

Figure 6. Passivation curves of 09Mn2Si steel with Zn coating (a), and after Zn+N diffusion saturation (b).

In contrast to the method of Zn+N diffusion saturation, the technique of nitriding of Zn coatings protects the steel both in salt and acid mediums. Table 4 shows that nitriding moves anodic dissolution potentials of Zn coatings E_c toward the higher negative values, and significantly decreases the anodic current density.

The corrosion resistance of nitrided zinc coatings depends on the parameters of the coatings: thickness, initial zinc concentration, and type of a binder. For example, Zn coating on the basis of silicate binder is more resistant than on the basis of organic binder; and double-layered coatings are more resistant than single-layered. Nitriding shades down these contradictions (Table 6).

Table 6. Electrochemical corrosion parameters of Zn coatings on 09Mn2Si steel in 0.5M solution of sulfuric acid H_2SO_4

Zn-coating parameters	E_c, mV		i_{max}, μA/sm^2	
	As-received	Nitrided	As-received	Nitrided
Silicate binder, 1-layered	-486	-627	11.32	2.89
Organic binder, 1-layered	-452	-690	26.23	3.14
Organic binder, 2-layered	-452	-630	16.21	2.48

3.3. Wear tests results
Wear tests have shown that combined processes of nitriding with Zn metallization increase wear resistance of steel. Both studied processes provide quite similar volumetric wear indicators W, which are about 5 times less compared to the wear rate of Zn coating without nitriding (Table 7). The process of Zn+N diffusion saturation improves the time profile of friction coefficient η and decreases η_{min} and η_{max} values (Fig. 7).

Table 7. Wear indicators of steel samples after different treatment modes

Treatment	η_{min}	η_{max}	V, mm^3	W, mm^3/h
Zn-filled coating	0.006	0.650	0.330	1.50
Nitriding of Zn coating	0.113	0.895	0.056	0.26
Zn-N diffusion saturation	0.004	0.320	0.066	0.30

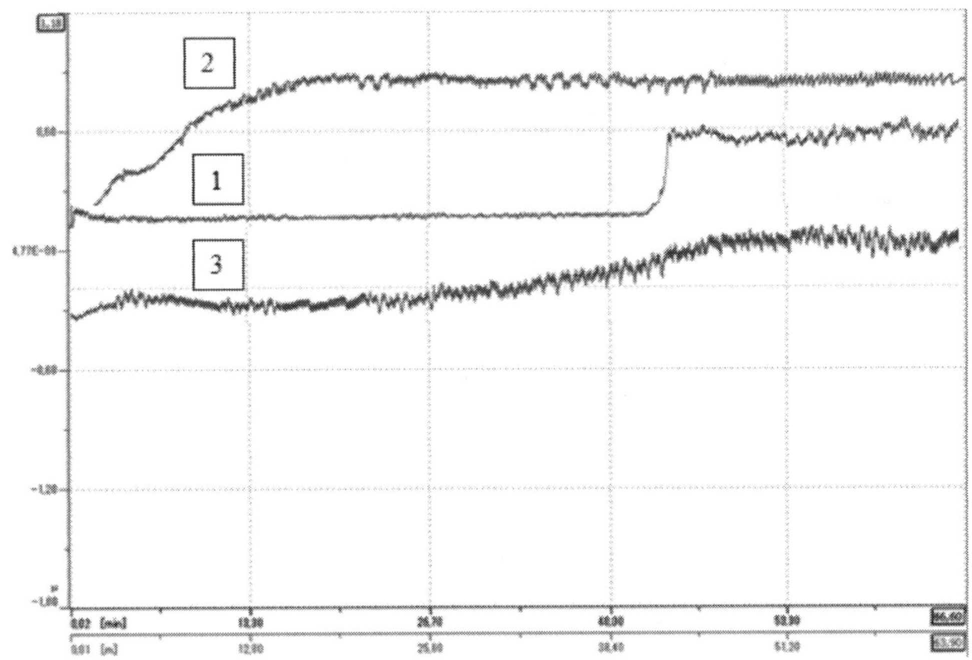

Figure 7. Time profiles of friction coefficient for steel samples: 1 – with Zn coating; 2 – with nitrided Zn coating; 3 – after Zn+N diffusion saturation.

4. Conclusions
Combined diffusion saturation of steel by zinc and nitrogen may be a good alternative to zinc-filled coatings for usage in conditions of corrosion and wear. Diffusion layers have a proper structure with transition zone/internal nitriding zone of significant thickness. This zone has increased zinc and

nitrogen concentration due to the diffusion of the elements under the surface layer. The diffusion layer with transition zone has a gradual microhardness profile that decreases residual stress at the boundary with the substrate steel and promotes adhesion improvement. Zn+N diffusion layers in low-carbon steel have increased wear resistance and reduced friction coefficients comparing to the parameters of Zn coatings either nitrided or not. Steel with such layers shows significant corrosion resistance increase in salt fog in comparison with untreated steel. On the other hand, for reasons of maximum improvement of electrochemical corrosion parameters Zn coatings remain more effective.

Acknowledgments

The material was prepared within the framework of scientific research under the project №FSFM-2020-0011 (2019-1342); experimental studies were carried out using the equipment of the MADI Centre of collective usage.

References

[1] Aleksandrov V A 2010 Combined TCT process for surface hardening of steels *Met. Sci. Heat Treat.* **52 (5-6)** 214-218
[2] Petrova L G, Aleksandrov V A, Malakhov A Yu et al. 2017 Improvement of operation lifetime of machine parts with composite coatings in wear conditions *Int. J. Mech. Eng. & Tech.* **8 (10)** 855-861
[3] Biryukov I, Batmanova V, Galin R G et al. 2019 The effect of the chemical composition of intermetallic phases on the corrosion of thermal diffusion zinc coatings *Surface and Coatings Technology* **372** 166-172
[4] Petrova L G, Demin P E, Kosachev A V and Barabanov S I 2017 Application of diffusion techniques for formation of zinc coatings to improve corrosion resistance of structural steels *Polymer Science, Series D* **10 (2)** 179-184
[5] Tseluikin V N, Koreshkova A A 2016 Corrosion resistance of composite coatings based on zinc *Chemical and Petroleum Engineering* **7-8(52)** 560-562
[6] Marchebois H, Touzain S, Joiret S et al. 2002 Zinc-rich powder coatings corrosion in sea water: influence of conductive pigments *Progress in Organic Coatings* **4(45)** 415-421
[7] Belashova I S, Shashkov A O 2012 Kinetics of growth of the diffusion layer in nitriding by the thermogasocyclic method *Met. Sci. Heat Treat.* **5-6(54)** 315-319
[8] Shestopalova L P 2016 Influence of preliminary oxidation on the subsequent nitriding *Russian metallurgy (Metally)* **13** 1241-1248

Application of digital modeling to determine design criteria for mobile frontal barriers

P S Mikheev[1], I A Karpov[1]

[1]Department of Structural Mechanics, Moscow Automobile and Road Construction State Technical University (MADI), 64, Leningradsky Prospect, Moscow, 125319, Russia

E-mail: psmikheev95@gmail.com

Abstract. The work is devoted to the issues of finite element modeling of mobile frontal barriers (MFB, truck mounted attenuators, mobile crash cushions), based on computer simulation of digital modeling of the behavior of elements in the MFB system, which includes a cover truck and a test vehicle designed to ensure the safety of road workers, when they are hit by cars in places of works on public roads that require temporary changes in the organization of traffic. It discusses various designs of trailer MFBs, the basic principles of damping devices, and defines the criteria for developing structures, based on digital mathematic modeling by the finite element method (FEM).

1. Introduction

Collision of vehicles (V), mainly cars, with a cover truck in the places of road works is one of the most dangerous types of road traffic accidents (RTA). A radical way to increase safety in such collisions is the installation of mobile frontal barriers (MFB) either directly on cover trucks or as a non-self-propelled trailer [1, 10] (figure 1). The use of this kind of security systems can reduce the number of victims from 44% to 35%, and the death rate from 19% to 4% [8]. Currently, active work is underway to create domestic MFB designs, but engineers are faced with a large number of difficulties due to the lack of knowledge of their mechanics. Digitalization is one of the main trends in the Russian economy, so digital models were the basis for the work. In this article, on the basis of digital modeling, the mechanics of the operation of the trailer MFB, one of the MFB types, is considered, typical problems of the structure's operation are highlighted and requirements for the safety of the structure and testing are formulated.

The vehicle and a structure collision process is fast (not more than 0.5 s), with complex boundary conditions in the form of contacts and restrictions on displacements, accompanied by large displacements and deformations in the system; therefore, to solve the problem, the finite element method was used in an explicit formulation. All calculations have been performed using the LS-DYNA software package [4].

Content from this work may be used under the terms of the Creative Commons Attribution 3.0 licence. Any further distribution of this work must maintain attribution to the author(s) and the title of the work, journal citation and DOI.
Published under licence by IOP Publishing Ltd

(a) (b)

Figure 1. Collision with a shadow vehicle, covering road works: a — without an impact attenuation system, b - with an impact attenuation system

2. MFB operating principle

The main task of impact attenuation systems like MFB is to absorb the kinetic energy of an oncoming vehicle, then stop or redirect it along a safe path.

Figure 2 shows some designs of foreign manufacturers [12, 13]. The oncoming vehicle impact attenuation in such structures is carried out by the work of deformation, friction forces or destruction of specially designed MFB structural elements [6, 7, 12, 13] (figure 3).

(a) (b)

Figure 2 Trailler MFB designs: (a) - Vorteq, (b) - Scorpion II

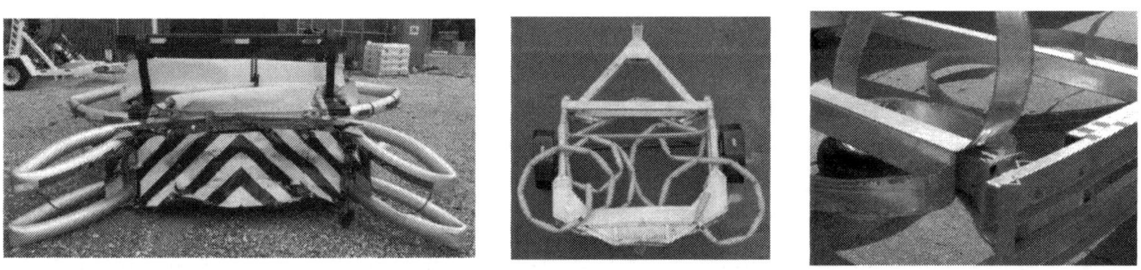

(a) (b) (c)

Figure 3. An example of energy attenuation at the impact to a MFB due to: (a) deformation, (b) friction, (c) destruction

The behavior of the vehicle and the risk of injury upon impact can be described by the following parameters: roll and pitch angles (rotation around the longitudinal and transverse axes of the vehicle), speed and the driver and passengers head overload, the center of gravity (CG) overload. These characteristics can be obtained by a number of tests [3, 5]: direct impact over the MFB axis, direct impact with displacement, and impact at a 10° angle to the MFB axis with displacement. However, in the real conditions it is impossible to predict the initial impact parameters, therefore, the general operation principle of the impact attenuation system should be maintained regardless of the vehicle collision conditions.

When designing, it is possible provide for an additional absorbing capacity, but there are restrictions on the mass-dimensional characteristics [11]: width, height and mass, so the design process becomes a non-trivial task.

IOP Conf. Series: Materials Science and Engineering 1159 (2021) 012092 doi:10.1088/1757-899X/1159/1/012092

3. Finite element model of MFB construction and virtual model of crash test

The investigated MFB (figure 4) is a structure of steel bent sections and sheets with a thickness of 2.8 - 5 mm, fastened to each other by bolted and welded joints, mounted on a uniaxial suspension. Energy absorption occurs due to deformation and the formation of plastic hinges.

The FE model was built using shell elements with the application of an elastoplastic material model MAT_024_PIECEWISE_LINEAR_ PLASTICITY with setting the true deformation curve and strain rate sensitivity constants according to the Cowper-Symonds model [4].

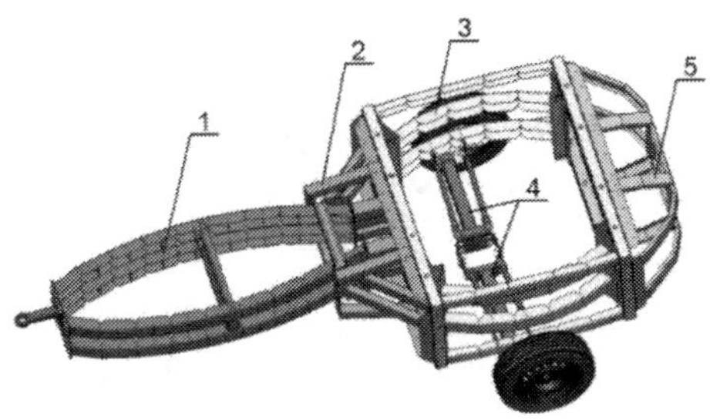

Figure 4. The main components of the MFB structure are: 1 - draw-bar, 2 - fairing frame, 3 - frame, 4 - wheels with axles, 5 - rear bumper reinforcement

Figure 5 shows the complete FE model of a virtual test (crush-test), consisting of a cover truck with a MFB attached to it, and an oncoming car weighing 1500 kg at a speed of 70 km/h.

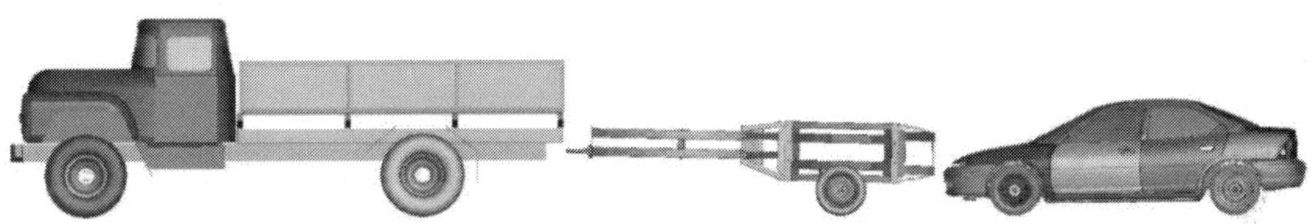

Figure 5. FE model of a virtual test with a collision along the axis of the MFB device

The after-collision view of the MFB obtained during the virtual test is shown in figure 6. The local area of impact is located below the center of gravity of the MFB, therefore, the resulting bending moment led in this case to the loss of stability of the draw-bar due to its insufficient rigidity. As a result, the longitudinal axis of the main impact attenuation part of the structure (frame) deflected from the initial position and turned out to be in a vertical position. Figure 7, a and b shows the process of deformation of the draw-bar, and figure 7, c - a general view of the energy-absorbing frame after impact.

Figure 6. The MFB after the impact

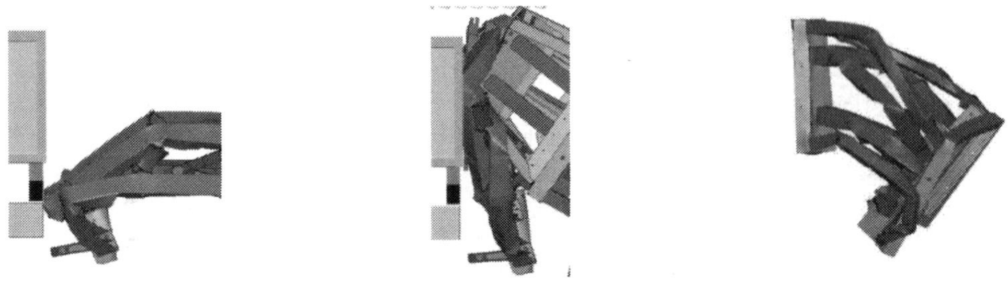

(a) (b) (c)

Figure 7. MFB structural elements deformation: (a), (b) draw-bar, (c) frame

For such damping systems as MFB, the sequential inclusion of structural elements in operation, starting from the point of impact, is the preferred case, while, in our example, the elements are deformed in a chaotic way (figure 8).

Figure 9 presents a chart of the change in the speed of the CG of an oncoming vehicle during an impact, showing an area without a decrease in speed, which indicates the ineffectiveness of the considered structure.

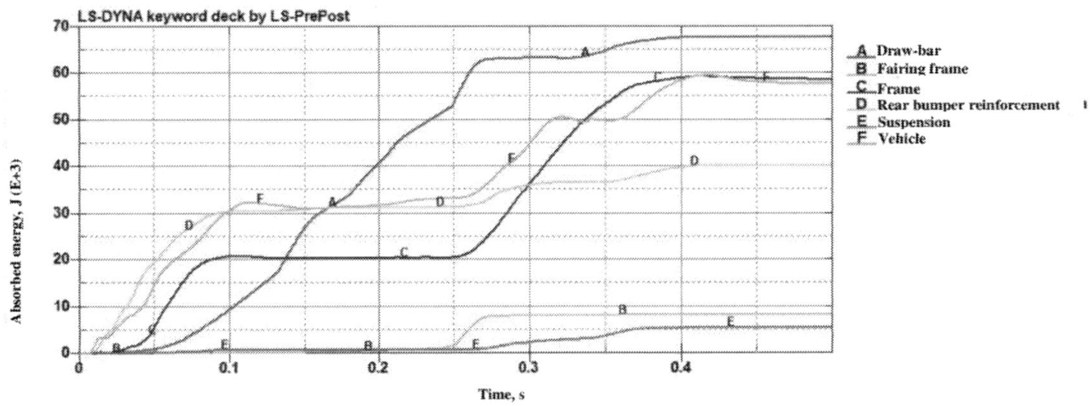

Figure 8. The time dependence chart of the energy absorbed by the MFB parts and the vehicle

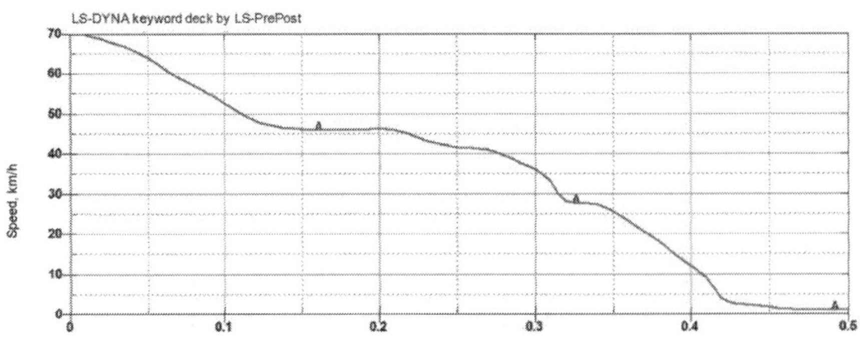

Figure 9. The chart of the vehicle speed change over time

Also, this MFB was tested with a displaced angle impact from a car (figure 10) with similar speed and mass, 70 km / h and 1500 kg, respectively. With such test parameters, the impact area is displaced relative to the CG of the MFB downward and sideways, and the force vector is at an angle, which leads to eccentric tension-compression and torsion of the structure.

This resulted in a negative test result: the vehicle lifted the guardrail and drove under it (figure 11). The speed of the approaching vehicle after contact with the MFB was 40 km/h, after which there was a collision with the rear of the cover truck. This result, as in the first case of impact, was due to the draw-bar low bending rigidity.

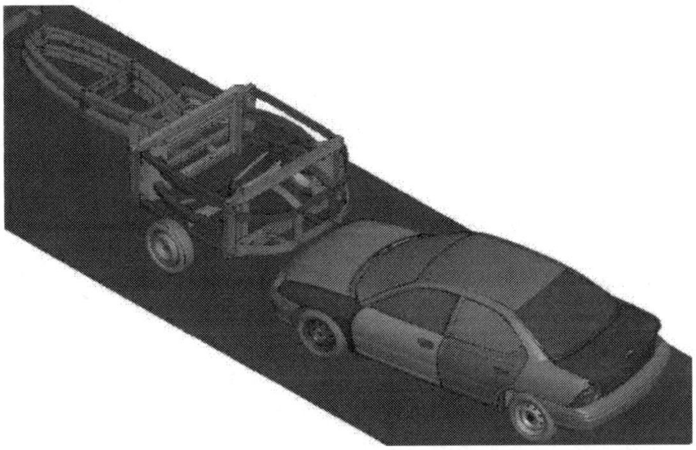

Figure 10. Testing the MFB at a 10° angle to the axis with an off-set

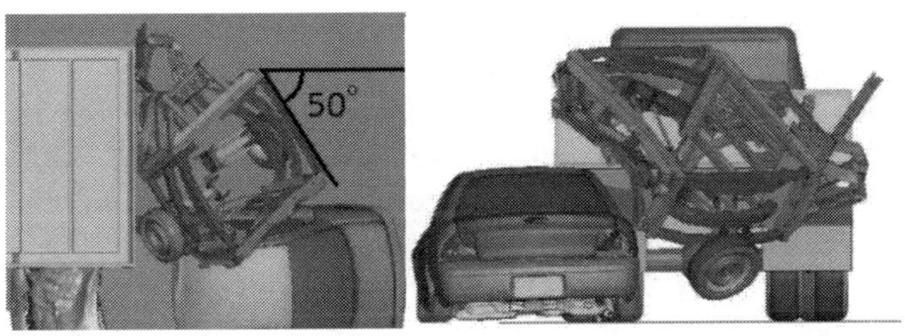

Figure 11. MFB after impact at an angle with an off-set

4. Formulation of principles for designing MFB

When designing MFB, it is necessary to structurally ensure longitudinal compression: the impact load must be evenly redistributed over the entire cross section.

In case of a direct impact, the uniformly retarded motion of the vehicle is achieved by selecting the rigidity of the main energy-absorbing elements and the geometry of the MFB. The height is selected in such a way so as to exclude oncoming vehicle driving under or onto the MFB during the impact, which can lead to serious consequences, and the CG of MFB should be at the impact level.

When the impact is produced with an off-set or at an angle, the MFB structure is subject to eccentric extension-compression, therefore, the bending rigidity must also be correctly selected. However, this will only partially solve the problem, since the rotational degree of freedom remains in the hinge of the trailer. Therefore, draw-bars with a telescopic structure [7] (figure 12, a, b) or hydraulic dampers [9] (figure 12, c) are used in the structures of the MFB, which allow preventing the MFB from turning during the impact.

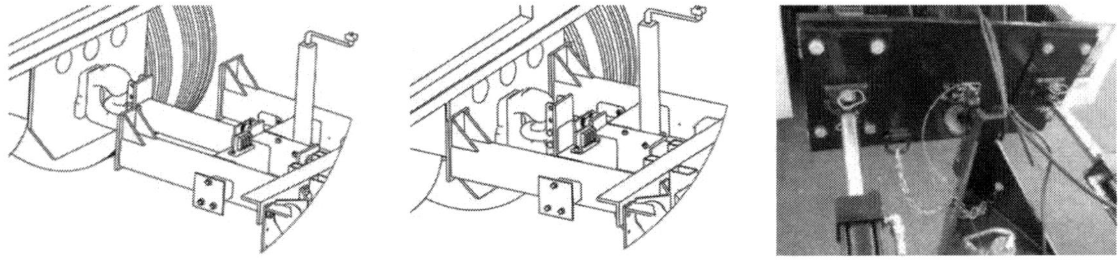

(a) (b) (c)

Figure 12. The MFB anti-rotation design of the draw-bar: (a), (b) telescopic before and after impact, (c) with a hydraulic damper

5. Analytical model for calculating the length of a trailer MFB

The design principles formulated above and the prerequisites for them made it possible to develop an analytical model for calculating the length of the MFB, developed for a certain impact energy. The resulting overloads upon impact should not exceed the permissible values, specified in regulatory documents [3, 5].

To determine the minimum distance S that the CG of the vehicle must pass upon impact, the equation of uniformly accelerated motion is used:

$$S = \frac{v^2 - v_0^2}{2 * \mathrm{a}}, \tag{1}$$

where:
v_0, v – initial and final vehicle speeds, m / s,
a – acceleration of the CG of the vehicle, m / s^2.

In case of an ideal impact, when the CG of a vehicle performs uniformly retarded motion along the longitudinal axis of the impact, only longitudinal accelerations occur N_x. They can be expressed in terms of the formula (2), which is used to find the injury severity index I [2]:

$$I = \left[\left(\frac{N_x}{12}\right)^2 + \left(\frac{N_y}{9}\right)^2 + \left(\frac{N_z}{10}\right)^2\right]^{0,5}, \tag{2}$$

where:
N_x, N_y, N_z – average values of a vehicle CG overloads in g fractions along X, Y, Z axes, m /s^2.

Next, the formula (2) is used and the overload along the longitudinal axis of the impact is expressed N_x :

$$N_x = I * 12g \tag{3}$$

Substitute N_x into formula (1):

$$S = \frac{v^2}{2 * N_x * g} = \frac{v^2}{24 * I * g}, \tag{4}$$

where:
v – vehicle speed before impact, m / s.

Thus, using formula (4), it is possible to determine the minimum distance that the CG of the vehicle must travel when driving over the MFB down to a standstill.

It is also necessary to take into account the margin for the deviation of the vehicle CG upon impact from the longitudinal axis, at which vertical and lateral accelerations occur, as well as the residual length of the deformed MFB elements. This distance is called the safety length $S_з$ and is found using the formula:

$$S_з = S * k,$$

where:
k – safety factor.

Figure 13 shows a diagram for determining the length of the l MFB draw-bar, determined by the formula:

$$l = l_1 + l_2 + l_3,$$

where:
l_1 - the distance from the trailer to the rear surface of the body of the cover truck,
l_3 - the distance from the back surface of the cover truck body to the main energy-absorbing part of the MFB,
l_2 - safety length when turning.
Distance from the back of the cover truck to the surface of the MFB - l_2 depends on the width of the cover truck. Also, it is necessary to take into consideration the trailer location l_1 and some margins l_3 required when turning. If the height of the MFB is lower than the height of the lower surface of the cover truck body, then the length of the draw-bar may be less, because, when turning, the MFB will pass under the truck body. It can also l be less if there is a limitation on the angle at which the MFB can turn when moving, or if the width of the MFB is reduced.

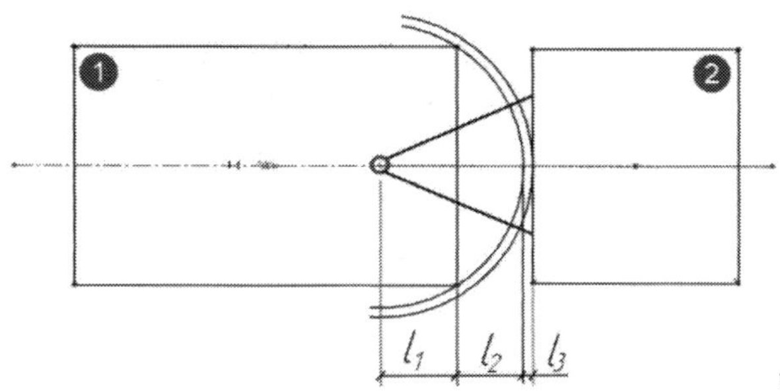

Figure 13. Calculation of the length of the MFB draw-bar: 1 - shadow vehicle, 2 - MFB

In total, the required full length of the MFB was obtained:

$$L = S_3 + l = \frac{v^2}{24 * I * g} * k + l_1 + l_2 + l_3$$

6. Conclusions

On the basis of the conducted virtual analysis using digital simulation modeling of the collision of a trailer MFB and a vehicle, the mechanics of work, the ongoing processes and design features of the MFB were studied. The possibility is shown using the example of the proposed energy method to assess the effectiveness of the MFB and its capacity to cope with its task which is the safe stop of an oncoming vehicle. This made it possible to formulate recommendations for the designs of trailer MFB and develop an analytical model for calculating their length for a given value of the injury index I [3].

7. Acknowledgment

The authors are grateful to their scientific advisor, Professor, Doctor of Technical Sciences I.V. Demyanushko for leadership and creative participation in research.

The work was facilitated by interaction with the MFB - developer enterprise - Prodmash Plant OJSC, as well as financing under the Development - Digital Technologies Program with the support of the Fund for Assistance to the Development of Small Forms of Enterprises in the Scientific and Technical Sphere (Fund for the Promotion of Innovations) (No. 47GRTSTS10-D5 / 56182), and

IOP Conf. Series: Materials Science and Engineering 1159 (2021) 012092 doi:10.1088/1757-899X/1159/1/012092

support of the FDA (ROSAVTODOR) for the creation of GOST R «Project. First revision. General-purpose highways. Mobile crash cushions. Specifications. Test methods».

References

[1] Demyanushko I V, Tavshavadze B T, Mikheev P S and Karpov I A 2020 Improving road safety during repair work *World of roads* **131** pp 102-107

[2] GOST 33128-2014 General-purpose highways. Road restraint systems. Specifications

[3] GOST R Project. First revision. General-purpose highways. Mobile crash cushions. Specifications. Test methods; Available at: http://niimech.ru/

[4] Hallquist John O 2006 LS-DYNA THEORY MANUAL, https://www.dynasupport.com/

[5] MASH 2016 2nd Edition, Available at: bookstore.transportation.org/Item_details.aspx?id=2707

[6] Rohde J R, Reid J D, Sicking D L and Mak K K 2008 Trailer truckmounted attenuator *International Journal of Crashworthiness* **13:1** pp 77-87 DOI: 10.1080/13588260701731708

[7] Scorpion II® Trailer Attenuator Assembly Manual and Mounting Instruction Guide, TrafFix Devices, Inc. 2018, Available at: www.traffixdevices.com

[8] Silyanov V V and Sidorova A S 2012 Traffic organization in the places of road works *Science and technology in the road industry* **2** pp 2-5

[9] SMT™, Product Description Assembly Manual, Trinity Highway Products, LLC, 2020, Available at: https://trinityhighway.com/

[10] Theiss LuAnn and Bligh Roger 2013 *Project Title: Compare Trailer-Mounted Attenuators vs Truck-Mounted Attenuators Protection for Workers* Available at: https://static.tti.tamu.edu/tti.tamu.edu/documents/0-6707-1.pdf

[11] TR CU 018/2011 Technical Regulations of the Customs Union "On the safety of wheeled vehicles", Available at: http://docs.cntd.ru/document/902320557

[12] TTMA-100, User's Manual, Gregory Industries, 2010 Available at: https://www.gregorycorp.com/

[13] Vorteq® Trailer, Product Description Assembly Manual, Trinity Highway Products, LLC, 2020, Available at: https://trinityhighway.com/

IOP Conf. Series: Materials Science and Engineering 1159 (2021) 012093 doi:10.1088/1757-899X/1159/1/012093

Analysis and mathematical modeling of cable barrier mechanics during car impact

I V Demiyanushko[1], I A Karpov[1]

[1] Moscow Automobile and Road Construction State Technical University (MADI)

E-mail: ilya.karpov@outlook.com

Abstract. The article discusses the mechanics of the response of cable barriers used in Russia. The first part of the article describes the general mechanics of the barriers. The second part describes the mechanics of the open-type posts, the most commonly used type worldwide. In the third part, accidents with closed-type posts are classified. The latter are unique structures that were developed in Russia to significantly reduce the deflection of barriers (by 35-50%). Understanding the principles of their response is not required for the further development of their structure only but also for the creation of their digital models since accurate models can only be developed if key physical processes are taken into account.

1. Introduction

Problems with passive road safety become clearer if we evaluate them in terms of fatalities per 1 billion kilometres traveled by vehicles. Such estimate gives more explicit information about the car - road infrastructure interaction period and the total number of cars than the death rate in road accidents per 100 thousand people. According to this indicator, in 2018 15.7 people per 1 billion km of car mileage were killed in Russia, 7.0 in the USA, 6.7 in Japan, 6.8 in Germany [1], which shows that the country falls significantly behind from developed countries.

The main way to reduce the severity of road accidents is road barriers, and one of the types are cable barriers (Figure 1). All cable barriers are similar in design and consist of cables, posts, anchor blocks, tension straps, and sleeves [2].

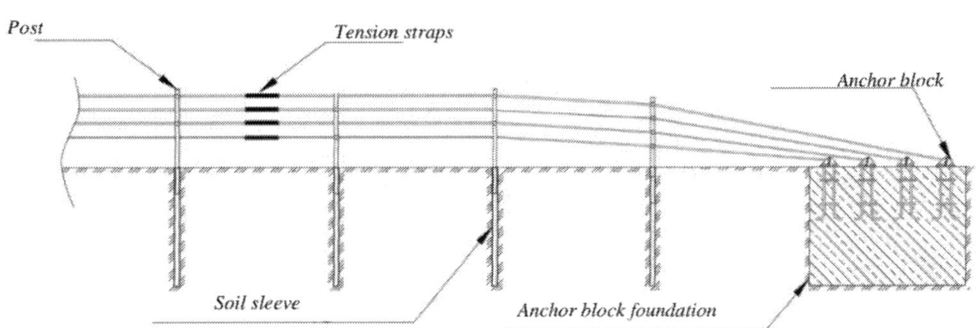

Figure 1. Cable road barriers

Content from this work may be used under the terms of the Creative Commons Attribution 3.0 licence. Any further distribution of this work must maintain attribution to the author(s) and the title of the work, journal citation and DOI.
Published under licence by IOP Publishing Ltd

Since real-life testing is quite expensive, computational modelling has become the main tool for the analysis of road barriers. The models are built on the basic equations of continuum mechanics and perform many operations automatically, nevertheless, none of them is fully automated yet. The most difficult part is the selection of kinematic constraints, as well as the choice of the equation of state, and the constitutive equations of the material; also the choice of models of the main structural elements of barriers (usually, these are finite element spatial models) is a complex task. This cannot be done without understanding the mechanics of the object and defining the most difficult issues for modelling.

2. Operating principles of cable barriers

The response of the barrier comprises three successive stages. At the first stage, the vehicle begins to deflect the cable of the barrier (Figure 2a), generating stress (T_r) in the cable, which is transmitted to the vehicle itself, as well as to the posts nearest to the active zone (Figure 2b). The friction forces of the wheels with the pavement F_{fw} and F_{rw} at the beginning of the impact prevent the vehicle from turning, hence, it continues to move in the initial direction, increasing the cable deflection.

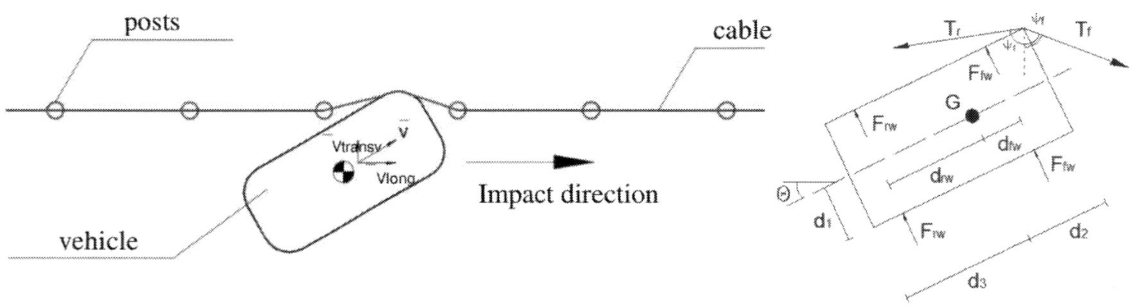

Figure 2. Interaction between the vehicle and the barrier: (a) the initial moment of impact, (b) free-body diagram of the vehicle

With an increase in the strain of the cable and overcoming of the friction forces, the second stage begins: the vehicle turns around the centre of gravity (G). The turn takes place until the vehicle body is parallel to the cable (Figure 3).

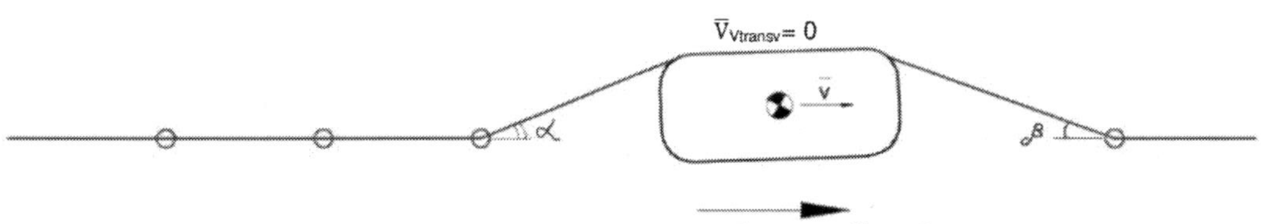

Figure 3. The second stage of vehicle – barrier interaction

As the vehicle turns, the stresses in the cable reach values sufficient to deform the posts of the barrier. At this stage, the further development of events is almost completely determined by the type of posts of cable barriers. Therefore, let us take a closer look at the mechanics of their response.

3. The response of the posts of cable barrier during the impact

The cable barriers are divided into two large groups: open-type and closed-type (For more details, see. ODM 218.6.018-2016). During the impact, the open-type posts only limit the transversal displacement of the cable, imposing almost no constraints to the vertical movement. The closed-type posts remain fully or partially connected to the barrier throughout the entire impact and cannot move vertically relative to the cable without destruction.

3.1. Open-type posts

3.1.1 Operating principle

When the vehicle hits the barrier, the forces in the cable reach values high enough to form a plastic hinge at the base of the post (Figure 4). The post is gradually pressed to the ground (Figure 4b) and, after complete deformation, ceases to bear any load or hinder the movement of the vehicle.

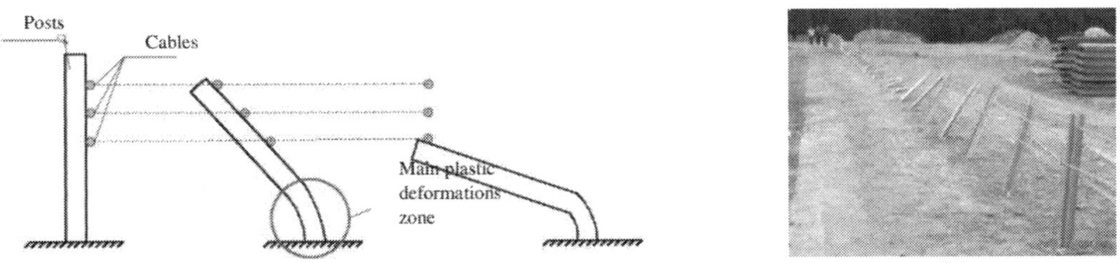

(a) (b)

Figure 4. Open-type posts operation mechanics: (a) diagram, (b) test results

The deformation of the posts leads to an increase in the active zone of the cable and a decrease in the angles between the cable and the posts of the barrier (angles α, β; Figure 3). A decreased angle results in a decreased stress impacting the next post, therefore it begins to limit the working zone of the cable until the deflection of the barrier and the increasing stress in the cable increase the angle enough to cause a complete deformation of the post. This process repeats until the vehicle turns in the direction opposite to the direction of impact, or the energy accumulated in the cable is high enough to overcome the frictional forces and force the vehicle back to its lane.

If the posts stiffness is low, when the vehicle gets closer to the post, the post has already partially or completely deformed. Therefore, there isn't any significant damping of speed in the longitudinal direction, deformations of side members and other load-bearing body elements, separation of parts (Figure 5). This ensures a constant geometry of the vehicle and the barrier in the contact area and stability of the collision process. After the interaction with these barriers, the vehicle is restorable, and the accelerations at the impact that determine the injury index are relatively small.

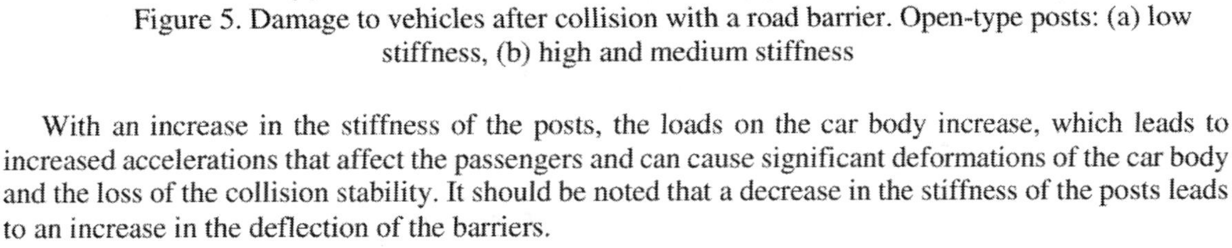

(a) (b) (c)

Figure 5. Damage to vehicles after collision with a road barrier. Open-type posts: (a) low stiffness, (b) high and medium stiffness

With an increase in the stiffness of the posts, the loads on the car body increase, which leads to increased accelerations that affect the passengers and can cause significant deformations of the car body and the loss of the collision stability. It should be noted that a decrease in the stiffness of the posts leads to an increase in the deflection of the barriers.

3.1.2 Energy absorption structure in barriers with open-type posts

From an energy point of view, the collision is a process of converting the kinetic energy of a vehicle into the internal energy of cables, the work of friction forces in contact pairs, and the work of forces causing plastic deformation of the posts, as well as losses due to internal friction in structural elements.

The structure of these responses and the components of the absorbed energy, as well as their influence on the behaviour of the vehicle and the barrier, were experimentally studied in [3], using 3 full-scale tests of a pickup-car collision. Table 1.1 shows the results of the study. Frictional energy losses were determined from the marks on the barrier and the pavement; since this determination method is not very accurate, the values are presented in ranges.

Table 1.1. Energy losses at the vehicle - cable impact [3]

Test No.	Vehicle energy, kJ		Posts deformation response, kJ	Friction forces, kJ		Energy in the cable, kJ
	Initial	Final		Tyres-road	Vehicle-barrier	
1	770.4	318.8	86.0	135.3 – 238.9	63.9-95.7	47.5
2	862.9	743.2	24.5	1.9-7.1	13.4-20.2	41.6
3	850.5	540.2	64.4	118.9-217.7	56.0-84.1	43.8

The results of the study showed that the barrier can absorb up to 60% of the kinetic energy of the vehicle, of which: up to 13% is transformed into plastic deformation of the posts, up to 31% is spent on friction in a pair tyre-road, up to 15% - on friction in a vehicle-barrier pair, up to 10% is transformed into the internal energy of the cable, of which, according to the authors' estimates, up to 3% is dissipated by the cable. It should be noted that the structure of energy absorption requires further research due to the small sample used in the study and the inaccuracy of estimation of some values. A more accurate energy absorption study can be carried out using computational modelling methods, which would allow not only to analyze various design options but also to estimate energy losses due to friction more accurately.

3.2. Closed-type posts

The characteristic feature of closed-type posts is that they do not disconnect from the cable during deformation, in part or completely. Closed-type posts act similarly to the open-type posts until they form a plastic hinge. After the hinge has formed, the post starts rotating around the hinge towards the road (Figure 3b).

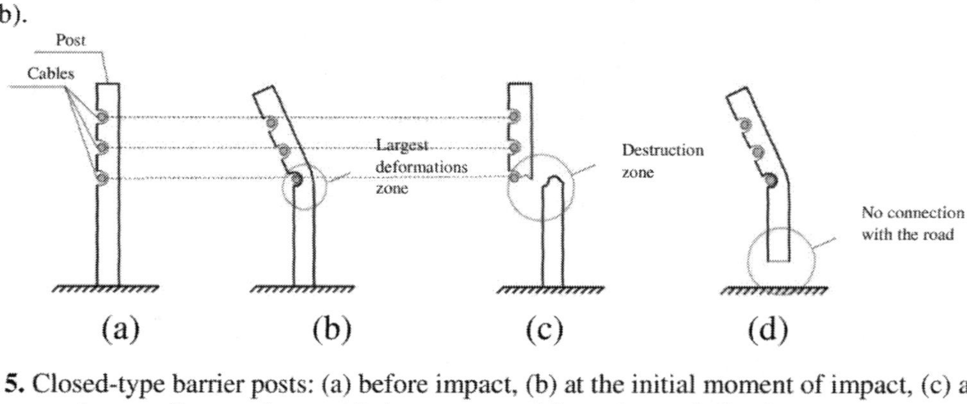

Figure 5. Closed-type barrier posts: (a) before impact, (b) at the initial moment of impact, (c) after impact. Damaged post, (d) A post severed from the road after impact.

Since the cables stay rigidly connected to the post, they, unlike the cables of the open structures, cannot maintain their vertical position and continue moving towards the surface of the road together with the post (Figure 5b). This generates a surface comprising cables and posts and inclined to the surface of the road that the vehicle can cross to drive onto the barrier. We will call this inclined surface of cables and posts that a vehicle can use to drive onto a barrier a "ramp". Some designs provide mechanisms for the release of the cables, such as zones of planned destruction (Figure 5c) or shallow positioning of the posts (Figure 5d), which allows the posts itself to move relative to the road without restricting the vertical movement of the cable, thus reducing the ramp effect. In all available designs, the vehicle comes into direct contact with the post, which can cause different consequences depending on the strength of the car body, the initial collision angle, and the collision speed.

If the car body and suspension are stiff enough not withstand deformation and destruction on contact with the post, the vehicle itself will press the cables against the road, forming a ramp that allows the vehicle to climb the barrier (Figure 6). In most cases, the vehicle's energy is not enough for a complete crossing of the barrier, and the vehicle ends up "hanging" on the cables without crossing to the wrong lane. This shall be considered a positive outcome, since staying on the oncoming lane significantly reduces the severity of an accident.

(a) (b) (c)

Figure 6. Partial crossing of the barrier (the vehicle is "hung" on the fence): (a) a passenger car, (b) an off-road vehicle, (c) a lorry

This happens most often with the vehicles with a high centre of gravity (off-road vehicles) and heavy vehicles (vans and trucks). In all cases the height of the barrier decreases and the posts are pressed to the road according to the mechanism described in Figure 5.

For passenger vehicles, the most important impact parameters are: angle and speed of collision, as well as the place of impact. If the approach angle is small and the impact occurred close to the post, the vehicle will not be able to deflect the cable enough to form a ramp. This will prevent crossing to the wrong lane, but will lead to a collision with the barrier post. If the initial speed of the vehicle was low, then the impact energy will not be enough for severing and deformation of the suspension. Therefore, the vehicle will start rotating relative to the contact zone formed by the deformed wing of the car, the suspension and the post (Figure 7a, b). After the vehicle turns 90 degrees, it suffers a second impact from the collision with the barrier with its front part, however, due to the high stiffness of the barrier, the vehicle only slightly crosses to the oncoming lane. If the speed of the vehicle and the energy are high enough, then the front suspension of the vehicle can completely detach from the vehicle, a partial turn occurs, and the vehicle continues sliding along the barrier, with repeated collisions of much lower energy (Figure 7a).

Comparing to other barrier types, a disproportionality of damage to the vehicle relative to the impact energy should be noted. For rail-type and parapet barriers, if the impact occurs with a low angle or energy, the vehicle will exit tangentially without significant damage. In cable barriers, even at very low angles and high speeds, a car wing will be crushed or the suspension detached, which significantly damages the vehicle.

(a) (b) (c)

Figure 7. A vehicle after a collision: (a) a vehicle turns without destruction of the suspension, (b) the suspension is completely destructed, (c) reduction of the bodywork area

At high angles and collision speeds, the two situations above are combined, because the vehicle has enough energy to create a ramp and partially run over the barrier, which reduces the load on the bodywork (Figure 8). The barrier is crossed when the allowed impact energy on the barrier is exceeded (Figure 8c).

(a) (b) (c)

Figure 8. Damage at large impact angles: (a) partial run over the barrier over the ramp, (b) partial crossing of the barrier, (c) complete crossing of the barrier.

The most dangerous collision cases are impacts when the contact area is in the area between the side members of the car, close to one of them (Figure 9a), as well as in the cases of roll cages with low stiffness. The consequences of such impacts are similar to the crash test of cars with a 25% displacement (Figure 9), which is considered one of the most difficult operational conditions for the bodywork because of a high asymmetry of the impact.

(a) (b) (c)

Figure 9. Vehicle after impact: (a) cable barriers, impact not to the side members, (b) cable barriers, impact to the side members, (c) crash test with 25% overlap

IOP Conf. Series: Materials Science and Engineering 1159 (2021) 012093 doi:10.1088/1757-899X/1159/1/012093

4. Conclusions

In most cases, cable barriers are effective for their purpose: they prevent frontal collisions of vehicles and reduce the severity of road accidents. The analysis of cable barriers with closed posts shows that the collision process follows different scenarios depending on the initial conditions of the impact. The initial conditions at the vehicle-barrier collision are largely random and circumstantial. During certification tests, only 1 passenger car collision is carried out and the latter - under rather limited conditions, which does not allow to fully understand whether other collision development scenarios are safe for drivers. Therefore, if one certification test with a passenger car is maintained, an understandable principle of the barrier action shall be achieved, which would be strictly followed during the entire collision process; otherwise, the safety of the barrier for the roads cannot be guaranteed.

References

[1] World Health Organization - Global status report on road safety 2018/1

[2] Demyanushko I V, Karpov I A, Storozhev S A 2013 Cable barriers: What you need to know [Ustrojstvo trosovyh dorozhnyh ograzhdenij: Chto nuzhno znat'] *Car roads* [Avtomobilnie dorogi] **10** 96-101 (in Russian)

[3] Schmidt J, Reid J D, Lechtenber K et al 2013 *Energy analysis of vehicle-to-cable barrier impacts* (Nebraska Transportation Center, New York)

IOP Conf. Series: Materials Science and Engineering 1159 (2021) 012094 doi:10.1088/1757-899X/1159/1/012094

The development of a dependent mathematical model of dielectric properties of an engine oil to the concentration of soot formation

A A Khaziev[1], N N Sugatov[1], A V Laushkin[1]

[1] Department of Operation and servicing of motor vehicles, Moscow Automobile and Road Construction State Technical University (MADI), 64, Leningradsky ave., Moscow, 125319, Russian Federation

E-mail: madi-chim@mail.ru, nsugatov@gmail.com, lav82@mail.ru

Abstract. The frequency of engine oil replacement reaches 150 thousand km for modern trucks, which increases the load on the lubrication system, leads to an increase in the level of contamination and to a greater level of oil degradation. The accumulation of dirt in the oil and, above all, soot leads accelerated wear of bearings, valves and camshafts, and increases the rate of formation of sludge and deposits on the piston rings. All this significantly reduces the engine life. The purpose of the work is to estimate the soot concentration in engine oil by its dielectric parameters measured by the waveguide method. To study the dependence of the dielectric parameters of engine oil on the soot concentration, the transmission line method was chosen and an installation was created that allows measuring the signal amplitude in a given frequency range. Tests performed on various engine oils at soot concentrations in the oil from 0 to 1.06% allowed us to identify the most optimal frequency regions where high sensitivity of the method is achieved. The nature of the change in the dielectric constant of engine oil with an increase in the concentration of soot in it is established. Diagnostic indicators such as the frequency of peak peaks and the signal amplitude at the optimal frequency are identified. Mathematical models of the dependence of the peaks amplitude and frequency on the soot concentration in engine oils are constructed.

1. Introduction

The aim in reducing the maintenance costs of vehicles, and the elimination of the negative impact on the surrounding environment over the years generates a tendency to an increase in the periodicity of engine oil change. The periodicity of engine oil change varies for modern vehicle makes, 60000 km for light weight vehicles, and 150000 km for heavy duty vehicles, which in keeping to this recommendation leads to the pollution of the environment, and the degradation of the engine oil.

In other to reduce the concentration of NOx in exhaust gases, an exhaust gas recirculation system (EGR) is used, where part of the exhaust gas is supplied to the intake manifold of the engine. The particles of soot in an engine with the EGR system is 20-30% denser, in comparison to soot particles in an engine without the EGR system. The combustion products that were supposed to be exhausted falls back into the cylinder as well as into the engine oil, leading to the increase in the accumulation rate pollution products in the engine oil. In other to keep the large impurities in suspension,

Content from this work may be used under the terms of the Creative Commons Attribution 3.0 licence. Any further distribution of this work must maintain attribution to the author(s) and the title of the work, journal citation and DOI.
Published under licence by IOP Publishing Ltd

manufacturers usually increase the dispersing forces of the engine oil, which adversely affects the corrosion of parts in the engine due to the presence of primary amines in the oil.

Purpose of studies is the estimation of the concentration of soot in an engine oil based on its dielectric properties, measured using the frequency method

2. The effect of the concentration of soot in an engine oil and the technical state of the engine

Of the soot formed by the combustion of diesel fuel in the engine, only 29% flows through the exhaust pipe, the rest remains in the engine oil, on the walls of the cylinder and on the piston base [1-5]. Soot in oil leads to accelerated wear of the cylinder-piston group and gas distribution mechanism (GDM).

The more the soot in an engine oil, the greater the possibilities of wear of the piston rings, the cams of the camshaft, the rocker arms, the guide valves and other components of the engine.

Soot's aggravates the wear of those contact surfaces with the least lubrication. The most intensive wear occurs with the GDM, as it is located in the upper part of the engine, which is not sufficiently lubricated, especially during cold starts, due to low pressure in the lubrication system and the high viscosity of the oil, which are normally observed during idling.

The concentration of soot in an engine oil can attain 3...4%. At same time, there is a common opinion that its content in an engine oil must not exceed 1%.

The high concentration of soot in an engine oil significantly increases the formation of sludge and sedimentation. Sedimentation on the piston rings is harmful, as this will lead to the ring losing its mobility and eventually ends up jamming. The limited freedom of motion of the piston rings will lead to the deterioration of the hermitization between the piston and the walls of the cylinder and the entrance of gases into the crankcase, which will lead to a further deposition of soot's. The analysis of fig.1 shows that an increase in the concentration of soots brings about an increase in the kinematic viscosity of the lubricants. It should be noted that, a 2% presence of soot in an engine oil will lead to an increase in viscosity of more than 12,5 mm^2/c [6].

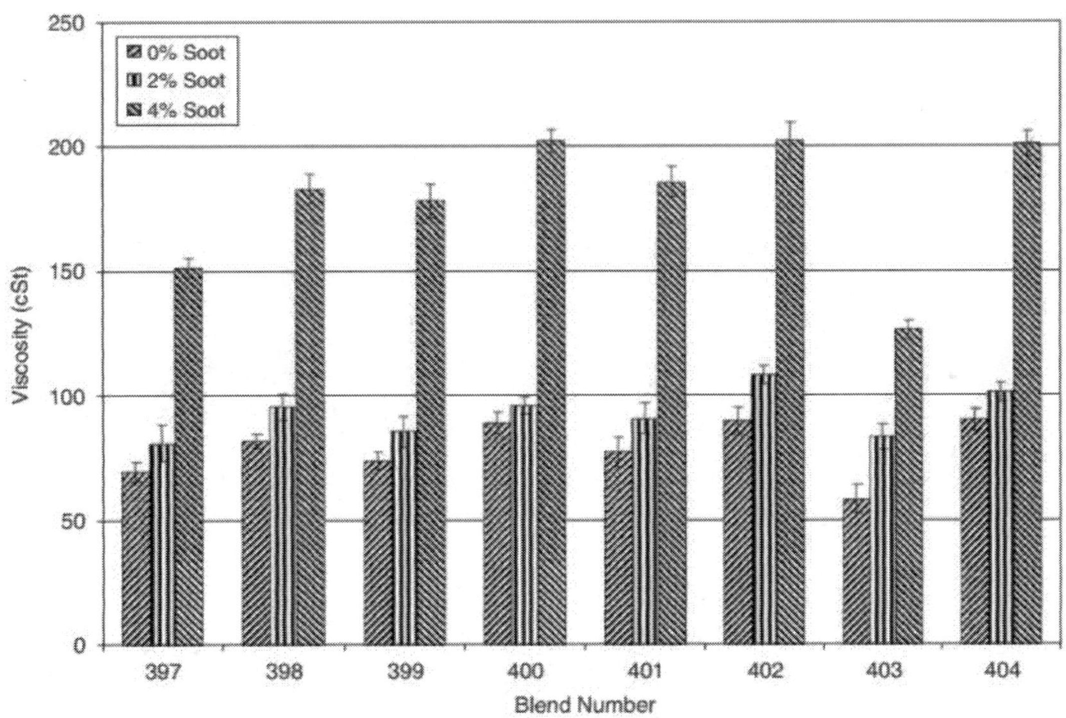

Figure 1. The change in the kinematic viscosity of an engine oil at 40°C depending on the concentration of soot

IOP Conf. Series: Materials Science and Engineering 1159 (2021) 012094 doi:10.1088/1757-899X/1159/1/012094

3. Experimental Research

Recently, there has been a trend in the use of small-size sensors for determining the dielectric characteristics of engine oil, based on which it is possible to judge the qualitative characteristics of engine oil [7-9]. Therefore, now it is actual to investigate and develop mathematical models of dependence of qualitative characteristics of motor oil on its dielectric parameters. Carbon black has a significant effect on the dielectric constant of motor oil [10-12].

To study the dependency of the dielectric properties of an engine oil to the concentration of soot, the transmission line method was chosen [13], of which a setup was developed consisting of a measuring cell representing a rectangular waveguide with dimensions 32 x 16 x 166 mm, and a scalar network analyzer assembled on the basis of a USB- TG124A generator, and an analyzer USB-SA124B manufactured by Signal Hound company. The setup allows measuring the amplitude of the signal transmitted through the measuring cell in the frequency range from 2 to 12 GHz. The measured characteristic is an indirect indicator characterizing the dielectric properties of the engine oil under studies.

The measurements were carried out on synthetic, semi-synthetic and mineral engine oils with the concentration of soot in the engine oil in the range of 0 to 1.06% by volume. For each type of oil, 9 measurements were taken.

The measured amplitude-frequency response (AFC) of a measuring cell filled with synthetic oil, depending on the concentration of soot, is shown in Fig. 2. The AFC with an increase in the concentration of soot shifts to the left, in the direction of decreasing frequency. This explains the fact that the relative permittivity of soot is 18.8, which is significantly different from the dielectric constant of fresh oil (~ 2) [14, 15], and with an increase in the dielectric permittivity of the aggregate, decreases the cut-off frequency.

The boundary frequency determines the maximum wavelength that can propagate along the frequencies. Four (4) peaks can be selected for each AFC. The frequencies of the peaks correspond to the modes of electromagnetic waves that can exist at these frequencies. It should be noted that peaks 1-3 have a steeper peak than peak 4.

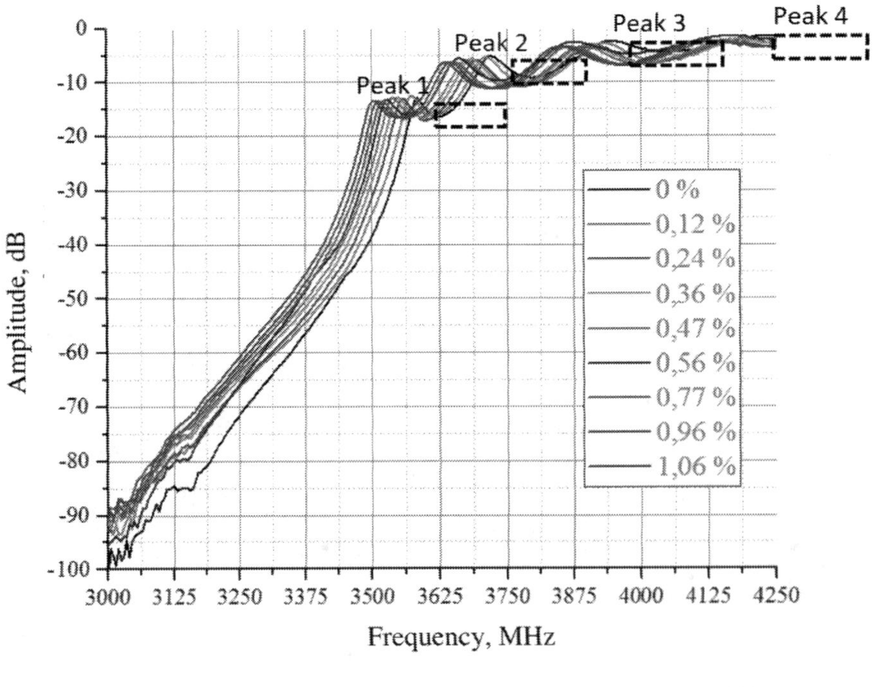

Figure 2. Frequency response characteristic of synthetic motor oil with different soot content

Fig. 3 shows an enlarged peak regions of 1 and 2 AFC in comparison to fig. 1. The graphical analysis shows practically no change in the amplitude of the peaks, but with a step by step increase in the frequencies of the peaks 1 and 2, due to an increase in the concentration of soot.

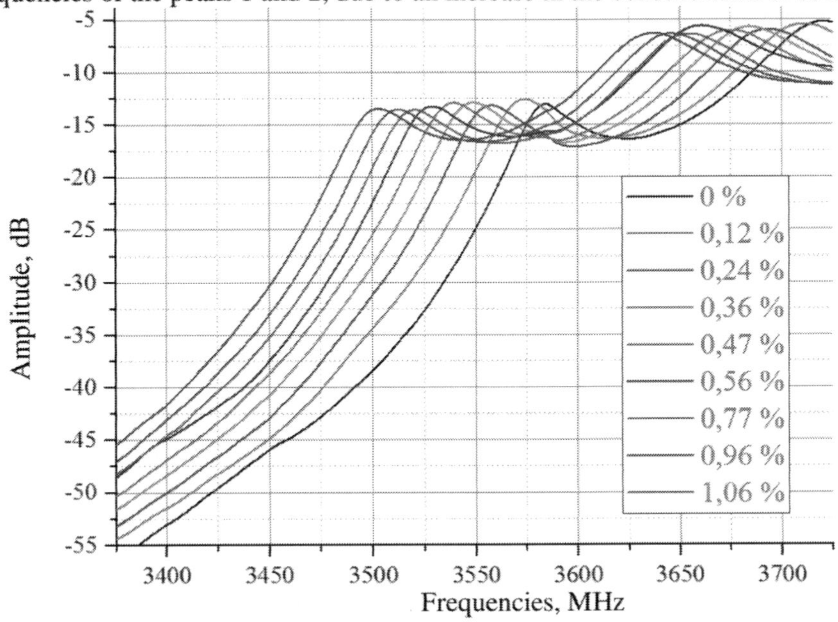

Figure 3. Frequency response characteristics of a synthetic motor oil with different soot contents at frequencies ranges from 3375 to 3725 MHz

4. The analysis of the results of experimental studies

The dependency of the coefficient of correlation of soot concentration to the frequency for each type of engine oil, shown in fig. 4, allows to identify the most optimal frequency regions, where high sensitivity of the method is achieved.

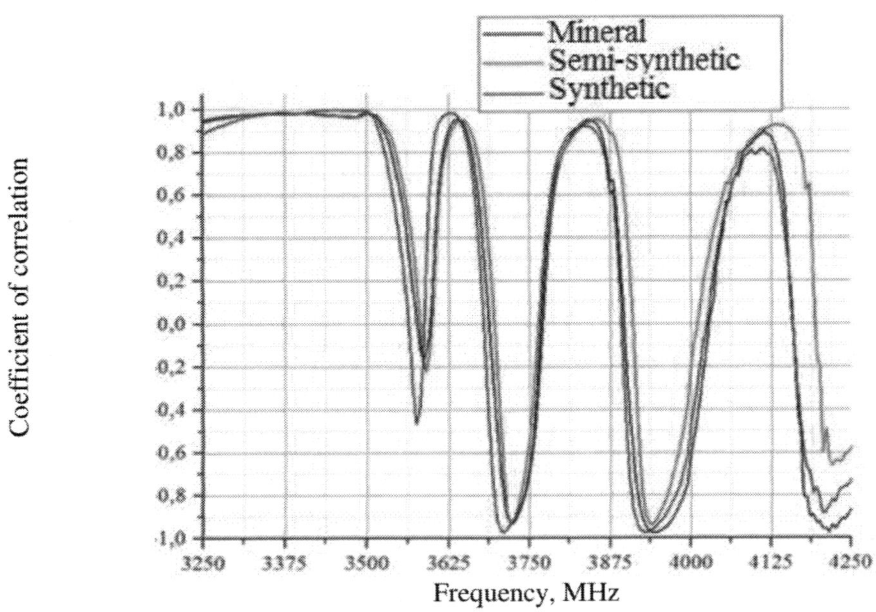

Figure 4. The dependency of coefficient of correlation of soot concentration of an engine oil to frequencies

IOP Conf. Series: Materials Science and Engineering 1159 (2021) 012094 doi:10.1088/1757-899X/1159/1/012094

The most optimal signal frequencies for each type of oil were found, at which the maximum values of the amplitude differences for soot of 0% and 1.06% were obtained. The dependency of the signal amplitude on the concentration of soot in the engine oil at the selected optimum frequency is shown in fig. 5.

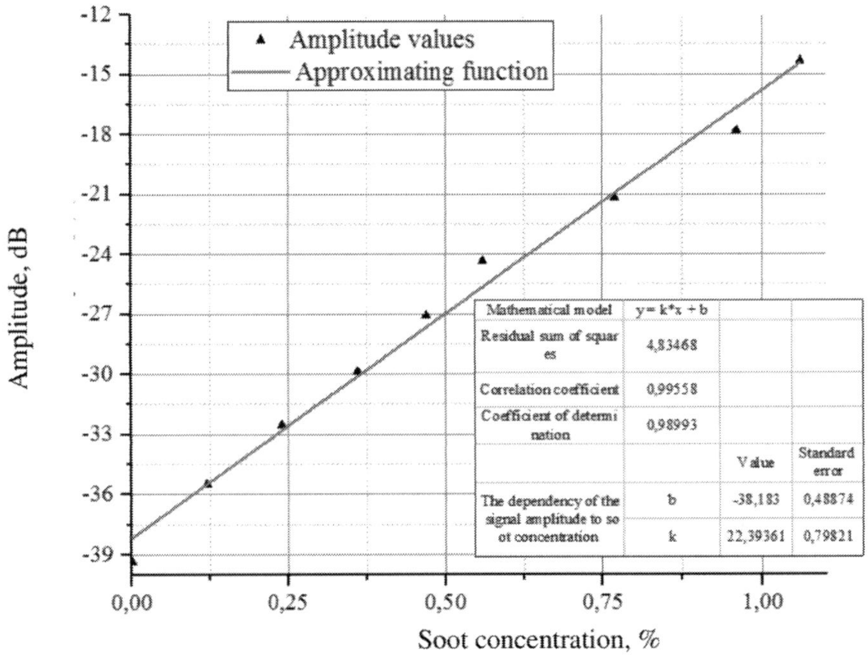

Figure 5. The dependency of the signal amplitude of a synthetic engine oil to soot concentration at a frequency of 3495 MHz

Table 1 shows the optimal frequencies, the results of the correlation analysis, and the approximating functions that determine the dependency of the signal amplitude to the concentration of soot in oil.

Table 1. The optimal frequencies, the results of the correlation analysis, and the approximating functions that determine the dependency of the signal amplitude to the concentration of soot in oil

Oil type	Optimal frequency, MHz	Coefficient of correlation	Determinant coefficient	Mathematical model, y = k*x + b	Standard error of the angular coefficient, k	Standard error of the free member, b
Synthetic	3495	0,99	0,99	22,39*x-38,18	0,79	0,49
Semi-synthetic	3495	0,97	0,93	30,79*x-46,22	1,82	2,98
Mineral	3485	0,96	0,91	30,32*x-46,63	3,24	1,98

Table 2 shows the results of the correlation analysis for peaks 1-4. An example of constructing an approximating function for the dependency of the frequency of peak 1 on the concentration of soot is shown in fig. 6.

Table 2. The analysis of the dependency of peak frequencies 1-4 on the concentration of soot

Oil type	Optimal frequency MHz	Coefficient of correlation	Determinant coefficient	Mathematical model, y = k*x + b	Standard error of the angular coefficient, k	Standard error of the free member, b
Synthetic	1	0,99	0,98	-78,08*x+3580,5	4,13	2,53
	2	0,98	0,95	-75,78*x+3714,34	5,9	3,6
	3	0,99	0,97	-89,82*x+3936,98	5,79	3,55
	4	0,93	0,84	-89,28*x+4218,37	8,34	13,63
Semi-synthetic	1	0,998	0,99	-97,65*x+3603,15	2,62	1,6
	2	0,995	0,99	-100,21*x+3738,9	3,87	2,37
	3	0,995	0,99	-115,75*x+3962,8	2,7	4,42
	4	0,97	0,92	-105,56*x+4236,6	6,74	11,01
Mineral	1	0,998	0,99	-97,91*x+3597,72	1,7	2,78
	2	0,995	0,99	-101,32*x+3734,4	2,45	4
	3	0,99	0,98	-111,58*x+3955,2	5,65	3,46
	4	0,7	0,41	-60,99*x+4207,43	23,71	14,52

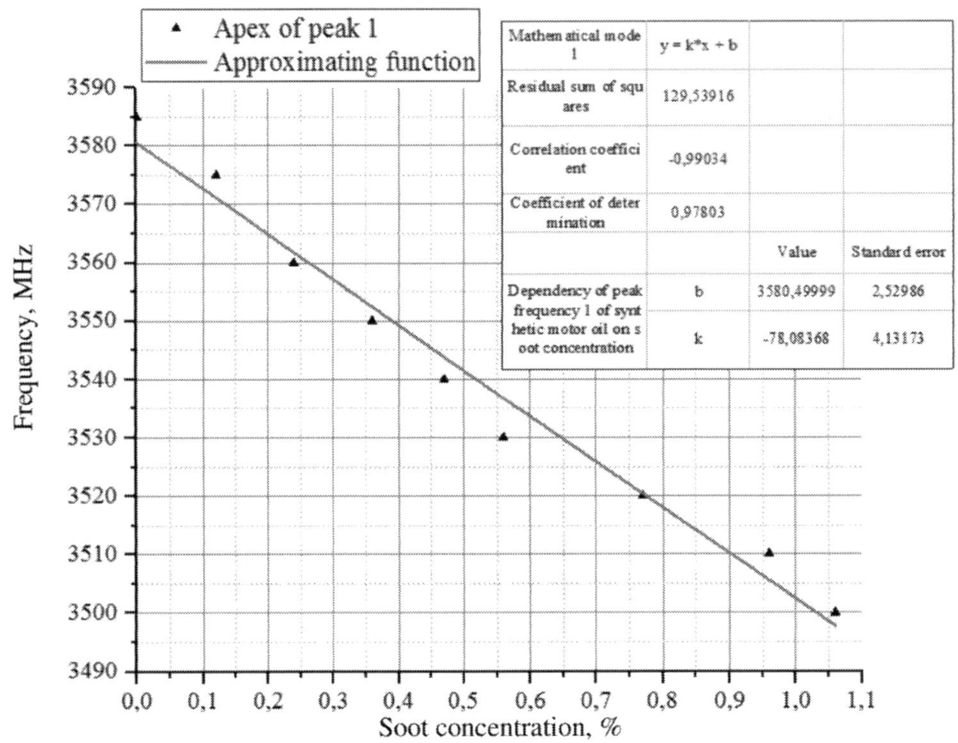

Figure 6. Dependency of peak frequency 1 of synthetic motor oil on soot concentration

The high correlation coefficients indicate a practically functional dependency of the amplitude on the selected optimal frequencies, and the peak frequencies on the concentration of soot. The determinant coefficients are high, and this means that the models constructed explain 93 ... 99% of the variation in values relative to their mean value.

5. Conclusion

The complex of experimental studies made it possible to establish the character of the change in the dielectric permitivitty of the engine oil with increasing soot concentration. Obtained are the diagnostic parameters, such as the peak frequencies, and the amplitude of the signal at the optimum frequency, where measurements most informatively describe the change in the state of the engine oil.

The mathematical model dependencies of peak frequencies and amplitude of various engine oil samples to the concentration of soot is being obtained. It should be noted that the obtained dependencies are almost completely identical for semi-synthetic and mineral oils, which can be explained by their relative chemical composition.

References

[1] Sam G, Santhosh B, Vishaal G, Mridul G 2007 Effect of diesel soot on lubricant oil viscosity *Tribology International* Vol 40 **5** 809-818 Available at: http://dx.doi.org/10.1016/j.triboint.2006.08.002

[2] Yudin A V 2014 *The method of controlling the effect of oxidation and thermal degradation processes on the antiwear properties of motor oils* Dissertation 05.11.13 (TPU, Tomsk - Krasnoyarsk)

[3] Runda M M 2014 *The method of monitoring the state of motor oils during long-term storage of machinery* Dissertation 05.11.13 (TPU, Tomsk – Krasnoyarsk)

IOP Conf. Series: Materials Science and Engineering 1159 (2021) 012094 doi:10.1088/1757-899X/1159/1/012094

[4] Belyaev S V 1993 *Motor oils and engine lubricatio* (Petrozavodsk: PetrSU)
[5] Grigoriev M A, Bunakov B M, Doletsky V A 1981 *Engine oil quality and engine reliability* (Moscow: Standards)
[6] Rigol S, Schenk M., Perryman R., Hosny W M 2009 Challenges in the onboard oil condition monitoring *Proceedings of Advances in Computing and Technology, (AC&T) The School of Computing and Technology 4th Annual Conference, University of East London* pp 26-34 http://citeseerx.ist.psu.edu/viewdoc/download?doi=10.1.1.426.9180&rep=rep1&type=pdf
[7] Mauntz M, Kuipers U, Gegner J 2011 New Electric Online Oil Condition Monitoring Sensor – an Innovation in Early Failure Detection of Industrial Gears *IMETI 2011 - 4th International Multi-Conference on Engineering and Technological Innovation, Proceedings*
[8] Perez A T, Hadfield M 2011 Low-cost oil quality sensor based on changes in complex permittivity *Sensors* **11** 10675-10690
[9] Bouaïcha A, Fofana I, Farzaneh M, Seytashmehr A, Borsi H, Gockenbach E, Béroual A, Aka N T 2009 On the Usability of Dielectric Spectroscopy Techniques as Quality Control Tool *IEEE Electr. Insul. Mag.* Vol 25 **1** 6-14
[10] Carey A A 1995 *The Dielectric Constant of Lubrication Oils* (USA, OH, Cleveland: CRC press)
[11] Rigol S 2011 *Monitoring Concept to Detect Engine Oil Condition Degradations to Support a Reliable Drive Operation. A thesis submitted in partial fulfilment of the requirements of the University of East London for the degree of Doctor of Philosophy*
[12] Zhu J, Yoon J M, He D, Qu Y, Bechhoefer E 2013 Lubrication oil condition monitoring and remaining useful life prediction with particle filtering *Int. J. Prognosis Health Manage* Vol 4 2-6
[13] Sugatov N N and Khaziev A A 2013 Selection of the optimal method for assessing the quality of a running engine oil by dielectric properties *Modernization and scientific research in the transport complex* (Perm: Perm National Research Polytechnic University) 343-354
[14] Green D A and Lewis R 2008 The effects of soot-contaminated engine oil on wear and friction: A review *Proceedings of the Institution of Mechanical Engineers Part D Journal of Automobile Engineering* **222** **(9)** 1669-1689 Available at: https://www.researchgate.net/publication/245391734_The_effects_of_soot-contaminated_engine_oil_on_wear_and_friction_A_review
[15] Khaziev A A, Postolit A V, Sugatov N N, Laushkin A V 2017 Investigation of the effect of water concentration on the dielectric properties of motor oil *Bulletin of the Moscow Automobile and Highway State Technical University (MADI)* **1 (48)** 21-27

IOP Conf. Series: Materials Science and Engineering 1159 (2021) 012095 doi:10.1088/1757-899X/1159/1/012095

Justification of efficiency of route fuel consumption norms for trucks operated in tropical conditions

A N Rementsov, V A Egorov and Fan An Vu

Moscow Automobile and Road Constraction State Technical University (MADI), Leningradsky prospect, 64, Moscow, 125319, Russia

E-mail: yegorov@tea.madi.ru, rementsov@yandex.ru, anhvumta@gmail.com

Abstract. Over the past decade, the economies of Southeast Asian countries, including Vietnam, have been developing dynamically, which leads to an increase in demand for the transportation of goods and passengers. At the same time, in a number of countries in this region, there is no state regulation in the field of rationing fuel consumption of trucks in real conditions. Therefore, improving the current fuel rationing system is an urgent task when performing cargo transportation in the tropics. The paper offers a justification for rationing fuel consumption of trucks performing transport work on permanent routes in tropical conditions on the example of Vietnam.

1. Introduction

In studies on fuel rationing in tropical conditions, experimental research was carried out in the Socialist Republic of Vietnam (SRV) for vehicles that perform cargo transportation on permanent routes. In the course of conducting research to analyze the possibility of using route fuel consumption regulation in Vietnam for trucks that perform transportation on established routes (often implemented when fulfilling transport orders in the construction industry), in accordance with the previously analyzed data [1,2,3] and the set goals [4,5], the tasks of choosing an experimental motor transport enterprise, collecting statistical information on fuel consumption on different routes, etc., were solved.

On the basis of the obtained approximations of average route fuel consumption, taking into account the data of the combined statistics on the variation of the studied random variables, , in order to conduct a more detailed study, the standards of hard (85% quantile) and soft (95% quantile) restrictions were calculated for the experimental truck models.

The main effect of the introduction of route rationing of fuel consumption into the practice of a motor transport enterprise is manifested in the discipline of drivers [6,7,8] when organizing more precise control over fuel consumption on the basis of reasonable norms. This makes it possible to make stricter demands on drivers without overestimating the standard. To assess the effectiveness of the study, an approximate calculation of the economic efficiency of using the calculated route fuel consumption standards in an experimental motor transport enterprise was also performed.

Content from this work may be used under the terms of the Creative Commons Attribution 3.0 licence. Any further distribution of this work must maintain attribution to the author(s) and the title of the work, journal citation and DOI.
Published under licence by IOP Publishing Ltd

IOP Conf. Series: Materials Science and Engineering 1159 (2021) 012095 doi:10.1088/1757-899X/1159/1/012095

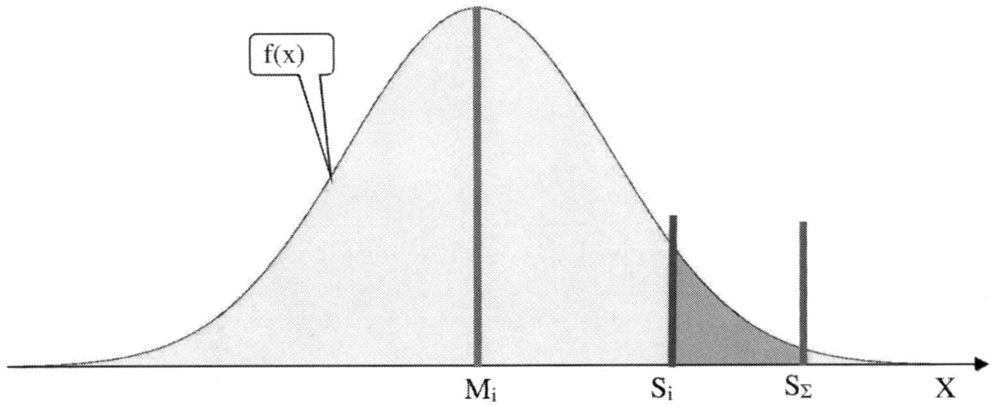

Figure 1. An illustration of the formation of the total fuel consumption by an enterprise under various assigned standards

2. General analysis of the efficiency of rationing fuel consumption for individual sub-groups (clusters) of vehicles

Let us assess the gain in fuel costs arising from the implementation of the fuel rationing system, at which the standard, i.e. the upper limit of the value of specific fuel consumption, within which the fuel costs are considered as acceptable, is assigned taking into account the properties and characteristics of the vehicle's route, in comparison with the standard norm assigned as a whole, for a generalized group.

Let the probability distribution of specific fuel consumption have the form shown in Fig. 1 with the parameters M - mathematical expectation and σ - standard deviation (In this estimated conclusion and further calculations, we rely on the form of the normal distribution, as the most common, and indeed manifested for the collected fuel consumption values separately along the routes. However, it should be noted that the conclusion formulas are valid for various types of probability distributions).

Figure 1 shows M_i – the mathematical expectation of fuel consumption on the i-th route, S_i and S_Σ - various assigned standards (individual and generalized), as well as the probability distribution density function of the true specific fuel consumption on the i-th route f(x) with color-marked zones: light green – that do not make any difference in the formation of total fuel consumption, dark green (swamp)- that make a difference.

Let's first determine the possible fuel consumption at a certain given permissible upper limit (standard norm) S, based on the fact that with a real consumption of a technically serviceable car less than S, with a probability k (in fact, it can be called the "temptation coefficient"), the consumption will be equal to S, - this is a fuel drain that is still widely implemented in real practice in Vietnam, and insufficient care of the vehicle, not optimal driving mode, poor quality control of technical condition, etc., - indeed, when the vehicle is operated on routes with low difficulty coefficients that require less fuel consumption, the driver may pay less attention to factors that lead to increased fuel consumption, especially if the standard is assigned for a generalized group of vehicles-operation on all routes of the enterprise as a whole, or even a national standard norm. According to preliminary estimates, the coefficient k can range from 0.3 to 0.8. If the actual fuel consumption is greater than S, the value of the specific fuel consumption added to the weighted average value does not change. This zone (x>S) is a zone of risk, a zone of social discomfort, tension, because at real consumption X, the driver is required to spend no more than S. As usual, however, when the standard is set correctly (95% quantile, and even in the case of particularly significant parameters, such as those affecting the safety of traffic, or when rationing strategically important resources, such as in certain cases fuel in the SRV, 85% quantile), this zone includes cases related to the reasons listed above, set by not quite efficient operation of the trucks.

Let's determine the weighted average value of the "additional, excess" value of the specific fuel consumption:

$$Q_s = k \cdot \int_{g_1}^{S} f(x, M, \sigma)(S - x)dx =$$

$$= k \cdot \int_{g_1}^{S} (f(x, M, \sigma) \cdot S)dx - k \cdot \int_{g_1}^{S} (f(x, M, \sigma) \cdot x)dx =$$

$$= k \cdot S \cdot F_0(\frac{S - M}{\sigma}) - k \cdot \int_{g_1}^{S} (f(x, M, \sigma) \cdot (x - M + M)dx$$

where g_1 is the lower limit of integration, in the general case (for a normal distribution, especially with small coefficients of variation of a random variable, for the convenience of integration, it is usually taken $-\infty$),

$F_0(z)$ - Cumulative function of the probability distribution law of the normalized random variable (for the standard norm selected in accordance with the methodology for a given random variable, $F_0((S-M)/\sigma)$ is equal to 0.85 under a hard constraint and 0.95 under a soft constraint).

Let's take $y = x-M$ and transform the last term:

$$\int_{g_1}^{S} (f(x, M, \sigma) \cdot (x - M + M)dx \xrightarrow[y=x-M]{} \int_{g_1-M}^{S-M} (f(y, 0, \sigma) \cdot y)dy +$$

$$+ \int_{g_1}^{S} (f(x, M, \sigma) \cdot M dx = - \int_{S-M}^{+\infty} (f(y, 0, \sigma) \cdot y)dy + M \cdot F_0(\frac{S - M}{\sigma}) =$$

$$= -\sigma \cdot \int_{\frac{S-M}{\sigma}}^{+\infty} (f(z, 0, 1) \cdot z)dz + M \cdot F_0(\frac{S - M}{\sigma})$$

Finally, we get:

$$Q_s = k \cdot \left[(S - M) \cdot F_0(\frac{S - M}{\sigma}) + \sigma \cdot \int_{\frac{S-M}{\sigma}}^{+\infty} (f(z, 0, 1) \cdot z)dz \right] \quad (1)$$

So, for values $S_\Sigma > S_i$ (taking into account that $S_i = M_i + r * \sigma_i > M_i$), we have for the difference in the values of excess specific fuel consumption:

$$Q_{S_\Sigma} - Q_{S_i} = k \cdot \left[(S_\Sigma - M_i) \cdot F_0(\frac{S_\Sigma - M_i}{\sigma_i}) + \sigma_i \cdot \int_{\frac{S_\Sigma - M_i}{\sigma_i}}^{+\infty} (f(z, 0, 1) \cdot z)dz \right] -$$

$$-k \cdot \left[(S_i - M_i) \cdot F_0(\frac{S_i - M_i}{\sigma_i}) + \sigma_i \cdot \int_{\frac{S_i - M_i}{\sigma_i}}^{+\infty} (f(z, 0, 1) \cdot z)dz \right] =$$

$$= k \cdot \left[(S_\Sigma - S_i) \cdot F_0(\frac{S_i - M_i}{\sigma_i}) + (S_\Sigma - M_i) \cdot (F_0(\frac{S_\Sigma - M_i}{\sigma_i}) - F_0(\frac{S_i - M_i}{\sigma_i})) \right] - \quad (2)$$

$$-k \cdot \sigma_i \cdot \int_{\frac{S_i - M_i}{\sigma_i}}^{\frac{S_\Sigma - M_i}{\sigma_i}} (f(z, 0, 1) \cdot z)dz$$

This expression characterizes the expected difference in fuel consumption for any one i-th route when assigning an individual norm compared to the assigned general group standard norm (for all routes at once), which in General can be either positive or negative.

Let us simplify the expressions presented above and consider the total difference for all routes of the group (enterprise) - based on the specific consumption.

Since $S_i = M_i+\sigma_i*r$, and $\sigma_i=M_i*\upsilon_i$ and $F_0(r)= \Theta$ – is the level of the specified limit of the standard (for example, $\Theta= 0.85$, and r is the corresponding quantile), we obtain from (1):

$$Q_{S_i} = k\cdot\left[\sigma_i\cdot r\cdot F_0(r)+\sigma_i\cdot\int_r^{+\infty}(f(z,0,1)\cdot z)dz\right] \qquad (3)$$

Table 1. Values of the elements of formula (3) for different values of the level of the specified norm limit.

Θ	r	$\int_r^{+\infty}(f(z,0,1)\cdot z)dz$
0,85	1,037	0,233
0,95	1,64	0,104

Assuming that the influence of variations in parameters that affect fuel consumption, but are not related to the characteristics of the routes (such as the technical condition of the truck, various driver qualities, etc.), on different routes served by the same automobile enterprise, can be assumed to be very close for this estimate we will accept the invariance of the coefficient of variation for the spread of values of specific fuel consumption by trucks of the same model on one (i-m) route: $\upsilon i= \upsilon \vee i$.: From this assumption we get:

$$Q_{S_i} = k\cdot\upsilon\cdot M_i\cdot[r\cdot\Theta+\tau] = C_k\cdot\upsilon\cdot M_i,$$

Let's take the notation: $C_k = [r\cdot\Theta+\tau]$.

Then, to combine routes into one group, we sum up the Qs_i for all I-th routes with weights – the probabilities of finding a truck on the i-th route, i.e. with the part of the i-th route in the total volume of traffic (for the truck model under consideration):

$$Q_\Sigma = \sum P_i\cdot Q_{S_i} = C_k\cdot\upsilon\cdot\bar{M}$$

In the case of calculating the considered value for the standard assigned for the entire combined group of routes, we have the following expressions:

$$S_2 = \bar{M}+r\cdot\sigma_U,$$

where σ_U - is the standard deviation for the combined route group, determined from:

$$\sigma_U^2 = \sum_{i=1}^n P_i\sigma_i^2 + \sum_{i=1}^n P_i(M_\Sigma - M_i)^2 = \upsilon\cdot\sum_{i=1}^n P_iM_i^2 + \sum_{i=1}^n P_i(M_\Sigma^2 + M_i^2 - 2\cdot M_\Sigma\cdot M_i)^2 =$$

$$= \sum_{i=1}^n P_i(M_i^2(1+\upsilon^2) + M_\Sigma^2 - 2\cdot M_\Sigma\cdot M_i)^2 = -M_\Sigma^2(1+\upsilon^2)+$$

$$+(1+\upsilon^2)\cdot\sum_{i=1}^n P_i\cdot M_i^2 + M_\Sigma^2\cdot\upsilon^2 = (1+\upsilon^2)\cdot(\sum_{i=1}^n P_i\cdot M_i^2 - M_\Sigma^2) + M_\Sigma^2\cdot\upsilon^2$$

Thus, the coefficient of variation of the combined group is determined by:

$$\upsilon_{\text{U}}^2 = \frac{\sigma_{\text{U}}^2}{M_\Sigma^2} = (1+\upsilon^2)\cdot\frac{D(M_i)}{M_\Sigma^2} + \upsilon^2 = (1+\upsilon^2)\cdot\upsilon(M_i)^2 + \upsilon^2$$

where $D(M_i)$, $\upsilon(M_i)$ are the operators of taking, respectively, the variance, the coefficient of variation for the sample consisting in this formula of the mathematical expectations of specific fuel consumption on the i-th route.

Therefore, the difference in expected fuel consumption can be estimated using the formula:

$$Q_{S_2} - Q_\Sigma = \bar{M}\cdot C_k \cdot (\sqrt{\upsilon^2 + (1+\upsilon^2)\cdot\upsilon(M_i)^2} - \upsilon)$$

and it is obviously positive, i.e. the expected fuel consumption in the case of setting the standard for the entire combined group at once will be higher. It should be noted that in any case of inequality of coefficients of variation for statistics on different routes, the excess of expected fuel consumption is also proved when assigning a combined standard using the same method.

Let $\upsilon i \neq \upsilon j$ be at least for some i,j. Then for the sum of individual routes we have the expression:

$$Q_\Sigma = \sum P_i \cdot \sigma_i \cdot C_k = M(\sigma_i)\cdot C_k$$

where M (..) is the operator for taking the mathematical expectation from the corresponding sample.

Let's consider expressions for a given generalized standard norm S_2. While:

$$D(\sigma_i) = -(M(\sigma_i))^2 + M(\sigma_i^2) \geq 0 \Rightarrow$$

$$(M(\sigma_i))^2 \leq M(\sigma_i^2) \Rightarrow$$

For the combined group we have:

$$Q_{S_2} = C_k \cdot \sigma_{\text{U}},$$

Compare the squares of values Q_Σ and Q_{S_2} :

$$Q_\Sigma^2 = C_k^2 \cdot M(\sigma_i)^2 \leq Q_{S_2}^2 = C_k^2 \cdot \sigma_{\text{U}}^2 = C_k^2 \cdot (M(\sigma_i^2) + D(M_i))$$

Consequently, the expected fuel consumption when assigning individual route standards will be less than when assigning the general group standard by the same method, and the difference will be the greater, the more the complexity of the routes differ.

The calculation of the difference in the total value of the specific fuel consumption when setting different normative boundaries, carried out according to formula (2), concerns the consumption of one model for any one route. To calculate what the effect will be for all routes of the same model, you need to find the average value of such differences for each route, and in the case when the load on the routes is not clearly defined, you can find a simple average value. If the company's work program is stable enough, and the route load is clearly set for a long time, it is more correct to take a weighted average value, where the average daily mileage of vehicles operated on the corresponding route is used as a weight.

An illustration of the relative positions of functions and compared regulatory boundaries for each route used for calculating the difference in total fuel consumption using formulas (2) is shown in figure 2 a,b.

Here in figure 2a shows a separate fragment of the entire field of functions with the notation and parameters consistent with figure 1, used to construct the output of formula (2), and with the shape and color representation consistent with figure 2b, and in Fig. 2b shows the complex formation of the difference in total consumption.

The density functions of the probability distribution of the specific fuel consumption separately for each route are shown in green, while the specific fuel consumption Q is given on the abscissa axis, and the route number is indicated on the ordinate axis. For each route, the standard S_1 calculated for this route is shown in blue. The red color indicates the limits of the S_2 standard defined for the entire set of

IOP Conf. Series: Materials Science and Engineering 1159 (2021) 012095 doi:10.1088/1757-899X/1159/1/012095

vehicles, based on the same principle as for individual routes, i.e. the corresponding quantile (the very distribution of probabilities for fuel consumption for the generalized group of trucks is shown in red at the position of route No. 11).

The zones between S_{1i} and S_2 are shown in different colors depending on whether the total flow consumption for a particular route decreases (red, $S_1 < S_2$) or increases (blue, $S_1 > S_2$). The data for constructing the illustration are taken from the results of calculations for the KAMAZ-55111 model, dry period, with a soft assignment of the standard, i.e., the cut-off probability is 0.95. It should be noted that when using vehicles in the tropics, it is necessary to consider separately the two seasons of operation - dry and rainy (wet).

To calculate the generalized effect from the use of individual route standards for each individual vehicles model, after calculating the averaged values of the effect both in absolute values (l / 100 km.) and in relative (%) for each season, these values were averaged over the entire year (i.e. for two seasons) to obtain the resulting performance indicators for each individual truck model.

To obtain the total final effect, the performance indicators for each model were added with weights equal to the part of individual truck models in the enterprise.

At the same time, a generalized standard can be assigned for each vehicles model as a whole throughout the whole year, as it happens now in Vietnam, and with division by seasons. In tropical climatic conditions (for example, in Vietnam), the division into dry and rainy seasons adopted in this work seems to be the most logical, as noted above. Further calculations, which were carried out using the formulas presented above, used the previously obtained values for the route norms for specific fuel consumption separately by seasons for each of the considered car models, together with the values of the mathematical expectation of consumption obtained after factor analysis, the standard deviation for the probability distributions of specific fuel consumption, obtained from the collected statistical data separately for routes and seasons. The corresponding values were also determined and used in the calculation for a generalized group of trucks (for each model separately) per season. For a complete comparison, we used data on currently existing single (model-based) standards in the SRV.

Although for traction and economic indicators, it is customary to use the assignment of standard norms for 85% quantile ("soft" standard), calculations according to a specially created program were also carried out for a more rigid option - 95% quantile. The effect of applying route rationing was also calculated in relation to the standards assigned in an individual motor transport enterprise for a generalized group of trucks (with a natural division by season and model), as well as in relation to the currently accepted industry model standards.

In accordance with the calculations, it can be concluded that even with the appointment of "soft" standards, the gain even in relation to the appointment of a norm for the generalized group of routes will be from 3.4% to 4.1%, depending on the redistribution of the volume of transport work along the routes in the enterprise, and in comparison, with the industry-wide standard, the gain reaches 4.4% - 5%. In absolute terms, this effect for the pilot enterprise is from 94 to 140 tons of fuel per year. When using more stringent rationing, the effect can range from 66 (2.4%) to 197 tons per year. Even assuming the above-described "temptation coefficient" of 0.5 in the current diesel prices in Vietnam, the minimum effect is 0.89 billion Vietnamese Dong or 2.7 million rubles.

IOP Conf. Series: Materials Science and Engineering 1159 (2021) 012095 doi:10.1088/1757-899X/1159/1/012095

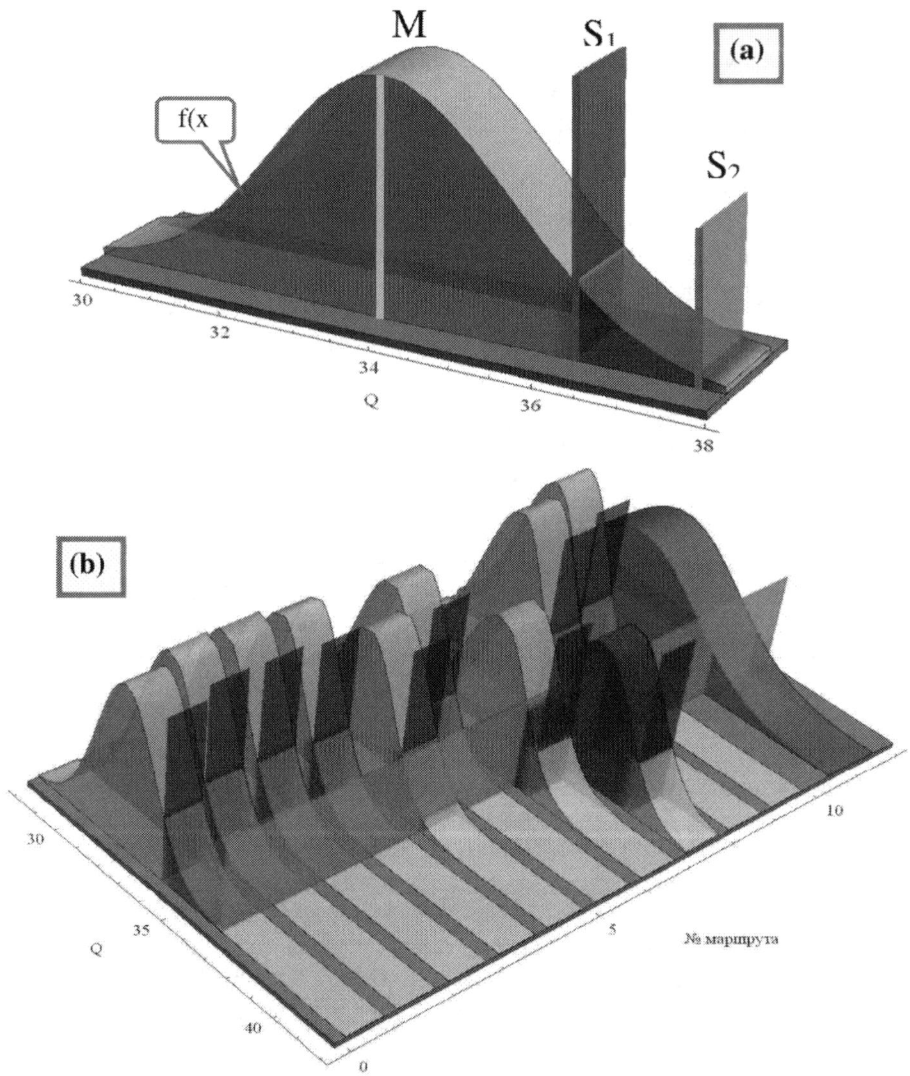

Figure 2. Illustration of finding the complex effect of applying separate route standards on the specific fuel consumption in an enterprise.
(a) – a separate element; (C) – a composition of jointly considered probability distribution densities for specific fuel consumption on individual routes of the enterprise.

3. Conclusion

Based on the above, it can be stated that in Vietnam, as in other countries of South-East Asia, the trend of increasing cargo turnover continues, with a corresponding increase in the number of vehicles in operation. This leads to an increase in fuel consumption in road transport companies. To reduce operating costs during cargo transportation, it is necessary to improve methods for calculating fuel consumption standard norms based on differentiated consideration of many factors that affect fuel consumption. The proposed justification for rationing fuel consumption by trucks will make it possible to more correctly assign fuel consumption standards for vehicles performing transport work in the tropics.

References

[1] Lasko V I, Ta Ch H 2014 The role of freight road transport in the development of the economy of the SRV *International Economics* 6 55-60
[2] Nguyễn Mạnh Hùng, Đinh Quang Toản 2016 Thực trạng loại hình vận tải đa phương thức trong ngành dịch vụ logistics tại Việt Nam *Giao Thông Vận Tải* **10/2016** 34-36
[3] Borisov G V, Leliovskiy K Y and Pachurin G V 2015 To the question of rationing the consumption of liquid fuels in road transport *Fundamental research* **3** 28-35
[4] Ispolatov B Yu 2005 *Development of a methodology for standardizing the route fuel consumption of the Li-AZ-6212 bus in operation.* Diss. Ph. D. (Moscow: MGIU)
[5] Rementsov A N, Fan An Vu, Egorov V A and Antonov D V 2020 Formation of fuel consumption norms for cargo transportation in Vietnam. *IOP Conference Series: Materials Science and Engineering* Vol 832 **1** Published under license by IOP Publishing Ltd. DO - 10.1088 / 1757-899x / 832/1/012067
[6] Kuzmin N A 2010 The problem of rationing the consumption of automobile fuels and lubricants in the Russian Federation *Motor transport enterprise* **8** 20-22
[7] Nguyễn Thị Thu Hà 2017 Phát triển hạ tầng giao thông nhằm nâng cao hiệu quả logistics tại Việt Nam *Giao Thông Vận Tải* **5/2017** 2017 87-89
[8] Rementsov A N, Egorov V A and Fan A V 2019 Rationing of fuel consumption of trucks in the conditions of Vietnam *Truck* **9** 36-40

IOP Conf. Series: Materials Science and Engineering 1159 (2021) 012096 doi:10.1088/1757-899X/1159/1/012096

Comparative characteristics of wear resistance of electroplated copper coating and composite electrolytic coating obtained with use of lead bronze electroerosive materials

E V Ageev[1], A N Novikov[2], M Yu Karelina[3]

[1]South-West State University, 94, 50 let Oktyabrya., Kursk, 305040, Russia
[2]Orel State University named after I.S. Turgenev, 77, Moskovskayastr., Orel, 302030, Russia
[3]Moscow Automobile and Road State Technical University (MADI), 64, Leningradskyave, Moscow, 125319, Russia

E-mail: karelinamu@mail.ru

Abstract. This paper presents the results of a study of the elemental composition of dispersed particles obtained from Brs30 waste, as well as their morphology. Samples of metal products with applied copper composite coatings with the obtained particles were experimentally made and their physical and mechanical properties were studied. It is noted that samples with composite coatings based on copper obtained from an electrolyte suspension with the addition of dispersed particles of electroerosive lead bronze of the Brs30 brand have the best wear resistance.

1. Introduction

Currently, the urgent problem is the processing of waste lead bronze. Of greatest interest among the existing technologies for processing metal waste is the technology of obtaining powder materials based on the phenomenon of electrical erosion - electroerosive dispersion[1-5]. Its main advantages are environmental friendliness and minimal energy consumption. Among the prospects for the repeated use of materials obtained by this method, it should be noted that there is a possibility of their use as materials for metallization and galvanic coatings with improved physical and mechanical properties. Currently, in the modern scientific and technical literature there is no complete information on the use of dispersed particles obtained from waste lead bronzes by electrical erosion, in the formation of galvanic coatings and their effect on wear resistance [6-9].

The purpose of this work was to study the characteristics of wear resistance of galvanic copper coatings and composite coatings based on copper obtained from a suspension electrolyte with the addition of dispersed particles of electroerosive lead bronze grade BrS30.

2. Materials and methods

Waste of BrS30 alloy in the form of shavings was chosen as a starting material for obtaining dispersed particles. Distilled water was used as a working liquid [10, 11]. The deposition of coatings of two types was carried out on an L1 DIGIT galvanic installation. Nickles made of steel grade 30 HGSA

Content from this work may be used under the terms of the Creative Commons Attribution 3.0 licence. Any further distribution of this work must maintain attribution to the author(s) and the title of the work, journal citation and DOI.
Published under licence by IOP Publishing Ltd

IOP Conf. Series: Materials Science and Engineering 1159 (2021) 012096 doi:10.1088/1757-899X/1159/1/012096

were used as a substrate for coating. To deposit a coating from a universal sulfuric acid electrolyte of copper plating, the following electrolyte composition was used: $CuSO_4 \cdot 5H_2O$ (200-250 g/l), H_2SO_4 (35-70 g/l), NaCl (0,01g/l), lemon acid (0,01 g/l) [5-6]. To prepare a suspension electrolyte suitable for obtaining a composite electrolytic coating, all these reagents must be introduced into a suspension of electroerosivelead bronze powder. The deposition of coatings was carried out with the following parameters: anode material - anode copper of the AMF brand; anode-cathode area ratio - 1: 1; bath material - chemical glass; working temperature – 18-25 °C; voltage – 2-5 V; current density range – 5-10 Aperdm2; deposition rate – from 1 µm /1 min. (at 5Aperdm2); stirring - obligatory.

The coefficient of friction and the rate of wear of the surface of the samples and counterbody were measured on an automated friction machine (Tribometer, CSM Instruments, Switzerland). The tests were carried out in air at a load of 2N and a linear velocity of 10 cm/s, with a radius of curvature of wear of 5-6 mm, the friction path was 200 meters.

3. The morphology and X-ray spectral microanalysis

Reasoning The results of the study of morphology and X-ray spectral microanalysis of the obtained dispersed particles are presented in Table 1 and Figure 1.

Table 1. Results of X-ray spectral microanalysis

Element	Massfraction, %	Atomicfraction, %
O	7,77	29,80
Sn	1,77	0,92
Fe	0,28	0,31
Cu	55,59	53,71
Zn	5,86	5,50
Pb	28,10	8,33
Al	0,63	1,44
Total	100,00	100,00

Figure 1. Microstructure and morphology of dispersed particles

Based on the results of the presented data, it was established that the main elements in the powder obtained by the method of electroerosive dispersion of lead bronze waste in distilled water were oxygen, copper, zinc and lead. The rest of the elements are distributed relatively evenly in the powder [7]. It has been experimentally established that the shape of the resulting particles is mainly spherical or elliptical, the average particle size is 9.73 μm, the arithmetic value is 9.731 μm [8-10].

The results of the study of the microstructure and morphology of the obtained standard and composite galvanic coatings are presented in Figure 2 and Table 2.

Table 2. Results of X-ray spectral microanalysis of the obtained coatings

Coating	Element, %					
	Al	Si	Fe	Ni	Cu	Pb
Standard	0,84	0,67	1,59	1,68	95,23	-
Composite	0,72	0,92	1,32	0,40	89,52	7,12

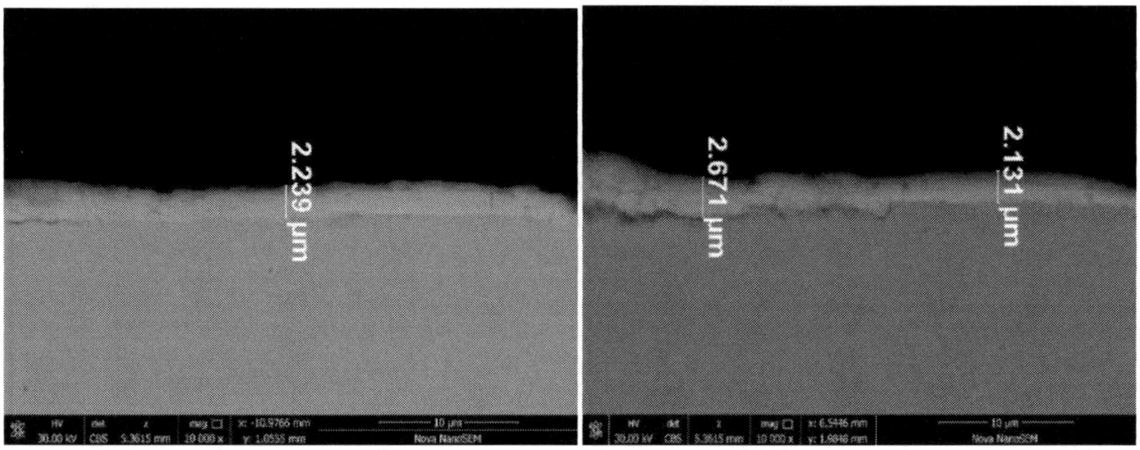

a) b)

Figure 2. Microstructure and morphology of the obtained coatings:a) standard; b) composite

The study of the microstructure of the transverse section of the samples showed that coatings with a thickness of 2-2.7 μm were obtained without visible defects at the "coating/substrate" interface. The coatings have a mixed structure with well-melted powder particles.

The small thickness of the coatings is due to the choice of the deposition time in accordance with the recommended thickness of the copper coatings.

The results of the study of the wear resistance of the experimental samples are shown in Table 3. It should be noted that the pores of the substrate are more uniformly filled with a composite coating with the addition of electroerosive lead bronze.

Table 3. Tribological characteristics of electroplated coatings

Sample	Statistical partner wear factor (ball of Al_2O_3 Ø6 mm), $mm^3{*}N^{-1}{*}m^{-1}$	Sample wear factor, $mm^3{*}N^{-1}{*}m^{-1}$
Standardcoating	3,753	407,9
Compositecoating	0,107	50,94

4. Research findings

It has been experimentally established that the wear factor of the statistical partner when testing a sample with a standard electroplated copper coating is higher than the wear factor of a statistical partner when testing a sample with a composite electrolytic coating based on copper obtained from a suspension electrolyte with the addition of electroerosive lead bronze particles. It is noted that the wear factor of a specimen when testing a specimen with a standard electroplated copper coating is higher than that of a specimen when testing a specimen with a composite electrolytic coating based on copper obtained from a suspension electrolyte with the addition of electroerosive lead bronze particles.

This is also confirmed by the study of the wear track on the surface of the coatings after testing on a Quanta200 3D scanning electron microscope (Figure 3).

a)

b)

Figure 3. Wear tracks on the surface of coatings after testing: a) standardcoating; b) compositecoating

5. Conclusion

It was found that the main elements in the powder obtained by the method of electroerosive dispersion of lead bronze waste in distilled water were oxygen, copper, zinc and lead. The rest of the elements are distributed relatively evenly in the powder. The particle shape of the resulting conglomerates is spherical and elliptical, the average size is 9.73 μm.

Samples of metal products with standard and composite copper-based galvanic coatings applied to them were made, their physical and mechanical properties were investigated:

– coatings with a thickness of 2-2.7 μm were obtained without visible defects at the "coating-substrate" interface;

– the main elements in the coatings are Cu and Pb;

– composite electrolytic coatings obtained from an electrolyte suspension with the addition of particles of electroerosive lead bronze have the best wear resistance.

6. Acknowledgments

The study was carried out with the financial support of the Russian Foundation for Basic Research within the framework of the scientific project № 19-38-90039

References

[1] Putintseva M N 2007 Analysis of the surface of particles obtained by electroerosive dispersion of alloys WC-CO *Surface. X-ray, synchrotron and neutron research* **5** 97-102

[2] Petridis A V 2006 Production of powders from wastes of hard alloys by the method of electroerosive dispersion *Repairs. Recovery.Modernization* **8** 29-35

[3] Putintseva M N 2007 Processes in wc-co hard alloys during electroerosive dispersion *Physics and chemistry of material handling* **5** 61-66

[4] Dvornik M I, Verkhoturov A D, Yershova T B, Metlitskaya L P, Bruy V N 2006 Preparation of nanostructured tungsten-cobalt powder by electroerosive dispersion of VK8 hard alloy *Promising materials* **3** 70-75

[5] Roik B O, Zubarev M A 2014 Development of a step-by-step diagram of the process of electroerosive dispersion of high-speed steel in distilled water *Theoretical and applied aspects of modern science* **4-1** 20-22

[6] Zvonarev D Yu, Kuznetsova L P 2005 Increasing the wear resistance of chromium electroplated coatings due to the use of dispersed powders obtained by electroerosive dispersion (EED) *New materials and technologies in mechanical engineering* **4** 58-60

[7] Novikov E P 2016 Historical aspects of obtaining powders from conductive industrial waste by the method of electroerosive dispersion *Modern materials, equipment and technologies* **1 (4)** 170-177

[8] Chaplygin V Yu 2016 Research methodology for metal powders obtained by electroerosive dispersion *Modern materials, equipment and technologies* **1 (4)** 225-231

[9] Putintseva M N 2016 Crushing of tungsten carbide grains during electroerosive dispersion of alloys WC-Co *Auditorium* **4 (12)** 57-66

[10] Zadorozhny R N, Romanov I V, Shemberev I A 2018 Study of the properties of metal powders obtained from wastes of hard alloys by electroerosive dispersion *Proceedings of GOSNITI* Vol 130 pp 208-213

[11] Ageev E V, Latypova G R, Davydov A A, Ageeva E V 2012 X-ray spectral microanalysis of hard-alloy electroerosive powders *Bulletin of the South-West State University* **5-2 (44)** 099-102

[12] Ageev E V, Semenikhin B A, Latypov R A 2010 A method for obtaining nanostructured powders based on the WC-Co system and a device for its implementation *Fundamental and applied problems of engineering and technology* **5 (283)** 39-42

A comprehensive mathematical model and the methodology for assessing technical excellence road trains with active links and all-wheel control

A N Sova[1], S A Eruslankin [1], O A Valyaev[1], I M Makeev[1], I O Sudarikov[1]

[1] Department of transport installations, Moscow automobile and road state technical University (MADI), 64, Leningradsky prospect, Moscow, 125319, Russia

E-mail: a.sova@madi.ru, slsova@mail.ru

Abstract. This article presents the main provisions of the new scientific and methodological approach, integrated mathematical model and methodics assessing the technical perfection of wheeled transport means of particularly large capacity based on road trains with active links and all wheeled steering. The results of the research allowed us to justify scientific and methodological approach to the assessment of technical excellence wheeled transport means of particularly large capacity based on road trains with active links and all wheeled steering, which includes implemented conceptual apparatus, scientific concepts and principles, theoretical and practical provisions. Results of justification and developing integrated mathematical model include the total mass equations, and methodics wheeled transport means of particularly large capacity based on road trains with active links and all wheeled steering. Introduced the concept of the integrated mathematical model, justified and developed by the integrated mathematical model wheeled transport means of particularly large capacity based on road trains with active links and all wheeled steering, in which includes the equation of the total mass of transport means in absolute and relative values, equations of dynamics of sprung and unsprung masses of links of wheeled transport means, equations of joint dynamics of wheeled transport means taking into account the applications of the saddle device and all-wheel control, models of interaction of wheel propellers with the support surface, models of the micro-profile and macro-profile of the support surface, and other component parts united comprehensive model. On the basis comprehensive mathematical model proposed metodics assessment of the technical excellence of the road train with active links and all-wheel control, recommendations on the choice of indicators and evaluation criteria are justified.

1. Introduction

Modern technologies of design and the choice of technical and technological solutions when building wheeled vehicles especially big load capacity based on the application of modeling techniques and algorithmic and software complex (software packages) for their implementation allow to justify design-layout solutions, providing implementation of technical requirements the technical requirements to experimental-design works on creation of the wheel transport means especially large load capacity on the basis of road trains with active links and all-wheel control [1, 2]. Making a decision on the creation and mass production of wheeled transport means especially large load capacity on the basis of road trains with active links and all-wheel control (WV) it requires a reliable and complete feasibility study with the use of indicators and criteria for evaluating the technical excellence of the products being

Content from this work may be used under the terms of the Creative Commons Attribution 3.0 licence. Any further distribution of this work must maintain attribution to the author(s) and the title of the work, journal citation and DOI.
Published under licence by IOP Publishing Ltd

created, taking into account the volume of serial production, as well as the costs of development, development of production, production, experienced working out, commissioning and service, including service station. [3, 4].

In the terms of reference for the implementation of experimental design work on the creation (development, manufacture, testing) and operation of wheeled transport means the purpose of the WV, the features and conditions of its operation, the general characteristics of the WV or WV complex and the necessary initial data must be specified. For the design, the following initial data are required: the mass of the main payload m_{gr}, kg; weight of technological equipment $m_{t.o}$, kg; dimensions, volume and centers of mass of cargo and technological equipment (dimensional drawing); minimum specific pressure on the ground q, MPa; the fuel consumption S_T, km; total warranty mileage for the entire service life S, km; distribution of road conditions in % from the total mileage; overall or road-legal restrictions; maximum speed of movement v_{max}; values of the minimum D_{min} maximum D_{max} dynamic factors [5-8].

The procedure for preliminary calculations and identification of the general structural and layout scheme of the WV and its base chassis on a specific example is considered in [9]. When developing structural and layout schemes of WV and special automotive base chassis (ABCH), it is necessary to take into account the experience of creating their analogues and prototypes, represented by statistical data on the completed structures, given in the [5, 9].

In the section «comprehensive mathematical model of a WV» the article presents the results of justification and development comprehensive mathematical model.

2. Comprehensive mathematical model of a WV

Mathematical model WV – this is the way (prototype) WV, represented by mathematical symbols and relations which allows to investigate the properties of WV, determining its technical excellence and degree of conformity of the achieved result of the application of WV desired result, specific technical requirements, different conditions. Comprehensiveness of the mathematical model of WV is determined by its unity and the number of factors that affect the technical perfection of WV, taking into account its structure, characteristics, modes and operating conditions. Comprehensive mathematical model wheelbase transport means especially large load capacity on the basis of road trains with active links and all-wheel control includes the equation of the total mass of WV in absolute and relative values, the equations of dynamics of sprung and unsprung masses of WV links, the equations of joint dynamics of WV links taking into account the use of a saddle device and all-wheel control, models of interaction of wheel propeller with seating surface, models of the micro profile and macro profile of the reference surface, and others component parts single comprehensive model [1, 2].

The total mass of any WV can be represented as an equation [5]:

$$m_0=m_r+m_{Ka}+m_a+m_{upr}+m_T+m_{op}+m_{dop}+m_{su}+m_{mp}+m_p,\ kg, \tag{1}$$

where m_0– gross weight vehicles (WV) with cargo, kg; m_r– weight of the frame (body), including load-bearing elements(spars), cross-links, bumpers and towing devices, frame brackets, control cabins and tail, kg; m_{Ka}– weight of WV wheel movers, taking into account the weight of tires, rims and hubs with bearing units, central pumping system and tire air,kg; m_a– weight of the springing system, taking into account the weight of the guides, elastic and damping elements, as well as the control and stabilization system of the housing, kg; m_{upr} – weight of chassis controls including power steering, wheel brakes, air receivers, etc.,kg; m_{op}– weight of hydraulic supports of the hanging system, kg; m_T– fuel weight including fuel tanks and fuel equipment located outside the engine, kg; m_{dop} – weight of additional equipment of the basic machine, including electrical equipment, spare parts, tools and accessories (SPTA), winches and other unrecorded mass, kg; m_{su} – weight of the power plant, taking into account the weight of the engine, its service systems, coolant and oil, kg; m_{mp} – the mass of the powertrain given the wheel gear and actuator control, kg; m_p– the mass of the payload taking into account the technological equipment and technical systems for cargo service, kg.

In relative (specific) parameters, the equation (1) it will have the following form [5]:

$$1 = \xi_r + \xi_{\kappa a} + \xi_a + \xi_{upr} + \xi_t + \xi_{op} + \xi_{dop} + \xi_{su} + \xi_{tp} + \xi_p, \quad (2)$$

where $\xi_i = \frac{mi}{m0}$ – relative masses of elements on the right side of the equation (1).

Based on the analysis of statistical data and design experience, all elements of the WV mass equation can be divided into three main groups. The first group includes elements of the chassis (frame, wheel propellers, etc.), the mass of which does not depend or practically does not depend on the WV energy capacity taken into account, but is determined mainly by the level of cross-country ability, permissible overloads, materials used and design quality. The second group includes elements (power plant and chassis transmission), the mass of which significantly depends on the energy capacity of the WV. The third group includes elements that make up the WV payload, including the transported cargo and technological equipment (TE) for its installation and maintenance.

For the first and second groups of elements, we introduce the notation [5]:

$$\xi k = \xi_r + \xi_{\kappa a} + \xi_a + \xi_{upr} + \xi_t + \xi_{op}. \quad (3)$$

$$\xi_{su} = m_{u.su} \cdot N_{u.e.} \quad (4)$$

$$\xi_{tr} = m_{u.tr.} \cdot N_{u.e.} \quad (5)$$

where $m_{u.su}$ и $m_{u.tr}$ – specific reduced masses of the power plant and transmission, respectively, kg/ kw: $m_{u.su} = m_{u.su} / N_e \eta_\Sigma$; $m_{u.tr.} = m_{u.tr} / N_e \eta_\Sigma$; N_e – rated power of the power plant, kW; $\eta_\Sigma = \eta_g \eta_y \eta_{tr}$ – total efficiency, taking into account the power consumption for the engine's own needs (η_g), on the steering wheel (η_y) and losses in the transmission ($\eta_{\text{тр}}$); $N_{u.e.}$ – total efficiency, taking into account the power consumption for the engine's own needs WV, kW/kg: $N_{u.e.} = \frac{N_e \eta_\Sigma}{m_o}$.

By solving the equations (2) relative mass of the payload (M_{PN}) taking into account the equations (3), (4) and (5), we get [5]:

$$\xi_п = 1 - \xi K - (m_{u.su} + m_{u.tr}) N_{u.e.} \quad (6)$$

When designing the base chassis for a transport or any other WV, the source data specifies the full M_{PN} When designing the base chassis for a transport or any other WV, the source data specifies the full M_{PN} *and* m_p is given, and there are dependencies for determining the components of the equation (3) ξ_c and the specific reduced masses in the equations (4) and (5), then at a given power level $N_{у.э.}$ can be found relative M_{PN} $\xi_п$ by the formula (6), after that, the total mass is determined WVm_o and all its components:

$$m_o = \frac{m_p}{\xi p}, \text{ kg}, \quad (7)$$

$$m_i = m_o \xi i, \text{ kg}, \quad (8)$$

Data on the accumulated statistical material and analytical expressions for calculating the relative masses of the elements of the base chassis are given [6-8].Consider the procedure for preliminary calculation of the total mass of the WV m_o for the case that M_{PN}, including the weight of the cargo being transported m_g и TE $m_{t.o.}$ given, i.e. the value of $m_p = m_g + m_{t.o}$ is known. In this case, to determine the mass m_o it is necessary to determine the mass of the base machine $m_{b.M.}$. Since the total mass WV m_o is determined as follows:

$$m_o = m_p + m_{b.M.} = m_g + m_{t.o} + m_{b.M.} \quad (9)$$

To determine the mass $m_{b.M.}$ it is customary to use the WV-specific parameter –load capacity factor K_p, equal to the ratio MPNm_p to the mass of the base machine $m_{b.M.}$:

$$K_p = \frac{m_p}{m_{b.M.}}, \quad (10)$$

Thus, summing up the mass of the actual cargo m_{gr}, technological equipment m_{TO}, including the mass of autonomous energy sources and fuel reserves for a given autonomy, the total mass of the payload is obtained m_p. If this transported mass is placed in the body, then the payload also includes the mass of the body (tare)m_{tar}. Then, given the coefficient K_p, from the formula (10) determine the mass of the base machine: $m_{b.M.} = \frac{m_p}{K_p}$ and find the full mass WV:

$$m_o = m_p + m_{b.M.} \quad (11)$$

The above method of approximate determination of the total mass can be applied only if all the data necessary for the determination are available or specified MPNm_p.However, in the technical specification for the development of WV, the weight of the cargo can be set m_{gr}, a mass of technological equipment m_{TO} unknown and must be determined in the process of calculating the total mass WVm_o. In this case, for the approximate calculation of the masses m_o,m_p и $m_{b.M.}$the method given in [5] can be used. Analysis of statistical data on the mass characteristics of various multiaxial WV showed that the stable dependences between the masses m_p,m_{gr} и m_{TO} missing. The most stable parameter for WV with different load factors (K_p=0,8...2,0) turned out the ratio of the mass of the base machine ($m_{b.M.} = m_o - m_p$) to weight WV without load (m_{agr}=$m_o - m_{gr}$). Denoting this relation byθ, based on the statistical data, we get [5]:

$$\theta = \frac{m_{b.M.}}{m_{agr}} = \frac{m_o - m_p}{m_o - m_{agr}} = \frac{1-\xi_p}{1-\xi_{gr}} = 0{,}62\ldots0{,}65. \quad (12)$$

$$\xi_{gr} = \frac{m_{gr}}{m_o} = \frac{\theta+\xi_p-1}{\theta}; m_o = \frac{\theta m_{gr}}{\theta+\xi_p-1}. \quad (13)$$

In the expression (12) weight m_{gr} specified, the valueθ accepted within the limits of θ=0,62...0,65, relative MPNξ_p it is determined by the load capacity factor [5]:

$$K_p = \frac{m_p}{m_{b.M.}} = \frac{m_p}{m_o - m_p} = \frac{\xi_p}{1-\xi_p}; \xi_p = \frac{K_p}{1+K_p}. \quad (14)$$

Therefore, there are data for an approximate calculation of the mass m_o the original masses:

payload $m_p = \xi_p \cdot m_o = \frac{K_p}{1+K_p} m_o$; technological equipment $m_{TO} = m_p - m_{gr}$
the basic machine $m_{b.M.} = m_o - m_p$; WV without cargo $m_{agr} = m_o - m_{gr} = m_{b.M.} + m_{TO}$. (15)

Next determined by the method, described in [6 – 8], the wheel diameter D_{sh} at a given specific pressure on the ground $q_{min.}$:

$$G_k = \frac{m_{os}}{2} q. \quad (16)$$

The relative weight of the frame is calculated after the design and layout scheme of the chassis or WV is developed. The method of determining the relative masses of WV and ABCH is sufficiently fully justified, developed, presented in [5] and includes the determination of the relative mass: 1) frames (with spar and tubular cross-sections); 2) WV wheel propellers; 3) springing systems; 4) machine controls; 5) fuel; 6) additional equipment (electrical equipment systems with batteries, SPTA and other unaccounted for equipment); 7) power plant; 8) transmissions (mechanical, hydrodynamic (hydro-mechanical), hydro-volume, electric on direct current, electric on alternating current with the use of valve-inductor electric machines, electric machines on permanent magnets).

The relative mass of the power plant and transmission depends on the accepted specific effective power of the machine $N_{u.e.}$, which, according to the power formula, determines the maximum speed v_{Max} of movement for a given dynamic factor $-D_{min}$:

$$N_{u.e.} = \frac{g \cdot D_{min} \cdot v_{\text{Max}}}{3600}, \frac{kW}{kg}, \tag{17}$$

where g– acceleration of free fall: g= 9,81 m/s^2.

Technological equipment (TE) is a component part of a machine or mobile unit (lifting, road-building, road-harvesting and other machines), which ensures that the machine or mobile unit created on the basis of a certain WV or ABCH, its purpose and performance of technological functions and operations defined by the technical task. WV or ABCH-based maintenance includes elements of metal structures, various mechanisms, their drives, power supply systems, communications, temperature and humidity maintenance, etc., as well as auxiliary equipment and SPTA. The method of determining the mass of elements of technological equipment are also adequately and reliably justified, developed, presented in [5, 9] and offers a lot of technological equipment in general, can be defined as relative mass of mechanical equipment, the masses of all the systems, that ensure the functioning of the WV and masses of accessories and spare parts WV, including masses: 1) supports (jacks and base plates); 2) pumping station; 3) Autonomous power source (AIP); 4) cooling-heating installation (henna); 5) accessories and spare parts; 5) mechanical equipment (cargo boom, front boom support, boom sub-axle beam, WV support brackets, fencing, boom lifting drive, power drive and pumps, hydraulic lines and other mechanisms and drives.

3. The methodology for assessing the technical excellence of wheeled vehicles of particularly high load capacity based on road trains with active links and all-wheel control.

The methodology for assessing the technical excellence of wheeled vehicles of particularly high load capacity based on road trains with active links and all-wheel control is based on the integrated WV model, indicators and criteria for evaluating its technical excellence.The technical level and efficiency of using any WV as vehicles is mainly determined by the specific energy capacity, that is, the specific power $N_{u.e.}$ which is implemented on the driving wheels of the units [5, 9]:

$$N_{u.e.} = \frac{N_e \eta_\Sigma}{m_o}, \frac{kW}{kg}, \tag{18}$$

where N_e – rated power of the running engine (power plant), kW; m_o – full weight of PA, kg; η_Σ – total efficiency, which takes into account the power consumption for the engine's own needs, steering drive and transmission losses.

Let's consider the influence of specific energy intensity on specific characteristics of WV:

$$v_{max} = \frac{3600 \cdot \text{Nu.e.}}{g \cdot D_{min}}, \text{ km/ч}, \tag{19}$$

where $N_{u.e.}$ – specific energy capacity, kW/kg;D_{min} – the minimum dynamic coefficient specified in the terms of reference; g– acceleration of free fall: g= 9,81 m/s^2.

For example, from formula (19) it can be seen that for a given minimum dynamic factor D_{min}value $N_{u.e.}$. largely determines the value of the maximum speed of the unit. On the other hand, if the required speed of the unit is set at a certain value of the dynamic factor, then the required specific effective power is uniquely determined. Value $N_{u.e.}$ it has a significant impact on the mass characteristics of the unit. In this regard, it is necessary to justify the choice of certain criteria for technical excellence, which will allow you to reasonably, fully and reliably assign the value $N_{u.e.}$. in accordance with each specific problem situation, in order to ensure the effective use of WV and / or ABCH, taking into account its technical level indicators that characterize the load capacity, performance, efficiency, etc. To assess the performance of WV and/or ABCH of different weights, it is advisable to use them Π_y, that is, the work of transporting the payload by mass m_n,per unit of total mass of the unitm_o:

$$\Pi_y = \frac{1-\xi_k}{D}\text{Nu.e} - \frac{(m\text{ту}.su + m\text{ту}.tr)}{D}N^2{}_{y.э}, \text{ kW/kg}, \tag{20}$$

where G_n — useful weight of the transported cargo, H;V — the speed of movement of the unit, km/ch; g – acceleration of free fall, m/s²;ξ_n — the relative mass of the payload.

Exploring the expression $\Pi_y = \Pi_y(N_{y.э.}, \xi_k, \text{D}, m\text{ту}.su + m\text{ту}.tr)$ to the maximum and determine the optimal value of the specific energy capacity$N_{y.э.opt}$:

$$N_{y.э.опт} = \frac{1-\xi_k}{2(m\text{ту}.su + m\text{ту}.tr)}, \frac{kW}{kg}. \tag{21}$$

Maximum value of the indicator Π_y:$\Pi_{y.max} = \frac{1-\xi_k}{4\,(m\text{ту}.su + m\text{ту}.tr)}, \frac{kW}{kg}$.(22)

The maximum specific performance depends on the road conditions, that is, the lower the required value D, the higher$\Pi_{y.max}$.The technical level of WV and / or ABCH can also be estimated by the specific effective fuel consumption – the fuel consumption per unit of transport useful work:

$$g_{y.э} = \frac{g_e N_{y.э}}{\eta_\Sigma \cdot \Pi_y} = \frac{g_e D}{\eta_\Sigma \cdot \xi_n} = \frac{g_e D}{\eta_\Sigma[1-(m\text{ту}.su + m\text{ту}.tr)N_{y.э}]}, \text{ kg·т/kW·ч} \tag{23}$$

where g_e– specific fuel consumption in the engine, kg.т/kW· ch;η_Σ – overall efficiency of the power plant and transmission.

As can be seen from the expression (23), the minimum value$g_{y.э}$, should be achieved when $\xi_{p\ max}$. However, the value of $\xi_{p\ max}$ meet, $N_{y.э}$=0. Therefore, the optimal value of energy capacity according to the criterion $g_{y.э}$ does not exist with growth $N_{y.э}$ value $g_{y.э}$ it will continuously increase. Given that the relation $\frac{\Pi_y}{g_{y.э}}$ it has an extremum (maximum) and characterizes both efficiency and productivity, let's take it as an indicator of the efficiency of using WV and / or ABCH and call it economic productivity:

$$Э = \frac{\Pi_y}{g_{y.э}} = \frac{\eta_\Sigma \xi^2{}_n N_{y.э}}{D^2\, g_e}, \frac{kW/kg}{kg.т/kW.ч}. \tag{24}$$

The study of the extremum of the expression (24) allows us to determine the optimal value of the specific energy intensity$N_{y.э.opt}$:

$$N_{y.э.опт} = \frac{1-\xi_k}{3(m\text{ту}.su + m\text{ту}.tr)}, \frac{kW}{kg}, \tag{25}$$

at which economic productivity takes the maximum value:

$$Э_{max} = \frac{4\,\eta_\Sigma\ (1-\xi_k)}{27\ g_e D^2\ (m\text{ту}.su + m\text{ту}.tr)}, \frac{kW/kg}{kg.т/kW.ч}. \tag{26}$$

The technical excellence of WV and / or ABCH is determined by the value of the load factor $К_п$, by which is meant the relation $M_{PN}m_п$ to the mass $m_{б.м.}$:

$$К_п = \frac{m\text{р.}}{m_{b.м.}} = \frac{\xi_п}{1-\xi_п} = \frac{1-\xi_к - (m\text{ту}.su + m\text{ту}.tr)N_{y.э}}{\xi_к + (m\text{ту}.su + m\text{ту}.tr)N_{y.э}}. \tag{27}$$

Based on equation (27), the expression of the maximum valueЭ it has the form:

$$N_{u.e.opt} \approx \frac{1-\sum_{i=1}^{n}\xi_{ki}}{3(m_{Tu.su} + m_{Tu.tr})}, \text{ kW/kg.} \quad (28)$$

$$Э_{max} = \frac{(1-\xi_{к})^4}{4{,}87(m_{Tu.su} + m_{Tu.tr})}, \frac{kW}{kg.т}. \quad (29)$$

As an indicator of the feasibility assessment of the unit may be taken unit cost of useful transport work, that is, the transport cost per unit of useful work, taking into account the cost of construction of the unit and the cost of fuel consumed for period of operation [10]. The total unit cost of a unit of transport useful work is also determined by the unit cost of maintenance and repair of the unit, its maintenance in parks and the costs associated with the organization and maintenance of normal operating conditions of WV and / or ABCH:

$$C_{u.agr} = C_{u.k}(1 + K_{\theta}.)\xi_{к} + C_{u.su.}(1 + K_{\theta}.)\xi_{su} + C_{u.tr.}(1 + K_{\theta}.)\xi_{tr} = C_{u.k}(1 + K_{\theta}.)\xi_{к} + (C_{u.su.}m_{u.su} + C_{u.tr.}m_{u.tr})(1 + K_{\theta}.)N_{y.э}, \text{ rub./kg,} \quad (30)$$

where $C_{u.k}$ – unit cost of the basic machine design, rub./kg;$C_{u.su.}$ – unit cost of the power plant, rub./kg;$C_{u.tr.}$ – unit cost of transmission, rub./kg;$C_{u.to}$ – unit cost of technological equipment of the unit, rub./kg;ξ_{TO} – the relative mass, equal to the ratio of its massm_{TO}, to the total weight of the unit m_o ;$\xi_{TO} = \frac{m_{TO}}{m_o}$.

Per-unit workA_y for the entire service life of T is determined as follows:

$$A_y = П_y\ T = \frac{T\,[1-\xi_{к} -(m_{u.su} + m_{u.tr})}{D}, \frac{kW{\cdot}ч}{kg}, \quad (31)$$

where T– motor mileage allowance unit for the service life, ch.

The unit cost of useful transport work due to the cost of fuel is expressed as

$$C_{y.т} = C_т\, g_{y.э} = \frac{C_т\, g_{e\ D}}{\eta_{\Sigma}[1\xi_{к} -(m_{u.su} + m_{u.tr})N_{y.э}]}, \frac{rub.}{kW.ч}, \quad (32)$$

where $C_т$ - the cost of one kilogram of fuel, rub/kg.

The total unit cost of useful transport work, taking into account (6), is written as:

$$C_y = \frac{C_{y.agr}}{A_y} + C_{y.т} = \frac{\eta_{\Sigma D}(1+K_{\theta}.)[C_{u.к}\xi_{к}(C_{u.su.}m_{u.su} + C_{u.tr.}m_{u.tr})N_{y.э})] + C_т\, g_{e\ D}TN_{y.э}}{\eta_{\Sigma т}\,[1\xi_{к} -(m_{u.su} + m_{u.tr})N_{y.э}]}, \text{ rub./kg.} \quad (33)$$

Thus, the unit costs of the structural elements of the unit depend on a large number of factors and therefore vary widely, they are significantly influenced by the scale of production (annual output) and wholesale prices [10]. Analysis of the results of studies of the technical perfection of wheeled vehicles of extra large carrying capacity based on road trains with active links and all-wheel control allowed to justify and develop a methodology for its assessment, which includes the following main stages:: 1) formation of databases of absolute and relative (specific) masses of WV and / or ABCH and their components; 2) preparation and input of initial data on WV and / or ABCH, their purpose, design and layout solutions, characteristics and operating conditions; 3) modeling of WV and/or ABCH using a complex mathematical model of a WV and / or ABCH; 4) evaluation of indicators of technical excellence of a WV and / or ABCH; 5) justification of the choice of design-layout solutions WV and/or ABCH, technical solutions, technical and operational characteristics of their components; 6) the synthesis of control algorithms for power distribution between driving wheels, the control algorithms for steering wheels, control algorithms for elastic and damping characteristics of the suspension system, algorithms, electronic brake control, etc.; 7) substantiation of general, technical, operational, cost and private (special) requirements.

4. Conclusions

In connection with the foregoing rationale for decisions about the establishment and further operation of WV and/or ABCH should be based on a detailed multivariate study of the competitiveness of the variants WV and/or ABCH by criterion «efficiency (compliance result in the creation and implementation (implementation) of WV and/or ABCH required (desired), technical excellence)- cost-feasibility».

The rationale for the decision on the feasibility of establishing WV and/or ABCH selection of comprehensive and private indicators and criteria for assessing the technical level and efficiency is determined by the appointment of WV and/or ABCH, the possibility of effective use of WV and/or ABCH taking into account the indicators characterizing economics and technical excellence for the lifting capacity, maneuverability, performance, speed, stability, smoothness, etc.

The results of the research allow us to formulate general and specific recommendations for the selection of indicators and criteria for technical excellence, for example: 1) the optimal energy capacity should be assigned in accordance with the criterion or set of criteria that most fully characterize the purpose of WV and / or ABCH; for example, if a WV and/or ABCH is created for which high dynamic qualities (performance) are most important, even at the expense of efficiency or load capacity, then as a criterion for $N_{у.э.opt}$ you should take the specific performance$П_у$; this class of vehicles includes passenger cars, off-road vehicles, tracked vehicles, etc.; 2) if a multi-purpose WV (ABCH) is designed, then the E indicator should be taken as a criterion for optimizing energy availability;3) if the transport of undivided large-mass cargo using WV (ABCH) - based vehicles plays a decisive role, then the indicator should be taken as the number of criteria $П_{кп}$ or $Э_{кп}$;4) assigned power capacity of the unit$N_{у.э.}$ it must provide: the required maximum speed of movement under the conditions of use of the unit for its intended purpose in given road conditions; high values of the indicator (s) of the technical level of the unit, which is achieved by choosing the optimal (or close to it) value $N_{у.э.}$, determined by the criterion that characterizes the most important properties of the unit at a given payload.

References

[1] Sova A N, Mazlumyan G S, Egorov O V, Shadrin S S, Eruslankin S A 2020 The results of modeling and evaluating the dynamics of the active semitrailer *IOP Conference Series: Materials Science and Engineering* **832**

[2] Sova A N, Mazlumyan G S, Egorov O V, Shadrin S S, Eruslankin S A, Sova V A 2020 The results of modeling and evaluating the dynamics of a road train with an active trailer link based on valve-inductor electric machines *IOP Conference Series: Materials Science and Engineering* **832**

[3] Decision of the Customs Union Commission on 09.12.2011 N 877 (ред. от 21.06.2019) " On the adoption of the Technical regulations of the Customs Union "On the Safety of Wheeled Vehicles "(together with «TP TS 018/2011. Technical regulations of the Customs Union. On the safety of wheeled vehicles»)

[4] GOST 33997-2016. Interstate standard. Wheeled vehicles. Requirements for safety in operation and methods of verification (put into effect by the Order of Rosstandart dated July 18, 2017 N 708-st).

[5] Nuzhdov Yu N 1995 *Comprehensive assessment of the technical level and effectiveness of the use of mobile equipment* (Moscow: Strategic Rocket Forces)

[6] Usikov V Yu 2015 *Increasing the cross-country ability of multipurpose vehicles by decentralizing the regulation of air pressure in tires [Text]:* dis. ... Cand. tech. Sciences: 05.05.03 (Chelyabinsk: FSBEI HPE "South Ural State University" (NRU))

[7] Filkin N M, Shaikhov R F, Buyanov I P 2016 *Theory of transport and transport-technological machines* (Perm: FGBOU VO Perm State Agricultural Academy)

[8] Chudakov O I 2017 *Development of the law of power distribution between the links in rectilinear movement of the road train based on the analysis of power factors in the coupling* [Text]: dis. ... Cand. tech. Sciences: 05.05.03 (Moscow: FGBOU VO MSTU named after N.E. Bauman)

IOP Conf. Series: Materials Science and Engineering 1159 (2021) 012097 doi:10.1088/1757-899X/1159/1/012097

[9] Stepanchenko E P, Falaleev P P 1986 *Technological equipment: Textbook for universities* (Moscow: Ministry of Defense of the USSR)
[10] Astafyeva O M, Broslavets P A, Budalin S V, Nekrasov D N, Sidorov B A 2013 Methods for assessing the technical and economic efficiency of trucks *Modern problems of science and education* **2**

Statistical means of the assessment of the passive safety of road vehicles

E Y Balaev[1] T V Konovalova[1] A E Litvinov[1] M P Mironova[1], I S Senin[1]

[1] Kuban State Technological University, 2 Moskovskaya St., Krasnodar, 350072, Russia

E-mail: artstyleone@mail.ru

Abstract. The article discusses the main criteria for assessing the design of vehicles in the field of passive safety. Road safety is influenced by a large number of factors, among the most negative factors are road accidents and their consequences. All categories of the population die and get injured in road accidents, which entails a decrease in the population of the country and the level of health, life expectancy, length of the working period, and distortion of the age pyramid. Existing methods for assessing measures to improve passive safety do not fully allow establishing a correspondence between the characteristics of vehicle structural elements that ensure its passive safety and the severity of the consequences in an accident. This situation is mainly due to the impossibility of assessing the passive safety of the car as a whole, by one indicator, since the applied assessment is based on the requirements of regulatory documents. In this case, the most optimal analysis will be the analysis of the consequences of real road accidents, which allows to identify the most dangerous conditions of road accidents and traumatic elements of the structure of vehicles. The result of the assessment of measures to improve passive safety is an accurate assessment of the severity of human injury in an accident and the risk of an accident. The existing and prospective criteria governing passive safety indicators cover a small number of all types of road accidents. Therefore, expanding the scope of application of safety criteria to the widest possible range of types of road accidents and taking into account the anthropological parameters of a person is the most significant potential way to reduce the number and severity of injuries in road accidents.

1. Introduction

The problem of ensuring the safety of modern passenger cars is of high social and economic importance and is one of the important problems of the country's motorization. Reducing the number of road accidents and the number of people injured in motorization is the main task in the road safety system. Nowadays, the term "Safe System Approach to Road Traffic" has become widespread, which includes the optimization of the transportation system for all road users by identifying and taking into account the vulnerability of road users to serious injuries in road accidents. One of the most important principles of the described approach is the safety of vehicles, aimed at reducing the severity of injury and reducing mortality as a result of road accidents. However, the existing methods for assessing measures to improve passive safety do not fully make it possible to establish a correspondence between the characteristics of vehicle structural elements that ensure its passive safety and the severity of the consequences in an accident. This situation is mainly due to the impossibility of assessing the passive safety of the car as a whole, by one indicator, since the applied assessment is based on the requirements of regulatory documents. The result of the assessment of measures to improve passive safety should be an indicator

Content from this work may be used under the terms of the Creative Commons Attribution 3.0 licence. Any further distribution of this work must maintain attribution to the author(s) and the title of the work, journal citation and DOI.
Published under licence by IOP Publishing Ltd

IOP Conf. Series: Materials Science and Engineering 1159 (2021) 012098 doi:10.1088/1757-899X/1159/1/012098

of the hazard of an accident. And an accurate assessment of the severity of injuries will determine the limit of the human body's endurance to the effects of shock loads at the time of an accident. Thus, it is possible to influence the evaluation criteria of the vehicle design and minimize injuries in road transport. Analysis of the consequences of real road accidents allows us to identify the most dangerous conditions of road accidents and traumatic elements of the structure of vehicles.[1,2]

In the Russian Federation, the national priority is the life and health of people. However, there are objective factors that affect the damage from the loss of a member of society or his temporary retirement from the sphere of production. Socio-economic damage from road accidents has a negative impact on the demographic indicators of Russia, such as: a decrease in the population of the country and the level of health, life expectancy, length of the working period, distortion of the age pyramid. All categories of the population die and get injured in road accidents. Young people between the ages of 15 and 29 account for about a third of road traffic victims. Road traffic crashes are the second leading cause of death among young people aged 5-29 and the third leading cause of death for people aged 30-44. The need for hospitalization of victims in subsequent road accidents is 7 times more likely and 6 times more likely to become disabled, including children. In Russia, the mortality rate of persons from road accidents is 14.0 per 100 victims of road traffic accidents, which is 5-7 times more than in developed countries The distribution of the number of road traffic victims by category of road traffic accident participants in the Russian Federation, for the same period of the accounting year is presented in Table 1.

Table 1. The distribution of the number of injured in automobile transport by category of accident participants in the Russian Federation

Participants in an accident	2019 November			2018 November			2017 November		
	total accident	died, people,	injured, people	Total accident	died, people	injured, people	Total accident	died, people	injured, people
1	**2**	**3**	**4**	**5**	**6**	**7**	**8**	**9**	**10**
People in cars	10230	1005	1359	10865	1067	14491	10482	1095	13945
People in trucks	880	138	1074	999	143	1247	961	181	1188
Pedestrians	4811	526	4509	5169	602	4808	5968	721	5532
Motorcyclists	85	13	83	53	10	49	69	19	69

Statistics show that about a thousand people die in road accidents in Russia every year due to technical malfunctions of vehicles. Consequently, the design safety of vehicles does not fully take into account the factors that can affect the severity of the consequences after an accident.

Vehicle safety as one of the sources of road accidents affects road safety in general. The continuously modernized design of the vehicle makes it possible to practically promptly implement measures to improve safety. An important role in this process is played by the constructive safety of the car and the methods of its testing, the perfection of these methods. Ensuring passive vehicle safety is part of one common problem - road safety.[3]

Passive vehicle safety is the property of a vehicle to reduce the severity of the consequences of an accident. Passive safety is manifested in a period when the driver, despite the safety measures taken, cannot change the nature of the car's movement and prevent a road traffic accident (the culmination phase of an accident). Distinguish between internal passive safety, which reduces injuries to passengers, the driver and ensures the safety of goods transported by the car, and external safety, which reduces the possibility of causing damage to other road users.

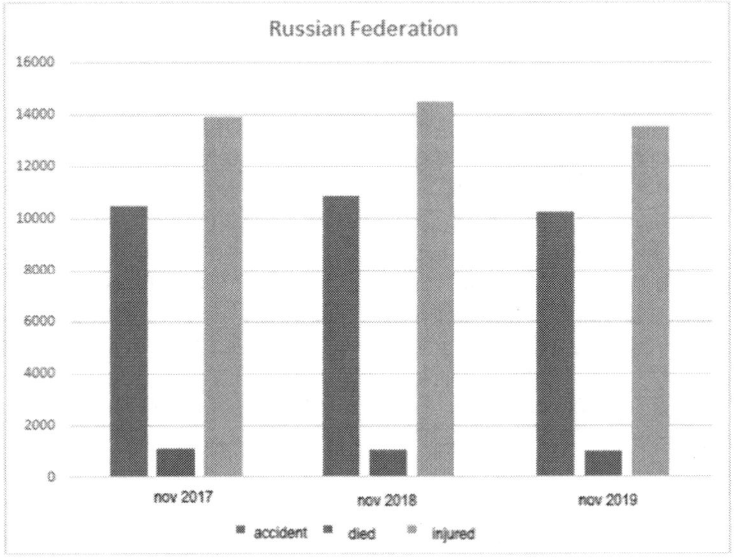

Figure 1. Absolute indicators of the state of road safety (drivers and passengers of passenger vehicles)

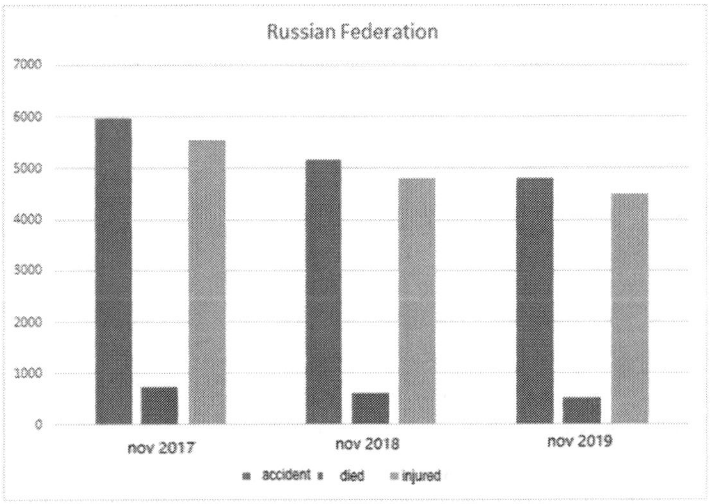

Figure 2. Absolute indicators of the state of road safety (accident involving pedestrians)

2. Materials and methods

Passive safety includes a number of factors that can contradict each other in various emergency situations. However, passive safety should provide a number of tasks in the event of an accident:

1. Ensure the possibility of saving human life inside the cabin;
2. Provide the ability to reduce the severity of injury (both a person inside the passenger compartment and outside the car, for example, a pedestrian);
3. Provide unhindered, emergency evacuation of people from the vehicle interior.

Passive safety is the main quality indicator of the system for ensuring the constructive safety of vehicles. Therefore, reducing the likelihood of injury to a person in the process of a road traffic accident is fundamental in the functioning of the passive safety system and its subsystems. In this case, the main indicators are the anthropological data of a person, and the ability of individual parts of the human body to withstand overload.[4]

The main ways to increase the passive safety of passenger cars are considered in the context of the main requirements for individual elements of the structure of passenger cars, to ensure the protection of drivers and passengers in an accident:

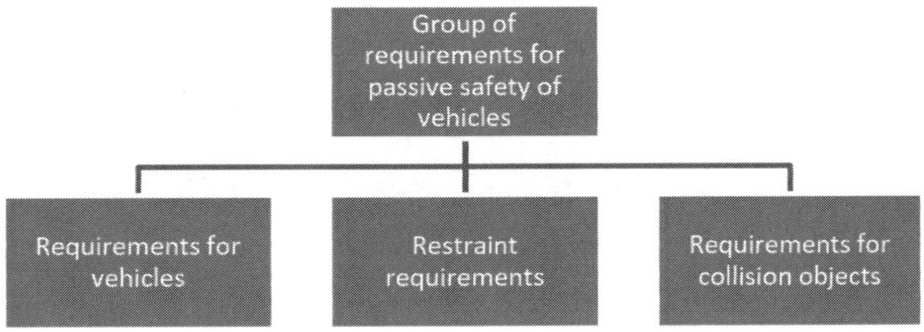

Figure 3. The main group of requirements for passive safety of vehicles

Light vehicle passive safety (PS) studies are a regulated procedure for the automotive industry. The main object under study is the many variations of various factors that affect the behavior of the car at the time of an accident.

Most reliably, the safety of a vehicle can be estimated by the consequences of an accident, which are expressed in the severity of a person's injury in an accident. The assessment of the severity of injury is based on data on the human tolerance of overloads at the time of an accident within a given time interval without receiving serious injuries.

The severity of injury to a person in an accident is determined by the maximum or average value of the following factors: overloads acting on a person, duration of exposure, rate of increase, direction of action and place of application of efforts.

The result of passive safety assessment is an accurate assessment of the severity of injury to a person in an accident and the risk of an accident.[5]

Determination of the hazard indicator of an accident makes it possible to determine the endurance limit of the human body to the impact of shock loads and how much a particular element of the car's structure is dangerous to humans, as a result of which the evaluation criteria of the vehicle design will be changed and the severity of the consequences of an accident will be minimized.

Based on research carried out by the Cornell Aeronautics Laboratory in accordance with the American Traffic Accident Severity Program, an analysis of injuries sustained by vehicle occupants by source was carried out, shown in Figure 4.

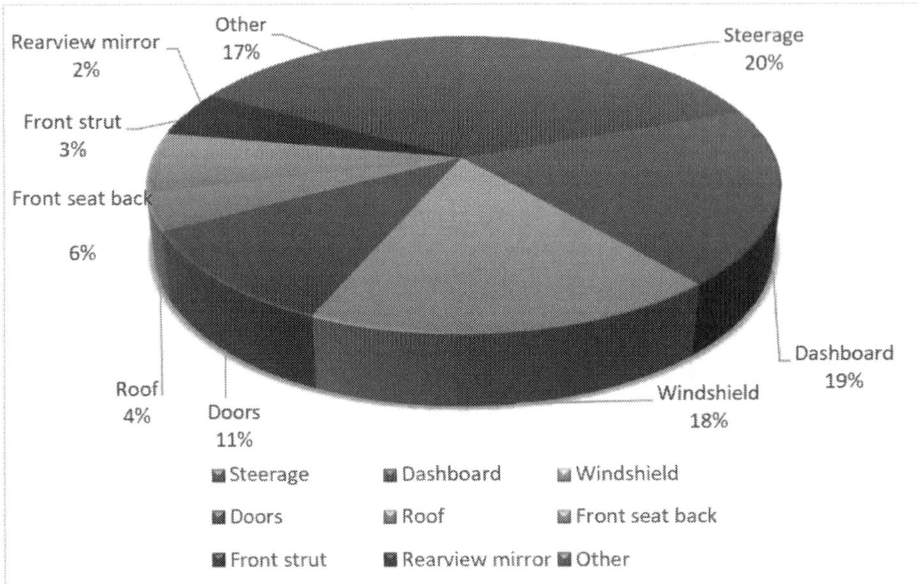

Figure 4. Analysis of the main sources of human injury

3. Analysis of Simulation Results and Experimental Data.

The analysis showed that the main cause of serious and fatal injuries was impacts on the front guard and steering column - 20.23%. In second place are impacts on the instrument panel, which accounted for 19.17% of serious injuries and deaths. In addition, windshield is the cause of 18.7% of injuries (skull trauma, concussion, etc.).

During an accident, the driver most often hits the dashboard or other structural element of the vehicle with his head (13%), and the front passenger - with his feet (11.3%). The use of seat belts reduced serious injuries to 7% and minor injuries to 34%. With the use of more effective restraints with an inertial device, as a result of road traffic accidents, only 5% of the victims were seriously injured and 29% light.

At the same time, the main causes of death in an accident of the driver and passengers of passenger cars are: ejection from the seat, impacts on the steering column, on the door, on the windshield and on the instrument panel.[5-6]

Also, studies have found that the vast majority of fatal accidents, namely 120 dead out of 136 people, were in the front seat during the accident. Therefore, the main focus should be on ensuring the safety of the driver and front passenger.[6]

The study of the statistical data of Italian, American and German researchers made it possible to identify the structural elements of the car interior, which most often injure a person. The distribution by groups of injury hazard in road accidents is shown in Figure 5.

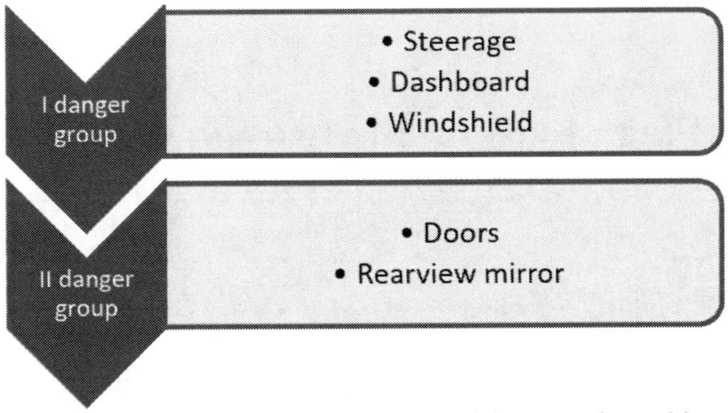

Figure 4. The distribution of injury risk groups in accidents

4. Conclusion

The existing and prospective criteria governing passive safety indicators cover a small number of all types of road accidents. The applied assessment principles are not quantitative, but qualitative, which implies compliance or non-compliance with established standards. There are practically no methods for a generalized quantitative assessment of the passive safety of a car. Therefore, expanding the scope of application of safety criteria to the widest possible range of types of road accidents and taking into account the anthropological parameters of a person is the most significant potential way to reduce the number and severity of injuries in road accidents.[7-9]

References

[1] Methodology for the integral assessment of passive safety of passenger cars deposited manuscript No. 1889-B2003 10/30/2003

[2] Muizemnek A Yu, Kartashova E D and Zemskov R A 2012 *Proceedings of higher educational institutions. Volga region. Technical science* **3 (23)** 110–119

[3] Agakhanov E K and Batmanov E Z 2012 *Bulletin of the Dagestan State Technical University. Technical science* **3 (26)** 67–72

[4] Zemskov R A and Parkina Yu I 2012 *Models, systems, networks in economics, technology, nature and society* **1 (2)** 138–141

[5] Komarov Yu Ya, Lemeshkin A V and Silchenkov D D 2014 *Bulletin of the Volgograd State Technical University Series: Ground transportation systems* **19 (146)** 53–55

[6] Razgovorov K I 2016 *New materials and technologies in mechanical engineering* **23** 80–83

[7] Orlov L N, Tumasov AV and Bagichev SA 2014 *Proceedings of NSTU im. R.E. Alekseeva* **3 (105)** 163–172

[8] Konovalova T V and Nadiryan S L 2013 *Science. Technics. Technologies (polytechnic bulletin).* **4** 78–79

[9] Konovalova T V and Kotenkova I N 2013 *Bulletin of the Saratov State Technical University.* **2 (71)** 275–279

IOP Conf. Series: Materials Science and Engineering 1159 (2021) 012099 doi:10.1088/1757-899X/1159/1/012099

The impact of lean tools on the growth of overall equipment efficiency (OEE) in vehicle manufacturing

V S Shmatkov[1], A V Shmatkova[2]

[1] Irkutsk Aviation Plant - a branch of PJSC "Corporation" Irkut ", Novatorov Street, 3, Irkutsk, 664020, Russia
[2] Irkutsk National Research Technical University, Lermontov Street, 83, Irkutsk, 664074, Russia

E-mail: shmatkov_vs@irkut.ru, annashmatkova@yandex.ru

Abstract. The paper considers the organizational activities allowing to improve the return from the equipment at the expense of reducing timely revealed losses. The analysis of the significance of the effect of the proposed measures on the growth of OEE is carried out. The statistics of losses during the operation of the fleet of metal-cutting high-performance equipment used in the production of vehicles for 10 years is given, an analysis is performed. A method for calculating the motivation fund for bonus payments to interested personnel is proposed.

1. Introduction

For the effective implementation of OEE accounting methodology at the vehicle manufacturing enterprise, the following sequence of actions at the enterprise (10 steps) is proposed:

1. The study of classical technique, proposed by the Institute for Complex Strategic Studies.
2. Adaptation of the classical methodology to the needs and specifics of your enterprise (industry).
3. Experimental testing of the developed methodology at pilot sites, development of a loss classifier.
4. Development of the company's own regulatory documents for accounting and analysis of OEE.
5. Implementation of information systems in order to increase the objectivity of the received data and reduce the time of their collection and processing.
6. Development of a methodology for planning OEE coefficients.
7. Development and implementation of measures aimed at increasing OEE indicators (mainly by reducing organizational losses).
8. Motivation of staff for increasing OEE, incl. through the introduction of progressive forms of labor.
9. Adjustment of regulatory documents (clarification of the classifier of losses).
10. Integration of OEE data into MES systems, etc.

At most enterprises, mastering OEE accounting and analysis methodology is limited to points 1, 5, 7 and partly 8. It should be borne in mind that the use of the classical methodology on one or another equipment, production line, etc. not always possible. For example, it may be necessary to replace the productivity factor, which is difficult to calculate in the production of various products, with different labor intensity and processing cycles in the absence of implemented MES systems at workplaces, this

Content from this work may be used under the terms of the Creative Commons Attribution 3.0 licence. Any further distribution of this work must maintain attribution to the author(s) and the title of the work, journal citation and DOI.
Published under licence by IOP Publishing Ltd

factor is applied on the conveyor with debugged cycles and a perfect planning system. Possible approaches to increase the objectivity of data on the operation of equipment are discussed in the article "Increasing the objectivity of data on the operation of CNC equipment using a log file (machine protocol) [1]. It is necessary to strive for online monitoring of OEE with the adoption of prompt corrective measures on a "here and now" basis in a standardized manner that is understandable for all involved personnel.

It is important to note that the very fact of accounting for the operation of the equipment allows at once, without any financial investments, to increase OEE by 5-10% by increasing production discipline and increasing personnel responsibility.

2. The main types of losses and corrective measures

The classifier of losses and their subsequent ranking according to their importance is usually compiled on the basis of the results of in-depth analysis when working on machines using a photograph of the working day, this method is more accurate, but very costly, it requires a large amount of labor of employees of the Bureau of Labor and Wages, especially if the equipment is working at the enterprise around the clock. It is possible to take as a basis the recommended classifier of losses with its subsequent updating for the specifics of production. In any case, the most significant losses that we can influence should be monitored, usually the list of such losses does not exceed 10. An example of a list of such losses is given in Table 1 with reference to corrective measures.

According to the information from table 1, it can be concluded that the main losses can be eliminated by the introduction of the main basic tools of Lean production, such as accounting for OEE, 5C, TPM, SMED, while the correction measures must be implemented mainly from left to right, as indicated in the table, thus achieving the maximum effect in increasing OEE. If in principle everything is clear with lean management, we will dwell a little on other corrective measures, they will not be considered in detail in this article. The main difficulty in calculating the load of equipment and scheduling its load, say for a month, is to determine the exact processing cycles, taking into account the performance of all the necessary intermediate operations, taking into account the control of the first part from the batch, especially on night shifts, lying, waiting and other force majeure. Shift is understood as the introduction of progressive forms of labor allowing to organize full-fledged round-the-clock operation of equipment, incl. on holidays and weekends, the most suitable mode is accounting of the timesheet according to the summarized hours worked for each operator separately, etc. taking into account all the requirements of the Labor Code of the Russian Federation. Staff motivation can be both direct and indirect. Direct motivation is understood as payments for a certain period of time (one-time, within a year, etc.) bonuses for achieving a certain level of OEE or reducing certain losses. At the same time, the motivational period may be due to the launch of a new product, a sharp increase in the production plan. Indirect motivation is understood as the share in wages, which depends on the achieved OEE indicators, but for this it is necessary to introduce new methods of calculating wages for machine operators. There is also another good motivator that allows you to increase OEE in certain areas - this is the brigade method of work. Today, there are few enterprises that can boast of a full-fledged transition to the brigade method of work when working on machine tools. Providing foremen with a certain set of functions is fraught with a number of legal and legal restrictions. With processing technology, not everything is so simple. Let's make a reservation right away that the replacement of equipment with an even more progressive one is not considered. Optimization of the processing technology of a part is a very labor-intensive process that requires fundamental knowledge and large resources of specialists. You also need to take into account a number of restrictions due to increased requirements for product quality, life tests, production certification, etc. [2] It will not be possible to obtain a large effect in increasing OEE due to the processing strategy, rather it will allow to achieve a higher stable quality, in some cases an increase in labor intensity is observed. Therefore, decisions must be very carefully weighed. The development of external cooperation in conditions of a decline in production strongly depends on the availability of guaranteed production plans for a long term of 5-10 years. Therefore, if there is a financial opportunity and confidence in promising launches, then it is better not to load equipment with non-core orders and external cooperation, with the exception of cooperation on a common manufactured product.

Table 1. An example of a list of such losses

		Corrective measures (corrective actions)								
		Lean management					Other			
Cipher	Name of losses	Organization of OEE accounting	5C	TPM	SMED (PPMWP)	Equipment load calculation	Shift organization	Motivation (demotivation)	Processing technology	External cooperation
		Start								finish
0.2	Manufacturing of a part according to the control program (CP) at low feeds								+	
0.3	Manufacturing of a part according to CP at high feed							(+)		
1	Preventive maintenance, equipment maintenance by specialists			+						
2	Losses due to emergency stops and repairs			+						
2.1	Equipment failure or repair			+				+		
2.2	Tool breakage				+				+	
2.3	Other emergency stops (no energy resources, etc.)			+						
3	Operator service	+	+	+						
4	Auxiliary time	+	+		+				+	
5	Preparatory and final time	+	+		+				+	
6	Organizational losses	+								
6.1	No operator					+				
6.2	Lack of download from the master				+			(+)		
6.3	Absence or incompleteness of the service station		+		+					
6.4	There is no blank				+					
6.5	Lack of support staff in the workplace				+	+				
6.6	Shift shifts, planning meeting					+				
6.7	There is no CP, technical process				+					
6.8	Solution of technical issues by services				+				+	
6.9	Downtime due to lack of orders									+
6.10	Unidentified losses				+			+ (+)		
7	Losses in the implementation of CP				+			+	+	
8	Losses during interoperational movements				+					
9	Regulated breaks							+		

3. Increasing OEE through Lean Management

5C is a basic tool; without it, the implementation of subsequent tools and the implementation of other organizational measures are impossible. As a rule, the organization of a workplace according to the 5C system allows, without special financial investments, to increase OEE by 3 - 5% by reducing the auxiliary time, preparatory and final time and other unproductive work. At the same time, the involvement of workers in kaizen is impossible without 5C. The 5C tool is simple and straightforward, at enterprises there are differences only in the assessment of the fifth stage - "Improvement". At our enterprise, we assess the fifth step if two conditions are met: regular audits of workplaces according to the approved methodology with filling in the "5C radar sheet", see Fig. 1. according to the schedules and participation of workers in the submission and implementation of kaizen proposals. If desired, this fifth level can be supplemented with two more evaluation criteria - labor protection (knowledge of labor protection instructions and their implementation) and fire safety (knowledge of fire safety rules and their implementation). There are two components of success with this tool - the will of the head of the enterprise and motivated staff without administrative "pressure".

Radar 5C					
Department / area code (workplace)		year			
Auditor					
Stage	Criteria	Assessment (score)			
		Week 1	Week 2	Week 3	Week 4
"1C" sorting	Absence in the workplace of items unused to perform work				
	Items are in the amount required for work				
"2C" creating your own places	Places for storing items needed for work have been identified and organized				
	Items required for work are in their "own" storage locations				
"3C" cleanliness	Regular cleaning of machines, equipment, office equipment				
	Keeping the workplace clean				
"4C" standardization	Application of standards, instructions, memos required for the performance of work				
	Original markings applied (P. 343.372)				
"5C" improvement	Regular 5C audits				
	Participation of employees in submitting kaizen proposals				
Total					

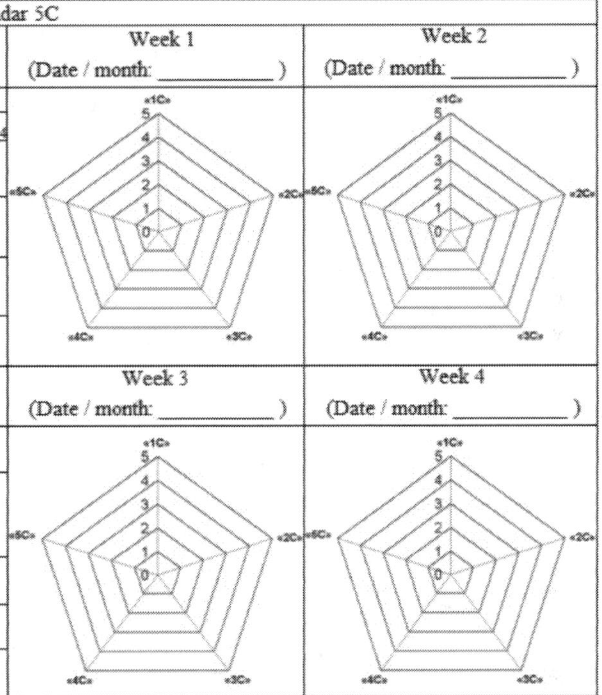

Figure 1. Radar 5C.

TPM - this tool is directly related to the equipment availability ratio. At the same time, this is the case when the growth of losses for equipment maintenance leads to either a significant reduction or prevention of losses associated with emergency downtime in the future. Therefore, this type of loss (codes 1 and 3, see Table 1) must be considered in combination with losses (codes 2.1 and 2.3). Correct implementation of the methods of scheduled maintenance and current repairs of high-performance equipment in modern conditions [3-8] can increase the equipment availability factor to 18%, in terms of OEE (with a loss factor of 0.65, a quality factor of 1.0) will be about 12%.

SMED tool is a quick changeover, it is not acceptable for metal-cutting equipment in the aviation industry in its pure form. This tool is well implemented in blank production when working on complex and massive dies, mainly in the automotive industry, where the changeover time is hundreds and thousands of times longer than the part stamping time. Nevertheless, such works on the study of SMED in 2014 on metal cutting machines were carried out. A video camera was used for observation. As a result of the subsequent viewing of the videos, the specialists filled out the protocols, the recommended form is shown in Table 2.

IOP Conf. Series: Materials Science and Engineering 1159 (2021) 012099 doi:10.1088/1757-899X/1159/1/012099

Table 2. Result of the subsequent viewing of the videos

Changeover Observation Protocol (SMED)							
Shop:			Plot:			The date:	
Machine:			Operator:			Observer:	
From detail:			For detail:			Start:	End (1st suitable):
1	2	3	4	5	6	7	8
No.	Action	Total time	Time of a separate stage	Internal	External	Changeover phase	Decision

The analysis of the existing changeover ended with the development of changeover improvements; the recommended form is shown in Table 3.

Table 3. The analysis of the existing changeover ended with the development of changeover improvements

No p/p.	Operation name	Actual time	Suggestions for improvement	Expected time	Actual received time

Based on the results of the work carried out, recommendations were made, the main ones are shown in Table 4.

When implementing the above recommendations, it is necessary to take into account:

- batch (seriality);
- relevance of the manufactured product for the future and the development strategy of the enterprise;
- financial feasibility (payback of expensive equipment, machines with interchangeable pallets), etc.

This tool, as applied to high-performance equipment, can also be supplemented with well-forgotten methods from the scientific organization of labor (STO) of the Soviet era, namely, elements of planned preventive maintenance of workplaces (PPMWP). An example of the time schedule for the PPMWP for machine tools is shown in Table 5.

IOP Conf. Series: Materials Science and Engineering 1159 (2021) 012099 doi:10.1088/1757-899X/1159/1/012099

Table 4. Results of the work carried out

No p/p.	RECOMMENDED ACTIONS	Executor
1.	Filming device in conjunction with the last item of the batch. The removal of the part from the device and its control should be carried out after starting the CP of the next batch.	Operator
2.	Fill in the operator's report on work with the machine running immediately before the end of the program, and immediately after starting a new one.	Operator, master
3.	Prepare fixtures, tools and fasteners in advance before the end of the work on the previous program.	Operator
4.	Use progressive wrenches (ratchet, ring, etc.).	Operator, technologist
5.	Do not unnecessarily clean the machine between changeovers.	Operator
6.	Monitor the safety of the equipment, its completeness. Ensure the safety of removable elements (identify them).	Operator, technologist
7.	Participate in the submission of Kaizen proposals for the unification of fasteners of the equipment, simplification of its design.	Operator, foreman, technologist
8.	Make the most of the quick-detachable elements of the service station (such as "plugs", etc.).	Constructor
9.	Provide the ability to install and remove the part on the device outside the machine.	Constructor
10	It is imperative to carry out external adjusting elements that allow you to configure the equipment on the machine with the installed and pressed part.	Constructor, programmer
11.	Indicate on sketches in the technological process and marks on the equipment of the exact location of the clamps, clamps, etc.	Programmer, constructor
12.	Provide in technological processes with complex equipment of a photo of a part installed and clamped in a fixture.	Technologist
13.	Make the most of typical lodgements - racks for attaching equipment. Create a catalog of existing overpasses (supplement if necessary).	Programmer, constructor
14.	Unify fasteners. Introduce restrictions on the use of standard sizes of fasteners.	Constructor
15.	Use, where possible, fastening the part to the cradle with bolts, screws, etc. into the holes of the part or technical admission.	Programmer, constructor
16.	Provide storage space for removable elements on the tooling plates. Provide spare parts and accessories - spare pins, nuts, washers.	Constructor
17.	Apply the same type of tool setups, minimize specialized tools.	Programmer
18.	The use of modern quick-detachable elements for fixing parts and accessories on the machine table. Availability of rotary tables (replaceable pallets).	Leading department

IOP Conf. Series: Materials Science and Engineering 1159 (2021) 012099 doi:10.1088/1757-899X/1159/1/012099

Table 5. Example of the time schedule for the PPMWP for machine tools

Organization of part launches

Subject of Labor	Launch		Charting	
	Responsible.	Ahead of time	Responsible	Ahead of time
Nomenclature of parts	Planning Technician	One month ahead	Master	5 working days before starting reporting month

Acquaintance with the production, shift task

Employee service	Formation		Familiarization of the employee	
	Responsible.	Ahead of time	Responsible.	Ahead of time
Production assignment	Master	2 days	Operator	2 days
Shift job	Master	2 days	Operator	2 days
Shift job (replacement option)	Master	1 day	Operator	1 day
Emergency deficit	Master	1 day	Operator	1 day

Completion of the tool, service station, TD

Subject of Labor	Request			**Completion**	
	Responsible No 1	Responsible No 2	Ahead of time	Responsible.	Ahead of time
Tool	Worker	Master	1 day before the start of work (taking into account the availability of the tool according to receipts and tool books)	Head instrumental distribution pantry (IDP)	Half a shift starting work (taking into account the availability of tools according to receipts and tool books)
Rig	Worker	Master	1 day before the start of work (taking into account the availability of a service station according to the service station issue card)	Head IDP	Half a shift before starting work (taking into account the availability of a service station according to the service station issue card)
TD (technical process, drawing)	Master	-	List for 1 week	Archivist	1 day before starting work
Acquaintance with TD	-	-	-	Worker	Half a shift before starting work
Blank	According to the Memo on the formation of request for cutting a workpiece				

Product control

Subject of Labor	Notification of employees of quality control department		Control	
	Responsible	Ahead of time	Responsible.	Control time, h no more
Control	Master (Worker)	1 hour	Employee of quality control department	0.2…1

Within the framework of the PPMWP, it is recommended that the procedures for providing workplaces with the required labor tools be accompanied by one-page instructions (leaflets), an example of such instructions is given in Table 6.

Table 6. Example of instructions

No p/p	On organizing the assembly and delivery of instrumental adjustments to the workplace of the malware operator within the framework of the implementation of PPMWP elements (typical for workshops)	Executor
1	Clarify the sequence and launch according to the schedules of manufacturing (implementation) of parts for the release of production equipment and in advance (one day) give a task to the distributor of work on completing the adjustments at a given time.	Master
2	Make an entry in the "Shift-daily assignments" journal for the collection of tool adjustments with reference to the details, indicating the number of start-ups in the batch, machine inventory number and indicate the time when the adjustments should be transferred to the operator.	Nork distributor
3	According to the designation of the part, print out from the PC a map of the set-ups acquisition (hereinafter MSA) or make an extract from the technical process, transfer the extract (including duplicates) to the IDP to the senior storekeeper.	Toolmaker
4	Not later than 30 min. select the necessary tool in the required quantity, while checking its compliance with the marking, the presence of chips, nicks, etc.	Storekeeper IDP
5	Get the instrument against signature in the MSA (or the journal for issuing the instrument to the workplace). Assemble the adjustment in compliance with the requirements of the technological instructions TI546.25000.00201, TI546.25000.00250.	Toolmaker
6	Deliver the adjustment (set of adjustments) to the machine within the specified timeframe according to the Journal and transfer it to the MSA (or the issuing journal ...) to the operator against signature. If there are used adjustments, take them also against signature.	Toolmaker
7	After manufacturing the part (batch of parts), remove the used adjustments from the tool magazine of the machine and place them in the tool trolley, load the newly delivered adjustments for the manufacture of the part.	Operator
8	Make a bypass of the entire malware (at 730, 930, 1130, 1330, 1530, 1730, 1930, 2230, 130, 430) collect used and unused adjustments against signature, receive applications from operators for preparing the next adjustments, if necessary. NOTE: check the integrity of the used adjustments and, if necessary, clarify with the operator the actual durability of the cutters, fix in the prescribed manner. Hand over the instrument on the same day against signature in the IDP. If defects are found, further actions on STP.	Toolmaker
9	Twice a day (in the morning after the 2nd shift and in the evening after the 1st shift), collect information on the durability of cutters and enter it into the technical bureau's database in the prescribed manner.	Tool engineer

As a result, the implementation of the PPMWP will increase OEE by 5 - 7% by reducing organizational losses. It should also be noted that the impetus for the introduction of this tool can serve as an increase in the workload of production and the formation of "bottlenecks". The tool does not work in low production conditions.

IOP Conf. Series: Materials Science and Engineering 1159 (2021) 012099 doi:10.1088/1757-899X/1159/1/012099

4. Loss statistics

Beginning in 2008, after receiving and launching the last batch of new high-performance equipment, a more in-depth analysis of losses and operational time was started. Statistical data on the achieved components for calculating the OEE coefficients are summarized in Table 7.

Table 7. Statistical data on the achieved components for calculating the OEE coefficients

Cipher	Description of works, losses	stage 1		stage 2			stage 3				
		2008	**2009**	**2010**	**2011**	**2012**	**2013**	**2014**	**2015**	**2016**	**2017**
0	Operational time	46	54	67.1	65.5	53.4	46.2	47.8	52.7	57.5	55.3
0.1	Manufacturing of a part according to CP			65.8	62.3	48.2	42.5	42.5	47.5	51.8	49.6
0.2	Manufacturing of a part according to CP at low feeds			1.3	3.2	5.2	3.4	4.6	4.6	4.9	5.2
0.3	Manufacturing of a part according to CP at high feed						0.3	0.7	0.6	0.8	0.5
1	Preventive maintenance, equipment maintenance by specialists			1.2	2.0	1.2	1.3	1.8	1.2	1.0	1.2
2	Losses due to emergency stops and repairs	7.0	11.0	6.0	4.6	5.4	3.2	5.0	6.0	4.7	6.4
3	Service			1.2	0.8	1.2	1.1	0.7	0.9	0.9	0.9
4	Auxiliary time	20.0	14.0	9.2	7.0	9.3	9.5	12.0	11.6	11.7	11.0
5	Preparatory and final time	2.0	2.0	1.6	1.0	1.0	1.3	1.5	1.7	1.7	2.5
6	Organizational losses	15.0	12.0	5.6	9.7	16.5	30.1	27.0	19.4	11.9	16.6
6.1	Absence of an operator (including on holidays)			0.3	3.1	2.9	3.6	4.0	3.8	3.0	2.8
6.2	Lack of download from the master			0.0	0.2	1.0	2.0	2.8	0.8	0.7	1.8
6.9	Downtime due to lack of orders			0.0	0.7	2.2	15.8	17.4	13.6	7.3	10.0
7	Losses in the implementation of CP			1.1	0.6	1.0	0.8	1.7	2.5	1.8	1.0
8	Losses during interoperational movements			0.2	0.1	0.1	0.1	0.1	0.1	0.1	0.1
9	Regulated breaks	10.0	7.0	6.6	7.0	5.9	4.0	2.5	2.8	2.6	2.6

The entire presented observation period can be divided into three stages:

- Stage 1 - manual collection and analysis of information according to our own developed methodology;
- Stage 2 - manual collection and analysis of information in accordance with generally accepted guidelines for accounting for OEE [9], development of standard documents of the enterprise;
- Stage 3 - automated collection and analysis of information to improve the objectivity of the data.

From the given data, we want to note some features:

- a sharp increase in accidents in 2009 (cipher 2) with the subsequent adoption of a large set of corrective actions;
- starting from 2011, it was decided to attribute losses not to the organization of work on official holidays, as a result of the increase in losses due to the lack of operators (cipher 6.1). To organize work on these days, the personal consent of the employee is required, which greatly complicates the formation of shift schedules for machine tools;

- starting from 2013, there has been a decrease in the load, see losses cipher 6.9 and, as a consequence, a slight increase in the preparatory and final time (cipher 5) due to small launch batches;
- since 2013, in order to prevent accidents, improve the quality of the surface layer of the part and reduce wear, a loss was introduced - the manufacture of a part according to CP at high feeds (cipher 0.3);
- the reduction in scheduled breaks (code 9) was mainly due to the reduction in lunch breaks.

It is important to note that with a constant, say OEE, or its slight decrease over the years, the volume of products removed from the machines increases due to:

- optimization and improvement of technology, namely mechanical processing from year to year due to new strategies, tools and modes, equipment characteristics, etc.;
- planned annual reduction in labor intensity;
- increasing the share of serial work instead of implementing control programs.

5. Development of activities

The recommended form of measures to reduce losses when working on high-performance equipment in the shop (on the site) is shown in Table 8.

Table 8. The recommended form of measures to reduce losses when working on high-performance equipment in the shop (on the site)

Loss cipher	Description of losses	The amount of losses for the previous period, %	Corrective Action	Performers	Period of execution	The type of document that ends the work	Expected reduction in losses, %
1	2	3	4	5	6	7	8
		TOTAL:					TOTAL:

Below are the most requested and most relevant actions.

Possible corrective measures:

- Organization of multi-station service if necessary;
- Provision of personnel for work in the 2nd and 3rd shifts - accelerated training and transfer of operators from other machines;
- Organization of work on weekends and holidays;
- Preparation and availability of a backup alternative loading of the machine in case of forced downtime, especially after hours and at night;
- Organization of photography of the working day, timing of auxiliary time and preparatory and final time, identification of reserves for revising labor standards;
- Organization of scheduled maintenance and repair of equipment during forced downtime;
- Organization of adjustments and implementation of control programs in the 2nd shift;
- Motivation of personnel for identifying and reducing losses.

Possible corrective actions:

- Application of new forms of payment for operators' labor (refusal from piecework payment).
- Increase in machine load due to the additional volume of transfer of the nomenclature from other equipment, incl. universal;
- Advance launches of promising product series;

- Organization of off-hours and night shifts for specialists of repair services;
- Implementation of procedures for training, testing and certification (recertification) of operators for the right to work;
- Implementation of procedures for the analysis of emergencies and prompt repair of equipment, incl. under warranty. Communication of information to all interested workers and operators. Creation and maintenance of a database on repairs;
- Development and implementation of regulations for scheduled preventive maintenance of equipment, procurement of spare parts and consumables, PPMWP, control of the first part from a batch during off-hours and at night, calculation of equipment load, scheduling of operators' shift, etc.
- The transition to the brigade method of labor.

6. Conclusion

Information on the specific gravity of the tools with which it is possible to increase OEE indicators is summarized in Table 9.

Table 9. Information on the specific gravity of the tools

No p/p	Tool (correction measures)	Growth OEE, %
1	Organization of OEE accounting	5 - 10
2	5C	3 - 5
3	TPM	5 - 12
4	SMED (PPMWP)	5 - 7
	TOTAL	**18 - 34**

In accordance with the data in Table 9, on average, the introduction of these Lean Manufacturing tools will increase OEE by 18 - 34% (on average by 25%), depending on the quality of the activities being carried out.

From practice, at the start of the introduction of lean technologies, OEE usually does not exceed 40%. Consequently, in a short time, it is possible to increase OEE up to 65% only through organizational measures. This value is an average indicator and applies to the entire fleet of equipment, so OEE on individual machines can reach 80% or more, and on machines with a specific unique technology, OEE may not exceed 30%. A further increase in OEE is possible due to:

- stable production load when entering a large-scale debugged production, ideally one product;
- technology optimization;
- the introduction of other forms of work organization, allowing to increase the internal motivation and consciousness of personnel, especially operators.

6.1. Motivation for the implementation of Lean manufacturing tools that increase OEE

We would like to give a simple method for calculating the motivation fund based on the cost of one machine-tool hour [10] using the example of one machine. The lower OEE, the higher the cost of the machine tool hour, and therefore the cost of losses (lost profits), also measured in hours.

Initial data:

1. Achieved or starting indicator OEE1 - 40%;
2. The planned indicator OEE2, to which we strive - 80%;
3. Conventional unit (1 cu) = 70 rubles;
4. The average cost of a machine-tool hour with OEE equal to 40% is 72.83 USD [];
5. Fund of the machine tool per month - 720 hours (with 30 days.

Thus, the calculation of losses per month is performed according to the formula:

$$\text{Losses} = ((\text{OEE2}-\text{OEE1})/100) \cdot \text{monthly fund of the machine} \cdot \text{machine hour cost} = ((80-40)/100) \cdot 720 \cdot 72{,}83 = 20\,975{,}04 \text{ USD/ month.}$$

20,975 USD losses per month from only one machine, with 10 machines this is already 200,000 USD. etc. It is more convenient to operate with a different value, namely, not achieving OEE by one% brings 524.36 USD losses per month from one machine.

Since some of the personnel are involved in the maintenance of more than one machine tool in the shop, I propose the following dependence of the calculation of the share of motivation on lost profits, the data are summarized in Table 10.

Table 10. The data summarized information

Number of machines	1 machine	10 machines	25 machines	50 machines	100 machines
% of the increase in OEE (avoidance of losses)	30%	25%	20%	15%	10%

Note: Achievable annual growth of OEE is approximately 10-12%. Thus, with properly planned work, you can reach the average planned OEE indicator of 65% in the third year of operation. And given that the enterprise bears the maximum losses at the very beginning of the journey, with a low OEE, then the share of motivation during this period will be greater. In the future, in the process of achieving the planned OEE, the motivation according to the proposed method decreases, since OEE will increase and the cost of the machine-tool hour will decrease. You should not completely abandon motivation, since the subsequent growth of OEE will be associated with the improvement of more complex processes, where each subsequent percentage of OEE will be given with great difficulty, and it is also not easy to maintain the achieved state.

References

[1] Shmatkov V S, Timokhin R S, Shmatkova A V 2017 Increasing the objectivity of data on the operation of CNC equipment using a log file (machine protocol) *Automation. Modern technologies* Vol. 71 **6** 254-259

[2] Karlina Yu I, Kargapoltsev S K, Gozbenko V E, Karlina A I, Leonovich D S 2021 Selection of tools and cutting modes for turning small-sized high-precision parts of micro-wave electronics from beryllium bronze *IOP Conference Series: Materials Science and Engineering* **1064** 012016

[3] Shmatkova A V, Shmatkov V 2018 Organization of scheduled maintenance and current repairs of high-performance equipment in modern conditions *Machines and tools* **5** 36-40

[4] Karlina Y I, Kargapoltsev S K, Gozbenko V E, Leonovich D S, Karlina A I 2021 Automation of preproduction processes for high-precision small-sized parts on CNC machines *IOP Conference Series: Materials Science and Engineering* **1064** 012017

[5] Konyuhov V Y, Gladkih A M, Zott R S 2021 Machine-building enterprise performance and quality improvement tools *IOP Conference Series: Materials Science and Engineering* **1064** 012021

[6] Gladkih A M, Konyuhov V Yu, Galyautdinov I I, Shchadova E I 2020 Suggestions for improving the efficiency of repair activity *IOP Conference Series: Materials Science and Engineering* **760** 012022

[7] Konyuhov V Y, Gladkih A M, Zott R S 2020 Accelerator as an effective replacement of a business incubator in the Irkutsk region *Journal of Physics: Conference Series* **1582** 012045

[8] Lysenko D A, Olentsevich V A, Vlasova N V, Konyukhov V Y 2021 Formation of new principles and models of operation of structural units of the industry under the conditions of implementation of digital technologies *IOP Conference Series: Materials Science and Engineering* **1064** 012025

[9] *Overall equipment efficiency* 2007 Translation from English (Moscow, Institute for Comprehensive Strategic Studies) p 120

[10] Shmatkov V S, Shmatkova A V 2017 Calculation of economic efficiency and cost of a machine-tool hour when working on high-performance equipment *Machine tools and tools* **12** 8-12

AUTHOR INDEX